High-Speed Monohull and Hydrofoil Craft

Zu-Shun Dong • Liang Yun • Raju Datla
Alan Bliault

High-Speed Monohull and Hydrofoil Craft

Performance, Technology, and Applications

Springer

Zu-Shun Dong
Department of Naval Architecture
and Ocean Engineering
Naval University of Engineering
Wuhan, Hubei, China

Raju Datla
School of Engineering and Science
Stevens Institute of Technology
Hoboken, NJ, USA

Liang Yun (deceased)
Professor Yun wrote this book
while at the Marine Design & Research
Institute of China (MARIC)

Shanghai, China

Alan Bliault
The Hovercraft Society
Sola, Norway

ISBN 978-3-031-62759-0 ISBN 978-3-031-62760-6 (eBook)
https://doi.org/10.1007/978-3-031-62760-6

This Springer imprint is published by the registered company Springer Nature Switzerland AG
The registered company address is: Gewerbestrasse 11, 6330 Cham, Switzerland

Preface

This volume is the sixth in a series on fast marine craft that incorporates knowledge and research from both China and Russia as well as the western world. The central figure in bringing this together has been Professor Liang Yun who has been dedicated and energetic in his encouragement and at times urging of other participating authors to achieve the objective that has now been reached.

The journey began for author Bliault with a rewrite and enlargement of a book he had prepared in the early 1990s on Air Cushion Craft for a Chinese publisher. Following much work this saw the light of day in 2000. This success and Liang's impending retirement from Marine Design and Research Institute of China (MARIC) in Shanghai led him to suggest a whole series of books. Somewhat reluctantly due to other professional commitments, Bliault agreed to a partnership and Liang began to assemble further materials. The series grew (see below) steadily. Each project has been a major challenge and has taken a significant period to complete as a joint project. Bliault additionally performed the role of technical editor on each volume.

Involvement in this last volume was initially to assist with some final editing and diagrams in 2020 rather than a joint project overall. Reviewing the text at that stage it became clear that rather more work was required. Following Liang's passing in December 2020 Bliault committed to support the completion of the book as such for his family. Subsequently a good deal of the content has been rewritten and revised, particularly that on model test series and analytical work carried out on semi-planing and planing vessels.

Both Liang and Bliault have had a focus on practical design and so it is hoped readers will find useful material both to assist in understanding the underlying physics, and for selection and analysis of a prospective high-speed monohull.

Consistent with earlier volumes Bliault has collated together a selection of resources that may assist in more detailed research and design. This is included as Appendix 2. Appendix 1 summarizes and discusses analytical models for resistance prediction for semi-planing and planing craft, respectively, for practical application that is covered in depth by Professor Radojcic.[1]

On the aspect of project management, readers are referred to Chap. 14 of our book *High-Speed Catamarans and Multihulls*, as the principles described there are applicable to any project for delivery of high-speed marine craft.

This book refers to CFD but only in passing. Since the beginning of the twenty-first century, there has been a revolution in computer capabilities such that designers now use CFD in the early stages of design to enable concept comparison and selection. This does not replace the need for model testing, particularly for high-speed craft where the physics of the air-water interface is involved. Bliault has attempted a commentary on this in the Chap. 12 "Postscript" to the volume which it is hoped can provide useful thoughts for the new generation of engineers and researchers working in this field.

The search for more speed—humanity has a tendency always to seek more—has been enabled through increasingly efficient and lightweight power plants such as high-speed diesels and gas turbine engines, and lighter and stronger structural materials (aluminum alloy, GRP, titanium alloy), that have enabled designers of fast boats, hydrofoils, and air cushion craft to develop performance close to physical limits of speed in a seaway.

In the last 30 years or so a revolution in electronics has given us the possibility for automated stabilization of motions that was simply not possible before. Together with big strides in power plant efficiency, not to mention satellite navigation, these have been important enablers to comfort at higher speed and high-speed vessel development.

Thus, a series of high-speed craft and high-performance marine vessels have entered service around the world after the Second World War, such as High-Speed semi-planing and planing mono-hull craft in and after the 1940s, Hydrofoil Craft in 1950s and 1960s, Air Cushion Vehicles and Surface Effect Ships in 1960s and 1970s, High-Speed Catamarans in 1980s, Small Water Plane Area Twin Hull vessels in 1980s, Wave Piercing Craft in 1990s, Trimarans at the beginning of the twenty-first century, and super high-speed Wing in Ground Effect Craft (WIG) from the 1970s.

High-speed semi-planing and planing craft (including Deep V craft) have high speed, together with simple structure and equipment (no skirt and lift system as with ACV/SES, no hydrofoils, no wing as with WIG) as well as lower price, so the concept has been vigorously developed, and is still a core military combat and civil transportation vehicle around the world.

[1] D Radojcic, M Kalajdzic, A Simic, 'Power Prediction Modelling of conventional high speed marine craft', Springer, 2019, ISBN 978-3-030-30603-6.

Regarding the hydrofoil, high speed and fine seaworthiness are available simultaneously, particularly with a fully submerged foil automated control system; however, its high construction and maintenance cost, speed limitation from the so-called cavitation barrier, as well as size limitation slowed its development in the past few decades. Nevertheless, some new medium size submerged hydrofoil vessels in service may show a new economy for the passenger routes in the future.

The two principal authors (Yun, Dong) of this volume have organized more than 20 domestic High-Performance Marine Vehicle (HPMV) Conferences, and 17 International HPMV conferences in China in the past three decades at the same time as working on HPMV R&D and design and do feel that High-Speed Monohull and Hydrofoil Craft have been increasingly valued by worldwide shipping companies, navy, and the passenger navigation community year by year in recent times. In addition, there are a lot of technical papers concerned with the HPMV performance, technology, and application presented at various worldwide conferences. However, fewer special technical books in such field have been published around the world so far!

This book is part of a series that includes five other books concerned with the high-performance marine vehicle (HPMV) instigated as a partnership two of the authors (Yun and Bliault):

- "Theory & Design of Air Cushion Craft", Liang Yun, Alan Bliault, Edward Arnold, part of Elsevier Publications, 2000, London, UK, ISBN 0 340 67650 7
- "WIG Craft and Ekranoplan, Ground Effect Craft Technology", Liang Yun, Alan Bliault, Johnny Doo, Springer, New York, US, 2009 ISBN 978-1-4419-0041-8; "Chinese version", National Defense Industry Press, Beijing, China, 2013; "Russian version", Technosphera Publishers, Moscow, Russia, 2014
- "High Performance Marine Vessels", Liang Yun, Alan Bliault, Springer, New York, US, 2012, ISBN 978-1-4614-0868-0
- "High Speed Catamarans and Multihulls, Performance, Technology, and Application", Liang Yun, Alan Bliault, Rong Huan Zhong, Springer, New York, US, 2018, ISBN 978-1-4939-7889-2
- "Air Lubricated and Air Cavity Ships", Gennadiy Alexeevitch Pavlov, Liang Yun, Alan Bliault, Shu-Long He, Springer, New York, US, 2020, ISBN 978-1-0716-0423-6

This is the sixth book of this series, describing the performance, technology, and application of High-Speed Monohull and Hydrofoil Craft. There are two parts with a total of 12 chapters in this book. The first part is a description of High-Speed Monohull Craft, and the second part is of Hydrofoil Craft. We start with a general survey, and description of development and application, of monohulls in Part I, Chap. 1. Chapter 2 describes the semi-planing mono-hull craft, model test programmes and characteristics. Chapters 3, 4, and 5 describe the basic principles of planing, resistance, stability, seakeeping quality, and maneuverability of planing hulls, and Chap. 6 describes the semi-displacement monohull "Deep V" craft which uses attributes from both semi-planing round bilge designs and hard chine designs.

Part II introduces Hydrofoil Craft. Through Chaps. 7, 8, 9, 10, and 11, the book provides a general overview in Chap. 7, fundamental theory of hydrofoils in Chap. 8, introduction to wing theory in a real fluid in Chap. 9, initial design and performance in Chap. 10, and introduction of some special hydrofoil craft types in Chap. 11.

We have introduced the power plant and transmission, as well as propulsion, hull material, and structures of both Air Cushion Craft (ACC), and Wing in Ground Effect Craft (WIG) in our first and second books respectively. The book covering multihull vessels covers propulsion systems, outfitting, and project execution that is equally applicable to fast monohulls, while the chapter on project management is applicable to all high-performance Marine Vessels. Chapters in the book series may thus be referenced separately if required, and this facility is available through digital download from the Springer site.

The intended audiences for this book are teachers, students, both at undergraduate and postgraduate level in naval architecture and marine engineering and other concerned faculties of universities and institutes, engineers, researchers, technicians, and operators who are involved in High-Performance Marine Vehicles research, design, construction, and operation, and all people, who are interested in HPMV in both military and civil applications.

Shanghai, China — Liang Yun
Wuhan, Hubei, China — Zushun Dong
Hoboken, NJ, USA — Raju Datla
Sola, Norway — Alan Bliault

Acknowledgments

The authors would like to express their sincere thanks to the leadership and colleagues of the PLA Naval University of Engineering, China; the Marine Design & Research Institute of China (MARIC); Stevens Institute of Technology, USA; the China Society of Naval Architecture & Marine Engineering (CSNAME); SNAME, USA; and the Shanghai Association of Shipbuilding Industry (SASI) of China, for their help in using their public papers and research during the writing of this book.

The authors would also like to show their appreciation toward Senior Researchers Hui-Ming Li and Cheng-Xia Zhang for providing their important public research and papers on the High-Speed Monohull and Deep V type craft, respectively.

The authors also wish to thank Ke-Zhi Li of China and Jeffrey Hu, a Senior Computer Software Engineer from the United States, for their help during the preparation of the manuscript of this book. The first and second authors of this book wish to thank their respective wives, Jia-fu Zhu and Li-hui Qiu, for their patience and continuous support during the writing of this book.

Author Alan Bliault extends personal thanks to Zu-Shun Dong and Liang for their preparation of the core material, to Raju Datla for the input of his and others' important research at Stevens Institute of Technology and his support together with his students in redrafting of many figures. Sincere thanks also go to Professor Radojcic for his advice on regression modeling, and to Prof. Sahoo for access to his regression work. Finally, this book could not have been published without the incredible support of the project and editorial team at Springer publications. Many thanks to you all!

Acknowledgments for Images and Data

This book could not have been written without the research programs carried out at Maritime Research Centers across the world. Reference is given to the papers and theses published by these authorities through the chapters of this book.

Following academic practice where diagrams or data are quoted in this book, reference is given directly, and the authors wish to reiterate their thanks and admiration for the technical achievements shared. Where possible diagrams used from reference material are reproduced as accurately as possible or redrawn for clarity. Readers are encouraged to also refer directly to the original references for their own verification where possible.

Many photographic images presented in this book are taken from the public domain, and from materials available to the authors. The following figures are published courtesy of the organizations listed below, with many thanks. Figure references are in parentheses.

Discovery Museum, Tyne and Wear, UK (1.1); Everglades Swamp Tours, Fort Lauderdale, Florida (1.3, 3.1); US Navy (1.4, 1.8a, 1.8b, 7.1, 7.3d, 7.5, 7.17, 7.18, 7.20, 10.45); Aries (1.9a); Cougar Marine (1.9b); Bluebird Supporters Club, Mike Varndell (1.11); Schnelker Marine (1.12a, 3.2); Vexus Boats (1.12b); Hardy Marine (1.13); Damen (1.14, 1.26, 6.2, 6.22); Boat International (1.15); Sunseeker Marine (1.19a); Windy Boats (1.19b); Swedeship (1.19c); Sea Doo Company (1.20); BAZAN (1.21); Fincantieri (1.22); Ares Shipyard (1.23), Bournemouth Echo/Cougar (1.24a); M.craft/Twindisc/Arnesen (1.24b); Strategic Marine (1.25); Imperial War Museum UK (2.2); Abeking & Rasmussen/Turkish Coastguard (6.1); Dodecanese Flying Dolphins/Turkish Sealines (7.7a), Supramar (7.13, 7.14 10.43); Rodriquez/SNAV (7.15, 7.16, 10.44), US Coastguard (7.20); Boeing Marine Systems (7.21, 7.22, 7.23, 10.39, 10.40, 10.47, 10.48); Far East Hydrofoils (7.26); Royal Canadian Navy (7.29); Grumman Marine (10.46); Kværner Fjellstrand (11.12); Mitsubishi (11.13); Daea Express Shipping (11.14); Westamarin (11.17).

Contents

Part I High Speed Mono-hull Craft

1 Introduction to High-Speed Monohull Craft 3
1.1 Introduction 3
1.2 Monohull Configurations 6
1.3 Recent Development of High-Speed Monohull Craft 17
1.4 Studying the Monohull Concept 28
References 32

2 Semi-Displacement Monohull Craft 33
2.1 Drag Components 37
2.2 Resistance Through Model Test Programs 38
2.2.1 NPL Model Test Series 42
2.2.2 SSPA Series 47
2.2.3 USA Series 63 49
2.2.4 USA Series 64 50
2.2.5 China Ship Science Research Center Series 52
2.2.6 MSC Series at CSRDC, China 54
2.2.7 Bazan 80 Series 1980 55
2.2.8 SKLAD Series 1980 58
2.2.9 NSMB HSDHF Test Series 59
2.2.10 AME CRC Series 1999 63
2.2.11 VTT Series 65
2.3 Resistance Estimation by Means of Regression Expressions . . . 68
2.3.1 Mercier-Savitsky Regression Expressions [27] 70
2.3.2 MARIC's Regression Expressions [30] 73
2.3.3 Regression Formula of Shanghai Jiao Tong University, China [31] 77
2.3.4 Regression Evaluation Summary 78
2.3.5 Appendage Drag for Semi-planing Craft 81
2.4 Effect of Craft Geometric Parameters on Drag 82

2.4.1 Slenderness ψ or Displacement/Length Ratio $\nabla/(L/\mathbf{10})^3$ 82
2.4.2 Beam-Draft Ratio 84
2.4.3 Sectional Area Curve 84
2.4.4 Prismatic Coefficient C_p 84
2.4.5 Profile of Sectional Area Curve and Longitudinal Position of C_B 86
2.4.6 Transom Configuration 86
2.4.7 Cross Section Profile 88
2.4.8 Half Entrance Angle 90
2.4.9 Conclusions 90
2.5 Transverse Stability of Fast Round Bilge Craft 91
2.6 Seakeeping Quality of Fast Round Bilge Craft 93
2.6.1 Motion Response of Fast Round Bilge Craft in Waves 93
2.6.2 Dynamic Responses in Waves 100
2.6.3 Influence of Hull Parameters of Semi-planing Craft on Seakeeping Quality 106
2.7 Maneuverability of Fast Round Bilge Craft 108
2.8 Appendages for Improving the Operational Performance of Round Bilge Craft 110
References 119

3 Basic Principles of Planing 123
3.1 Force Acting on a Planing Surface in Ideal Flow 125
3.1.1 Thin Wing Analogy of Two-Dimensional Planing Surface 125
3.1.2 Basic Equations of Motion 126
3.1.3 Boundary Condition at Body Surface 127
3.1.4 Boundary Condition for Free Surface 127
3.1.5 Simplified Boundary Condition for "High-Speed Small Disturbance" 128
3.2 Approximate Solution of L. Y. Sedov with Consideration of Gravity Effect 131
3.2.1 Nonlinear Solution of Two-Dimensional Planing Plate in Ideal Flow—High-Speed Large Disturbance Problem 133
3.3 Wing Simulation Method for the Planing Surface with Finite Width 134
3.3.1 Planing Surface with High Aspect Ratio 134
3.3.2 Short Wing Modeling for the Planing Surface with Lower Aspect Ratio 135

3.4 Theory for Slender Planing Surface—The Slender Wing Analogy . . . 138
3.4.1 Basic Idea of Monk-Jones' Slender Wing Theory . . . 138
3.4.2 Virtual Mass Theory on Slender Planing Surface—Slender Wing Analogy on Planing Surface . . . 142
3.5 Force Acting on Planing Surface in Viscous Flow . . . 147
3.5.1 Viscous Effect on Flow . . . 147
3.5.2 Effect of Gravity and Correction of Static Buoyancy . . . 149
3.6 Influence of Planing Surface Shape on Its Hydrodynamic Performance . . . 151
3.6.1 Influence of Beam of Planing Plate . . . 152
3.6.2 Influence of Varying Beam in Longitudinal Direction . . . 154
3.6.3 Influence of Longitudinal Curvature . . . 155
3.6.4 Influence of Transverse Deadrise on Bottom . . . 157
3.6.5 Influence of Transverse Curvature . . . 158
References . . . 159

4 Planing Hull Resistance . . . 161
4.1 Introduction . . . 161
4.2 Initial Power Estimation for Planing Hulls Using Coefficients . . . 163
4.2.1 Admiralty Coefficient Method . . . 163
4.2.2 K_η Method . . . 164
4.2.3 Niwa Method . . . 165
4.3 Semi-theoretical and Empirical Methods for Planing Hull Resistance . . . 167
4.3.1 CAHI Method . . . 168
4.3.2 Daniel Savitsky's Method (USA) . . . 172
4.3.3 Shuford-Brown Method . . . 174
4.4 Estimation of Resistance from Model Test Series . . . 179
4.4.1 EMB Series 50 . . . 180
4.4.2 "Series 62" Model Tests (USA and Holland) . . . 182
4.4.3 "CSSRC" Series . . . 184
4.4.4 Series 65 Tests . . . 187
4.4.5 US Naval Academy Series . . . 188
4.4.6 US Coast Guard Planing Hull Series . . . 190
4.4.7 Technical University of Nova Scotia (TUNS) Series . . . 193
4.4.8 Southampton University Series . . . 196
4.4.9 Naples Warped Hard Chine Series . . . 200
4.4.10 Delft Systematic Deadrise Series . . . 203

4.5 Analytical Equations Derived from Model Testing 207
4.6 Model Test Issues for Planing Hulls 209
4.7 Planing Hulls with Steps 216
4.7.1 Characteristics of Planing Hulls with Steps 216
4.7.2 Estimation of Drag of Stepped Planing Hulls 217
4.7.3 Arrangement of Step for Stepped Planing Hulls 222
4.8 Appendage and Air Resistance 226
4.8.1 Estimation of Appendage Drag Components 227
4.8.2 Air Drag 231
4.9 Influence of Appendages and Propellers on Running Trim 231
4.9.1 Influence of Appendages 231
4.9.2 Influence of Propellers 232
4.10 Resistance of Planing Craft in Shallow Water 241
References 247

5 Planing Craft Stability, Maneuverability, and Seakeeping 251
5.1 Transverse Stability of a Planing Hull 251
5.2 Longitudinal Stability and Porpoising Motion of Planing Hulls 255
5.2.1 Experimental Investigation of Porpoising Motion . . . 256
5.2.2 Theoretical Assessment of Longitudinal Stability of Planing Hulls 263
5.3 Maneuverability of a Planing Hull 268
5.3.1 Course Keeping Stability of Planing Hull 269
5.3.2 Turning Characteristic of a Planing Hull 270
5.3.3 Features of Rudders on a Planing Hull 272
5.4 Seakeeping Quality of Planing Hulls: Impact Forces 273
5.4.1 Spray on Planing Hulls 274
5.4.2 Planing Hull Wave Slamming 276
5.5 Seakeeping Quality of Planing Hulls: Motions and Speed Loss 285
5.5.1 Heaving, Pitching, and Speed Loss of Planing Craft in Waves 285
5.5.2 Broaching of Planing Hulls in Following Waves 289
5.6 Prediction of Seakeeping Quality of Planing Hulls 292
5.6.1 Prediction from Model Series Experimental Results 294
5.6.2 Theoretical Calculation for Predicting the Seakeeping Quality of Planing Hulls 301
5.7 Influence of Planing Hull Geometry on Seaworthiness 304
5.7.1 Geometric Parameters of Craft Configuration 304
5.8 Special Measures for Improving Planing Hull Performance . . . 311
References 312

6 Deep Vee Semi-planing Craft . . . 315
6.1 Basic Concepts and Features . . . 315
6.2 Drag of Deep Vee Semi-planing Craft in Calm Water . . . 318
6.2.1 Effect of Length Beam Ratio L/B . . . 320
6.2.2 Effect of Deadrise Angle β . . . 320
6.2.3 Effect of Longitudinal Position of C_B (X_C) . . . 321
6.3 Seakeeping Quality of Deep Vee Semi-planing Craft . . . 325
6.3.1 Heaving, Pitching, and Vertical Acceleration of Deep Vee Craft in Waves . . . 325
6.3.2 Drag Increment of Craft in Waves . . . 326
6.3.3 Slamming in Waves . . . 327
6.3.4 Green Water and Surface Emergence of Propellers . . . 330
6.3.5 Roll and Rolling Acceleration in Waves . . . 331
6.3.6 Influence of Size and Profile on the Seakeeping Quality of Deep Vee Craft . . . 332
6.4 Maneuverability of Deep Vee Craft . . . 333
6.5 Stability of Deep Vee Craft . . . 334
6.6 Semi-submerged Bow for Deep Vee Craft . . . 336
6.7 Development of the Axe Bow . . . 338
References . . . 341

Part II Hydrofoil Craft

7 Introduction to Hydrofoil Craft . . . 345
7.1 Introduction . . . 345
7.2 Transverse Arrangement of Foils . . . 347
7.3 Plan Arrangement of Foils . . . 352
7.4 Hydrofoil Development . . . 354
7.5 Development Opportunities for Hydrofoils . . . 375
7.6 Summary Data for Sample Hydrofoils (Tables 7.1, 7.2, 7.3, 7.4, and 7.5) . . . 378
References . . . 387

8 Introduction to Hydrofoil Wing Theory . . . 389
8.1 Theory of Two-Dimensional Thin Wing . . . 389
8.1.1 Basic Assumptions . . . 389
8.1.2 The Boundary Condition and Its Solution . . . 390
8.1.3 Vortex Model of Two-Dimensional Thin Wing . . . 394
8.1.4 Solution of Flow Around the Two-Dimensional Thin Wing . . . 396
8.1.5 Forces and Moments Acting on a Thin Wing . . . 399
8.1.6 Simplified Lifting Line Model . . . 403

8.2 Influence of Limited Span, the Three-Dimensional Thin Wing 404
8.2.1 Flow Situation and Lifting Line Model of Wing with Limited Span 404
8.2.2 Downwash Caused by Free Vortices 406
8.2.3 Lift on a Limited Span Wing and Equation for Induced Drag 408
8.2.4 Explanations 411
8.3 Effect of Wing Configuration on Hydrodynamic Performance 414
8.3.1 Calculation of Hydrodynamic Performance of Swept-Back Wing 414
8.3.2 Effect of Wing Dihedral 419
8.4 Experimental Reference Data for Wing Sections 421
8.4.1 Some Test References for Airfoils Frequently Used for Hydrofoils 422
8.4.2 Correction to Lift Coefficient and Approximate Estimation of Viscous Drag 427
References 435

9 Hydrofoil Wing Theory Influence of Free Surface and Cavitation 437
9.1 Influence of the Free Surface 437
9.2 Mirror Image Vortex Model on Free Surface and Its Use 440
9.2.1 Using a Mirror Image Vortex Model to Calculate Free Surface Effect 441
9.2.2 Velocity Induced by Attached Vortex 442
9.2.3 Velocity Induced by Mirror Image Vortex 444
9.2.4 Induced Velocity from Free Vortex 445
9.2.5 Induced Velocity from Mirror Image Free Vortex 445
9.3 Estimation of Free Water Surface Effect 452
9.3.1 Estimation of Free Water Surface Influence for a Surface Piercing Hydrofoil 457
9.3.2 Average Submerged Depth of V Profile Foil 457
9.3.3 Correction to Angle of Attack 458
9.3.4 Hydrofoil Performance from Model Tests Under Free Water Surface 459
9.4 Hydrofoil Cavitation and Its Avoidance 462
9.4.1 The Formation of Cavitation and Its Effects 462
9.4.2 Cavitation Inception of Hydrofoil 464
9.4.3 Avoiding Hydrofoil Cavitation: The Features of Sub-Cavitating Hydrofoils 466
9.4.4 Hydrofoil with Swept Plan Form 467
9.4.5 Hydrofoil Air Ventilation 468

9.5 Hydrodynamic Characteristics of Cavitating Hydrofoils 469
9.5.1 Cavitation Flow and Its Simplified Model 469
9.5.2 Hydrodynamics of Two Dimensional Fully Cavitating Hydrofoil (at $\sigma = 0$) 474
9.5.3 Flat Foil at $\sigma = 0$ 474
9.5.4 Linearized Theoretical Solution for Camber Foil at $\sigma = 0$ 476
9.5.5 Two-Dimensional Fully Cavitating Foil at $\sigma > 0$ . . . 480
9.5.6 Betz's Zero Approximation Method 480
9.5.7 Richardt's Zero Approximation 481
9.5.8 Numerical Calculation of Cavitating Hydrofoil at $\sigma > 0$ 482
9.6 Influence of Free Water Surface and Correction for Limited Span 483
9.7 Section Profiles for Super-Cavitating Foils 486
9.7.1 Foil Profile with Low Drag at Zero Cavitation Number 486
9.7.2 Foil Profile of Super-Cavitating Hydrofoil 491
References 492

10 Hydrofoil Craft Initial Design and Performance 493
10.1 The Arrangement, Type, and Definition of Hydrofoil Size 493
10.1.1 Arrangement 493
10.1.2 Geometry and Sizing 494
10.1.3 Initial Definition of Hydrofoil Area 495
10.1.4 Hydrofoil Dimensions and Plan Profile 495
10.1.5 Foil Section Profile 499
10.1.6 Hydrofoil Takeoff Aids and Struts 500
10.2 Hydrofoil Drag 502
10.2.1 Determination of Craft Running Attitude in Foil Borne Operation 503
10.2.2 Interaction Between Bow and Stern Foils 506
10.2.3 Estimation of Drag 507
10.3 Stability of Hydrofoil Craft 516
10.3.1 Vertical Stability of Hydrofoil Craft 517
10.3.2 Longitudinal Stability of Hydrofoil Craft 518
10.3.3 Transverse Stability of Hydrofoil Craft 524
10.4 Maneuverability of Hydrofoil Craft 529
10.5 Seaworthiness of Hydrofoil Craft 533
10.6 Platforming Operation 533
10.7 Motion Following the Wave Profile 534
10.8 Intermediate Response Motion 534
10.9 Automated Control of Hydrofoil Motions 549
10.10 Review of Foil Arrangements on Hydrofoil Craft 557
References 564

11 Some Special Hydrofoil Craft . . . 567
11.1 Introduction . . . 567
11.2 Foil-Assisted Catamaran . . . 568
11.3 Foilcat . . . 577
References . . . 585

12 Postscript . . . 587
References . . . 590

13 Appendix 1: Analytical Models . . . 591
13.1 Introduction . . . 591
13.2 Round Bilge and Semi Planing Craft . . . 593
13.2.1 S C Fung . . . 593
13.2.2 Mercier and Savitsky . . . 596
13.2.3 MARIC . . . 597
13.2.4 Shanghai University . . . 598
13.2.5 VTT . . . 598
13.2.6 Sklad . . . 600
13.2.7 HSDHF/AMECRC . . . 601
13.2.8 NPL . . . 601
13.2.9 Southampton University—High L/B NPL Form . . . 602
13.2.10 US Naval Academy Series (Round Bilge) . . . 603
13.3 Hard Chine Planing Vessels . . . 604
13.3.1 Mercier and Savitsky . . . 604
13.3.2 Series 62/Series 65 . . . 604
13.3.3 Delft DSDS . . . 604
13.3.4 US Naval Academy Series (Hard Chine) . . . 605
13.3.5 VTT . . . 605
13.3.6 USCG and TUN . . . 605
13.3.7 Series 50 . . . 605
13.3.8 Naples University Warped Hard Chine Series . . . 606
13.4 General Discussion . . . 606
13.5 Spreadsheet Models Provided . . . 607
References . . . 609

14 Appendix 2: Resources and General Reference Material . . . 611
14.1 Internet Resources . . . 611
14.2 General References and Internet Links . . . 627

Index . . . 637

List of Figures

Fig. 1.1 (**a**) *Turbinia* general arrangement. (**b**) *Turbinia* at speed 4
Fig. 1.2 (**a**) Planning craft body plan and profile. (**b**) Semi planning craft body plan and profile 5
Fig. 1.3 Everglades airboat 7
Fig. 1.4 Monohull LCS example. (Courtesy U.S. Navy photo by Mass Communication Specialist 2nd Class Nathan Laird/Released) from Wikimedia commons. Note: ATLANTIC OCEAN (Sept. 28, 2009) The littoral combat ship USS Freedom (LCS 1) conducts flight deck certification with an MH-60S Sea Hawk helicopter assigned to the Sea Knights of Helicopter Sea Combat Squadron (HSC) 22) 9
Fig. 1.5 Series 64, developed by DTMB in USA early 1960s 10
Fig. 1.6 A typical planing hull profile, with hard chine—wireframe model 10
Fig. 1.7 Running attitude and drafts at bow and stern change with volumetric Froude Number, where I denotes displacement mode; II transitional mode; III planing mode 11
Fig. 1.8 (**a**) A typical medium speed planing hull. Type VI patrol boat from USN. (**b**) A typical high speed planing craft, the 17m Sentinel-class patrol vessel 13
Fig. 1.9 Typical stepped planing hulls. (**a**) Cruiser. (**b**) Racing craft with multiple steps 14
Fig. 1.10 (**a**) Spray rail configuration. (**b**) Chine and spray rails examples 15
Fig. 1.11 Three-point hydroplane—Bluebird K7 16
Fig. 1.12 (**a**) Schnelker marine tracker. (**b**) Vexus boats DS20 high speed fisherman shown at speed riding on spray rails 18
Fig. 1.13 Semi-planing cabin cruiser-Hardy Marine 40DS 19

Fig. 1.14 Offshore patrol vessel (Damen Stan 4207 Royal Bahamas defense network). (Courtesy Chuck Oldham April 1, 2015, Defense media network) 20
Fig. 1.15 Semi-displacement Superyacht Silver Fast 77m, 27 knots and global range, courtesy Boat International 20
Fig. 1.16 Typical Torpedo boat. (From WIKI at https://en.wikipedia.org/wiki/Motor_torpedo_boat) 21
Fig. 1.17 Vosper MTB firing torpedoes in English Channel. (Courtesy BAE Systems Maritime) 21
Fig. 1.18 Guided missile craft. (Komar class boat courtesy Wikipedia, at https://en.wikipedia.org/wiki/Komar-class_missile_boat) 22
Fig. 1.19 (**a**) Fast day boat (Sunseeker Hawk 38, 2 × 400 hp, multi step, 60 knots plus. Courtesy www.sunseeker.no). (**b**) Fast tender craft for superyacht (Windy SR26 tender b1 and b2 side a quarter view showing lines and seating layout. Courtesy Windy, https://www.windyboats.com/range/sr-line/sr26/). (**c**) Fast Rescue craft—Tornado semi-inflatable 9m? (Swedeship 24m fast rescue craft 38 knots. https://swedeship.se/vessel-production/22-m-fast-rescue-craft/) (**d**) Fast MOB boat (Fast mob boat – Courtesy WIKIcommons. https://commons.wikimedia.org/wiki/File:MOB_Boat_(man_overboard_rescue_boat).jpg. Craft is being lowered from Construction vessel Silja in Baltic and has self-righting at stern) 24
Fig. 1.20 Sea doo craft. (Courtesy Sea Doo Inc.) 25
Fig. 1.21 HSC Silvia Ana built by BAZAN. (Courtesy WIKI commons) 25
Fig. 1.22 MDV 1200 ferry, 37 knots hard chine design. (Courtesy WIKI commons at https://commons.wikimedia.org/wiki/File:SSC3-2.jpg and https://en.wikipedia.org/wiki/MDV_1200-class_fast_ferry) 26
Fig. 1.23 Modern Fast Interdiction craft—Ares 24 fast interceptor for Abu Dhabi. (Courtesy Ares Marine) 26
Fig. 1.24 (**a**) Racing deep vee craft—Cougar craft Fortido (Courtesy Bournemouth Echo June 7, 2018 at https://www.bournemouthecho.co.uk/sport/16277328.powerboat-racers-aim-glory-poole-bay-100/). (**b**) Surface drive 27
Fig. 1.25 Fast offshore logistics craft Strategic Marine Centus 42m, 30 knot fast crew boat. (Courtesy Strategic Marine at https://www.strategicmarine.com/vessel/42m-gen-4-fast-crew-boat/) 28

Fig. 1.26 Damen offshore fast craft with wave slicer bow, the Damen FCS-7011 40 knots, 73.6m. (Courtesy Damen Shipbuilding, at https://www.damen.com/insights-center/news/damen-revolutionary-fcs-7011-completes-sea-trials and https://products.damen.com/-/media/products/images/clusters-groups/high-speed-crafts/fast-crew-supplier/fcs-7011/documents/damen_fcs_7011_brochure_04_2019.pdf?rev=559f18fa102746e487a9dd27997c1e7a) 29
Fig. 1.27 Design flow chart for fast monohull vessels 31

Fig. 2.1 Bow/stern draft of the craft versus speed 34
Fig. 2.2 German E Boat. (German E-Boat S204 surrenders at Felixstowe on 13 May 1945, By Lt. J.E. Russell, Royal Navy official photographer—This is photograph A 28558 from the collections of the Imperial War Museum, it is Public Domain, https://commons.wikimedia.org/w/index.php?curid=734851) 36
Fig. 2.3 Proportion of residual drag to friction drag of fast displacement monohulls versus Fn_L 37
Fig. 2.4 Slenderness range for key test series 41
Fig. 2.5 Body plan of NPL series prototype 42
Fig. 2.6 (**a**) Parameter range of NPL series models (**b**) NPL series models wetted area calculation 43
Fig. 2.7 Sample residual drag graph for NPL series models 45
Fig. 2.8 Comparison of effective power between craft with round bilge and hard chine types 46
Fig. 2.9 Drag difference between round bilge and hard chine craft at different speeds 46
Fig. 2.10 Range of Application for round bilge and hard chine craft 47
Fig. 2.11 Body plan of SSPA prototype 47
Fig. 2.12 C_R curve versus different $L/\nabla^{1/3}$ and B/T 48
Fig. 2.13 Body plan of USA Series 63 parent hull form 49
Fig. 2.14 Body plan of USA Series 64 parent hull form model 4783-1. (From original paper by Yeh MT 1965) 51
Fig. 2.15 Body plan of CSSRC series models 52
Fig. 2.16 Effective power coefficient of CSSRC series 53
Fig. 2.17 Typical body plan of MSC series 54
Fig. 2.18 Residual drag coefficient curves for MSC series 55
Fig. 2.19 Body plan BAZAN 80 model 100 57
Fig. 2.20 Body plan BAZAN 80 series model 200 57
Fig. 2.21 Calm water resistance data, BAZAN 80 prototypes 58
Fig. 2.22 SKLAD parent body plan 58
Fig. 2.23 Body Plans NSMB Series (**a**) Model Matrix (**b**) Models 1, 2 and 3 61

Fig. 2.24 NSMB adjusted model test dimensions "magic cube" diagram from Robson 1988 fig. 8 showing the base cube, the C_p tests, and the frigate series ... 63
Fig. 2.25 Typical body plan AMECRC ... 64
Fig. 2.26 AMECRC Testing parameter cube. (Note is from Sahoo HPMV paper 2012 or FAST 99 paper) ... 64
Fig. 2.27 VTT body plan ... 67
Fig. 2.28 (**a**) Residual resistance of NOVA models (from Figs. 9 and 10 of paper). (**b**) Total resistance of NOVA models. Note total drag in fig. 2.28b is scaled to 45.36 Tonnes (100,000 lbs) vessels, while residual drag in Fig. 2.28a is the residual drag coefficient ... 68
Fig. 2.29 $F_1 = f(C_\nabla,\ F_n)$ ... 74
Fig. 2.30 $F_2 = f\left(A_T/A_X, F_n\right)$ ... 75
Fig. 2.31 (**a**) $F_3 = f(C_P, F_n)$ (**b**) $F_3 = f(C_P, F_n)$ ... 76
Fig. 2.32 $F_4 = f(i_S, F_n)$ ... 77
Fig. 2.33 $F_5 = f(X_C, F_n)$ ... 77
Fig. 2.34 Nordstrom's $C_R = f\left[\frac{\Delta}{(L/10)^3}, V/\sqrt{L}\right]$ curves ... 83
Fig. 2.35 NSMB's $C_R = f\left[\frac{\Delta}{(L/10)^3}, V/\sqrt{L}\right]$ curves ... 83
Fig. 2.36 Range of C_P for fast displacement ships ... 85
Fig. 2.37 Sectional area curve versus F_n ... 86
Fig. 2.38 Favored range of longitudinal position of CB ... 86
Fig. 2.39 Effects of transom width to the drag of fast destroyers ... 87
Fig. 2.40 Comparison of residual drag coefficient for different lines ... 89
Fig. 2.41 Comparison of relative total drag Rt/Δ of MSC series with other series models ... 89
Fig. 2.42 Change of transverse stability of a round bilge craft versus speed ... 91
Fig. 2.43 Hydrodynamic pressure acting on the round bilge craft in heeling ... 92
Fig. 2.44 Comparison of calculation prediction of longitudinal motion response of high-speed craft in regular head sea by means of 2.5D method with test results for a model at $Fn = 1.14$ and trim angle 1.62° ... 94
Fig. 2.45 Significant values of pitch and heave of models in irregular head sea waves from reference 12, fig. 56 ... 96
Fig. 2.46 Significant values of vertical acceleration of models in irregular head sea waves, from ref. 12 fig. 57 ... 97
Fig. 2.47 Wave response for different beam draft ratio ... 98

Fig. 2.48 Influence of longitudinal radius of gyration on motion response. (Figures alongside reorganize as in PowerPoint file) 99
Fig. 2.49 Distribution of impact pressure on fast round bulge craft in irregular waves 103
Fig. 2.50 An example of revised bow section 105
Fig. 2.51 Forces acting on the hull during turning 109
Fig. 2.52 Schematic of stern plate and wedge installation 111
Fig. 2.53 Schematic of stern interceptor 112
Fig. 2.54 Controllable interceptor equipment example from NAIAD 113
Fig. 2.55 Wave making generated by hydrofoil near the water surface 114
Fig. 2.56 Bow mounted hydrofoil with negative dihedral 115
Fig. 2.57 Efficiency of wave making elimination by foil with different foil span 116
Fig. 2.58 Drag curve of craft models with bow anti-spray rails 118

Fig. 3.1 Florida Everglades air boat 124
Fig. 3.2 Planing craft with deadrise 124
Fig. 3.3 Flow around a two-dimensional planing plate, force acting on the plate and boundary condition 125
Fig. 3.4 Flow condition for fully immersed two-dimensional plate wing 126
Fig. 3.5 Comparison of both lift (*N*) and moment coefficient (*M*) for $F_n \geq 2.8$ with correspondent value at $F_n = \infty$ 132
Fig. 3.6 Comparison of Sedov's solution with numerical precise solution 133
Fig. 3.7 Flow around a wing with lower aspect ratio 136
Fig. 3.8 Flow around the slender wing 138
Fig. 3.9 Plate planing condition 142
Fig. 3.10 Comparison of flow around the whole wing and planing surface sections 143
Fig. 3.11 Trailing section on both hard chine wetted and non-wetted situations 145
Fig. 3.12 Normal force comparison between Wadlin and McGehoe experiments and theoretical results 146
Fig. 3.13 Normal force from rectangular plate by Shuford 146
Fig. 3.14 Force acting on planing plate 148
Fig. 3.15 Planing plate and displaced volume 149
Fig. 3.16 The calculation of submerged volume of section with deadrise 150
Fig. 3.17 Wave making resistance proportion against speed 151
Fig. 3.18 Influence of beam 153
Fig. 3.19 Resistance of planing surface changes with angle attack ... 154
Fig. 3.20 Influence of stern beam taper on lift efficiency 155

Fig. 3.21 Influence of longitudinal curvature ... 156
Fig. 3.22 Longitudinal curvature versus pressure center ... 156
Fig. 3.23 Influence of β ... 157
Fig. 3.24 Optimum angle of attack versus β ... 158
Fig. 3.26 Several alternative chine plates ... 159
Fig. 3.25 Test results for transverse sections with different curvature ... 159

Fig. 4.1 Statistical curve for K_η ... 165
Fig. 4.2 Shinishi Niwa Ne/Δ vs. V_s curves for planing hulls ... 166
Fig. 4.3 Shinichi Niwa curves for assessment of effect of fineness on powering ... 166
Fig. 4.4 Forces acting on planing hulls ... 167
Fig. 4.5 Curve for $C_b/_\alpha = f(\lambda, F_b)$ ... 170
Fig. 4.6 Curve for $C_{M\Delta} = f(\lambda, F_b)$ ... 171
Fig. 4.7 Lift curves for Savitsky's method ... 173
Fig. 4.8 Moment coefficient curves for Savitsky's method ... 174
Flow Chart 4.1 Procedure to determine equilibrium for a planing craft by iterative calculation following Savitsky's approach ... 175
Fig. 4.9 Centre of lift for Shuford Brown method ... 177
Fig. 4.10 Body plans of BK and MBK planing hull parent forms ... 180
Fig. 4.11 Body plans and lines plans EMB Series 50 models. (**a**) Parent hull body plan. (**b**) Body plans for the series. (**c**) Lines plan for parent hull ... 181
Fig. 4.12 Body plan of base case "Series 62" model 4667-1. (From reference [18]) ... 183
Fig. 4.13 Example resistance cross curve of Series 62, with lift coefficient as parameter ... 183
Fig. 4.14 Body plan of "CSSRC" series, with transverse deadrise, curved rocker, longitudinal warp profile ... 184
Fig. 4.15 Three groups of curves at $Fn_y = 3.0$ ... 185
Fig. 4.16 Correction coefficients of CG ... 186
Fig. 4.17 Statistical data of $\Delta = f(l_s)$ ... 186
Fig. 4.18 Series 65 Hull forms for A and B series ... 187
Fig. 4.19 US Naval Academy Series Hull form YP81-5 ... 188
Fig. 4.20 US Naval Academy Series YP81-5 Resistance Curves example. (**a**) Residual resistance coefficient as a function of speed and displacement for YP81-5 with LCG/Lpp = 0.244. (**b**) Residual resistance coefficient as a function of displacement for YP81-5 with LCG/Lpp = 0.244 ... 190
Fig. 4.21 US Coast Guard Series Hull form of parent 5628 ... 191
Fig. 4.22 US Coast Guard Series Hull Model 5628 example plots ... 192
Fig. 4.23 TUNS Series parent Hull form ... 194

Fig. 4.24 TUNS model test results example for model 3018 with thrust through CG, from reference 25, figure B25 TUNS plots of CT × 10^{-3}, RT/Δ, trim τ, heave at CG/BPX, and wetted area coefficient Cswph, all against Cv The data is from model 3018, the parent hull, and is plotted for the range of loadings CDL with LCG at 35%L forward of transom 195
Fig. 4.25 Southampton University Series Hull forms 197
Fig. 4.26 Southampton Model C results from calm water testing. (**a**) Resistance. (**b**) Trim. (**c**) CG heave. (**d**) Wetted area ... 198
Fig. 4.27 Resistance vs slenderness for the four models 199
Fig. 4.28 Resistance for C, C1, C2 200
Fig. 4.29 Resistance component assessment for model C 201
Fig. 4.30 Naples Series Hull form 201
Fig. 4.31 Naples Series results. Hull form C1 RT/W and tau vs Fr_v for both without and with interceptors 203
Fig. 4.32 Delft Series Hull forms 204
Fig. 4.33 Delft Series Hull forms twist and rocker 205
Fig. 4.34 Example resistance and trim graphs from the twist and rocker tests 207
Fig. 4.35 Model tests programmes parameter coverage 208
Fig. 4.36 Different moment due to different CG height of planing hull in operation 210
Fig. 4.37 Towing test method 211
Fig. 4.38 Friction coefficients of planing flat plate with small angle of attack 212
Fig. 4.39 Friction coefficients of planing flat plate with large angle of attack 213
Fig. 4.40 Average flow speed under the bottom of planing surface. (Fig. 14 from Savitsky 1964—take new image) 214
Fig. 4.41 Wetted area of planing surface with transverse deadrise angle 215
Fig. 4.42 Increment of length/beam ratio due to spray friction 215
Fig. 4.43 Profile of planing hull with single step 216
Fig. 4.44 Comparison of both drag and trim curves of planing hull with step and conventional planing hull without a step (**a**) on the left, and a stepped hull (**b**) on the right 217
Flow Chart 4.2 Resistance and trim analysis of a stepped planing craft 219
Fig. 4.45 Stepped planing craft geometry. (From [40], Chapter 5, Fig. 5.12) 220
Fig. 4.46 (**a**) Longitudinal central profile of free water surface after the transom of planing hull at speed (after [40], Fig. 9.14). (**b**) Water surface geometry behind transom as it varies with speed. (From reference [41]) 223

Fig. 4.48 Planing surface on bottom of stepped craft with transverse deadrise ... 225
Fig. 4.47 Longitudinal central profile of free surface profile after step of planing hull with prismatic configuration and square stern, based on Savitsky's empirical equation ... 225
Fig. 4.49 Planing surface on bottom of stepped craft with "Reverse V" (**a**), "Reverse U" plane profile (**b**) ... 226
Fig. 4.50 Drag curves of appendage and bare planing hull ... 227
Fig. 4.51 Velocity diagram and so vertical force acting on propeller in inclined flow ... 233
Fig. 4.52 Axial-induced velocity before and after the propeller disc ... 235
Fig. 4.53 Integral values of pressure coefficients at any point before and after propeller disc, at various r/R ... 237
Fig. 4.54 y/R diagram ... 237
Fig. 4.55 Integral values of pressure moment coefficients along the X direction, at various r/R ... 238
Fig. 4.56 Average axial induced velocity in propeller wake at different longitudinal position ... 239
Fig. 4.58 Test results for the drag of model No. 1174 of 'Series 62' in shallow water, where R is model drag is in lbs ... 242
Fig. 4.57 Vortex system diagram ... 242
Fig. 4.59 Effect of CG moving forward of model No. 1174 "series 62" on the drag in shallow water ... 243
Fig. 4.60 Geometry figure ... 245

Fig. 5.1 Restoring moment of planing hull with flat bottom in heeling ... 252
Fig. 5.2 Restoring moment of craft with deadrise. (**a**) Pressure profile for a craft with deadrise. (**b**) pressure profile with positive restoring moment when rolling ... 253
Fig. 5.3 Heeling angle versus speed curves ... 254
Fig. 5.4 Stable boundary limit for planing flat plate ... 256
Fig. 5.5 Stable boundary limit, with λ and C_b as coordinates ... 257
Fig. 5.6 Stability boundary limit for a flat plate and prismatic section. Black dots are plate. Open dots are prismatic section with 22 degrees deadrise ... 258
Fig. 5.7 The calculation graphics for third stability boundary limit ... 259
Fig. 5.8 The porpoising motion boundary limit of planing hulls ... 260
Fig. 5.9 Porpoising motion boundary limit for the planing hull with prismatic profile, suggested by Savitsky ... 261
Fig. 5.10 Physical explanation of porpoising motion ... 262
Fig. 5.11 Coordinate systems ... 265
Fig. 5.12 Swaying and heeling due to lateral wind ... 269

Fig. 5.13 Swaying phenomenon of planing hull 270
Fig. 5.14 Turning circle of planing hull 271
Fig. 5.15 Heeling moments acting on planing hull during turning: (**a**) swaying stage and (**b**) stable turning 271
Fig. 5.16 Trapezoid and peach-like rudder 272
Fig. 5.17 Spray of craft planing in the calm water 274
Fig. 5.18 C_Δ influence on spray height ($L_G/b = 2.5$, $F_b = 3.55$) 275
Fig. 5.19 Influence of trim angle on spray height ($C_\Delta = 0.625$, $F_b = 3.55$) 275
Fig. 5.20 Longitudinal anti-spray rail at bottom of craft 277
Fig. 5.21 Distribution of overload coefficient of craft without step along the craft length, $n + 1 = 3.4$ at CG ($C_\Delta = 0.56$, $F_B = 2.5$, Sea state 6). Note n is the acceleration, the plus 1 is to account for gravity. Normalized acceleration at CG 278
Fig. 5.22 Distribution of overload coefficient of craft with step along the craft length, $n + 1 = 5.4$ at step ($C_\Delta = 0.63$, $F_B = 2.85$, ss $5 \sim 6$) 278
Fig. 5.23 Distribution of impact load along the craft length versus speed 278
Fig. 5.24 Vertical acceleration time trace at CG of a patrol boat 279
Fig. 5.25 Distribution of impact force on the bottom of bigger craft along the craft length 280
Fig. 5.26 Impact acceleration of planing hulls in irregular waves: (**a**) bow waves and (**b**) following waves 280
Fig. 5.27 $n \sim C_\Delta$ at different β 282
Fig. 5.28 Impact load of various hull cross sections with $\beta = 15°$, where $b = 300$ mm, $C_\Delta = 0.585$ 283
Fig. 5.29 Four models of Vosper Co, UK. (**a**) A type, (**b**) B, (**c**) C, (**d**) D. (From ref. [2] in ch 4) 284
Fig. 5.30 Max acceleration versus average speed of planing hulls at bow in irregular heading waves 284
Fig. 5.31 Typical section of W craft 285
Fig. 5.32 Heaving and pitching motion of planing craft in heading regular waves 287
Fig. 5.33 Heaving and pitching motion of planing hulls in regular heading waves with two peaks 287
Fig. 5.34 drag increment of planing hull in regular waves: (**a**) head seas and (**b**) following seas 288
Fig. 5.35 Effects of following seas and trajectory circular motion of waves 289
Fig. 5.36 Test results for speed range of a craft losing handling 290
Fig. 5.37 Test results for handling loss by three models 291
Fig. 5.38 The external bow section recommended by D. Savitsky ... 292

Fig. 5.39 Double hard chine bow section and stern section with large deadrise 292
Fig. 5.40 Comparison of calculation values of pitch and heave motions of planing hull by means of strip method with test results $Fr = 0.5$ and 1.2 293
Fig. 5.41 Heaving amplitude at $Vs/\ L^{0.5} = 4$, $\tau = 4^0$ 295
Fig. 5.42 Pitching amplitude at $Vs/L^{0.5} = 4$, $\tau = 4^0$ 296
Fig. 5.43 Max additional drag and its corresponding speed $C_\Delta = 0.6$, $L/B = 5$ 297
Fig. 5.44 Generalized additional drag curve 297
Fig. 5.45 Average acceleration at CG at $V/\sqrt{L}=4$ $(\tau = 4^\circ, \beta = 20^\circ)$ 298
Fig. 5.46 Average acceleration at bow at $V/\sqrt{L}=4$ $(\tau = 4^\circ, \beta = 20^\circ)$ 299
Fig. 5.47 Statistical relationship between the max deck beam and displacement of planing hulls 305
Fig. 5.48 Statistical data for length displacement relationship of planing craft 306
Fig. 5.49 Statistical data for depth versus length of craft 307
Fig. 5.50 Statistical data for craft depth versus deck beam 307
Fig. 5.51 Resistance of planing craft at various area load coefficients 309
Fig. 5.52 Drag for varying CG 309
Fig. 5.53 Various transverse profiles 310
Fig. 5.54 β changed longitudinally 311

Fig. 6.1 SAR33 GA. (**a**) General arrangement. (**b**) Photo. (**c**) Stability curve. (**d**)Seakeeping comparison—ability to maintain speed 316
Fig. 6.2 Damen Stan 2205 Patrol Craft, general arrangement, and 3D model viewed from underside forward of the bow to show underwater shape, fins, and propulsion 317
Fig. 6.3 Effective power of Deep Vee NHP48 and round bilge craft PR72. (From Serter [2]) 319
Fig. 6.4 Effect of L/B and β to drag 321
Fig. 6.5 Effect of longitudinal position of CB to drag 322
Fig. 6.6 Running trimming angle versus speed 322
Fig. 6.7 Residual drag coefficient of Deep Vee and round bilge craft 323
Fig. 6.8 Body plan of round bilge craft 902 and Deep Vee craft 904 326
Fig. 6.9 Drag increments of 2400 t Deep Vee ship and round bilge ship in waves. Dash is round bilge. Line is Vee 326
Fig. 6.10 Slamming of Deep "Vee" 2400t vessel compared with round bilge. (Ref. fig. 4 in Kehoe alt ref. 72) 329

Fig. 6.11 Maximum speed for 2400t frigate with Deep Vee and round bilge against sea states. (Fig. 7 in Kehoe al ref. 72) 329
Fig. 6.12 Comparison of slamming frequency of 1700 t frigate with Deep Vee and round bilge 330
Fig. 6.13 Frequency of wetted deck for Deep Vee and round bilge in various sea states 330
Fig. 6.14 Emergence of propeller at different sea states for Deep Vee and round bilge 331
Fig. 6.15 Rolling and transverse acceleration in sea state for 1000t Deep V and round bilge 332
Fig. 6.16 Body plan with SSB 337
Fig. 6.17 Vertical acceleration of craft with different SSB in regular waves 337
Fig. 6.18 Arrangement of SSB with fins 338
Fig. 6.19 Lines of EPC, revised EPC at bow (TUD), and Axe. (From Keuning et al. [12] Fig. 2) 339
Fig. 6.20 Probability distribution of bow vertical accelerations of three ship types. (From Keuning et al. ref. [12] Fig. 3) 340
Fig. 6.21 Lines of Axe ship. (From Keuning et al. ref. [12] Fig. 12) 340
Fig. 6.22 Profiles of Axe ships—Damen FCS 7011 and 4008 340
Fig. 7.1 PGH-2 *Tucumcari* 346
Fig. 7.2 Foil cross-section diagrams and annotation 347
Fig. 7.3 Forces and moments acting on hydrofoils when rolling and turning. (**a**) and (**b**), rolling moments. (**c**) Forces on a submerged hydrofoil in a turn. (**d**) Tucumcari during a coordinated turn 348
Fig. 7.4 PHM foil arrangement 349
Fig. 7.5 *Tucumcari* with retracted foils 350
Fig. 7.6 Early ladder foil craft—Bell Hydrodome HD-4 350
Fig. 7.7 (**a**) Photo of Aegean Prince. A Kolkhida craft in service from Rhodes to Kos with sister Flying Poseidon. (**b**) Kolkhida general arrangement 351
Fig. 7.8 Foil plan configurations 352
Fig. 7.9 Meteor 356
Fig. 7.10 250-passenger Cyclone 358
Fig. 7.11 (**a**) Turya class vessel at speed. (**b**) Matka 359
Fig. 7.12 Mukha 360
Fig. 7.13 PT 20 Frechia del sole 361
Fig. 7.14 (**a**) PT 50 front foil system. (**b**) PT 50 at speed 362
Fig. 7.15 (**a**) Rodriquez RHS-150. (**b**) Rodriquez RHS 200 364
Fig. 7.16 FSH 38 365
Fig. 7.17 PCH-1 *High Point* at speed and (a) launching a Harpoon missile 366

Fig. 7.18 AGEH-1 *Plainview* 367
Fig. 7.19 PGH-1 Flagstaff 368
Fig. 7.20 PHM fleet at speed 369
Fig. 7.21 PHM power plant 370
Fig. 7.22 Jetfoil Power plant layout 371
Fig. 7.23 Jetfoil 929 at speed 372
Fig. 7.24 Chinese Hu Chwan bow hydrofoil patrol boat 372
Fig. 7.25 Chinese Chang Jiang Hydrofoil (Russian Raketa design built under licence) 373
Fig. 7.27 Chinese river hydrofoil ferry *Mountain Ge Lu* 374
Fig. 7.26 PS-30 Far East Hydrofoil 374
Fig. 7.28 GA of FHE-400 376
Fig. 7.29 FHE-400 *Bras d'Or* 377

Fig. 8.1 Geometrical profile and nomenclature of wing section 391
Fig. 8.2 Coordinate system of two-dimensional thin wing 391
Fig. 8.3 The normal direction of any point on wing surface 392
Fig. 8.4 Separation of boundary conditions for a thin wing 393
Fig. 8.5 Vortex model for a thin wing 395
Fig. 8.6 Geometric relations relating to circulation around a wing section 398
Fig. 8.7 Nomenclature for plate wing 401
Fig. 8.8 Wing with camber line profile 401
Fig. 8.9 Camber line wing—zero lift angle diagram 402
Fig. 8.10 Flow nearby the surface of limited span wing 405
Fig. 8.11 Lifting line "Π" model 406
Fig. 8.12 Downwash caused by free vortices 407
Fig. 8.13 Correction coefficients for rectangular wing 410
Fig. 8.14 Value of τ for the trapezium profile wing, as shown in Fig. 8.15 410
Fig. 8.15 Value of δ' for trapezium profile wing 411
Fig. 8.16 Lift coefficient curves 411
Fig. 8.17 Simplified "Π" vortex model 413
Fig. 8.18 Wing tip fences 413
Fig. 8.19 $K_e = f(h_e/l)$ curve 414
Fig. 8.20 Free vortex series on swept back wing 415
Fig. 8.21 Lift coefficient slope of swept back wing versus aspect ratio 416
Fig. 8.22 Vortex lines distribution of wings with typical plan profiles 416
Fig. 8.23 Vortex geometry for swept back wing 417
Fig. 8.24 Hydrofoil "dihedral" 419
Fig. 8.25a NACA64-208 wing section characteristics lift 424
Fig. 8.25b NACA64-208 wing section characteristics drag 425
Fig. 8.26 E836 wing profile and its cavitation characteristics 428

Fig. 8.27 E838 wing profile and its cavitation characteristics 428
Fig. 8.28 E874 wing profile and its cavitation characteristics 430
Fig. 8.29 YS-920 cavitation characteristics compare with NACA 16 and 66 431
Fig. 8.30 RAF-6 wing profile 431
Fig. 8.31 Coordinates of wing profile chord 432

Fig. 9.1 Hydrofoil wave making close to free surface 438
Fig. 9.2 Lift versus submerged depth of foil at high F_n 438
Fig. 9.3 Pressure distribution at the lower/upper surface of a hydrofoil at high $F_{n.}$ (The foil has a camber profile and cross section with 7.5% thickness) 439
Fig. 9.4 Mirror image vortex of hydrofoil under the free surface ... 441
Fig. 9.5 Simplified mirror vortex model of free water surface 442
Fig. 9.6 velocity induced by attached vortex 443
Fig. 9.7 Velocity induced by attached vortex 443
Fig. 9.8 Relationship between various vortices 444
Fig. 9.9 Effect of free water surface to lift coefficient 450
Fig. 9.10 Additional drag coefficient K_I induced by free water surface 450
Fig. 9.11 Transverse flow velocity on foil and induced velocity on foil surface induced by its mirror image free vortex 453
Fig. 9.12 Nomenclature for calculation of K_e 454
Fig. 9.13 Downwash correction factor $\psi = f\left(\frac{h}{l}, \frac{a}{l}\right)$ for hydrofoil with struts 454
Fig. 9.14 Coefficient $m = f\left(\overline{h}\right)$ 455
Fig. 9.15 $\gamma = f\left(\overline{h}\right)$ from 9.33 457
Fig. 9.16 $\zeta = f\left(\frac{\overline{h}}{A}\right)$ from 9.35 457
Fig. 9.17 average submerged depth of the surface piercing foil 458
Fig. 9.18 Geometry and test results for trapezoid hydrofoil. (**a**) geometry diagrams, (**b**) $C_L = f(\alpha)$, (**c**) $K = f(C_L)$ 460
Fig. 9.19 Geometry and test results for rectangular hydrofoil. (**a**) geometry diagrams, (**b**) $C_L = f(\alpha)$, (**c**) $K = f(C_L)$ 461
Fig. 9.20 Cavitation curves (**a**) vs angle of attack (**b**) vs lift coefficient and speed 465
Fig. 9.21 Camber section with a small downward section at leading edge 467
Fig. 9.22 Cavitation flow sketch 470
Fig. 9.23 Simplified flow field of fully cavitating hydrofoil 470
Fig. 9.24 various models for two dimensional cavitation flows 471
Fig. 9.25 Kirchhoff's flow model around a vertical flat plate 474
Fig. 9.26 separated flow around the inclined flat plate from Rayleigh 475
Fig. 9.27 circular curve foil diagram 477

Fig. 9.28 definition diagram for α_n 479
Fig. 9.29 Diagram for definition of Betz's method 480
Fig. 9.30 Various super-cavitating foil profiles at $\sigma = 0$ 489
Fig. 9.31 Section profile of super-cavitating hydrofoil and its agitating means (**a**) agitating thread; (**b**) agitating wedge; (**c**) air ventilation hole 491

Fig. 10.1 Relation between hydrofoil unit loading and craft takeoff and service speed 496
Fig. 10.2 Example relationship of vertical bending moment of hydrofoil craft operating on calm water with changing position of foils 497
Fig. 10.3 Relationship of relative thickness of foil and sweep angle with speed for avoiding cavitation 499
Fig. 10.4 Takeoff aid additional foils 501
Fig. 10.5 Lateral profile of strut 501
Fig. 10.6 Installation of struts and interceptor fences on the strut of surface piercing hydrofoils 502
Fig. 10.7 Running attitude of hydrofoil craft at foil borne operation 504
Fig. 10.8 $m_p = f(C_L)$, where the blue region shows the range of m_p which may vary up to 10% at any given C_L 508
Fig. 10.9 K value for local protrusion drag—welds and surface imperfections 510
Fig. 10.10 K value for local protrusion drag—bolts and rivets 511
Fig. 10.11 Interference drag coefficient of strut and foil interface 514
Fig. 10.12 Longitudinal balance of foil loading 519
Fig. 10.13 Relationship of stiffness with submerged depth 521
Fig. 10.14 Aquavion general arrangement 523
Fig. 10.15 Transverse stability of shallow submerged flat foil 525
Fig. 10.16 Raketa bow foils with takeoff aids and planing stabilizer foils 527
Fig. 10.17 Transverse stability of surface piercing foil 527
Fig. 10.18 Forces acting on the rudders during turning of hydrofoil craft 530
Fig. 10.19 Foil velocity vectors in a turn 531
Fig. 10.20 Operation of flaps for coordinated turning 532
Fig. 10.21 Coordinated turning of PHM craft 532
Fig. 10.22 Motion response of hydrofoil craft in waves 533
Fig. 10.23 Influence of vertical velocity 535
Fig. 10.24 Velocity caused by wave circular oscillatory motion 535
Fig. 10.25 Hydrofoil craft running in head and following seas 537
Fig. 10.26 Main and bow foils of hydrofoil craft Bras D'or 539
Fig. 10.27 Platforming motion range for various craft displacements 541

Fig. 10.28a Sea state in North Atlantic Ocean 542
Fig. 10.28b Average annual wave height occurrence for the sea regions around Russia 542
Fig. 10.29 Vertical acceleration of hydrofoil craft at CG 543
Fig. 10.30 Comparison of comfort between hydrofoil catamaran with automated ride control system with conventional catamarans in various sea states 544
Fig. 10.31 Vertical acceleration a_g/g versus wave height for the hydrofoil *Dolphin* 544
Fig. 10.32 Influence of wavelength on vertical acceleration according to research by Supramar for the PT-50 vessel (LBP = 20 m, *V* 46 knots, in regular waves) 545
Fig. 10.33 Comparison of vertical acceleration for various craft 546
Fig. 10.34 Drag increment of hydrofoil craft in waves versus craft speed 547
Fig. 10.35 Comparison of speed loss in waves for various craft 547
Fig. 10.36 Speed loss in waves of various high-speed craft 548
Fig. 10.37 Power requirements for various craft in waves 548
Fig. 10.38 Various lift control types (and required powers) 552
Fig. 10.39 Logic block chart for automated controls system 554
Fig. 10.40 Control system components and arrangement of *Jetfoil* 555
Fig. 10.41 Foil arrangement of hydrofoil craft type *Volga* 558
Fig. 10.42 Hydrofoil arrangement of Russian hydrofoil *Meteor* 559
Fig. 10.43 General arrangement of PT-50 MK-II 560
Fig. 10.44 General arrangement of RHS-150 560
Fig. 10.45 Foil arrangement of hydrofoil craft AGEH-1 "Plainview" 561
Fig. 10.46 Foil arrangement of hydrofoil craft PGH-1 *Flagstaff* 561
Fig. 10.47 PCH-1 "Highpoint" 563
Fig. 10.48 PGH-2 "Tucumcari" foil arrangement 564

Fig. 11.1 (**a**) Superfoil-40 at speed (**b**) Superfoil-40 bow view afloat with raised bow foils (**c**) Geometry of Superfoil-40 foil system deployed and raised (**d**) Stern view of one demi hull showing one fixed and one steerable propulsion waterjet 569
Fig. 11.2 FACAT 40 general arrangement 570
Fig. 11.3 FACAT 40 performance 572
Fig. 11.4 (**a**) FACAT 40 vertical acceleration performance in a seaway (**b**) FACAT 40 motion comfort limits 573
Fig. 11.5 FACAT 40 foil control system 574
Fig. 11.6 The HYSUCAT foil assisted catamaran (**a**) The principle (**b**) Typical general arrangement (**c**) Nordblitz 115 passenger fast ferry 576
Fig. 11.7 Kværner Fjellstrand Foilcat FC40 profile 577

Fig. 11.8 Foilcat foil arrangement 578
Fig. 11.9 Foilcat propulsion system 578
Fig. 11.10 Foilcat resistance components from hull and foils with speed 579
Fig. 11.11 Foilcat instrumentation to allow active automated ride control 580
Fig. 11.12 Turbojet Hong Kong ferries Fjellstrand foilcat at speed 580
Fig. 11.13 Mitsubishi Super-Shuttle 400 "Rainbow" 581
Fig. 11.14 Daea Express Shipping Super Shuttle 400 "Dream" 582
Fig. 11.15 Westamarin Foilcat FC2900 general arrangement 582
Fig. 11.16 Foil and propulsor details Westamarin FC2900 583
Fig. 11.17 Westamarin 2900 in service 583
Fig. 11.18 Motions response data for Catamaran and Westamarin FC2900 584

Fig. 13.1 Suggested Bx/Tx and IE 595
Fig. 13.2 Suggested Cp, Cx, TA, and TW 596
Fig. 13.3 C_R Comparative plot 608
Fig. 13.4 RT/Δ Comparative plot 608

List of Tables

Table 2.1 Summary of the range of test parameters used in the model test programmes reviewed in this chapter 40
Table 2.2 Characteristics of the prototype BAZAN 80 models 56
Table 2.3 Typical vessel data 60
Table 2.4 NSMB model testing parameters 62
Table 2.5 Residuary resistance comparison normalized on model 2 62
Table 2.6 AMECRC model parameters 65
Table 2.7a VTT Series main dimensions (design draft) 66
Table 2.7b VTT Series main coefficients (design draft) 66
Table 2.8 Coefficients A_i in (2.7) versus volumetric Froude number 72
Table 2.9 Regression coefficients of relation (2.14) 79
Table 2.10 Regression coefficients of relation (2.15) 80
Table 2.11 Taylor's suggested Cp values (in %) for $B/T = 3.75$ 85

Table 3.1 Transverse flow resistance coefficient with deadrise angle 145

Table 4.1 Hard Chine Model Test Series key parameters 180
Table 4.2 Particulars of the DTMB Model Series 62 182
Table 4.3 Summary parameters for NAHL Model Series 188
Table 4.4 Coefficients for NAHL regression equation 189
Table 4.5 USCG test series model parameters 191
Table 4.6 TUNS Model series 194
Table 4.7 Southampton University Fast Planing Series 197
Table 4.8 Naples Warped chine model series summary 202
Table 4.9 DSDS Model data 206

Table 5.1 Application range from Fridsma 295
Table 5.2 Correction factor E for C_Δ and L/B 298

Table 6.1 Increment of drag of NHP48 vs PR72 in calm water 319
Table 6.2 Drag results of Deep Vee craft in calm water(model test results with round bilge model 902 as reference) 324
Table 6.3 Motion response of Deep Vee craft and round bilge craft in waves $\Delta = 519.4\text{t}$, $H_{1/3} = 2.4\text{m}$, $T_1 = 6\text{s}$ 325

Table 6.4 Comparison of drag increments of Deep Vee craft and round bilge craft in waves ... 328
Table 6.5 Significant values of single amplitude of motion response per unit wave height ($\Delta = 620\text{t}, T_1 = 6\text{s}$) ... 333
Table 6.6 Maneuverability comparison of Deep Vee and round bilge craft ... 335

Table 7.1 Key data for early commercial Russian and Supramar PT craft ... 379
Table 7.2 Leading particulars for later commercial hydrofoils ... 381
Table 7.3 Example commercial hydrofoils ... 382
Table 7.4 Example military hydrofoils ... 384
Table 7.5 Example key data for some Russian military hydrofoils ... 386

Table 8.1 Properties of the NACA 64-208 wing section ... 423
Table 8.2 Offsets of E836 and E838 wing profiles ... 427
Table 8.3 Offsets of E874 and YS-920 wing profile ... 429
Table 8.4 offsets of RAF-6 airfoil section ... 431
Table 8.5 Offset of symmetric profile RAE101 ... 431
Table 8.6 Gottingen wing profile ... 432

Table 9.1 Vapor pressure versus water temperature ... 462

Table 10.1 Automatic control systems in several hydrofoil craft of US ... 513

About the Authors

Liang Yun has more than 40 years' experience at the Marine Design and Research Institute of China, Shanghai (MARIC). He graduated from the Shipbuilding Engineering Faculty of Da-Lian Polytechnic University in 1953 and completed a postgraduate diploma at the Military Engineering Academy of China in 1955. He has been involved in ACV development in China since the first prototypes were constructed in Harbin in the 1950s. He was involved in the design and prototype construction of WIG craft in the 1990s and the development of high-speed catamarans and air cavity vessels through the millennium. He was director of the HPMV division, MARIC, from 1983 to 1987, and Deputy Chief Naval Architect of MARIC from 1980 to 1997. He has been a Guest Professor supporting HPMV postgraduate students at Harbin Engineering University and Wu Han Water Transportation University in the early 1990s. Prof. Yun has been Chairman of HPMV Design subcommittee of the China Society of Naval Architecture and Marine Engineering, CSNAME, over the last 20 years, and Vice Chairman of the organizing committee of the annual International HPMV Conference, Shanghai, China, since 1996. He continued to play an active role in promoting and developing HPMV technology in China through his association with the industry and Chinese Universities. Professor Yun has partnered with Alan Bliault on five textbooks covering ACV, WIG, and HPMV technology before the current volume on monohull and hydrofoil vessels. Liang Yun died in Shanghai on December 26, 2020.

Zu-shun Dong graduated from Naval Engineering University (NEU), Peoples Liberation Army (PLA), China, in 1959. After graduation, he was appointed as an Assistant Professor, Associate Professor, and Professor in the NEU, working on the teaching, research, and design of hydrodynamics of High-Performance Marine Vessels (HPMV), such as high-speed monohull craft, hydrofoils, air cushion craft, wing in ground effect craft, hydroplanes, catamarans, and multihulls. He was appointed as a member of the Hydrodynamics Committee of the General Armament Department (GAD), PLA, from 1987 to 2007, and the first specialist on hydrodynamics of the HPMV Committee of GAD, PLA. He was a Professor at Harbin Industrial University from 2011 to 2014 and Qian-Dao Science and Technology University from 2010 to 2015. He was a technical consultant to the Special Airplane Research Institute from 1995 to 1996; a member of the Academic Committee on Key Hydrodynamics Laboratory of National Defense Technology from 2001 to 2009; and a member of the Academic Committee of National Key Laboratory on Multihull in Harbin Engineering University, from 2009 to 2012. He has also been the Editor of *China Shipbuilding* and Deputy Chief Director of the Yacht Design and Production Academic Committee of the China Society of Naval Architects and Marine Engineers.

Raju Datla is a Research Associate Professor in Naval and Ocean Engineering at Stevens Institute of Technology. He manages the experimental marine hydrodynamics research at the Davidson Laboratory. His expertise is in the areas of high-speed craft hydrodynamics and ocean engineering, in which he has so far prepared more than 50 technical articles. He teaches courses to undergraduate and graduate students in naval engineering and ocean engineering at Stevens. He has served as principal advisor to six Ph.D. dissertations and more than 50 master's and undergraduate theses and design projects so far. He received his Ph.D. in Ocean Engineering from Stevens Institute of Technology. He is a fellow of the Society of Naval Architects and Marine Engineers. Dr. Datla has previously participated in editing a compendium of the first 20 years of papers presented to the International HPMV Conference held regularly in Shanghai by CSNAME, which the RINA also supports.[2] During the 2000s, he worked on hydrodynamic planing research together with Daniel Savitsky at the Davidson Laboratory, publishing original papers on the spray boundary in partnership with him.

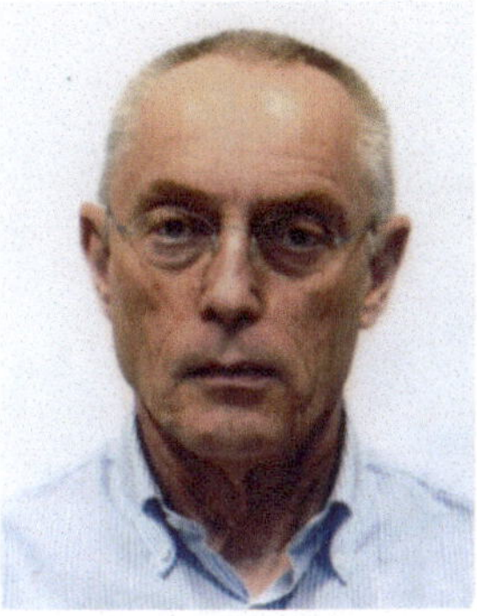

Alan Bliault Naval Architect and Offshore Engineer, Alan Bliault is a Fellow of the Royal Institution of Naval Architects and graduated from the University of Newcastle upon Tyne. His initial career was at Vosper Thornycroft, working with the design and operation of hovercraft and air cushion platforms. Subsequently, he worked in the offshore industry, developing new offshore loading systems. He was responsible for hydrodynamic design for the Conoco Hutton field Tension Leg Platform in the UK and subsequently at Norske Shell for Draugen Platform substructure mechanical outfitting, hydrodynamics, tow-out, and installation in Haltenbanken. In the mid-1990s, he led the development of new API and ISO standards for subsea flexible flowlines and risers based in Holland. Throughout the millennium, he led Shell International's development of the Floating LNG production system for remote gas fields. Since then, he has held various project management, construction, and R&D roles. He has worked as a senior auditor in Shell's central Internal Audit group evaluating risk and management controls at major projects and operating companies worldwide from 2013 to 2016. He has maintained a close interest in high-speed marine craft throughout his career, leading to his partnership with Liang Yun on maritime engineering textbooks, including the present volume on High-Speed Monohull Vessels and Hydrofoils. This work has, since the late 1980s, been a continuous research project alongside his "day job"!

[2] 'Performance, Technology and Application of High-Performance Marine Vessels', Volume One, Cambridge Scholars Publishing, 2018, edited by L Yun, R Datla, X Yang, ISBN 978-1-5275-0356-4

Part I
High Speed Mono-hull Craft

Chapter 1
Introduction to High-Speed Monohull Craft

1.1 Introduction

The description "high speed" attached to a vessel has evolved over time. In earlier times, rowed triremes were fast craft traveling at between four and perhaps ten knots or more when wind assisted with a large sail. These craft were what might be referred to today as high length to beam ratio or slender craft, and the sharp bow ram also formed a relatively slender bow entrance. In more recent times, while still wind powered, the clipper and schooner hull forms were also relatively slender, with a fine bow entry and similarly fine stern geometry. These vessels raised the speed envelope to the upper teens of knots.

On a smaller scale, there are many examples of very slender craft throughout history, though the main intent of these was to minimize drag rather than a focus on extreme speed.

To run a craft at high speed, minimizing drag is clearly important, whether the motive power is wind or, as came on the scene in the nineteenth century, the steam engine. The introduction of mechanical motive power demanded a means to drive a vessel forward from a rotating shaft, and so after experiments with paddle wheels, the screw propeller proved to be more convenient to install, driven by a shaft from a mid-mounted engine and boiler, and less affected by ship rolling motions.

Steam engines were refined and became rapidly more efficient and powerful, and once boilers were also able to be fired with liquid fuel rather than coal, another boost was given to vessel speed as well as size. It was in this era of the late nineteenth century that inventors and indeed shipbuilders began to study how to design vessels with much higher operating speed. A significant breakthrough came in 1894 when Parsons designed the *Turbinia* Fig. 1.1, a slender craft specifically aimed at demonstrating extreme speed to the British Navy [1]. This vessel was powered by a steam turbine rather than a reciprocating engine and achieved a speed of 34 knots, unheard of in those days.

Z.-S. Dong et al., *High-Speed Monohull and Hydrofoil Craft*,
https://doi.org/10.1007/978-3-031-62760-6_1

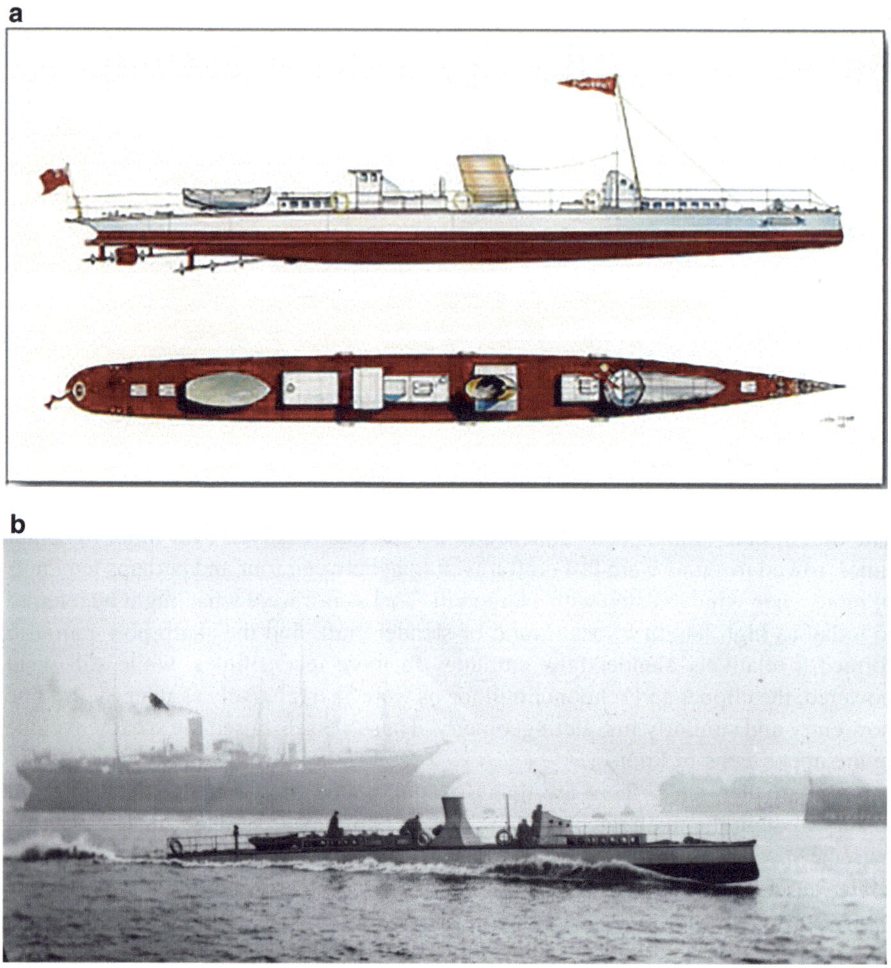

Fig. 1.1 (**a**) *Turbinia* general arrangement. (**b**) *Turbinia* at speed

The Turbinia had a hull with a rounded cross section (round bilge configuration), and while it would have experienced dynamic lift forces at top speed, it was not intended to ride on these.

Prior to this, many naval architects and inventors had looked at the possibility of using dynamic forces to lift a boat hull out of the water and skim it like a flat stone skimmed over a pond with much reduced drag. One of these inventors, Rev. C. M. Ramus [2, 3], had the idea of a hull with a pair of angled flat surfaces forming a stepped hydroplane that could be launched with rocket power as a means of sinking an enemy ship two miles away. Prototype models of this craft were tested and

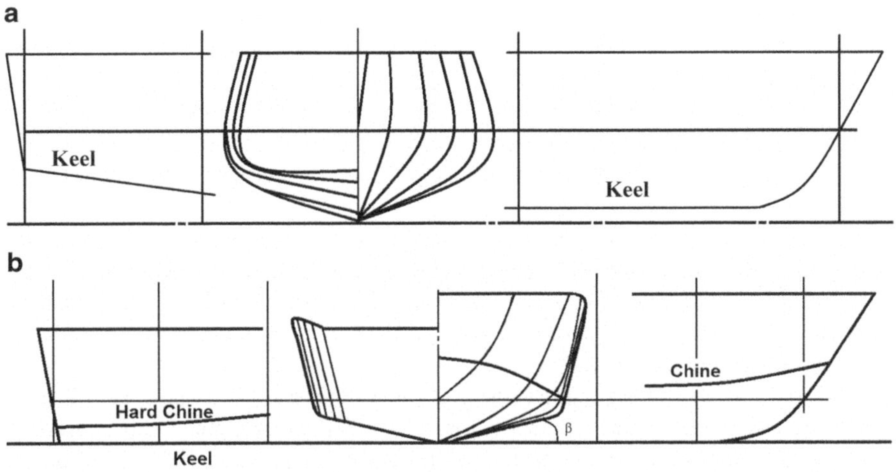

Fig. 1.2 (**a**) Planning craft body plan and profile. (**b**) Semi planning craft body plan and profile

reached speeds in excess of 30 miles per hour. Later, Froude carried out a series of model tests in his tank in Torquay to investigate the proposal. Due to the rounded corner of the stern, the drag was found to be higher than a simple angled planing surface with a sharp transom, so the Royal Navy was not interested in the concept. Morabito [4] carried out a re-analysis including model tests that confirmed Froude's original findings.

Once smaller steam engines became available and later internal combustion engines, faster boats could be built, which interested inventors and people with sufficient personal wealth to wish to own one and compete against others in races [1, chap 1]. Therefore, toward the later part of the nineteenth century, there arose two drivers for high-speed craft—private leisure and naval operations. The latter were craft that could outrun larger vessels and attempt to sink them or alternatively to board and take control if the target was a commercial "pirate."

The competition for faster speed affected the hull shape adopted. For small craft, the use of a flat or slightly canted bottom in cross section and near vertical sides leading down to a small bilge radius or even a sharp corner was convenient for construction so long as the waves to be faced were small, Fig. 1.2a. For close inshore, river, or lake vessels, this was not a problem, and racing or leisure does not naturally fit with extreme conditions.

Naval vessels needed to be able to face open seas as well as travel faster than commercial shipping. It was necessary to pass through waves rather than to skate across their surface with an attendant bumpy ride, and this led to the hull form that has a fine bow entry, rounded bilges, and canted almost flat bottom surfaces aft of amidships Fig. 1.2b. The after-hull form generates lift at high speed, while the forward geometry can slice through waves in a seaway.

Competition boats from early in the twentieth century began to take varied forms based on the idea of the planing surface, including hydroplanes with two or three planing sponsons [1].

The monohull form also developed with much experimentation of the canted lower surface angle (deadrise), the use of steps cut across the planing surface, strakes along the surfaces to guide flow and improve lift, and the curves of the hull along its length.

Most of these innovations have been aimed at reducing drag while also minimizing motions and accelerations in a seaway. Such a combination could allow offshore races to be won with improved comfort for the crew. For high-speed small military craft, this also meant that operations could be completed as fast as possible with minimum fatigue to the crew.

Since the early twentieth century, propulsion systems have become available with increasing power rating and efficiency, lower installed weight, and smaller physical envelope. Initially, this has been through the development of the high-speed diesel engine, then the gas turbine as motive power. Delivery of thrust through high-speed propellers has been refined, and where boat speeds higher than 40 knots or so are involved, the use of supercavitating or ventilated propellers has been adopted. Since the late twentieth century, waterjet design has advanced to the point where for commercial vessels such as high-speed ferries, they are the base case due to their higher efficiency than propellers in the speed range of 30–60 knots.

Turning back to the planing plate, airboats, as used in the Florida Everglades, are simple flat-bottomed punts that plane across the shallow waters powered by air propellers to avoid underwater appendages that might snag on roots, etc. in that swampy environment, Fig. 1.3 [1].

So, what do we mean by a "high-speed" monohull craft that we will study in these chapters? One can perhaps see from this introduction that there are several different directions that the hull configuration may follow, depending on the mission requirements—primarily the service speed and sea state, though also influenced by desired endurance. Clearly, a vessel aimed at river, lake, or swampy glade operation can take different shape to one needing to face the open sea.

In terms of speed, we consider here vessels that will experience a significant dynamic lift rather than being simply displacement vessels. Dynamic forces exerted by water on a shaped hull begin to be significant when speed is above 25 knots. The actual force depends on the geometry and the angle of attack. Let us look at the alternatives.

1.2 Monohull Configurations

Conventional ships float on the water surface with the aid of static buoyancy to support the weight of the ship. For these craft, when ship speed V increases, the wave-making drag of the ship will increase at a rate proportional to its displacement and to V^2, while the power absorbed will be proportional to V^3. Wavemaking drag is

Fig. 1.3 Everglades airboat

minimized if a vessel has a high length-to-beam ratio, so one way to design a high-speed vessel is to use this geometry, as referenced above.

A long, thin vessel will have a substantial wetted surface and, therefore, drag from friction with the water flow. To enhance speed by reducing this immersed area, one has to reduce static buoyancy and support the vessel weight by other means. Some quite radical possibilities are the air cushion, hydrofoil, and air foil lift, so as to form air cushion craft (ACV), hydrofoil craft, and wing-in-ground effect (WIG) craft, respectively [1, 5, 6].

We are concerned in Chaps. 1–6 of this book with the least complex approach—that of hydrodynamic lift applied to a monohull vessel underside shaped appropriately. In Chaps. 7–11 we will study the hydrofoil configuration.

The high-speed monohull can be considered in the following categories:

- Displacement craft (slender hull, high L/B)
- Semi-displacement craft (with partial dynamic lift)
- Planing craft
  - Flat punt (airboats)
  - Low deadrise with hard chine
  - Stepped planing hull
  - High deadrise hull (Deep Vee craft)

It should be noted that a planing craft traveling at low speed will operate as a displacement craft. As it accelerates, the dynamic lifting forces will increase until, depending on the precise geometry, most of the weight will be supported. There will always be a small proportion of the weight supported by buoyancy since a 'wedge' has to protrude into the water to create the wetted lifting surface. Full planing is usually achieved at Froude Number $Fn_L = 1.0$ and higher or $Fn_\nabla = 2.5$ or higher.[1]

1. *Displacement Craft*:

A vessel with fine lines, that is, a bow with a sharp entry, and a stern form that is also fine, will have minimum self-induced drag via wavemaking. With careful design of the curved shape, the waves generated at the bow can be designed to partially cancel those generated in the stern part of the vessel at the intended operating speed. The use of a bulbous forefoot can enhance this.

The higher the length-to-breadth ratio, the greater the immersed surface area, and so viscous drag will increase, offsetting some of the gains made in reduced wavemaking drag. Wavemaking or inertial drag is the predominant force at low speed, while friction forces, being proportional to V^2, rapidly increase as a craft accelerates.

In addition, while stability and, therefore, stiffness in pitch will increase with greater length, a very slender hull will have reduced transverse stability.

Depending on the proportions, and particularly the hull cross section aft of amidships and the lines to the stern, whether a transom or a canoe stern, the hull may generate dynamic lifting forces as Fn_L approaches 1.0. These lifting forces will affect the drag.

If configured favorably, the dynamic forces may help to minimize drag. If they are significant, these forces may also affect the vessel trim and particularly the transverse stability. If we are considering a vessel where the dynamic forces are rather small, this should not be an issue. So, stability assessed for the static vessel should be accurate.

It may be noted that a slender vessel with fine lines at the bow will slice through sea waves and so will have low pitch, heave motions, and added drag. Very slender vessels may be rather tender in roll when in oblique seas since the righting moment will be lower than a fatter vessel.

2. *Semi-displacement Craft*

For craft running at medium speed, $Fn_\nabla = 1 - 2$, i.e., running speed where a small portion of hull weight will be supported by buoyancy, the hull configuration will be similar to that shown in Fig. 1.2b, i.e., with a fine round bilge body plan forward, and with canted flat surfaces toward the stern, as well as high length/beam ratio.

[1] Froude Number based on vessel length $(V/(gL)^{0.5})$ or Froude Number based on vessel displaced volume $(V/(g\nabla^{1/3})^{0.5})$.

Fig. 1.4 Monohull LCS example. (Courtesy U.S. Navy photo by Mass Communication Specialist 2nd Class Nathan Laird/Released) from Wikimedia commons. Note: ATLANTIC OCEAN (Sept. 28, 2009) The littoral combat ship USS Freedom (LCS 1) conducts flight deck certification with an MH-60S Sea Hawk helicopter assigned to the Sea Knights of Helicopter Sea Combat Squadron (HSC) 22)

By giving the craft high length/beam ratio, round bilge form ahead of amidships, and fine bow entry, slamming in waves will be minimized and seaworthiness maximized. Such craft can be developed for displacements up to several thousand tons based on gas turbine power plants currently available, as applied for the monohull version of the US Navy LCS (littoral combat ship) Fig. 1.4.

In the 1960s the David Taylor Model Basin in the USA developed three hull cross-section shapes, i.e., U-shape, V-shape, and intermediate type, as shown in Fig. 1.5 [7] in a research series designated "series 64" that was tested in its high-speed model test tank to evaluate resistance and seakeeping and compare their attributes. The U-shaped hull featured an almost flat lower surface to substantially forward amidships with a flat keel line. The V-shaped body plan had a deep vee to the hull underside that extended toward the stern, while the intermediate shape had a faster-reducing deadrise to flat at the stern.

This series was followed by a sequence of model testing programs in the UK [8] and other countries investigating the effect of geometric parameters on resistance and seakeeping. We will discuss the results of these investigations in Chapter 2.

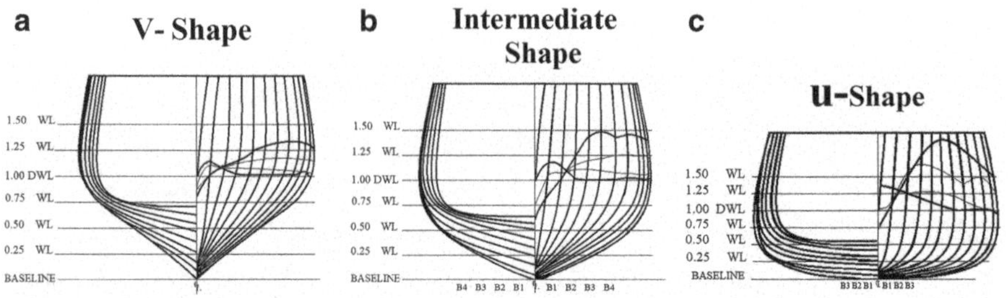

Fig. 1.5 Series 64, developed by DTMB in USA early 1960s

Fig. 1.6 A typical planing hull profile, with hard chine—wireframe model

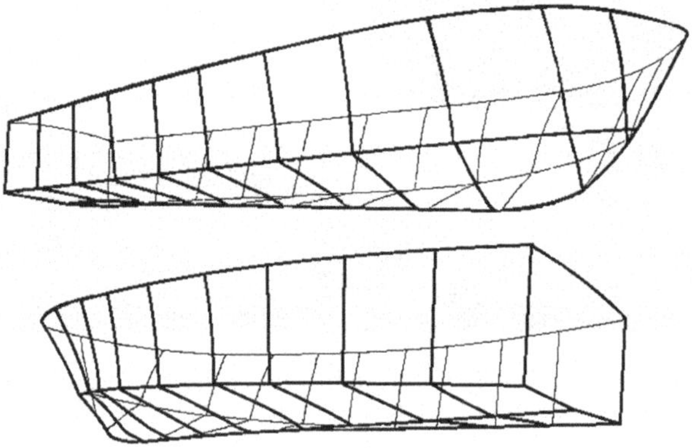

3. *Planing Craft*

To lift a high-speed hull out of the water, dynamic forces are required. A planing craft therefore needs a flat underside to the hull. If the surface has an "angle of attack" against the oncoming water flow, there will be a lift vector and drag vector of the generated force tangential to the surface. Since water is dense, the lift force is substantial.

If the hull is literally a punt shape as in Fig. 1.3, then the lift force per surface area is maximized. On the other hand, if a turning moment is applied—in the case of an airboat by an air rudder—the boat will slide out of the turn until sufficient thrust is applied to counterbalance the centrifugal force. In a seaway, the sliding can be dangerous and so seagoing craft have a form such as that shown in the wireframe model, Fig. 1.6 [9].

One can see that the hull has undersides that cant upwards from the keel line at an angle referred to as the deadrise angle. This may remain steady, forming a prismatic surface, or gradually aft of the craft amidships, forming a "warped" planing surface. The advantage of a warped surface is that hydrodynamic lift is higher and drag is lower at a lower average deadrise angle.

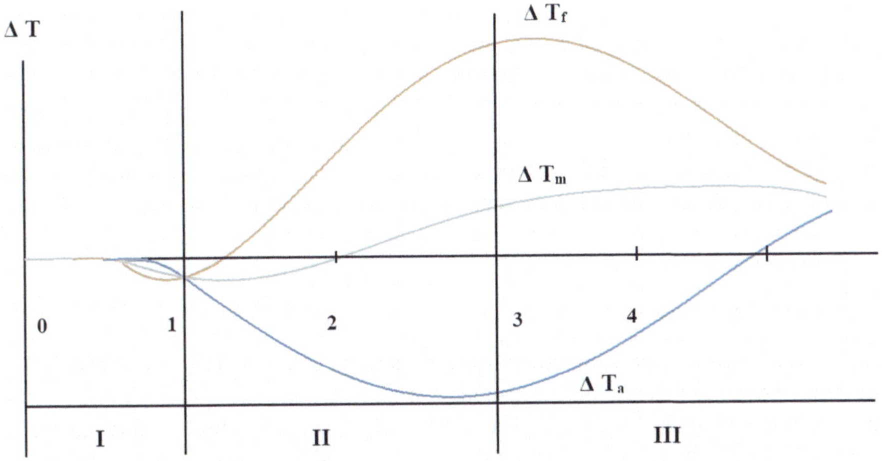

Fig. 1.7 Running attitude and drafts at bow and stern change with volumetric Froude Number, where I denotes displacement mode; II transitional mode; III planing mode

This latter is the form commonly adopted in the mid-twentieth century, having a deep forefoot and reducing deadrise toward the stern. Longitudinally, the form may be slightly convex (referred to as a positive "rocker") along the whole length, flat from the middle area, or even slightly concave if the deadrise is flattened out completely at the stern. The stern is always a "transom" or vertical cutoff.

In order to avoid water wetting the hull sides and minimize friction drag, a hard chine (sharp corner) is incorporated at the base of the hull sides. This encourages the water to separate from the side hull and form spray rather than wetting the hull side area below the running waterline.

The deadrise angle at the forward part of the craft is increased to give a fine entry, which will improve drag at low speeds and lower the drag hump experienced prior to achieving full planing. It will also improve the seakeeping quality.

Since the wetted length changes with speed for planing hulls, using a length-based Froude number is not practical, and a displacement-based volumetric Froude number, i.e., Fn_∇, is typically used to characterize the speed, where ∇ represents the volume displacement, normally in m^3. Figure 1.7 shows an example of the change in draft, ΔT, at the forward, mid, and aft locations of a typical planing craft with speed represented by Fn_∇. From the figure, one can see the draft stays unchanged at bow, stern, and amidships while at $Fn_\nabla < 1$, the displacement mode I. The craft trims as its self-generated waveform increases, such that somewhere between $Fn_\nabla = 2$ and 3, the wavelength is twice the craft length, and so maximum trim is reached in mode II. At the same time, dynamic forces increase sufficiently to lift the hull relative to the water surface shape. As speed is increased beyond $Fn_\nabla = 3$, in mode III, the wavelength continues to increase, and as the dynamic forces further lift the hull, the generated wave height will decay. The result is that trim diminishes, as can be seen on the right side of the figure.

Figure 1.8a shows a typical modern planing patrol craft with a low deadrise in the after section, somewhat like the wireframe in Fig. 1.6, the US Navy Mk VI. Figure 1.8b shows a faster design, the 17 m Sentinel-class patrol vessel.

The features of these two craft are high deadrise angle at the bow for improving both takeoff and seakeeping, and lower deadrise toward the stern. They have a hard chine at the intersection of the bottom and sides for clean separation of spray, and the deadrise angle from amidships to stern is almost unchanged, forming a prismatic planing surface.

There are some derivative types of planing hull as introduced in Sect. 1.1:

- Stepped Planing Hull

As a craft accelerates from reaching planing speed at between $Fn_\nabla = 2$ and 3, the lift on the bottom surface will increase in relation to V^2, so a flat planing surface will rise and the wetted area decrease. The same will happen for a hull with deadrise.

In both cases, the angle of attack will decrease, and the wetted length will adjust to balance the moment from center of gravity. When traveling in waves, the dynamic movement of center of lift and pitch moment can accentuate motions and even cause instability.

One configuration that minimizes this is to include one or more "steps" across the hull bottom and have the surface aft of the step at a slightly higher angle of attack (see Fig. 1.9a). In effect, the craft then rides on two tandem planing surfaces, and the rise at higher speeds changes the center of overall lift only marginally.

This design approach has been adopted successfully for many high-speed designs. Recently, high-speed small craft designers have adopted alternative approaches to the integration of steps into a hull bottom. If the step is actually a cut out in the hull bottom, air will still ventilate the cut-out as the hull rises with speed. With a suitable geometry of the back of the cut-out air, it will ventilate the running surface aft at least with a bubble sheet and so reduce drag. Lift aft of the cut-out will also reduce, and the bow-down pitching moment with increased speed will also reduce.

The one downside of the ventilated rear part of the hull will be a tendency for the stern to slide out of a strong turn maneuver. Some craft have experienced complete rotation around the vertical axis. In order to lessen this tendency, spray rails along the planing surfaces are very important.

Racing craft have been built including more than one step or cut-out, see Fig. 1.9b.

The angle of attack of the planing surfaces can be more closely optimized for running in calm water with a stepped configuration, since craft trim will stay almost constant above full planing speed, and so drag can be minimized. In a seaway, the response of the craft will depend on the deadrise of the planing surfaces and the shape of the bow. Commonly, very high-speed craft have a long slender "nose" above water that, when wetted, can provide strong pitch stability. The downside to this approach is heavier accelerations in proportion to the additional area wetted.

Fig. 1.8 (**a**) A typical medium speed planing hull. Type VI patrol boat from USN. (**b**) A typical high speed planing craft, the 17m Sentinel-class patrol vessel

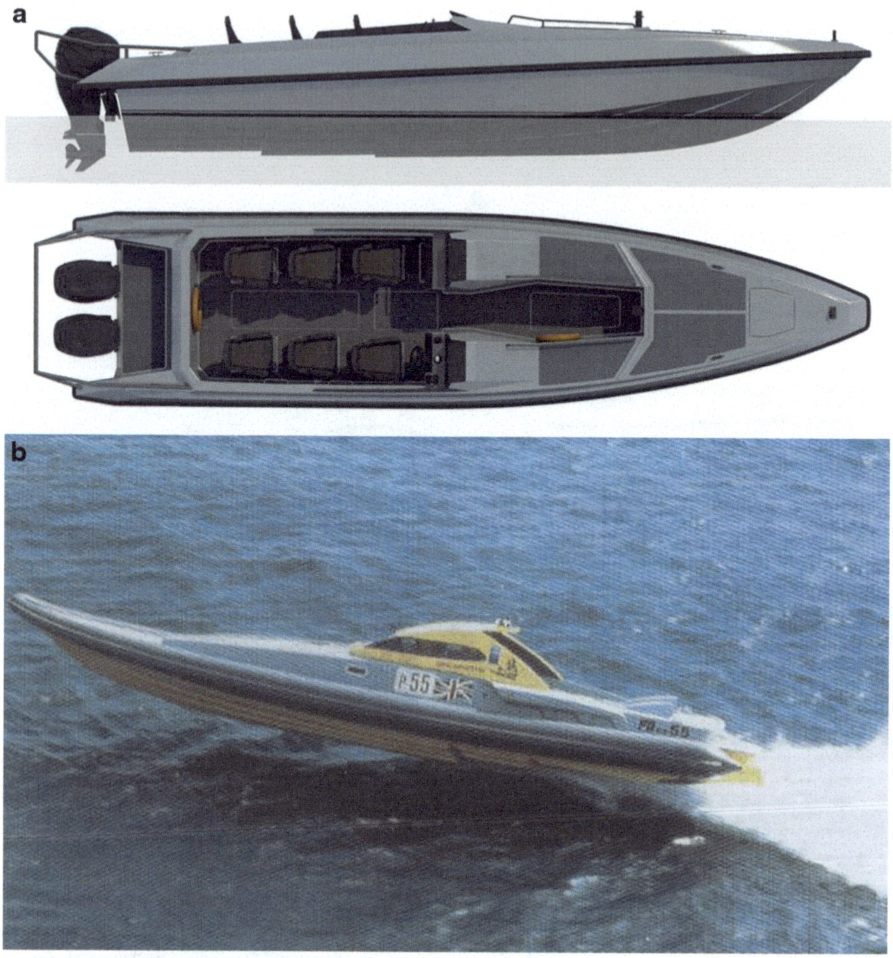

Fig. 1.9 Typical stepped planing hulls. (a) Cruiser. (b) Racing craft with multiple steps

This can be adjusted a little by having a flared shape to the sides of the bow form rather than simple deadrise. Nevertheless, there are limitations on the reduction in acceleration forces for high-speed planing monohulls, and this leads to a natural limiting envelope for practical operations in terms of sea state.

Crews running a racing craft will have installed seating that absorbs acceleration forces, and in addition, they will not expect operation in extreme sea states and durations. In contrast, crews operating an offshore patrol vessel may need endurance in terms of days, and the ability to loiter comfortably at low speed. A sport or racing craft does not have that kind of wide design envelope, with its performance focused around its speed and environmental limitations.

- Planing hull with spray rails

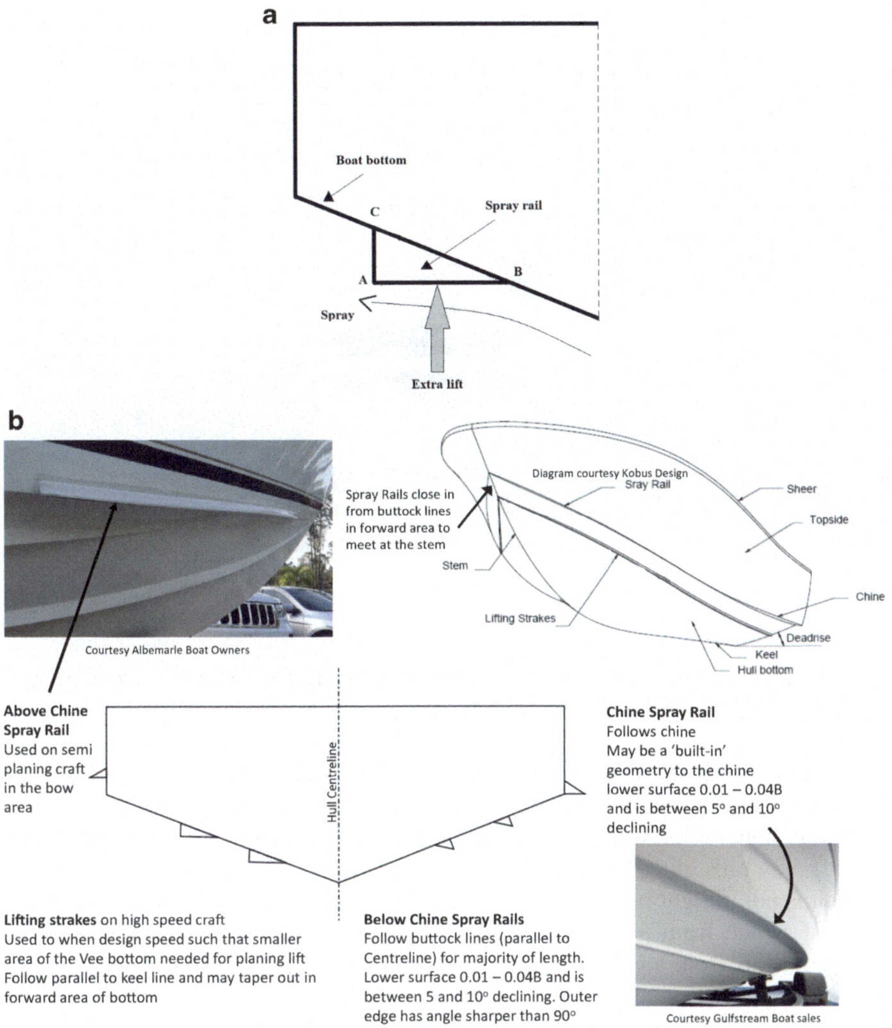

Fig. 1.10 (**a**) Spray rail configuration. (**b**) Chine and spray rails examples

If the hull bottom has a deadrise, there will be a tendency for water surface flow to be deflected outward and upward, reducing the lift effect. If small section rails, Fig. 1.10a, are fitted to the surface that itself has a level or even slightly downward-angled surface, the flow can be constrained more lengthwise along the surface and so provide a higher lift.

Typically, two or three such rails may be fitted to the underside of the hull, as shown in Fig. 1.10b, as it has been found by experience that this gives good results. Additionally, a strake may be installed at the chines to "sharpen the corner" so spray is directed downward.

Fig. 1.11 Three-point hydroplane—Bluebird K7

While there have been model tests and analytical studies to optimize positioning and sizing of spray rails and chine strakes, this aspect of design remains somewhat of an art.

One key aspect, though, is that spray rails also increase damping during craft motions and resistance to a craft sliding outward in turning maneuvers.

At very high speed, where the planing surface can reduce inward from the chines toward spray rails, the transverse stability may reduce, such that the stable running position is not upright but lolled to one side by some degrees. The stable condition can then switch from port to starboard as oblique waves pass. This effect is not common, apart from some high-speed deep vee craft. It is also much lessened by planing surface spray rails.

- Three-Point Hydroplane

In order to minimize the wetted surface and drag, the planing hull can be designed to be supported on three separate planing surfaces, as shown in Fig. 1.11, i.e., two planing "sponsons" ports and starboard of the central hull respectively. The central hull is designed so that the central planing surface is close to the stern. This is the configuration adopted by many water speed record craft [1]. By its nature, such a craft is intended for calm water operations.

In addition, the above water shape of the sponsons and main hull must be arranged so that aerodynamic lift does not catastrophically affect the craft trim, since at speeds above 100 knots if trim rises above 5° or so the aerodynamic lift could simply pitch the craft until it turns right over. Such craft are thus a very fine balance of hydrodynamic and aerodynamic forces.

4. *Deep Vee Craft*

The Vee shape of the hull bottom with deadrise can be developed into a Deep Vee configuration using an angle of between 20° and 35° instead of the 10–15° used in low deadrise planing craft. The design aim of a deep vee hull is to significantly improve seaworthiness.

This deep vee configuration can also be applied to high-speed semi-planing craft for the after part of the hull (see Chap. 6).

Deep Vee planing craft are characterized by:

- Vee-shaped body plan, with hard chine, and large deadrise angle forward, as well as straight line planing surfaces in transverse section
- The transverse section aft of amidships also retains a significant deadrise angle ($\beta = 20$–$35°$), so as to decrease the wave slamming forces on the flat surfaces
- Multiple spray rails on the deadrise surface (typically 3 off), and chine spray deflection rails

The advantages of this configuration are:

- Smaller pitching and heaving displacements, vertical acceleration, slamming, and smaller probability of bow submergence in waves, and emergence of the propellers at stern as waves pass, compared to round bilge displacement forms
- Fine maneuverability, and course stability
- Lower speed loss in waves
- Simple hull structure using single plane curves, reducing production costs

Two examples are shown in Fig. 1.12a, b.

1.3 Recent Development of High-Speed Monohull Craft

We introduce here some of the designs developed in the last half century or so as a background to our study of the hydrodynamics of the different configurations in the next chapters.

Semi-Planing Craft

Semi-planing craft have found their main market as small utility and pleasure craft, as well as for coastal patrol vessels with national coastguard services and navies for protection of their maritime Exclusive Economic Zone (EEZ). With EEZ extending to 200 nautical miles from shore, the performance while deployed for days at a time often loitering is as important as the high-speed performance.

Smaller utility boats may be used for offshore fishing, for example, as the hull form will give smoother motions than a planing form at low speeds in a seaway.

Most "cabin cruiser" craft will have a displacement or semi-displacement form, depending on the intended cruising range. This form then allows for cruising speeds in the range of 10–20 knots. When at design speed, a semi-displacement geometry will give some additional damping to roll and pitch motions.

Fig. 1.12 (**a**) Schnelker marine tracker. (**b**) Vexus boats DS20 high speed fisherman shown at speed riding on spray rails

Fig. 1.13 Semi-planing cabin cruiser-Hardy Marine 40DS

Recently, some "superyachts" have also adopted this hull form, which allows a designer to deploy a form that will have efficient and motion-kind long distance transit at medium speed, with comfortable performance at higher speed for transit between locations used for entertaining clientele on board.

We give examples of a cabin cruiser, an offshore patrol vessel, and a fast superyacht below (Figs. 1.13, 1.14, and 1.15).

Planing Craft

Monohull planing craft developed rapidly during World War I (WWI) as motor gun boats and attack craft to launch torpedoes [3]. This was enabled by the appearance of high-speed, lightweight engines developed for aircraft. During WWII, the high-speed planing boat developed rapidly again, supported by further improvements in high-speed engine technology enabling speeds in excess of 40 knots, Figs. 1.16 and 1.17.

Particularly for coastal and strait patrol, motor gun boats and fast torpedo boats could be operated in conjunction with airplanes to attack enemy maritime assets. For instance, the US and British alliance navy torpedo boats took part in 800 attacks and sank 400 German ships; the former Soviet Union's torpedo boats attacked German vessels 383 times, shot 539 torpedoes, and sank 209 German ships, respectively. A selection of reference books about the developments, including WWII, are included in the Resources section at the back of this book. During this war, fast boats also became important as rescue craft to retrieve flight crews from aircraft shot down over the sea.

Post WWII, because of the rapid development of radar and other surveillance equipment as well as aviation, the development of even faster patrol boats to enable stealth and enhanced armament was attempted by a number of builders. For

Fig. 1.14 Offshore patrol vessel (Damen Stan 4207 Royal Bahamas defense network). (Courtesy Chuck Oldham April 1, 2015, Defense media network)

Fig. 1.15 Semi-displacement Superyacht Silver Fast 77m, 27 knots and global range, courtesy Boat International

Fig. 1.16 Typical Torpedo boat. (From WIKI at https://en.wikipedia.org/wiki/Motor_torpedo_boat)

Fig. 1.17 Vosper MTB firing torpedoes in English Channel. (Courtesy BAE Systems Maritime)

Fig. 1.18 Guided missile craft. (Komar class boat courtesy Wikipedia, at https://en.wikipedia.org/wiki/Komar-class_missile_boat)

example, the UK developed the "Brave" class at the beginning of the 1960s. Gas turbine power plant was installed on the craft, allowing it to reach speeds of up to 50 knots.

In addition, with the rapid development of guided missiles after WWII, the development of high-speed boats to deploy these weapons, Fig. 1.18, was the next step for many navies. A typical example was the Egyptian guided missile craft type "Komar," made in former USSR, which attacked and sank Israel's destroyer *Eilat* from a distance of 12 nautical miles from the destroyer on October 21, 1967. With the same type of craft and guided missiles, the Indian Navy attacked and sank one of Pakistan's destroyers in December 1971.

While in the 1960s radar surveillance was improving and protection of wider areas was being developed by the use of airborne radar with AWACS aircraft, accurate monitoring of potential hostile objects was still difficult, particularly if fast moving. That was a key impetus behind continued development of military high-speed craft.

The strategic concept of using small craft to sink a large naval ship continued to be attractive due to the relative low cost involved, [10]. This generated a very active period for research, design, and construction of high-speed guided missile craft in the late 1970s in countries such as the US, former USSR, UK, and Germany, and an export market to other countries as well.

From the 1990s onward, satellite surveillance became ubiquitous, both for global positioning (GPS) and for military surveillance. While extreme speed can, in certain circumstances, be a key asset for military defense, other missions have become more important, and so the use of fast marine craft has focused more on EEZ patrol and smuggling interdiction.

Planing craft are nevertheless built in very large numbers around the world for leisure, utility, and rescue purposes. Typical examples are fast day boats and tenders, Fig. 1.19a, b, fast-rescue craft and man-overboard boats, Fig. 1.19c, d. FRC and MOB boats often have solid foam or inflated tubular buoyancy around the gunwale, as shown in Fig. 1.19. All these craft are designed for relatively short deployments, in terms of an hour to a few hours. Most are powered by outboard motors or stern mounted inboard motors on a transom-mounted z-drive, while some rescue craft are now powered by water jet units.

At the smallest end of the size range, there are the Sea-doo craft that have seating for one or two persons and water jet propulsion, as shown in Fig. 1.20. These are also mass manufactured and used for leisure the world over. Since the 1980s when they were first introduced, they have developed from a simple fun machine to activities such as those depicted, such as fishing offshore (with care!).

At the other end of the range, on a large scale, a number of shipyards in Europe have developed high speed monohull ferries for both passengers and vehicles, including trucks. An example is the vessel shown in Fig. 1.21; the Bazan Shipyard built Silvia Ana, 100 m × 18.7 m × 2.5 m, which has a maximum speed close to 40 knots, a displacement approximately 3500 tons, carrying 1200 passengers and 202 cars, and is powered by six diesels with total of 33,900 kW of driving waterjets for propulsion. It currently operates from Buenos Aires, Argentina to Piriapolis, Uruguay.

Fincantieri in Italy have also built a series of six high speed RoPax ferries, the MDV1200 class. These ships, 100 m × 17.1 m × 2.6 m also with a service speed of approximately 40 knots are somewhat smaller than the Silvia Ana, at approx. 2000 tons displacement, carrying 600 passengers and 147 cars powered by four Ruston RK270 diesels totaling 19,200 kW. One of the vessels is shown in Fig. 1.22.

Deep Vee Craft

Deep Vee craft generally target both high speed and open seaway environments. Due to the power required to achieve this, they tend to be much smaller craft than the high-speed ferries shown in Figs. 1.21 and 1.22.

During the last half of the twentieth century, through practical experience as well as analysis, the hull form has settled into one using a relatively sharp bow that is wave riding rather than wave piercing and parallel cross section aft of amidships. A number of small shipbuilders have developed designs for fast interdiction missions, such as that shown in Fig. 1.23 from ARES. This hull is a simple, deep vee with spray strakes and sharp chine. Such craft may be used, for example, by marine police forces or paramilitary groups to deliver teams to remote locations for missions ashore.

Outside the paramilitary market, the main application has been for pleasure and racing, where speeds have risen to above 100 knots for the fastest craft, for example, the craft in Fig. 1.24a powered by surface drive propellers that are located behind the transom, Fig. 1.24b [1]. Such craft may include steps as well as spray strakes along the planing surface.

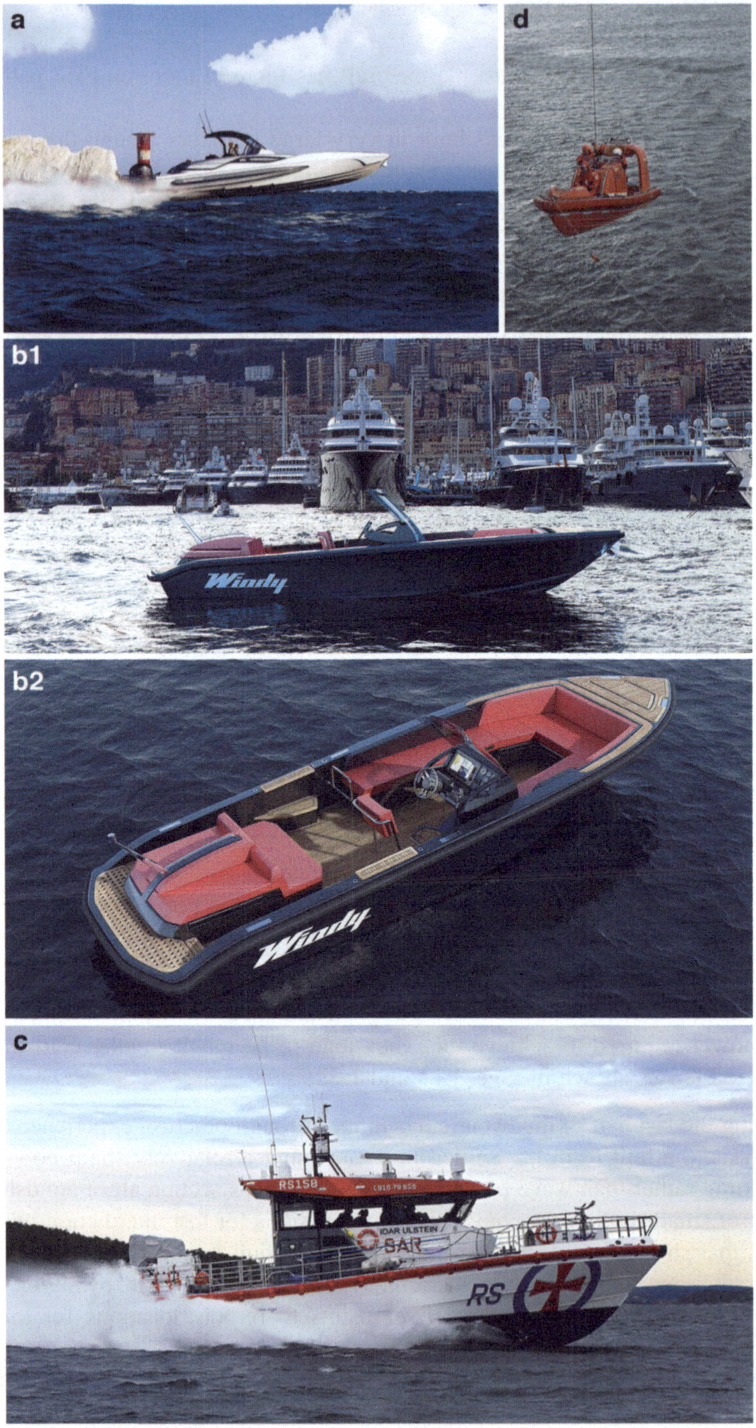

Fig. 1.19 (**a**) Fast day boat (Sunseeker Hawk 38, 2 × 400 hp, multi step, 60 knots plus. Courtesy www.sunseeker.no). (**b**) Fast tender craft for superyacht (Windy SR26 tender b1 and b2 side a quarter view showing lines and seating layout. Courtesy Windy, https://www.windyboats.com/range/sr-line/sr26/). (**c**) Fast Rescue craft—Tornado semi-inflatable 9m? (Swedeship 24m fast

Fig. 1.20 Sea doo craft. (Courtesy Sea Doo Inc.)

Fig. 1.21 HSC Silvia Ana built by BAZAN. (Courtesy WIKI commons)

Fig. 1.19 (continued) rescue craft 38 knots. https://swedeship.se/vessel-production/22-m-fast-rescue-craft/) (**d**) Fast MOB boat (Fast mob boat – Courtesy WIKIcommons. https://commons.wikimedia.org/wiki/File:MOB_Boat_(man_overboard_rescue_boat).jpg. Craft is being lowered from Construction vessel Silja in Baltic and has self-righting at stern)

Fig. 1.22 MDV 1200 ferry, 37 knots hard chine design. (Courtesy WIKI commons at https://commons.wikimedia.org/wiki/File:SSC3-2.jpg and https://en.wikipedia.org/wiki/MDV_1200-class_fast_ferry)

Fig. 1.23 Modern Fast Interdiction craft—Ares 24 fast interceptor for Abu Dhabi. (Courtesy Ares Marine)

Fig. 1.24 (**a**) Racing deep vee craft—Cougar craft Fortido (Courtesy Bournemouth Echo June 7, 2018 at https://www.bournemouthecho.co.uk/sport/16277328.powerboat-racers-aim-glory-poole-bay-100/). (**b**) Surface drive

Fig. 1.25 Fast offshore logistics craft Strategic Marine Centus 42m, 30 knot fast crew boat. (Courtesy Strategic Marine at https://www.strategicmarine.com/vessel/42m-gen-4-fast-crew-boat/)

A significant market has also developed for larger deep vee craft operating at relatively slower speed in the 35–45 knot range, powered by waterjets, for offshore logistics to oilfield installations. These craft are deployed in regions such as the Caribbean and South East Asia. An example is shown in Fig. 1.25. A wave-slicing bow form is now used on some of these craft, as shown in Fig. 1.26, rather than the overhung fine but wave-riding bow form used in older designs. The advancing market for offshore wind power will offer these craft another market going forward.

As for planing craft in the previous subsection, there is a plethora of deep vee high-speed boats built for private use that continue to be built around the world. Many boat builders do not have access to model test tanks, and so until recently, hull design has been largely passed down through experience and tradition.

Since the beginning of the twenty-first century, computer programs have become available for fluid dynamic analysis (CFD) that can enable more "science" to be injected, but the physics is not yet accurately modeled for the interface and spray, so a combination of analysis and model or prototype testing is still needed to optimize a design. We will discuss this a little further as we progress in the following chapters and provide some general reference materials in the resources section at the back of the book.

1.4 Studying the Monohull Concept

In the last two sections, we have introduced the categories of hull shape for vessels to run at speeds exceeding $Fn_L = 1.0$ or $Fn_\nabla = 2.5$. In the next chapters, we will first look at the semi-planing hull form (Chap. 2), and then study the basic principles of a

Fig. 1.26 Damen offshore fast craft with wave slicer bow, the Damen FCS-7011 40 knots, 73.6m. (Courtesy Damen Shipbuilding, at https://www.damen.com/insights-center/news/damen-revolutionary-fcs-7011-completes-sea-trials and https://products.damen.com/-/media/products/images/clusters-groups/high-speed-crafts/fast-crew-supplier/fcs-7011/documents/damen_fcs_7011_brochure_04_2019.pdf?rev=559f18fa102746e487a9dd27997c1e7a)

planing surface (Chap. 3) before applying this to the geometry of a planing hull, first for drag and then looking at stability, seakeeping, and maneuverability (Chaps. 4 and 5). Finally, in Chap. 6, we look at the Deep Vee form for semi-planing speeds and its performance in waves.

All these forms need to possess stability while being static in the normal way. In addition, each of them will experience increasing dynamic forces as they accelerate to service speed. Once at service speed, a significant proportion, or indeed, almost all support for vessels operating above $Fn_\nabla = 4$ or so, will be dynamic rather than the buoyant stability when static. This has consequences for the vessel when operating in a seaway. As waves pass along the hull surface, the local velocity will vary, as will the forces acting on it. In addition, a wave passing a semi-planing or planing hull will still apply inertial forces in a similar way to a displacement vessel.

Both model-scale testing and full-scale vessels have demonstrated motions in a seaway that confirmed these variable forces could effectively reduce stability in roll and pitch. The consequence could be the high-speed vessel lolling to one side, having unsteady motions, or increasing motions.

The selection of vessel geometry to avoid these unwanted characteristics is not a simple matter, but once an understanding of the causes is available, steps can be taken by a designer to mitigate them. We first address dynamic stability for semi-planing vessels. We will then look at it again when studying seakeeping of planing vessels, and finally extend this to the deep vee form.

The influence of dynamic forces as speed increases is related to V^2, i.e., response is nonlinear. Predicting changes in response when a parameter is changed is therefore more difficult. Full-scale prototyping whether electronically or physically therefore becomes more important to the fast vessel designer.

The trends can be assessed, but solving analytical models requires an iterative approach to determine equilibrium for fully planing vessels, as we shall see in Chap. 4.

So, what will be our objective, thinking as a designer? Given a mission brief for the vessel, we will be aiming at minimized powering, and the most comfortable possible ride, since low motions and accelerations also minimize structural loading and so weight of the structure itself.

To do this for the monohull vessel, we have the following parameters that can be utilized:

- Slenderness
- Cross section form and its variation
- Flow guidance such as spray rails and chines
- Transverse steps on fully planing forms
- Stern tabs and interceptors

Figure 1.27 shows a design flow chart for fast monohull vessels. This chart covers the complete project delivery. Our focus in this book is on the tasks labeled 1–7. It should be noted that normally a vessel design is based upon mission requirements that are set by a client operator or owner. The designer will use this to prepare a set of input data by liaising with the client before carrying out preliminary sizing and selection (1). This may be done for a series of vessels within a parameter space of hull form so that more than one potential design is taken through a first design cycle. At this stage, statistical or other guidance may be used for preliminary sizing, form, weight estimate, etc.

In stage (2), hull lines will allow hydrostatics and static stability, including checks against any regulations to be completed. Calm water resistance and trim follow in stage (3), and resistance and motion analysis are added in (4). At this point, the parameter options can be compared (5). Depending on the results, it may be possible to select and continue with the concept design. Alternatively, there may be a case for an adjustment to be studied. In this case, it is necessary to return to re-estimate vessel weights and centers (6) and repeat stages (2) to (5). Once the key parameters give satisfactory hydrodynamic characteristics, the concept design stage (7) can be started. During this stage, model tests and/or detailed CFD studies will be completed to verify the performance of the selected design while machinery is selected, and a more detailed weight estimate is prepared. That will then provide the starting data for the detailed design of the vessel.

Depending on the combination of required speed, sea state capability, and endurance, it may be that the motions and accelerations of a monohull are just too high. In that case, a designer may consider deploying the hydrofoil form. This option is discussed in the second half of this book. Hydrofoils, with a very much lower motion response but at the cost of considerable structural and control system complexity, can meet such mission requirements for passenger transport.

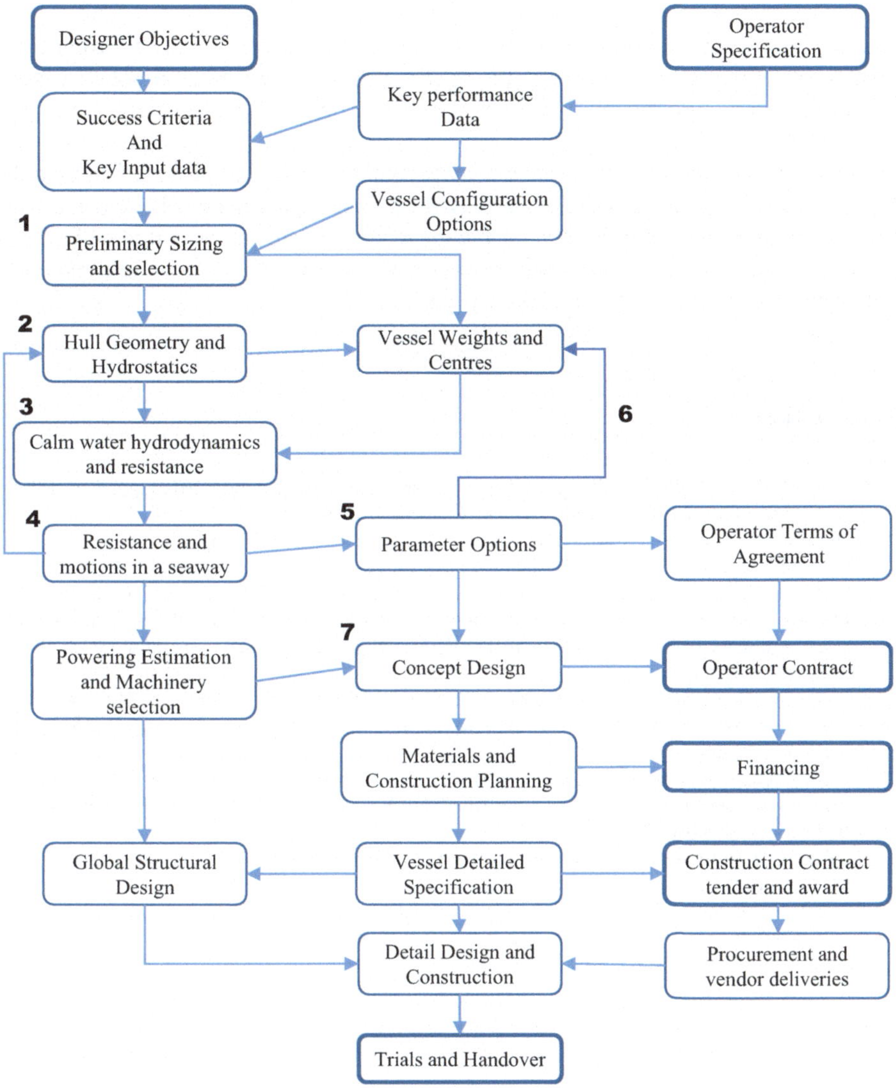

Fig. 1.27 Design flow chart for fast monohull vessels

The basic advantages of a high-speed monohull are its simple structure and machinery. If using some combination of the above parameters can result in a vessel that has satisfactory motion and acceleration levels in the intended service, then it will tend to be the product of choice. At present, the high-speed marine market has borne this out, as alternatives such as hydrofoils have developed within a limited market.

This brings us to one other ability possessed by a monohull—that of scalability. A displacement vessel can be successfully built at very large dimensions. If we consider a slender semi-planing vessel, dimensions beyond 150 m length and carrying capacities in thousands of tons are practical with today's propulsion machinery. If we wish to consider speeds above 50 knots where a monohull vessel would need to be fully planing, then there are lower limits to practical size due to the installed power required.

As we work our way through the hydrodynamics of this group of monohull forms, a picture should emerge of the way the different parameters above can affect the performance. This should then enable the reader to make choices that will give a satisfactory monohull form for their own mission requirement, or to understand how a designer would carry out selection and design.

References

1. A. Bliault, L. Yun, *High Performance Marine Vessels* (Springer Science, 2012), pp. 5–8., ISBN 978-1-4514-0868-0. https://doi.org/10.1007/978-1-4614-0869-7
2. H.F. King, *Aeromarine Origins* (Putnam and Co Ltd, London, 1966), pp. 66–68
3. See 'Spitfires of the Sea' internet site: A Quest for Speed: The History of the Hydroplane – Spitfires of the Sea. The site also covers Motor Gun Boats, and Motor Torpedo Boats.
4. M.G. Morabito, Reanalysis of William Froude's studies of planing craft. Royal Institution of Naval Architects Transactions **157**(part B1) (2015., ISSN 1740-0694). https://doi.org/10.3940/rina.ijsct.2015.b1.173
5. L. Yun, A. Bliault, *Theory and Design of Air Cushion Craft* (Arnold Publisher Ltd, 2000)
6. L. Yun, A. Bliault, J. Doo, *WIG & Ekranoplan: Ground Effect Technology* (Springer, New York, 2009)
7. H.Y.H. Yeh, Series 64 Resistance experiments on high-speed displacement forms. Marine Technology (1965), pp. 248–272, published by Society of Naval Architects and Marine Engineers, New York
8. D. Bailey, Performance Prediction of Fast Craft. UK National Physical Laboratory, Report No 181, October 1974
9. P. Du Cane, High Speed Small Craft, 3rd & 4th edition, 1964, 1972
10. D. Savitsky, Planing Craft, Chapter IV. Naval Eng. J. **97**(2) (1985)

Chapter 2
Semi-Displacement Monohull Craft

In this chapter, we will focus on craft that operate in the speed range where kinetic forces are limited. Forces on a hull will then be primarily inertial due to variation in buoyancy (vertical forces) and the horizontal component of pressure forces, including incident and self-generated surface waves. Kinetic forces due to surface friction and boundary layer turbulence are smaller components. Kinetic vertical force components are small.

Presently, computer software (CFD) is available to simulate these forces in the time domain, using linear superposition for waves comprising a sea spectrum to generate statistics for craft motions, accelerations, and drag forces. Since the kinetic forces are limited, their nonlinear nature does not create too much of a problem for computer predictions or analytical methods. The results from CFD still require calibration with physical tests before they can be relied on for projections suitable for optimization of hull parameters.

Traditionally, naval architects have used model tests in calm water to determine total drag from a hull and compared them with analytical calculations made separately for the inertial wave making drag and surface friction drag. The results then have to be scaled to the full-scale dimensions of the craft, bearing in mind the different factors that will apply to inertial and kinetic forces.

Tests in regular waves can be used to determine the additional drag due to incident waves, and this factor is identified by the extraction of the calm water elements. On the basis of linearity, a series of regular wave tests can be combined using a sea wave energy spectrum (for example, the Pierson Moskovitz spectrum)[1] to predict motion and drag statistics for a projected operational sea state and added to a projection separately of the calm water inertial and friction components. Model tests in waves do depend on the moments of inertia for the craft to be correctly modeled to obtain reliable results for the motions.

---

[1] https://wikiwaves.org/Ocean-Wave_Spectra#Pierson-Moskowitz_Spectrum

Z.-S. Dong et al., *High-Speed Monohull and Hydrofoil Craft*,
https://doi.org/10.1007/978-3-031-62760-6_2

If a hull is tested with its appendages modeled, the drag due to these is an integral part of the friction drag component. An alternative is to calculate these elements separately at full scale and add to the calculated hull skin friction component.

Since the 1970s model test basins have been able to simulate a natural sea spectrum increasingly accurately, using fast Fourier transformation to determine the response spectra directly rather than using a discrete set of regular wave tests. The overall approach is nevertheless the same, to identify the additional drag for a seaway, or the craft motion response statistics.

We have three main components to evaluate skin friction drag, induced wave drag, and additional drag from waves in a seaway. Induced waves will affect the mean vessel trim as speed rises. For large displacement vessels, this change is small, while for high-speed vessels, it becomes increasingly significant.

It may also be useful to note here that the induced waveform is the result at the surface of the hull form of physically displacing the water as it moves forward, creating a pressure field around the hull. It is this total pressure field that results in the forces acting on the hull. The surface profile changes the buoyancy distribution so that the hull will trim to reach an equilibrium.

Figure 2.1 shows a typical variation in the running attitude versus speed of a high-speed craft. The abscissa shows volumetric Froude number $Fn_\nabla = (V/(g\nabla^{1/3})^{0.5}$, where V is craft speed (m/s), and ∇ is the displaced volume in m^3. The ordinate shows the change in draft ΔT from the still water value at bow, T_f, amidships T_m, and stern T_a

At increasing speed, the running attitude can be divided into three stages as follows:

1st stage: $Fn_\nabla \leq 1.0$ is displacement operation mode; the variation of bow/stern draft is small; the running attitude changes a little from static buoyancy

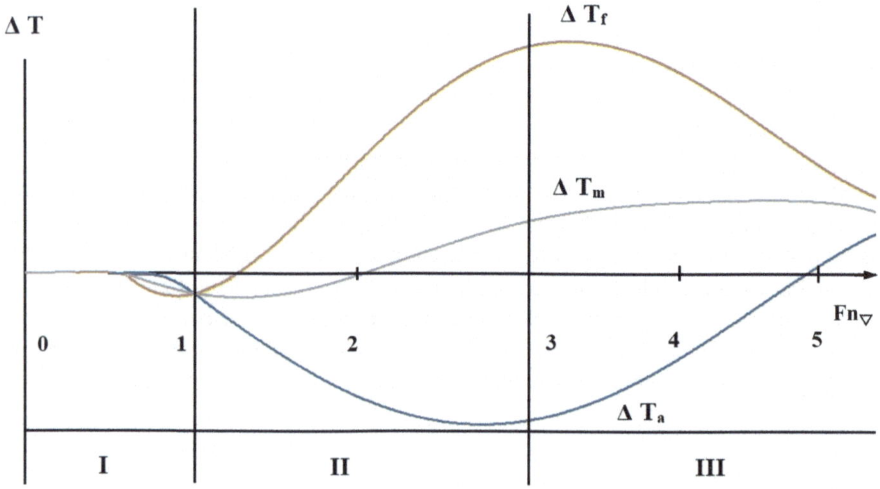

Fig. 2.1 Bow/stern draft of the craft versus speed

condition with the bow going down a little, but with no change at stern draft similar to conventional ships.

2nd stage: $1.0 < Fn_\nabla < 3.0$ is transition mode with the bow rising and stern sinking with speed, so that the trim angle increases with speed. This is due to induced wave height increasing and length increasing steadily to 2× craft length.

3rd stage: when $Fn_\nabla \geq 3.0$, the craft is in planing mode. Dynamic lift is the main force to support the craft weight, and so the mean draft relative to the static waterline decreases as the area required to support the weight decreases with increasing speed. The induced wavelength increases beyond 2× craft length, and the surface slope flattens, decreasing trim angle. At the same time, the induced wave height decreases.

It may be noted that we refer here to Fn_∇ rather than $Fn_L = (V/(gL)^{0.5}$ as we are considering speeds where dynamic lift is important and when planing the wetted length is not constant. For semi-planing craft, where the bow is not lifted clear of the water surface, Fn_L can be more appropriate, as comparisons can be made directly to other displacement craft. Related to Fn_L, the 1st stage is below $Fn_L = 0.65$ and third stage is above 1.0–1.2.

Figure 2.1 shows an example variation for the running attitude of fast craft. The actual running attitude will be different for varying hull forms, and the critical points for the three stages will be different. For example, some well-designed hard chine craft can be planing at $Fn_\nabla = 2.5\sim2.8$, while some round bilge craft will remain in transition mode at $Fn_\nabla > 3.0$. Similarly, some slender round bilge craft remain in displacement mode of operation with small dynamic lift even at $Fn_L = 1.2\sim1.5$.

In this chapter, the second stage of craft operation—the transition mode— will be discussed. In this operation mode, the dynamic lift cannot be neglected, so such craft are generally referred to as semi-displacement or semi-planing craft. We first present the approach to the evaluation of drag from several model test series carried out in different countries, and then look at evaluation of motions.

In the first half of the twentieth century, the design of fast military craft was focused on high *L*/*B* round bilge designs, with fine bow entry. Typical of such craft was the German E Boat design [1] built in large numbers during the World War II, Fig. 2.2.

At their design speed of over 40 knots, such craft could be tender (sensitive in roll) in a seaway, and after the war, the main shipbuilder for the E-boats, Lürsen, developed a new design, the Jaguar class Fast Attack Craft [2]. These craft had a hard chine form transverse section for about 20% of LWL from the stern and the remaining part was round. This achieved improved seaworthiness with minimized slamming.

This led to the development of fast displacement monohulls with various transverse sections along the craft length, i.e., mixed lines craft with hard chine toward the stern and round bilge at mid and forward part of the craft during the 1960s and 70 s.

In this same period, the focus for fast patrol boats was on planing hulls for service speeds in the 50-knot range. These craft had acceptable seaworthiness for coastal

Fig. 2.2 German E Boat. (German E-Boat S204 surrenders at Felixstowe on 13 May 1945, By Lt. J.E. Russell, Royal Navy official photographer—This is photograph A 28558 from the collections of the Imperial War Museum, it is Public Domain, https://commons.wikimedia.org/w/index.php?curid=734851)

conditions but were not suited to endurance missions due to their harder motions. In this period, nations became more sensitive to their "Economic Zones" extending up to 200 nautical miles from the coast and so attention returned to fast vessels that also had seakeeping aligned with long endurance missions, and thus turned back to the more rounded semi-displacement form.

At this time, powering was considered the most important parameter to optimize in the performance of high-speed craft.

Several model test series were carried out to evaluate the effect of varying hull shapes on power, and later on, to looking at motion response. The aim of the various test series was to develop analytical expressions or graphs that could be interpreted to determine the required power for a given hull form. This was an extension of various model test series carried out from the 1930s onwards on round bilge hull forms [3].

We will summarize some of the later model test work referred to in [3] and its interpretation below, together with additional work carried out in China.

More recently, with increasing demands for better seaworthiness, displacement craft with aft hard chine sections were adapted with a deeper vee cross section aft, including craft to the Combattante II design of CMN in France [4].

Thus, the modern semi-displacement monohull cannot be considered simply as a round bilge craft with small bilge radii or a hard chine form, but as a mix with different lengths of round bilge or hard chine and possibly also deeper Vee aft. In this chapter, we mainly discuss craft with forward round bilge and aft planing sections with medium deadrise, while more extreme Deep Vee geometries will be discussed separately in Chap. 6.

2.1 Drag Components

The fast displacement monohull is an extension of displacement craft. Its main drag components are wave making, friction, and viscous pressure drag. When $Fn_L > 0.5$, the wave making drag coefficient decreases with increase in speed as the generated wavelength increases and the trim of the vessel also reduces. Meanwhile, the friction and viscous pressure drag increase in proportion to V^2. The component of "residual drag" in total drag therefore starts to reduce after the hump speed, $Fn_L = 0.5$.

Figure 2.3 shows the change in residual drag with the total drag of a fast displacement monohull. For $Fn_L < 0.5$ the "residual drag" increases with speed, reaching a hump value at $Fn_L = 0.5$. For $Fn_L > 0.5$ the "residual drag" decreases with speed, and after about $Fn_L = 1.0$, the "residual drag" will be smaller than friction drag.

The value of total drag relates to the craft profile, especially to its slenderness, as shown in the figure, where the slenderness $\psi = L/(\nabla^{1/3})$ values are 6.934 and 7.94 respectively. As slenderness is increased, the residual or wave making drag component is reduced, and friction drag component is increased in relative terms. This hull geometry, being round bilge form develops little dynamic lift above $Fn_L = 0.5$ and

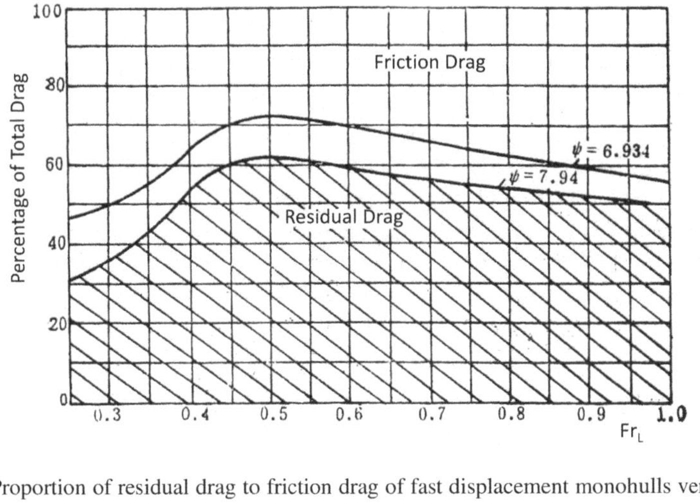

Fig. 2.3 Proportion of residual drag to friction drag of fast displacement monohulls versus Fn_L

so there is still a significant wave making drag at $Fn_L = 1.0$ where a hard chine vessel would be planing, and the viscous drag would be above 90%.

Once speeds exceed $Fn_L > 0.6\text{~}0.7$, wave making at the bow will create a breaking waveform that wets the hull above the still water line. Depending on the bow shape, it will curl outward as a breaking wave. The additional drag caused by this is often referred to as spray drag, as it is not solid water. If a vessel is fitted with deflecting strakes above the static water line extending around amidships, the rising water can be deflected to minimize the area wetted and the additional drag.

These drag components may be termed the "bare hull" drag components. The total resistance of a hull will comprise the following components:

$$R_T = R_R + R_f + R_{SP} + R_{AP} + R_A \tag{2.1}$$

Where

R_T Total drag of craft

R_R Residual drag, the sum of wave making and viscous pressure drag i.e., $R_R = R_W + R_{VP}$

R_f friction drag, the sum of flat plate friction R_{f0} and "roughness" additional friction drag ΔR_f, so $R_f = R_{f0} + \Delta R_f$

R_{SP} spray drag

R_{AP} appendage drag

R_A air drag

The residual drag is strongly affected by the shape of the hull, including the surface curvature for both the forward and aft parts of the hull, similarly to a normal ship form without the center parallel section typical of large ships. The friction drag is related to the wetted surface area of the craft.

2.2 Resistance Through Model Test Programs

Given a design requirement for a vessel to support a particular payload and mission, the key question is how to shape the hull to optimize the geometry and requirements for the propulsion system and supporting equipment.

Starting with the research by Froude and his contemporaries, an approach was developed to vary a limited number of form coefficients in a geometric series to be able to interpolate between results from model tests within a specific range. For displacement ships, this proved a powerful approach, enabling optimization from an initial base geometry using relatively small changes to the parameters. In the mid-twentieth century, refinements such as the "bulbous bow" enabled even finer optimization of calm water resistance to be achieved.

If we turn to the smaller craft, as discussed in the introduction, the main approach to minimizing drag was the use of high L/B and a fine bow form. In order to develop a model series with varying geometry, L/B, B/T, and C_B define the primary space, with ψ as a secondary variable.

Model Test series with parametric variation from a base geometry looking at fast craft were begun by H F Nordström in Sweden at the Swedish Ship Model Basin (now SSPA, Göteborg). His test series carried out beginning in 1936 used displacement hull forms run at higher speeds. His work was followed by a more extensive series reported by De Groot at the Netherlands Ship Model Basin, Wageningen at the beginning of the 1950s.

In the 1960s, substantial programs were run in Great Britain, the USA, and Sweden, as well as in China. Subsequently, the work at National Physical Laboratory in Great Britain was extended, and NSMB completed another series in the 1980s as a joint program with both the Australian and US Navy Authorities (HSDHF) to look at both resistance and seakeeping. The HSDHF program was later extended with the same participants by the AME CRC Series. Meanwhile, through the 1980s BAZAN in Spain tested its own series, and the University of Belgrade tested the SKLAD series.

A summary of the range of form parameters in these test series is given in Table 2.1 for reference.

Where

$\psi = L/\nabla^{1/3}$	slenderness coefficient
C_B	Block coefficient
C_p	longitudinal prismatic coefficient
L	craft length
B	craft beam
A_T	sectional area at transom under design water line
A_M	Max sectional area under the design water line
X_c	distance of center of buoyancy aft of amidships T draft
$Fn_L = V/(gL)^{0.5}$	Froude number with respect to craft length

These series, as listed in Table 2.1 above, have subsequently been further compared and analyzed to study the effect of key parameters on performance in calm water and in waves. Until the 1990s output from the tests continued to be presented in graphic form, while regression analysis and associated expressions for projection began in the 1970s. Key reference material for the test series can be found as follows:

(a) Nordström, Royal Institute of Technology, Sweden, 1951, [5]
(b) DeGroot Series, Netherlands, NSMB 1951, [6]
(c) Marwood and Silverleaf, Great Britain, NPL, 1959 (1960 paper), [7]
(d) USA Series 63, Davidson Laboratory, Steven Institute of Technology, 1963, [8]
(e) USA Series 64, David W Taylor, Naval Ship Research and Development Centre, 1964 reported 1965, [9]

Table 2.1 Summary of the range of test parameters used in the model test programmes reviewed in this chapter

Series	ψ	C_B	C_p	L/B	B/T	A_T/A_M	X_c/L % aft	Fn_L	Fnv	No models
Sweden Nordström	5.65–7.72	0.373–0.41		4.83–6.94	3.16–3.57				0.9–2.0	12
NSMB DeGroot	5.23–7.76	0.421–0.457	0.65–0.677	4.55–7.39	3.16–3.57	0.17	1.6–2.64	0.2–1.10	0.8–2.7	31
Sweden SSPA	6–8	0.40	0.68	4.62–8.21	3–4	0.42	4.15	0.4–1.20	1–2	9
Great Britain NPL	4.47–8.5	0.397	0.693	3.3–7.5	1.69–9.83	0.573	6.4	0.3–1.19		19
USA Series 63	4.5–6.4	0.6–0.78	0.38–0.64	2.5–6.0	2.89–9.47	0–0.7	0–5.8	0.2–1.0		5
USA Series 64	8–12.4	0.35–0.55	0.63	8.454–18.264	2–4	0.41	5.66	0.1–1.5		27
NSMB HSDHF	4.3–13.1	0.35–0.55	0.626	4–12	2.5–5.5	0.296	5.4	0.1–1.2		40
China CSSRC	6.3–8.74	0.355–0.50	0.63	6–9	2.54–5.03	0.174	3.2	0.278–0.97		n/a
CSSRC Series extended	5.42–7.62	0.355–0.50	0.63	3.6–6.0	3.6–10	0.174	3.2	0.278–0.97		n/a
China MSC series	5.45–7.62	0.440–0.452	0.613–0.634	4.49–8.03	2.98–4.00	0.17–0.32	2.64–4.41	0.40–0.85		8
Spain Bazan80	5.7–7.9	0.35–0.55	0.54–0.70	4.5–8	2.5–4.5	0.2–0.4 approx.	6.9	0.35–1.1		2 + 24
Serbia SKLAD	4.5–8.5	0.35–0.55	0.715	4–8	3–5	0.621	8.8–9.2		1–3	27
AME CRC Series	4.3–8.7	0.396–0.50	0.626	4–8	2.5–4	0.296	5.4	0.1–1.0		14
VTT Series	6.586	0.45–0.60	0.693	5.41–6.25	4.39–6.90	0.52	6.4		0.6–3.8	5

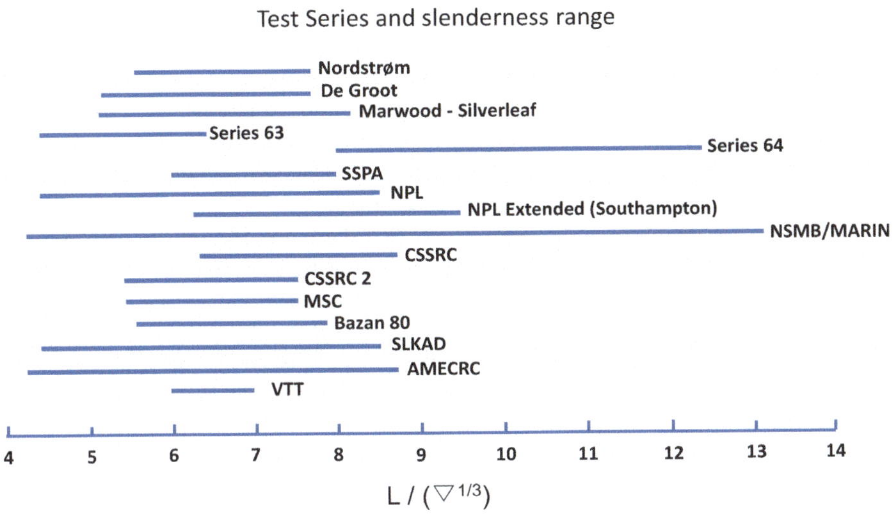

Fig. 2.4 Slenderness range for key test series

(f) Sweden SSPA Series, 1968, [10]
(g) Great Britain, NPL Series, 1969, 1976 extension, [11, 12]
(h) The Netherlands, NSMB Series 1988, [13, 14, 20]
(i) China, CSSRC Series, 1964, [15, 16]
(j) China, MSC Series of CSRDC, [17]
(k) Spain, Bazan 80 series 1980 [18]
(l) Serbia (Yugoslavia) SKLAD series 1980 [19]
(m) US/NL/Australia AME CRC series 1999 [3, 21, 22]
(n) Finland, VTT Series, 1990 [23]

Figure 2.4 gives a diagrammatic impression of the $L/\nabla^{1/3}$ ranges for the different test series. This is one of the key starting parameters a designer may work with and may suggest the initial source of test data to use for comparing predictions.

It may be noted that the common characteristics shared by all these model sets are as follows:

- Fine lines forward of amidships with Cp typically at around 0.6 and half angle of entrance at the bow between 10° and 15°
- Transom stern with relative immersed area at the stern varying widely, resulting in a wide range of rise of the keel from amidships to the transom
- Buttock lines generally are straight aft of amidships with reducing bilge radius and transverse deadrise of the bottom that varies with *B/T*. Base case National Physical Laboratory (NPL) model had 12° deadrise in the after part of the hull

If the variation in geometry for the various programs listed in Table 2.1 is considered, it becomes clear that developing an empirical equation that covers all the tested hulls is not appropriate. The best approach is to look at each program to develop a relation to predict $R_R/g\nabla$ and then use this to see what may change if the hull form uses attributes from other series.

In this chapter, an outline of the work is presented to give an overview. Additional information, including empirical resistance prediction equations, is presented in Appendix 1, together with a commentary on their use. We will focus our attention on the later programs before we move on to the analyses, developing empirical parameters that can be applied for design assessment.

It will be seen from our summaries below that there are differences between the series that may be useful to check. Examples would be the effect of increasing or reducing the transom area, varying the bottom curvature aft, and adding a knuckle or spray rails forward.

The NSMB/HSDSF/AMECRC series have also looked at the interaction of parameters for resistance compared with seakeeping and come to some useful conclusions.

2.2.1 NPL Model Test Series

For about 10 years, the UK National Physical Laboratory (NPL) systematically carried out model tests of high-speed round bilge craft models, including drag, propulsion, stability, and seakeeping quality [7, 11, 12]. Figure 2.5 shows the body plan of the parent hull of the NPL series.

There were 19 models, the parameters of which are shown in Fig. 2.6, where Ⓜ is slenderness, referred to here as $\psi = \mathrm{Lwl}/\nabla^{1/3}$. A set of graphs for the residual drag coefficient with different L/B is presented in the reports. Figure 2.7 shows one of such graphs, with $L/B = 5.14$ where abscissa is Ⓜ, the slenderness ratio, the ordinate is residual drag for unit displacement, and the speed parameter is volumetric Froude number Fn_∇.

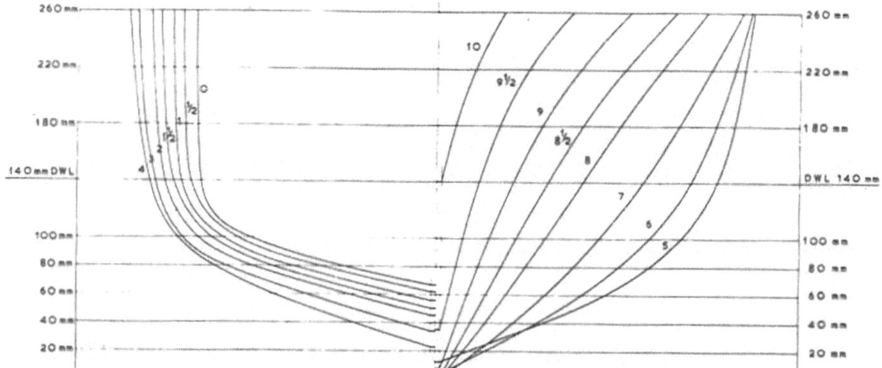

Fig. 2.5 Body plan of NPL series prototype

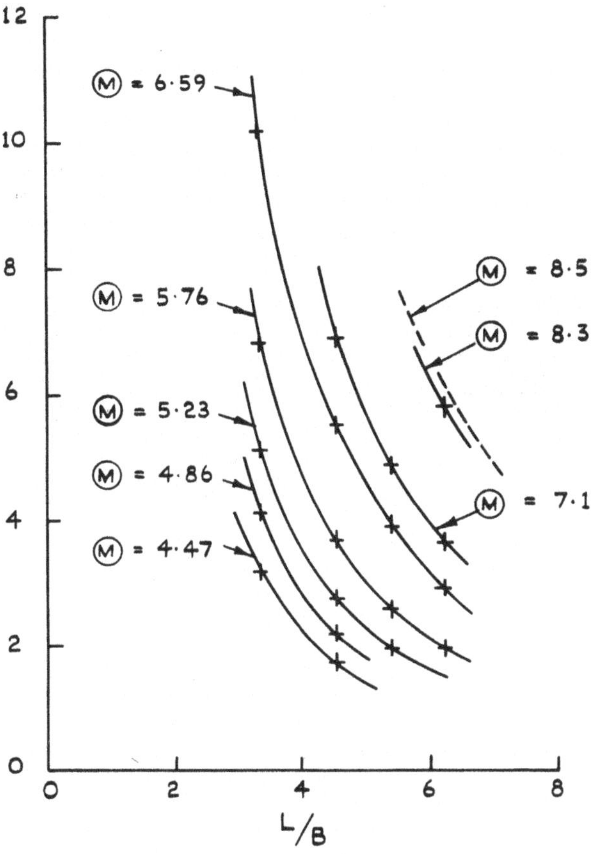

Fig. 2.6 (**a**) Parameter range of NPL series models (**b**) NPL series models wetted area calculation

Calculation of model and full-scale wetted area is done by using a plot, as shown in Fig. 2.6a. The graphs give curves for calculating the wetted surface area of a hull, and so calculate the friction drag using the 1957 ITTC expression with roughness allowance of $\Delta C_f = 0.2 \times 10^{-3}$. Full calculation is presented in [12]. Additional graphs also give curves for calculating the effective power of full-scale craft, to obtain the effective power directly.

References [11, 12] give full details of the parameter space that the models cover, together with sample graphs and a worked example. The body plans show the hull form has a fine bow profile and rounded amidships area, with the bilge radius reducing steadily to the stern and a flat lower hull aft from amidships with deadrise diminishing. The transom lower edge also has significant deadrise. As breadth is increased in models b, c, and d of each displacement the deadrise reduces and so the lower surface produces higher dynamic lift.

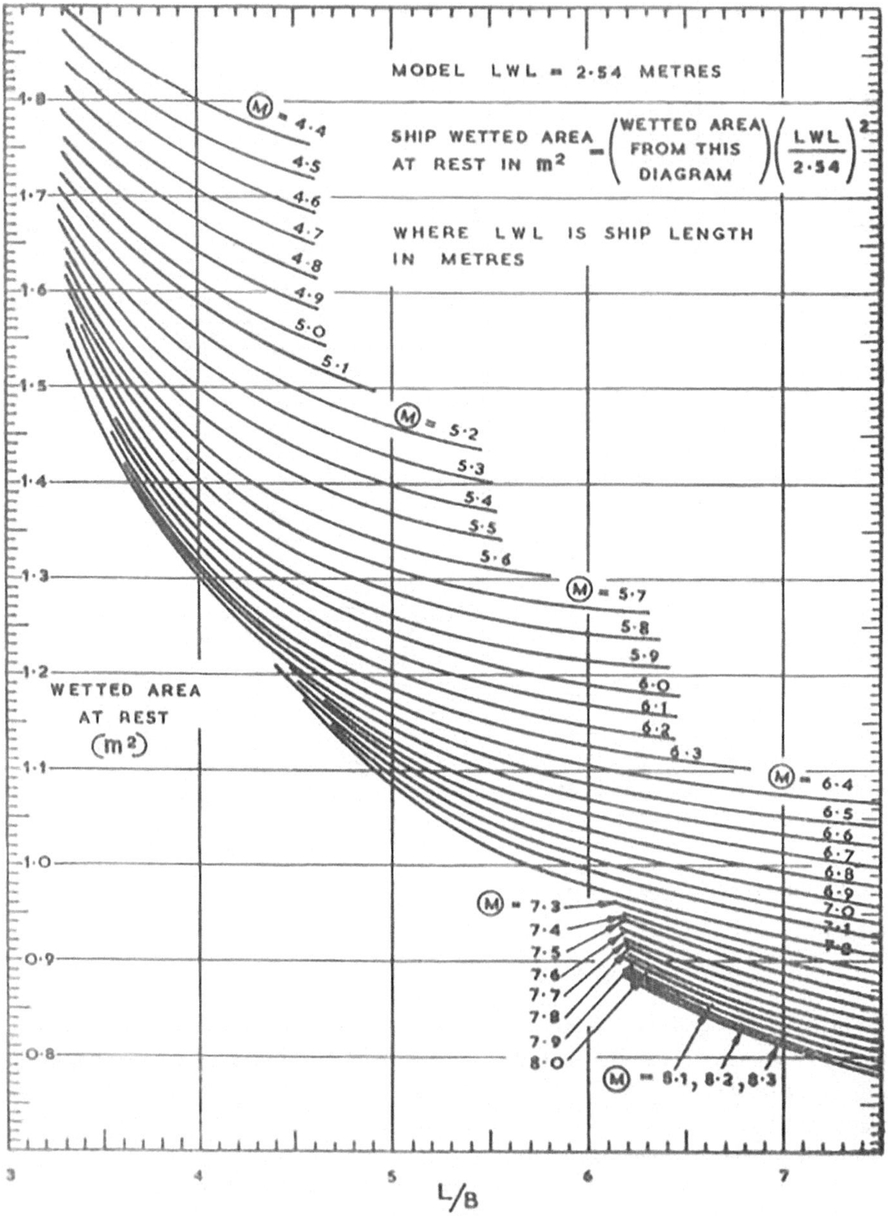

Fig. 2.6 (continued)

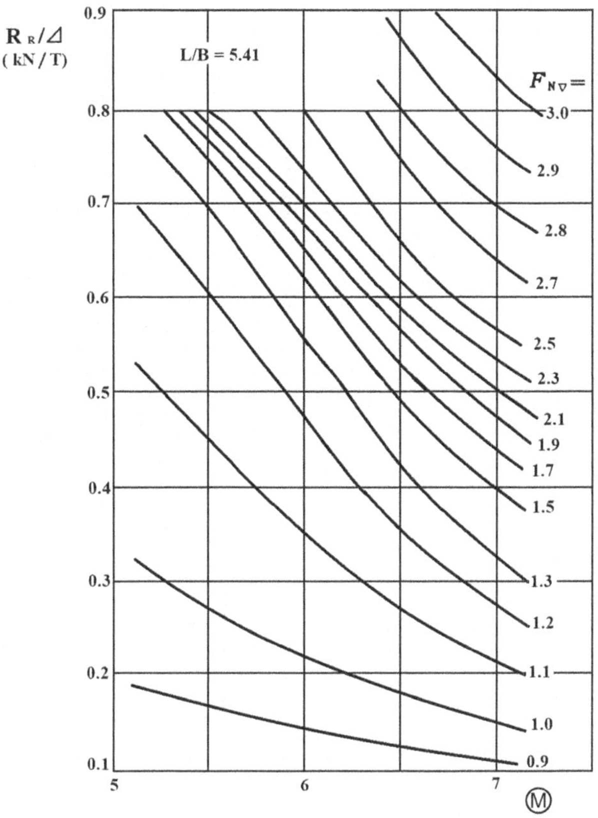

Fig. 2.7 Sample residual drag graph for NPL series models

An extension to the test series was carried out in the early 1970s considering also hard chine geometry variations for the same size and displacement. A sample effective power comparison from this series is shown in Fig. 2.8 [24]. From the figure, the intersection point of the effective power curve for two equivalent craft is located at $Fn_L = 1.05$ (denoted F_n in the diagram). Figure 2.9 shows the increment of resistance with Fn_L where $\Delta R\%$ represents the increment of resistance of hard chine craft over the round bilge craft, positive value represents a relative drag increment compared to the round bilge form. The data shows that in case of $Fn_L \leq 1.05$, the advantage of round bilge craft is apparent as resistance is lower; however, when $Fn_L > 1.05$, the hard chine performance improves as speed increases, as it comes fully on the plane. Based upon such results, the author suggests the range of application of both types of craft as shown in Fig. 2.10.

The application range mentioned by Bailey, was obtained in relation to craft with same size and displacement. However, in general, round bilge craft may be designed with increased length to decrease drag, so the practical application range will be a bit wider, corresponding to $Fn_L = 1.0{\sim}1.2$. Here we discuss the performance of craft at speeds $0.6 \leq Fn_L \leq 1.2$. For $Fn_L \leq 0.6$ a craft will perform as a displacement vessel.

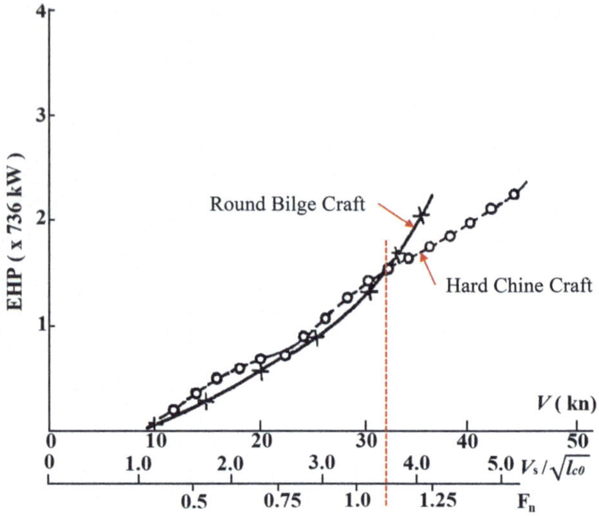

Fig. 2.8 Comparison of effective power between craft with round bilge and hard chine types

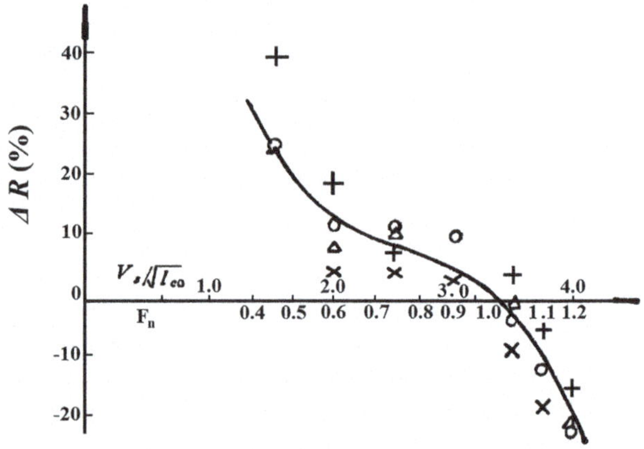

Fig. 2.9 Drag difference between round bilge and hard chine craft at different speeds

The NPL series was developed from a parent hull geometry for 100 ft LWL and beam varying from 16 to 30 ft. The hull depth was adjusted so that four models had the same displacement of 100 tons. The dimensions were then further varied to create models representing craft with displacements of 50, 80, 150, 200, 250 and 320 tons.

The parent hull characteristics were chosen by experience from model tests of similar geometry that demonstrated low resistance and reasonable seakeeping performance. No spray strips were fitted, and so wetted area increased with speed. Wetted areas were estimated from photographs using judgment for determining the line between solid water and spray.

The authors noted that Saunders [25] suggested that C_p, C_M, and LCB had significant effect on resistance and suggested optima. The values chosen for the

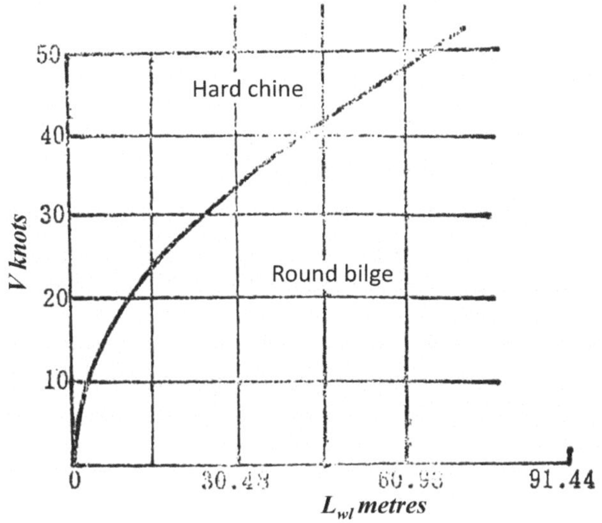

Fig. 2.10 Range of Application for round bilge and hard chine craft

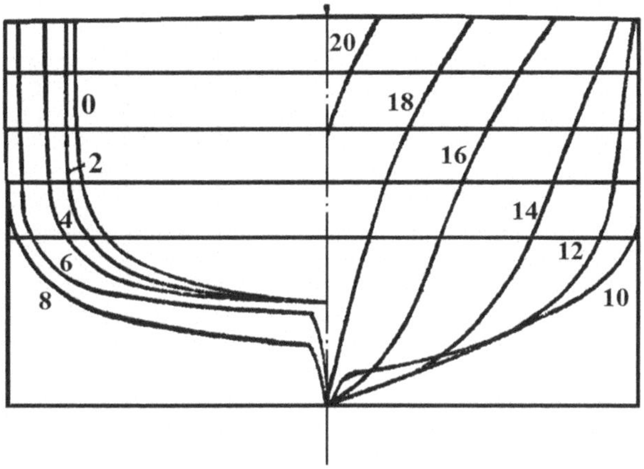

Fig. 2.11 Body plan of SSPA prototype

test series (and maintained constant) of 0.693, 0.397, and 6.4% aft of amidships were thought to be close to the optimum. Review of the results from the test series at NPL suggested that sensitivity was low to adjustments of LCB. The key parameters that needed careful selection were C_p and Ⓜ (referred to as ψ here).

2.2.2 SSPA Series

Figure 2.11 shows the body plan of the SSPA series prototype [10], with constant block coefficient C_B, maximum sectional area coefficient C_M, and longitudinal location of center of buoyancy LCB, with three values of B/T and ψ. They constructed nine models for testing. The parameters were as follows.

Ψ	6, 7, 8
B/T	3.0, 3.5, 4.0
C_B	0.4
C_M	0.59 (referred to as C_x by Fung)
LCB	4.15% aft of amidships

It can be seen from the body plan of the parent hull that it has a somewhat fuller shape, with C_M at 0.59 compared to NPL of 0.397, and the sections aft of amidships show an almost flat underside, larger bilge radii, and a stabilizing keel aft, so that the hull baseline is straight.

The results of the test series for Residual Resistance coefficient C_R are shown in Fig. 2.12 plotted against Fn_L. The higher *B/T* forms show higher resistance above Fn_L 0.8, and as ψ is increased from 6 to 8, the residual resistance is significantly reduced.

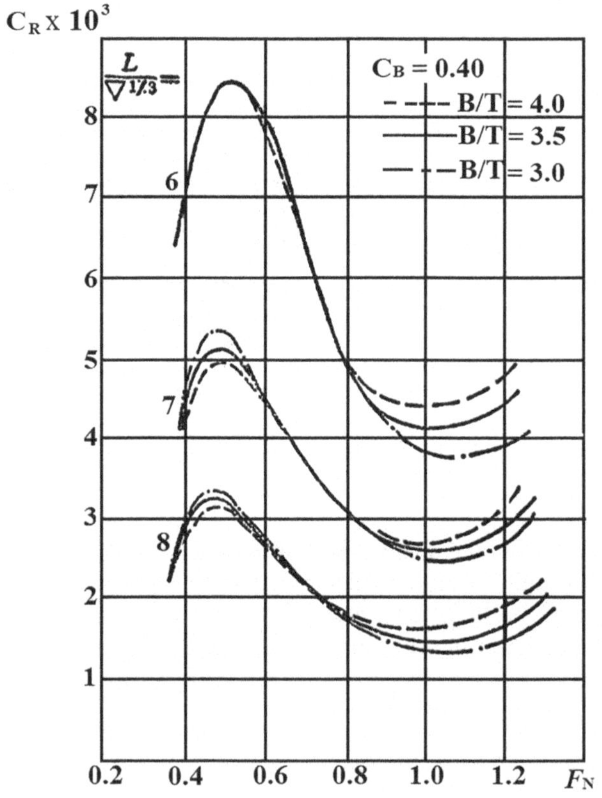

Fig. 2.12 C_R curve versus different $L/\nabla^{1/3}$ and B/T

The wetted area coefficient can be calculated by expression (2.1)

$$C_S = 4\left[0.77 - 0.045\frac{B}{T} + 0.01\left(\frac{B}{T}\right)^2\right] \tag{2.1}$$

The definition of wetted area coefficient is:

$$C_S = \frac{S_S}{\sqrt{\nabla L_{WL}}} \tag{2.2}$$

Where

S_s hull wetted area, m^2
∇ displacement volume, m^3
LWL length at design waterline, m

Friction drag coefficient C_f is based on the 1957 ITTC friction line with additional roughness allowance $\Delta C_f = 0.4 \times 10^{-3}$.

2.2.3 USA Series 63

The USA Series 63 hull has similarities to the NPL round bilge series, while having shallower draft and flatter bottom geometry. Aft of amidships the waterline breadth

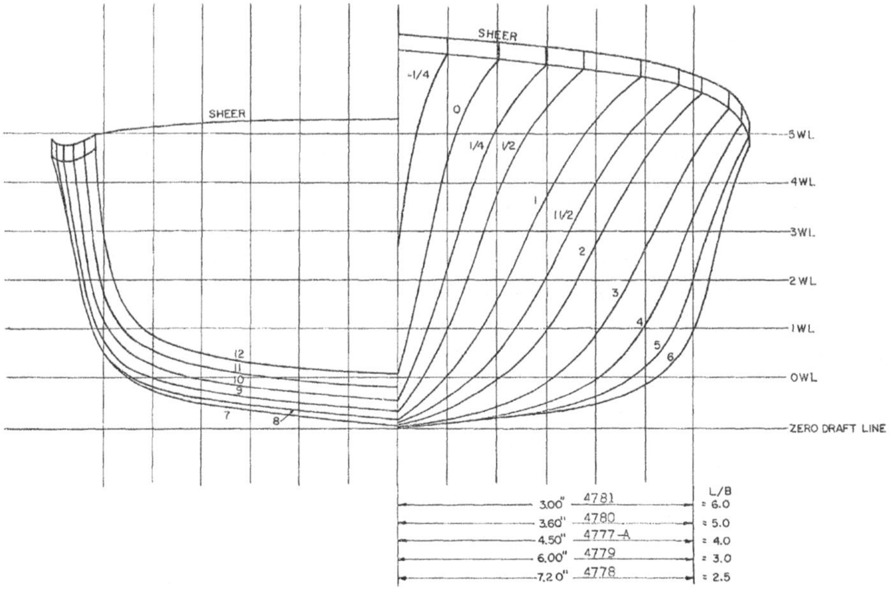

Fig. 2.13 Body plan of USA Series 63 parent hull form

tapers and bottom flattens from a low deadrise at amidships. The parent hull body plan is shown in Fig. 2.13 below.

There were 5 models each with length 3 ft (modeling 48 ft at scale 1:16). The *L*/*B* proportions were 2.5, 3, 4, 5, and 6. The parent hull had *L*/*B* 4. As *L*/*B* increased from 2.5 to 6, T was increased to maintain displacement, while L was held constant. This gave a range of parameters as follows:

ψ 4.5–6.4
L/*B* 2.5, 3, 4, 5, 6
B/*T* 2.89–9.47
C_B 0.6–0.78
C_M 0.8 approx.
LCB up to 5.8% aft of amidships

The friction drag was determined using the 1947 ATTC friction line with $C_A = 0$.

While observing that the models with higher *L*/*B* and thus higher ψ had lower resistance, the report [8] does not discuss optimization per se. Results are presented in graphical form for interpolation.

2.2.4 USA Series 64

The USA series 64 comprised 27 models covering a wide variation of geometric parameters. *L*/*B* proportions were from 8.454 to 18.264. The parent hull had *L*/*B* 10. The series was carried out at David Taylor Model Basin with speeds up to Fn_L 1.5

ψ	8–12.4
L/*B*	8.454–18.264
B/*T*	2, 3, 4
C_B	0.35, 0.45, 0.55
C_M	0.556–0.873 (referred to as C_x by DTMB, where $C_x = C_B/C_p$)
LCB	5.66% aft of amidships
I_{half}	3.7–7.8°, half entrance angle at bow

A_T/A_M was 0.41 compared with the NPL series 0.573. One can see from the body plan in Fig. 2.14 [9] below that deadrise is reduced to zero at the transom. Meanwhile though the body plan suggests a more rounded shape toward the bow, the much higher *L*/*B* results in C_p being lower for Series 64 than the NPL series, 0.63 vs 0.693. The LCB was maintained at 5.66% aft compared with 6.4% in the NPL series.

The friction drag was determined using the 1947 ATTC friction line with $C_A = 0$.

The program began with comparative tests of two existing models at DTMB; models 3474 and 3584 that both had very good characteristics at high speed. Two new models were constructed, 4783 and 4783-1 guided by the earlier models' lines. The new models were differentiated by lines aft of amidships, 4783 having a hard chine aft, and 4783–1 having a rounded chine. The latter proved to have best performance and so was chosen. The parent model characteristics are as follows:

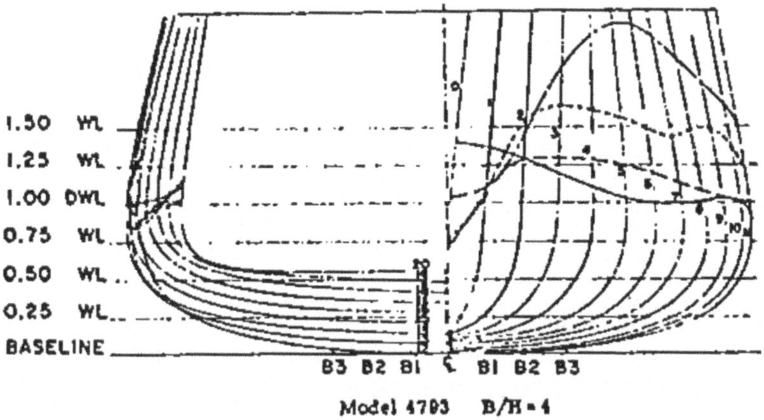

Fig. 2.14 Body plan of USA Series 64 parent hull form model 4783-1. (From original paper by Yeh MT 1965)

L/B	10
B/T	3
C_B	0.45
Cp 0.63	(This was held constant through the series)
C_M	0.714
I_{half}	7°

It may be noted that the maximum section in this series is at station 12 from the bow (of 20 stations).

The test program was focused on variation of three parameters, ψ, *B/T*, and C_B as shown at the beginning of this section. Waterline profiles were recorded by photography to assess the change in wetted area from static waterline.

Test results from this series is again presented in graphical form and data tables in [9]. The residuary resistance in general increases with wider and shallower hull forms in the series.

The wetted area reduces by about 30% between the lowest and highest *L/B* for a given *B/T*, and 5–7% as *B/T* is reduced. It was noted that in the Fn_L range between 1.0 and 1.5 that frictional resistance dominated and that the slenderer forms had lower residual resistance. Frictional resistance would be minimized through the lowest wetted area.

The series 64 model tests referred to here were calm water tests. The hull forms were aimed at high-speed frigates or destroyers. Comments were made concerning seakeeping performance and whether this would be optimal, and dynamic stability for the extreme slender forms.

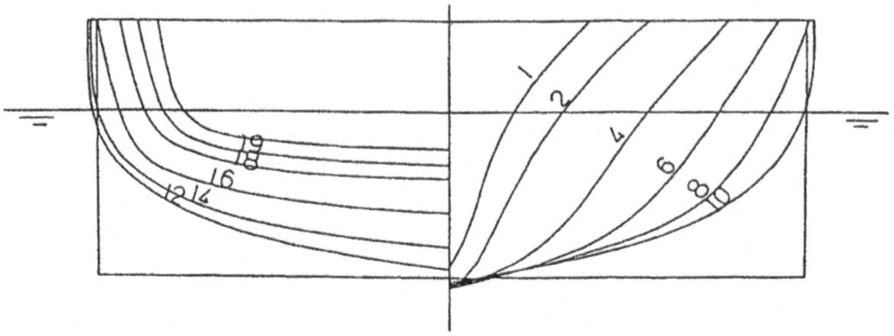

Fig. 2.15 Body plan of CSSRC series models

2.2.5 *China Ship Science Research Center Series*

Several model experimental series have been carried out in China. There is the 64–031 series experimental data and calculation graphs compiled by Q Jiang and C Gao in China Ship Science Research Center (CSSRC) [15], an extension of which is presented by S Liu in [16]. China Ship Research and Design Center (CSRDC) also carried out the MSC series model tests for the craft with short stern hard chine profile and presented equivalent calculation graphs [17] that we summarize in Sect. 2.2.6 below.

CSSRC in China completed their series testing of round bilge craft models in 1964 and added tests of models with low length/beam ratio, from the original range of 6.9–9.0 extended to 3.6–9.0.

Figure 2.15 shows the parent body plan of the model series. One can see that the beam /draft ratio is relatively high. The hull form has a fine entry forward, and aft of amidships, the bilge radius tightens toward the transom, which is relatively shallow. The aft hull underside is shaped and close to flat.

The test results for drag were converted into effective power coefficient R_0 with standard craft length L = 27 m, where R_0 is defined as follows

$$R_0 = 1000 \times \frac{P_e}{V_s^3 \Delta} \tag{2.3}$$

Where

R_0 Effective power coefficient with standard craft length 27 m
P_e Effective power coefficient, ps
V_s speed, knots
Δ displacement, *t*

The 1957 ITTC expression was used to calculate the friction drag with roughness allowance $\Delta C_f = 0.4 \times 10^{-3}$.

Figure 2.16 shows a multivariable carpet plot for R_0 based on the following variables for a speed of 26 knots and hull with *L*/*B* of 6.0

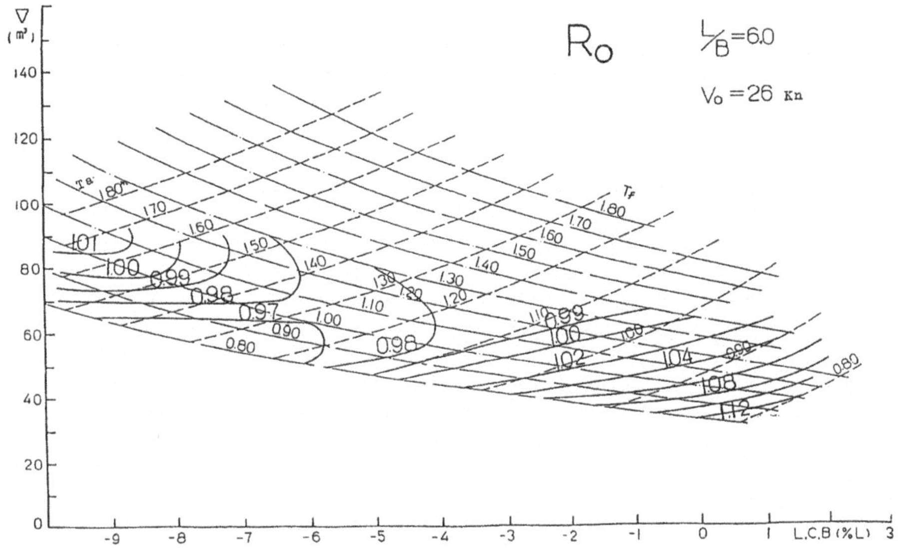

Fig. 2.16 Effective power coefficient of CSSRC series

Where:

∇ volumetric displacement, m^3
X_g distance of CG from mid ship section (% craft length)
T_f bow draft, m
T_a stern draft, m

In case of full-scale craft length not equal to 27 m, the correction of effective power can be expressed as follows.

① for sea craft, effective power is

$$p_e = \lambda^{3.5} p_{eo} + AS \tag{2.4}$$

Where	P_e	effective power of real craft (hp)
	Λ	scale ratio with craft of L=27m
	P_{eo}	effective power for L=27m craft (hp)
	S	wetted surface area of real craft(m^2)
	A	length correction factor
		When L > 27m, A < 0, when L < 27m, A > 0
		$A = 0.09497\ (C_{fs} - C_{fs0})V_s^{\,3}$ hp/m^2
Or		$A = 0.0698\ (C_{fs} - C_{fs0})V_s^{\,3}$ kW/m^2
	C_{fs}	friction drag coefficient of real craft at Vs
	C_{fsO}	friction drag coefficient of standard craft
		With L= 27m at correspondent $V_{s0} = V_s/\lambda^{0.5}$

② for inland watercraft the effective power is

$$P_e = 0.9738\lambda^{3.5} P_{eo} + BS \tag{2.5}$$

Where

B	length correction factor for L < 26.3 m, B > 0
	$B = 0.09248\,(C'_{fs} - C'_{fs0})V_s^3$ hp/m^2
or	$B = 0.06797\,(C'_{fs} - C'_{fs0})V_s^3$ kW/m^2
C'_{fs}	friction drag coefficient of real craft at 15 ° C in fresh water
C'_{fs0}	friction drag coefficient of standard craft with L=27m at speed

Other units are the same as before.

2.2.6 MSC Series at CSRDC, China

This series of models are of short hard chine configuration, 20% of craft length from the stern is hard chine while the other 80% craft length to the bow is of round bilge. A typical body plan is shown in Fig. 2.17.

The series was composed of 8 models with different slenderness ψ. The drafts were changed ±10% from base condition to create 26 test conditions. Residual drag coefficient curves were determined from the test results, as shown in Fig. 2.18. The

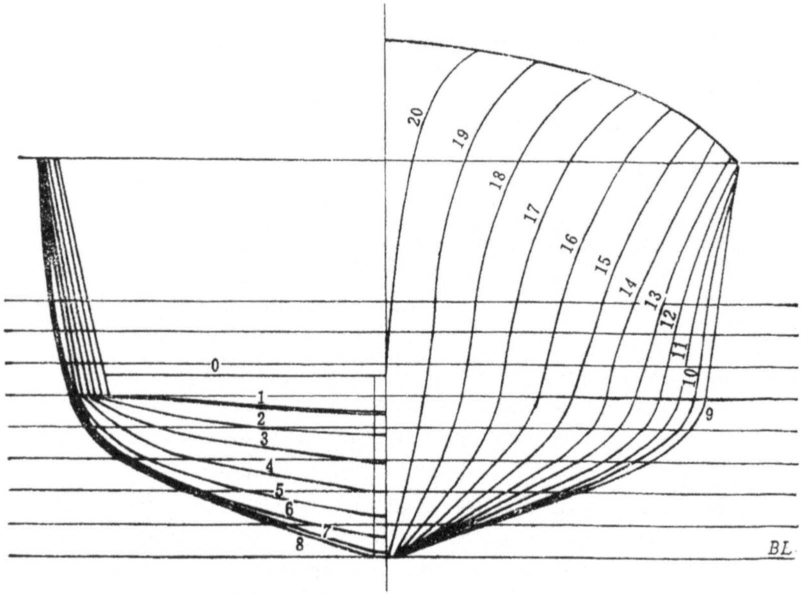

Fig. 2.17 Typical body plan of MSC series

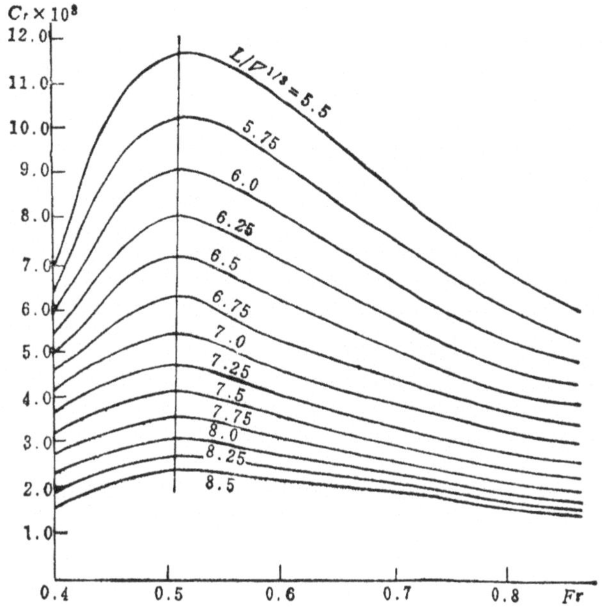

Fig. 2.18 Residual drag coefficient curves for MSC series

test results demonstrate that slenderness ψ and Fn_L are the main factors influencing the residual drag coefficient. Since other parameters such as L/B and B/T influence weakly on the residual drag coefficient, the authors used $C_R = f(\psi, Fn_L)$ as main graphical representation and neglected the influence of L/B, B/T.

The Prandtl-Schlichting formula for friction drag coefficient and a roughness allowance $\Delta C_f = 0.4 \times 10^{-3}$ were used to calculate frictional drag using static wetted surface area of the craft. The following approximate expression is suggested for calculating the wetted surface area of this series:

$$S = 1.83\,(0.49 + C_B)L\ \sqrt{(B/2)^2 + T^2} \tag{2.6}$$

where

T craft draft, m
B max beam of water line at draft T, m
L length of water line at draft T, m
C_B block coefficient at draft T

2.2.7 *Bazan 80 Series 1980*

The BAZAN test series [18] was carried out to develop data to allow design of semi-planing vessels for the Spanish Navy. BAZAN carried out a survey of data available from the UK NPL series and the US Series 64 before selecting an envelope of characteristics for their program. They chose a range for the main variables to be:

L/B	4.5 to 8.
B/T	2.5, 3.5, 4.5.
C_B	0.35, 0.45, 0.55
$L/\nabla^{1/3}$	5.7 to 7.9
LCB	6.9% aft of amidships (approx.)
C_p	0.540, 0.647, 0.700
C_M	0.5 to 0.8 (note $C_M = C_B/C_p$)
C_x	0.648, 0.696, 0.786
i_E	11.0° approx.
Fn	0.35–1.1
b_t/B 0.9	approx.

Prior to building the complete series of 24 models to explore the parameter envelope, two prototype models were built, with key data as in Table 2.2.

Comparing with the NPL characteristics in Table 2.1 these models are toward the slender end of the range, and with slender C_p and fuller C_B, these hulls are more shaped around the midships. Comparing between the two models, see Figs. 2.19 and 2.20, one can see model 200 has a little more volume forward and sharper bilge toward the transom, which on these vessels is flatter than an equivalent NPL hull.

Calm water testing of these two prototypes showed model 200 to have resistance profile a little higher than NPL or SSPA, while Model 100 showed an improving trend, as in Fig. 2.21, and therefore became the parent to develop the full model test series of 24 models. The main test series models were all 3.5 m length after review of Reynolds number issues for turbulence and laminar flow.

Table 2.2 Characteristics of the prototype BAZAN 80 models

Item		BAZAN model 100	BAZAN model 200	NPL baseline
Lwl	m	1.433		2.53
B	m	0.220		0.405
T	m	0.063		0.14
L/B		6.5		6.26
B/T		3.5		2.89
$L/\nabla^{1/3}$		7.0		6.585
C_B		0.435	0.433	0.397
C_p		0.627	0.642	0.693
C_M		0.694	0.674	0.573
X_B/L		6.6% aft	6.9% aft	6.4% aft
$S/(\nabla L)^{0.5}$		2.68	2.74	

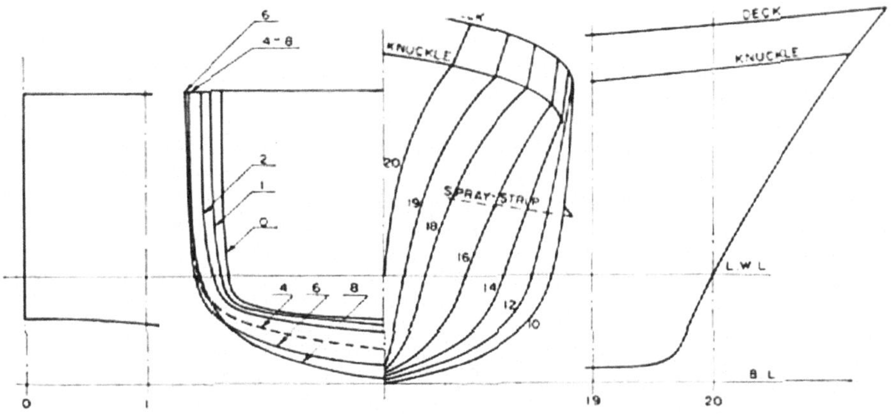

Fig. 2.19 Body plan BAZAN 80 model 100

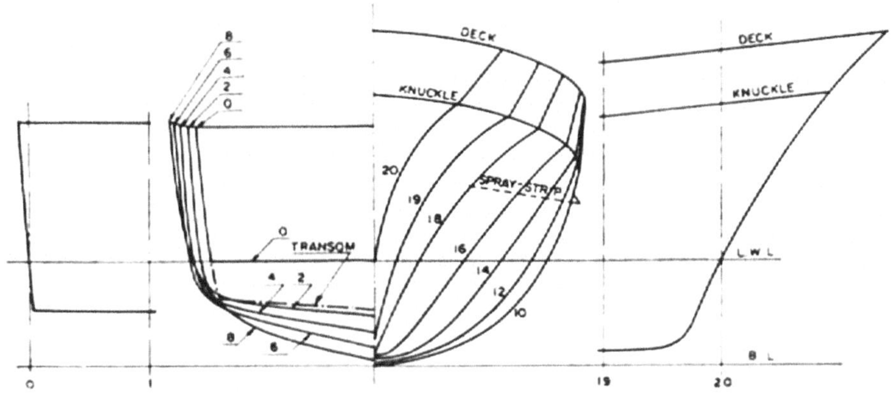

Fig. 2.20 Body plan BAZAN 80 series model 200

Following the prototype testing, analytical predictions were made for motions and added resistance in waves. The projections in Pierson Moskowitz spectra of 2.0, 2.5, and 3.0 meters significant sea state. Comparing results, the added resistance was least for model 100, while the motions results favored model 200. This appears consistent with the NSMB series where hulls with fuller forward shape showed improved seakeeping. Perhaps it should be pointed out here though that this is relative, as all these hull shapes have relatively fine lines.

It may be noted that BAZAN included spray deflector strips above the waterline forward and a knuckle at main deck level.

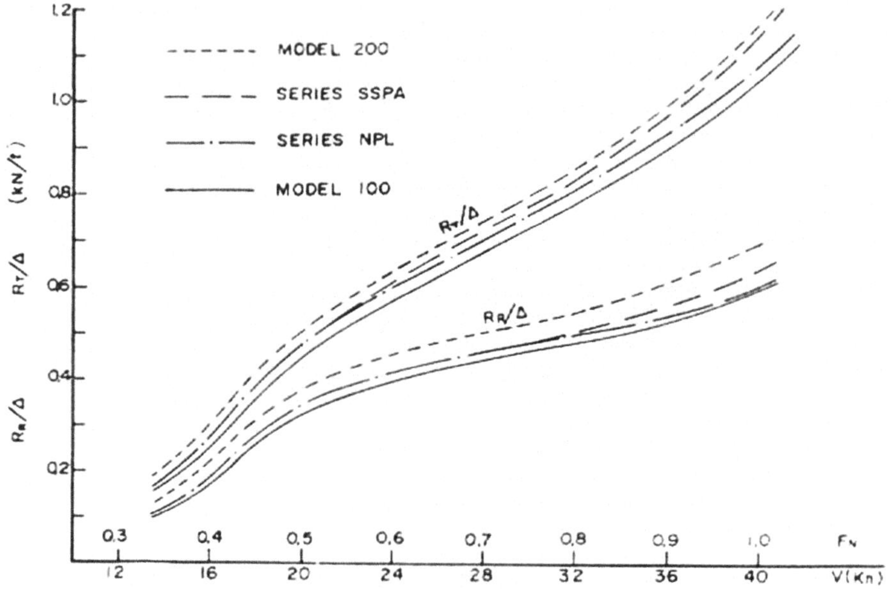

Fig. 2.21 Calm water resistance data, BAZAN 80 prototypes

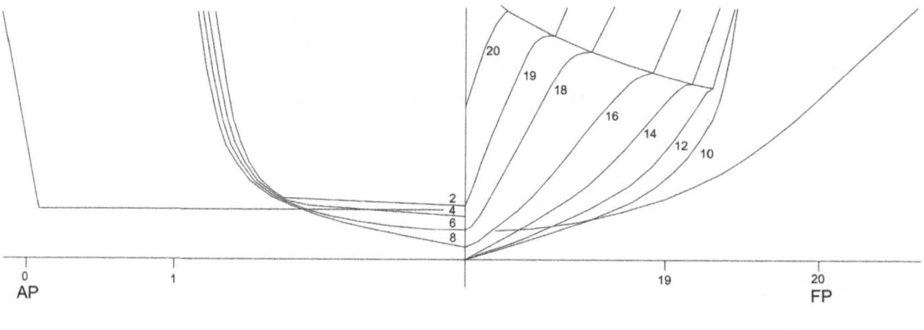

Fig. 2.22 SKLAD parent body plan

2.2.8 SKLAD Series 1980

This is a model test series carried out at the Brodarski Institute in Belgrade from 1972 to 1980 [19]. The body plan had a bilge radius reducing to almost a chine for the last 20% of Lwl to the transom with deadrise of 12°. Forward of amidships there was a spray deflecting knuckle for about 40% of Lwl. The models were divided into three groups based on their C_B. There were 27 models in this series based on the parent hull shown in Fig. 2.22. Parameters for the parent hull and the variations in *L/B*, *B/T*, and C_B are shown below.

$L/\nabla^{1/3}$	6.5	4.5, 6.5, 8.5
L		2.7 to 6 m with displacement constant at 0.23 m^3
L/B	6	4, 6, 8
B/T	4	3, 4, 5
C_B	0.45	0.35, 0.45, 0.55
LCB	9.3	8.8, 9.3, 9.2% aft of amidships for each C_B (or is this LCF)
C_p	0.715	
C_M	0.621	0.49, 0.621, 0.77
i_E	6.5–11.0°	
Fn_∇	1–3 (seakeeping series tested 0.285–1.14)	
Bt/B	Constant approx. 0.9	

This series had some parameters similar to the NPL series, while the bottom surface at the stern was flatter. The LCB, on the other hand, was set 2.5% further aft. At higher speeds one might expect the chined stern area to develop more lift so reducing trim. Meanwhile the knuckle form from the bow would deflect upwelling water.

Radojcic completed a detailed regression analysis of the test results [19] to define equations and parameter tables for both drag and hull wetted surface area. We discuss the analysis in Sect. 2.3 and give details of the expressions in Appendix 1.

2.2.9 NSMB HSDHF Test Series

The NSMB/MARIN series program [13, 20] was carried out through the 1980s jointly supported by the Office of Naval Research in USA, the Royal Netherlands Navy in the Netherlands and later also the Department of Defence in Australia. It was aimed at developing a hull form series that could be usefully applied to a patrol boat, a frigate, or a cruiser. A key aim was a series with low resistance at high speed while also having low motion response. On a patrol boat one may expect to have L/B at the lower end of the range and the larger vessels to have a slimmer profile.

When selecting the parameters to vary, it was noted that L/B and C_B have a primary effect on resistance, together with C_p and C_M. The variables B/T and C_{WP} together with secondary variables C_{VP} and L/T are key factors controlling transverse stability and seakeeping response.

NSMB chose to fix Cp, C_{WP}, C_{VP}, LCB, and LCF, B_T/B and i_E. It can be seen therefore that draft would increase as L/B is increased. LCB was selected at 5% aft of amidships, as experience had shown this to be close to optimum over the parameter range considered. Cp was chosen at 0.63 in the initial phase of testing to determine a suitable parent hull for a main series of tests.

Table 2.3 Typical vessel data

Type	Lwl m	Displacement m³	Speed knots	Fn_L	Fn_∇	$L/\nabla^{1/3}$
Patrol boat	40	300	40	1.04	2.54	6.0
Corvette	85	1800	30	0.53	1.41	7.0
Frigate	125	3800	30	0.44	1.25	8.0
Destroyer	155	7300	35	0.46	1.31	8.0
Cruiser	210	18,000	40	0.45	1.28	8.0

For information, reference [20] gives typical dimensions for the type of vessels that the test program was intended to cover, as follows (Table 2.3):

Due to uncertainty on the optimum for LCF (center of floatation at waterplane) three base models were drawn up as shown in Fig. 2.23 with LCF varied from 6% to 9% aft of amidships. It was not possible to keep C_{WP} constant, and it reduces as the hull shape is changed between forward narrow and aft wide, to forward wide and aft narrow. Model testing in waves did later show significant differences in response between these geometries.

Results with the first three models (models 1–3 in Fig. 2.23) suggested the possibility of further optimization and so three further combinations of forward and after geometry were created from the initial models—medium/wide, wide/wide, and wide medium. The models built typically had length of 5 m and a displacement around 0.2 m^3.

Model testing initially for calm water resistance and then in waves was carried out on these six models to select a best possible geometry for a "parent hull" for the main series of tests with 27 models following the parameter cube formed by *L/B*, *B/T*, and C_B shown in Table 2.4.

The six initial models had the same displacement and closely the same wetted area, and so comparisons could be made based on the residuary drag. The results were normalized based on results found for model 2 (see Fig. 2.23) which was the "parent" for this initial series. This is shown in Table 2.5 below extracted from [13].

To evaluate the initial subseries of six models for motion response, model tests were carried out in regular bow waves ranging from 0.6 to 2.8 ship lengths with a height of 2% of ship length. These results were then evaluated, and a figure of merit determined, based on that suggested by Bales [26], listed as FOM in Table 2.5. From this work, it was concluded that Model 5 should become the parent hull for the main set of model tests as mentioned above.

Model testing was begun on the main series, and after testing 15 models in this program two additional models were constructed to check whether the chosen Cp of 0.626 was optimum. These models were as the parent (model 5) with Cp altered to 0.561 and 0.685 and labeled as subseries 2. In 1984, a second subseries was introduced following studies in the US by DTNSRDC showing that seakeeping

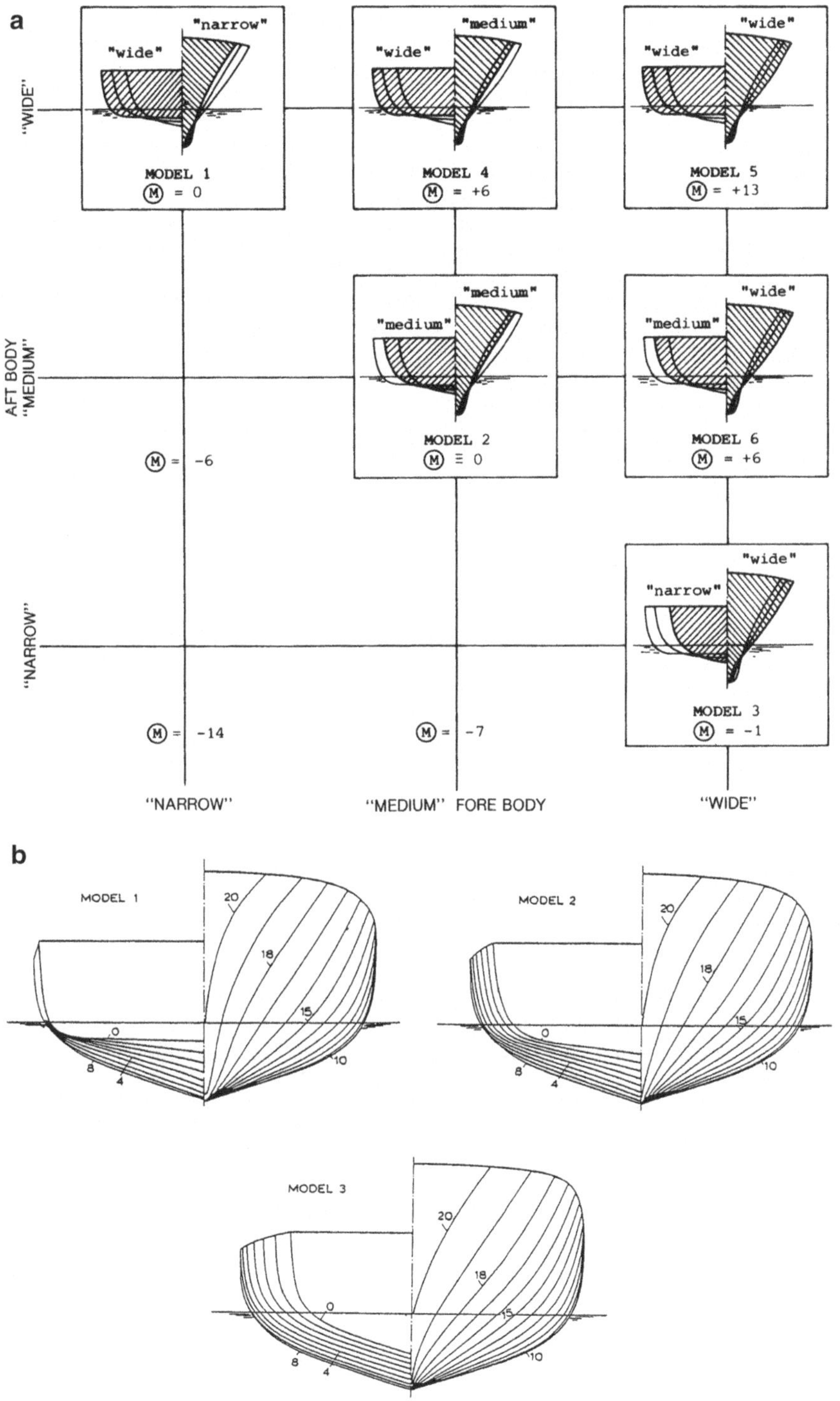

Fig. 2.23 Body Plans NSMB Series (**a**) Model Matrix (**b**) Models 1, 2 and 3

Table 2.4 NSMB model testing parameters

	Series parameters	Model 5
$L/\nabla^{1/3}$	4.3–13.1	8.65
L/B	4, 8, 12	8
B/T	2.5, 4, 5.5	4
C_B	0.35, 0.396, 0.5 (0.55)	0.396
LCB	5% aft of amidships (approx.)	5.11
C_p	0.626	0.626
C_M	0.633 0.559, 0.633, 0.798	0.633
i_E	6.5–11.0°	11.0
b_t/B	0.59, 0.74, 0.89	0.89
Fn_L	0.1–1.2 (seakeeping Fn_∇ 1.0–3.5)	
Freeboard	6% Lwl	

Table 2.5 Residuary resistance comparison normalized on model 2

	Model number					
Fore	Narrow	Medium	Wide	Medium	Wide	Wide
Aft	Wide	Medium	Narrow	Wide	Wide	Medium
Fn_∇	1	2	3	4	5	6
1.0	95.2	100	100	120.8	96.2	120.6
1.5	97.6	100	91.0	96.0	91.6	88.4
2.0	100.3	100	93.8	99.2	89.4	91.1
2.5	100.2	100	93.2	99.5	94.9	90.9
3.0	104.1	100	89.9	100.0	98.0	88.7
3.5	103.5	100	89.8	101.1	98.0	90.4
FOM	0	0	−1	+6	+13	+6

performance of the parent was better than constructed designs to date in the frigate to destroyer size range.

A further eight models were constructed with C_B 0.45 and 0.55, *L*/*B* 6 and 8, and *B*/*T* 2.5 and 4, as can be seen in Fig. 2.24 to test this possibility.

Key results from the model tests showed that increasing *L*/*B* reduced resistance and improved seakeeping, and further that increased *B*/*T* showed the same tendency. A lower C_p also reduced calm water resistance.

Testing data from this series was digitized and regression analysis carried out to produce predictive models for residual drag and other parameters of interest in selecting parameters for vessels within the tested envelope. Subsequent analysis documented in [3] has resulted in predictive equations for drag and wetted surface area that can be used to compare with results from other series.

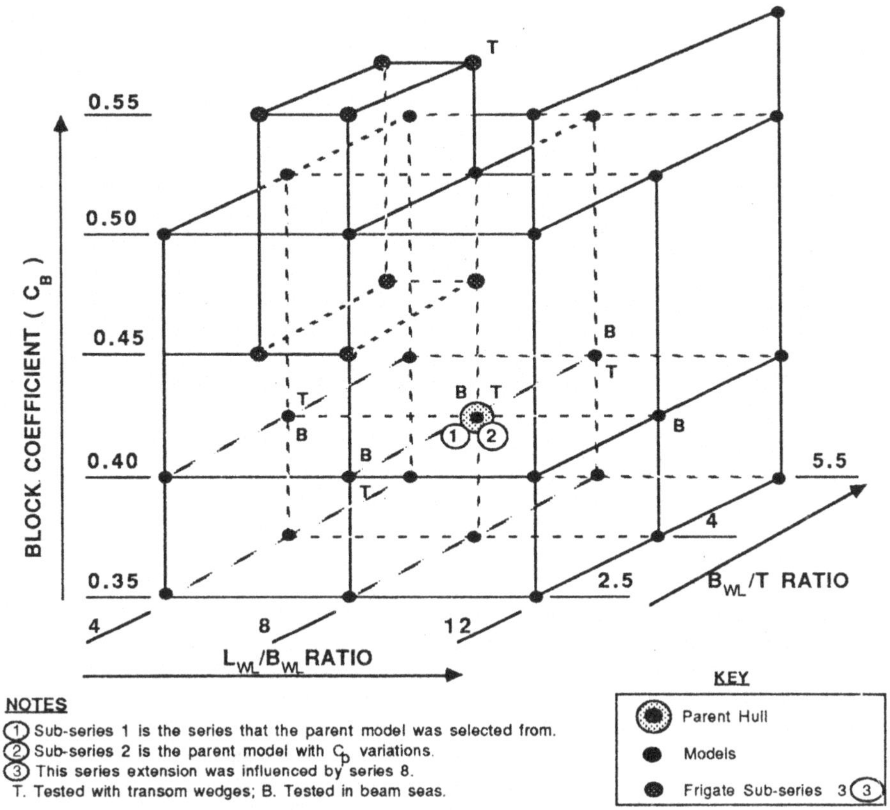

Fig. 2.24 NSMB adjusted model test dimensions "magic cube" diagram from Robson 1988 fig. 8 showing the base cube, the C_p tests, and the frigate series

2.2.10 AME CRC Series 1999

Following on from the HSDHF series, the same sponsors completed a further series of model tests aimed at further improvements particularly in seakeeping [3, 21, 22].

The same parent model hull form was used. Fourteen models were built with the same length of 1.6 m to variations in dimensions as shown in Fig. 2.25. The parent lies in one corner of the parameter cube defined by *L/B* 4–8, *B/T* 2.5–4, and C_B 0.396–0.50. It may be seen that this envelope falls within the HSDHF envelope and extends the frigate subseries to lower C_B and lower *L/B*.

The parent hull body plan is shown in Fig. 2.25 for reference. Related to the initial selection process at NSMB this was termed "wide-wide" fore and aft. This geometry formed the center of the variation in parameters for the NSMB series.

Having determined from NSMB Frigate subseries that for smaller vessels there were advantages to lower *L/B* and higher C_B for seakeeping the AMECRC series focused on this area. While the parent and some other points duplicated models in the earlier series, the remainder provided finer granularity within this smaller envelope as can be seen in Fig. 2.26.

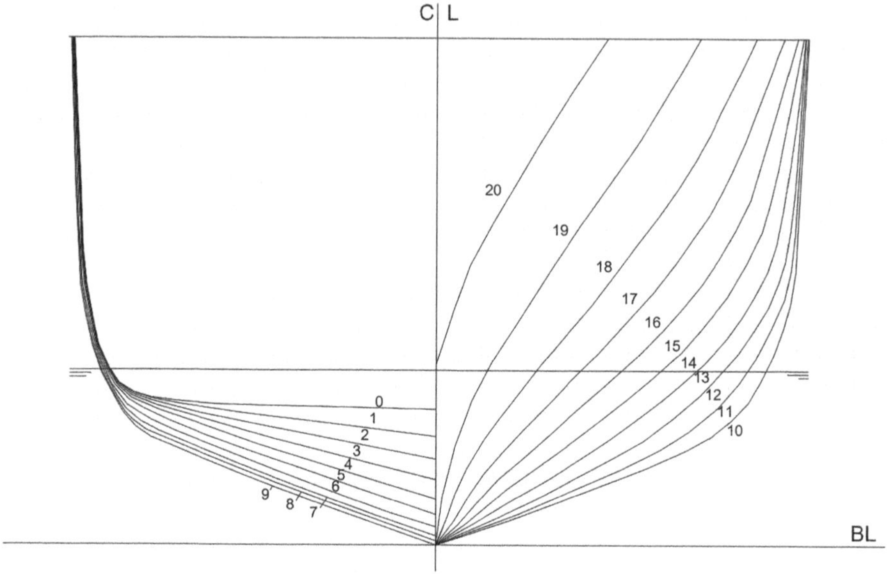

Fig. 2.25 Typical body plan AMECRC

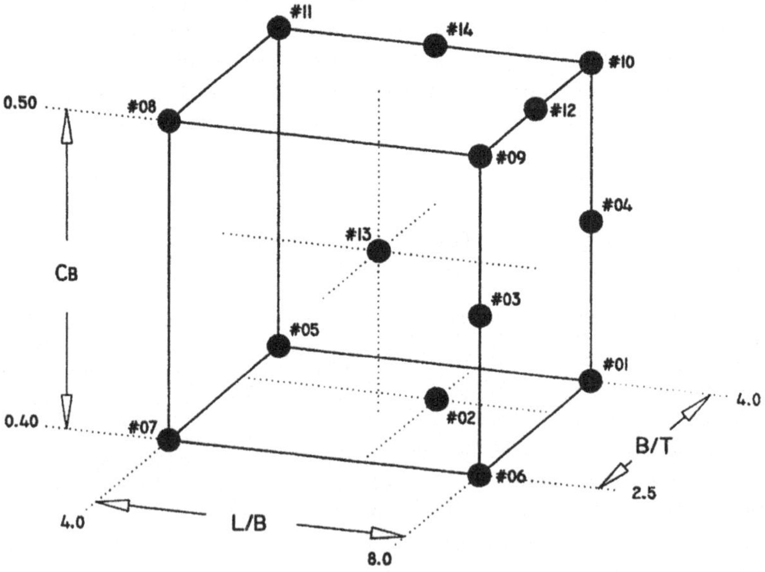

Fig. 2.26 AMECRC Testing parameter cube. (Note is from Sahoo HPMV paper 2012 or FAST 99 paper)

Table 2.6 AMECRC model parameters

Model	L/B	B/T	C_B	Model Δ kg	$L/\nabla^{1/3}$
1	8	4	0.396	6.321	8.653
2	6.512	3.51	0.395	11.455	7.098
3	8	2.5	0.447	11.454	7.098
4	8	4	0.447	7.158	8.302
5	4	4	0.395	23.344	5.447
6	8	2.5	0.395	10.123	7.396
7	4	2,5	0.396	40.523	4.658
8	4	2.5	0.5	51.197	4.308
9	8	2.5	0.5	12.804	6.839
10	8	4	0.5	8.002	7.998
11	4	4	0.5	32.006	5.039
12	8	3.25	0.497	9.846	7.464
13	6	3.25	0.45	15.784	6.379
14	6	4	0.5	14.204	6.606

The models were all tested in the Ship Hydromechanics center at the Australian Maritime College (AMC) over a speed range of Fn_L between 0.1 and 1.0. By constraining the models to the same length, displacement varied from 6.321 kg to 51.197 kg and so $L/\nabla^{1/3}$ ranging from 8.653 down to 4.308, see Table 2.6.

In addition to the model tests, computer-based analysis for each model was carried out to predict resistance. Two software programs were used, both developed at AMC, called Shipflow, and Hydros/4, the results of this work are presented in [28]. The authors note that while MARIN developed a computer prediction program from the HSDHF series, the test data themselves have not been published in detail.

Following up on the AMECRC test program regression analysis has been able to be carried out on both HSDHF and AMECRC data, so that predictions are able to be made for hull geometries within this envelope [21]. We discuss this later in the chapter.

2.2.11 VTT Series

The VTT series [23] is a subset of the earlier NPL series [12]. The aim was for a shallow draft vessel suitable as a patrol craft in the coastal waters surrounding Finland. A total of five models were constructed, with characteristics as in Table 2.7a. The parent model NOVA I was based on NPL model 100A with lower L/B, higher B/T, and higher C_B.

The length/displacement ratio at design waterline for all models was kept constant at the same value as NPL model 100A, but tests were carried out at lighter and heavier drafts resulting in this ratio varying between 6.0 (85% load) and 7.0 (130% load).

Table 2.7a VTT Series main dimensions (design draft)

Model	L (m)	B (m)	T (m)	∇(m^3)	S (m^2)
NPL	2.8	0.4480	0.1543	0.076831	1.296
NOVA I	2.8	0.5176	0.1178	0.076831	1.343
II	2.8	0.6154	0.0892	0.076831	1.492
II	2.8	0.5176	0.0964	0.076831	1.365
IV	2.8	0.4480	0.1021	0.076831	1.279
V	1.2	0.4364	0.0726	0.01510	0.444

Table 2.7b VTT Series main coefficients (design draft)

Model	$L/\nabla^{1/3}$	L/B	B/T	C_B	C_X
NPL	6.586	6.25	2.9	0.397	0.5730
NOVA I	6.586	5.41	4.39	0.45	0.6494
II	6.586	4.55	6.90	0.50	0.7215
II	6.586	5.41	5.37	0.55	0.7937
IV	6.586	6.25	4.39	0.60	0.8658
V (hard chine)	4.855	2.75	6.00	0.397	0.5730

Constant LCB 6.4% aft amidships, B_T/B 0.815, T_T/T 0.513, A_T/A_X 0.52

Model V was a hard chine hull for comparison purposes. After the main tests, model NOVA I was converted to hard chine form increasing its wetted area slightly to 1.381 m^2. Other characteristics were held to same constants as NPL 100A, as noted below in the table (Table 2.7b).

A view of the body plan and lines plan of NOVA I is shown in Fig. 2.27. The bow and keel profile also follow that of the NPL hull giving a raked fine forefoot.

Calm water resistance tests were carried out from Fn_∇ 0.6 to 3.8 for the round bilge models, whereas for NOVA V, they were carried out from Fn_∇ 1.2 to 5.0.

Analysis of the resistance data followed normal practice, based on determination of the frictional drag coefficient C_F using the ITTC-57 ship-model correlation, thus

$$R_F = 0.5\ C_F\ \rho\ V^2\ S$$
$$C_F = 0.075/(\log_{10} \mathrm{Rn}-2)^2$$

Where

Rn Reynolds number $V\,L\,/\,\nu$
ν kinematic viscosity of water

The total resistance coefficient C_T is defined as

$$C_T = C_F + C_R + C_A$$

Where C_R is the residual resistance and C_A is a correlation coefficient for model to ship correlation. C_R is assumed the same for model and ship. A value for C_A of around 0.00025 is generally accepted for semi displacement vessels.

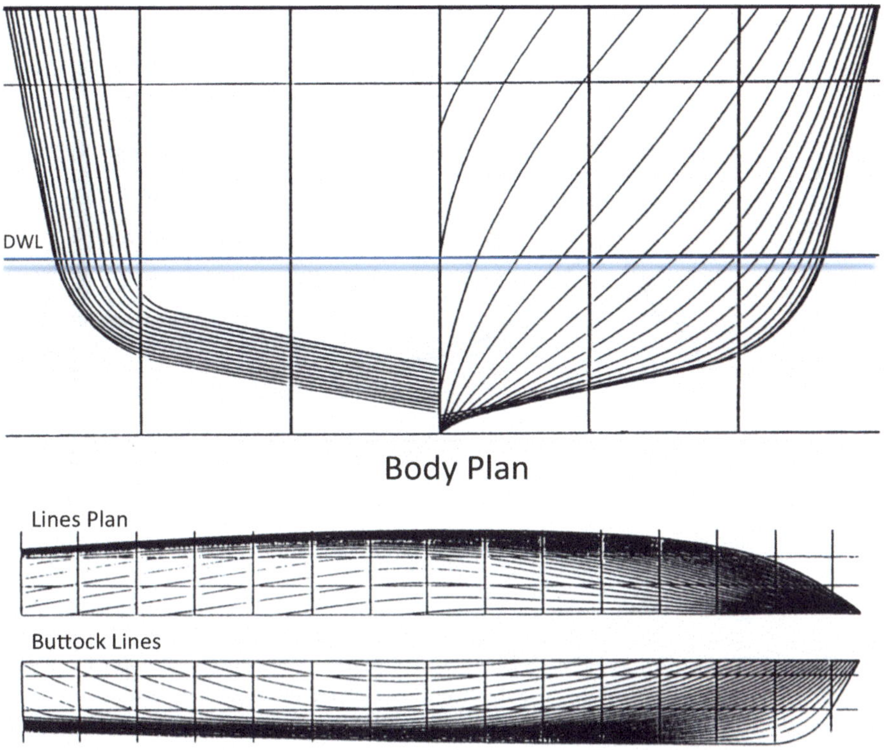

Fig. 2.27 VTT body plan

From the model testing therefore, C_R is extracted from the measured total drag by determining the frictional drag component. VTT used the static waterline wetted area for this.

Results from the resistance model tests showed those with higher C_B had higher resistance up to $Fn_\nabla = 2.5$. Above this there was no difference. At speeds above $Fn_\nabla = 2.5$ the form of the bottom aft affected the ability to generate planing lift and so while model IV was narrow, it had a larger effective lifting surface and so relative reduced drag at high speed. It can be seen though that model V with a hard chine form has dramatically lower resistance compared with the others above $Fn_\nabla = 2.75$, Fig. 2.28.

Following the test program for calm water resistance VTT carried out regression analysis of the data and developed equations in similar form to the work by Mercier and Savitsky [27] for prediction purposes. Bearing in mind that this earlier work was valid in the Fn_∇ range of 1 to 2, the new analysis was carried out for Fn_∇ up to 3.2. Data from NPL, SSPA, and the VTT series was used, yielding 65 sets of model data. Correlation testing was carried out, resulting in conclusions that the original Mercier

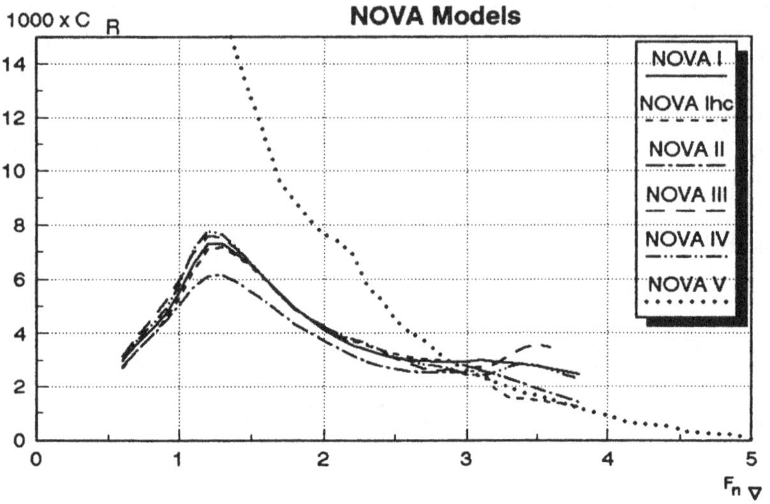

Fig. 2.28 (**a**) Residual resistance of NOVA models (from Figs. 9 and 10 of paper). (**b**) Total resistance of NOVA models. Note total drag in fig. 2.28b is scaled to 45.36 Tonnes (100,000 lbs) vessels, while residual drag in Fig. 2.28a is the residual drag coefficient

and Savitsky model was reliable below Fn_∇ 1.8 while a new equation was developed for the speed range up to 3.2.[2] We discuss this further in Sect. 2.3 below.

VTT also carried out seakeeping tests of two models, NOVA II and NOVA IV, that represented the widest and narrowest hull forms. We discuss this work in Sect. 2.6 below.

2.3 Resistance Estimation by Means of Regression Expressions

At the initial design stage, while the lines of a craft have not been developed, it is necessary to estimate drag from the main parameters to select a start point for a hull form. In addition, a designer will probably already have an idea of desired form to accommodate the main machinery and propulsion system.

We have presented in Sect. 2.2 a range of the test series carried out allowing interpolation within a hull series either graphically or by means of regression expressions derived from the tests.

Early work such as that at NPL and CSSC presented graphs. Later, regression analysis has been carried out on both individual series and collections of model data. Probably the widest to date is that carried out by Mercier and Savitsky [27]. Because

[2] Savitsky noted in discussion of [23] that with a correction to coefficient A6 from the published +0.20152 to −0.20152 the equations should give reliable results up to Fn_∇ of 2.0

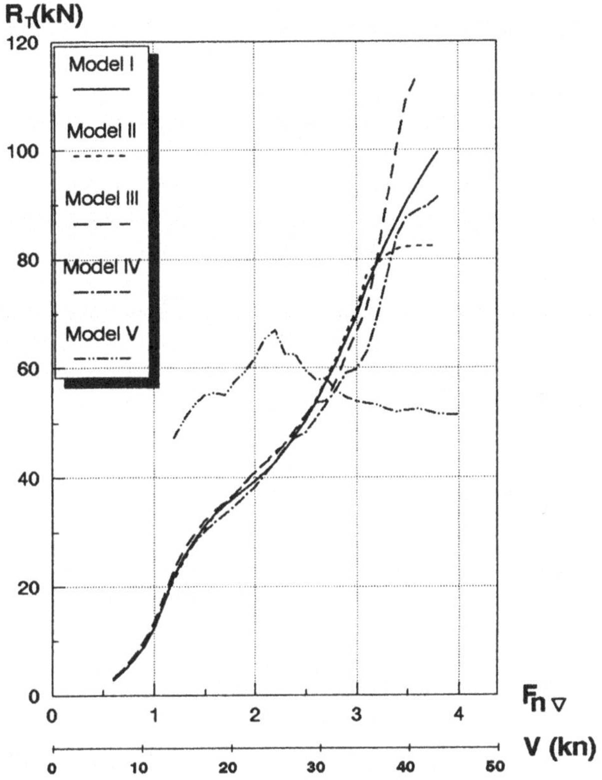

Fig. 2.28 (continued)

the regression expressions from this work are obtained from experimental results of many model series, including both round bilge and planing hulls, the precision is not high. The intent behind the work was to estimate drag through the hump speed prior to a craft achieving planing rather than to optimize a hull for operation in the intermediate speed regime. VTT regression analysis mentioned above gave a model calibrated by their own model testing that was reliable in the pre-planing and transition regime.

Radojcic et al. [19] has prepared regression analysis of the NPL series and the SKLAD series, while Bogovic [21] and Sahoo et al. [3, 28] have completed regression analysis for the AMECRC series, which encompasses the main characteristics of the earlier NSMB series.

In China MARIC has analyzed a collection of model tests, including the round bilge model data used by Mercier and Savitsky, to generate a more refined regression model for round bilge geometries, while Shanghai Jiao Tong University have carried out another analysis using data from 38 of their own models. We summarize these analyses below together with the work by Mercier and Savitsky.

First, a little background on this approach, though. Regression analysis of collections of model test data from hull forms with systematic varying parameters has a history going back to Froude. S C Fung [29] gives a summary of the history and its transition from generation of residuary resistance prediction charts to production of analytical expressions. A key issue discussed by S C Fung is the importance of primary and secondary parameters, and the regression techniques used by the different researchers. Clearly, the wider and more random with respect to form parameters the collection of data points is, the more general and less accurate the prediction will be.

S C Fung carried out his own analysis to develop a preliminary design prediction equation for use at US NAVSEA for fast combatants and discussed the effect of the variables used in his predictions. We will refer to these in our discussion of the effect of different parameters in Sect. 2.4 below.

In Appendix 1 we present a discussion of material, including that referred to above, that might be used for prediction, allowing some evaluation of drag components in relation to primary parameters that can be chosen at an initial stage. The secondary parameters may well need more detailed study, including CFD and eventual model testing. Employing the approach based on a particular model test series, it should be possible for a designer to check initial estimates against a realistic variation of hull form.

The general approach recommended by the present authors is for a designer to consider their proposed main parameters, compare with the hull forms used by NPL, SKLAD, AMECRC, CSSC, CSDC, VTT etc. and check predictions with the formulation(s) closest to the preferred parameters as a starting point. This is what we discuss in more detail in Appendix 1. Once a lines plan is available, secondary parameters might be considered and varied to identify potential effects. It has also to be noted that there will have to be a balance between minimizing drag and providing a "seaworthy" hull at the intended service speed and sea state.

We begin our discussion here with J A Mercier and D Savitsky's analyses.

2.3.1 Mercier-Savitsky Regression Expressions [27]

Mercier and Savitsky aimed initially to predict the drag of planing craft in displacement and transition operation modes. They selected a wide series of full hard chine craft, round bilge bows, and hard chine stern craft from almost all of the famous series, including the NPL series, De Groot series, Nordstrom series, SSPA series, 62 series, 63 series, and 64 series, for a total of 118 models where 62 of them are hard chine craft. The range of speeds considered was volumetric Froude Number $Fn_\nabla = 1.0\sim 2.0$.

The resistance was expressed as a function of four parameters:

- slenderness ratio $L/\nabla^{1/3}$
- half entrance angle of water line i_x
- static load coefficient ∇/B_x^3
- and the ratio transom area to maximum section area A_T/A_X

Slenderness and water line entrance angle are the main parameters. Relative transom area influences mainly the flow separation at stern, and the static load coefficient seems only to affect the planing performance but apparently not the drag of craft at displacement mode at lower speeds. They give the regression expressions for the total drag per unit displacement, as:

$$\begin{aligned} R_T/\Delta = A_1 + A_2X + A_4U + A_5W + A_6XZ + A_7XU + A_8XW + A_9ZU \\ + A_{10}ZW + A_{15}W^2 + A_{18}XW^2 + A_{19}ZX^2 + A_{24}UW^2 + A_{27}WU^2 \end{aligned} \tag{2.7}$$

Where:

$X = \nabla^{1/3}/L$, $Z = \nabla/B_x^3$, $U = (2i_x)^{0.5}$, $W = A_T/A_x$ and

∇ volumetric displacement, m^3
L length of water line, m
B_x max water line beam, m
A_x Max sectional area, m^3
A_T Transom area, m^2

Table 2.8 below gives the various coefficients in the equation for the range of volumetric Froude numbers between 1.0 and 2.0.

The coefficients in Table 2.8 are for a craft weight of 100,000 pounds (45,630 kg), at sea water temperature 15 ° C, and using Schoenherr friction drag coefficient with roughness allowance $\Delta C_f = 0$. In case of craft with different dimensions, the drag has to be corrected using the following expression:

$$\frac{R_T}{\Delta} = \left(\frac{R_T}{\Delta}\right)_0 + \left[\left(C_f' + \Delta C_f\right) - C_{fo}\right]\frac{1}{2}S/\nabla^{3/2} \times F_V^2 \tag{2.8}$$

where

R_T/Δ total drag of calculated craft, per unit displacement
$(R_T/\Delta)_0$ total drag per unit displacement at standard condition
Cf' Friction drag coefficient for revised craft data
ΔC_f roughness allowance (normally set at 0.004)
C_{f0} Friction drag coefficient using Schoenherr Formula
∇ volumetric displacement
Fn_∇ volumetric Froude Number
S wetted surface area

Table 2.8 Coefficients A_i in (2.7) versus volumetric Froude number

Fn_∇Coefficient	1.0	1.1	1.2	1.3	1.4	1.5	1.6	1.7	1.8	1.9	2.0
A_1	0.0617	0.1078	0.0948	0.0348	0.0301	0.0316	0.0319	0.0434	0.0504	0.0561	0.0597
A_2	−0.4868	−0.8879	−0.6372	0.0	0.0	0.0	0.0	0.0	0.0	0.0	0.0
A_4	−0.0103	−0.0163	−0.0154	−0.0098	−0.0064	0.0	0.0	0.0	0.0	0.0	0.0
A_5	−0.0649	−0.1344	−0.1353	−0.0510	−0.0554	−0.1051	−0.0860	−0.1329	−0.1560	−0.1866	−0.1976
A_6	0.0	0.0	−0.1605	−0.2188	−0.1936	−0.2054	−0.1944	−0.1806	−0.1781	−0.1829	−0.2015
A_7	0.1063	0.1819	0.1680	0.1043	0.0961	0.0601	0.0619	0.0549	0.0510	0.0474	0.0465
A_8	0.9731	1.8308	1.5597	0.4351	0.5182	0.5823	0.5205	0.7820	0.9286	1.1857	1.3003
A_9	−0.0027	−0.0039	−0.0031	−0.0020	−0.0022	−0.0037	−0.0036	−0.0033	−0.0031	−0.0024	−0.0021
A_{10}	0.0109	0.0147	0.0348	0.0411	0.0390	0.0479	0.0443	0.0419	0.0411	0.0412	0.0434
A_{15}	0.0	0.0	0.0	0.0	0.0	0.0832	0.0737	0.1215	0.1493	0.1809	0.1977
A_{18}	−1.4096	−2.4670	−2.1556	−0.9266	−0.9528	−0.7090	−0.7206	−0.9593	−1.1218	−1.3864	−1.5513
A_{19}	0.2914	0.4731	1.0299	1.0639	0.9776	1.1974	1.1812	1.0156	0.9314	0.7841	0.7828
A_{24}	0.0297	0.0588	0.0520	0.0221	0.0241	0.0	0.0	0.0	0.0	0.0	0.0
A_{27}	−0.0015	−0.0036	−0.0030	−0.0011	−0.0014	0.0	0.0	0.0	0.0	0.0	0.0

Wetted surface area can be calculated by the following approximate formula:

$$S/\nabla^{2/3} = 2.262\sqrt{\frac{L_{WL}}{\nabla^{1/3}}}\left[1 + 0.046\frac{B_X}{T} + 0.00287\left(\frac{B_X}{T}\right)^2\right] \tag{2.9}$$

Or

$$S/\nabla^{2/3} = \left(\frac{L_{WL}}{\nabla^{1/3}}\right)^2\left(1.7\frac{B_X}{L_{WL}}\cdot\frac{T}{B_X} + \frac{B_X}{L_{WL}}\cdot C_B\right) \tag{2.10}$$

The error for this formula is expected to be within 10%.

2.3.2 MARIC's Regression Expressions [30]

The Marine Design and Research Institute of China (MARIC) considered that the geometric variation of the models used in Mercier-Savitsky's regression expressions was very large and so reduced the calculation precision. Thus, MARIC deleted the planing hulls of 62 series, the very slender craft of 64 series, craft with vertical keel as SSPA series, and only kept the NPL series, De Groot series, Nordstrom series and 63 series. They then added some test data from Chinese round bilge craft to formulate a regression equation.

The residual drag coefficients were given in the regression formula as a function of volume displacement/length coefficient $C_\nabla = \nabla/(L/10)^3$, the prismatic coefficient Cp, the relative transom area A_T/A_X, the longitudinal location of C_B, X_C, the half entrance angle of the bow water line i_S, and the length Froude number Fn_L (Fn). The equation for the drag coefficient can be expressed as follows:

$$C_R = C_\nabla(F_1 + F_2 + F_3F_4 + F_5) \times 10^{-3} \tag{2.11}$$

Where:

$F_1 = f(C_\nabla, F_n)$ coefficient related to displacement/length coefficient
$F_2 = f(A_T/A_X, F_n)$ coefficient related to relative transom area
$F_3 = f(C_P, F_n)$ coefficient related to prismatic coefficient
$F_4 = f(i_S, F_n)$ coefficient related to half entrance angle of water line
$F_5 = f(X_C, F_n)$ coefficient related to longitudinal location of CB
which are presented in Figs. 2.29, 2.30, 2.31, 2.32 and 2.33 respectively.

The applied ranges of (2.11) are:

C_∇ = 1.5 – 5.7 for Fn = 0.9–1.0
C_∇ = 1.5 – 11.0 for Fn = 0.4–0.9
A_T/A_X = 0.0 – 0.74

Fig. 2.29 $F_1 = f(C_\nabla, F_n)$

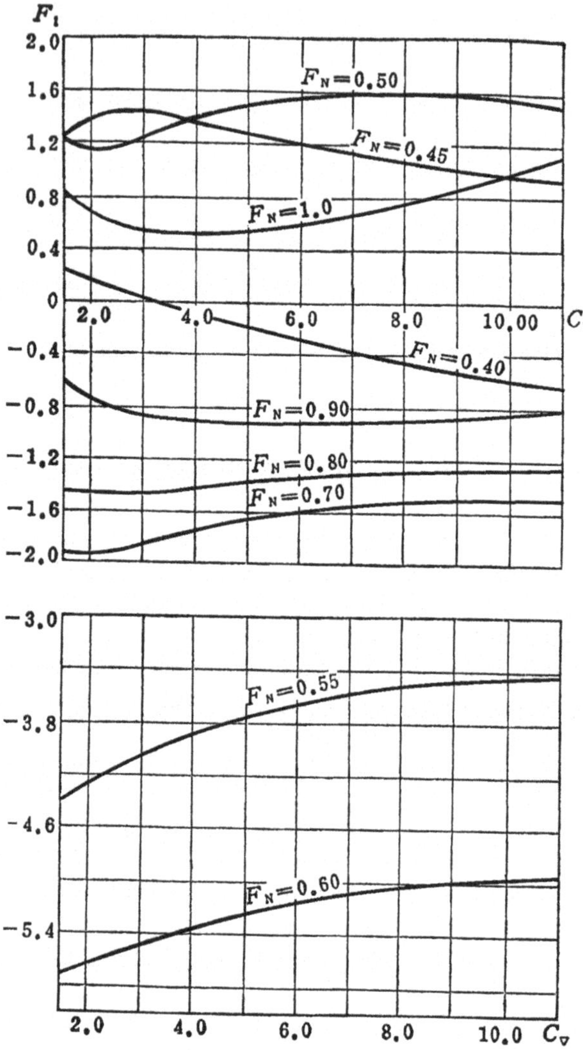

C_P = 0.573 – 0.764
X_C = 0.0 – 6.4% aft of amidships
i_S = between 10 and 20°

Considering the wetted surface area is related to B/T, C_P, and C_B, the following regression expressions for the wetted surface area were obtained.

$$C_S = 6.6912 - 0.6742\left(\frac{B}{T}\cdot C_P\right)^{0.5} + 2.8961\frac{B}{T}\cdot C_P - 0.1156 \cdot C_B\left(\frac{B}{T}\cdot C_P\right)^{2.5} - 0.1591\,C_B \tag{2.12}$$

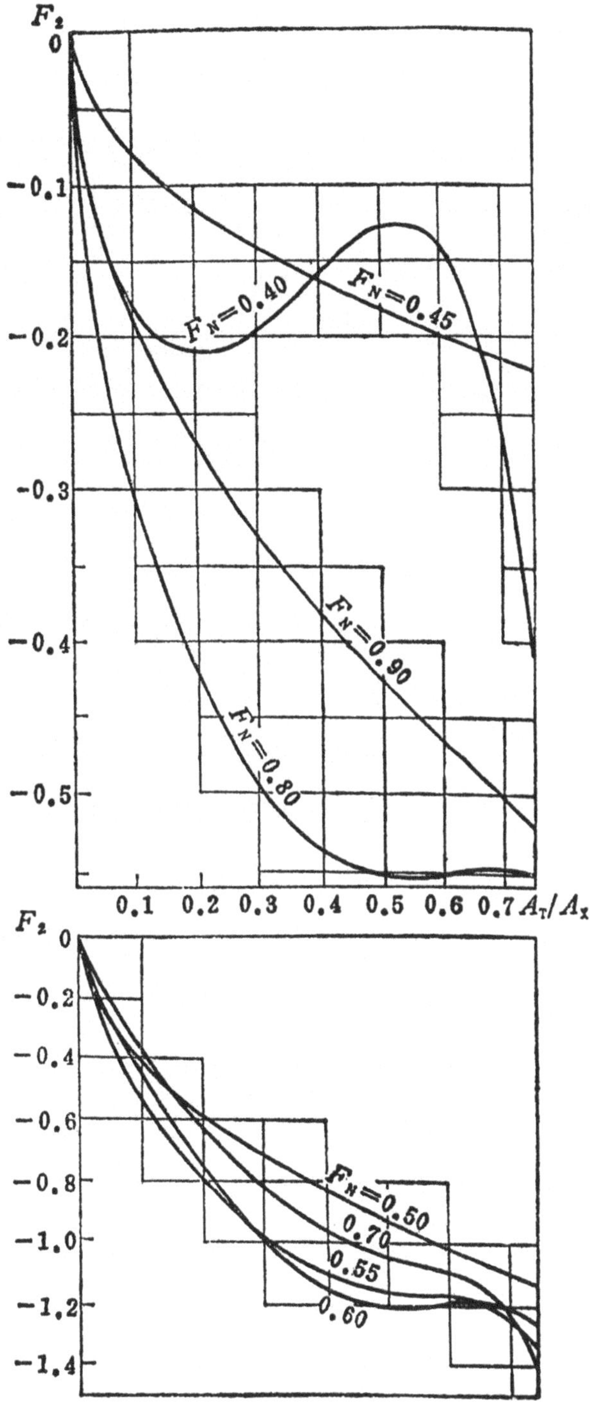

Fig. 2.30 $F_2 = f\left(A_T/A_X, F_n\right)$

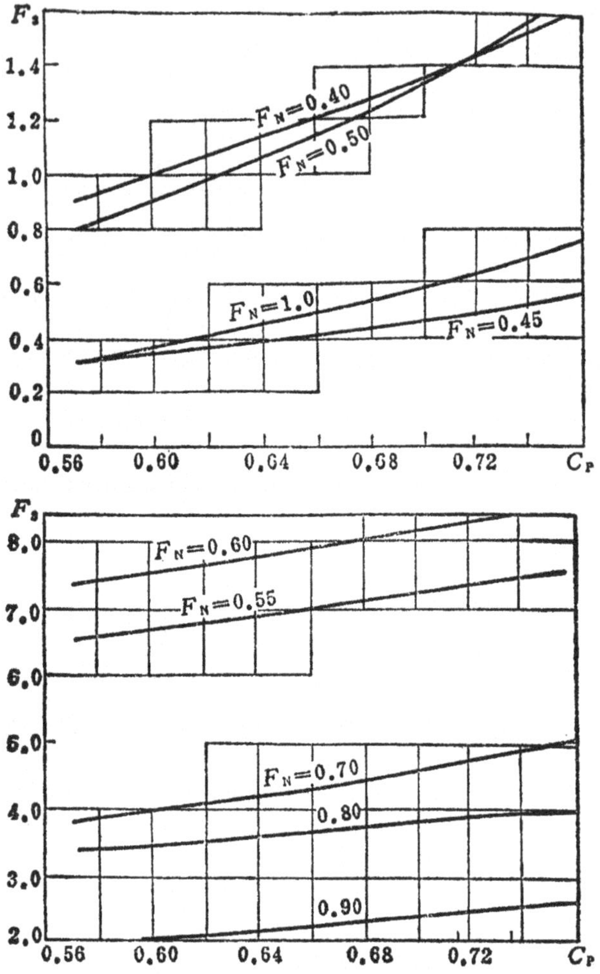

Fig. 2.31 (**a**) $F_3 = f(C_P, F_n)$ (**b**) $F_3 = f(C_P, F_n)$

The definition of wetted surface area coefficient C_S is:

$$C_S = \frac{S}{\sqrt{\nabla L}} \tag{2.13}$$

The error in calculated residual drag from (2.11) is mostly under 8%, and for wetted surface area from (2.12) is under 9%, and for total drag using both formulae is under 6%.

Fig. 2.32 $F_4 = f(i_S, F_n)$

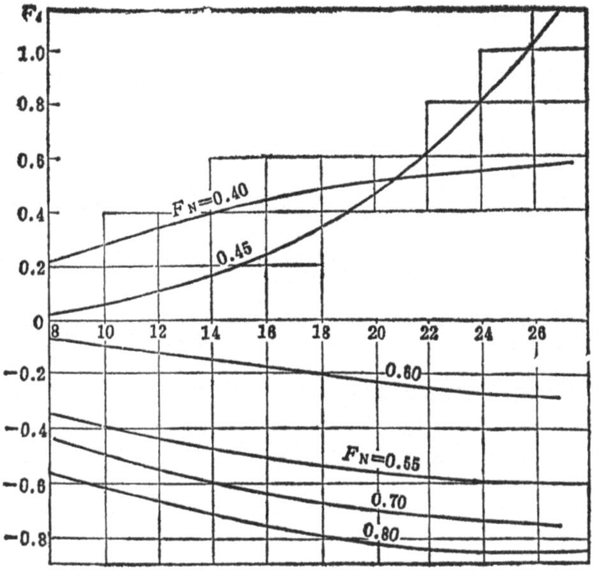

Fig. 2.33 $F_5 = f(X_C, F_n)$

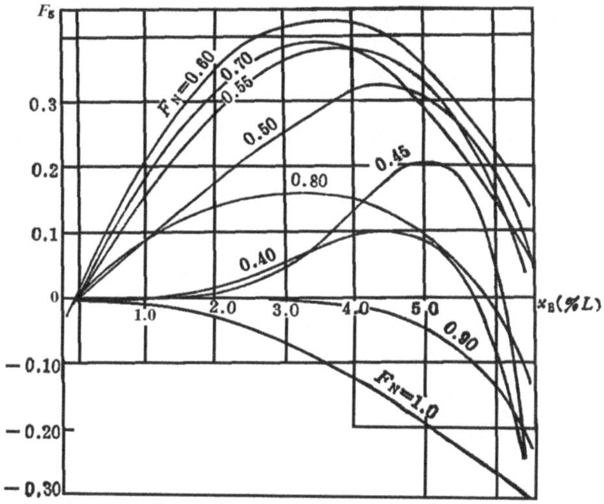

2.3.3 *Regression Formula of Shanghai Jiao Tong University, China [31]*

Researchers at Shanghai Jiao Tong University collected the drag data of 38 round bilge craft and performed regression analysis in terms of slenderness ratio, longitudinal position of CB, and block coefficient to obtain the drag estimation formula as follows:

$$C_R\, x\, 10^3 = A_0 + A_1X + A_2Y + A_3Z + A_4X^2 + A_5XY + A_6XZ + A_8X^2Y \quad (2.14)$$

Which is applied for Froude number based on craft length, $Fn = 0.4$ to 0.95 and also:

$$C_R\, x\, 10^3 = A_0 + A_1X + A_2Y + A_3Z + A_4X^2 + A_5XY + A_6XZ + A_8X^2Y + A_{11}X^2YZ \quad (2.15)$$

which is applied for Froude number based upon volumetric displacement, $Fn_\nabla = 1.0 - 2.2$.

where

$X = L/\nabla^{1/3}$ the slenderness
$Y = -X_c/L$ the longitudinal position of CB (%)
$Z = C_B$ block coefficient

The coefficients of formulas (2.14) and (2.15) are listed in Tables 2.9 and 2.10 respectively.

Application ranges of (2.14) and (2.15) are

Slenderness Coefficient	$L/\nabla^{1/3} = 5.16$ to 8.15
Longitudinal position of CB:	$X_c/L = 1.25$ to 6.4% aft of midship section
Block coefficient	$C_B = 0.39$ to 0.51
Beam draft ratio	$B/T = 3.00$ to 4.26

Wetted surface area coefficient $C_S = \frac{S}{\sqrt{\nabla L}}$ can be calculated from:

$$C_S = 2.77 + 0.065\frac{B}{T} + 0.008\frac{L}{\nabla^{1/3}} - 0.67C_B$$

2.3.4 Regression Evaluation Summary

S C Fung prepared a regression model based upon the existing work and model test data available in the late 1970s including NPL, Series 63, etc. His regression model was incorporated into a spreadsheet program for initial assessment of potential US Navy fast naval combatants as described above.

VTT took the approach of Mercier and Savitsky to develop a regression equation building on the NPL test series data added to their own. Testing the model, they felt it adequately predicted resistance for the hull forms they had tested. The main difference between VTT and original NPL series was the flatter aft hull geometry for VTT.

Radojcic completed a regression analysis of the SKLAD test series and came to similar conclusions within the form parameters he had tested.

Table 2.9 Regression coefficients of relation (2.14)

$Fn_L \rightarrow$ Coefficients ↓	0.40	0.45	0.50	0.55	0.60	0.65	0.70	0.75	0.80	0.85	0.90	0.95
A_0	35.1740	69.7017	88.6147	83.5466	58.4763	51.2576	45.1331	40.5358	40.9690	37.3348	31.5225	27.5969
A_1	−7.0758	−15.4303	−20.4756	−19.2147	−16.2014	−14.3125	– 121,302	−10.2175	−8.9232	−8.1314	−6.7029	−5.8773
A_2	0.0	−0.2048	0.0	0.0	0.0	0.0	0.0	0.0	−0.1632	−0.1785	0.1621	−0.1533
A_3	0.0	0.0	0.0	0.0	49.6175	45.0454	33.0458	21.9512	0.0	0.0	0.0	0.0
A_4	0.3821	0.9025	1.2443	1.1626	1.1430	1.0193	0.8431	0.6775	0.5176	0.4742	0.3846	0.3349
A_5	0.0	0.0	0.0	0.0	0.0	0.0	0.0	−0.0282	0.0	0.0	0.0	0.0
A_6	0.0	0.0	0.0	0.0	−5.5690	−5.0832	−3.6976	−2.4689	0.0	0.0	0.0	0.0
A_8	0.0	0.0	−0.0046	−0.0042	−0.0050	−0.0048	−0.0043	0.0	0.0	0.0	0.0	0.0

Table 2.10 Regression coefficients of relation (2.15)

F_V A_{ij}	1.0	1.1	1.2	1.3	1.4	1.5	1.6	1.7	1.8	1.9	2.0	2.1	2.2
A_0	67.0743	84.2097	83.8999	70.5845	64.0225	41.3855	37.5895	30.5925	28.3251	29.8611	26.9594	24.4377	20.4846
A_1	−15.4060	−19.3666	−18.9541	−15.3765	−15.3052	−10.9187	−9.8194	−7.6497	−6.5740	−6.1575	−5.4528	−4.7110	−3.5004
A_2	0.0	0.0	0.0	0.0	0.0	0.0	0.0	0.0	−0.1569	−0.1474	−0.1434	−0.1405	−0.2961
A_3	0.0	0.0	0.0	0.0	9.8746	36.2846	29.7697	27.1753	18.8334	2.6156	1.8630	0.0	0.0
A_4	0.9137	1.1462	1.1052	0.8710	0.9169	0.7333	0.6548	0.5128	0.4150	0.3368	0.2962	0.2531	0.1567
A_5	0.0	0.0	0.0	0.0	0.0	0.0	0.0	−0.0233	0.0	0.0	0.0	0.0	0.0
A_6	0.0	0.0	0.0	0.0	0.0	−3.6030	−2.7821	−2.9325	−2.0064	0.0	0.0	0.0	0.0
A_8	0.0	0.0	0.0	0.0	0.0	0.0	0.0	0.0	0.0	0.0	0.0	0.0	0.0034
A_{11}	0.0	0.0	0.0	0.0	−0.0097	−0.0070	−0.0082	0.0	0.0	0.0	0.0	0.0	0.0

Bojovic completed regression analysis of the AMECRC data. This work was subsequently incorporated into a preliminary assessment tool for the Australian Navy. Both the US Navy and the Royal Netherlands Navy have used the same database to develop tools for themselves.

MARIC completed a regression analysis of test data culled from the NPL series, De Groot series, Nordstrom series, and 63 series added to the test data they had also generated.

Shanghai Jiao Tong University carried out regression analysis with drag data of 38 round bilge craft that they had tested.

Finally, Sahoo et al. made a revised regression analysis of several of these test series. Their initial intent had been to prepare a unified model but found it more appropriate to create separate models based around NPL test series, SKLAD, and AMECRC. They also prepared a model based on the test program carried out at Southampton University for catamarans where slenderness was extended to $\psi = 9.5$.

If we collate this material together, we have nine models plus those referred by S C Fung that might be used as tools to make first estimates for drag. These are all summarized in Appendix 1 with a discussion of their use. Further work has been carried out by Radojcic and an overall presentation and spreadsheet models given in [32].

When selecting the main parameters, it is helpful to understand their effect on performance, so a summary is presented in the next section.

The next step after making predictions based only on form parameters is the generation of lines plans and completion of hydrostatics and CFD-based predictions of drag. At that stage refinement of the hull form can be considered. It will be useful to recheck the regression model prediction using the chosen model, for eventual comparison with the CFD and eventual model tests and/or full-scale measurements.

2.3.5 Appendage Drag for Semi-planing Craft

The appendage drag, which is usually included as a percent of bare hull drag in displacement ships, cannot be included in the calculations of fast round bilge craft due to higher speeds compared to the general displacement ships and the appendage drag being proportional to V^2.

There is a misunderstanding that the percent of appendage drag of fast round bilge craft should be higher than the general displacement craft. For fast round bilge craft, wave making increases with sixth power of speed, however, viscous drag only with second power. Therefore, the increase in bare hull drag is significantly larger than appendage drag, which is mostly due to friction. Thus, the ratio of bare hull drag to appendage drag typically decreases with speed.

However, unfortunately, many designers do not consider such facts on the estimation of appendage drag of the round bilge craft. Z S Dong, one of the authors, has carried out both model and full-scale tests of a 37 m length round bilge craft with four rudders and four propellers at $Fn_L = 0.81$, and found that the appendage drag is only 13% of bare hull drag. Thus, the ratio of appendage drag with bare hull drag is smaller than that on displacement craft. Therefore, the authors do not suggest that estimating appendage drag should be in terms of number of appendages and a percentage of bare hull drag but suggest that the appendage drag should be determined in terms of appendage drag data of a reference prototype with the same speed. Of course, it can also be calculated separately for individual appendages as shown in Sect. 4.8 in Chap. 4.

2.4 Effect of Craft Geometric Parameters on Drag

In Sect. 2.2, we have summarized the main model test series with parametric variation of a number of parameters, specifically ψ, C_B, C_p, L/B, B/T, A_T/A_M, and X_c. The different series were aimed at understanding the variation of drag as these parameters were varied within specific limits.

In Sect. 2.3, we have presented an overview of regression analyses carried out by the hydrodynamicists that have either attempted to develop relations for prediction of drag within the same envelope, or to use data from multiple test series to develop more general prediction equations.

There are two subjects of interest related to the regression analyses. First there is the issue of which parameters are most influential individually, and second whether parameters interact rather than being independent, and are significant.

To address these questions, we will go back to S C Fung [29] who used test data from several of the series in Sect. 2.2 and then we will present additional views on a number of these parameters. Finally, we will conclude with an approach to parameter selection and its testing through use of relations documented in Appendix 1.

2.4.1 Slenderness ψ or Displacement/Length Ratio $\nabla/(L/10)^3$

S C Fung showed with his correlation analysis that this is the most important parameter for resistance. While being the most important at all speeds in the range, its contribution reduces at higher speeds. There is no optimum at a given speed within the range, simply higher slenderness (lower displacement/length (D/L) coefficient) gives lower drag. Nevertheless, there is a link to L/B_x and wetted areas. If higher slenderness results in an increased wetted areas, then total drag will increase.

The slenderness, or DL coefficient, is a major factor that affects the drag of round bilge craft as it directly influences the wave making as well as wetted surface area, and so friction drag. In general, a low DL or high ψ is suitable for fast craft for reducing wave making drag. However, for small displacement fast round bilge craft and their requirements for general arrangement, the slenderness ratio can't be as high as that of the destroyers. Additionally, at high speed and associated dynamic lift, it is

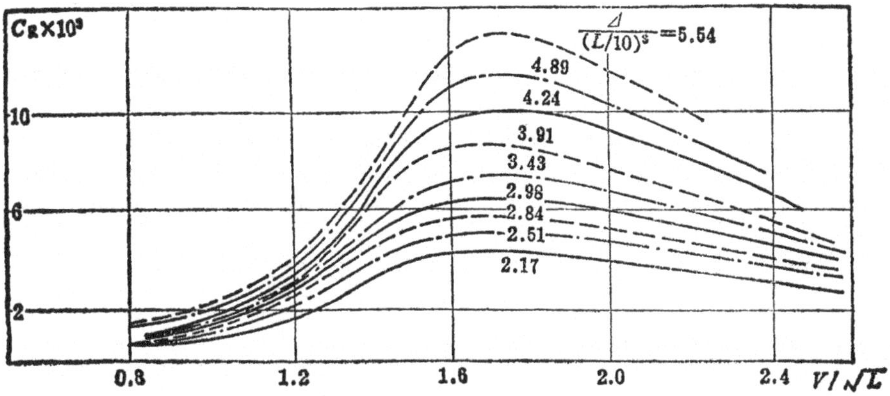

Fig. 2.34 Nordstrom's $C_R = f\left[\frac{\Delta}{(L/10)^3}, V/\sqrt{L}\right]$ curves

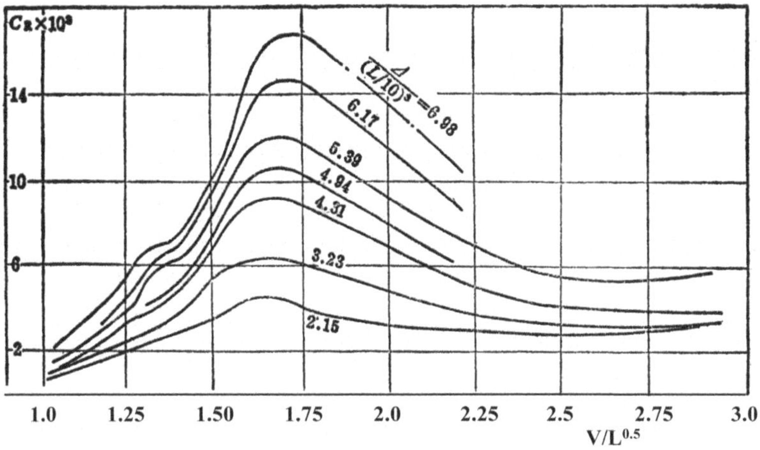

Fig. 2.35 NSMB's $C_R = f\left[\frac{\Delta}{(L/10)^3}, V/\sqrt{L}\right]$ curves

unnecessary to use a high slenderness coefficient. For example, ψ is typically 8–9 for destroyers, but 6–8 or even lower for round bilge fast craft, especially for the Deep Vee craft and craft with mixed craft profiles.

Nordström [5] presented the residual drag coefficient versus displacement-length coefficient and speed as shown in Figs. 2.34 and 2.35 shows the same curves given by Netherlands model basin. It must be noted that the speed parameter is in English units.

From the curves, it can be seen that residual drag decreases with DL, especially at high speed, with hump at $V/L^{0.5} = 1.7$ ($Fn = 0.5$). Therefore, a small value of DL (high slenderness) should be taken for high-speed craft, because in this case the decrease in wave making drag may be larger than the increase of friction drag.

To sum up, high slenderness is favorable for fast round bilge craft where $Fn > 0.4$ for reducing drag. Nevertheless, the general arrangement, especially arrangement of power plant should be considered carefully in relation to the required beam at a fixed displacement. In addition, the increase of structure weight for high slenderness should also be considered to arrive at compromise solution that best meets the requirements.

2.4.2 Beam-Draft Ratio

The effect of this ratio on resistance was found to be low by Fung, thus allowing flexibility in choice of form. Its importance did increase with speed. This is logical since at higher speeds a wider aft hull and underside enables improved dynamic lift.

2.4.3 Sectional Area Curve

The sectional area curve reflects the distribution of displacement along the craft length, which is the second important parameter after the slenderness of the craft drag. The section area curve links directly with prismatic coefficient C_p, the longitudinal position of C_B, X_c, the transom geometry and so wetted area.

From the point of view of drag of fast round bilge craft, it is required that the forward part of the sectional area curve gives a small entrance angle i_E for reducing wave making, with maximum sectional area to be located a little after the midship section for keeping good running attitude, and with a square transom geometry. We look at each of these parameters in turn.

2.4.4 Prismatic Coefficient C_p

The classical Taylor's resistance graphs gave the favorable range of C_p for displacement ships, and one of which is shown in Table 2.11, in which the displacement length ratio DL (C_∇ in Table 2.5) and $V/L^{0.5}$ (English units) were taken as parameters.

From the table it can be seen that for fast ships, C_p should be taken a little bit larger, and a little bit smaller for slender ships with same speed. In view of the speed of model tests being lower, i.e., $V/L^{0.5} \leq 2.0$, these are not suitable for fast round bilge craft.

Table 2.11 Taylor's suggested Cp values (in %) for $B/T = 3.75$

C_∇ V/L$^{0.5}$	2	3	4	5	6	7	8	9
1.0	51.0	51.7	52.2	52.6	53.0	53.1	53.2	53.3
1.1	52.6	53.3	54.0	54.2	54.6	54.9	55.0	55.0
1.2	56.2	56.0	56.2	57.0	57.9	58.5	59.0	59.6
1.3	60.5	60.5	60.5	60.5	60.5	60.5	–	–
1.4	62.5	62.5	62.5	–	–	–	–	–
1.5	66.0	66.0	66.0	–	–	–	–	–
1.6	68.0	70.0	–	–	–	–	–	–
1.8	68.4	–	–	–	–	–	–	–
2.0	69.0	–	–	–	–	–	–	–

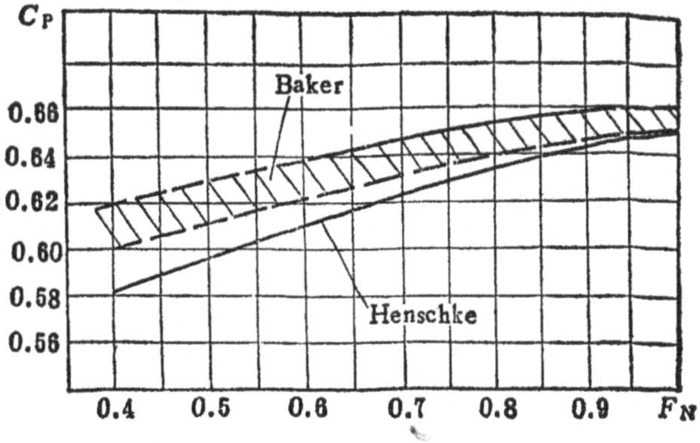

Fig. 2.36 Range of C_P for fast displacement ships

According to the test data of destroyers, where $Fn > 0.5$ it is better to take $C_p = 0.61 - 0.65$ and Fig. 2.36 shows Baker and Henschke's suggestion for the C_p range of fast displacement ships in the speed range Fn_L 0.4 to 1.0.

In view of high wave making for fast ships, it is unhelpful to make the sectional area curve with a "protruding shoulder," but rather use a flat and straight fore part with small entrance angle to give a larger C_p.

S C Fung found that C_p influence on residuary resistance was weak except at the wave resistance hump region $Fn_L = 0.5$. At high speeds, he found no correlation, which is consistent with increased effect from dynamic lift on a hull.

The recommended range for C_p is 0.64–0.68 for fast round bilge craft based on the authors' experience. Of course, the general arrangements should be considered carefully as well.

2.4.5 *Profile of Sectional Area Curve and Longitudinal Position of* C_B

Figure 2.37 shows the basic profile of sectional area curve at different speeds suggested by Henschke [33]. They basically reflect the features of sectional area curves for fast round bilge craft for reducing wave making drag, i.e., flat, and straight forward hull with small entrance angle and full aft hull, as well as C_B located aft of amidships and a submerged transom area larger than zero. With increasing Fn_L, the fullness of sectional area and C_B should be moving afterward, with the ratio of X_c and craft length 1.5–5.5% aft of the midship section. Figure 2.38 shows the favorable X_c versus Fn_L suggested by Henschke and Pyszka.

2.4.6 *Transom Configuration*

The transom area ratio TA reflects the profile of sectional area curve aft of amidships and its fullness, which influence the running trim angle, drag, seakeeping, and maneuverability.

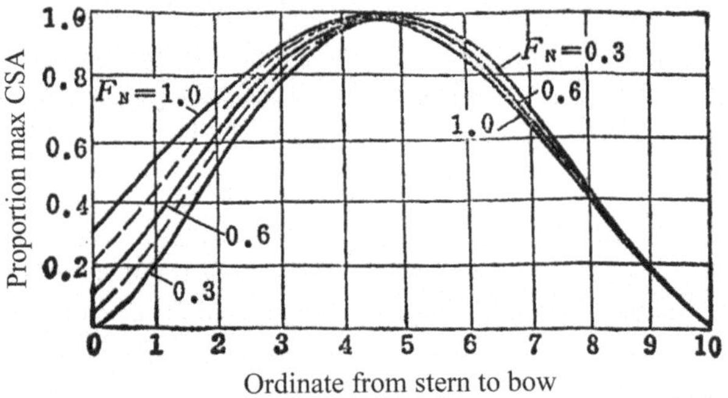

Fig. 2.37 Sectional area curve versus F_n

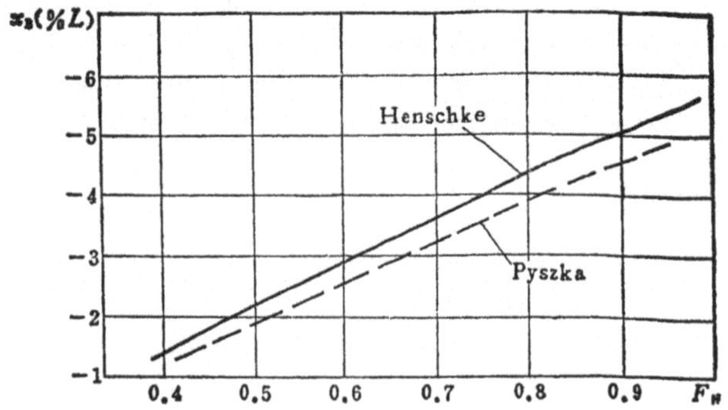

Fig. 2.38 Favored range of longitudinal position of CB

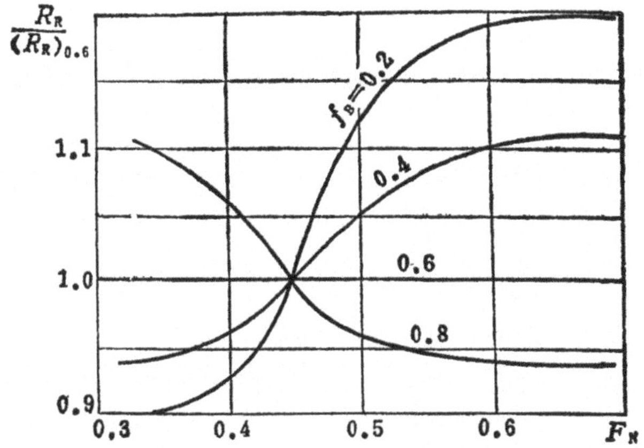

Fig. 2.39 Effects of transom width to the drag of fast destroyers

The transom area ratio also correlates with the transom width, which significantly effects the powering performance of round bilge craft. In general, when $Fn_L > 0.5$ the flow is separated from transom lower edge. The transom is then dry and the water surface forms a parabolic shape behind it returning to still water level. At planing speeds a "rooster tail" is formed beyond where the cavity ends. The cavity extending from the stern forms a "virtual length" increasing the slenderness of the craft and thus reducing wave making drag. At speeds from $Fn_L = 0.3$ to 0.5, the transom wetted area decreases to zero with attendant reducing drag component from the reentrant flow behind the transom.

Wider transom beam and square stern creates longer virtual length and so square sterns are present on high-speed destroyers and frigates. Vorogin of Russia made some plots to show the effect of transom width on drag of fast destroyers as shown in Fig. 2.39. It shows drag versus F_n for three different ratios of width of transom to maximum beam where $f_B = B_T/B_X$ and taking the drag at $f_B = 0.6$ as the reference. It can be seen that when $Fn < 0.45$ a narrow transom is favorable. However, when $Fn \geq 0.45$ a wide transom is favorable. This is since the flow behind a transom recirculates and wets the transom at low speed causing larger viscous pressure drag.

At a speed above $Fn = 0.45$ to 0.5, the flow cleanly separates from the transom reducing viscous drag, extending the hull virtual length, and consequently reducing the wave making drag. This conclusion of Vorogin can also be used for fast round bilge craft, thus a "square stern" used on fast round bilge craft is a positive selection.

Although the experimental investigation of Vorogin relates to fast displacement ships, it still can be applied for fast round bilge craft, even though the test speed for former is only limited to $Fn < 0.7$. According to the present authors' experience, when $Fn \geq 0.8$ the transom beam ratio may be taken as $f_B = 0.75$ to 0.9, i.e., using a hull with almost constant beam in the after part. This will result in the hull bottom having a shape adapted to generate higher dynamic lift.

So far as inclusion of transom parameters in a regression equation are concerned, S C Fung's analysis suggested that the three parameters of Transom width TW, depth TT, and area TA are highly correlated as far as effect on drag, and therefore should not all be incorporated individually. He included all three in his analysis, and the result in a sense validated the use of just TA by Mercier and Savitsky. Fung's recommendation is to consider the geometry at the stern as a whole. As we have discussed above, if the design vessel speed is below $Fn_L = 0.5$ then a smaller TA will be appropriate. As design speed increases above $Fn_L = 0.5$ increasing attention on the effect of transom geometry on aft hull lower surface geometry is required trending to constant beam aft of amidships and cross section area.

2.4.7 Cross Section Profile

The performance comparison of hard chine with round bilge craft was illustrated in the first part of this chapter, and the speed range of application of both types of craft was also discussed. Since the 1970s, in addition to aiming at minimum drag, improved seaworthiness was increasingly demanded by both designers and users of semi-planing craft. This led to designs based on the round bilge configuration with a short hard chine section profile at the stern, semi-hard chine lines, or long hard chine lines.

An advantage is that roll and heave damping are increased for a hard chine craft, which can improve the seakeeping performance so long as the bow profile is slender. Nevertheless, the drag of a hard chine craft is higher than round bilge forms in the speed region between $Fn_L = 0.5$ and 1.0, so designers developed the semi hard chine lines and short hard chine lines for higher speed craft and achieved a nice powering performance.

A drag comparison for both long and short hard chine and round bilge craft was carried out in Shanghai Jiao Tong University in the 1980s [34]. Residual drag curves for three different types of craft at the same slenderness coefficient $\psi = 6.95$ as shown in Fig. 2.40 were obtained.

From the figure, it can be seen that, at low speed, the round bilge craft drag is the smallest, and largest for the model with long part of hard chine. However, at high speed, the drag of models with both short and long part of hard chine models are the same and smaller than round bilge craft. It must be noted that the drag mentioned above is the residuary drag coefficient, and in general the friction drag of hard chine craft is larger than that of round bilge craft, particularly for the craft with long hard chine due to the larger wetted surface area.

The MSC test series of the China Ship Research and Design Center (CSRDC), demonstrated that nice powering performance is possible on well-designed short hard chine craft. Figure 2.41 shows the comparison curves of total drag of MSC series with three well known series models at the same displacement, slenderness coefficient, and length/beam ratio [17].

It is apparent that an effective design using short hard chine lines at the stern with round bilge lines forward is suitable for improving both seakeeping quality and powering performance. Of course, using full hard chine lines to improve seakeeping quality is also possible, particularly using the Deep Vee form, though this form

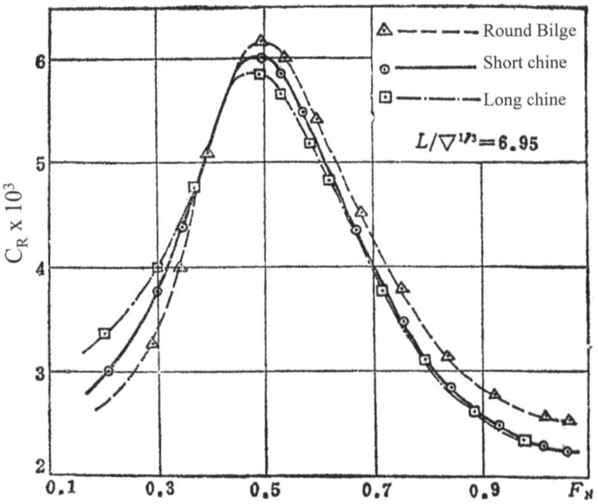

Fig. 2.40 Comparison of residual drag coefficient for different lines

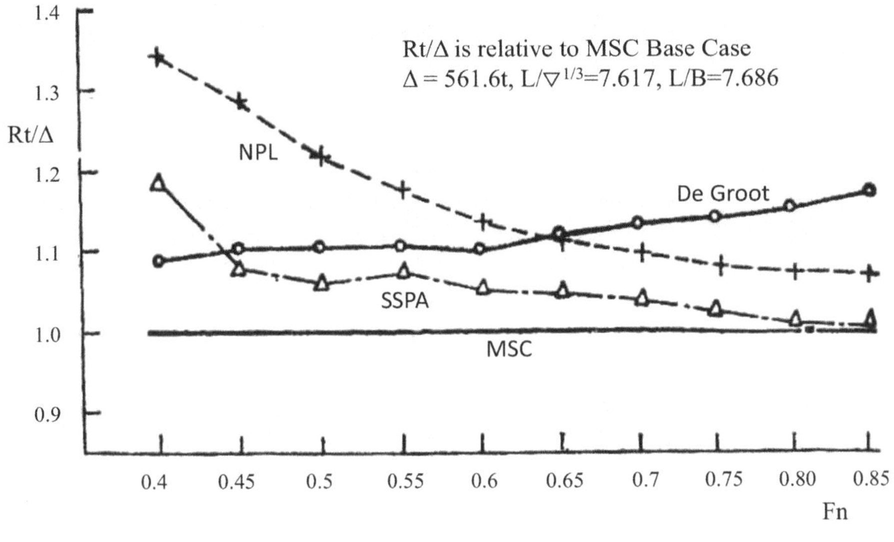

Fig. 2.41 Comparison of relative total drag *Rt*/Δ of MSC series with other series models

provides a more optimum performance at planing speeds. This will be discussed in Chap. 6.

It is interesting that the comparison shown in Fig. 2.41 compares with performance found by S C Fung, in that craft with high C_x (maximum section area coefficient) and transom stern showed reduced residual drag at low speeds compared to designs with rounder cross section. Examples quoted were the AE-36, T-AGS

45, and FDL vessels. These had transoms with low immersion, which was also the case for the MSC models, see Fig. 2.17.

2.4.8 *Half Entrance Angle*

Fung found that the half entrance angle was second only to the slenderness in correlating with residual resistance, and consistent through the speed range. Thus, at speeds up to $V/L^{0.5}$ about 1.4 increasing i_E gave higher resistance. Above this the opposite trend started. This was due to the dynamic effects increasing and the bow wave crest moving aft as it lengthened with increasing speed. Both effects are also influenced by other hull parameters.

Half entrance angle i_E is therefore important as a parameter in the regression equation. The problem is what should a designer select as a starting point? S C Fung suggested the following relation as a starter. It should be remembered that this will be for a semi-planing hull where the bow remains immersed at design speed.

$$\begin{aligned} I_E = {} & 120.0192 + 69.2808/\mathrm{DL} + 0.1131\, L_{WL}/B_x + 104.8640/(L_{WL}/B_x) \\ & - 28.5611\, C_p - 26.5287/C_p - 24.8166\, C_x - 16.2033\, C_x \\ & + 2.6392\, \mathrm{TW} - 49.6570\, L_{cb}/L_{WL} \end{aligned} \tag{2.16}$$

In (2.16) DL is displacement to length ratio, and TW is the transom width ratio. The values of I_E used in the model test series summarized in Table 2.1 varied as follows. The NPL base model had I_E of 11°, and this was also the case for the BAZAN and VTT models. NSMB and SKLAD series used I_E in the range between 6.5° and 11° while the Series 64 models, which were slenderer, had I_E in range 3.7–7.8°.

2.4.9 *Conclusions*

We have discussed the key parameters affecting drag for a semi displacement craft from two points of view, first their correlation with drag and so relevance to a regression model, and second from the point of view of performance generally.

There are several models developed from the different test series. They all use the key parameters discussed in somewhat varying formulations. In Appendix 1 we present some of these and some examples of their use.

To select a starting point a Naval Architect needs to have a clear idea of the vessel configuration and operating envelope, including its design speed. A preliminary data set for the parameters ψ, C_B, C_p, L/B, B/T, A_T/A_M, and X_c can be set, and a first pass estimate performed using the regression models. Variation of some parameters can be made within a given model such as ψ, *L/B*, and *B/T*, or the differences between models such as NPL and AMECRC or VTT can be used to assess the effect of changing the aft hull form.

2.5 Transverse Stability of Fast Round Bilge Craft

The static transverse stability of fast round bilge craft is the same as that of ordinary displacement ships, and it can be judged in terms of initial metacenter height and static stability curves of craft at large heeling angle. However, when at speed, particularly at high speed, hydrodynamic pressure acting on the craft bottom influences the stability. In case of craft running in waves, the stability may reduce and even lose the positive stability due to the influence of wave perturbation. The "broaching" phenomenon may occur in following waves leading even to overturning.

Even without the influence of wave action, the static transverse stability of fast round bilge craft running in calm water may reduce with speed, as shown in Fig. 2.42, which shows the test results of restoring moment of a round bilge craft versus speed, carried out in Central Aero-Hydrodynamic Institute, former USSR. The results show that the dynamic stability is the same as when static for $Fn < 0.5$ however, the stability starts to deteriorate with further increase in speed and may be dangerously low when $Fn \geq 1.2$.

Calculation of static stability from measured waterline from the model tests demonstrates that even though the transverse stability of round bilge craft running at high speed is lower, the change of static stability in terms of static buoyancy is not apparent, which means the deterioration of stability is not due to the static buoyancy but caused by hydrodynamic pressure.

In general, in round bilge craft, especially those with ordinary propellers, the stern keel line and longitudinal buttocks are raised up at the stern in order to reduce the

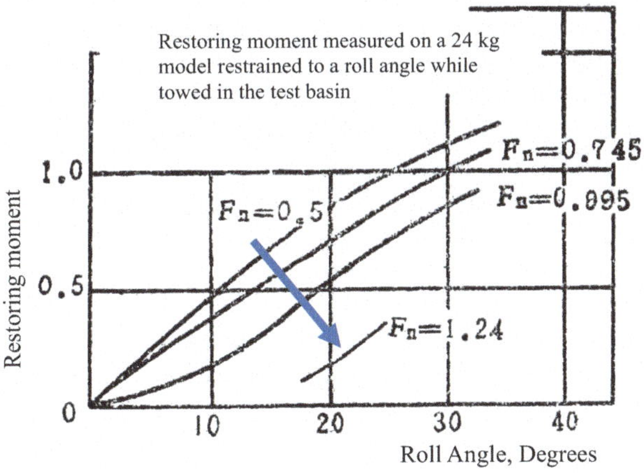

Fig. 2.42 Change of transverse stability of a round bilge craft versus speed

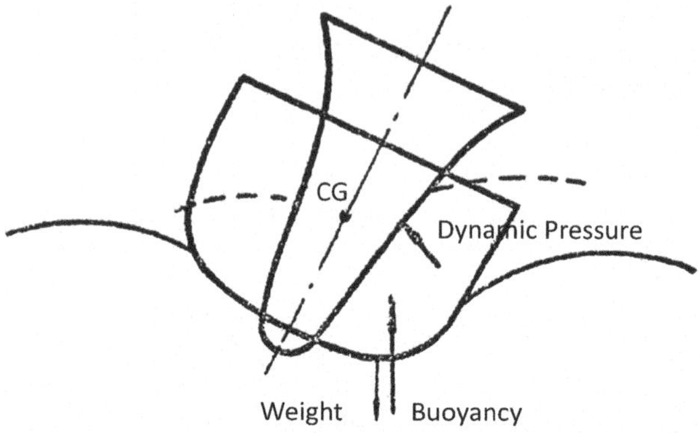

Fig. 2.43 Hydrodynamic pressure acting on the round bilge craft in heeling

inclination angle of propeller shaft and to avoid part of propeller disc extending below the baseline. This causes keel and buttocks at large inclination angles with the baseline. However, the trim angle should not be high for lower drag and even the positive angles of attack of sections at bow can't compensate for the negative moment caused by negative angle of attack of sections at stern. This, combined with the torque from free propellers can cause an offset negative pressure zone at the stern of craft running at high speed may be formed so as to make a heeling moment as shown in Fig. 2.43.

Therefore, more attention to transverse stability has to be paid during the design of the lines of fast round bilge craft in trying to reduce the negative angle of attack of stern buttocks as far as possible so as to make a positive or at least zero angle of attack with the trim angle added during running at high speeds. In the case of such requirement can't be met, a stern plate or wedge plate can be added to increase the hydrodynamic force and avoid negative pressures acting on the stern bottom.

The short hard chine profile at stern is also favorable for improving the transverse stability of craft at high speed.

A positive angle of attack on the buttocks is generally available, which provides hydrodynamic lift on craft in heeling, so as to provide the restoring moment and compensate partly the negative restoring moment caused by the stern part. However, too big an inclination angle of the buttocks may cause an increase in drag and increase in the motions of craft in waves. The anti-spray rails mounted near the waterline can also improve the transverse stability, and the width of which should be larger than 3% of craft beam.

Other measures for improving the transverse stability include lowering the CG, increasing beam, etc. It is suggested that the metacenter of fast round bilge craft should be larger than 0.8 m for $Fn > 0.6$ and use of 1.0–1.5 m may be more reasonable. Of course, it should be determined according to the specific condition for stern lines, including the inclination angle of buttocks at stern, the propulsion type, and the suction of propellers, not only from the negative pressure before the disc of propeller, but also from the pressure drop caused by cavitation.

For specific craft, the decrease in transverse stability of round bilge craft at high speed can be obtained by model tests in a towing tank, similar to that for planing craft, i.e., moving a certain load transversely to make a small heeling angle and measuring the heeling angle at various speeds. Different from planing hulls, the transverse stability of round bilge craft will be deteriorated with increasing speed, and more attention has to be paid.

Similar to displacement craft, the loss of stability of fast round bilge craft in waves is due to change in the draft of sections, and thus the transverse stability of craft, in waves. In general, when the midship section is located at wave hump and stern on trough, the loss of static transverse stability is most serious; however, when midship section is located on trough, the transverse static stability is best. Since the length of round bilge craft is small, the wavelength closed to craft length should be with small wave height, so the condition mentioned above will not be serious.

The broaching of fast round bilge craft will be discussed later in the section of seakeeping quality.

2.6 Seakeeping Quality of Fast Round Bilge Craft

Semi-planing round bilge craft dynamic lift isn't large enough to support the craft fully and so plane on the water surface. The motion response of these craft in waves will be between a conventional displacement hull and a planing hull. In general, the running attitude of round bilge craft is close to a conventional displacement ship.

The trim of semi-planing craft at speed is rather different from that in slower speed displacement mode, but the change of draft and trim is not so significant as that on the planing hulls, and so the response to waves is close to that of the conventional displacement ships.

2.6.1 *Motion Response of Fast Round Bilge Craft in Waves*

When a fast round bilge craft runs at low and medium speed, the heave and pitch motions in waves can be predicted by conventional linear methods, i.e., using strip theory, and the prediction in irregular waves also can be determined by linear superposition of components. The prerequisite for using linear theory is waves with small wave slope and non-breaking, as well as the craft not running at very high speed.

The strip theory method is only valid at low speeds. Bertram and Iwashita [35] deemed it suitable only for speeds $Fn \leq 0.4$, however, Dong has carried out calculations of wave response for round bilge craft and found the strip method usable as high as $Fn \leq 0.7$ [36–38] and suggested it can be used for the prediction of motion response in waves below such Froude number. As we illustrate below, work at NPL, MARIN, and VTT has shown that with care strip theory methods can give useful results at speeds of *Fn* above 1.0 for finer vessels where bow emergence and slamming are not significant.

When $Fn > 0.7$, using the two-dimensional strip theoretical methods will not be accurate, and modern three-dimensional methods such as Rankine source element method, etc., still can't improve the accuracy of the predictions due to assumption of non-viscous flow. Faltinsen has suggested to use 2.5D or 2D+t method to obtain good seakeeping predictions for fast displacement ships [39].

The so-called 2.5D method assumes that the flow on each section is only influenced by upstream incoming flow but has no relation to the downstream flow and also assumes that the flow ahead of the ship bow is not influenced by the hull. When carrying out the numerical solution of flow around the hull, the calculation can be started from the bow moving aft while applying the free surface boundary condition to obtain two-dimensional flow on each section.

This method combines the two-dimensional flow on a section and longitudinal three-dimensional free surface condition, and so is called two and half dimensional method (2.5D)or 2D+t method. It requires that the change of longitudinal flow between various sections is small compared with the flow on the section plane, and the errors could be substantial if this is not met. Therefore, it can only be applied for craft at higher speed where the transverse wave produced is very long and the change of longitudinal flow is relatively small.

Ohkusu and Faltinsen [40] showed that the 2.5D method considers the function of diverging waves but neglects the effect of transverse waves, so it was only applicable for high-speed craft in the speed gap where the standard 2D method cannot be applied. Figure 2.44 shows the comparison of the calculation of motions of a high-speed craft in regular head sea waves performed by Faltinsen and Zhao using 2.5D method with model test results, [39]. It shows that the predictions agree rather well with model test results for heave, pitch, and vertical acceleration.

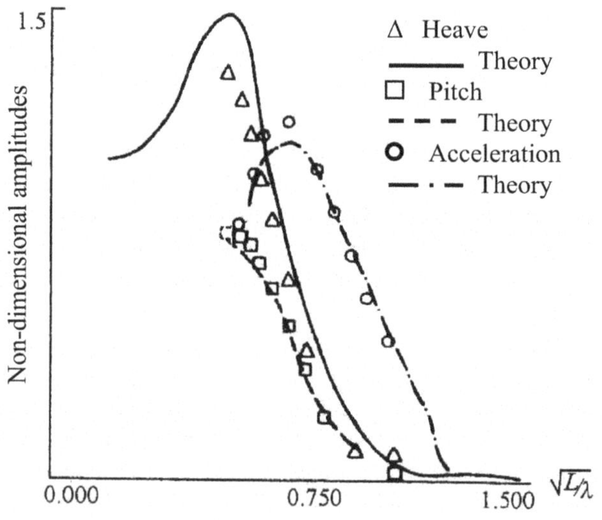

Fig. 2.44 Comparison of calculation prediction of longitudinal motion response of high-speed craft in regular head sea by means of 2.5D method with test results for a model at $Fn = 1.14$ and trim angle 1.62°

The roll of a fast round bilge craft in waves is greater than that of hard chine craft due to its lower roll damping coefficient, but this can be compensated by mounting bilge keels for increasing roll damping and thus decreasing the roll motion.

Using model tests to analyze seakeeping quality is still an important method, particularly in head seas generated in a towing tank. Many standard series test results, including two dimensional regular waves and irregular waves, have been carried out. With respect to towing test results in different headings, limited data is available due to limitation of maneuvering basin test facilities suitable for high speed. Towing tank test results from NPL and SSPA series are summarized below, with some commentary taken from [41]. Further results from testing by MARIN and VTT are also discussed.

1. *Test Results with Fast Round Bilge Craft at NPL, UK*

Experimental investigations have been carried out at NPL, UK, for seakeeping, in which five models were selected for tests, the parameters of which are listed as follows:

Length/beam ratio:	L/B	$= 4.54 - 6.25$
Length displacement coefficient:	$L/\nabla^{1/3}$	$= 5.76 - 7.10$
Radius of gyration of models in pitching:	K_y	$= 0.25$ L

No appendages were included on the models, and the tests were carried out in two-dimensional irregular waves in towing tank.

In the tests, it was found that there was no obvious influence of slenderness, including length/beam ratio and length/displacement coefficient, on the motion response and vertical acceleration of models in head sea waves; thus, the test results averaged over different slenderness are shown in Figs. 2.45 and 2.46, in which the influence of speed and wave height are shown.

The figures show that heave and pitch increase with wave height and decrease with craft speed, especially for pitch, due to the increase in encounter frequency of waves with craft speed. The vertical acceleration increases with wave height and craft speed almost linearly, and the vertical acceleration at bow is significantly larger than that at stern.

While the range of model parameters in the NPL series is limited, the test results can be useful for designers to estimate the motion response of round bilge craft in waves at the form selection stage. Ref [12] is available via the internet.

2. *Test Results of SSPA Series Models in Regular Waves*

Three models of the SSPA series [10] were selected for towing tests in head and following waves. The models have the same slenderness coefficient of $L/\nabla^{1/3} = 7.0$, but with different beam/draft ratios of $B/T = 3.0$, 3.5, and 4.0. The radius of gyration of the models is $K_y = 0.32$ L.

Using different wave lengths and corresponding wave heights to keep the same wave slope, the wavelength to craft length ratio is taken as $\lambda/L = 0.50$, 0.75, 1.00,

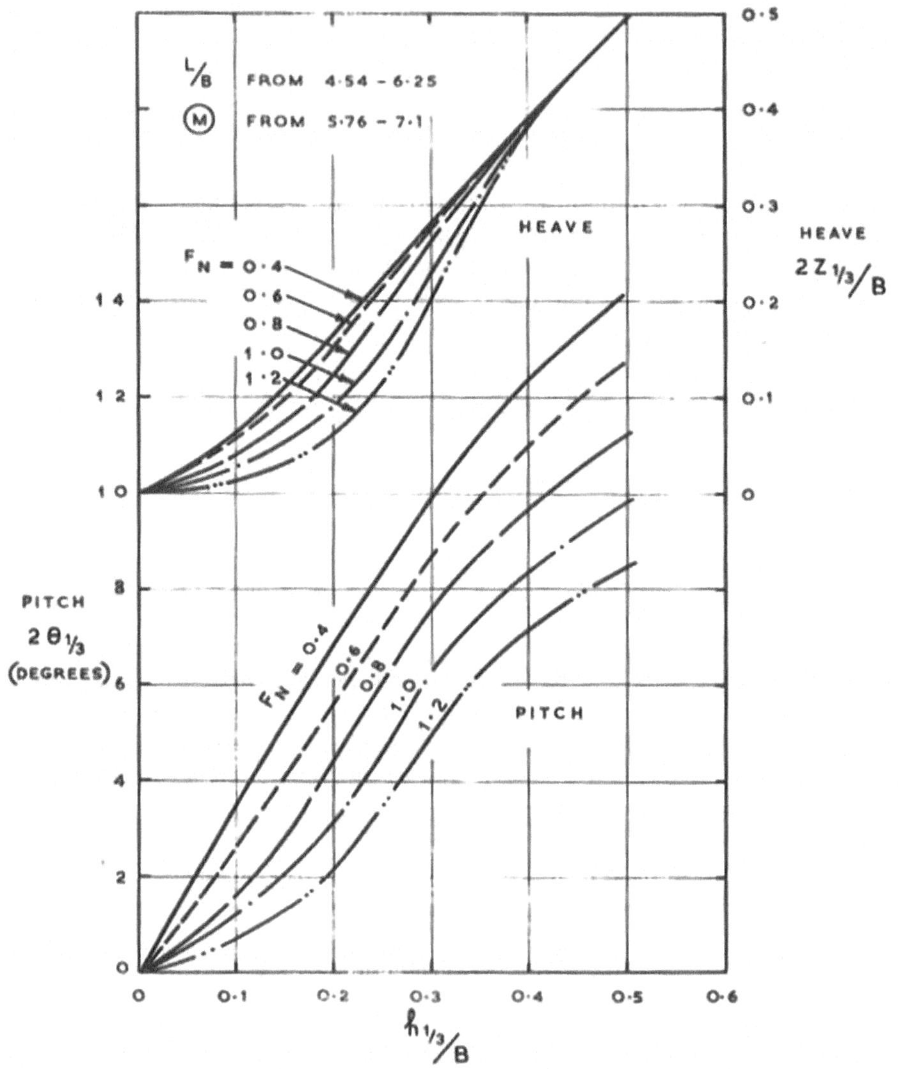

Fig. 2.45 Significant values of pitch and heave of models in irregular head sea waves from reference 12, fig. 56

1.25, 1.50, and 2.00 while $H/\lambda = 1/40$ in each case. The test speed range was $Fn = 0.5 - 1.0$.

Figure 2.47 shows the motion response of models with different beam/draft ratios and showing that at low speed $Fn = 0.66$ the influence of B/T on motion response is small, but at high speed $Fn = 0.925$, it is larger, though not very significantly possibly due to the small range of B/T in these tests.

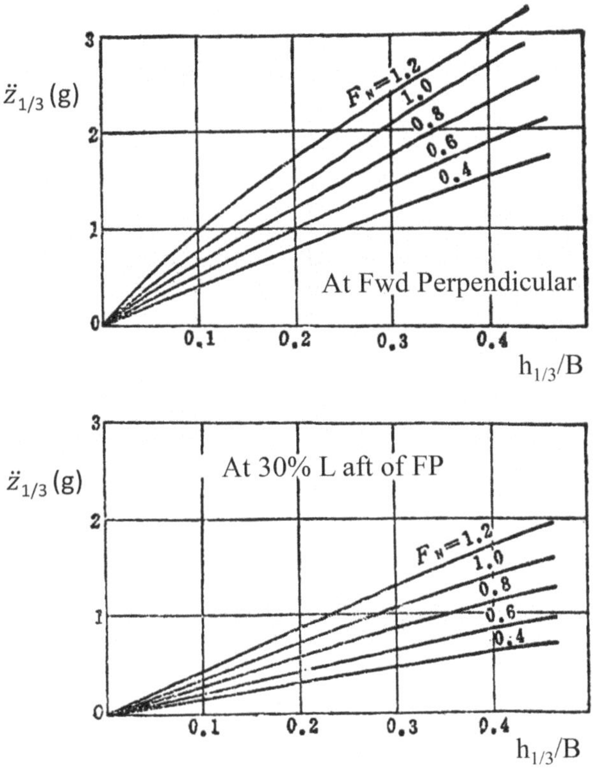

Fig. 2.46 Significant values of vertical acceleration of models in irregular head sea waves, from ref. 12 fig. 57

The influence of pitch radius of gyration (k_y) on motion response is shown in Fig. 2.48, which shows that motions might be reduced with greater pitch radius of gyration.

The tests show that the motion response of models in following waves is lower than in head sea waves, especially for vertical acceleration. Therefore, for fast displacement craft, the motion responses of craft in head sea waves are more important than that in following waves; consequently, model tests are mostly limited to head seas.

In view of the small effect of viscosity on responses in seakeeping tests, small models can be conveniently used without introducing large scale effects.

3. *Test Results from VTT*

VTT carried out a series of tests in both regular and irregular head waves following their work on calm water resistance discussed above [23], using the NOVA II and NOVA IV models.

The start of this part of their work was to prepare analytical predictions. Initially they reviewed the linear strip theory methods developed by Korvin-Kroukovsky and adjusted models by Salvesen et al. and other researchers. Key differences between the models were coefficients in the motion equations relating to added mass and

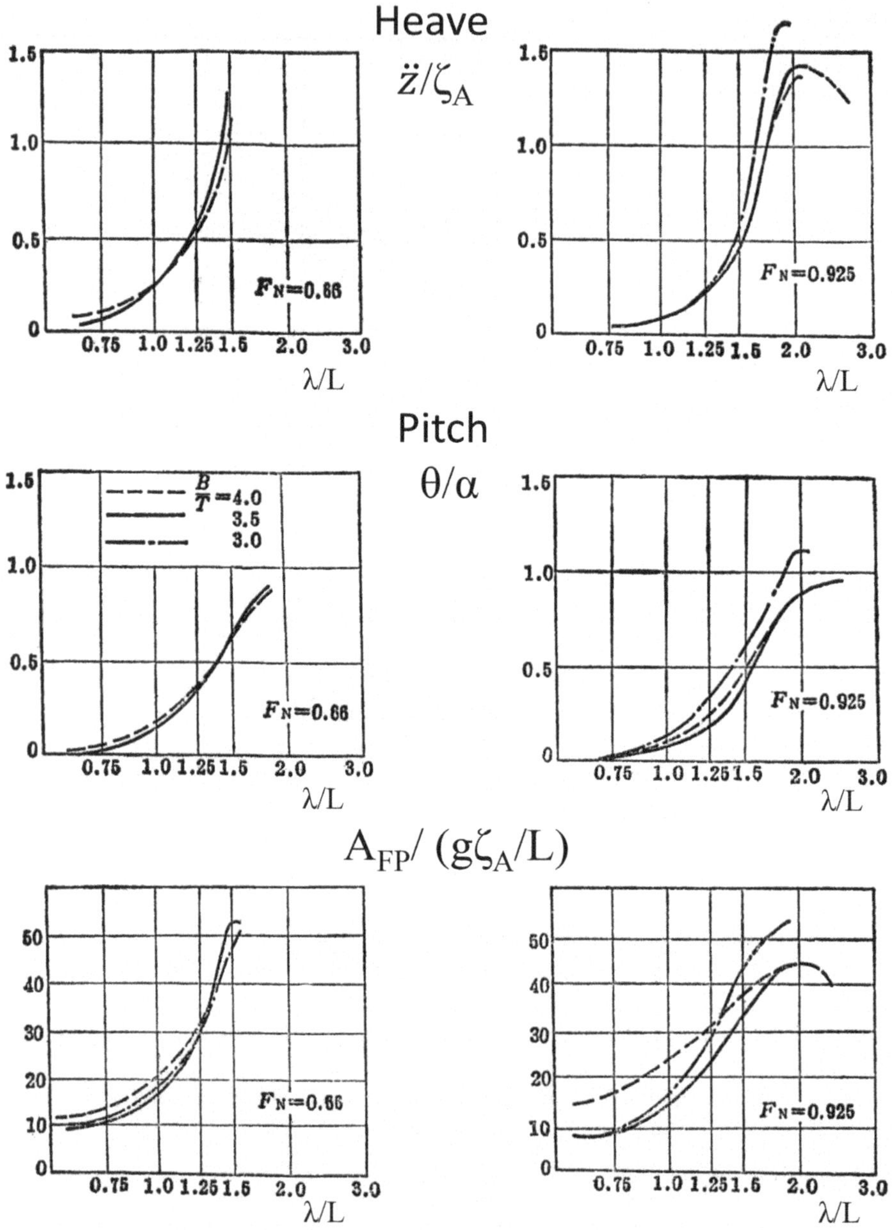

Fig. 2.47 Wave response for different beam draft ratio

damping in the stern area, and treatment of the pitch damping, added inertia and cross coupling coefficients. The original Korvin-Kroukovsky model and coefficients are based on thin ship ideal fluid theory and close to symmetrical fore and aft, while a fast craft is highly asymmetrical.

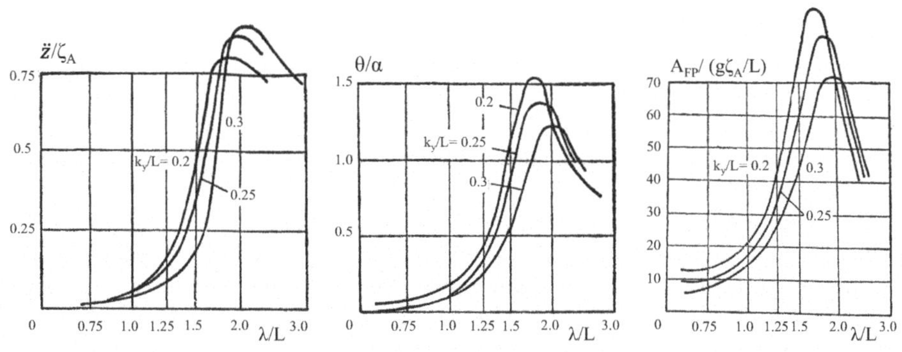

Fig. 2.48 Influence of longitudinal radius of gyration on motion response. (Figures alongside reorganize as in PowerPoint file)

The later models included damping coefficients based on forced oscillation testing. VTT compared calculations of motion transfer functions for NOVA II and NOVA IV using the original and the modified strip theory. These were compared with results from model tests in regular waves for heave, pitch, and acceleration. The responses were calculated for speeds between $Fn_L = 0.0$ to 1.202. This covered the regime up to the start of planing for a hard chine hull. They found that the two theories started to diverge at Fn_L 0.38 with the modified theory showing much higher response at Fn_L 0.635 close to resonance. At Fn_L 1.202 the modified theory showed further amplified pitch and heave response, while the original model did not show this. Model tests showed that at this speed the bow was emerging and wave slamming occurring.

Predictions of the response amplitudes at Fn_L 0.635 for NOVA II and IV compared with model test data from regular waves showed that the original strip theory gave results that were closest to the finer NOVA II, and the adjusted theory closest to the wider NOVA IV.

A further comparison was also made between irregular wave testing using the ISSC spectrum and predictions made using the RAO's from strip theory. The results showed that although at Fn_L 0.635 the heave response was close, pitch and particularly vertical acceleration at the bow were under predicted. Since bow emergence had started at this speed this may be understandable.

Based on this work, transom stern round bilge craft motions may be predicted using strip theory and RAO linear superposition to predict motions in a seaway below resonance and below around Fn_L 0.635. Above this time domain analysis or testing in irregular waves is required.

It may be observed that this speed is significantly below planing, while for drag predictions it is possible to use a regression model reliably up to planing speed so long as the hull parameters are close to those used for the regression.

4. *Test Results from Wageningen (MARIN) HSDHF Series*

As part of the model test series described in Sect. 2.2.9 above [13] the initial six models shown in Table 2.5 were tested in a range of regular bow waves ranging in length from 0.6 to 2.8 ship lengths with a height of 2% of ship length. The influence of hull geometry on motion response within the range bounded by the six models was evaluated.

By inspection the authors concluded that higher waterplane area coefficient will reduce heaving (for constant block coefficient), and similarly low vertical prismatic coefficient would lead to reduced pitch response. While the overall result is also influenced by the magnitude of coupling terms, these trends were demonstrated in their tests and were accentuated as speed increased.

The authors also used strip theory to calculate motions, accelerations and added resistance of the six models. Two versions of the strip theory model were used, the first being that of Korvin-Kroukovsky and the second of Salvesen et al. Response functions were calculated for the models and regular waves using a computer program developed at Delft University of Technology to compare with the model test data. Further the responses to an irregular sea used in the model basin were calculated and the dimensionless values of motions, accelerations and added drag were compared between the calculated and model test derived responses.

It was concluded that model test and analytical results compared favorably, even at the higher speed of $Fn = 1.14$. The subsequent evaluation of a figure of merit was thus felt to be reliable within the range of hull parameters to enable a choice based on a balance between optimized resistance and motions response. This enabled selection of a parent model for the larger main test program initially focused on resistance, as outlined in Sect. 2.2.9.

It may be noted that these initial model tests were for $L/B = 8$. If we look at the ranges in Table 2.1 it can be seen that the NPL, Bazan and VTT series are all in the range lower than this, that one may expect for a patrol vessel rather than a larger craft such as a frigate or destroyer. This combined with the effect of the transom stern form may indicate why VTT found rather higher response particularly in pitch at higher speeds and where wave lengths were close to resonance.

We discuss the effect of hull form on wave response further in Sects. 2.6.2 and 2.6.3 below. We consider first the drag increment, the problem of slamming and green water on deck and of broaching in following seas. In Sect. 2.6.3 we return to motions and accelerations.

2.6.2 Dynamic Responses in Waves

Dynamic responses of craft in waves apart from motions and accelerations include drag increment, slamming and deck wetness, and broaching. We discuss each in turn.

1. *Drag Increment of Fast Craft in Waves*

We assume to start with that the drag increment of craft, no matter they are displacement ships or fast round bilge craft, is caused by vessel pitching and heaving in waves. The drag increment in regular head sea waves as shown in [42] is given as:

$$\frac{R_{AW}}{\rho g \varsigma_a^2 (B^2/L)} = \frac{L^2}{32B^2}\left[\left(\frac{Z_a}{\varsigma_a}\right)^2 P_1 + \frac{\pi^2 l^2}{L_W^2}\left(\frac{\theta_a L_w}{2\pi\varsigma_a}\right)^2 P_3 - \frac{2\pi L}{L_W}\left(\frac{L_W\theta_a}{2\pi\varsigma_a}\right)\left(\frac{Z_a}{\varsigma_a}\right) P_2 \cos\varepsilon\right] \tag{2.17}$$

Where:

$$P_1 = \frac{\omega_e^3}{\rho g^2}\frac{2}{L}\rho\nabla\sqrt{\frac{g}{L}}\times B_{33}{}^{\prime} P_2 = \frac{\omega_e^3}{\rho g^2}\frac{2}{L}\rho\nabla\sqrt{\frac{g}{L}}\times (B_{35}+B_{53})^{\prime} P_3$$
$$= \frac{\omega_e^3}{\rho g^2}\frac{2}{L}\rho\nabla\sqrt{\frac{g}{L}}\times 4B_{55}$$

And $\varepsilon = |\varepsilon_Z - \varepsilon_\theta|$

ε_Z phase lag between interference force and heaving
ε_θ phase lag between interference moment and pitching
Heaving amplitude: $Z = Z_a \cos \omega_e t$
Pitching amplitude: $\theta = \theta_a \cos(\omega_e t + \varepsilon)$
Heaving damping coefficient: $B_{33} = \frac{1}{\omega_e \nabla}\sqrt{\frac{g}{L}}\int \xi\, b_n d\xi$
Pitching damping coefficient: $B_{55} = \frac{1}{\omega_e \nabla}\sqrt{\frac{g}{L}}\int \xi^2 b_n d\xi$

Correlation coefficients for heaving and pitching:

$$B_{35} + B_{53} = \frac{2}{\omega_e \nabla}\sqrt{\frac{g}{L}}\int b_n d\xi$$

Where

L	wavelength
ω_e	encounter frequency
∇	volumetric displacement
b_n	damping coefficient of sections
$\xi = 2X/L$	non-dimensional longitudinal ordinate

The heaving, pitching and section damping coefficients can be calculated theoretically by means of the strip method or can be obtained by model towing tests in head sea waves.

Both calculations and tests demonstrate that:

- increment of drag increases with second power of wave height.
- drag increment is linearly proportional to heaving and pitching amplitude and will be maximum at resonance.
- The drag increment is caused by craft inertia and wave diffraction. Viscosity is assumed not relevant, so incremental drag can be scaled directly using cubic power of scale ratio consistent with Froude similarity.

Apart from this, both calculation and tests demonstrate that the average value of drag increment in waves is not large compared with the drag in calm water and is not the main factor leading to speed loss of craft in waves.

Vessel speed loss is mainly caused by heavy slamming, green water shipment, emergence of propellers and the resultant voluntary speed reduction by operating crew due to poor working conditions, seasickness, etc. On fast round bilge craft with small displacement, the slamming forces and motions will be highest and are thus the most important parameters to determine so as to define operating limits for a fast craft.

Similar to evaluation of motions, the prediction of drag increment can be carried out using model test results in regular waves with the aid of linear superposition and sea spectra to determine significant response.

It should be noted that the nonlinear effect on drag increment in waves may be greater than that on motion response at speeds approaching planing and the accuracy will be poor compared with that on the prediction of heave and pitch; however, it may be useful for the evaluation of seakeeping quality during initial design.

2. *Slamming and Green Water Shipment*

Although the slamming of fast round bilge craft is not so serious compared with that of planing craft, it is more serious than for conventional displacement ships due to higher speed, so attention has to be paid to this.

The impact force is generally concentrated in the bow region. NPL measured the impact force at different locations at the bow of a craft [12] and demonstrated the impact force at bow keel was significantly greater than other places at the bow as shown in Fig. 2.49. The craft was 100 ft. long with 91 t displacement, and the figure shows hydrodynamic pressure at various points under the keel and in the bow area in waves with significant wave height 4 ft versus craft speed.

From the figure, it can be seen that where $Fn \leq 0.4$ the change of pressure at points both fore and aft is not great, but when $\mathrm{Fn} > 0.4$ the impact force at bow keel point ① increases significantly with speed, but the change of pressure at other points is rather flat, so the impact on bottom of craft at bow has to be paid much attention.

However, the difficulty is that the impact force is strongly non-linear, and the round bilge craft is rather different from a planing hull running in waves with heave and pitch motions coupled with roll motion. Therefore, it is difficult to predict the impact force at full scale either by theoretical calculation or model tests.

Since the running trim of round bilge craft is rather different from that on planing hulls, the maximum impact load is not located at 1/3 length from bow as on the

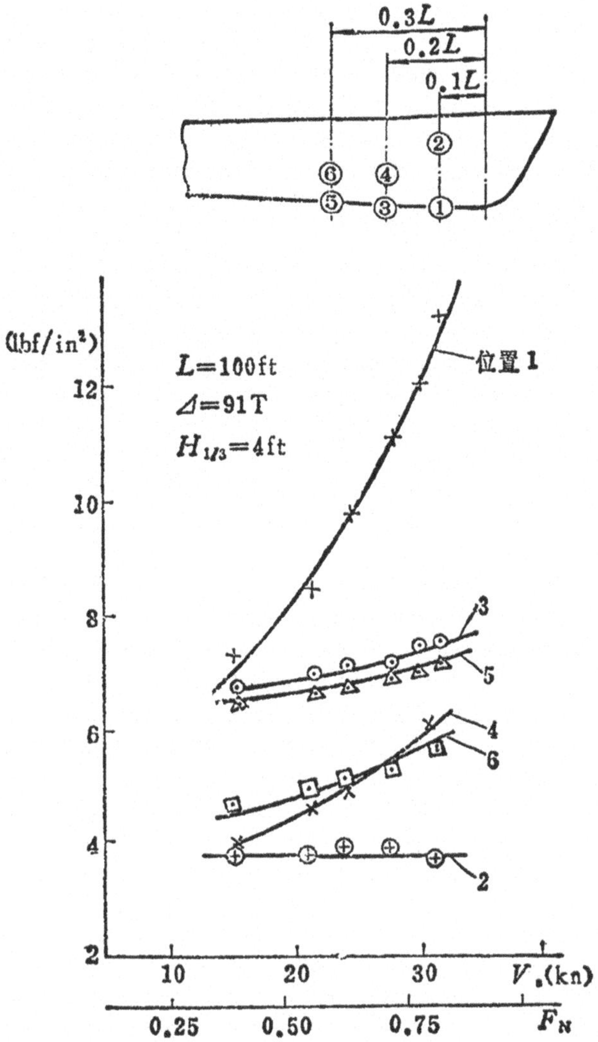

Fig. 2.49 Distribution of impact pressure on fast round bulge craft in irregular waves

planing hulls but sometimes is located on the bottom as far back as the mid-ship section. Also, only a small portion of the bow is lifted out of water and attention should be paid to the impact load on bow keel and bow side plating due to the roll motion coupling with heave and pitch motions.

In general, some measures can be taken for avoiding excess impact load in waves, such as increasing deadrise at the bow, decreasing running trim angle etc., which can be done on lines and by adding appendages as will be discussed later.

The deck wetness due to green water shipment is still difficult to predict, and it concerns wave breaking which is difficult to simulate in model seakeeping tests due to scale effects for surface tension. Thus, model tests cannot accurately simulate the green water shipment.

Many model tests have demonstrated green water shipment increases with wave height and craft speed; however, many full scale craft tests have demonstrated that green water problem is not as severe as shown in model tests.

At the design stage, the impact load and green water shipment can be evaluated qualitatively by means of model tests; the impact force can be measured at bow keel, and green water shipment can be recorded as number of occurrences per unit time.

3. *Broaching in Following Waves*

In Chap. 4, the conditions of "broaching" and loss control phenomenon are discussed, and it is found that the speed range of round bilge craft in operation is exactly located where the loss of control is easiest to happen, i.e., $0.2 < Fn < 1.04$ and therefore much attention has to be paid to this issue.

To minimize wave making drag, the bow lines of round bilge craft are often sharp and slender, and the stern is of square transom shape. The slender and sharp bow reduces the buoyancy forward and square stern reduces draft at the stern particularly when using conventional propellers. To avoid the propeller extending below the base line, the keel lines are raised up creating a stern with small lateral projection area.

These contribute to poor course stability; in addition, in following waves, since the speed of oncoming flow is reduced, the rudder efficiency will also be reduced, weakening course stability further. All these factors make "broaching" of high-speed round bilge craft very likely at high speed in following waves.

It is easiest to theoretically investigate the broaching problem through the coupled equations of yawing and swaying. Strictly speaking, the rolling motion should also be included into the equations.

Compared with the maneuvering motion equations of craft in calm water, the difference is the determination of various coefficients, i.e., hydrodynamic derivatives in the equations. Considering the special case where craft speed is nearly equal to wave speed and the encounter wave frequency is equal to zero, it is clear that when running in following waves, the hydrodynamic derivatives are different due to different drafts at each section and different phases between the craft and wave. In addition, calculation of the hydrodynamic force induced by waves is complicated, and simplified calculations may not reflect the real situation of the fast round bilge craft running in following seas. Therefore, there is still a lot of work to be done on theoretical investigation of "broaching" of round bilge craft in following waves.

Regarding the model experimental investigation of "broaching" many towing tanks have carried out experimental investigations such as NPL in UK, Davidson Laboratory in US, and CSRDC in China.

Many test methods are also used for response components, such as towing model tests in calm water for measuring the pitching and yawing moment of models under the conditions of restricted swaying and yawing. Negative pitching (plough-in) and large yawing moments show the tendency of "broaching." If this is the case towing model tests in following waves should be carried out with the same condition of restricted yawing and swaying to measure pitching and yawing moment at different phase lags between the craft and wave.

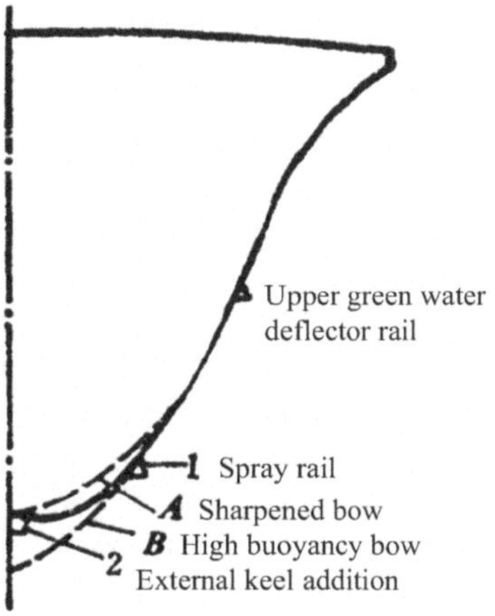

Fig. 2.50 An example of revised bow section

Large yawing moment and serious plough-in shows serious "broaching" tendency, and negative yawing moment shows positive course stability. A more ideal method is carrying out free running model tests in following waves in which the "broaching" tendency can be demonstrated directly; however, in such case, the simulation of propulsion is important as this greatly influences the course stability of craft in waves.

During the design of fast round bilge craft, the important measures to take for avoiding broaching include not too slender bow for providing enough buoyancy forward, and not to use a "round bottom" bow to avoid the asymmetric transverse flow which could lead to transverse force and yawing moment. Additionally, too deep draft should also be avoided as this causes the lateral motion center of hull to move forward. Figure 2.50 gives a typical example of adjusted bow section, in which the dotted line shows the revised profile that is more favorable to avoid broaching than the original profile with solid line.

In the figure, two revised designs of A and B are provided for avoiding the round bottom bow lines. Type A is more favorable to avoid changing draught. Mounting an bow anti-spray rail (item 1) can compensate the low fullness of the bow, and a positive angle of attack can provide some hydrodynamic lift at high speed for preventing plough-in and excess roll in turning, which is favorable for avoiding "broaching." A small keel plate (2) can help avoid transverse flow on an existing vessel with broaching issues.

For the stern section, increasing the deadrise to move the vertical center of area below the waterline aft is favorable, as well as mounting a "Skeg," if necessary. Increasing rudder area, in the wake of propeller, for enhancing the rudder efficiency is also an effective measure for avoiding "broaching." Generally, frequent handling of rudder of fast round bilge craft in following waves is necessary for keeping course and avoiding "broaching." For this reason, enough power of rudder machinery and high rudder machine frequency response are also necessary, particularly for the craft at high speed of $Fn = 1.0$ and higher. Using water jet propulsion or surface piercing propeller can provide high rudder efficiency and is favorable for avoiding "broaching."

Increasing the moment of inertia of yawing is also favorable to avoid "broaching" of craft in following waves. However, in general, it is difficult to alter the yawing moment of inertia once the general arrangements have been fixed. In order to avoid plough-in of craft in following waves, maintaining large running trim angle is preferred. In general, the trim angle of craft in waves is larger than in calm water; however, larger trim angle will lead to larger drag and larger motion responses and wave drag increment of craft in head sea waves, which is unfavorable to powering performance and seakeeping quality of the craft and is contrary to the requirement for the craft running in following waves.

The ideal measure is using trim controllable installations such as controllable stern flap or controllable interceptor. Since the motion response and drag increment of craft in following waves is significantly smaller than that in head sea waves, the increase of trim angle in following waves will not significantly influence the drag increment and motion response, but it will reduce the risk of "broaching."

2.6.3 Influence of Hull Parameters of Semi-planing Craft on Seakeeping Quality

At the early stage of fast round bilge craft design, the designer uses the calm water drag requirements in selecting the hull profile main parameters such as slenderness ψ or displacement length ratio DL, the transverse section area including profile and distribution, prismatic coefficient, position of CB, and transom geometry. This has been discussed in Sect. 2.4.

The requirements for seakeeping quality will be higher for larger craft operated in a high sea state compared to smaller craft. However, in view of complicated issues related to seakeeping quality, the investigation of seaworthiness for fast round bilge vessels is less well proven than powering performance, with some conclusions on the seaworthiness even being contradictory. The influence of hull profile parameters on seakeeping must be paid more attention, especially when using various modern types of weapons, electronic equipment, as well as powerful power plants. Therefore, in this section we will describe some general trends influencing the seakeeping quality, but not introduce the conclusions of various model test results as these may be best interpreted in relation to the tested model parameter ranges.

1. *Effect of Slenderness* ψ

   Slenderness ratio $\psi = L/\nabla^{1/3}$ is a primary parameter influencing the drag of craft in calm water as has been described in previous section, and most experimental investigations, including NPL, used a typical slenderness ratio in the range $\psi = 6 - 8$. The influence of slenderness ratio on seakeeping is not obvious however, the range of this ψ is not large. Keeping displacement constant with large ψ, the seakeeping quality should be improved due to the wavelength relative to craft length being large, which has been validated by tests. Delft University of The Netherlands with Damen Shipbuilding Group presented the "Enlarged Ship Concept—ESP" with an improvement of seakeeping quality by adding 25% and 50% length on a craft with length of 26 m, showing that the longer craft length has much better seakeeping quality so as to demonstrate that higher ψ provides better seakeeping quality. Of course, it may make some difficulties for general arrangement and could also lead to structural weight penalty.

2. *Effect of Sectional Area Curve*

   The sectional area curve showing the distribution of displacement longitudinally reflects not only the longitudinal prismatic coefficient and longitudinal center of buoyancy, but also the sharpness of hull in the bow and stern areas of the hull.

   The effect of transom beam on seakeeping quality still requires further analysis. In general, larger transom area ratio gives larger buoyancy at the stern and is favorable for avoiding "broaching" in following waves.

3. *Effect of Beam/Draft Ratio*

   The model experimental investigation of SSPA has demonstrated that the effect of *B/T* is not clear and has little influence at high speed. To date studies with a large range of *B/T* are limited for fast round bilge craft, so the performance of such designs is not clear compared to more traditional forms. The focus has been on *L/B*, slenderness as above and block coefficient or body plan as below.

4. *Effect of Body Plan*

   The body plan shape influences operational performance significantly, such as the hard chine and round bilge body plan. In the semi-planing regime, differently from planing mode, the hard chine craft style generally implies a "Deep Vee" style, which will be discussed in Chap. 6. It is a mixture craft type in the transitional speed region, including short, medium, long hard chine craft types.

   The pure hard chine form is suitable to reduce drag at high speed, however, for transitional mode craft this geometry causes higher drag, as has been validated by many researchers. Nevertheless, the motion response of round bilge craft is greater than that on hard chine craft, a fact that confuses a lot of designers. Then the mixture type of craft with partly hard chine and partly round bilge was introduced, which reasonably met both powering performance and seakeeping quality.

   The initial successful example was the fast craft type "Jaguar" of West Germany referred to in the introduction to this chapter [1]. It had small speed loss in waves and same drag in calm water as round bilge craft. Its success promoted the mixture type craft on the application of high-speed craft, including short, medium, and long hard chine profiles.

   A lot of tests have shown that the well-designed mixture type craft improves the seakeeping quality while also providing satisfactory powering performance, which has been validated not only on "Jaguar," but also on craft "684" and MSC-1 series of China, with short hard chine stern and long round bilge fore part providing nice powering performance and seakeeping quality compared with other craft.

5. *Effect of Pitch Inertia*

   SSPA, Sweden carried out seakeeping tests of three models in regular waves with different pitch inertia. The test demonstrates that larger pitch inertia lowers motion response as shown in Fig. 2.48. Since the moment of inertia about the transverse axis Y and vertical axis Z are very close, larger pitch inertia is also favorable for avoiding "broaching" of craft in following waves.

   The hull pitch inertia depends on many factors such as general arrangement, longitudinal overall strength, hull structure weight, etc., and is difficult to be specified only in terms of the requirement of seakeeping quality.

   Apart from this, wave direction also greatly influences the seakeeping quality, with head sea waves causing the highest motion and dynamic response, while following and stern quartering waves can cause serious "broaching" as discussed here. Such phenomena are important for instructing the craft handling and suggesting additional controls such as stern flaps and interceptors touched on below.

2.7 Maneuverability of Fast Round Bilge Craft

The maneuverability of fast round bilge craft is similar to that of conventional displacement ships, which can be dealt with by means of the method used in displacement ships. However, due to the high speed and other features there are some differences between the two craft.

The high speed and high efficiency of a rudder located in the wake of a high-speed propeller, gives a semi displacement craft nice course stability and the ability to keep course with a little adjustment with small rudder angle in calm water.

The turning characteristic of fast round bilge craft is similar to that of displacement ships. At the initial stage of turning, the heeling moment due to the rotating rudder heels the craft inward for a short period of time, then under the action of sideslip and drift angle an outward moment will be formed on the underwater part of the hull which is greater than the inward moment caused by the rudder and causes the outward heeling of the craft. When the turn is stable, the outward heeling angle will be kept constant, and when the rudder angle is brought back to zero, the outward heeling disappears, and the craft runs in normal straight course.

A smaller high-speed craft may attain a greater drift angle while turning under the action of turning moment on the hull and rudder due to high speed, and the resulting ratio of turning radius to craft length may be about $R/L = 4 - 6$, somewhat smaller than conventional displacement ships.

However, the heeling angle of fast round bilge craft during turning is larger than that of conventional displacement ships, and sometimes more than 10°, which may scare people on board. Figure 2.51 shows the force acting on the hull during stable turning course. Differential application of stern flaps or use of canted rudders can assist to minimize the heeling.

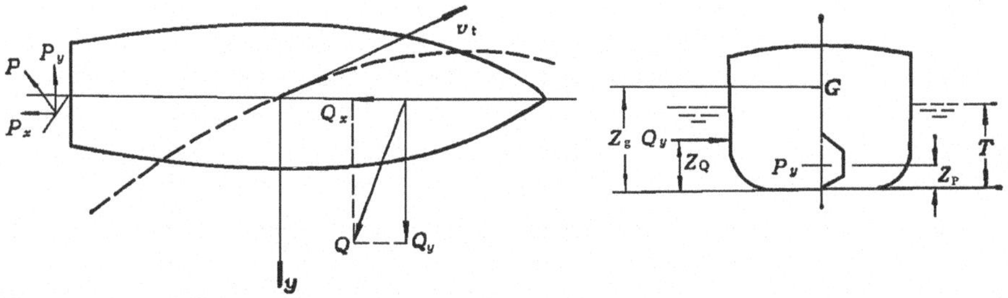

Fig. 2.51 Forces acting on the hull during turning

There is a centrifugal force $FP = (\Delta/g).\ V_t^2/R$ acting on the CG, where Δ is displacement, V_t, the tangential speed of craft during turning, and R turning radius. Large turning speed generates high centrifugal force. The hydrodynamic force Q acts on the hull during turning and its component at transverse section is Q_y and the component of rudder force P in the y direction is P_y, so the total moment generating the heeling of craft is:

$$M_x = F_p(Z_g - Z_Q) - P_y(Z_Q - Z_P) = \frac{\Delta}{g}\frac{V_t^2}{R}(Z_g - Z_Q) - P_y(Z_Q - Z_P) \quad (2.18)$$

Where Z_g is height of CG from the base line, Q_y is hydrodynamic transverse force, Z_Q is the height of center of transverse force from the base line, and Z_p is height of rudder force acting point from base line.

Since P_y is a small value compared with F_p in (2.18), and $(Z_Q - Z_p)$ is also a small value, so the second item of the equation can be neglected safely for predicting the heeling angle. Then (2.18)can be rewritten as follows

$$M_x = \frac{\Delta}{g}\frac{V_t^2}{R}(Z_g - Z_Q) \quad (2.19)$$

When heeling angle is small, it can be judged from initial stability, and then

$$M_Q = \Delta h\ sin\ \theta \approx \Delta h\theta = M_x = \frac{\Delta}{g}\frac{V_t^2}{R}(Z_g - Z_Q)$$

Where

h static metacentric height, GM
Δ vessel displacement

then

$$\theta = \frac{V_t^2}{ghR}(Z_g - Z_Q) \quad (2.20)$$

From (2.20) it can be seen that high speed generates a large outward heeling angle. A large distance between CG and transverse hydrodynamic force also generates a large heeling angle. For the high-speed round bilge craft, in order to avoid excess heeling angle during turning, limiting the craft speed or rudder angle during turning is necessary, however, it will result in larger turning radius. Since the heeling angle is proportional to V^2, limiting the speed will obtain much better effect.

During turning, the height of the center of hydrodynamic force on the hull Z_Q also influences the heeling angle directly. On conventional ships, $Z_Q \approx T/2$, but tests of fast round bilge craft demonstrate that the center of hydrodynamic force during turning relates to transverse deadrise angle β and B/T. Large deadrise angle causes the center of hydrodynamic pressure to descend due to the suction acting on the inward bottom during turning. Large B/T causes the hydrodynamic pressure center to descend when turning as well, which is also an important factor in generating high heeling of fast round bilge craft during turning, and so more attention being needed in design.

The static metacentric height GM is an important parameter for heeling as well. The appropriate metacentric height used to be considered as $GM = 0.8$–1.0 m, however, this has been enhanced gradually to $GM = 1.2$–1.5 m and higher especially for Deep Vee craft where it reaches $GM = 1.8$–2.0 m. It is an important parameter to reduce heeling angle during turning, especially in waves, avoiding "broaching," and maintaining craft stability in operation.

2.8 Appendages for Improving the Operational Performance of Round Bilge Craft

The effect of various appendages mounted on a high-speed craft is strong, and a reasonable design of such appendages can significantly improve the performance of the craft. We give an overview of the key equipment options and their use.

1. *Stern Flap, Wedge, and Interceptor*

Stern flaps have been mounted on planing hulls since the 1940s. This is a generally effective measure for improving performance by means of adjusting the running trim of planing hulls. Initially fixed stern flaps were used and later the adjustable stern flap to maintain different running trim for different applications including reducing the hump drag, keeping optimum trim angle at cruising speed, and maintaining small trim angle in waves for improving seakeeping quality.

The technology of mounting adjustable stern flaps on fast round bilge craft started in the 1960s and has been developed continuously since then [43–45].

A stern flap or wedge mounted on a fast round bilge craft can reduce the drag in calm water due to lengthening the "virtual length" of the craft, both extending the craft length by the stern flap itself and moving the "rooster tail" aft. Such "virtual length" corresponds to increasing slenderness which reduces the wave making drag while keeping friction drag unchanged. In addition, decreasing the trim angle also decreases the wave making drag. Apart from these, the stern flap can lower the height of "rooster tail" flow by the so-called "pressing wave" action.

However, from a physical point of view, the extension of a stern wave afterward or the increase of "slenderness" are difficult to justify as the reason for drag reduction because the craft profile is still not changed and the length of stern wave after the transom is not very relevant to the force acting on the hull. The main reason for the drag reduction is the change of trim angle which reduces wave making drag.

Secondly, the wedge plate deployment leads to an increase of pressure at the stern, so as to reduce the pressure difference between the bow and stern and consequently reduction of drag. Actually, in case of a well-designed stern profile, the stern wedge cannot reduce the drag. This has been validated by many test results, in spite of visual observation of increase in the "virtual wavelength."

The stern wedge cannot be designed to reduce drag at all speeds. For given hull lines, an appropriate trim angle at the design speed for minimum drag should be determined. In case of well-designed lines, such running trim can be obtained from the lines optimization, and it is unnecessary to mount an additional stern wedge. However, in view of the difficulty predicting hydrodynamic forces acting on the hull precisely to get such ideal running trim during the design of lines, the application of a stern wedge to adjust the running attitude becomes quite effective.

With respect to the drag in calm water, the optimum trim angle is specific to the lines especially at the bow. A trim angle of 1–2° for minimized wave making drag is normally used as the optimum trim angle for powering performance for most fast round bilge craft.

Regarding the application range of speed for a stern wedge, it can be applied on most craft with operational speeds where there is transom separation of the flow ($Fn > 0.5$). Actually, the application of stern flap has extended from high-speed craft to medium speed craft, from planing craft to round bilge craft, from $Fn = 0.9$ to 0.7 and even to 0.5 for a frigate and obtained a good reduction in drag.

There is no optimum size and installation angle for a stern wedge. The experimental results of Naval University of Engineering, China demonstrate that the influence of length on the drag is not apparent, and it can be taken as 1–2% of the length. The key problem is with installation angle of the stern wedge to get optimum trim angle, i.e., use small installation angle with hull bottom for a longer stern wedge, and vice versa. The installation angle is in the range 0–5°, as shown in Fig. 2.52.

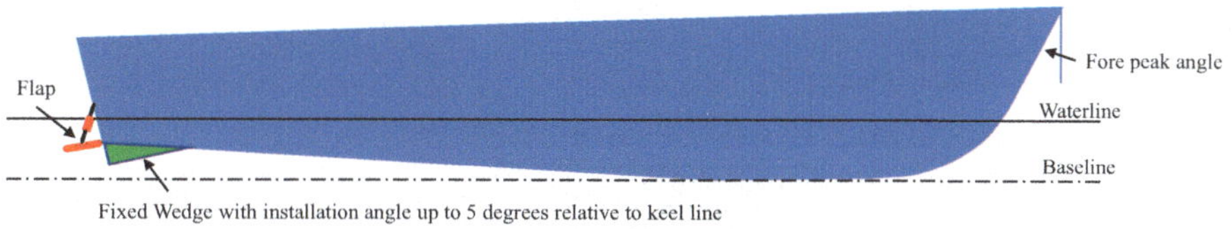

Fig. 2.52 Schematic of stern plate and wedge installation

A stern wedge is installed at the bottom of the transom with wedge form to make the stern bottom inwardly concave and with the same function as a stern flap to locally increase the stern bottom pressure so as to adjust the operational trim angle and decrease the pressure difference of bow and stern and consequently reduce the drag. Figure 2.52 shows the installation schematic of stern wedge.

Compared with stern flap, the advantages of fixed stern wedge are that it does not protrude beyond the transom, is simplified configuration and favorable to avoid local vibration, which may happen at a stern flap. However, the major disadvantage is it is not as easily adjustable as the stern flap. In addition, protruding under the bottom can create some interference for the propeller. The stern flap being a little further rearward can extend the "rooster tail" rearward and cause more uniform flow into the propellers [46]. In addition, a stern flap can be used easily to adjust the trim angle in operation, as a manual control system on luxury boats and automatic control system in some high-speed craft.

It is very convenient to install a fixed stern plate and wedge on hard chine craft, including the hybrid craft with hard chine stern. For round bilge craft, the stern plate and wedge can be mounted between two side bilges corners; however, in case of controllable stern plate, the installation will be of some difficulty with discontinuities between the stern plate and stern with changing inclination angle, which leads to spray blown from the gaps.

A stern interceptor is a rather simple appendage on the transom, being a small panel protruding below the transom and perpendicular to the stern bottom and its function is similar as the stern plate and wedge, to enhance the pressure on the stern bottom, as shown in Fig. 2.53. Due to that the stern interceptor is perpendicular to the oncoming flow the increase of pressure is significantly larger than the stern plate and wedge. The height of stern interceptor must be kept within the boundary layer so as not to destroy the oncoming flow; a typical value is about 0.1% of craft length L.

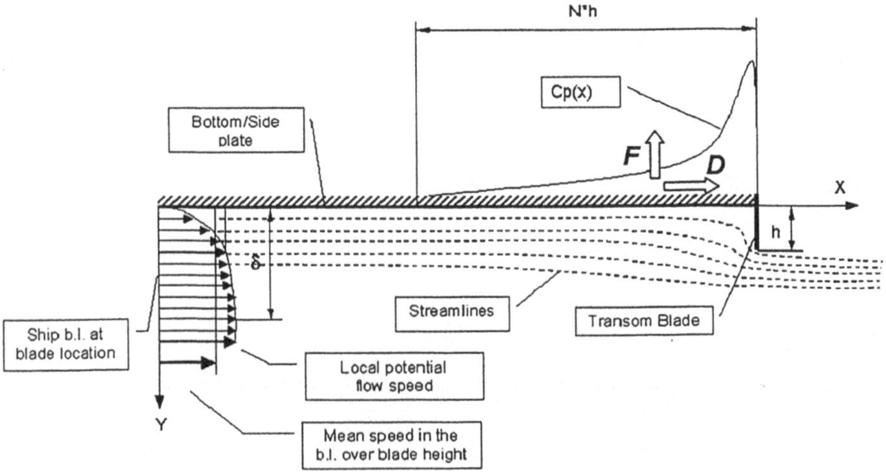

Fig. 2.53 Schematic of stern interceptor

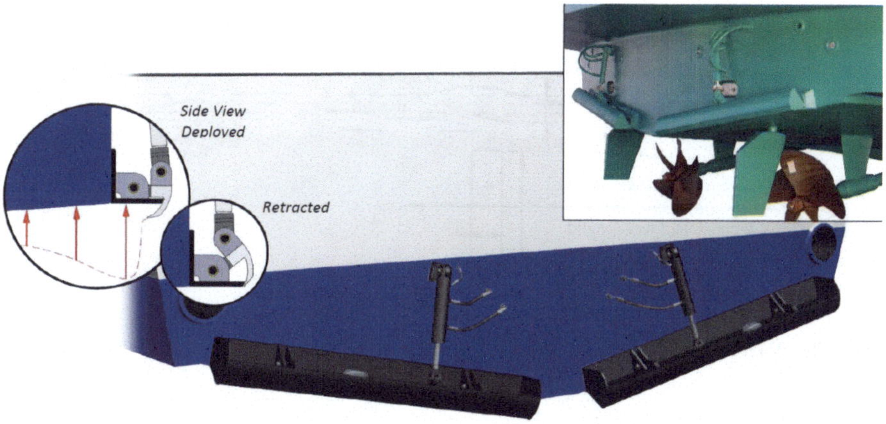

NAIAD stern intercepters: These operate by rotation downwards by hydraulic cylinder. The system can be controlled actively in a seaway, and the two interupters below can be coordinated so as to trim the vessel in roll and give a 'coordinated' turn with vessel rolled inwards.

Fig. 2.54 Controllable interceptor equipment example from NAIAD

The size and weight of interceptors are smaller than stern plates and wedges and can be suitably mounted on the sterns of both hard chine and round bilge craft.

For a round bilge craft, they are installed across the "flat" bottom area. Interceptors with automatic control systems are widely used on craft [45]. Figure 2.54 shows an example of the system supplied by NAIAD Dynamics Inc. of the USA.

The controllable stern interceptor and stern plate can improve the drag and seakeeping quality as the running attitude directly influences the motion responses, vertical acceleration, slamming, and speed loss of the craft in waves.

In general, it is expected that low running trim is suitable for a craft running in head sea waves, but larger trim angle is needed in following waves for avoiding the "broaching" phenomenon. The running trim angle of craft can be regulated by means of a controllable interceptor according to the sea sate to decrease the pitching motion, especially combining its action with bow hydrofoils or fins.

The investigation of controllable flaps on a fast round bilge craft for improving the motion response has been carried out by Naval University of Engineering, PLA, China, [47]. The model tests validated that the pitching of a 150 t craft running in head sea waves of medium sea state could be reduced by up to 50% using controllable stern plate. By controlling the left/right stern plates and interceptors separately, improving the roll motion is also possible.

2. *Hydrofoils and Fins*

It is well known that a foil has the most efficient hydrodynamic geometry, so the performance of a fast craft should be improved by means of these. Many types of foils are used on the fast round bilge craft for various purposes, drag reduction, for reduction of wave making, for improving the seakeeping quality, as damping fins, and as anti-rolling fins.

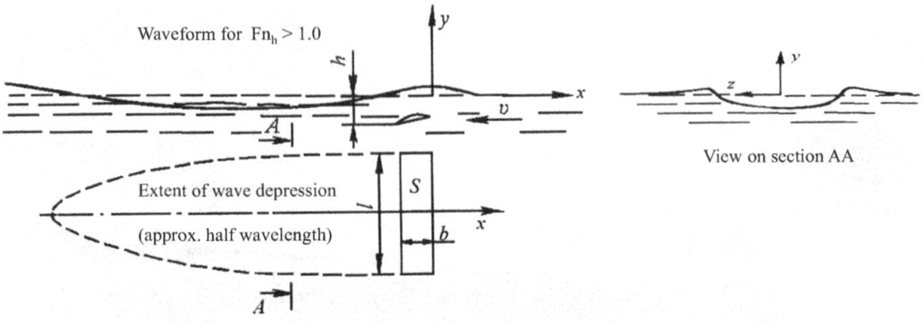

Fig. 2.55 Wave making generated by hydrofoil near the water surface

Foil for Elimination of Wave Making

It is very important to reduce the wave making of fast round bilge craft and using the Eliminating Wave Making Foil (EWMF) is an important method similar to the effect from a bulbous bow on medium speed ships. The wave generated by the EWMF fin interferes with the ship waves so as to reduce the wave making drag.

A hydrofoil moving near the free surface generates a trough behind the trailing edge as shown in Fig. 2.55. The wave making drag can be reduced so long as the trough superposes with the wave peak generated by the hull, therefore, the mounted position and size of hydrofoil has to be selected carefully.

The position of an EWMF relative to the surface is a key factor to reduce the drag, i.e., superposing the wave trough generated by the foil with the wave peak generated by the vessel bow. The position of the bow wave peak changes with speed so the longitudinal position of EWMF must be determined according to design condition for the vessel. The experimental investigation [48] demonstrated that for most fast round bilge craft the bow wave peak is located at about station #4 to #6 from the bow where there are 20 stations for total craft length, therefore generally the EWMF may be mounted at station #3 to #4.

The depth of mounted foil also influences the drag reduction significantly, and too deep a foil makes less reduction in drag and a too shallow EWMF makes the foil takeoff easily from the water surface, particularly in waves. According to our experience, the relative submersible depth of foil root (where the foil attaches to the hull) is about $h/c = 0.5$ where h is the foil submersed depth and c is the foil chord length.

Considering that the bow rises up at high speed, decreasing the submersible depth of the hydrofoil to $h < 0.5c$ is unreasonable. To decrease the probability of foil taking off from the water surface, it is recommended to use a foil with negative dihedral, take the submersible depth of foil root as minimum submersible depth and take the submersible depth of foil tip as maximum submersible depth as shown in Fig. 2.56. The negative dihedral can be taken as $\theta = 10$ to $20°$ according to the sea state of the operated craft, with larger value for high sea state requirements.

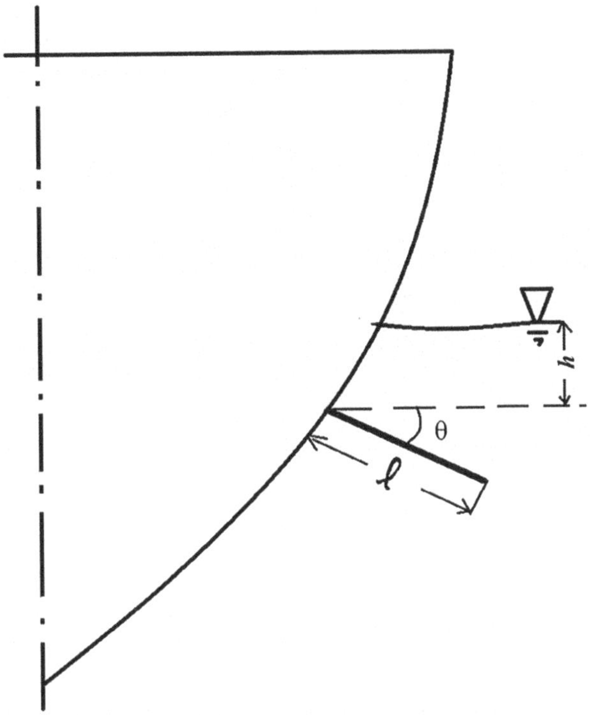

Fig. 2.56 Bow mounted hydrofoil with negative dihedral

Geometrical Parameters of EWMF
The geometrical parameters also influence the effect on wave making, and the most important parameter is foil span. The longer the foil span, the wider and longer the depression after the foil, which provides higher efficiency of wave making elimination as shown in Fig. 2.57, where *b* is Foil Span, B maximum craft beam, *b*/*B* is relative span.

From the figure, with increasing span the drag reduces significantly, and for foil with small span, where $b/B \leq 0.25$ drag reduction is not available; so largest possible span should be used from the point of view of drag reduction.

The EWMF chord length must be determined after the determination of span based on the requirement of foil strength; a typical value $b/c = 2.0$ to 2.5 may be used. Regarding the plan profile, the trapezoidal shape is better with the ratio of chord length from root to tip of about 1.5–2.0. The foil thickness can be determined in terms of strength, and generally the relative thickness t/c is in the region of 10–15%. Model tests of various foil sections and angles of attack for investigating the influence of circulation on reducing the wave making of craft have been carried out and found that the circulation didn't have much influence on the efficiency of the elimination of wave making, probably as the larger circulation leads to deeper but shorter trough and the small circulation makes shallower and longer trough. Based upon such an idea, the foil section can be taken as an airplane wing profile or circular camber profile with zero mounted angle of attack; and in case of the later, the leading

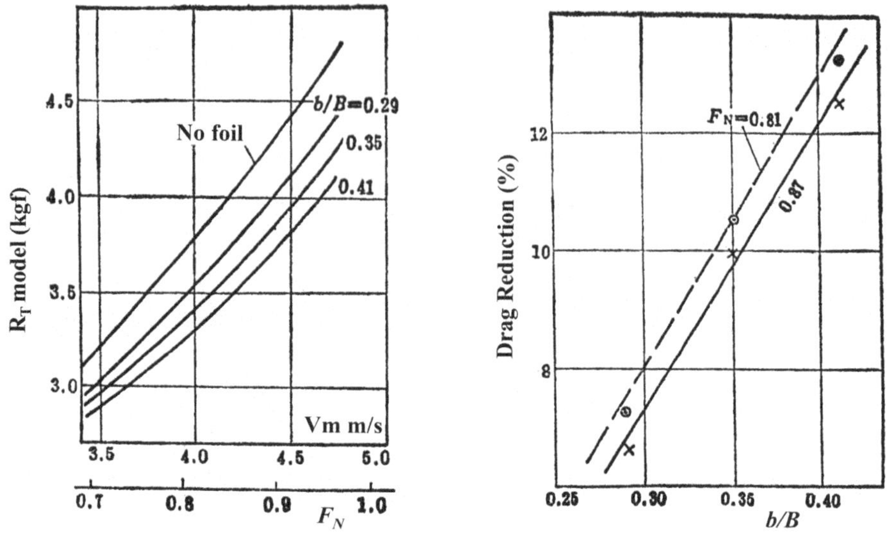

Fig. 2.57 Efficiency of wave making elimination by foil with different foil span

edge should have a circular profile to avoid flow separation at the leading edge in waves. Model tests show that a tip plate to a foil and circular correction at the foil end in the plan profile do not influence the efficiency of wave making elimination.

The theoretical calculation of EWMF was carried out in [49], and further investigations, including the study of the influence of stern plate on the running attitude and craft drag, were carried out in [50]. These investigations have demonstrated that the theoretical calculation of hydrodynamics of foils by means of low aspect ratio short wing theory and high aspect ratio planing plate theory agrees with the test results.

The EWMF also provides hydrodynamic lift, which may cause excess running trim angle and large wave making drag of hull, so a stern plate must be mounted along with EWMF. Additionally, enlarging the size of stern plate or providing negative dihedral angle may also be necessary to increase the stern hydrodynamic pressure and to obtain the optimum trim angle. Actually, raising up the hull a little bit could also be favorable for drag reduction, and a real craft test of a fast round bilge craft weighing more than 100 tons demonstrated that for a ratio of foil span and craft beam $b/B = 0.35$, the craft speed could be increased by up to 2 knots, by properly enlarging the stern plate for getting optimum running trim angle of approximately 1°.

Ship trials have also demonstrated that the heave and pitch responses in waves decreased significantly, apparently due to increased motion damping of the craft by means of the EWMF. For craft weighing one hundred tons plus in medium sea state with significant waves of $H_{1/3} \leq 2.0$m, the EWMF can significantly improve the seakeeping quality. In the case of crafting running in heavy waves, serious impacts may occur once the hydrofoil takes off from the water surface, and this may be a limiting condition for using the EWMF.

For fast round bilge craft with high requirements of seakeeping quality, the EWMF with negative lift, which is a foil with downward camber profile, can be considered due to the fact that in case of craft riding a wave peak, the local static buoyancy will be increased, and meanwhile the negative lift of a shallow submerged foil is also increased to compensate the change of static buoyancy, so as to decrease the wave interference force. Therefore, the shallow submersible foil with negative lift can reduce the motion response and vertical acceleration of craft while also reducing the probability of foil takeoff from the water surface. Tests show that this foil can also decrease the wake. Since the wave trough generated by the shallow submerged foil with negative lift is before the foil and wave peak behind the foil, the foil should be mounted behind the wave peak generated by hull bow to "eliminate the wave." These places the foil close to the mid ship, which is also favorable to avoid "take off" of the foil.

The efficiency of drag reduction by means of foil with negative, lift is less than the foil with positive lift. First, the negative lift of foil makes the craft sink and increase the draft of craft running in calm water; second, the trough generated is shallow as it is ahead of the foil and effected by the oncoming flow, resulting in lower drag reduction efficiency.

In case of using EWMF with controllable angle of attack, the powering performance and seakeeping quality can be improved simultaneously. In calm water, the foil is at positive lift condition to create nice powering performance, and when in heavy waves, the foil can be rotated to negative lift and improve seakeeping quality without consideration of the efficiency of elimination of waves.

T Type Hydrofoil Under the Craft Bottom

A T type foil mounted under the craft bottom can significantly improve the seakeeping quality of fast craft. Since the foil is located far from the free water surface, it is unable to lift off from the water surface, and the motion response can be reduced significantly by means of its damping effect; however, it may increase the craft drag.

In general, the T foil is mounted at bow under the keel, and deeper draft is favored to avoid the air ingestion and cavitation. However, cavitation may happen at higher speeds more than 50 knots. Due to high vertical motion at the craft bow, the foil is usually mounted there to provide significant damping effect. A controllable flap can be installed on a foil to regulate the phase relative to vertical velocity to increase the heaving and pitching damping, and this will increase the motion reduction efficiency even more.

To decrease the drag generated by the T foil, it can be designed to be retractable into the hull bottom, however, with a penalty of higher cost for small craft.

3. *Application of Anti-spray Rails*

Generally, bilge keels are installed for improving the roll motion of the fast round bilge craft, and no anti-spray rails are placed aft of amidships when mounting bilge keels. On the mixture type of craft with stern hard chine, anti-spray rails are possible at the hard chine part, with horizontally, vertically, and inclined mounting.

Generally, the bilge keel must be mounted on a craft with short hard chine and without anti-spray rails. For the mixed-type craft with medium and long hard chine, the bilge anti-spray rails are mounted but no bilge keels. Since the bilge keel may increase the drag, the drag of mixed-type craft with medium and long hard chine mounted with anti-spray rail is usually lower than round bilge craft with bilge keel, especially at high speed and with same rolling performance.

Aside from this, anti-spray rails mounted at the bow of round bilge craft can also improve the drag and decrease the spray blown on to the deck. The bow anti-spray rail can be mounded at 10%T (T = draft) above the water line. In the case of heavy spray at bow, especially in waves, larger width and length of the spray rail can be taken, with width of $b \geq 0.02B$ (B = craft beam) and length of 20% L (L = craft length) from the bow or even more, and with horizontal mounting. An anti-spray rail together with a stern plate or flap can regulate the running attitude and decrease the drag. Meanwhile, the anti-spray rail also increases the motion damping in waves and improves the heaving and pitching motion of the craft.

Model tests have been carried out, [34] and the test results are shown in Fig. 2.58. The drag reduction from using bow anti-spray rail alone is only at high speed ($Fn > 1.04$) and using a stern plate alone increases drag at high speed; however,

Fig. 2.58 Drag curve of craft models with bow anti-spray rails

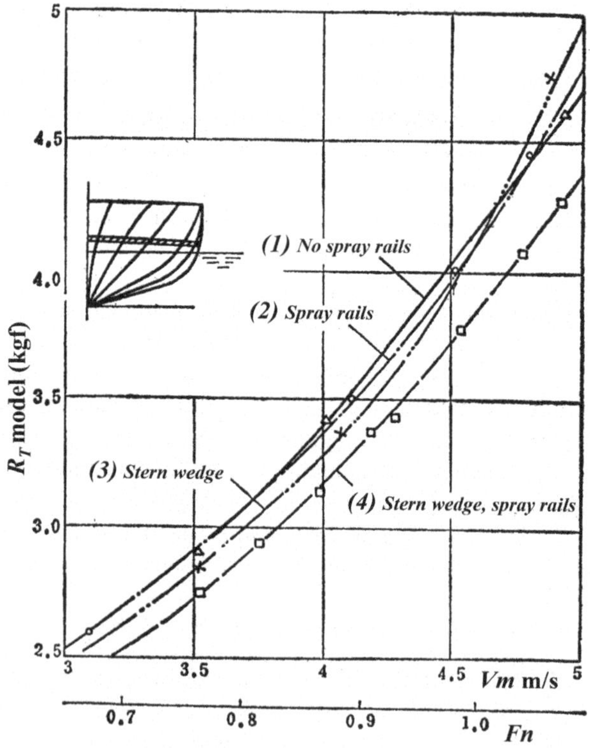

when both the stern plate and anti-spray rails are used together, nice drag reduction is available. Tests have also demonstrated that the bow anti-spray rail can decrease the pitching and heaving up to 8% and 20%, respectively.

References

1. German Jaguar Class Fast attack craft.: https://en.wikipedia.org/wiki/Jaguar-class_fast_attack_craft
2. German Eboats.: https://en.wikipedia.org/wiki/E-boat
3. P.K. Sahoo, H. Peng, J. Won, D. Sangarasigamany, Re-evaluation of resistance prediction for high-speed round bilge hull forms, in *11th International Conference on Fast Sea Transportation, FAST 2011, Honolulu, Hawaii, USA September 2011*, (2011)
4. The French La Combattante class fast attack craft.: https://en.wikipedia.org/wiki/La_Combattante_class_fast_attack_craft
5. H.F. Nordström, *Some Tests with Models of Small Vessels* (Teknisk Tidskrift, Skeppsbyggnadskonst, 1936). Published in English as publication No.19 of the Swedish State Shipbuilding Experiment Tank, Göteborg, 1951
6. D. de Groot, *Weerstand en voortstuwing van motorboten* (Schip en Werf, 1951). Published in English in International Shipbuilding Progress, Vol. 2, No.6, 1955
7. W.J. Marwood, A. Silverleaf, Design data for high-speed displacement hulls and a comparison with hydrofoil craft, in *3rd Symposium on Naval Hydrodynamics – High Performance Ships, ONR ARC-65*, (1960)
8. P.M. Beys, *Series 63 Round Bottom Boats*, Report 949 (Davidson Laboratory, Stevens Institute of Technology, 1963)
9. H.Y.H. Yeh, *Series 64 Resistance Experiments on High-Speed Displacement Forms* (SNAME Marine Technology, 1965)
10. H. Lindgren, A. Williams, *Systematic Test with Small Fast Displacement Vessel, Including a Study of the Influence of Spray Strips*, No. 65 (Publications of the Swedish State Shipbuilding Experiment Tank, 1969)
11. W.J. Marwood, D. Bailey, *Design Data for High-Speed Displacement Hulls of Round Bilge Form* (National Physical Laboratory Ship Report No. 99, 1969)
12. D. Bailey, *The NPL High Speed Round Bilge Displacement Hull Series* (Maritime Technology Monograph No.4, 1976)
13. J.J. Blok, W. Beukelman, The high–speed displacement ship systematic series hull forms—Seakeeping characteristics. SNAME Trans. **92**, 125–150 (1984)., ISSN 0081-1661
14. P. van Oossanen, J.B. Peifers, NSMB-systematic series of high-speed displacement ship hull forms, in *NSMB Workshop on Development in Hull Form Design, Wageningen, The Netherlands, October 22–24*, (1985)
15. J. Qianwei, G. Caisheng, *Systematic Tests with Round Bilge Displacement Craft* (CSSRC Shipbuilding of China No. 4, 1991) (in Chinese)
16. L. Shen, *Lines Design and Power Calculation of Defence Boat* (CSSRC Report 64-031, 1965) (in Chinese)
17. L. Guopei, W. Xieren, *Methodical Series Resistance Experiments on Medium Speed Craft Forms* (CSRDC Shipbuilding of China No. 3, 1991)., (in Chinese)
18. R. Martin, H. Sierra, *Form Selection of Developing a High-Speed Round Bilge Displacement Hull Series BAZAN80*, Ingeniera Naval, Numero 546 (1980), pp. 424–430, 444. (in Spanish)
19. D. Radojcic, T. Rodic, N. Kostic, Resistance and trim prediction for the SKLAD semi displacement hull series. Ocean Eng. Int. **3**(1), 34–50 (1999) see download
20. B.L. Robson, *Systematic series of high-speed displacement hull forms for Naval Combatants*, vol 130 (Transactions of Royal Institution of Naval Architects, 1988), pp. 241–259. ISSN

0035-8967. Paper presented 2nd December 1987 in London. Earlier presented to the Australian Branch of RINA in August 1987
21. P. Bojovic, Resistance of AMECRC systematic series of high-speed displacement hull forms, in *4th Symposium on High-Speed Marine Vehicles, 18–21 March 1997, Naples, Italy*, (1997). 7667.pdf available from https://www.repository.tudelft.nl
22. P.K. Sahoo, L.J. Doctors, *Hydrodynamics of the AMECRC Systematic Series—High-Speed Displacement Monohull Forms* (Australian Maritime College, 1999)
23. E. Lahtiharju, T. Karppinen, M. Hellevaara, T. Aitta, Resistance and seakeeping characteristics of fast transom stern hulls with systematically varied form. SNAME Trans. **99**, 85–118 (1991). ISBN 0-939773-07-04
24. D. Bailey, *Performance Prediction of Fast Craft* (NPL Report No.181, 1974)
25. H E Saunders, 'Hydrodynamics in Ship Design', 2, Chapters 76 and 77, SNAME, New York, 1957
26. N.K. Bales, Optimising the seakeeping performance of destroyer-type hulls, in *Proceedings, 13th Symposium on Naval Hydrodynamics, Japan*, (1980)
27. J.A. Mercier, D. Savitsky, *Resistance of Transom-Stern Craft in Pre-Planing Regime* (Report SIT-DL 73-1667, 1973)
28. P.K. Sahoo, L.J. Doctors, M.R. Renilson, *Theoretical and Experimental Investigation of Resistance of High-Speed Round Bilge Hull Forms*, FAST 99
29. S.C. Fung, Resistance and powering prediction for transom stern forms during early-stage ship design. SNAME Trans. **99**, 29–84 (1991)., ISBN 0-939773-07-04
30. Z.K. Tan, *Resistance Estimation of Round Bilge Craft by Means of Regression Method* (China Shipbuilding 4, 1979)., (in Chinese)
31. S.M. Shao, Y.C. Wang, M.Y. Zheng, Resistance estimation of high-speed displacement ship in calm water. J. Shanghai Jiao Tong Univ. **2** (1987) (in Chinese)
32. D. Radojcic, M. Kalajdzic, A. Simic, *Power Prediction Modelling of Conventional High-Speed Craft* (Springer Nature, 2019). ISBN 978-3-030-30606-0
33. W. Henschke, G. Hahnelm, K.H. Labes, *Widerstand von Rundspant-Motorbooten Schiffbautechnik 14, 4/1964 Also Contained in 'Resistance of Round Bilge Craft'*, Translation papers on shipbuilding, China, 1965
34. S.M. Shao, Y.C. Wang, B. Chen, S.X. Tu, *Experimental Investigation of Lines of High-Speed Displacement Craft* (China Shipbuilding, 1, 1985) (in Chinese)
35. V. Bertram, H. Iwashita, *Comparative Evaluation of Various Methods to Predict Seakeeping of Fast Ships* (Schiff & Hafen 48.6.54-8, 1996)
36. Z. Dong, X. Lu, *The Criterion of Water Depth without Shallow Water Effect for High-Speed Round Bilge Hull* (Shipbuilding of China No.2, 1993) (in Chinese)
37. Z. Dong, X. Lu, *Prediction of Shallow Water Resistance for High-Speed Round Bilge Hull*, (in Chinese)
38. P. Yinsheng, D. Zushun, *Initial Investigation of Seaworthiness on Large, Guided Missile Craft*, Research report (Naval Academy of Engineering, 1978) (in Chinese)
39. O.M. Faltinsen, *Hydrodynamics of High-Speed Marine Vehicles* (Cambridge University Press, 2005)., ISBN 978-0-521-84568-7
40. M. Ohkusu, O.M. Faltinsen, Prediction of radiation force on a catamaran at high froude number, in *International Proceedings 18th Symposium on Naval Hydrodynamics*, (National Academy Press, Washington, DC, 1990), pp. 5–19
41. S.M. Shao, Y.C. Wang, *Dynamics of High-Speed Craft* (Shanghai Jiao Tong University Press, 1990) (in Chinese)
42. R. Bhattacharya, *Dynamics of Marine Vehicles* (Wiley Publishing Co, 1978)
43. S. Shiming, Y. Wang, *The Effect of Stern Trim Flap on Resistance of High-Speed Craft*, No. 1 (Shipbuilding of China, 1981) (in Chinese)
44. Z. Surong, H. Wulin, *The Effect of Stern Trim Flap on the Resistance and Propulsion of High-Speed Displacement Craft*, No.5 (Ship Engineering, China, 1983) (in Chinese)

45. D. Zushun, S. Zhongli, Effect of stern trim plate on speed performance of semi-displacement craft, in *Shanghai International HPV Conference Proceedings, China, November 2–5, 1988*, (1988)., (in Chinese)
46. A.V. Ponomarev, Application of automatically controlled bottom interceptors for high-speed craft of various types, in *HPV'92, CHINA Proceedings, November 12–15*, (Krylov Shipbuilding Research Institute, 1992)
47. L. Wujiang, D. Zushun, Reducing Longitudinal Motion of High-Speed Round Bilge Craft with Controllable Flaps in Waves, in *5th Ship Seakeeping Symposium of China, Wushan, China*, (1991) (in Chinese)
48. S.M. Shao, A.K. Yang, Experimental investigation of the influence of bow hydrofoil on the operational performance of fast round bilge craft. Shanghai Jiao Tong Univ. J. **1** (1981)., (in Chinese)
49. X.P. Lu, D.J. Xue, Z.K. She, On the hydrofoil of fast round bilge craft nearby water surface for eliminating the wave making. Sci. J. Central Univ. Technol. **7** (1998) (in Chinese)
50. X.P. Lu, X.S. Zhu, Z.S. Dong, D.L. Tang, Z.S. She, *Influence of Stern Plates and Hydrofoils on the Running Attitude and Resistance of Fast Round Bilge Craft*, No. 3 (China Shipbuilding, 1999) (in Chinese)

Chapter 3
Basic Principles of Planing

Following our review of semiplaning craft where dynamic forces are small relative to inertial forces, we now move on to consider higher speeds where dynamic forces predominate. In this chapter, we will discuss primarily the generation of lift acting on the hull bottom of a craft at high speed. In Chap. 4, we will look at drag, and finally in Chap. 5, we will discuss dynamic stability and seakeeping.

A planing vessel needs a somewhat different form to a displacement or semiplaning craft to generate lifting force efficiently. Essentially the hull lower surface needs to present a plane area at a small angle of attack of sufficient area to lift the mass of the craft.

One configuration would be a simple plate, and indeed this is the geometry adopted by airboats that operate over the marshes and shallow waters of the Florida everglades (Fig. 3.1). In an open sea rather than a flat plate, craft have surfaces with dihedral about the longitudinal keel, as shown in Fig. 3.2. This has two advantages, giving directional stability and reduced response to waves. In both cases, the lifting surface form has low aspect ratio and so has lower lifting efficiency compared to a submerged wing. A submerged lifting wing—a hydrofoil—will have pressure applied on the lower surface and suction applied on the upper surface. A planing craft has just a lower surface generating lift from the dynamic pressure of the oncoming water velocity.

The pressure profile will vary depending on the exact shape of the planing surface. While many smaller planing craft have a simple prismatic form to their lower surface from midships aft, the forward hull will be shaped. The aft part of the hull may also have a warped dihedral flattening to the stern. The flow across the hull may also wet the sides toward the stern as well as the deadrise causing transverse flow in the same way as an aircraft wing with dihedral.

The flow may therefore be complex. In addition, since the support to the hull is now essentially dynamic, it will vary in proportion to V^2 and so is nonlinear. An iterative approach is required to determine equilibrium of forces and moments at any given speed and the associated "angle of attack" to the oncoming flow.

Z.-S. Dong et al., *High-Speed Monohull and Hydrofoil Craft*,
https://doi.org/10.1007/978-3-031-62760-6_3

Fig. 3.1 Florida Everglades air boat

Fig. 3.2 Planing craft with deadrise

Similar to semiplaning craft, model testing has formed a basis for analysis and design, with analytical projections derived from empirical formulae. These have been developed by relating to the classical hydrodynamics for the influencing parameters and analysis of various model test series. To clarify the basic hydrodynamics of a planing hull, one has to study the principles of the planing phenomenon and the consequences of geometrical variations.

In this chapter, first we discuss the planing principle in ideal flow, from two-dimensional to three-dimensional flows, then in real flow and correction of various factors to allow realistic determination of equilibrium.

3.1 Force Acting on a Planing Surface in Ideal Flow

3.1.1 Thin Wing Analogy of Two-Dimensional Planing Surface

We begin with discussion of a planing plate without camber, i.e., a two-dimensional flat plate moving with angle of attack α, at speed V_∞. If we consider the plat as fixed relative to the free water surface, the fluid flow with uniform velocity V_∞, and angle between the flow toward the plate at angle α, the flow around the plate will be as shown in Fig. 3.3, i.e., the flow moving under the plate being deflected downward.

Where the free surface intersects the underside of the plate there will be a point of inflexion, a stagnation point. Ahead of it the flow is directed forward, and rising above the free surface it will turn into spray. This stagnation point is a separation point between the forward and afterward flow at which the velocity of flow to the plate is equal to zero with maximum static pressure.

Aft of the trailing edge fluid flow is directed down, while static pressure in the fluid will return it to the static water level. A depression will be seen in the free surface.

If we consider the same plate fully immersed in fluid, Fig. 3.4, instead of the up welling and spray forward of the stagnation point, fluid will flow around the leading edge to join the accelerated flow over the upper surface. Instead, for a planing plate, the local pressure at the upper surface is atmospheric pressure.

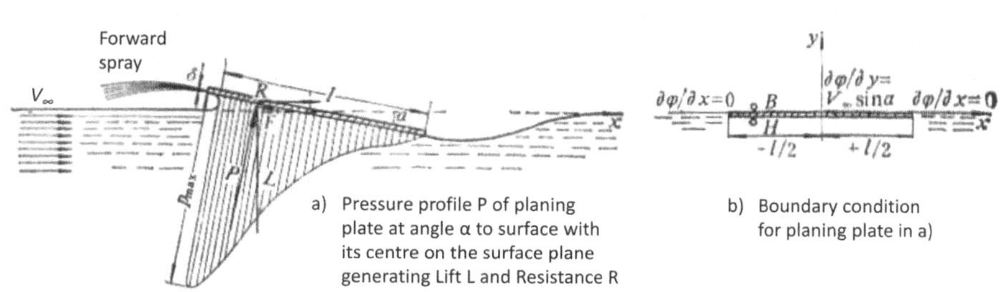

Fig. 3.3 Flow around a two-dimensional planing plate, force acting on the plate and boundary condition

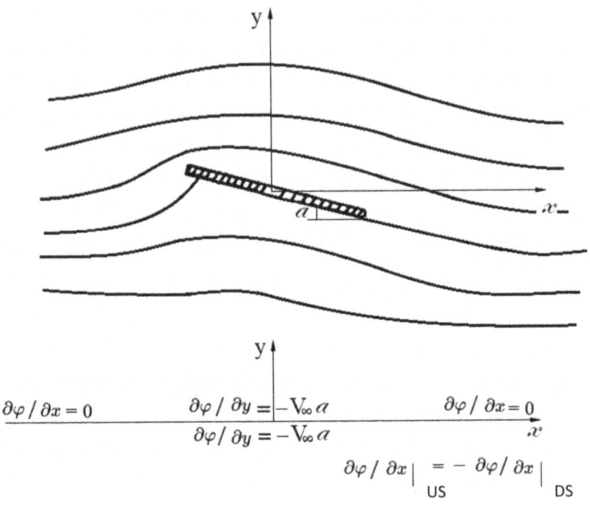

Fig. 3.4 Flow condition for fully immersed two-dimensional plate wing

To develop the equations of motion and potential flow, we need to define the boundary conditions, and to transform to a planing surface, rather than a lifting surface, we need to include the boundary conditions for the free surface.

3.1.2 Basic Equations of Motion

For the potential flow, the continuity equation can be written as a Laplace equation:

$$\nabla^2 \varphi = 0 \tag{3.1}$$

And the dynamic equation, i.e., Euler equation, as:

$$\frac{d\vec{V}}{dt} = \vec{F} - \frac{1}{\rho}\nabla_{\rho} \tag{3.2}$$

The Euler equation can also be presented as an integral equation following Lamb's form[1]:

$$\frac{\partial \varphi}{\partial t} + \frac{1}{2}\left(V_x^2 + V_y^2 + V_z^2\right) + U + \frac{p}{\rho} = f(t) \tag{3.3}$$

[1] https://en.wikipedia.org/wiki/Euler_equations_(fluid_dynamics)#Incompressible_case_and_Lamb's_form

where

φ is the velocity potential
$\vec{V}$ is the flow velocity
$\vec{F} = -\nabla U$
U is the potential of mass force; in gravity field $U = \text{gy}$.
p/ρ is the flow force or mass inertia force.

The velocity potential φ in the continuity Eq. (3.1) can be obtained uniquely in terms of boundary conditions, as follows:

3.1.3 Boundary Condition at Body Surface

It is a condition that flow cannot penetrate across the body surface, i.e., the normal velocity of a point on the body is equal to that of the fluid at that point, so:

$$V_n = \left[\frac{\partial \varphi}{\partial n}\right]_B = -\frac{\partial F}{\partial t} \Big/ \sqrt{\left[\frac{\partial F}{\partial x}\right]^2 + \left[\frac{\partial F}{\partial y}\right]^2 + \left[\frac{\partial F}{\partial Z}\right]^2} \tag{3.4}$$

Where $f(x, y, z, t) = 0$ is the equation of the body surface.

3.1.4 Boundary Condition for Free Surface

The induced velocity is equal to 0 at infinite distance from the plate, and the pressure at the free surface is atmospheric pressure, i.e., $p = p_a$. When we place the coordinate origin at the undisturbed steady water surface, then on the free surface:

$$\frac{\partial \varphi}{\partial t} + \frac{1}{2}\left(V_x^2 + V_y^3 + V_z^2\right) + g\eta_s(x, z) = 0 \tag{3.5}$$

Where $\eta_s(x, z)$ is the free surface coordinate in the y direction.

In the case of limited water depth, i.e., another solid but permeable boundary, such as a riverbed or bank, the boundary conditions will change; however, we consider here simply the water surface and infinite water depth.

Based on the above boundary conditions, a unique solution of the motion equation, i.e., the flow velocity field, can be obtained. Then, the pressure distribution in the flow field can be obtained from the Euler equation. In general, due to the unsteady and nonlinear nature of the basic equations, it is very difficult to obtain the solutions. However, for a planing surface, due to high speed and small angle of attack, the equations can be simplified.

3.1.5 Simplified Boundary Condition for "High-Speed Small Disturbance"

At high planing speed, i.e., Froude number $F_{n\nabla} > 4$–5, the kinetic force will be greater than that due to gravity, and the gravity term in the basic Eq. (3.3) and in boundary condition (3.5) can be ignored.

When the angle of attack is small ($\alpha < 5$ to 6 degrees), the disturbance velocity in the y direction induced by the planing surface is substantially smaller than the planing speed, and the square of disturbance velocity, i.e., second term of left side of Eqs. (3.3), and (3.5) can be neglected, which is referred to as "high-speed small disturbance" simplification of motion.

In general, putting the coordinate system on the planing surface and considering the plate stationary and the flow with uniform speed v moving to the planing surface, the motion can be regarded as steady.

In addition, due to small angle of attack, the boundary condition at the plate surface can be met approximately, and Eqs. (3.1), and (3.3) can be simplified as follows:

$$\nabla^2 \varphi = 0$$
$$V_\infty \frac{\partial \varphi}{\partial x} = \frac{p - p_0}{\rho} \tag{3.6}$$

where

φ is the disturbance velocity potential
V_∞ is the uniform velocity of incoming flow
p_0 is the atmospheric pressure
P is the pressure in flow field

The boundary conditions (3.4) and (3.5) can be simplified as follows:
At the plate, the boundary when $y \approx 0$, and $-½ \leq x \leq +½$

$$\frac{\partial \varphi}{\partial n} \approx -\frac{\partial \varphi}{\partial y} = V_\infty \sin \alpha \approx V_\infty \alpha \tag{3.7}$$

For the free surface boundary when $y = 0$, and $x \leq -½$ or $x \geq +½$

$$\frac{\partial \varphi}{\partial n} = 0 \tag{3.8}$$

It can be shown that the approximation method discussed above is similar to that of thin wing theory. Comparing the boundary condition for thin wing with this (see Fig. 3.4), one can find that the difference between them only lies in that the potential function of a thin wing can be an analytical expression on the whole surface; however, it is only valid on the lower plane for the planing surface.

Since the planing plate can meet Eq. (3.8) on the physical plane, i.e., correspondent to the virtual axis of the complex potential function, it may be used to apply to the upper plane of physical plane and define complex velocity as an analytical function on fully physical plane with separation along the plane surface. Then the velocity of the points symmetrically on both up and down surface is available with the following relationship:

$$\begin{aligned} \left(\frac{\partial \varphi}{\partial y}\right)_U &= \left(\frac{\partial \varphi}{\partial y}\right)_D \\ \left(\frac{\partial \varphi}{\partial x}\right)_U &= -\left(\frac{\partial \varphi}{\partial x}\right)_D \end{aligned} \tag{3.9}$$

Equation (3.9) is just the same as the boundary condition of the thin wing; therefore, the flow and force acting on the lower surface of a two-dimensional planing plate is similar to that on the two-dimensional thin wing, enabling us to apply the results of thin wing theory to planing plate directly, i.e., the so-called the "thin wing simulation."

To obtain the force acting on the planing plate directly from the thin wing equation, the pressure distribution on both upper and lower surfaces of the thin wing needs to be obtained first; thus, the Bernoulli equation can be applied on any point of upper surface on thin wing at infinite far distance, as:

$$p_0 + \frac{\rho V_\infty^2}{2} = p_0 + \frac{\rho}{2}\left[\left(V_\infty + \left(\frac{\partial \varphi}{\partial x}\right)_0\right)^2 + \left(\left(\frac{\partial \varphi}{\partial y}\right)_0\right)^2\right] \tag{3.10}$$

Subscript "*u*" denotes physical value on upper surface and "*d*" on lower surface; for simplification, take:

$$\begin{aligned} \left.\frac{\partial \varphi}{\partial x}\right|_u &= -\left.\frac{\partial \varphi}{\partial x}\right|_d = v_x \quad \text{and} \\ \left.\frac{\partial \varphi}{\partial y}\right|_u &= \left.\frac{\partial \varphi}{\partial y}\right|_d = v_y \end{aligned} \tag{3.11}$$

The pressure at any point on upper surface can then be written as:

$$p_u - p_0 = -\rho V_\infty v_x - \frac{\rho}{2}\left(v_x^2 + v_y^2\right) = -\rho V_\infty v_x \tag{3.12}$$

Where $(p - p_a)$ or $p_u - p_0$ is the "gage" pressure. Similarly, the gage pressure of the point symmetrically at the lower surface is:

$$p_d - p_0 \approx +\rho V_\infty v_x \tag{3.13}$$

Therefore, the gage pressure on both upper and lower surfaces of a thin wing is equal, but with opposite sign, and so suction on upper surface, pressure on lower. Since the surface situation of both planing surface and thin wing underside is the same, hydrodynamic pressure on a planing plate is half of thin wing. The lift coefficient of the submerged plate is $C_L = 2\pi\alpha$. Thus, the planing plate total pressure coefficient and total pressure per unit width are:

$$C_p = \pi\alpha,$$
$$P = \frac{1}{2} C_p \rho V_\infty^2 l = \frac{\pi}{2} \rho V_\infty^2 \alpha l \tag{3.14}$$

The position of pressure center of a planing plate is the same as on a thin wing, i.e., at 3 $l/4$ from leading edge, where l is the wetted length of the planing plate. Note that for a wing the length l will be the chord, notmally represented by C (see chapter 8). Here we will continue to refer to the plate length in x direction as l.

The unique difference of flow between the planing plate and plate-wing with fully submerged flow is at the leading edge, see Fig. 3.3. With respect to the flat plate wing in fully immersed flow, the forward stagnation point is located at a point on the lower surface aft of the leading edge, and the flow around the leading edge at high speed creates a "suction force" at leading edge, which generates a lift perpendicular to incoming flow, but not perpendicular to the wing surface.

Thus, in ideal flow, a fully immersed thin wing does not induce resistance. In the case of planing plate, flow does not go around the leading edge. Forward of the stagnation point, the flow is reversed and this flow follows a circular path losing energy and with entrained air generates spray. No suction happens at the leading edge and the total pressure p exerted by the water wetting forward of the stagnation point is perpendicular to the plate, and generates a resistance, $p \sin \alpha$, in the direction of motion.

It should be noted that the fluid flung forward above the static waterline mixes with air. The spray generated will extend the wetted area significantly from that due to the "solid" water. The air content grades from zero and so the true wetted area is uncertain. Nevertheless, even in ideal flow, a planing plate generates resistance from the flow forward of the stagnation point as well as aft of it. In a real fluid, the leading edge of the wetted length will be above still water level. At present, we are considering a two-dimensional thin plate at small angle of attack and so the wetted length does not need to be adjusted. We will discuss the real fluid case later. The resistance can therefore be expressed as:

$$R_\delta = p \sin \alpha = p\alpha = \frac{\pi}{2} \rho V_\infty^2 \alpha^2 l$$

This can be expressed as a resistance coefficient:

$$C_R = \frac{R_\delta}{1/2\rho V_\infty^2 l} = \pi\alpha^2 \tag{3.15}$$

In the case of a curved planing plate, which corresponds to a thin wing configuration with camber the lift coefficient of the thin wing is:

$$C_L = 2\pi \sin(\alpha - \alpha_0)$$

where

$\alpha_0 = -2f/c$
f is the camber of thin wing
c is the chord length of the thin wing

The total pressure coefficient of a planing surface is still equal to half of thin wing, i.e.,

$$C_p = \pi \sin(\alpha - \alpha_0) \approx \pi(\alpha - \alpha_0) \tag{3.16}$$

To summarize, for a planing plate with high speed and small disturbance, the modeling method of a thin wing can be used taking the planing plate as half of a thin wing, and then full development of the theory of the thin wing is used to predict the calculation of planing plate, but with deduction of the suction on the leading edge.

3.2 Approximate Solution of L. Y. Sedov with Consideration of Gravity Effect

When $F_n > 4$ or 5, the gravity effect can be neglected. However, in general, the speed of a planing hull is less than that, so the gravity effect has to be considered, and the simple "thin wing model" method cannot be applied. The gravity parameter cannot be neglected in the equations and boundary condition. However, the parameter of square of disturbance velocity still can be neglected to form a linear equation. When considering the gravity effect, waves will be generated, and the boundary condition on the wave surface must be satisfied in the equations, complicating the approach to solve the equations.

Figure 3.5 shows the variation of both lift and moment coefficients of a planing plate with varying F_n [1]. In the figure, N_∞ and M_∞ are equal values at $F_n = \infty$ but neglecting the gravity effect. N and M are the values considering the gravity effect. From the figure, at $F_n = 6$, one can find a difference of about 5% between considering or not considering the gravity effect, 10% for F_n =4.6, and 15% for F_n =3.8; therefore, the gravity effect is clearly present when Fn is not large.

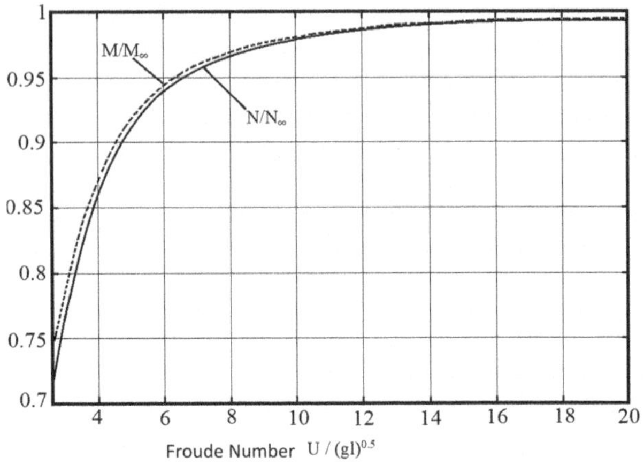

Fig. 3.5 Comparison of both lift (N) and moment coefficient (M) for $F_n \geq 2.8$ with correspondent value at $F_n = \infty$

The calculation discussed above is for a small disturbance; however, for a large angle of attack, the assumption is invalid; thus, the equations and boundary conditions will be nonlinear.

Sedov carried out extensive investigations [2] and determined approximate equations for the resistance and lift of a planing plate at high Fn and without considering the gravity effect as follows:

$$C_L = \pi\alpha\left[1 - \frac{1}{F_n^2}\left(\frac{\pi^2 + 4}{2\pi}\right)\right] \tag{3.17}$$

$$C_D = \pi\alpha^2\left[1 - \frac{1}{F_n^2}\left(\frac{\pi^2 + 4}{2\pi}\right)\right] \tag{3.18}$$

The moment coefficient about the trailing edge is

$$C_M = \frac{\pi}{4}\alpha\left[1 - \frac{1}{F_n^2}\left(\frac{\pi^2 + 8/3}{2\pi}\right)\right] \tag{3.19}$$

The second term of the equations above is the gravity effect. When $F_n \rightarrow \infty$, the gravity effect disappears, and the equations match the simplified thin wing approximation. These equations can be applied to $F_n > 2.8$.

Figure 3.6 shows the comparison of Sedov's approximate solution with Y. S. Chaplegen's precise solution at the lower speed and small disturbance condition. In the figure, the solid line I shows the precise numerical solution, while the line II shows the approximate solution in terms of (3.17). From the figure, one can see that (3.17) can be applied when $F_n > 2.8$, but with significant error when $F_n < 2.8$.

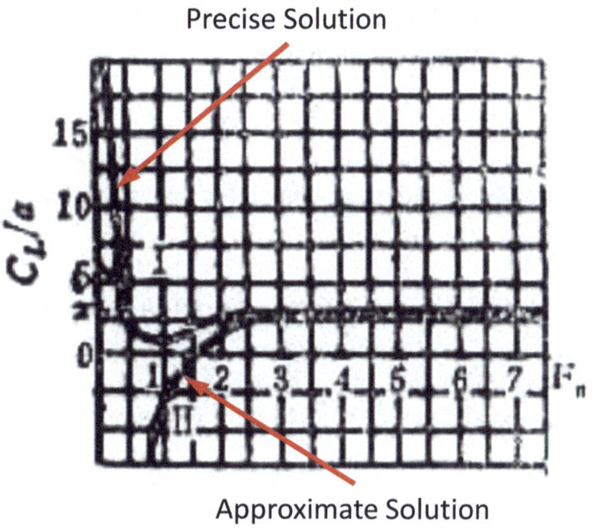

Fig. 3.6 Comparison of Sedov's solution with numerical precise solution

3.2.1 Nonlinear Solution of Two-Dimensional Planing Plate in Ideal Flow—High-Speed Large Disturbance Problem

For a planing plate running at large angle of attack, the disturbance on the water surface may be significant, the gravity effect can be neglected if speed is high, and the question can be solved with aid of jet theory. Using two-dimensional flow, the complex potential can be expressed as $W = \varphi + i\psi$, i.e., an analytical function on the flow plane.

Using conformal transformation, the transform function between the physical plane and complex potential plane can be obtained, i.e., obtaining the expression of complex potential, in order to avoid solving the nonlinear equations directly.

References [3, 4] provide such method for solving planing plate problems with any angle of attack by means of jet theory, and the pressure coefficient is obtained as

$$C_p = \frac{2\pi}{\cot\frac{\alpha}{2} + \pi + \tan\frac{\pi}{2}\ln\left(\cot^2\frac{\alpha}{2} - 1\right)} \tag{3.20}$$

The position of the relative pressure center from the leading edge can be expressed as:

$$\frac{l_p}{l} = \frac{1 + 1/2\cos\alpha + 2(1 - \cos\alpha)\ln 2 + \pi/2\sin\alpha}{(1 - \cos\alpha)\ln\dfrac{2\cos\alpha}{1 - \cos\alpha} + 1 + \cos\alpha + \pi\sin\alpha} \tag{3.21}$$

For a small angle of attack, the higher order terms in the equations can be neglected and the equations can be solved.

3.3 Wing Simulation Method for the Planing Surface with Finite Width

When the width of planing surface is finite, the flow becomes three dimensional. Using the wing modeling method, the wing theory for a wing with finite span can be applied directly into the solution for a planing surface.

3.3.1 *Planing Surface with High Aspect Ratio*

For the thin wing with finite span and high aspect ratio, by using the down wash effect induced by free vortices at the wing tips, the lift coefficient can be obtained as:

$$C_L = \frac{2\pi \sin \alpha_e}{1 + 2/A} \approx \frac{2\pi\alpha_e}{1 + 2/A} \tag{3.22}$$

A planing surface at high speed and small angle of attack equals half of a plate wing, in addition, it is customary to use the wetted length ratio λ as the specific wetted size, rather than aspect ratio A (the inverse to λ); therefore, the total pressure coefficient of planing plate is:

$$C_p = \frac{\pi\alpha}{1 + 2\lambda} \tag{3.23}$$

Since a planing plate is different from a wing with surrounding flow, the suction at leading edge does not exist on a planing plate, the total pressure for the three-dimensional plate is also perpendicular to the plate surface. Then the total pressure can be separated to parts, i.e., lift perpendicular to incoming flow, and resistance with same direction of incoming flow, with coefficients as:

$$\begin{aligned} C_L &= \frac{\pi\alpha}{1 + 2\lambda} \cos \alpha \approx \frac{\pi\alpha}{1 + 2\lambda} \\ C_D &= \frac{\pi\alpha}{1 + 2\lambda} \sin \alpha \approx \frac{\pi\alpha^2}{1 + 2\lambda} \end{aligned} \tag{3.24}$$

If the planing plate has camber, then the camber effect can be calculated into angle attack with zero lift, as $\alpha_0 = -2f/l$, then:

$$C_p = \frac{\pi}{1 + 2\lambda}(\alpha - \alpha_0) \tag{3.25}$$

It must be noted that even though the effect of limited span of a planing surface can be obtained by means of the thin wing modeling method, the physical meaning is rather different between them.

For a wing with limited span, the vortex layer will reduce along the span as a free vortex, particularly at the wing tip, to create the downwash effect from the free vortex. Consequently lift is decreased, inducing resistance. However, there is no vortex layer exiting the planing surface, thus no free vortex, and water rises to atmosphere at the plate tip, losing pressure, which generates the spray blown out at planing surface sides. This is similar to the free vortex flow on a wing, but does not lead to change of the lift direction while nevertheless it induces a component of resistance.

Experimental results demonstrate that there are errors in calculating the hydrodynamics of a planing surface by means of thin wing modeling method. This is due to Eq. (3.17) being only suitable to for a thin wing with high aspect ratio (in general, $A > 3$) and with elliptical profile for the wing plane. However, in practice, the aspect ratio for a planing hull is rather lower, and the speed is also lower, in which the gravity effect cannot be neglected.

So, for calculating hydrodynamics of a planing surface by means of such method, a correction is needed that will be introduced in the next chapter.

With respect to the three-dimensional planing surface with a large angle of attack, it is difficult to resolve such problems by simplified analytical methods discussed so far, even for a flat plate. Fortunately, in practice, the operational trimming angle of a planing hull is rather small; thus, it can be simplified using the small disturbance method.

3.3.2 *Short Wing Modeling for the Planing Surface with Lower Aspect Ratio*

For a real planing hull, the aspect ratio is rather small, say $A \leq 3$, and large errors will be introduced by using "down wash" for correction of limited span effect (limited beam). It is necessary to use an empirical method for correction.

In principle, serious disadvantages exist with the empirical method. Nevertheless, similar to a planing surface with high aspect ratio, a method can also be applied for calculating the hydrodynamics of a planing surface with lower aspect ratio by similitude to low aspect ratio wings.

There are many methods for a wing with small aspect ratio and small angle of attack. The lift coefficient for a short wing with $A \geq 1$ can be obtained by means of R.T. Jones's "Corrected lifting line theory" [5], as follows:

$$C_L = \frac{\pi A}{1 + \frac{A}{2}E} \sin \alpha \tag{3.26}$$

For $A < 1$, according to K. Krienes' Lifting Surface Theory:

Fig. 3.7 Flow around a wing with lower aspect ratio

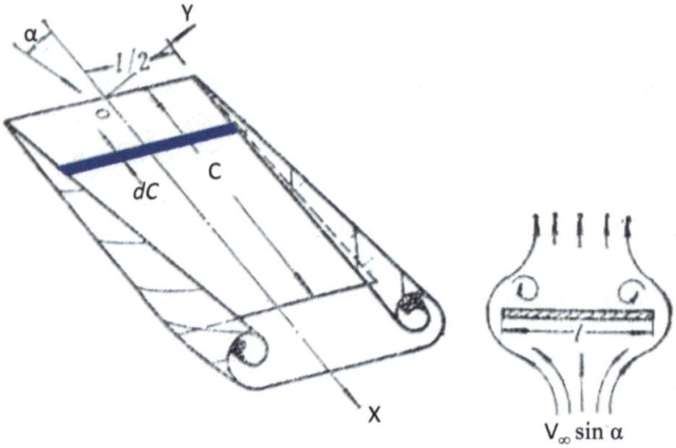

$$C_L = \frac{\pi A}{1+A}\sin\alpha \tag{3.27}$$

Both Eqs. (3.26) and (3.27) can be suitable for small angle of attack, neglecting the effect of transverse flow. Since the velocity of transverse flow is proportional to sin α, the effect of small angle of attack is a small value with second order and can be neglected. However, it cannot be neglected for a large angle of attack. The velocity of transverse flow, perpendicular to wing plane, is V_∞ sin α, which generates "transverse resistance" on the wing transverse section.

Considering the flow circumstances around a wing with low aspect ratio, a strong tip vortex which can be considered as a vortex induced by transverse velocity V_∞ sin α is generated, flowing perpendicularly to the wing section as shown in Fig. 3.7, and the projection perpendicular to the "transverse resistance" generates a lift of "transverse resistance."

In the direction of chord *dc*, transverse resistance can be written as:

$$dD_c = C_{DC}\frac{1}{2}\rho V_\infty^2 \sin^2\alpha l(x)dc$$

Where *l*(*x*) is the chord length at *x*.

Neglecting camber, the transverse resistance on the whole wing can be written as:

$$D_c = C_{DC}\int_0^C \frac{1}{2}\rho V_\infty^2 \sin^2\alpha l(x)dc = C_{DC}\frac{1}{2}\rho V_\infty^2 S\sin^2\alpha \tag{3.28}$$

"Lift induced by transverse resistance" can be written as:

$$L_{DC} = C_{DC}\frac{1}{2}\rho V_\infty^2 S\sin^2\alpha\cos\alpha \tag{3.29}$$

Thus, the total lift coefficient acting on the short wing without camber:

$$C_L = \frac{\pi A}{1 + \frac{A}{2}E} \sin\alpha + C_{DC} \cos\alpha \sin^2\alpha \quad (A \geq 1) \tag{3.30}$$

Or

$$C_L = \frac{\pi A}{1 + A} \sin\alpha + C_{DC} \cos\alpha \sin^2\alpha \quad (A < 1) \tag{3.31}$$

According to Kirchhoff's "splitting around flow theory for perpendicular plate," the drag coefficient is:

$$C_{DC} = \frac{2\pi}{\pi + 4} \approx 0.88$$

However, the experimental value is significantly greater than the calculations mentioned above, and $C_{DC} \approx 2.0$ for a plate wing with sharp tips and 1.0 with round tips. The difference between Kirchhoff's drag coefficient and experimental results is due to Kirchhoff's assumption that the flow after the flow splitting point is a dead water area with zero flow speed. However, in practice, there are a lot of vortices aft of the plate. Kirchhoff did not consider the energy loss induced by the vortex; therefore, the calculation results should be significantly lower than real conditions as verified by experiment.

With respect to the planing surface, the first term of right side of (3.30), or (3.31), i.e., the lift generated by longitudinal flow is half of the total lift of a fully immersed wing, but the transverse resistance of a planing plate is equal to that of a wing. However, suction at the leading edge exists on a wing and generates lift with a linear term perpendicular to incoming flow. According to wing theory, suction at the leading edge is $\sin^2\alpha$ of lift. Since there is no suction on the planing plate, this element of lift should be deducted in the linear term of lift; thus, the lift coefficient of a planing plate without camber can be written as:

$$C_L = \frac{0.5\pi A \cos^2\alpha \sin\alpha}{1 + \frac{A}{2}E} + C_{DC} \cos\alpha \sin^2\alpha \quad (A \geq 1) \tag{3.32}$$

$$C_L = \frac{0.5\pi A \cos^2\alpha \sin\alpha}{1 + A} + C_{DC} \cos\alpha \sin^2\alpha \quad (A < 1) \tag{3.33}$$

Or it can be expressed in terms of λ, then:

$$C_L = \frac{0.5\pi \cos^2\alpha \sin\alpha}{\lambda + E/2} + C_{DC} \cos\alpha \sin^2\alpha \quad (\lambda \leq 1) \tag{3.34}$$

$$C_L = \frac{0.5\pi \cos^2 \alpha \sin \alpha}{\lambda + 1} + C_{DC} \cos \alpha \sin^2 \alpha \ (\lambda > 1) \tag{3.35}$$

3.4 Theory for Slender Planing Surface—The Slender Wing Analogy

In general, length/beam ratio is larger than 1 for a planing hull (L/B > 1); therefore, the calculation for slender planing surface can be taken as that for slender wing, thus, the so-called slender wing simulation, or theory of slender planing surface.

This idea for processing the slender wing was introduced in M. M. Monk's "Theory of slender body," following which R. T. Jones developed this to "slender wing theory," and used it for calculation of a planing surface with high wetted length/beam ratio.

3.4.1 Basic Idea of Monk-Jones' Slender Wing Theory

With respect to a wing with infinite span, there is longitudinal flow and no transverse flow around the wing. For a wing with infinite chord and limited span, only transverse flow and no longitudinal flow exists. Therefore, the so-called "slender wing" means a wing mainly with transverse flow, i.e., low aspect ratio, with small change of span along the longitudinal direction of wing.

To simplify the processing for small disturbance, the angle of attack for Monk-Jones' slender wing theory has to be small.

Regarding flow around a slender wing, assume that a slender triangular wing is moving with speed V_∞ and angle of attack α, and the reference coordinate system is placed in the perpendicular plane *N* as shown in Fig. 3.8. The motion of the wing moving ahead corresponds to the wing section going continuously through the *N* plane, as shown in Fig 3.8a. For an observer on plane *N*, this corresponds to the wing section moving down continuously as shown in Fig 3.8b. The flow at any

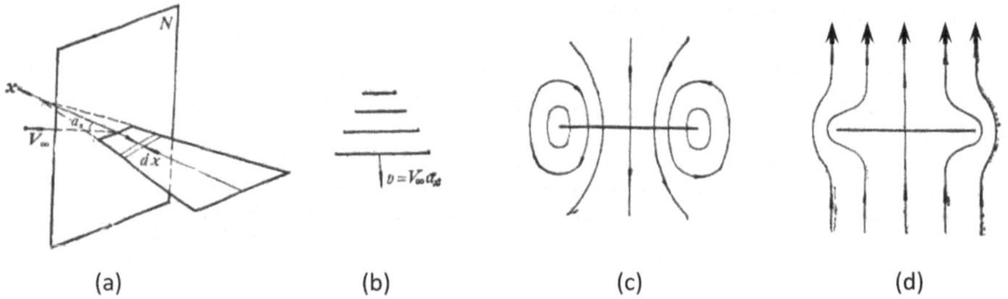

Fig. 3.8 Flow around the slender wing

instant can be illustrated as shown in Fig 3.8c. For a stationary wing and moving flow, the flow field can be shown as Fig 3.8d. When angle of attack is small, the disturbance velocity is small, and longitudinal disturbance velocity being second order is even smaller; consequently, the flow can be taken as two-dimensional transverse flow.

If only resultant force and moment acting on the wing are needed, and the pressure distribution on the wing is not needed, then the force can be obtained in terms of change of virtual mass, the so-called "theory of virtual mass", as follows:

- Assume a fixed perpendicular plane in space *N*, the triangular wing is moving through *N* increasingly at different time period, and the section moving down with vertical velocity $v = \alpha_x V_\infty$, as shown in Fig 3.8b, where α_x is the local angle of attack at the section. For a triangular plate wing, the α_x = constant = α.
- Assume force *dP* acting on the length of *dx*, then force acting on the unit length on chord is *dP/dx*, equal to momentum change rate of virtual mass *m'*, as:

$$\frac{dp}{dx} = \frac{d}{dt}(m'v) \tag{3.36}$$

It should be stressed that the motion is observed on fixed plane *N*; thus, the motion is changing. In ideal flow with non-rotating motion, force can exert on the body only in the case of body in changing motion. Such change can be due to acceleration or change of virtual mass due to change of body profile. Equation (3.36) can be written as two items, for the two parts, as:

$$\frac{dP}{dx} = m'\frac{dv}{dt} + v\frac{dm'}{dt} = m'\frac{dv}{dt} + v\frac{dm'}{dx}\frac{dx}{dt} \quad \text{and} \quad \frac{dx}{dt} = V_\infty \cos\alpha_x \approx V_\infty$$

Thus,

$$\frac{dP}{dx} = m'\frac{dv}{dt} + V_\infty v\frac{dm'}{dx} \tag{3.37}$$

In case of wing camber equal to zero (flat plate triangular wing), and wing moving at constant speed, then the first term on the right side of the above equation disappears, and $v = \alpha V_\infty$. The integral of Eq. (3.37), then becomes

$$P = V_\infty^2 \alpha \int_0^x \frac{dm'}{dx} dx = V_\infty^2 \alpha m'_x \tag{3.38}$$

In case of wing with camber, then

$$\frac{dP}{dx} = m' V_\infty \frac{dv}{dx} + V_\infty^2 \frac{dm'}{dx} = V_\infty^2 \left[m'\frac{d\alpha_x}{dx} + \alpha_x \frac{dm'}{dx} \right] \tag{3.39}$$

In this case, the direction of force in the normal direction changes with the section considered. If the force is projected at normal direction at each section to the chord, and assume the angle between the chord and moving direction is α, then one obtains

$$P = V_\infty^2 \int_0^x \left[m' \frac{d\alpha_x}{dx} + \alpha_x \frac{dm'}{dx} \right] \cos(\alpha_x - \alpha) dx \tag{3.40}$$

In case of small camber, cos $(\alpha_x - \alpha) \approx 1$, then above equation can be simplified to:

$$P = V_\infty^2 \int_0^x \left[m' \frac{d\alpha_x}{dx} + \frac{dm'}{dx} \right] dx \tag{3.41}$$

And

$$P = V_\infty^2 \alpha_x m'_x$$

The equation shows that the normal force only relates to angle of attack at rear edge and virtual mass, and no relation with medium section, and it can be shown that due to camber, $\alpha_x > \alpha$ and the force in the normal direction is increased.

With respect to the plate in span direction, its virtual mass is:

$$m' = \frac{1}{4} \pi \mu b^2 = \pi \rho \left(\frac{b}{2} \right)^2 \tag{3.42}$$

This equation is suitable to the rear section of the wing, and substituting into (3.36), the force in the normal direction is obtained as:

$$P = \rho \pi \left(\frac{b_x}{2} \right)^2 V_\infty^2 \alpha_x \tag{3.43}$$

For a triangular plate wing with small angle of attack, if the lift is equal to the normal force, then the lift coefficient can be obtained as:

$$C_L = \frac{1/4 \pi \rho b^2 V_\infty^2 \alpha}{1/2 \rho V_\infty^2 S} = \frac{\pi}{2} A \alpha \tag{3.44}$$

Or

$$C_{Lb} = \frac{P}{1/2 \rho V_\infty^2 b^2} = \frac{\pi}{2} \alpha \tag{3.45}$$

In the case of a wing not being of triangular profile, then there are two conditions.

The *first condition* is where the maximum wing beam is not located at trailing edge. In such case, the force in normal direction at the place before the maximum beam is positive, but negative at the maximum beam.

This is due to the section after the maximum beam moving downward in the N plane has a decreased virtual mass, i.e. dm'/dx <0; therefore if the section at trailing edge has zero beam, then the total force in the normal direction on the wing is also equal to zero.

However, this is incorrect in practice. A trailing vortex will be formed at maximum span making the wing sections after the maximum beam located in trailing vortex. Thus, Jones assumed, "After maximum beam, the wing will be not subject to force".

In the case of calculating force in the normal direction, it can be calculated before maximum span, and the condition of wing after the max beam is not related to normal force. Consequently, the theoretical calculation agreed with experimental result.

The *second condition* is where the leading edge is not sharp, e.g., rectangular wing. In such case, with sudden change of beam at leading edge, dm'/dx at leading edge is also changed suddenly, which does not satisfy the assumption for the slender wing that change of wing beam in longitudinal direction should be small, and the slender wing theory can't be applied.

However, the experimental results agree with Eqs. (3.39) and (3.40). Therefore, for a slender wing with blunt leading edge, the method for calculation of force in normal direction, as mentioned above can be applied, so long as adding a small delta wing on the leading edge and driving the wing section of this triangular wing to be zero gradually.

Therefore, for slender wing, the force in normal direction can be calculated so long as angle of attack α_x and virtual mass m'_x are known. Since the angle of attack is known, the key problem is obtaining the virtual mass, in terms of profile of transverse sections.

The *final* issue is that the direction of the normal force using slender wing theory is based on small angle of attack and flow disturbance. Thus, the direction of normal force does not influence the lift; however, it influences the resistance significantly, and it cannot be solved by means of the virtual mass method.

When calculating the velocity potential in terms of boundary conditions, the pressure distribution on a triangular wing can be obtained. Then the integral of the pressure along the wing chord, and thus the lift distribution along the wingspan, can be obtained as well, with elliptical profile. Therefore, the trailing vortex distribution after plane delta wing is also obtained with elliptical profile.

In the case of circulation distributing on an elliptical shape, the trailing vortex distribution after plate delta wing should be elliptical. When the distribution of circulation (i.e., trailing vortex) along the span is elliptical, the induced resistance on a wing with limited span is:

$C_{Di} = \frac{C_L^2}{\pi A}$ For slender plate delta wing, since $C_L = \frac{\pi}{2} A\alpha$, then

$$C_{Di} = C_L \frac{\alpha}{2} \tag{3.46}$$

It can be seen that the angle between the resultant force and perpendicular plane is $\alpha/2$. It can be understood that owing to suction on the sharp leading edge, the angle between resultant force (normal force and suction) and perpendicular plane is $\alpha/2$. In case of the planing plate, the normal force is perpendicular to the plate surface, since suction is not available.

3.4.2 Virtual Mass Theory on Slender Planing Surface—Slender Wing Analogy on Planing Surface

For a slender planing surface, the normal force can be obtained by means of the Monk-Jones method together with considering the difference between the planing surface and a fully submerged slender wing.

The first difference is the virtual mass. The virtual mass of a planing surface skimming on the water surface should be half of that on fully submerged wing section. For a plane section with no transverse curvature:

$$m' = \frac{1}{2}\pi\rho\left(\frac{b}{2}\right)^2 \tag{3.47}$$

Besides, in the case of planing, whole sections do not skim on the water surface, but some sections submerge under the water surface, as shown in Fig. 3.9, and then the virtual mass should be the summation of section virtual mass and the virtual mass induced by aft water depression; however, the latter is very difficult to calculate. Reference [6] suggested that virtual mass for a rectangular section on unit length in an infinite medium can be written as:

$$m' = \left[1 + \frac{1}{2}\sqrt{\frac{h}{b}}\right]\pi\rho\left[\frac{b}{2}\right]^2 \tag{3.48}$$

Here h is the height of rectangular plate (see Fig. 3.9), and considering the relation between submerged depth y, angle attack, and the distance between the section and stagnation line x, we have:

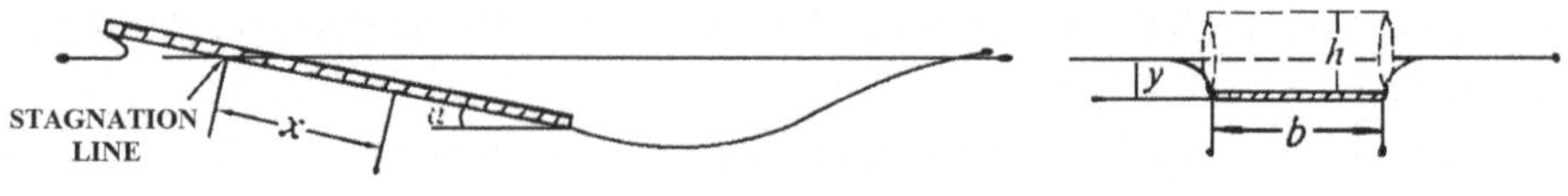

Fig. 3.9 Plate planing condition

$$y = x \sin \alpha = x\alpha$$

The stagnation line is shown along the intersection of the planing surface with the water surface, and the virtual mass of the transverse section on the planing surface can be written as:

$$m' = \frac{1}{2}\pi\rho(b/2)^2\left[1 + 1/2\sqrt{\frac{2x\alpha}{b}}\right] \tag{3.49}$$

In the case of the distance between the trailing edge and "stagnation line" equal to x_s, then the virtual mass at trailing edge:

$$m_s' = 1/2\pi\rho\left(\frac{b_x}{2}\right)^2\left[1 + \frac{1}{2}\sqrt{\frac{2x_s\alpha}{b_s}}\right] \tag{3.50}$$

Substituting (3.50) into (3.41), the normal force induced by virtual mass on the planing surface can be written as:

$$P_{m'} = \frac{1}{2}\pi\rho\left(\frac{b_s}{2}\right)^2 V_\infty^2 \alpha_s\left[1 + \frac{1}{2}\sqrt{\frac{2x_s\alpha}{b_s}}\right] \tag{3.51}$$

The first term on the right is the linear expression related to the angle of attack, equal to the half of normal force acting on the thin, slender wing moving in the infinite medium. The second term, which does not appear in the infinite medium, denotes the nonlinear expression due to the surface depression effect upon the planing surface. For a planing plate, it is proportional to $\alpha^{3/2}$.

The second difference is that the "transverse flow resistance" was not counted in Jones' "thin, slender theory," as it is a small value in the slender wing in the case of small perturbation and can be neglected.

Since the upper surface of a planing plate is not in contact with water, even in case of small perturbation, the flow is asymmetric, and neglecting the transverse flow resistance is unsuitable theoretically. Figure 3.10 shows a comparison of the flow

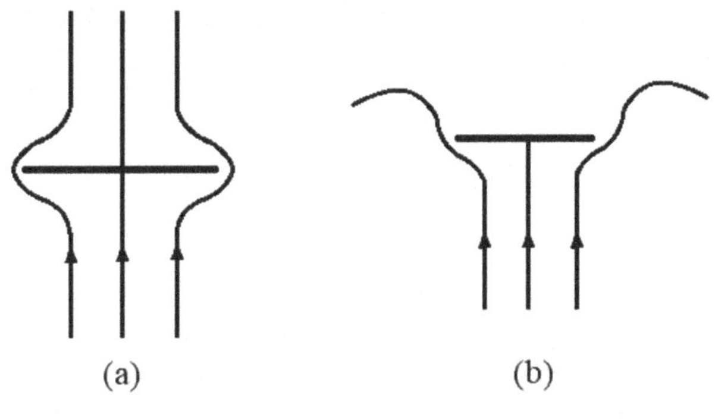

Fig. 3.10 Comparison of flow around the whole wing and planing surface sections

between a slender wing and planing surface. Lift induced by transverse flow resistance should be added on the slender planing surface, which leads to the lift calculation method for slender wing being suitable for limited angles of attack.

For a deeply submerged section, the transverse flow resistance can be calculated theoretically and directly by means of Kirchhoff's equation[2] for splitting flow around the perpendicular plate as $C_{DC} = 2\pi/(4 + \pi) \approx 0.88$, and it should be a suitable approximate value for shallow submergence.

Substituting the "transverse flow resistance" term into (3.51), total normal force on the planing surface is obtained as:

$$P = \frac{1}{2}\pi\rho\left(\frac{b_x}{2}\right)^2 V_\infty^2 \alpha_s\left[1 + \frac{1}{2}\sqrt{\frac{2x_s\alpha}{b_s}}\right] + \frac{1}{2}\rho(V_\infty \sin\alpha_s)^2 b_s x_s C_{DC}.$$

Where x_s denotes the distance between the trailing edge and stagnation line, i.e., wetted length theoretically. For a rectangular planing plate where $\alpha_s = \alpha$, $b_s = b$, then the non-dimensional normal force coefficient is:

$$C_p = \frac{P}{1/2\rho V_\infty^2 S} = \frac{\pi\alpha}{2x_s}\left(\frac{b}{2}\right)\left[1 + \frac{1}{2}\sqrt{\frac{2x_s\alpha}{b_s}}\right] + C_{DC}\sin^2\alpha \qquad (3.52)$$

For a prismatic planing surface with transverse deadrise angle β, where the hard chine does not submerge into water, the virtual mass of the section can be taken as half of displacement mass for the prismatic section. In case of small angle of attack, half of water line at trailing edge can be defined as (see Fig 3.11a):

$$\frac{b_1}{2} = \frac{x_s\alpha}{\tan\beta} \qquad (3.53)$$

The virtual mass of the trailing section is:

$$m_s' = 1/2\pi\rho\left(\frac{b_1}{2}\right)^2 f(\beta) = 1/2\pi\rho\left[\frac{x_s\alpha}{\tan\beta}\right] f(\beta) \qquad (3.54)$$

Where $f(\beta) = 1\text{-}\beta/\pi$, which is the correction of virtual mass induced by β.

Substituting (3.54) into (3.43), the normal force induced by virtual mass is obtained as:

[2] Kirchhoff's equations https://en.wikipedia.org/wiki/Kirchhoff_equations can also be found in H Lamb, Hydrodynamics.

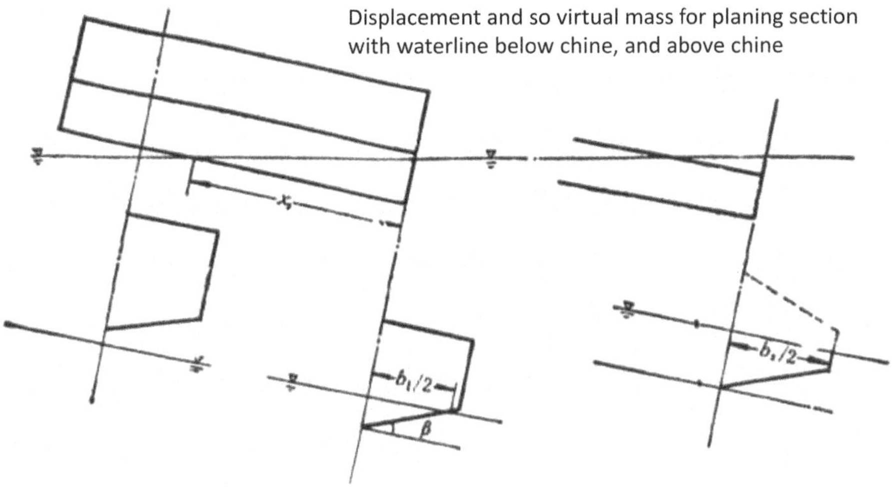

Fig. 3.11 Trailing section on both hard chine wetted and non-wetted situations

Table 3.1 Transverse flow resistance coefficient with deadrise angle

β	10°	15°	20°
C_{DC}	0.818	0.796	0.732

$$P_{m'} = \frac{1}{2}\pi\rho V_\infty^2 \alpha \left(\frac{x_s \alpha}{\tan\beta}\right)^2 f(\beta)$$

Since the wetted surface is triangular:

$$S = \frac{1}{2} x_s b_1$$

Thus, total normal force coefficient

$$C_P = \frac{P}{\frac{1}{2}\rho V_\infty^2 S} = \frac{\pi f(\beta)\alpha^2}{\tan\beta} + C_{DC}\sin^2\alpha \tag{3.55}$$

Korvin-Kroukovsky and Chabrow [7] calculated the C_{DC} with deadrise angle β by means of free streamline theory, and the values are shown in Table 3.1:

For a hard chine surface submerged into the water line, as shown in Fig 3.11b, the virtual mass of the surface depression behind the trailing edge should be added, which can be taken as half of virtual mass generated by a prismatic body fully immersed.

To demonstrate the correctness of the calculation method for a slender planing surface, a comparison with experimental results is made in [8]. Figure 3.12 shows a

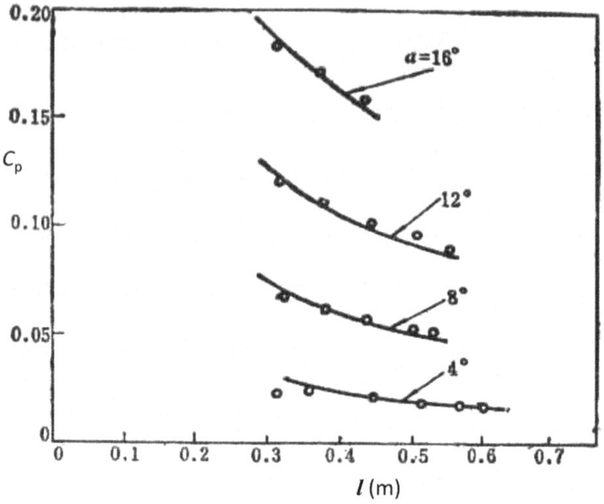

Fig. 3.12 Normal force comparison between Wadlin and McGehoe experiments and theoretical results

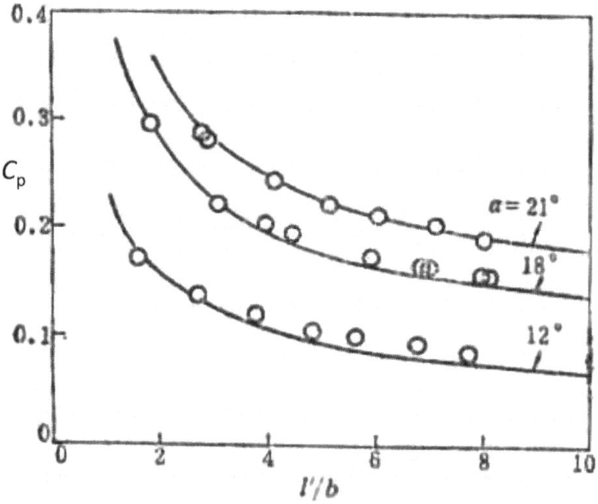

Fig. 3.13 Normal force from rectangular plate by Shuford

comparison of results between Wadlin and McGehoe's test results and theoretical results. The tests are for a planing surface with triangular profile and sharp edge (transom) at trailing edge. The ordinate is normal force coefficient $C_p = P/(0.5\rho\, V^2\, S)$ based on wetted surface S, and the abscissa is a nominal wetted length.

The "nominal" refers to wetted length calculated theoretically based upon the interception of the horizontal plane with the planing plate and does not consider the water surface deformation. From the comparison, one can see the theoretical calculation agreeing well with the test results.

Figure 3.13 presents the test results from a rectangular plate by Shuford compared with calculation of (3.55). In the Shuford method, the real wetted length and area, i.e., considering the deformation of water surface, has been considered. In the figure, the ordinate denotes non-dimensional normal force coefficient based on the real wetted length/beam ratio C_p, and the abscissa is the real wetted length and beam ratio, l'/b. For high wetted length/beam ratio, the effect of water surface deformation on wetted length is small, for both small and large trimming angle. The calculation agrees nicely with test results. The requirement of small angle of attack for theoretical calculation of a slender planing surface is not as strict as for slender wing.

Finally, we would like to explain the term of "transverse resistance." In the case of using a wing with low aspect ratio to simulate a planing surface in 3–2, the "transverse resistance" coefficient $C_{DC} = 2\pi/(4 + \pi) \approx 0.88$ had been obtained by Kirchhoff's flow theory, with too small value. However, for a planing surface, the vortex region after the planing surface being nonexistent, no significant error should be present in Kirchhoff's result, but the coefficient is still smaller compared with test results. However, in case of a slender planing surface, theoretical results agree perfectly with test results and can be considered reasonable theoretically. The error mentioned above may be due to greater error in the linear term of transverse resistance.

3.5 Force Acting on Planing Surface in Viscous Flow

The motion of a planing surface discussed above is based upon the assumption of ideal flow; however, in practice the flow is viscous. Simultaneously, the assumption of high speed and influence of gravity was made, that is, high Froude number $F_n > 4$, or $F_n > 2.8$ in Sedov's approximate expressions (3.18), (3.19), in which gravity force on the flow is neglected. In general, such high speed is not the normal case for planing hulls.

For a conventional planing hull, volumetric Froude number $F_{n\nabla} \geq 3$ corresponding to length Froude number $F_{n\mathrm{L}} \geq 1.2$ is where craft take off and enter the planing regime. Some well-designed planing hulls can be planing at $F_{\nabla} > 2.5$ or $F_n > 1.0$. Therefore, viscosity and gravity effects should be considered in practical hydrodynamic calculations.

3.5.1 *Viscous Effect on Flow*

A viscous fluid generates a boundary layer at the surface of the body.[3] At the surface, the fluid is at rest related to the body and through shear the velocity increases to the same as the free stream at a distance away from the surface. The distance is controlled by the viscosity of the fluid. Surface roughness of the body will cause

[3] A summary of viscosity and viscous flow is at https://en.wikipedia.org/wiki/Viscosity

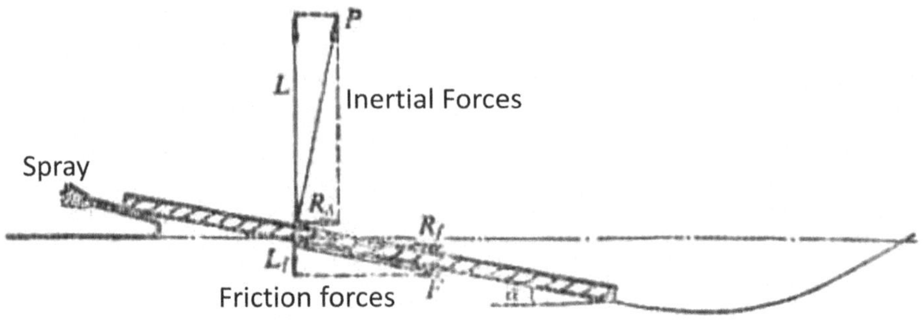

Fig. 3.14 Force acting on planing plate

unsteady flow phenomena. Unsteady flow causes eddies or vortices where locally the fluid will change direction. The effect of viscosity results in:

- Viscous resistance, including friction composed of the tangential force of flow around the body surface, and form resistance caused by separation of the boundary layer of trailing flow or viscous-pressure resistance induced by expansion of boundary layer.
- Change of dynamic lift around the body due to the boundary layer.

Figure 3.14 shows additional forces acting on a 2D planing plate due to the existence of viscosity. With respect to the viscous pressure resistance of planing surface, it can be included in the "drag due to pressure difference" term due to its small value. The friction drag can be included as "equivalent flat plate friction," which will be described in the next section, and the change of dynamic lift caused by viscosity is discussed below.

Based on the wing approximation for calculating the dynamic lift on a planing surface discussed previously, the effect of viscosity on lift can also be included in the slope rate of lift coefficient. The viscous effect on a two-dimensional thin wing can be written as:

$$C_L = 2\pi\eta \sin(\alpha - \alpha_0)$$

Where η is a correction coefficient for viscosity and $\eta = 0.8$–0.9.

For a planing surface, then

$$C_p = \pi\eta \sin(\alpha - \alpha_0) \tag{3.56}$$

For a three-dimensional planing surface:

$$C_p = \frac{\pi\eta}{1 + 2\eta\lambda} \sin(\alpha - \alpha_0) \tag{3.57}$$

Where $\lambda = l/B$, the wetted length to beam ratio of the planing surface.

For a wing with low aspect ratio, the effect of viscosity can also be treated similarly. However, since no significant test results are available for a wing with low aspect ratio and the effect of viscosity on transverse flow lift is still unclear, the influence of viscosity on the lift of a wing with low aspect ratio and a planing surface still needs further investigation.

3.5.2 Effect of Gravity and Correction of Static Buoyancy

To include a gravity term into the basic condition of the boundary equations, the solution will be quite complicated. However, is it possible to simplify the influence of gravity? Regarding physical attribution, gravity mainly relates to the static pressure, which can be seen clearly in the Lagrange[4] integral expressions. Not counting the interference between the static and dynamic pressure, the gravity influence on flow can be considered as adding a static pressure distribution to the dynamic pressure distribution for which gravity effect is not considered.

Precise calculation of the static buoyancy acting on the planing surface is not easy, due to which the water surface will be deformed by planing surface. However, an approximate calculation can be acceptable, i.e., not counting the distribution of static pressure, but calculating the resultant force caused by static pressure via the displaced volume of water and including an empirically based correction factor.

For a planing plate, if not counting the influence of water deformation aft of the trailing edge, the displaced volume is a wedge-shaped volume OEJ as shown in Fig. 3.15. However, this results in too large an error for the calculated buoyancy. If considering water surface deformation, the rise at leading edge should be considered. Thus, the static buoyancy should cover the positive buoyancy of the wedge-shaped volume OEJ and negative buoyancy of forward wedge-shaped volume OKG, and the resultant force can be written as:

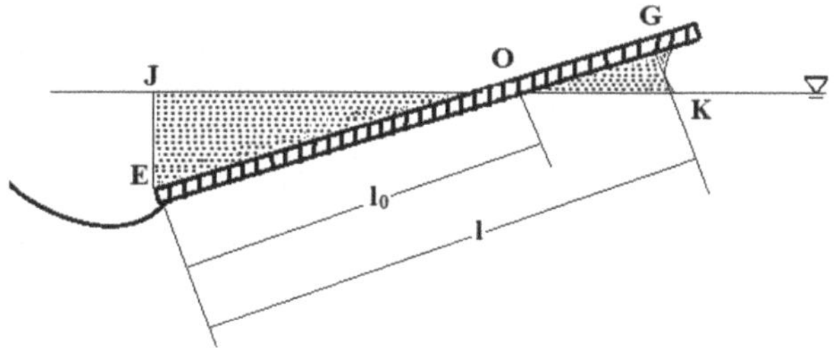

Fig. 3.15 Planing plate and displaced volume

[4]Refer to https://en.wikipedia.org/wiki/Lagrangian_mechanics for Lagrange equations, also see Euler Lagrange equations and Hamilton's principle

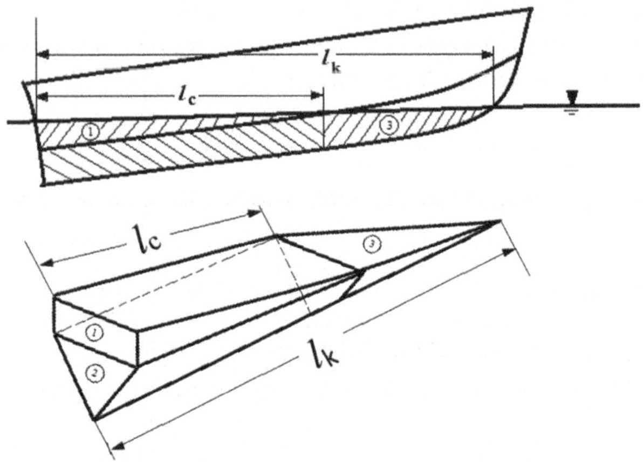

Fig. 3.16 The calculation of submerged volume of section with deadrise.

$$L_o = \rho g b l^2 \left(\frac{l_0}{l} - 0.5 \right) \sin \alpha \tag{3.58}$$

Where l_O is the distance between intersection line of the plate plane with the static water line and trailing edge. Its non-dimensional expression can be written as:

$$C_{\Delta s} = \frac{L_0}{\rho g b^3} = \lambda^2 \left(\frac{l_0}{l} - 0.5 \right) \sin \alpha \tag{3.59}$$

It must be noted that the consideration of the effect of water surface deformation mentioned above is not strictly due to the pressure at E, which is atmospheric, but the static pressure head of JE. Therefore, the calculation result should be over large.

For a transverse section with deadrise, the calculation may be more complicated, as the summation of (1), (2), (3) as shown in Fig. 3.16 is necessary, if not counting the water surface deformation. The total volume based on this assumption is as follows:

$$\nabla = \frac{b^3}{4} \lambda^2 \sin 2\alpha \left[1 + \frac{1}{12\lambda^2} (\lambda_k - \lambda_c)^2 \right] \tag{3.60}$$

The non-dimensional expression can be written as

$$C_{\Delta x} = \frac{\nabla}{b^3} = 0.25\lambda^2 \sin 2\alpha \left[1 + \frac{1}{12\lambda^2} (\lambda_k - \lambda_c)^2 \right] \tag{3.61}$$

Where $\lambda_k = l_k/b$ wetted length to beam ratio at keel and $\lambda_c = l_c/b$ at the hard chine.

The influence of the water surface rising at the leading edge and the reduction of static pressure caused by atmospheric ambient pressure are not calculated in (3.55) and (3.56); therefore, the results should be conservative. Some specialists suggested

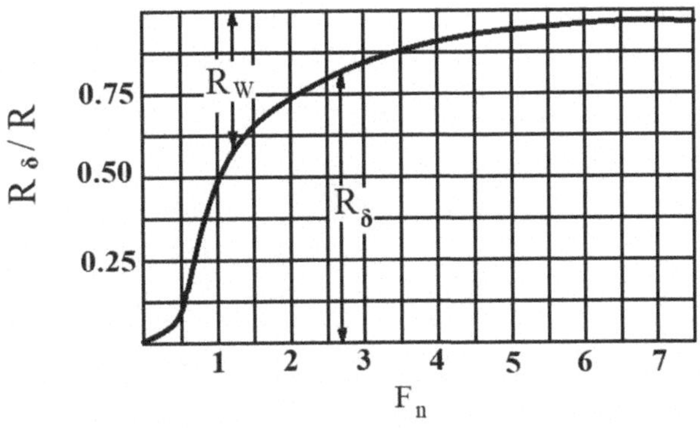

Fig. 3.17 Wave making resistance proportion against speed

$C_{\Delta x} = 0.25\lambda^2 \sin 2\alpha$, which also gave larger values compared with test results. Reference [8] mentions that according to the test results, that is, in case of $F_n > 0.5$, changing the factor from 0.25 to 0.156 makes the calculation agree better with test results, i.e., as follows:

$$C_{\Delta x} = 0.156\lambda^2 \sin 2\alpha \tag{3.62}$$

Strictly speaking, the "static buoyancy" is resultant of static pressure, the perpendicular component of which should be static buoyancy; however, due to the small value of angle of attack, such error can be neglected.

The horizontal component of static pressure will generate wave-making resistance. As speed increases, the proportion of static buoyancy decreases and the proportion of wave-making resistance in total resistance decreases. Figure 3.17 shows the change of proportion of wave-making resistance (below static water level) with spray resistance (above static water level, forward) in pressure resistance versus speed.

3.6 Influence of Planing Surface Shape on Its Hydrodynamic Performance

The calculation mentioned above is only concerned with planing surfaces with simple shape, e.g., flat plate or prismatic section with deadrise angle; however, in practice, the shape of real craft is not so simple, and here we introduce the influence of various shapes on the hydrodynamic performance of planing hulls.

In view of experimental data being presented in non-dimensional form to enable scaling, the non-dimensional forms for expressing the hydrodynamic performance should be introduced firstly as follows.

- *Froude number* characterizing relative speed.

In calculation of the hydrodynamics of planing hulls, in general, we use F_∇ and F_b and less Fn_L due to wetted length changing with speed.

- *Static load factor* C_Δ and dynamic load factor C_b

$$C_\Delta = \frac{\Delta}{\rho g b^3} = \frac{\nabla}{b^3} \tag{3.63}$$

$$C_b = \frac{\Delta}{\frac{1}{2}\rho V^2 b^2} \tag{3.64}$$

The relation between static and dynamic load coefficient is

$$C_b = \frac{2C_\Delta}{F_b^2} \tag{3.65}$$

- *Moment coefficient*

$$C_{M\Delta} = \frac{M}{b\Delta} \tag{3.66}$$

3.6.1 *Influence of Beam of Planing Plate*

Figure 3.18 shows the test results for a rectangular plate with various beam dimensions, at test speed $F_b = 3.5$. When Δ is fixed, different C_Δ indicates a plate with different beam, and three curves are shown in the figure, as follows:

(a) Drag to Lift ratio $\varepsilon = \frac{R}{\Delta} = f(a, C_\Delta)$.
(b) Wetted length ratio: $\lambda = f(a, C_\Delta)$.
(c) Moment coefficient about trailing edge:$C_{M\Delta} = f(a, C_\Delta)$.

From the figure, it can be shown that in the case of a fixed angle of attack, ε, λ, and $C_{M\Delta}$ are inversely proportional to beam (C_Δ increases). In case of the load being fixed, reducing beam and keeping angle attack unchanged, then wetted length increases, i.e., λ increases, and aspect ratio decreases, hydrodynamic property decreases, and resistance to lift ratio ε increases.

It should be stressed that the dynamic lift is always perpendicular to plate surface, so no induced resistance is present, as on a fully immersed wing surface. Therefore, in case of not considering viscosity, the resistance to lift ratio of a planing surface, $\varepsilon = R/\Delta = \tan\alpha$. When the angle of attack is fixed, $\varepsilon =$ constant, and so there is no change with aspect ratio. However, following a decrease of aspect ratio (increase of wetted length to beam ratio), loss of bottom pressure increases; therefore, in case of

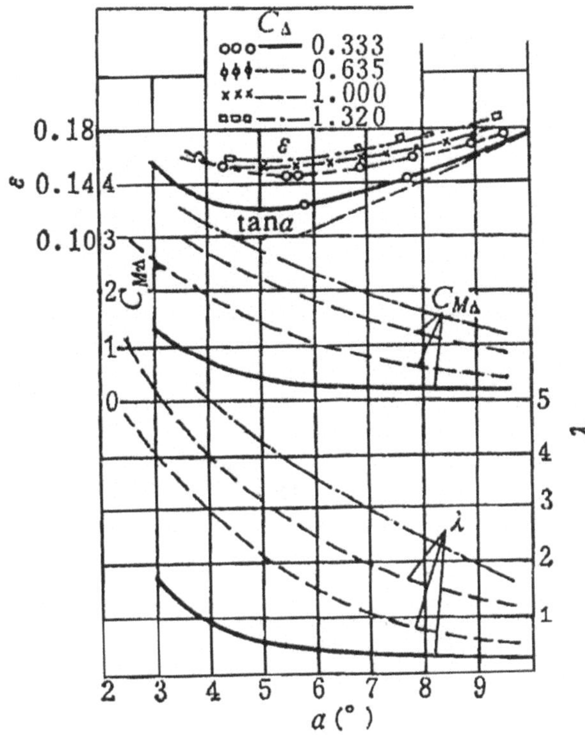

Fig. 3.18 Influence of beam

fixed load, the wetted length should be increased. Thus, the main reason for efficiency reducing is an increase in wetted surface area leading to an increase in friction resistance.

Increased wetted length to beam ratio causes the center of pressure to move forward, thus increasing the moment about the trailing edge. *An important conclusion from this is that increased beam leads to increased efficiency.* This is consistent with a wing where increased aspect ratio leads to improved efficiency.

In the case of ε changed by changing angle of attack α, for a certain C_Δ, an optimized angle attack α_{opt} is available, corresponding to minimum ε, as the resistance of the planing surface comprises the projection of the normal force in the moving direction (pressure difference resistance R_p) and friction resistance R_f, i.e., $R = R_p + R_f = \Delta \tan \alpha + R_f$.

When load Δ is unchanged, pressure difference resistance $\Delta \tan \alpha$ changes with angle of attack and with linear relation in case of small angle attack. When speed is fixed, to keep lift unchanged the wetted length should be decreased if the angle of attack is increased, thus decreasing wetted surface and decreasing R_f. However, a decrease in R_f with wetted length is not a linear relationship due to the Reynold's effect, which makes the coefficient of friction resistance increase inversely with wetted surface.

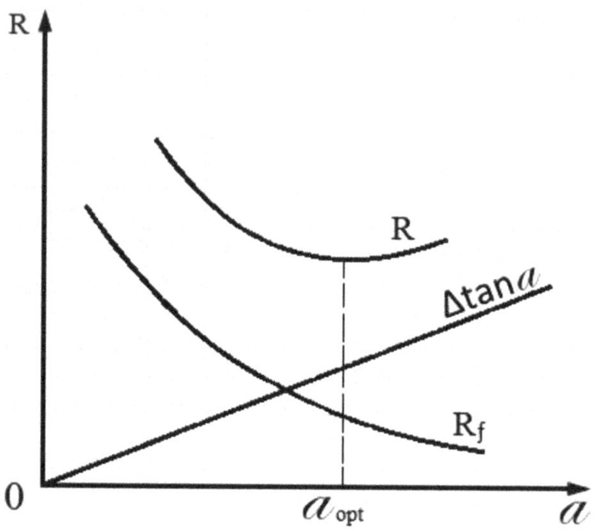

Fig. 3.19 Resistance of planing surface changes with angle attack

Pressure difference resistance and friction resistance are changed with angle of attack, as shown in Fig. 3.19. In the figure, one can see that minimum total resistance $R = f(\alpha)$ is available at an optimal angle of attack. Thus, we have an important conclusion: *an optimum angle of attack is available in the case of certain load, which is located around 4 to 6 degrees.*

These important conclusions above are also applicable for planing surfaces with non-plane profile on the bottom. For a planing hull, hopefully in general, a large beam and proper angle of attack are practical, i.e., small length to beam ratio compared with a conventional displacement ship. This is one important reason that longitudinal center of gravity significantly affects the powering performance, since it directly affects trim and so angle of attack and wetted length, which then controls the dynamic L/B.

3.6.2 Influence of Varying Beam in Longitudinal Direction

Figure 3.20 shows the test results of four plates with varying beam in longitudinal direction, all starting with same beam b forward, however, with different beams at stern of b (28), 0.75b (27), 0.5b (26), and 0 (pointed stern) respectively. Test speed is $F_b = 3.5$, $C_b = 0.109$.

From the figure, it can be seen that a narrower stern gives higher efficiency, which is due to the stagnation line moving forward at small angle of attack with high pressure forward and lower at rear as shown in profile in Fig. 3.3. Thus, decreasing stern beam corresponds to decrease in the area with lower efficiency, which corresponds to increase of total efficiency.

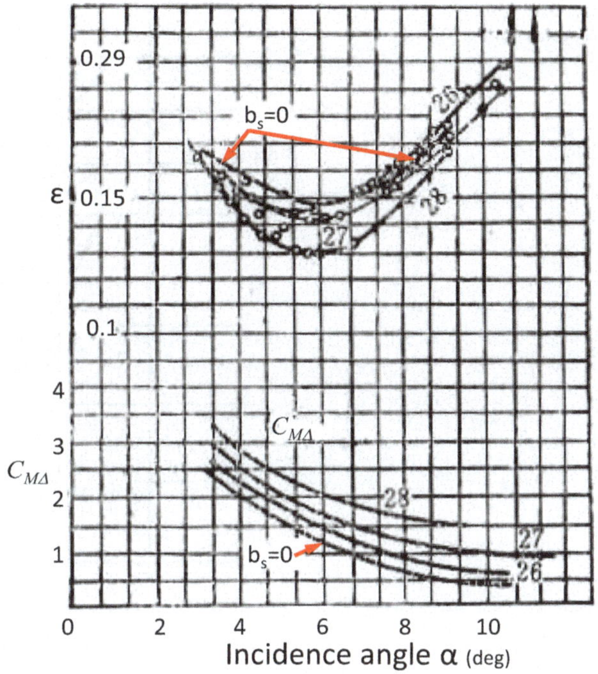

Fig. 3.20 Influence of stern beam taper on lift efficiency

Thus, a narrow stern might be preferable on a planing hull in general. However, considering the requirement for general arrangement, the beam at stern can't be too narrow. In addition, a too narrow stern may cause too large trimming angle so as to take it out of optimum angle of attack range.

3.6.3 Influence of Longitudinal Curvature

Using the wing analogy, longitudinal curvature corresponds to wing camber. Figure 3.21 presents the ε, λ, and $C_{M\Delta}$ of a planing plate with three different longitudinal curvatures. The beam $B = 300$ mm for all three plates. There are flat plate with no curvature, No. 16; a longitudinal circular plate protruding upward with curvature radius $r = 11.5$ m, No. 17; and a plate with same profile but $r = 6$ m. Test speed is $F_b = 3.5$, $C_b = 0.109$.

From the figure, one can see that planing efficiency is enhanced with longitudinal curvature, but wetted length and moment about the trailing edge increase inversely with longitudinal curvature. That means the center of pressure moves afterward.

Figure 3.22 shows the position of the pressure center versus longitudinal curvature; the ordinate is the ratio of pressure center distance from stern edge l_p and with wetted length l.

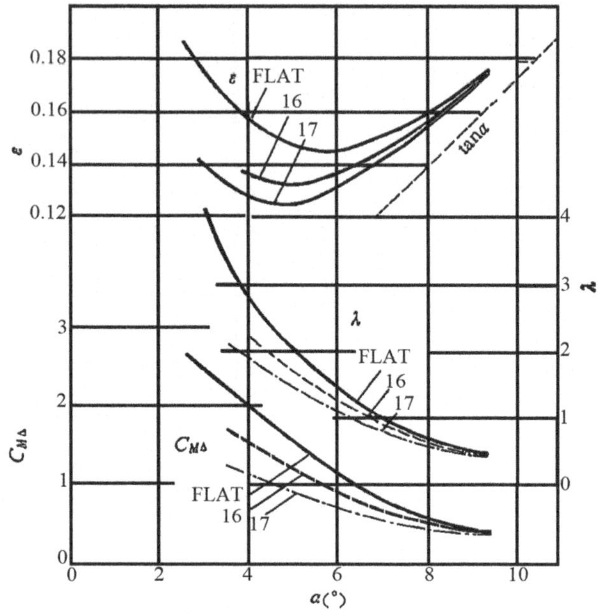

Fig. 3.21 Influence of longitudinal curvature

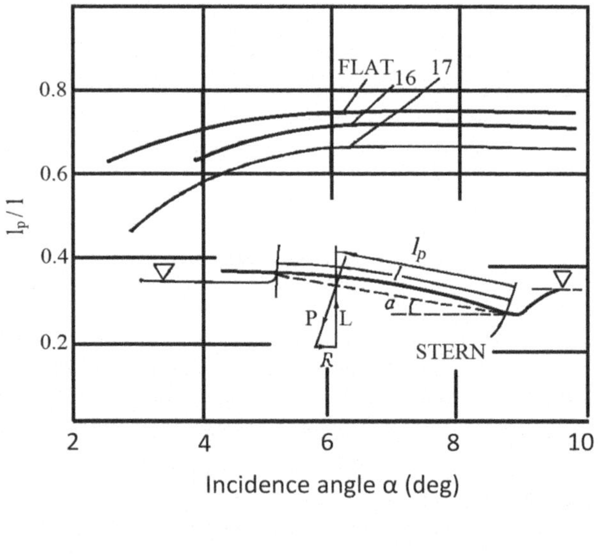

Fig. 3.22 Longitudinal curvature versus pressure center

Longitudinal curvature increases the pressure per unit planing surface area, particularly at the stern, thus enhancing the planing efficiency significantly as shown in the figure. For the same load, drag decreases by 10.3% and 15.6% for Nos. 16 and 17, respectively, compared to the flat plate. For practical planing hulls, the raising of specific pressure at the rear may reduce flow velocity there to improve the working condition for the propeller and defer or avoid the cavitation. An inverse effect will happen for the planing plate with longitudinal circular curvature protruding downward.

Some planing hulls have been designed with a hooked curve near the stern, but an alternative approach using either flaps or an "interceptor" provides a controllable curvature that can be varied with speed to provide optimum trim [9].

3.6.4 Influence of Transverse Deadrise on Bottom

Figure 3.23 shows test results for a planing surface with four different deadrise angles, of 10, 15, 24, and 40 degrees. Test speed is $F_b = 3.5$, $C_b = 0.109$. The planing efficiency is inversely proportional to β, which causes pressure loss on bottom due to the transverse flow, which can be verified by the spray blown out from both sides of the bottom. In addition, the planing surface area increases with deadrise angle. Therefore, in view of planing efficiency, small β and even negative

Fig. 3.23 Influence of β

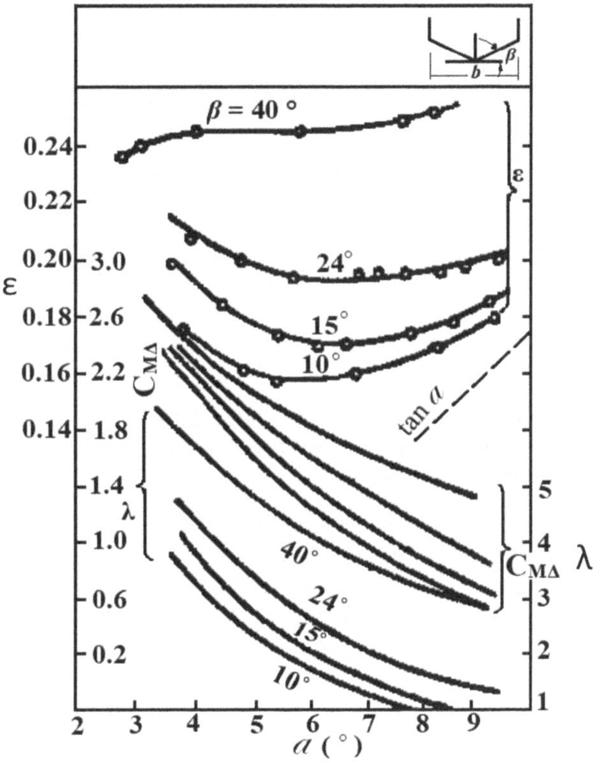

$F_b = 3.5$; $C_b = 0.109$

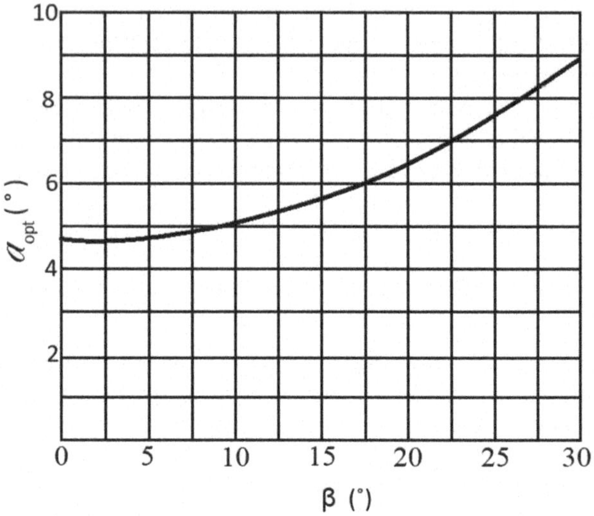

Fig. 3.24 Optimum angle of attack versus β

deadrise could be used (inverse V profile). However, high impact loads on the bottom will occur in waves in case of excessively small deadrise angles, particularly at the bow.

Experimental investigation demonstrates that for a bottom profile with a deadrise angle, the optimum angle of attack α_{opt} should be increased, and Fig. 3.24 shows the optimum angle attack curve verses deadrise angle obtained by Shoemaker.

3.6.5 Influence of Transverse Curvature

Change of bottom transverse section from straight line to curve, i.e., transverse curvature, can reduce transverse spray and maintain higher dynamic pressure on the bottom, so as to give higher planing efficiency. Figure 3.25 shows a comparison of various curvatures of transverse section with straight line deadrise, and from the figure, it can be seen that:

1. It is advantageous to change the deadrise angle at keel from sharp angle to circular angle for keeping higher dynamic pressure at keel.
2. Keeping integrated smooth curvature is better than local sharp curvature.
3. Having transverse curvature is better, nevertheless, excess curvature will generate higher local impact load and makes construction difficult.

Adding anti-spray rails or chine plates to the transverse section can give beneficial results, similar to using transverse curvature. The chine plate can be installed perpendicularly or horizontally, or with inclination angle, as shown in Fig. 3.26. We discuss this further in Chap. 4.

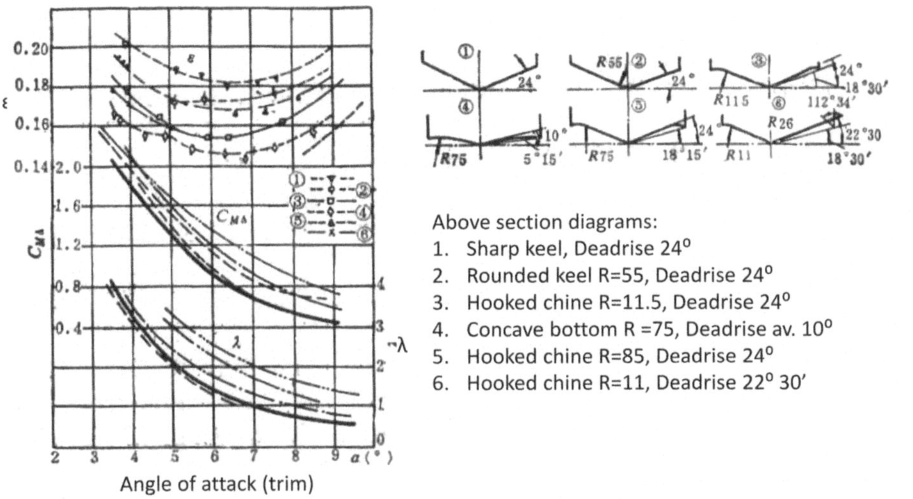

Fig. 3.25 Test results for transverse sections with different curvature

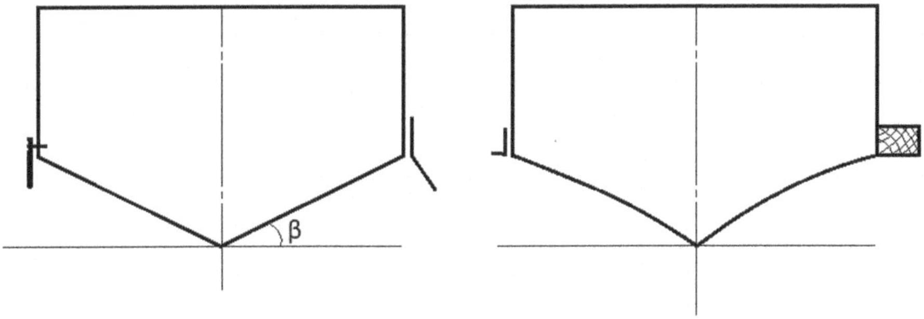

Fig. 3.26 Several alternative chine plates

References

1. Z.S. Dong, *High Speed Craft Dynamics* (Press of Central China University of Technology, China, April 1991) (in Chinese)
2. P.I. Sedov, *Plane problem of Hydrodynamics and Aerodynamics* (Science, Moscow, 1966., (in Russian)), p. 237
3. L.M. Milne-Thomson, *Theoretical Hydrodynamics*, 4th edn. (Dover Publications, Released April, 2013) ISBN 9780486318691
4. M.E. Gorivin, Theory of ideal fluid. Moscow (1961) in Russian
5. R.T. Jones, *Correction of the Lifting Line Theory for the Effect of the Chord* (National Advisory Committee for Aeronautics, Technical Note No 817, Washington, July 1941) Available from the following location, https://apps.dtic.mil/sti/pdfs/ADB814348.pdf
6. P.R. Payne, The normal force on a planing surface. J. Hydrodyn. **15**(1–4), 39 (1981)

7. B. Korvin-Kroukovsky, F.R. Chabrow, *'The Discontinuous Fluid Flow Past an Immersed Wedge', Technical Report, Stevens Institute of Technology*, vol 48 (Hoboken New Yersey, Davidson Laboratory, Experimental Towing Tank, Report SITDL-48-9-334, Oct. 1948), p. 334
8. J.B. Hadler, The prediction of power performance on Planing craft. Transactions SNAME **74** (1966)
9. P.W. Brown, *An Experimental and Theoretical Study of Planing Surface with Trim Flaps* (Davidson Laboratory Report SIT-DL-71-1463, Apr. 1971)

Chapter 4
Planing Hull Resistance

4.1 Introduction

In Chap. 3, we have addressed the fluid forces on a planing surface. To incorporate this knowledge into the design of a planing hull we need to determine equilibrium and identify the vertical and horizontal components, the lift and drag. A displacement craft is generally analyzed considering the drag forces independently. The waveform generated by the hull creates a buoyant moment that trims the vessel while generating a drag force that can be determined independently of friction drag. The semi-displacement hull forms studied in Chap. 2 depart from this assumption, but the geometry of such vessels makes an analytical approach difficult, and so Naval Architects have used model testing and regression analysis to relate trends to geometrical parameters.

Fully planing hull forms can have a simpler geometry as discussed in Chap. 3; nevertheless, there are three issues to face concerning the determination of resistance. The first is that both lift and drag are directly related to the planing surface angle of attack. Secondly, the kinetic forces are nonlinear, proportional to V^2. Finally, a planing hull operates in the displacement, and semi-displacement regimes as it accelerates up to planing speed.

Taking the last issue first, similar to semi-planing craft, model test series have been carried out on a range of geometries, and regression analyses completed to provide equations for drag prediction for planing craft at low speed. See Appendix 1 for more on this aspect.

When a craft is planing, it is necessary to carry out stepwise analyses of forces and moments at a particular speed to identify equilibrium. These analyses build on the knowledge of the forces described in Chap. 3.

Supplementary Information The online version contains supplementary material available at https://doi.org/10.1007/978-3-031-62760-6_4.

Z.-S. Dong et al., *High-Speed Monohull and Hydrofoil Craft*,
https://doi.org/10.1007/978-3-031-62760-6_4

The resistance of a planing hull comprises viscous and nonviscous parts similar to displacement hulls. The nonviscous resistance of planing hulls is divided into two parts, the hydrodynamic pressure resistance and static pressure resistance. The former can also be called "Spray resistance" for the above waterline part of the hull. Static water pressure can be considered similar to wave-making resistance of a conventional displacement ship, as both are due to the pressure difference between bow and stern of the ship. Of course, such division is not precise in physical meaning; however, for engineering application, such division is acceptable. The viscous resistance includes frictional resistance and viscous-pressure resistance, or form resistance. For planing hulls, there is little pressure difference between bow and stern. In addition, no vortex flow is generated at the stern and associated expansion of the boundary layer so form resistance can usually be ignored.

At present, there are many methods to determine the resistance of planing hulls and they can be divided according to the required level of accuracy into four groups as follows:

1. *Estimation of Resistance Using Statistical Data and Coefficients*

   This method is simple, nevertheless, rather approximate, and just suitable for predicting the initial powering performance of craft at the concept design stage. Since the method uses statistical data from existing craft with no division into components of resistance, so only the total resistance prediction is available with this method.
2. *Semi-theoretical and Semi-empirical Method*

   This method is based on the theory of planing and adds correction coefficients from experimental results. The method is characterized by clear physical concepts combined with semi-empirical approaches to various factors affecting the resistance and gives credible results as well as a significantly better understanding of the concepts.

   Compared to direct estimation from model testing, the calculation is more complex. The calculation error is different for specific hull configurations due to the limitation of correction coefficients derived from different test conditions. However, it has applicability to a wider geometric range compared with direct model test analysis.

   In this method the viscous and nonviscous resistances are calculated separately, the latter based on the basic theory of planing discussed in Chap. 3. The frictional resistance is treated similar to that of a conventional ship, calculating the frictional resistance of the wetted ship surface in terms of the "Equivalent flat plate."
3. *Use of Data from Results of Experimental Test Series of Planing Hulls*

   Here data is based on specific model series test results, so it is usually very precise. However, the configuration of the design craft should be close to and within the bounds of the model series parameters, otherwise the predictions could be inaccurate. In this method also, individual viscous and nonviscous components of resistance may not be available and only the total resistance calculated. Similar to semi-planing model test series, regression analyses have been carried out to gain more insight into the influence of design parameters.

4. *Determination of Resistance Directly from Model Tests*
   This is a method with high accuracy as well as a most general and credible method in engineering practice. It is generally not used on an individual craft until the initial design assessment has been made, using methods 1, 2, or 3 above.

Methods 2 and 3 above are most widely used at an early phase in planing craft design. We will discuss this approach and the model test programs that have been carried out to support it in Sect. 4.3.

4.2 Initial Power Estimation for Planing Hulls Using Coefficients

There are a number of simplified coefficient models available for estimation of planing hull required power from more general statistical data than the test series described in 4.3. The advantage of these coefficients is simplified usage, but they do have a limited range of application and limited accuracy of the results. Additionally, only limited parameters can be estimated. Three examples are summarized here. They can be helpful to identify potential power requirements, which in turn would allow main engines and their mass to be checked as a first pass. Analysis can then move to an empirical approach as discussed in Sect. 4.3 to determine equilibrium and a drag curve. The analyses discussed here, in Sects. 4.3 and 4.4 should be considered complimentary to each other.

4.2.1 Admiralty Coefficient Method

This is a well-known method for initial estimation of required installed power for displacement hulls [1] where the main parameters are given as:

$$C_e = 0.736 \times \frac{\Delta^{2/3} V_S^3}{\text{EHP}} \quad \text{or} \quad C_N = 0.736 \times \frac{\Delta^{2/3} V_S^3}{\text{Ne}} \tag{4.1}$$

The Admiralty coefficients for planing hulls may be expressed as:

$$C_e = 0.736 \times \frac{\Delta V_S^2}{\text{EHP}} \quad \text{or} \quad C_N = 0.736 \times \frac{\Delta V_S^2}{\text{Ne}} \tag{4.2}$$

where:

C_e effective power coefficient
C_{N} engine power coefficient
Δ displacement of craft, tonnes

V_s speed, knots
EHP effective power, kW
Ne engine power, kW

The Admiralty coefficients, which are not dimensionless, can be considered the same for the same Froude Number $Fn_L$and similar craft shape, enabling the power to be estimated.

The definition of Admiralty coefficient of a planing hull is rather different from that of a conventional displacement ship, due to the different resistance characteristics of planing hulls. On planing hulls, residual resistance (i.e., pressure difference resistance) is proportional to displacement, and frictional resistance is proportional to 2/3 power of displacement at speeds lower than planing. However, in planing mode, for a planing hull at high speed, residual drag can be considered as a minor drag component, so that the total drag is directly proportional to displacement.

Since the trimming angle of planing hulls decreases a little after take-off, residual drag decreases a little bit with speed while friction drag is nominally proportional to the square power of speed. However, the wetted area is inversely proportional to speed, therefore, the increase of friction drag is less than that mentioned above. Therefore total drag can be taken roughly as in proportion to speed, and power in proportion to square power of speed. The admiralty coefficient of hydrofoils can also be defined in a similar way. They are a little higher due to the presence of the foils as can be seen below.

It can be deduced that to get a more precise result, the displacement, speed, and prototype configuration have to be close to the reference craft when using such methods. In general, the admiralty coefficient of planing hulls can be taken as:

Planing hulls with large size:	$C_N = 30 - 35$
Small planing hulls	$C_N = 33 - 37$
Hydrofoils with large size:	$C_N = 34 - 41$
Small Hydrofoil	$C_N = 38 - 44$

Note that these coefficient values relate to the dimensions used above. The lower number in the ranges above applies to large vessels and the higher value to smaller vessels in a length range from say 30 m down to 15 m. Smaller vessels than this really need to use the semi-empirical methods discussed in the next subheading.

A caveat to the use of the Admiralty coefficient for power estimation is that it is best used to make an estimate based on a Froude number at or less than 1.1, that is, at or below full planing since as mentioned above in the range perhaps between Fn_L 1.1 and 2.0 or so the resistance curve in calm water will be almost flat.

4.2.2 K_η *Method*

The statistical data in the $K_\eta$method is often used for estimating the drag and power of a planing hull, where $K_\eta = \Delta/R$, that is, the lift/drag ratio that is also called the

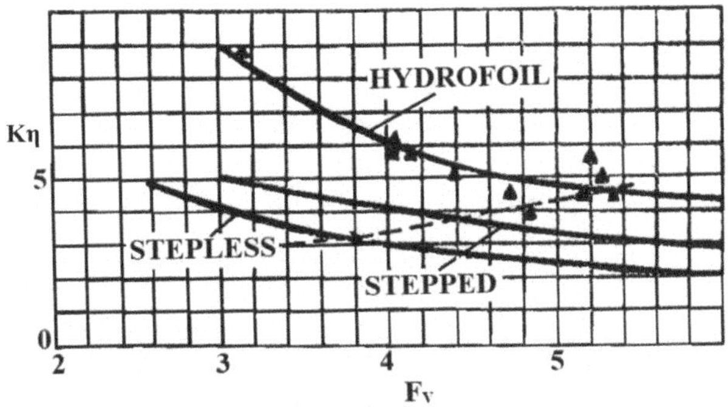

Fig. 4.1 Statistical curve for K_η

planing property coefficient, and is a measure of planing efficiency that characterizes the hydrodynamics. The symbol η represents total propulsive efficiency, and K_η represents the efficiency of planing powering performance, where higher is better.

A prototype with Froude number, Reynolds number, and configuration similar to the reference craft can be used for estimating the drag and power by means of the K_η method. The K_η of the prototype is compared with the reference craft, and engine power is estimated using the following equation:

$$\mathrm{Ne} = \frac{RV}{75\eta} = \frac{\Delta VR}{75\eta\Delta} = \frac{\Delta V}{75K\eta} \tag{4.3}$$

Figure 4.1 shows example data for stepless, stepped planing hulls and hydrofoils against volume Froude number, Fn_∇.

4.2.3 Niwa Method

Dr Shinishi Niwa (丹羽城) collected characteristic data of about 300 hard chine planing hulls and prepared curves by regression analysis, as shown in Figs. 4.2, and 4.3, [2].

Figure 4.2 shows the relation Ne/Δ vs V_s from two data sets that can be used to give an upper and lower estimate for a new prototype for a given design speed.

Figure 4.3 shows the relation $\mathrm{Ne}/\Delta \cdot V_S = f(V_S/\Delta^{1/6}, L_{\mathrm{OA}}/\Delta^{1/3})$ curves for planing hulls that can be used to look at the effect of $\mathrm{Loa}/\Delta^{1/3}$ on power requirement.

where:

Ne	engine power	(ps)
Δ	craft displacement,	(t)
V_s	speed	(knots)
L_{OA}	craft overall length	(m)

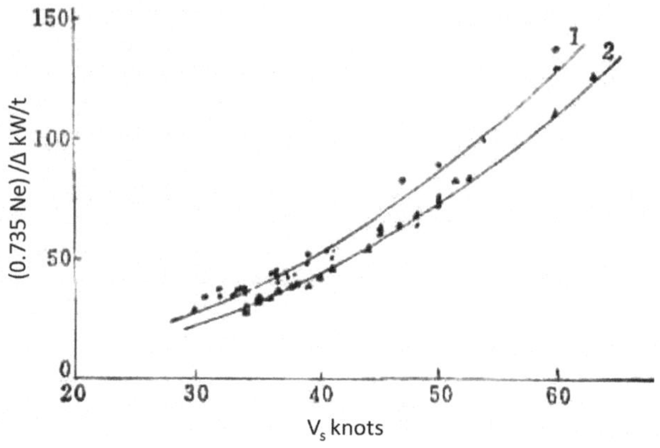

Fig. 4.2 Shinishi Niwa Ne/Δ vs. V_s curves for planing hulls

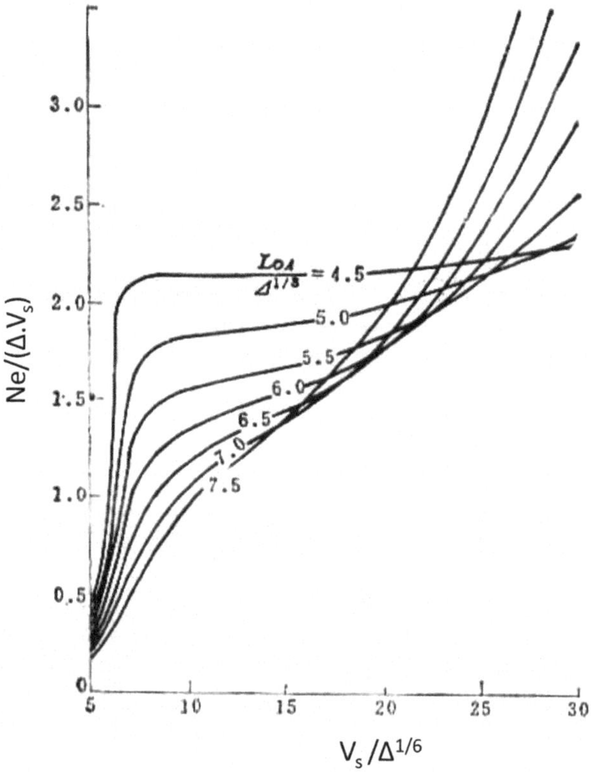

Fig. 4.3 Shinichi Niwa curves for assessment of effect of fineness on powering

Engine power can be obtained simply by using Fig. 4.2 if displacement and length L_{OA} are given. In addition, optimum length can be found from Fig. 4.3 if displacement, speed, and engine power are given.

These methods mentioned above are based on statistical data of existing craft to define specific coefficients. The craft may have varied profiles and configurations; however, only the main factors that influence the engine power are included.

4.3 Semi-theoretical and Empirical Methods for Planing Hull Resistance

Firstly, we analyze forces acting on the planing hull operating at uniform linear velocity, as shown in Fig. 4.4. For a craft running at trim angle α, the angle between craft bottom and incoming flow, both hydrodynamic lift P and static pressure P_0 are acting perpendicular to the craft bottom. The center of gravity of craft weight Δ is at G, the center of action of hydrodynamic lift P is typically located forward of CG, and that of the static pressure P_0aft of CG, as shown in Fig. 4.4.

The frictional force F parallel to the bottom and propulsion force T are also shown in the figure, but the air resistance and appendage resistance are not included.

The nonviscous pressure difference resistance is the component of summation of hydrodynamic and static pressure in the direction of motion, that is, $(P + P_0) \sin \alpha$.

Neglecting pressure resistance due to viscosity (form resistance), viscous resistance is the component of friction F in the direction of motion, $F \cos \alpha$, thus:

$$R = R_r + R_f = (P + P_0) \sin \alpha + F \cos \alpha \tag{4.4}$$

To calculate the nonviscous resistance, the hydrodynamic resultant P, and static resultant P_0, and trim angle α of a planing hull has to be obtained. When using an "equivalent flat plate" for calculating friction, the wetted surface area S and running trim angle also must be calculated. Since the running attitude of planing hulls is rather different from that in static condition, the position of the running water line and so wetted length must be obtained. Thus, first, we have to calculate the trimming angle and wetted length, that is, equilibrium position of the planing hull, and force balance equations for the planing hull.

As a first approximation, not counting the propulsion force and vertical component of friction and the moments due to these forces about the trailing edge (the transom); and considering a small angle of attack, the equilibrium equation of forces in the vertical direction, and moment equation about the trailing edge, can be expressed as:

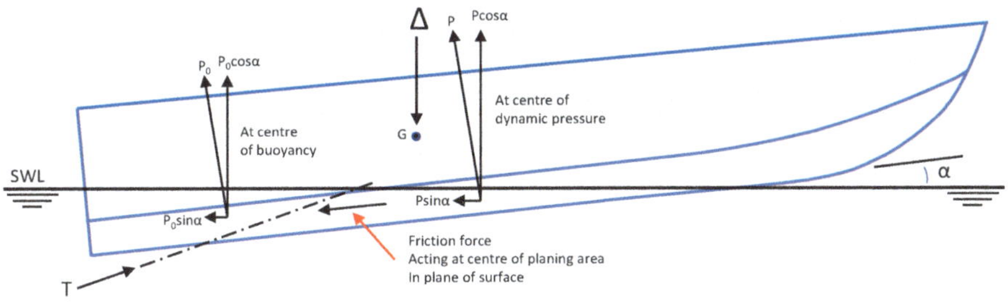

Fig. 4.4 Forces acting on planing hulls

$$\begin{cases} \Delta = (P + P_0)\cos\alpha \approx P + P_0 \\ \Delta l_G = M_P + M_{P0} \end{cases}$$

where l_G, M_p, and M_{p0} represent the distance of CG to the trailing edge at the keel, the moment of dynamic pressure resultant about the trailing edge, and static pressure resultant about the trailing edge, respectively.

In principle, the two unknown values l_G and α can be obtained using the above equations and thus obtain the resistance. Since lift and drag are related to V^2 an iterative method is normally used to determine equilibrium. Based upon the approximate values from the first iteration and assuming propulsion force equal to resistance, add the moments due to the friction and propulsion forces about the trailing edge to the moment equation and calculate l_G and α as second approximation values. This can be continued until the result converges to the desired level of accuracy.

This forms the basic idea of analytically determining the resistance of planing hulls. The differences between the various methods described below reflect the different methods adopted in calculating the hydrodynamic pressure.

4.3.1 CAHI Method

The Central Aerodynamic and Hydrodynamic Research Institute of Russia (CAHI) has also developed a method based on the iteration principle. This method calculates the hydrodynamic pressure and adds some correction factors to two-dimensional theory to apply a limited beam, thin wing to model the hull planing surface [3]. For a planing plate with a small angle of attack, the hydrodynamic lift and static buoyancy can be obtained according to (3.19), (3.53), as:

$$L + L_0 = \frac{\pi\alpha}{1 + 2\lambda}\frac{1}{2}\rho V^2 S + \rho g b l^2\left(\frac{l_0}{l} - 0.5\right)\alpha = \Delta \tag{4.5}$$

where

- l wetted length
- l_0 distance, hull bottom static water plane intersect to trailing edge
- b planing surface breadth
- S wetted surface
- λ *l/b,* inverse to aspect ratio for a wing
- L hydrodynamic lift
- L_0 static buoyancy

In nondimensional coefficient form:

$$C_b = \frac{\pi\lambda\alpha}{1 + 2\lambda} + \frac{2}{F_b^2}\lambda^2\alpha\left(\frac{l_0}{l} - 0.5\right) \tag{4.6}$$

where

F_b Froude number based on planing surface breadth
C_b $(L + L_0)/\Delta$

According to results from tests carried out by CAHI, empirical correction factors to (4.5) and (4.6) can be obtained as follows:

1. The first term of Equation 4.5, that is, hydrodynamic lift should have a correction coefficient K added, equal to 0.7, to account for the viscous effect on the slope of the lift coefficient, the effect of transverse flow, and effect of plate configuration
2. When $\lambda > 0.4$

$$l_0/l = \frac{\lambda}{\lambda + 0.4}$$

Thus, the equilibrium equation for perpendicular forces in coefficient form is:

$$C_b/\alpha = \frac{0.7\pi\lambda}{1 + 1.4\lambda} + \lambda^2 \frac{\lambda - 0.4}{\lambda + 0.4} \frac{1}{F_b^2} \tag{4.7}$$

With respect to moment equilibrium, according to thin wing theory, the dynamic center of pressure is located at $0.75l$ from the trailing edge. However, the high-pressure region at the high-pressure surface extends forward of the stagnation point at the lower surface due to the reverse flow, but the negative pressure distributes uniformly on the suction surface, so the pressure center for the whole "wing" should be located forward of the $0.75l$. According to CAHI's experimental data, the center of dynamic pressure should be at a distance of $0.751 + 0.08\ (\lambda^{0.865}/F_b^{0.5})l$ from the trailing edge.

The center of buoyancy should be at the volume center, and according to Fig. 4.4, it is not difficult to obtain the moment of buoyancy and dynamic pressure about the trailing edge, as:

$$M_{P0} = \frac{1}{2}\rho g b l^3 \alpha \frac{\lambda - 0.8}{3\lambda + 1.2} \tag{4.8}$$

Thus, giving the moment equilibrium equation as:

$$C_{M\Delta} = \frac{l_G}{b} = \frac{\frac{0.7\pi\lambda}{1 + 1.4\lambda}\left(0.75 + 0.08\frac{\lambda^{0.865}}{\sqrt{F_b}}\right) + \lambda^2 \frac{\lambda - 0.8}{3\lambda + 1.2} \times \frac{1}{F_b^2}}{\frac{0.7\pi}{1 + 1.4\lambda} + \lambda \frac{\lambda - 0.4}{\lambda + 0.4} \times \frac{1}{F_b^2}} \tag{4.9}$$

where α and λ can be obtained by using (4.7), (4.9). To use them easily, the two equations above can be expressed in graphical form, as shown in Figs. 4.5 and 4.6.

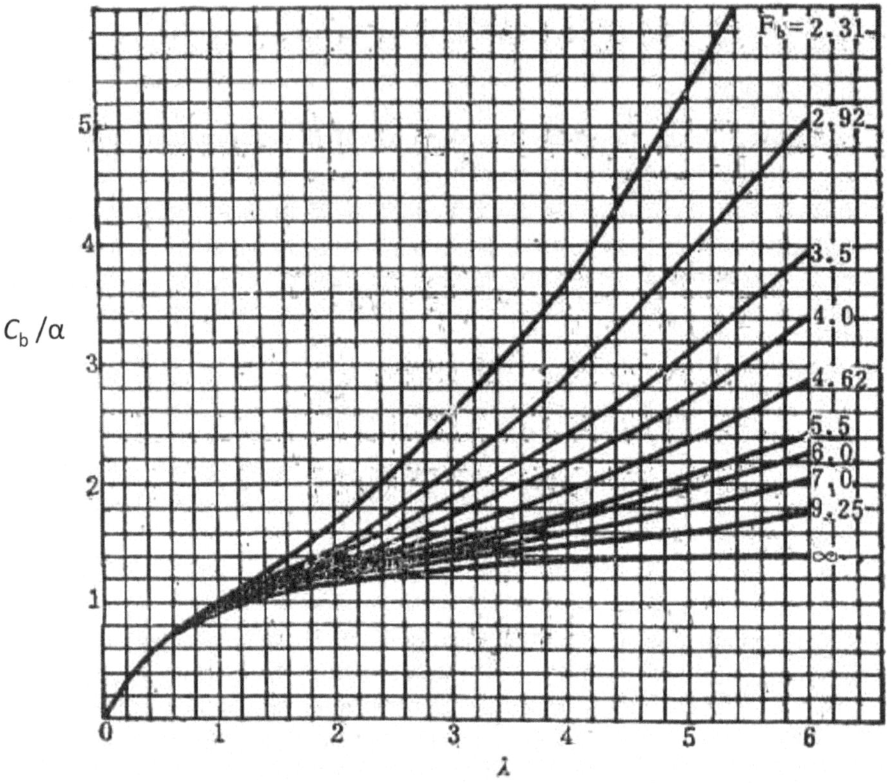

Fig. 4.5 Curve for $C_b/\alpha = f(\lambda, F_b)$

For a planing surface with transverse deadrise, a correction can be added to the equation for a flat plate surface. Since the planing lift will be decreased due to the deadrise angle, to maintain the original hydrodynamic lift the angle of attack should be increased from that of a flat plate as follows:

$$\alpha_\beta = \alpha_0 + \frac{0.15(\sin\beta)^{0.8}}{(F_b)^{0.3}} \frac{1 - 0.17\sqrt{\lambda_\beta \cos\beta}}{\sqrt{\lambda_\beta \cos\beta}} \tag{4.10}$$

where λ_β is the corresponding wetted length beam ratio, for increased wetted surface S_β due to deadrise angle β.

The corresponding angle of attack and wetted length/beam ratio, respectively, of a flat plate with the same lift, $\lambda_\beta = S_\beta/b^2$ can be obtained according to the following empirical equation:

$$\lambda_\beta = \lambda_0^{0.8} \frac{1}{\cos\beta}\left[1 - 0.29(\sin\beta)^{0.28}\right]\left[1 + 1.35(\sin\beta)^{0.44}\frac{C_{M\Delta}}{\sqrt{F_b}}\right] \tag{4.11}$$

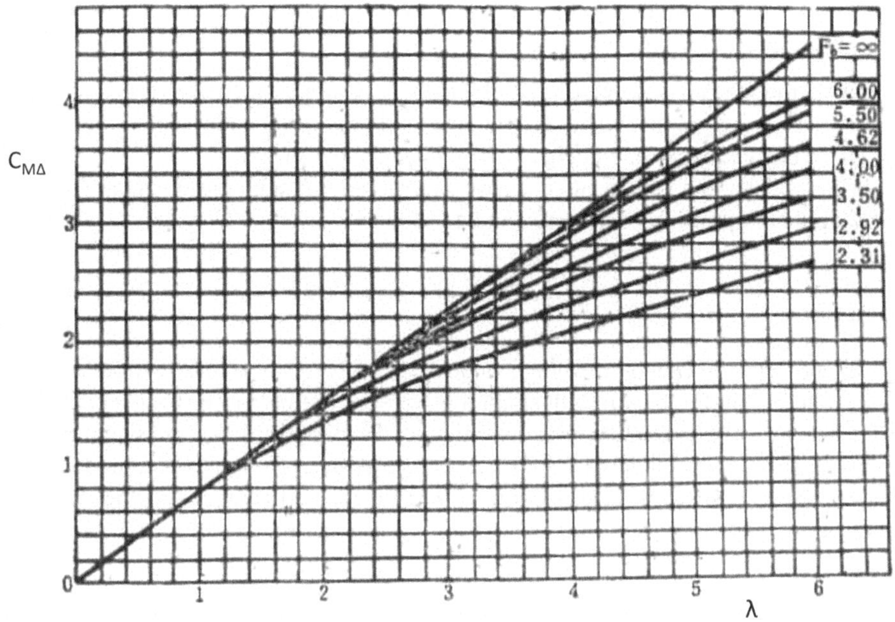

Fig. 4.6 Curve for $C_{M\Delta} = f(\lambda, F_b)$

The calculation can be carried out in the following order:

First, one calculates α_0 and λ_0, then determines α_β and λ_β according to (4.10), (4.11), and, finally, calculates the resistance as outlined below.

1. For known craft weight Δ, beam b, CG of craft l_G, and speed V, calculate the following nondimensional values: $F_b = V/\sqrt{gb}$; $C_{M\Delta} = l_G/b$; $C_b = \Delta/\frac{1}{2}\rho V^2 b^2$
2. From Fig. 4.6, obtain the wetted length/beam ratio for the corresponding wetted length/beam ratio $\lambda_0 = f(C_{M\Delta}, F_b)$
3. From Fig. 4.5, obtain the corresponding angle of attack of the planing flat plate: $C_b/\alpha = f(\lambda, F_b)$, and $\alpha_0 = C_b/(C_b/\alpha)$
4. From (4.10), (4.11), obtain α_β and λ_β
5. Calculating wetted surface S_β, wetted length l, and Reynolds number Rn

$$S_\beta = \lambda_\beta b^2$$

$$l = S_\beta \cos\beta / b - \lambda_\beta b \cos\beta$$

$$Rn = Vl/v$$

6. Calculating resistance

$$R = \Delta \tan\alpha_\beta + \left(C_f + \Delta C_f\right)\frac{1}{2}\rho V^2 S_\beta$$

where ΔC_f is a correction for surface roughness, which is taken as 0.4×10^{-3}

7. Calculate the second approximate value for the additional moment due to propulsion force, if necessary

The iterations can be done directly by using a computer program or spreadsheet without using the graphs; however, the graphical approach can be more convenient for quick estimation by engineers at an initial stage of design.

In case of changing beam and deadrise angle longitudinally, the average of values between that at midships and stern can be used, as:

$$b_m = \frac{b_{\mathrm{ms}} + b_s}{2} \quad \text{and} \quad \beta_m = \frac{\beta_{\mathrm{ms}} + \beta_s}{2}$$

where subscript "m" denotes average value, "ms" is the value at mid-section, and "s" value at the stern. For a craft with small transverse variation to the cross-section and deadrise angle, the hydrodynamic properties are close to flat plate values, and can be approximated as a flat plate in CAHI's method; however, the real wetted surface for transverse section with deadrise angle should be calculated for the friction calculation.

4.3.2 Daniel Savitsky's Method (USA)

This method was developed by Daniel Savitsky at the Davidson Laboratory of Steven's Institute of Technology, New Jersey, USA [4], based upon analysis of a large number of model test results.

In this method, it was considered that the relation between dynamic lift and angle of attack is not linear, and the planing plate lift coefficient can be written as:

$$C_{\mathrm{bo}} = \alpha^{1.1}\left(0.0120\lambda^{\frac{1}{2}} + 0.0055\frac{\lambda^{2.5}}{F_b^2}\right) \tag{4.12}$$

The first right-hand term is a nondimensional dynamic lift and the second a nondimensional static lift, while the angle of attack is expressed in "degrees." For a craft with transverse section having a deadrise angle, the lift coefficient is smaller, given as:

$$C_{b\beta} = C_{b0} - 0.0065\beta C_{b0}^{0.6} \tag{4.13}$$

where deadrise angle is expressed in "degrees"

For a surface with deadrise, the position of lift resultant from studies by Sotorf is located at:

$$l_p/l = K\ \lambda^n \tag{4.14}$$

where

$K = (0.84 + 0.015\ \beta)/\alpha^m$
$m = 0.125 + 0.0042\ \beta$
$n = -(0.05 + 0.01\ \beta)$

Savitsky, based on his own testing concluded that the center of lift varied little with deadrise and trim once the mean wetted length l_m is used, and proposed the following expression:

$$C_p = \frac{l_p}{l_m} = 0.75 - 1/\left[\frac{5.21 C_v^2}{\lambda^2} + 2.39\right] \tag{4.14a}$$

where $C_V = F_b = V/(gb)^{0.5}$ Expressions (4.12) and (4.13) are shown in Figs. 4.7 and 4.8, respectively, so the resistance can be obtained either by calculation on a computer or by use of the graphs.

The calculation procedure for equilibrium is as follows:

1. Given displacement Δ, beam b, speed v, and deadrise angle β, calculate Fn_b and $C_{b\beta}$, where the latter is lift or loading coefficient, that is, $C_{b\beta} = \Delta/(0.5\ \rho\ V^2\ b^2)$. The dynamic load coefficient of a flat plate at a deadrise angle $C_{b\beta}$ can be obtained in Fig. 4.7. Alternatively, C_{b0} and $C_{b\beta}$ can be calculated using Equations 4.12 and 4.13
2. Taking a series of angle of attack α_i, where i_1,2,3 ...n, calculate the corresponding $C_{b0}/\alpha_i^{1.1}$ values and obtain λ_i from Fig. 4.7
3. Obtain $\lambda_i^n = f(\lambda, \beta)$ and $K_i = f(\alpha_i, \beta)$ and calculate the center of pressure at the new angle of attack l_{pi} at α degrees $= \left(K_i\ \lambda_i^{n+1}\right).l$ from Fig. 4.8a, b, or use (4.14a)
4. Plot curves $l_{pi} = f(\alpha_i)$ and $\lambda_i = f(\alpha_i)$ and with CG at longitudinal position l_G, put $l_G = l_p$, and obtain equilibrium α and λ

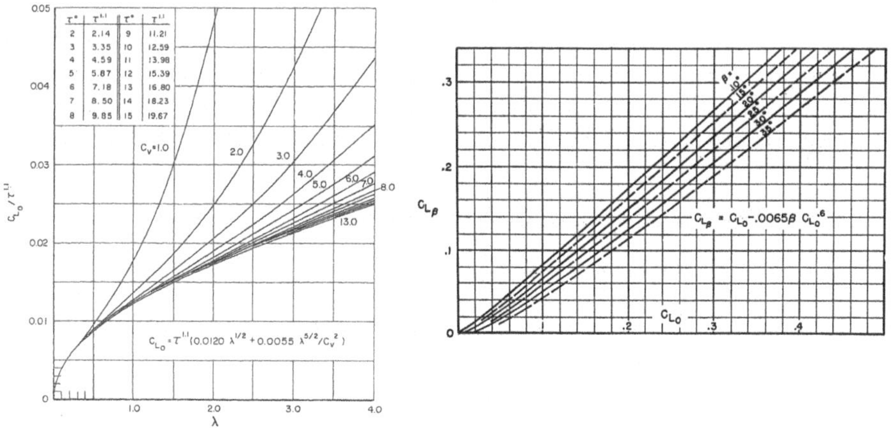

Fig. 4.7 Lift curves for Savitsky's method

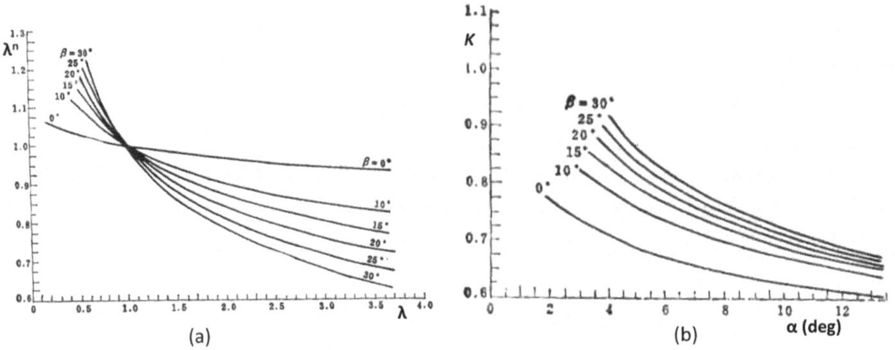

Fig. 4.8 Moment coefficient curves for Savitsky's method

5. From α and λ, calculate resistance:

$$R = \Delta \tan \alpha + \left(C_f + \Delta C_f\right) \frac{\rho}{2} V^2 S^{\beta}$$

where $S_\beta = l.\ b/\cos \beta$ and the roughness addition $\Delta C_f = 0.4 \times 10^{-3}$

Where b and β are variables through the length the average values can be taken. This procedure can be carried out as an iterative process with a computer program or spreadsheet calculation for a speed range. The procedure is shown in Flow Chart 4.1.

4.3.3 Shuford-Brown Method

Since the wetted length/beam ratio of a planing hull is large in general, it is unreasonable in principle to use the "downwash" for correcting the effect of limited beam. Shuford proposed using simulation of a wing with a lower aspect ratio, that is, "transverse flow resistance," to calculate the hydrodynamics of a planing surface [5]. Later, Brown corrected the Shuford equations and formed the "Shuford-Brown" method.

1. *Original Shuford Method and Clement Graphs*

In Chap. 3, lift coefficient equations for a low aspect ratio planing surface were given by means of "transverse flow resistance" theory. Shuford worked with Eq. (3.26) for a planing surface preparing experimental results with various aspect ratios.

It was noted that in spite of the suction at the leading edge of a wing relating only theoretically with lift linear term and has no relationship with transverse flow

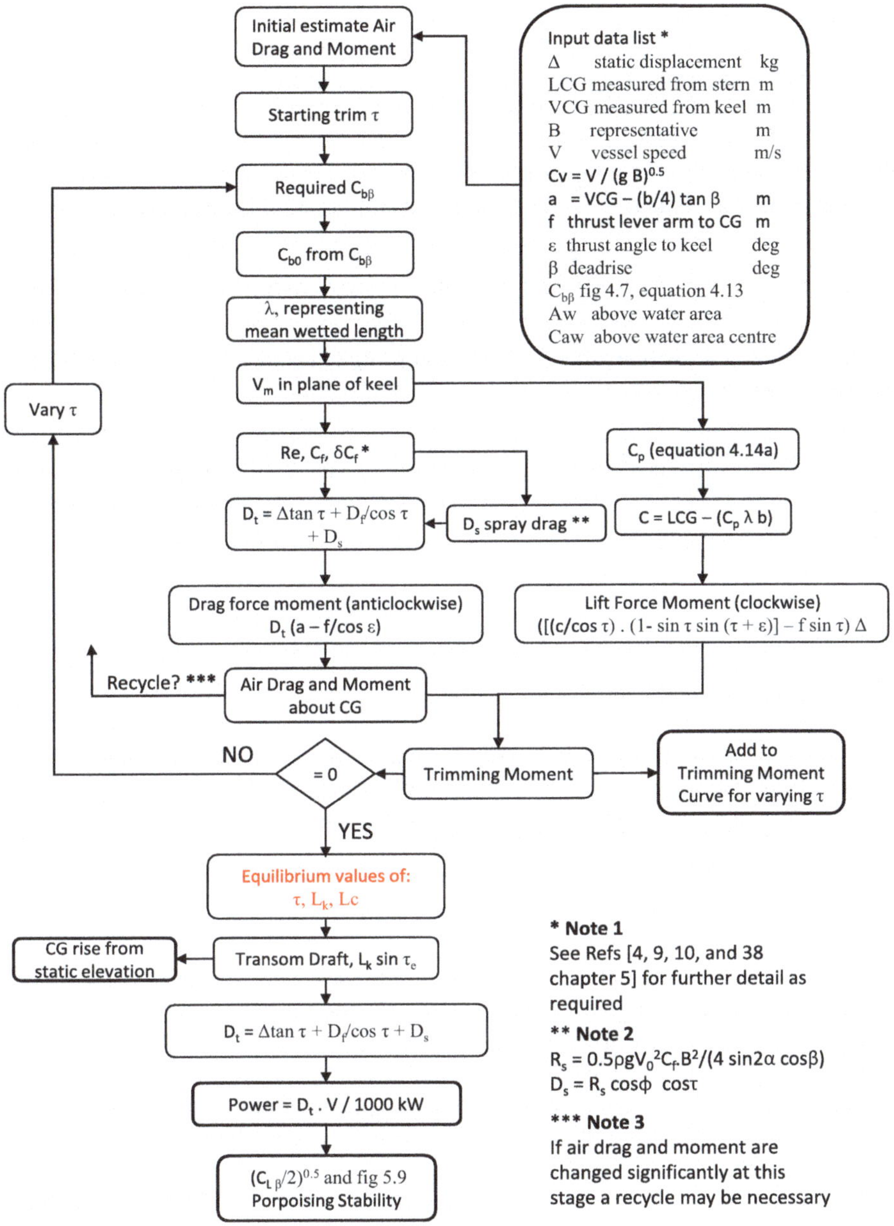

Flow Chart 4.1 Procedure to determine equilibrium for a planing craft by iterative calculation following Savitsky's approach resistance, the calculated lift of a planing surface agreed better with test results when considering the suction at the leading edge not only in terms of $\sin^2 \alpha$ of lift in linear term, but also in terms of $\sin^2 \alpha$ of total lift and deducting it.

With respect to the transverse flow resistance coefficient of a flat planing surface, he obtained 2/3 of that on flat wing, and with $C_{DC} = 2.0$ for flat plane wing, $C_{DC} = 4/3$ is used for planing surface.

Then Shuford's equation for lift coefficient of planing flat surface is written as:

$$C_L = \frac{0.5\pi A}{1+A}\cos^2\alpha \sin\alpha + \frac{4}{3}\cos^2\alpha \sin^2\alpha \tag{4.15}$$

For a transverse section with deadrise, according to test results, the lift coefficient can be written as:

$$C_L = \frac{0.5\pi A}{1+A}\cos^2\alpha \sin\alpha(1 - \sin\beta) + C_{DC}\cos^3\alpha \sin^2\alpha \cos\beta \tag{4.16}$$

where the transverse flow coefficient C_{DC} can be obtained from the following:

Section type	C_{DC}
Straight deadrise at any β value	4/3
Twisted deadrise with horizontal expansion	$4/3 + 0.93 \sin\beta$
Straight deadrise with perpendicular anti-spray rail	$5/3 + 0.93 \sin\beta$

With respect to the hydrodynamic lift center, Shuford used a distance of 7/8 forward from the trailing edge (transom) based on average wetted length.

The lift center due to transverse flow resistance is located at ½ of the average wetted length, so that the center of pressure is at:

$$l_p/l_m = (0.875\,C_{LL} + 0.5\,C_{LC})/C_L \tag{4.17}$$

where l_m average wetted length, and C_{LL} linear lift coefficient, that is, the first term of the right side in (4.16), C_{LC} lift coefficient due to transverse flow resistance, that is, second term of right side in (4.16);

Since buoyancy corrections are not counted in (4.16) and (4.17), it is only suitable for a pure planing situation at sufficiently high speed that the gravity effect can be neglected. In general, this will be at $F_{n\nabla} \geq 4$, which may be relevant for racing boats and hydroplanes.

E. P. Clement and J. D. Pope plotted graphs from (4.16), (4.17) [6], for calculating hydrodynamics of a pure planing surface and demonstrated that the calculation agreed nicely with the test results, so long as the planing hull speed is higher than the speed where there is a hollow in the resistance curve. Nevertheless, since they didn't count buoyancy, this conclusion still needs to be investigated.

In Clement's graphs, lift coefficients, center of pressure, and resistance per unit displacement versus angle of attack $\alpha = 1° \sim 10°$, wetted length beam ratio $b/l_m = 0.3\sim2.3$, transverse deadrise angle $\beta = 5, 10, 15°$, are presented and include the resistance of planing hulls with displacement of 0.454~45.4 t, for direct application.

Later, Clement expanded the graphs again to beam/wetted length ratio 0.1~0.3 [7], for the calculation of Twin Hull Planing Hulls.

2. *Shuford-Brown Method*

Buoyancy was not included in Clement's graphs. Shuford deemed that if taking the displacement volume based on the static waterline as buoyancy (i.e., not counting the elevating water surface at leading edge), this will be an overestimate, thus proposing ½ of this value as a correction to allow for the rise of the hull as dynamic lift increased. However, this calculation is only suitable for large angle of attack, $\alpha > 8°$. At lower angles of attack the buoyancy at equilibrium will be less as the hull will rise further. Based on many test results, Brown improved the original method of Shuford calculation with a method for buoyancy, and also for determining the center of lift [8].

Brown deemed buoyancy acted at 1/3 of wetted length l_m from the trailing edge. Since no force is acting on the wing surface outside the maximum beam, according to slender wing theory, linear lift only acts on a triangular area of $(l_k - l_c)$, as shown in Fig. 4.9, where l_k wetted length at keel, l_c wetted length at the hard chine. Assuming uniform distribution, the force center of action should be as follows, (from the trailing edge)

$$l_{p1} = l_c + \frac{1}{3}(l_k - l_c) = l_m - \frac{1}{6}(l_k - l_c)$$

Because $l_k - l_c$ is proportional to $b\frac{\tan\beta}{\tan\alpha}$, then $l_{p1} = \varepsilon l_m - \eta\frac{\tan\beta}{\tan\alpha}$

According to test results, $\varepsilon = 7/8$ and $\eta = 0.08$, then

$$l_{p1} = \frac{7}{8}l_m - 0.08\frac{b\tan\beta}{\tan\alpha} \tag{4.18}$$

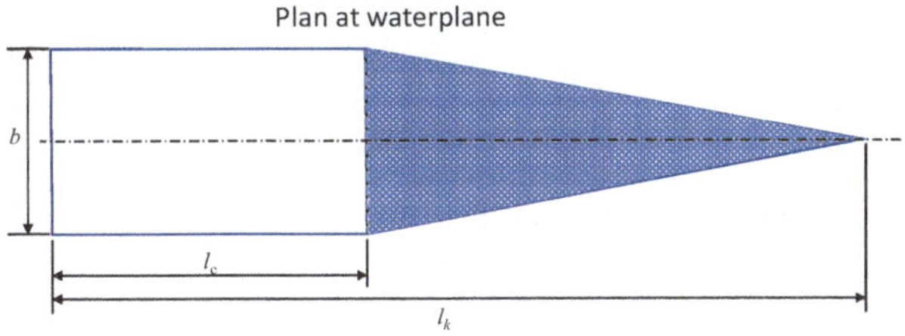

Fig. 4.9 Centre of lift for Shuford Brown method

Taking the centre of action of lift, due to transverse flow resistance, at $l_m/2$ again, and changing the lift coefficient based upon beam b, and including Brown's correction, we obtain:

$$C_{Lb} = \frac{\pi}{4}\sin 2\alpha\cos\alpha\left[(1-\sin\beta)\frac{\lambda}{1+\lambda} + \frac{C_{DC}}{\pi}\lambda\sin 2\alpha\cos\beta + 0.4\sec\alpha\left(\frac{\lambda}{F_b}\right)^2\right] \tag{4.19}$$

At force equilibrium, we have $C_{Lb} = C_b$
Then the moment coefficient can be written as:

$$C_M = \frac{\pi}{4}\lambda\sin 2\alpha\left[\left(\frac{7}{8}\lambda - 0.08\frac{\tan\beta}{\tan\alpha}\right)\frac{1-\sin\beta}{1+\lambda} + \frac{C_{DC}}{2\pi}\lambda\sin 2\alpha\cos\beta + 0.133\sec\alpha\left(\frac{\lambda}{F_b}\right)^2\right] \tag{4.20}$$

According to moment equilibrium $C_M = C_b(l_G/b)$. Then λ and α can be obtained by using the equilibrium equations of force and moment above, with (4.19) and (4.20).

In principle, the Shuford-Brown method is more reasonable than CAHI and Savitsky methods. Someone compared the calculation results of this method with Savitsky's and found greater precision; however, the calculated angle of attack was lower than that from model tests. The present authors have found that the calculation agreed with some test results at $Fn_\nabla = 3.0$. The industry has nevertheless found the Savitsky empirical equations to be reliable for analysis of the planing phase, and to a lesser extent also the CAHI procedure.

Following the development of the initial empirical relations presented in [4], Savitsky and Brown presented revised relations including buoyancy and the effect of side wetting in [9]. Further work was carried out on whisker spray [10] allowing adjustment of the lifting force and drag calculations. Work was also completed on the surface geometry behind the transom or a "step"' in the hull planing surface. We discuss that aspect in Sect. 4.6 below and present spreadsheets for these calculations in the electronic materials to this Chapter.

The Savitsky formulae have been used for comparison with many of the model test series discussed in Sect. 4.3 and it has been found that the semi-empirical formulae gives accurate values of trim and "bare hull" drag, used as an iterative procedure to determine equilibrium for the planing regime (i.e., above $Fn_\nabla = 2.5 - 3.0$ or $Fn_L = 1.1 - 1.2$).

It is important to note that this approach is based on the planing surface being a trapezoid (as does the CAHI or Shuford-Brown method). Thus, if the hull includes rocker (positive or negative curve longitudinally, twist (varying deadrise), or transverse curvature, the longitudinal flow will be disturbed in a way that is unique to the particular geometry. While varying deadrise, within reasonable limits, may be substituted by an average for estimation, the curved shapes can generate vortex flow that requires dedicated physical testing, or RANS-based CFD with a large number of elements (10^6 or greater).

At planing speeds any deflection of the fluid surface from the direction of flow will generate forces and the tendency to additional vorticity. This is why the trapezoid has proven a useful model. The challenge for a Naval Architect is that when not planing, a hull does need a shape with curves to minimize pressure drag/ wavemaking. Hence the complications and the needs for model tests to explore the consequences.

4.4 Estimation of Resistance from Model Test Series

Altmeter [11] has summarized a number of model test series carried out for planing craft in the USA and Europe since the 1960s. Prior to that a number of test series were carried out in the first part of the twentieth century partly aimed at design of floatplanes and flying boats. A B Murray gave a summary of this work in 1950 [12]. There have been further series since Altmeter' s review. We give an overview below and refer readers to the references for direct use. Our purpose here is to introduce the range of planing hull shapes studied and the analytical methods employed.

Some of the test series have also been subjected to regression analysis so that hulls of very similar geometrical parameters can be estimated. We present key materials available in Appendix 1 for readers' use.

The different series all use a hard chine hull form with variations to the geometry. Some series have focused on speeds well below planing, while others, particularly the Series 62, have been tested to high planing speeds, see Table 4.1 for a summary. At the lower speeds, clearly, the same approach to analyzing drag as for round bilge craft is appropriate, and hence the regression analysis.

At first sight, one may think that a planing hull has less potential of variation of form compared to a round bilge or short chine/round bilge semi planing hull. The problem is that rather small variations from a basic trapezoidal shape can generate significant change in performance at planing speeds. Indeed, curvature or simply too low loading can cause dynamic instability as we will discuss in Chap. 5. The key variables available for a designer are the deadrise, the L/B ratio, and the use of spray rails and chine rails to "direct" the high-speed flow.

Two early test series for planing craft were the BK/MBK series in Russia used to support development of the CAHI semi-empirical procedure discussed above [13], and the EMB 50 series in the USA [14]. This latter studied some quite extreme L/B.

Table 4.1 Hard Chine Model Test Series key parameters

Test Series	Fn∇	Lpp/Bmax	LCG%	βm (degrees)
BK	1.00–4.5	3.75–7.00	35–45	21, twist to 12
MBK	1.00–4.5	2.50–3.75	35–45	18, twist to 7
EMB Series 50	1.25–5.5	2.20–8.51	16–44	5.2–20, twist
DTMB Series 62	0.20–6.0	2.00–7.00	36–48	12.5
Series 65	1.00–4.0	2.32–9.28	46–63	16.3–30.4
CSSC	1.50–3.5	4.54–6.04	38–42	Varied, bifurcated
US Naval Academy	0.30–1.5	3.86–5.17	35–45	12.4.21.0
US Coastguard	0.50–6.0	3.24–4.47	38, 42	16.6, 20.0
TUNS Series (Canada)	0.55–3.5	2.5,3.0, 3.5	25, 30, 35	12,18,24
Southampton University	2.30–8.02	3.77–6.25	33	22.5
Naples Series	1.10–4.20	3.45–6.25	35, 40	22.3, twist to 8
Delft University (DSDS)	0.75–3.0	2.00–7.00	38–50	19,25,30

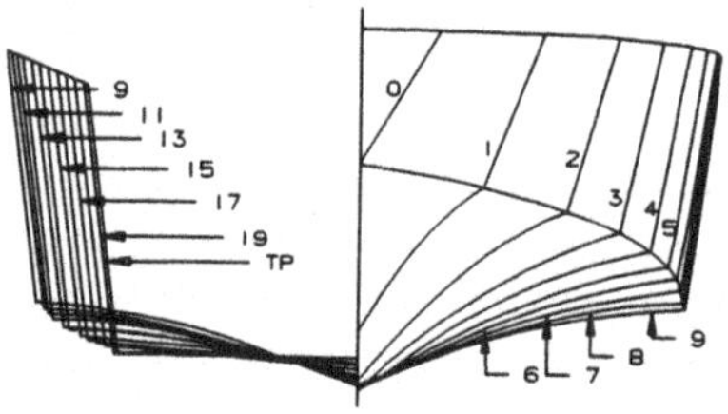

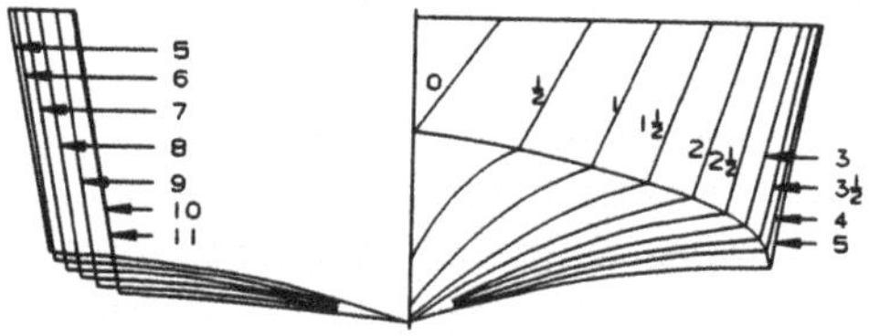

Fig. 4.10 Body plans of BK and MBK planing hull parent forms

Limited information is available in the public domain on the BK/MBK series. The hull forms are shown in Fig. 4.10 for reference. The planing surface is somewhat curved transversely with a fairly uniform reduction of deadrise from 21° to 12° and 18° to 7° for BK and MBK, respectively.

Note ref [15] reviews regression analyses carried out on many of the test series in Table 4.1 and discusses the approach in detail.

4.4.1 EMB Series 50

This was a series of 20 hulls including very wide and shallow craft, with concave, low deadrise planing surfaces, see reference [14]. Figure 4.11 shows an overview of the forms tested. The test series was reanalyzed by M Morabito [16] reported in 2015. The hull forms followed the approach of designers in the 1960s with a bluff bow form, transversely curved planing surfaces, and deadrise reducing to zero at the

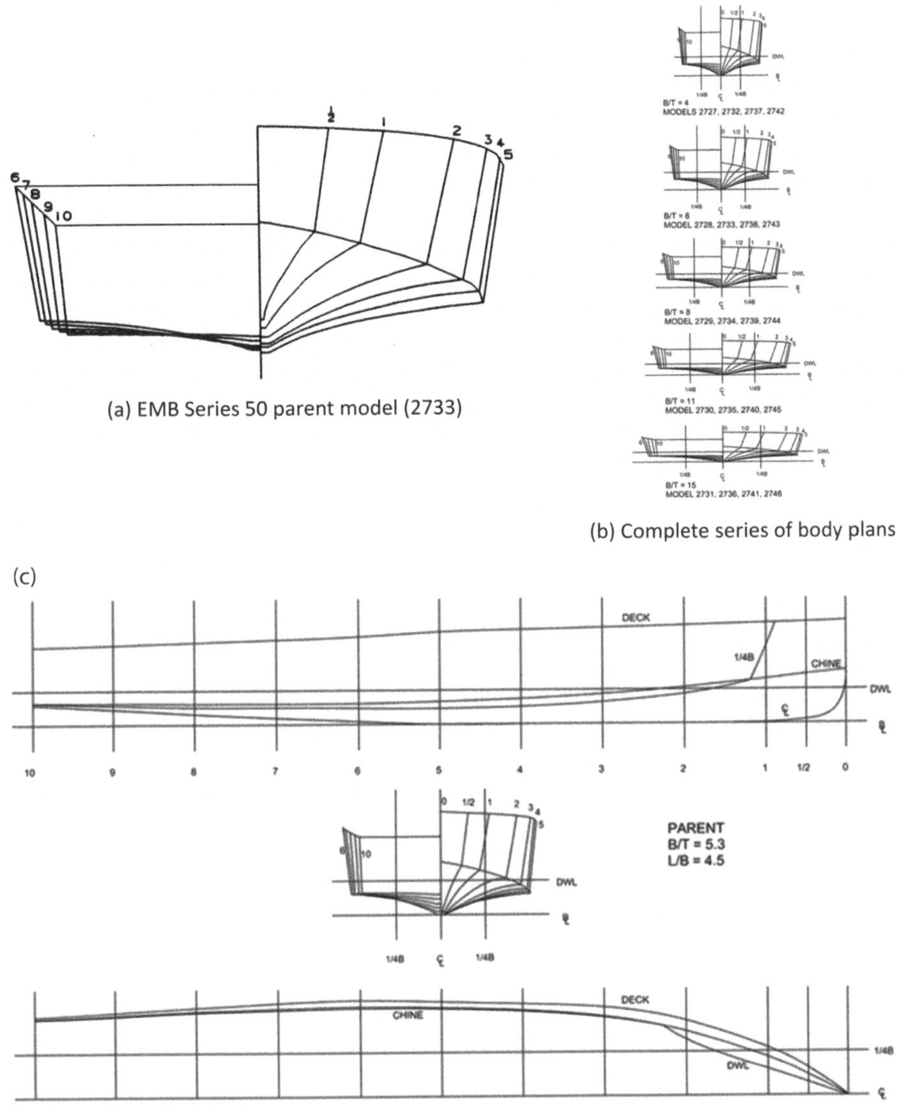

Fig. 4.11 Body plans and lines plans EMB Series 50 models. (**a**) Parent hull body plan. (**b**) Body plans for the series. (**c**) Lines plan for parent hull

transom. Morabito presented results in tabular and graphical form for interpolation and gave examples. The principal variables evaluated were beam/draft b/t, and slenderness or length/displacement ratio L/(Δ/100), where L was in feet and displacement in tons.

4.4.2 "Series 62" Model Tests (USA and Holland)

The well-known "series 62" test program was established in David Taylor Ship Model Basin, USA, using a planing hull form with straight dead-rise transverse sections, medium prismatic coefficient forward [17, 18], and includes five models with different length/beam ratio, see Table 4.2. The length/beam ratios are: $l_{co}/b_{cmax} = 2.0$, 3.06, 4.09, 5.50, and 7.00, where l_{co} is the projected length of the hard chine (in ref. [17], it is L_p), and b_{cmax} is the maximum beam at the hard chine, (in ref [17], it is B_{px}). Deadrise angle $\beta = 12.5$ °, constant after the midship section, and where the model with $l_{co}/b_{cmax} = 4.09$ is a base model. The model tests for resistance of every model with different displacement and CG were carried out, with the speed of $0.5 < Fn_\nabla < 6.0$. Finally, the test result was transformed to the resistance of full-scale craft with a displacement of 454 to 45400 kg.

The resistance per unit displacement R/Δ versus dynamic load coefficient C_{Lb} and displacement was plotted in graphs, where b_G is the hard chine width at CG (in ref. [17] it is b, and corresponding longitudinal CG is l_{op}, and all the units in ref. [17] are English units).

Figure 4.12 shows the body plan of the Series 62 base model. Figure 4.13 shows one of the resistance cross curves derived from the test series, with a relative CG longitudinal position at $l_G/b_G = 1.6$. Curves for various CG positions can be found in [17].

When using these "Series 62" graphs for estimating the resistance with known static displacement and leading dimensions, at first one has to find corresponding figures according to l_G/b_G, then obtain R/Δ at different C_{Lb} and certain Δ, and the resistance curve can be obtained according to speed V and R knowing the static

Table 4.2 Particulars of the DTMB Model Series 62

Series 62	Model				
	4665	4666	4667-1	4668	4669
A_p, ft^2	6.469	9.715	12.8	9.518	7.749
L_p, ft	3.912	5.987	8.0	8.0	8.0
B_{PA}, ft	1.654	1.623	1.6	1.19	0.935
B_{PX}, ft	1.956	1.956	1.956	1.455	1.143
B_{PT}, ft	1.565	1.386	1.25	0.934	0.734
L_p/B_{PA}	2.365	3.69	5.0	6.72	8.56
L_p/B_{PX}	2.00	3.06	4.09	5.5	7.0
B_{PX}/B_{PA}	1.18	1.21	1.22	1.22	1.22
B_{PT}/B_{PX}	0.80	0.71	0.64	0.64	0.64
A_p centroid[a]	47.5	48.2	48.8	48.8	48.8
β a-b, deg	5.0	5.0	5.0	3.7	2.9
i_a, deg[b]	58.0	49.0	46.0	39.0	37.0

Notes: [a]Centre of A_p forward of transom, % L_p. [b]Half angle of entrance at waterline
B_{PA}, B_{PX}, and B_{PT} are average, maximum, and transom breadth, respectively, across chines

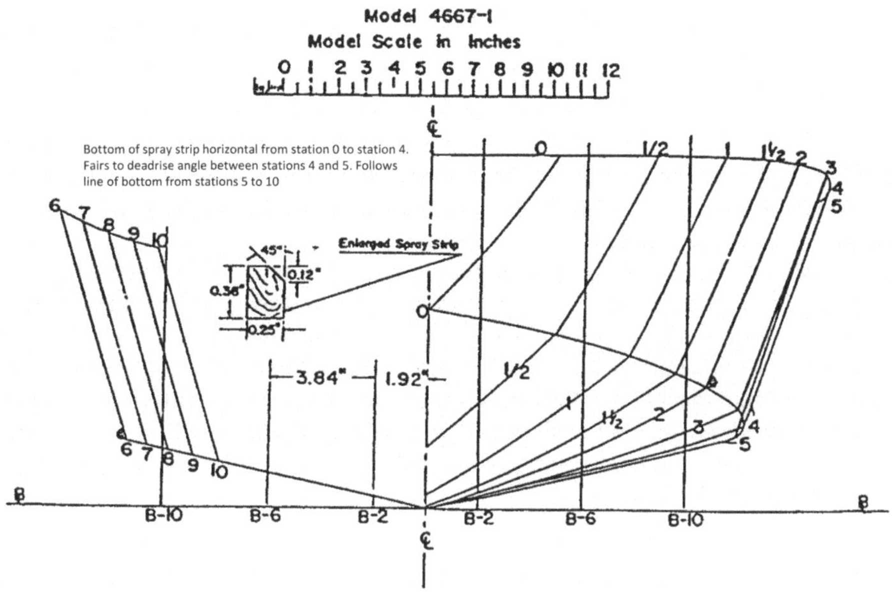

Fig. 4.12 Body plan of base case "Series 62" model 4667-1. (From reference [18])

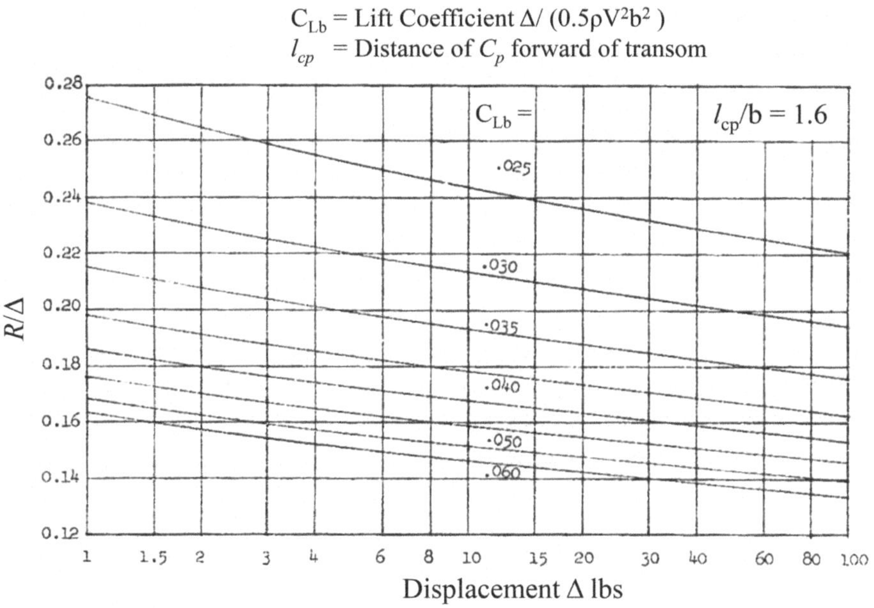

Fig. 4.13 Example resistance cross curve of Series 62, with lift coefficient as parameter

displacement. Interpolation can be used if L_G/b_G is rather different from the series model.

The "Series 62" is a planing hull with good performance. The test series results include the effects of various parameters, however, with fixed dead-rise. Following the increasing requirements for seakeeping quality, the importance of deadrise angle in design has increased. Therefore, J.A. Keuning and J. Gerritsma carried out a series of model tests with $\beta = 25°$ [19], with the same parameters as "Series 62," except β. The test speeds were limited to $0.75 < Fn_\nabla < 3.0$.

Section 4.4.10 below describes the further Delft test series (DSDS) that include 19 °, 25 °, and 30 °, followed by regression analysis to provide polynomial models for resistance up to $Fn_\nabla = 3.0$ including the original model range.

The test results of the initial series for 12.5° deadrise demonstrated that due to the increase in length/beam ratio, the resistance increment decreased with an increase in β, and showed no relation with area load coefficient $A_p/\Delta^{2/3}$. So, this can be considered as an additional result for the "Series 62" database, but with a lower range of test speed.

4.4.3 "CSSRC" Series

The "CSSRC" series model tests for estimating the resistance of planing hulls were carried out by the China Ship Scientific Research Center (CSSRC).

In China, many planing hulls have transverse deadrise cross section and longitudinal warped deadrise profile aft of amidships. The basic body plan of the series is shown in Fig. 4.14, which is rather different from "Series 62" having also a bifurcated deadrise forward. The hull has closer relation in cross section to the EMB Series 50 forms in Fig. 4.8 while having a slimmer forebody shape than either Series 50 or Series 62. CSSRC accumulated a large number of model test results for resistance and provided a method for estimating the resistance of their planing hulls using this data [20].

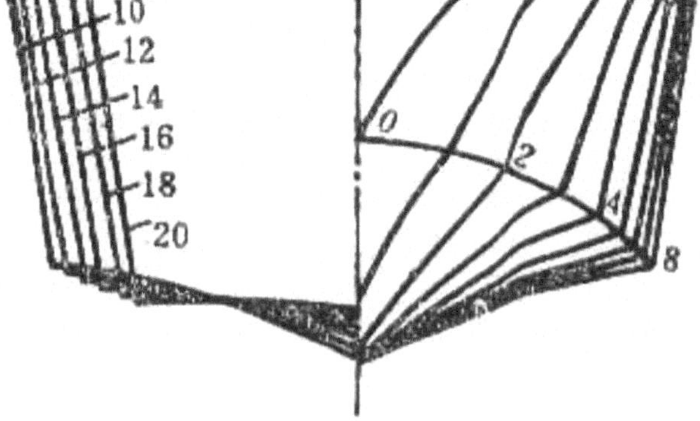

Fig. 4.14 Body plan of "CSSRC" series, with transverse deadrise, curved rocker, longitudinal warp profile

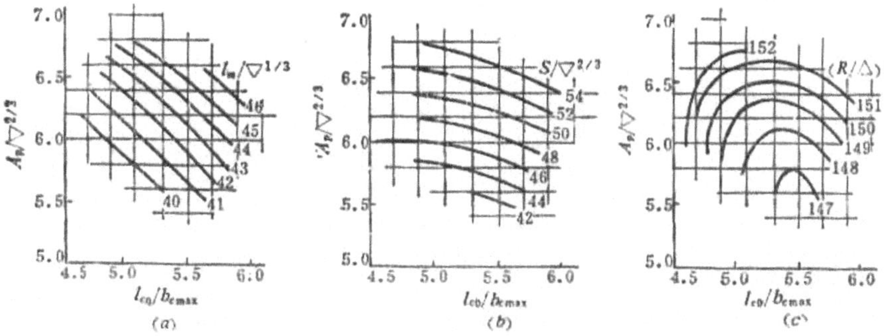

Fig. 4.15 Three groups of curves at $\mathrm{Fn}_y = 3.0$

The CSSRC estimation method assumes that resistance only relates to four factors: the volumetric Froude number Fn_∇, the area load coefficient $A_p/\nabla^{2/3}$, the length/beam ratio at the hard chine l_{co}/b_{cmax}, and the distance of CG from trailing edge l_G (i.e., the transom).

Initially, three sets of test results of a bare model were generated, see Fig. 4.15, at standard water temperature $t = 20\ ^\circ\mathrm{C}$, with $l_{co} = 1.987\mathrm{m}$, $l_G/l_{co} = 40.5\ \%$, and various shapes, were generated as follows:

$$\frac{1_m}{\nabla^{1/3}} = f\left(F_V, \frac{1_{c0}}{b_{c\max}}, \frac{A_p}{\nabla^{2/3}}\right), \frac{S}{\nabla^{2/3}} = f\left(F_V, \frac{1_{c0}}{b_{c\max}}, \frac{A_p}{\nabla^{2/3}}\right) \text{ and}$$

$$\frac{R}{\nabla} = f\left(F_V, \frac{1_{c0}}{b_{c\max}}, \frac{A_p}{\nabla^{2/3}}\right)$$

Since the CG is fixed, the effect of the CG is not shown in the Fig. 4.15. In case the position of CG is different from standard, a correction can be made by means of Fig. 4.16, where C_1, C_2, and C_3 are the correction coefficients for wetted length $l_m/\nabla^{1/3}$, the wetted area $S/\nabla^{2/3}$, and resistance fraction R/Δ, respectively. The correction values for the CG can be obtained by matching these to the corresponding values in the graphs.

Thus, the model resistance and corresponding values for full-scale craft, with given craft sizes and speed, can be obtained by means of graphs with $l_{c0} = 1.987\mathrm{m}$, $S/\nabla^{2/3}$, and R/Δ of model.

To estimate the power and speed of craft similar to the CSSC series in preliminary design with unknown main dimensions and lines, the initial main dimensions can be obtained using statistical data, and the relation of overall length with displacement can be estimated from Fig. 4.17, and the following approximate relationships used:

Length of hard chine is:

$$l_{c0} = 0.945\ l_s \tag{4.21}$$

where the static waterline length l_s may be in the range 10–50 m.

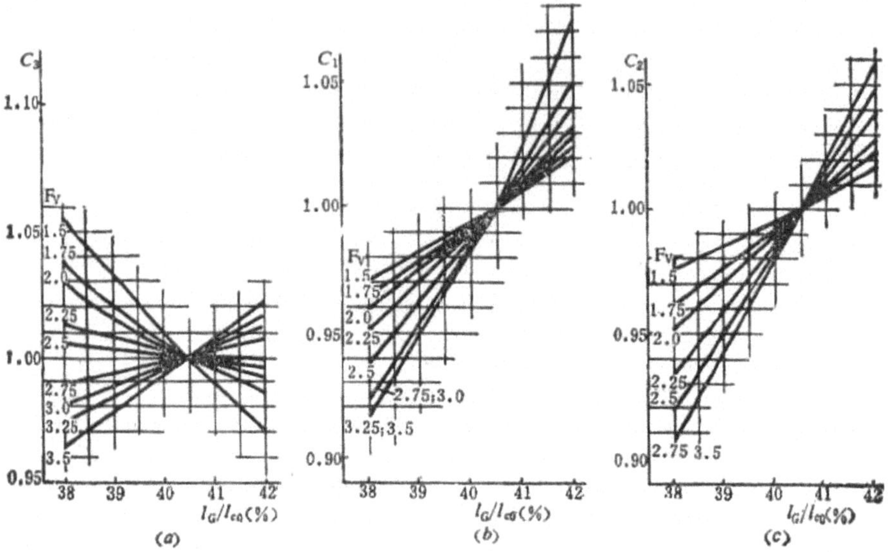

Fig. 4.16 Correction coefficients of CG

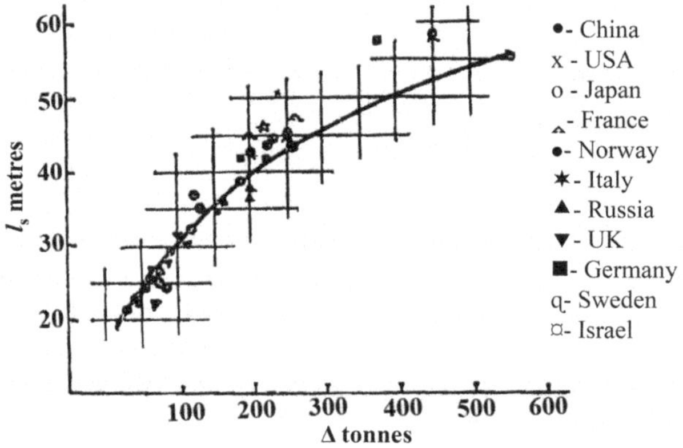

Fig. 4.17 Statistical data of $\Delta = f(l_s)$

The approximate plan area at the hard chine for area loading is:

$$A_p = 0.835\ l_{c0}\ b_{cmax} \tag{4.22}$$

The applicable parameter range for this method is:

$$Fn_\nabla = 1.50 \text{ to } 3.50$$
$$l_{c0}/b_{cmax} = 4.54 \text{ to } 6.04$$
$$A/\Delta^{2/3} = 5.40 \text{ to } 6.84$$
$$l_G/l_{c0} = 0.38 \text{ to } 0.42$$

Following estimation of drag and the other parameters above using the CSSC data in Figs. 4.12 and 4.13, it can be helpful to use the CAHI or Savitsky method to determine equilibrium trim, powering, etc and compare with the initial estimate, bearing in mind the effect of warped deadrise aft.

4.4.4 Series 65 Tests

The model test "Series 65" test program at the Naval Ship Research and Development Center, Bethesda, Maryland was carried out with two different hard chine hull form series, as shown in Fig. 4.18, aimed at planing hulls suitable for submerged hydrofoil craft using either airplane configuration foils, or canard configuration foils. The report was delivered in May 1974. Series 65-B aimed at the canard foil craft is perhaps most useful as the cross section aft of amidships is parallel with constant deadrise and so applicable to modern planing hulls. Series 65-A had a significant beam taper to the stern together with a rising keel line and so is less useful. The bow lines for both sub-series was fine as their performance investigation was aimed primarily at the preplaning speeds during which a hydrofoil would take off.

In line with practice at that time the models were towed at fixed trim, so interpolation of results presented in graphical and tabular form in [21] is necessary.

There were a total of 16 models in the series, with 9 in series 65-B covering a range of L/Bx from 2.35 to 9.38. The deadrise angles aft for series 65-B were 16.3° for (L/Bx 2.35–4.69), 22.5° (L/Bx 3.32–6.64), and 30.4° (L/Bx 4.69–9.38). The data thus provides a significant resource for estimation. The original report is available on the internet for reference, as shown in [21]. This series is included in regression analysis [24] discussed below for resistance up to $Fn_\nabla = 3.0$.

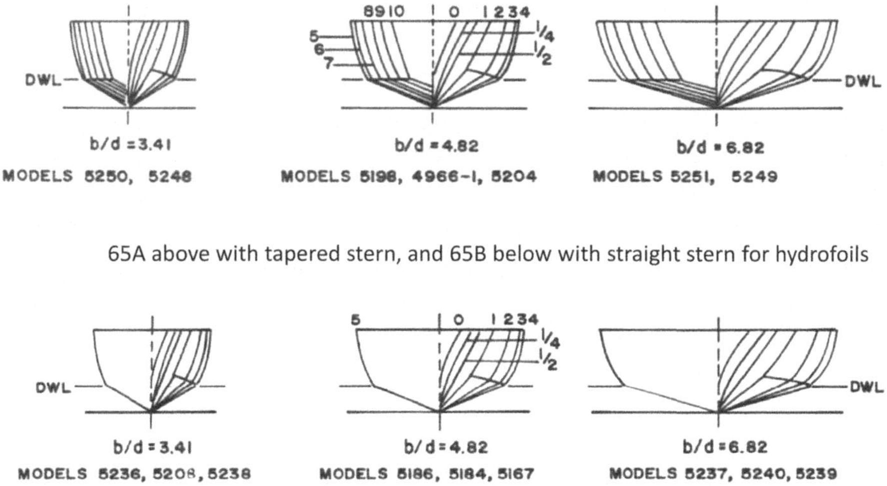

Fig. 4.18 Series 65 Hull forms for A and B series

4.4.5 US Naval Academy Series

The US Naval Academy Hydromechanics Laboratory (NAHL) at Carderock Division, Naval Surface Warfare Center, West Bethesda, Maryland, carried out a test series with three round bilge and three hard chine hull geometries in 1984.5 [22]. The application objective was coastal patrol, training, or recreational powerboats. The model lines are shown in Fig. 4.19, and a summary of the characteristics is shown in Table 4.3.

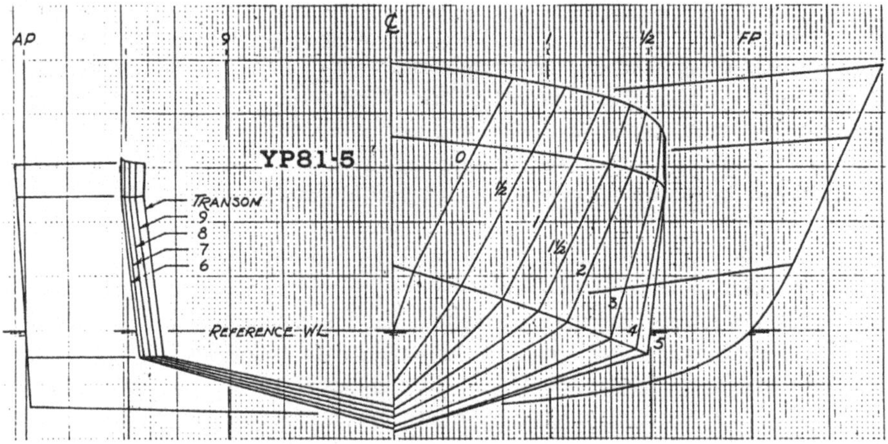

Fig. 4.19 US Naval Academy Series Hull form YP81-5

Table 4.3 Summary parameters for NAHL Model Series

	YP81-2	YP81-1	YP81-3	YP81-5	YP81-4	YP81-6
Type	Round chine	Round chine	Round chine	Hard chine	Hard chine	Hard chine
Lpp/Bref	3.97	4.59	5.17	3.86	4.49	5.17
Bref/T	5.05	3.84	3.06	5.25	3.90	3.03
Cb	0.432	0.437	0.432	0.452	0.437	0.437
Cp	0.724	0.720	0.716	0.720	0.721	0.719
Cwp	0.798	0.807	0.799	0.084	0.802	0.809
LCB/Lpp	−0.082	−0.081	−0.086	−0.086	−0.086	−0.084
Kmt/Bref	0.815	0.686	0.594	0.817	0.684	0.598
BML/Lpp	2.294	2.020	1.807	2.246	1.986	1.798
ie deg	20.9	17.5	16.5	21.5	17.7	17.3
β deg	12.4	16.1	20.8	12.4	16.4	21.0
Δ/L^3(ton/ft^3)	155.2	154.7	151.0	164.7	158.3	154.8
$S/\nabla^{2/3}$	7.222	6.695	6.412	7.345	6.818	6.513

Table 4.4 Coefficients for NAHL regression equation

	Soft Chine				Hard Chine			
Fn_L	**A**	**B**	**C**	**D**	**A**	**B**	**C**	**D**
0.1	−3.1551	0.11734	0.03036	0.24272	−4.0091	0.38551	0.02891	0.21182
0.15	−2.9756	0.13670	0.02882	0.23790	−3.8616	0.36559	0.02838	0.21086
0.2	−2.5305	0.14597	0.02780	0.24179	−3.4016	0.30585	0.02864	0.22727
0.25	−2.3637	0.16082	0.03099	0.27135	−3.8165	0.32552	0.03506	0.26631
0.3	−3.2734	0.28838	0.03994	0.31552	−4.9268	0.49059	0.04371	0.31399
0.35	−3.3070	0.34376	0.05040	0.31348	−5.0339	0.45322	0.05459	0.31524
0.4	−2.5114	0.19296	0.05704	0.31572	−3.3313	0.33365	0.06064	0.30777
0.45	−6.3063	0.85295	0.07999	0.35185	−7.4907	0.97457	0.08270	0.33763
0.5	−13.3525	1.71032	0.11094	0.34211	−13.4033	1.55367	0.11335	0.33905
0.55	−12.8032	1.66136	0.10600	0.27149	−13.4605	1.73983	0.10502	0.27315
0.6	−10.9613	1.46348	0.09407	0.119962	−11.7117	1.71066	0.08773	0.19874

Note: Long tons, ft, and fps are dimensions used in this table and the NAHL Model

The bow forms are relatively fine for all models, as one may expect for a patrol craft. The "soft chine" hulls had a quite small bilge radius in the after part, and both round bilge and hard chine models had straight deadrise flattening slightly toward the stern. The hard chine hulls have the chine continued right forward to the stem.

Each model was tested with LCG set at three different locations, equivalent to 5, 10, and 15% of Lpp aft of amidships. Testing was carried out at speeds between $Fn_L = 0.1$ and 0.67 thus in the displacement and semi-displacement speed regimes.

Friction resistance was determined for each test case in order to reduce the primary results to residual drag. Model test results were presented in a series of graphs. In addition, a regression analysis was carried out for each model based on a regression equation, as described in Table 4.4 and equation shown below the table.

$$C_R \times 10^3 = \mathbf{A} + \mathbf{B}\,(\mathrm{Lpp}/\mathbf{B}) + \mathbf{C}\,(\Delta/\mathrm{Lpp}) + \mathbf{D}\,(100.\mathrm{LCG}/\mathrm{Lpp})$$

A Sample curve of residual resistance, for model YP81-5 is shown in Fig. 4.20.

Overall, the tests suggested the following trends, within the bounds of the parameters tested for Residual Resistance:

- Residual resistance is highest for the aftmost LCG positions tested and the heaviest displacements tested.
- Resistance is less affected by LCG position at higher speeds.
- The primary resistance hump is close to $Fn_L = 0.5$ or $Fn_\nabla = 1.2$.
- The hard chine hulls had slightly lower resistance for the same L/B, Δ/Lpp, and LCG/Lpp at a given speed, this difference increases as speed is increased.
- The slender hulls had higher resistance than the wider models.

When frictional resistance is included, the slender models should have lower overall resistance as their specific wetted area is lower. Equally, the round chine

(a) Residual resistance coefficient as a function of speed and displacement for YP81-5 with LCG/Lpp = 0.244

(b) Residual resistance coefficient as a function of displacement for YP81-5 with LCG/Lpp = 0.244

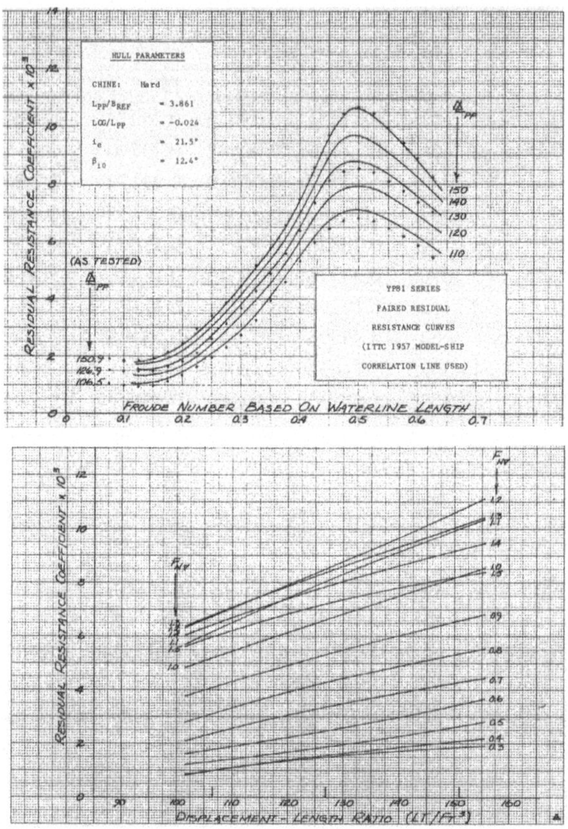

Fig. 4.20 US Naval Academy Series YP81-5 Resistance Curves example. (**a**) Residual resistance coefficient as a function of speed and displacement for YP81-5 with LCG/Lpp = 0.244. (**b**) Residual resistance coefficient as a function of displacement for YP81-5 with LCG/Lpp = 0.244

models had slightly lower wetted area than the hard chine models. Nevertheless, it is important to note that the use of a hard chine form can be advantageous at speeds down to the displacement region.

4.4.6 US Coast Guard Planing Hull Series

This test series was conducted some 20 years after the Naval Academy series was focused on planing hulls for fast intervention vessels [23]. Hull parameters were similar to the Naval Academy series, while the speeds evaluated were in the range between Fn_∇ 0 and 5.6 at model scale or between 10 and 48 knots at full scale for the 47 ft Motor Lifeboat (MLB).

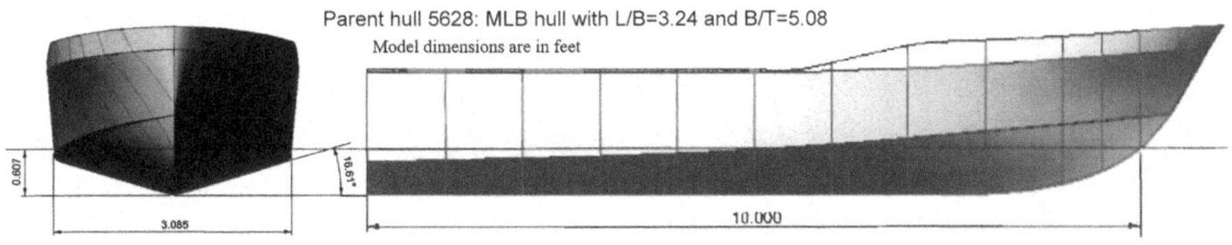

Fig. 4.21 US Coast Guard Series Hull form of parent 5628

Table 4.5 USCG test series model parameters

Model	LBP (ft)	B (ft)	T (ft)	L/B	B/T	β (deg)	∇ (ft^3)	AP (ft^2)
5628	10	3.08	0.608	3.24	5.08	16.61	8.05	25.88
5629	10	2.50	0.492	4.00	5.08	16.61	5.29	20.97
5630	10	2.24	0.441	4.47	5.08	16.61	4.24	18.76
5631	10	2.24	0.510	4.47	4.39	20.0	4.88	18.76

The hull parent was derived from the USCG 47ft MLB self-righting planing hull. This craft is used for inshore search and rescue. The hull form had proved successful in terms both of resistance and seakeeping at speed.

The objectives of the test series were to study the influence on L/B, transom deadrise angle, LCG position, and displacement on the resistance and the trim angle of the range of forms in the systematic series.

Four models were constructed at scale 1:4.3, and each was evaluated at three displacements and two positions of LCG (38% and 42% Lwl from aft perpendicular). The test condition range simulated vessels of 55–100 ft length, and 23–221 long tons displacement at full scale.

The hull lines for model 5628 are shown in Fig. 4.21. The form particulars are given in Table 4.5. It may be noted that at the transom the lines were flattened to the horizontal. The original 47 ft MLB had a wedge at its stern.

Test results are presented as a set of graphs and data tables documenting R_T/Δ, The dynamic wetted surface, residuary resistance, CG heave displacement, and pitch angle, all against Fn_∇. The dynamic wetted surface was determined at each speed by measuring with wave profile along the hull at 4 locations along the model—the keel-to-water intersection at the bow, the forward intersection between chine and spray sheet, the chine reattachment point, and the water height at the transom.

Further figures show the influence of deadrise and L/B on R_T. In general, it was found that high L/B gave the lowest drag, while model 5631 with higher deadrise had higher drag than model 5630.

Example plots are shown in Fig. 4.22a–d. For (a) R_T/Δ for all models, (b) wetted surface area model 5628, (c) derived residuary resistance of model 5628 with LCG at 38% forward from transom and (d) the influence of L/B and deadrise on R_T for models at 375 lbs displacement

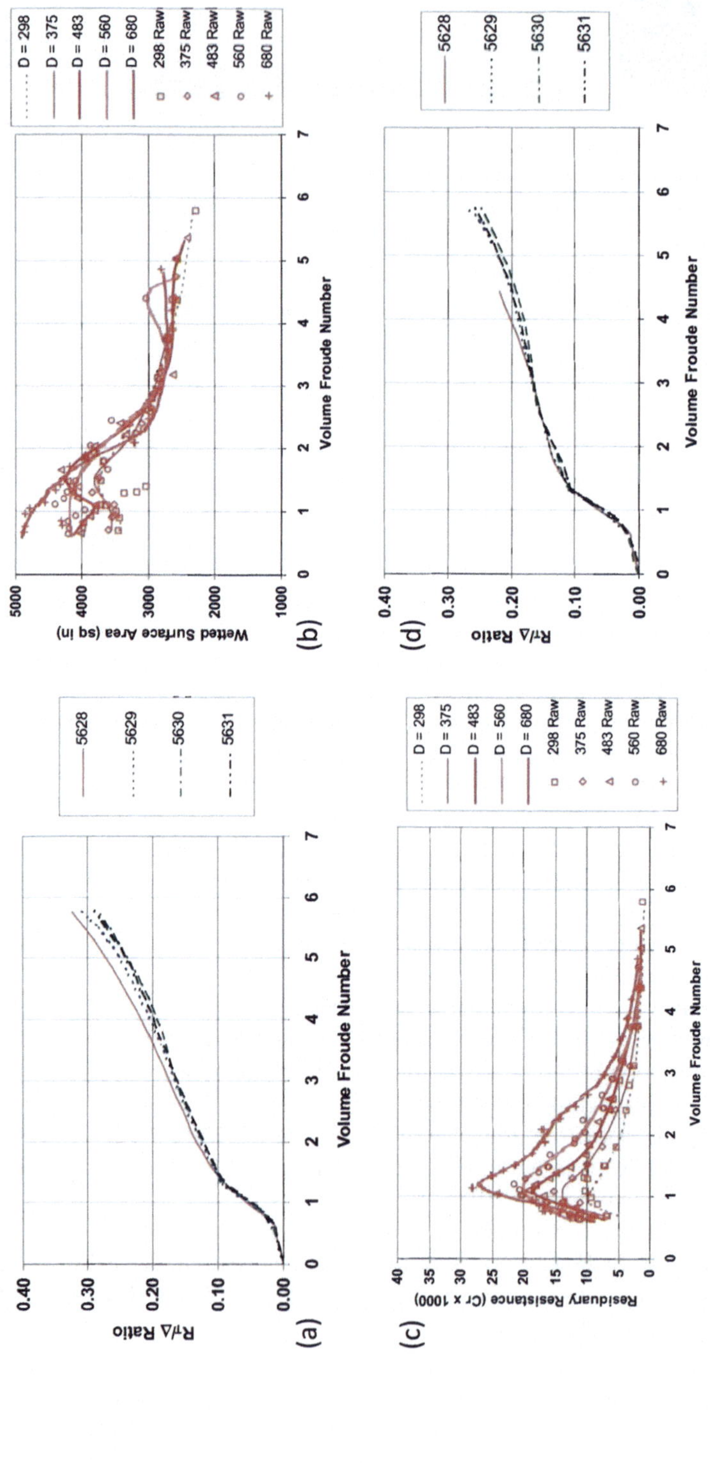

Fig. 4.22 US Coast Guard Series Hull Model 5628 example plots

Conclusions from the program were as follows:

- R_T/Δ was more favorable for the high displacements at high speed, and low displacement at low speed.
- R_T/Δ is favorable for the aft LCG at higher speeds, and forward LCG at low speeds.
- The dynamic wetted surface area for a particular hull form and LCG converged to the same value at high speeds above $Fn_\nabla = 3.6$ for all the displacements tested. The forward LCG location caused an increased dynamic wetted surface.

Added to these conclusions one may add the finding that higher deadrise did come at a cost of higher drag in calm water. In a seaway the performance would be improved for the higher deadrise compare with the flatter geometry.

A further observation is the characteristic drag peak around $Fn_\nabla = 1.5$ ($Fn_L = 0.5$) shown by the residuary resistance curves that subsequently decay as planing is achieved.

Readers are referred to the formal report [23] for full data. Meanwhile, results from this program have been included in regression analysis for the lower part of the speed range tested [24]. We discuss this in Sect. 4.5 below. The characteristics above $Rn_\nabla = 3.5$ can be checked against evaluation using the Savitsky or CAHI method summarized in Sect. 4.3.

4.4.7 Technical University of Nova Scotia (TUNS) Series

In 1992/3 Gabriel Delgado-Saldivar carried out an experimental investigation of a new series of planing hulls as his MSc project at the Technical University of Nova Scotia, Halifax, Nova Scotia, Canada [25]. The aim of this series was to provide design data for planing hull geometries applicable to modern designs. It was noted that many earlier series used forms with tapering shape toward the stern and had other characteristics different from modern craft which tended to have trapezoidal shape consistent with the evaluations by Savitsky [4, 9]. This is a series of nine hull forms where the after form is prismatic and having deadrise angle from 12° to 24°. The L/B ratio varied between 2.5 and 3.5, see Table 4.6. Parent model parameters are outlined in bold for reference

To select hull form parameters data was compiled from 368 designs of contemporary craft at that time and statistics prepared before choosing representative parameters and ranges. Statistical analysis was carried out and the series generated with three L/B ratios and three deadrise angles. Lcg was tested for each model at three positions, and the loading parameter $Cd_L(\Delta/(\rho B_x^3))$ was varied in steps of 0.05 in steps as shown below. The models were then tested at ten velocities ranging from 0.72 m/s to 3.93 m/s equivalent to Cv between 0.5 and 2.75 ($Fn_\nabla \approx 0.5$–3.5).

Hull bow shape was derived from plots of design data from the database, selections for the parent hull were bow stem angle 42.5°, chine length 0.912, constant beam lengths on deck and at chine 0.512 and 0.41, bow length 0.37,

Table 4.6 TUNS Model series

β deg	**12**								
L/B	2.5			3.0			3.5		
Lcg%	35	30	25	35	30	25	35	30	25
Cd_L	0.075	0.084	0.10	0.10	0.10	0.11	0.10	0.10	0.12
In steps	to	to	to	to	to	to	to	to	to
of 0.05	0.2	0.2	0.2	0.25	0.25	0.25	0.30	0.3	0.25
β deg	**18**								
L/B	2.5			3.0			3.5		
Lcg%	35	30	25	35	30	25	35	30	25
Cd_L	0.086	0.086	0.086	0.114	0.114	0.114	0.12	0.12	0.12
In steps	to	to	to	to	to	to	to	to	to
of 0.05	0.15	0.15	0.15	0.20	0.20	0.20	0.30	0.3	0.25
β deg	**24**								
L/B	2.5			3.0			3.5		
Lcg%	35	30	25	35	30	25	35	30	25
Cd_L	0.075	0.10	0.10	0.10	0.10	0.11	0.10	0.10	0.117
In steps	to	to	to	to	to	to	to	to	to
of 0.05	0.2	0.2	0.2	0.25	0.25	0.25	0.30	0.3	0.25

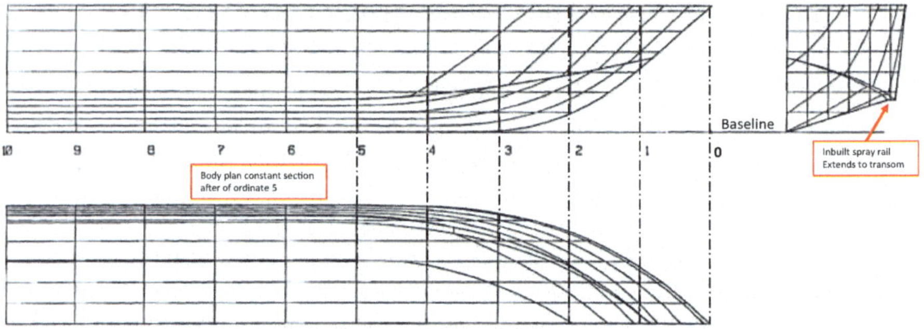

Fig. 4.23 TUNS Series parent Hull form

depth at forward end 0.182, and chine height 0.107, all as proportions of Loa. The parent hull had Loa 690 mm with deadrise β at 18°.

The final lines after fairing are shown in Fig. 4.23. Spray strakes were incorporated along the chine with a width of 7% of the maximum beam reducing to the bow forward of the constant deadrise section. Turbulence stimulation was provided by Hama strips at station 4.

To ensure the data produced by the program was reliable, the test tank was initially calibrated by using a model previously tested by Fridsma at Davidson

Laboratory, Stevens Institute of Technology. Reproducibility testing was also carried out, and error analysis of the results.

The test series data is presented in a comprehensive set of plots and data tables. A set of example plots for C_T, R_T/Δ, trim, CG rise, and Wetted area coefficient are shown in Fig. 4.24.

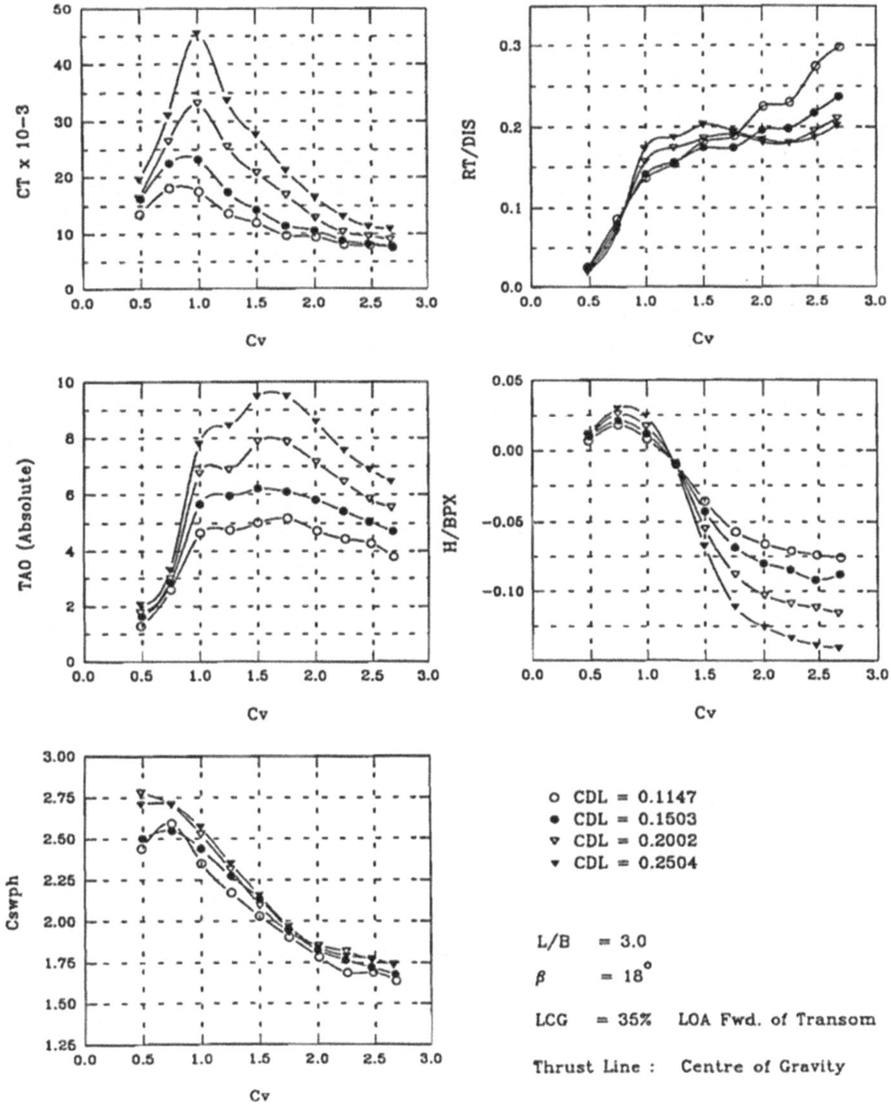

Fig. 4.24 TUNS model test results example for model 3018 with thrust through CG, from reference 25, figure B25 TUNS plots of CT × 10^{-3}, RT/Δ, trim τ, heave at CG/BPX, and wetted area coefficient Cswph, all against Cv The data is from model 3018, the parent hull, and is plotted for the range of loadings CDL with LCG at 35% L forward of transom

On many runs a video camera at the rear of the tow carriage recorded the wetting of the transom. Analysis showed that the transom ventilated completely at Cv between 0.75 and 1.0 compared to the value of 0.5 suggested by Savitsky and Brown [9]. Photographic data was also used to estimate the hull-wetted area.

In the planing regime above Cv $\approx$ 1.75, the resistance coefficient curves trended to the same value regardless of displacement. In contrast, the curves are distinct up through the drag hump and proportional to displacement.

It was noted that at very low loading coefficient C_D values the R_T/W curve is unstable.

The trim angle curves formed a uniform family for most conditions tested. Maximum trim is close to Cv = 1.5 and occurrence of maximum drag. In some combinations of LCG and L/B (35% and 3.0/3.5) a double hump appeared.

Sinkage of VCG was observed at low speeds and this changed to a rise at Cv between 1.0 and 1.25 as speed approached planing and settled to a stable value around Cv = 2.0

Porpoising was observed for some tests at 25% LCG, mostly at 12 deg deadrise, and L/B 2.5, thus the wider flatter geometry and higher static trim. The drag coefficient C_T was also highest for this aftmost LCG.

When LCG, deadrise, and C_D are constant, and L/B is varied it was found that slenderer models had lower C_T.

The effect of thrust line was studied by running tests with the thrust modelled along the keel line as well as through the CG. Initial analysis seemed to indicate that thrust line through CG gave lower C_T below planing speed and then the results switched above planing speed. Further work was concluded to be needed on this issue.

Gabriel Delgado-Saldivar did not carry out checks against the Savitsky scheme for evaluation of equilibrium and drag for the fully planing regime so that is an exercise that would add value to this test program. For the preplaning regime Radojcic has carried out regression analysis of the US Coastguard and TUN series and this is discussed in the next section.

4.4.8 Southampton University Series

In 2010, D J Taunton, D A Hudson, and R A Shenoi from the University of Southampton published results from calm water tests on a series of planing [26].[1] hulls representative of modern high-speed military, paramilitary, and offshore racing craft. Key parameters were selected from a survey of 24 such craft that varied in length from 10.1 to 24 m and L/B 3.7 to 5.8. Four models were constructed, as

[1] The calm water testing was followed by a test series for the same models in waves [27]. This work is discussed in Chap. 5.

Table 4.7 Southampton University Fast Planing Series

Model		A	B	C	D
L	m	2.0			
B	m	0.32	0.39	0.46	0.53
T	m	0.06	0.08	0.09	0.11
Δ	N	119.25	175.83	243.4	321.95
$L/\nabla^{1/3}$		8.70	7.64	6.86	6.25
L/B		6.25	5.13	4.35	3.77
β	deg	22.5			
LCG	%	0.33			

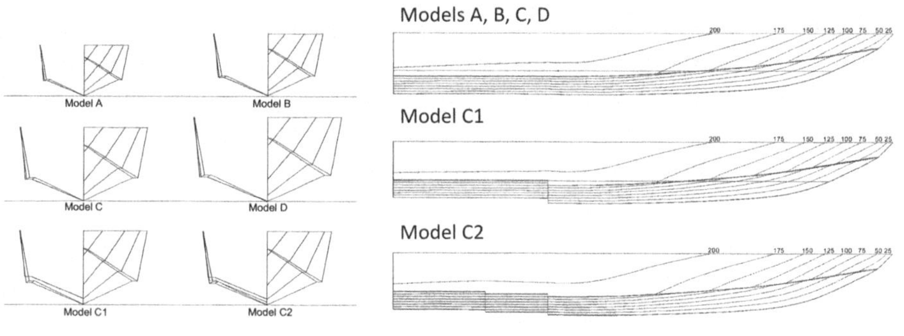

Fig. 4.25 Southampton University Series Hull forms

shown in Table 4.7. In addition, two hulls based on model C were constructed, having a single and a double step designated C1 and C2.

Figure 4.25 shows the body plans and profiles of the hull forms where model C is the parent hull form. Compared with the TUN hull form the forward lines are finer and the bow profile rather sharper. The form has an in-built spray rail at the chine that continues right to the bow, but no lower surface spray rails as are common on modern planing craft.

Models C1 and C2 had transverse steps incorporated at 18.5 and 30% of LOA from the stern so that LCG was located just forward of the main step.

Testing was carried out at the GKN Westland Aerospace No 3 Test Tank in Cowes, IOW. This facility has dimensions L/B/D 198/4.57/1.68 m and a maximum carriage speed of 15 m/s. In earlier decades, this facility was used by Saunders Roe for flying boat testing, high-speed vessels, and hovercraft development. During the testing of this planing monohull series speeds ranged from 4 to 12 m/s.

It may be noted that models in this series are approximately three times as large as the TUN models at 0.69 m LOA as the test tank at TUN is L/B/D 27/0.9/1.0 having a maximum carriage speed of 4.0 m/s

The capability at GKN No 3 Tank allowed testing to be carried out at speeds up to Cv = 6 ($Fn_\nabla = 7$). The lowest speeds tested were in the range Cv = 2 ($Fn_\nabla = 2.5$) thus at the start of the fully planing regime. In contrast, the TUN tests were in the range Cv = 0.5 to 2.75, thus investigating performance through the semi-planing regime and into the initial planing speed range.

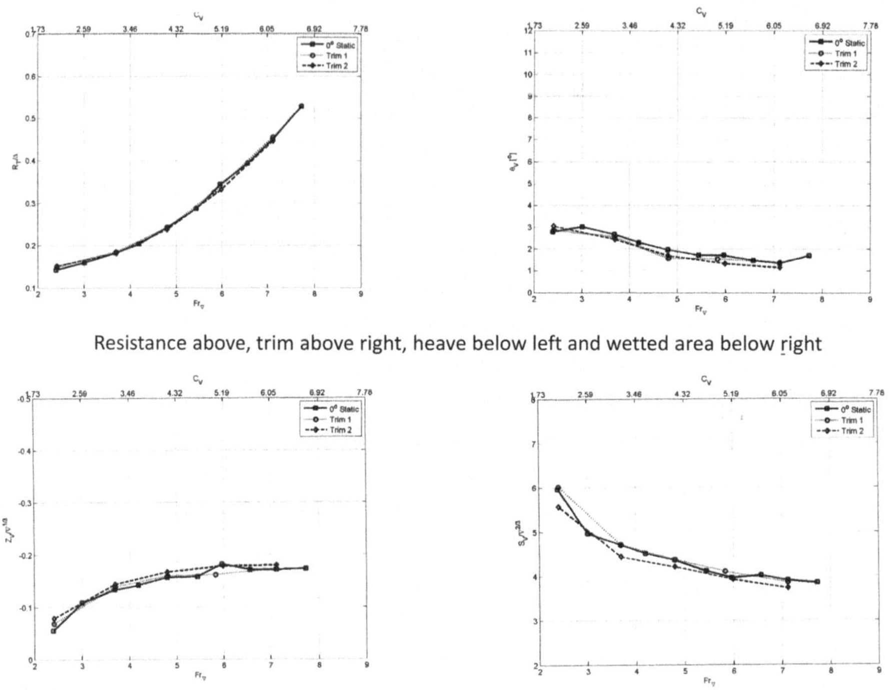

Fig. 4.26 Southampton Model C results from calm water testing. (**a**) Resistance. (**b**) Trim. (**c**) CG heave. (**d**) Wetted area

The Southampton University program measured resistance, dynamic trim and sinkage, and wetted surface area through observations of the running waterline. The models were also ballasted to give level trim and two further conditions on static bow-up trim which effectively simulated a movement of the LCG back from the 33% base condition.

Results are presented in tabular fashion, and plots against Fn_∇. Figure 4.26a–d show example results from the parent model C.

Analysis was also carried out to look at the influence of slenderness, $L/\nabla^{1/3}$, on dynamic trim and wetted surface area; and load coefficient on the four parameters. It was found that wetted surface area increased with slenderness, but at the same time, the CG rise increased and the dynamic trim reduced, so that the resistance curve had just a small trend toward lower resistance for lower slenderness, see Fig. 4.27.

The effect of the steps was found to be significant on resistance, Fig. 4.28, while having little effect on dynamic trim, and seemed to actually show increased wetted surface area from measurements. The trim may be linked to the step configuration adopted, and it is difficult of course to evaluate the area offloaded between a step and where the flow reattaches.

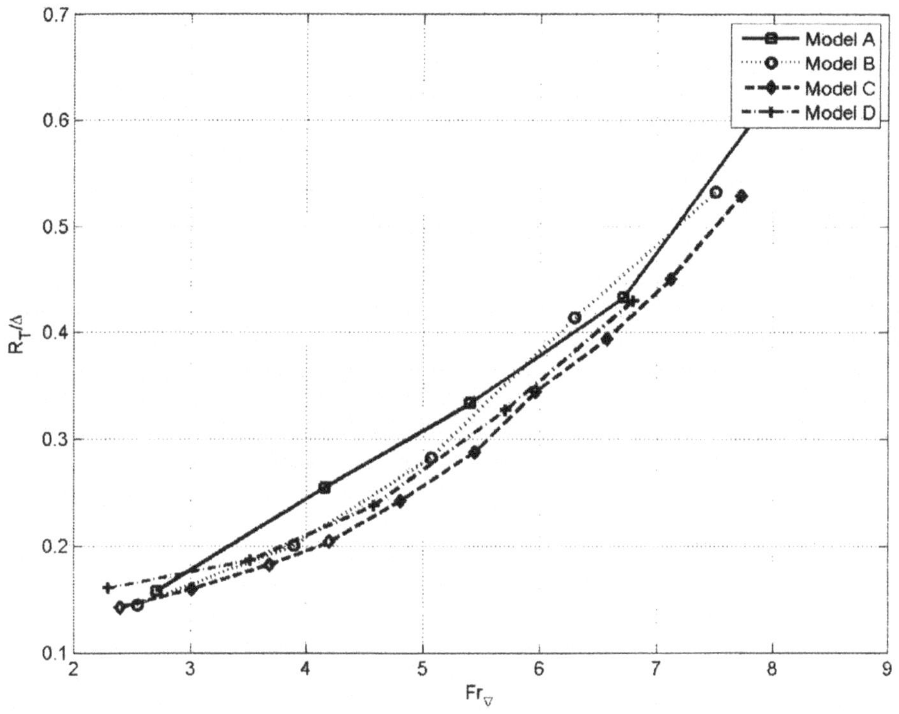

Fig. 4.27 Resistance vs slenderness for the four models

The authors also carried out an evaluation of the resistance components for model C using the methods of Savitsky et al detailed in [28], where the incorporation of "whisker spray" was discussed. The result for Model C was as shown in Fig. 4.29, where it can be seen that "spray drag" appears to increase steadily above $Fn_{\nabla} = 5$. In the subsequent evaluation of total drag, including air drag at full scale rather than model scale, there were found to be discrepancies suggesting that the scaled wetted area may be too high as the inclusion of spray drag then creates an error. It is suggested that more work is needed regarding scaling of planing data.

While not explicit in the discussion the issue here may well be the behavior of water droplets, which do not scale according to Froude or Reynolds. There is continuing discussion around spray and wetting, and we touch on that in the next subheading. The key result from this test series is a set of data for modern planing hulls with mid-range deadrise of 22.5° for speeds through the high-speed regime. While not a specific calibration between testing and the Savitsky empirical method, the work done to generate Fig. 4.29 does suggest that above full planing, with trapezoidal planing surface form, the method can be applied with confidence.

The work was continued with testing of the same models in irregular waves [27] to look at motions and accelerations. This is summarized in Chap. 5.

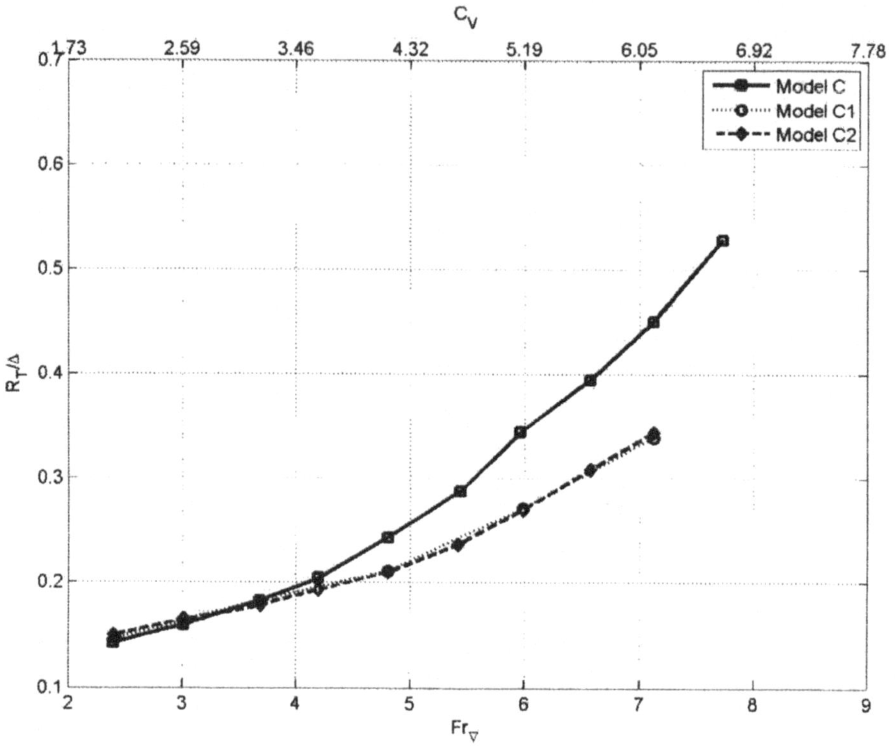

Fig. 4.28 Resistance for C, C1, C2

4.4.9 Naples Warped Hard Chine Series

Fabio De Luca and C Pensa from published results from a planing model test series carried out at the University of Naples "Federico II," Italy, in 2017 [29]. This was a series of five models having reducing deadrise from amidships to the transom and a bow form more like the TUN and USCG series rather than the sharp bow of the Southampton University Series. Tests were carried out in the speed range Rn_{∇} from 0.5 to 3.5, covering the lower part of the planing speeds and the semi-planing regime. The deadrise on these models increased from 13.2° at the transom, to 22.3° at amidships, and to 38.5° at 75% of length from the transom. Following the testing regression analysis was carried out to generate polynomial equations for prediction of resistance.

In addition to the basic resistance testing, tests were also carried out with interceptors modelled at the transom to assess the response of the hulls used.

Figure 4.30 shows body plan and profile lines of the parent model designated C1. Inset is a diagram showing the two versions of interceptor installed, either across the complete transom, or half the breadth centered on the keel. Table 4.8 summarizes the model key parameters.

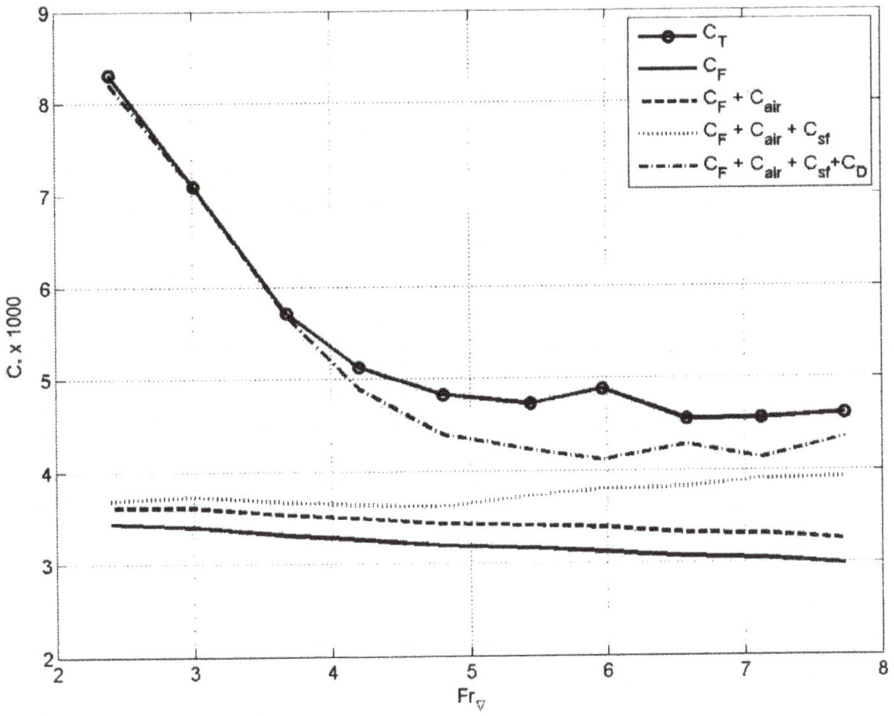

Fig. 4.29 Resistance component assessment for model C

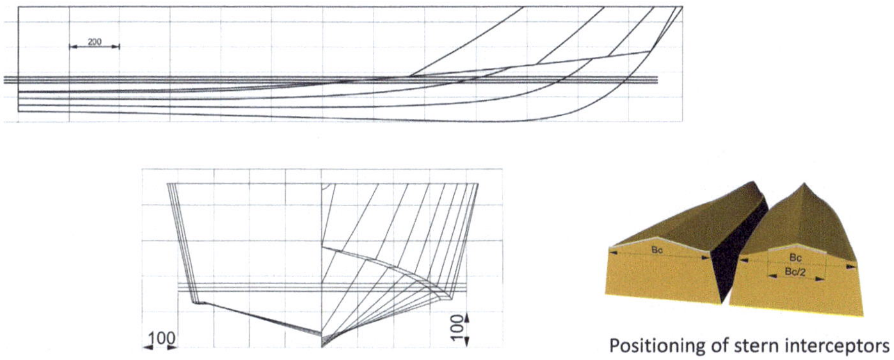

Fig. 4.30 Naples Series Hull form

The testing was completed in the towing tank of the Naval Division of the university which has dimensions 136 × 9.0 × 4.5 m (LxBxD). No turbulence stimulators were installed as for the tests Re > 3.5 × 10^6. Similar to the test series described above, towing force was applied at L_{CB} (measured from the transom) at 0.945 m and at the estimated VCG. The models were free to heave and pitch, but restrained in surge, sway, yaw, and roll.

Table 4.8 Naples Warped chine model series summary

Model		C1	C2	C3	C4	C5
L_{OA}	m	2.611				
L_{WL}	m	2.374				
B_{WL}	m	0.733	0.651	0.569	0.487	0.407
Interceptor		B_{WL}			$B_{WL}/2$	
$L_{CB}(\tau_s = 0)$	m	0.944	0.944	0.944	0.944	0.943
B/T		4.667	4.667	4.667	4.667	4.397
L_P/B_C		3.45	3.89	4.45	5.19	6.25
Δ	kg	92.25	72.74	55.54	40.66	32.31
$L/\nabla^{1/3}$		4.83–5.25	5.23–5.69	5.47–6.22	6.06–6.90	6.86–7.49
S_w	m^2	1.530	1.360	1.180	1.000	0.870
Fn_L		0.5–1.6				
Fn_∇		1.1–3.6	1.2–3.7	1.2–3.8	1.3–4.1	1.4.4.2
i_E	deg	23.7	21.1	18.9	16.3	13.7
τ_s, static trim	deg	0.0, 1.0				
T_H (est VCG)	m	0.191	0.171	0.154	0.145	0.134

Note: Models were tested at three displacements, the above base case, plus 15% and 33% higher, leading to increasing L_{WL} and S_W, and reducing B/T. i_E is given for $L_{WL} = 2.387$ m. Interceptor depth was approx. 2 mm.

For each model, test runs were made at speed intervals of 0.5 m/s, for C1 for example from 2.5 to 7.5 m/s while measuring the trim and resistance. This was repeated for the range of displacements for each hull, the two static trim values. The dynamic wetted length at the keel was documented and the wetted area was derived for each run. From these data, the coefficients of total resistance and residual resistance were determined, both for the bare hull and for the hull with interceptors. It is of interest that the wetted area includes reattached flow above the chines, but the whisker spray area is not evaluated due to the uncertainty of its effect.

In the reference paper, full tables of the recorded data are provided. The authors have also carried out regression analysis to derive polynomial equations to predict C_R, S_{WD}, and L_{WLD}. We provide more information on this in Appendix 1. Checks were carried out for accuracy and within the envelope of hull parameters the polynomial was shown to predict within around 1%.

Figure 4.31 shows sample results from hull C1 bare and with interceptors. It can be seen that the interceptors depress the trim, which also reduces the resistance up to Fn_∇ around 2.5. Above this speed, without interceptors being reduced in height the resistance is increased compared to the bare hull. It is of interest therefore that the impact of interceptors is high in the preplaning region, and once planing need to be retracted to a height that simply optimizes the trim. An example resistance plot for hull C1 is also shown in Fig. 4.31 together with the dynamic trim without and with interceptors. It is difficult to clearly see common trends though the zero static trim appears to show minimized resistance. The addition of the interceptor changes this balance.

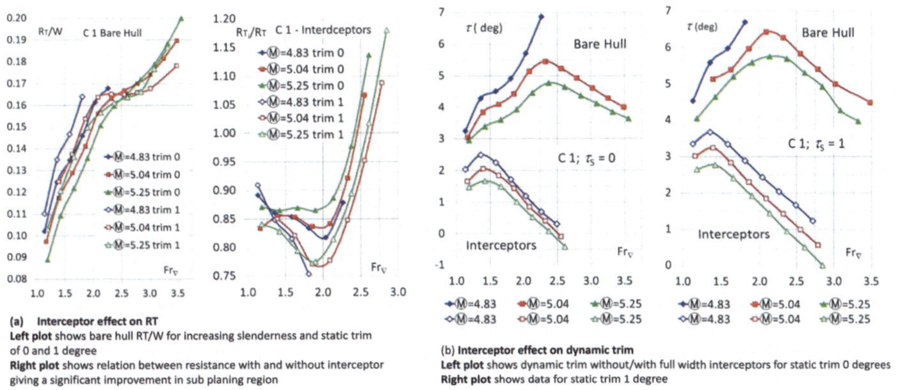

Fig. 4.31 Naples Series results. Hull form C1 RT/W and tau vs Fr_∇ for both without and with interceptors

This series thus gives information on potential resistance and trim characteristics for a planing craft with varying deadrises to compare with the data from TUN and Southampton university, where the deadrise is constant. The TUN hull has bow form similar to the Naples series while the Southampton University series has bow form more consistent with normal operations at rather higher planing speeds.

It may be noted that the Naples test series was extended to tests in waves. We discuss that in the next chapter.

4.4.10 Delft Systematic Deadrise Series

J A Keuning and L Hillege from the Ship Hydromechanics department of Delft University of Technology, The Netherlands, summarized the test programs carried out with the Delft Systematic Deadrise Series (DSDS) in a paper at the FAST 2017 conference [30]. This covers work extending over several decades based on the original work by Clement and Blount [17], the Series 62 program discussed above.

The Series 62 program comprised 5 different hulls with constant deadrise aft of amidships of 12.5°, see Tables 4.1 and 4.2 above. From testing in Delft of a model similar to Series 62 but with a high deadrise by JJ van den Bosch in 1970 it was shown that this gave significant improvements in seakeeping. The problem was that resistance in calm water was increased. Keuning and Hillege began their work with a series of five models similar to the Series 62 but with 25° deadrise rather than 12.5 °, tested to the same series of conditions and parameters as the original DTMB tests, see Fig. 4.32. Model 46671 is the Series 62 parent, while Model 188 is the DSDS parent with 25° deadrise.

The Delft University Towing tank is smaller than that at DTMB and so speeds tested were limited to Fn_∇ from 0.75 to 3.0. This work was published in 1982 by a

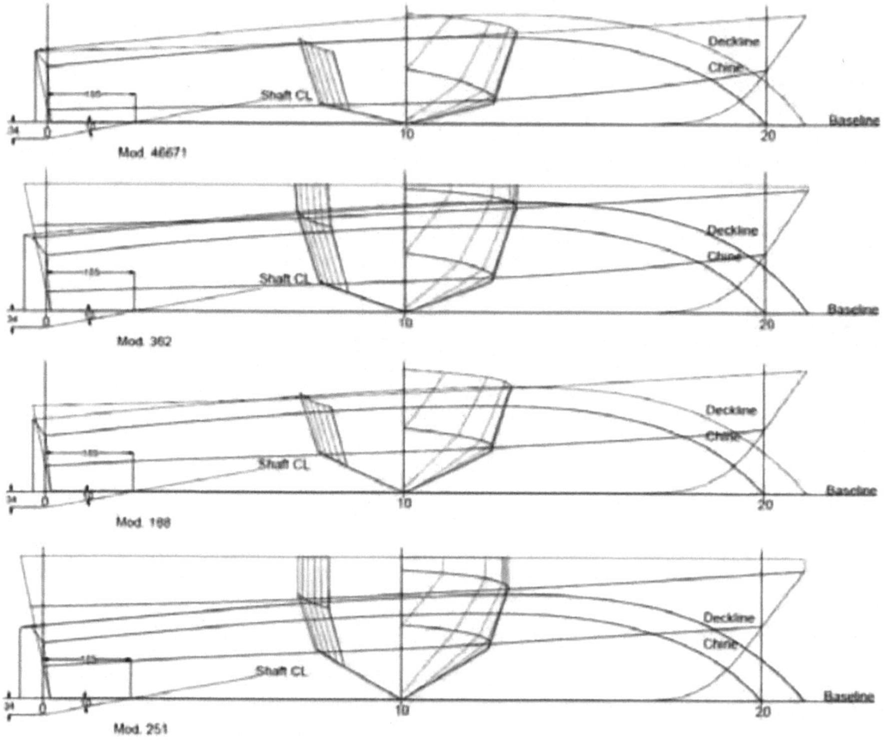

Fig. 4.32 Delft Series Hull forms

Delft University report and reported in an article in International Shipbuilding Progress [31].

The problem for craft with parallel aft deadrise was known to be relatively high resistance in the preplaning speed range. Most craft spend much of their operational time in this range rather than at top speed. During the 1980s, the fast boat industry experimented with variable deadrise (flattening to the stern as for the Naples hull form, also called deadrise twist), and "rocker" or curvature applied to the average surface. Both of these, combined with a rising keel line aft could significantly reduce resistance at semi-planing speed and improve seakeeping throughout the speed range.

These design elements also allowed larger propeller diameter or/and reduced shaft inclination for traditional machinery installations. It may be noted here that application of water jet propulsion, or transom mounted Z drive or outboard motor may adjust the preference and indeed the craft performance. The testing at Naples University with the two different thrust lines gives some information on influences here.

In 1986, Keuning extended the DSDS series with two models based on their parent Lp/Bpx = 4.09, and 25-degree deadrise hull applying a twist angle to the

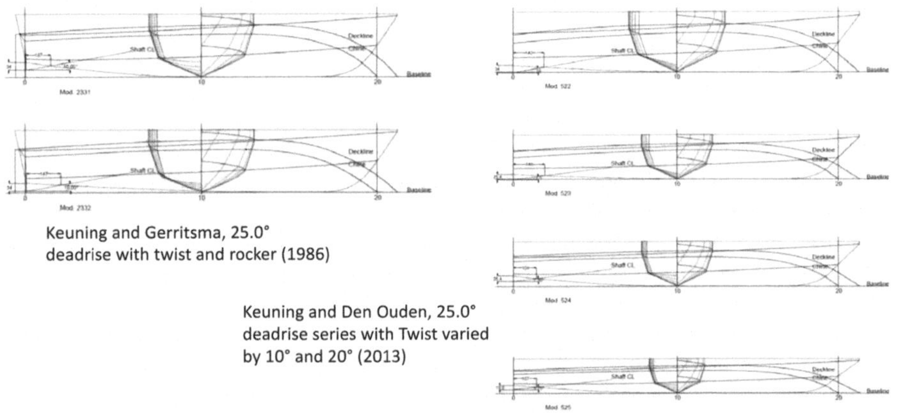

Fig. 4.33 Delft Series Hull forms twist and rocker

deadrise (ε) and a buttock angle (γ) measured at the transom, see Fig. 4.33. Carrying out tests to measure the effect of these parameters on resistance, heave, and trim compared to the initial DSDS tests it was possible to develop a polynomial equation to adjust the basic hull results for the effect of twist and rocker.

Later, in 1986, four new models were built, this time using 30-degree deadrise, and an identical series of tests was carried out. This work thus generated a database encompassing 12.5-, 25-, and 30-degree deadrise, together with the additional tests for twist and rocker.

During the late 1980s, this data was used to develop polynomial expressions for resistance, heave, and trim initially known as the "Planing Hull Forms" or PHF program. This was published by Keuning, Gerritsma, and Terwisga in 1993 [32].

In 1996, this work was followed with a further series of four models, this time with deadrise of 19°, thus filling the gap between 12.5° and 25°.

In 2013, the work on twist and rocker was extended by another series of four models covering the full range of L/B tested for the baseline data sets.

Table 4.8 below presents key data for the parent hulls in the series for each deadrise together with that for the twist and rocker work. For comparison, the parent for Series 62 with 12.5° deadrise is included (model 46671). Reference [30] tabulates the data for all the 24 different models tested.

In the table, the L/B one may consider appropriate for planing calculation following the Savitsky approach at the high-speed planing range needs careful attention. Using the maximum beam at the chines, Bpx, would be appropriate if the aft hull was prismatic. This series has a reduced transom beam, Bpt, and so an average breadth for the planing surface is needed, hence calculation of Bpa.

The Delft towing tank has dimensions 142 × 4.25 × 2.5 m length, breadth, and water depth. Maximum carriage speed is 7 m/s. Models were towed at their LCG and a height based on the intersection of the assumed shaft line and the LCG. Models were constructed in GRP without gel coat so as to be transparent for photography of the dynamic waterline and wetted area determination. Models were constrained in

Table 4.9 DSDS Model data

Model		46671	362	188	251	2331	522
β	deg	12.5	19.0	25.0	30.0	25.0	25.0
ε, twist	deg	0.0	0.0	0.0	0.0	20.0	10.0
γ, rocker	deg	0.0	0.0	0.0	0.0	−4.93	−2.2
Shaft Angle	deg	10.0	10.0	10.0	10.0	10.0	10.0
Lp	m	2.438	1.5	1.5	1.5	1.5	1.5
Bpx	m	0.596	0.299	0.299	0.367	0.367	0.367
Lp/Bpx	–	4.09	4.09	4.09	4.09	4.09	4.09
Bpt	m	0.381	0.236	0.236	0.236	0.320	0.276
Bpt/Bpx	–	0.640	0.643	0.640	0.640	0.872	0.752
Bpa (Ap/Lp)	m	0.487	0.299	0.30	0.30	0.306	0.312
Lp/Bpa	–	5.006	5.011	5.00	5.00	4.90	4.808
Ap	m^2	1.189	0.449	0.450	0.450	0.459	0.468
Cap	%Lp	48.80	48.773	48.80	48.80	48.80	47.467
$Ap/\nabla^{2/3}$	–	4.0–8.5	4.0–8.5	4.0–8.5	4.0–8.5	4.0–8.5	4.0–8.5
LCG	%Lp	38–50	42–50	38–50	42–50	38–50	38–50
Fn_∇		0–6	0–3	0–3	0–3	0–3	0–3

roll, yaw, surge, and sway, and allowed freedom in pitch and heave. Resistance and the two motions were measured during the runs. Using the wetted area data and the ITTC 1957 procedure, frictional resistance was calculated and so "residual" resistance determined that included wavemaking drag (Table 4.9).

Figure 4.34 shows example resistance and trim graphs from the twist and rocker series of tests.

Using ITTC 1957 and Froude scaling, estimation of the resistance, trim, and CG heave can be predicted for full-scale vessels. The authors used the polynomials to test such prediction against direct model test data for a 20 m search and rescue craft with parallel deadrise, and a larger 35 m patrol vessel having 15° of twist and 3° of rocker. Within the speed regime of Fn_∇ 1–3 the results correlated well.

The authors have updated the PHF program to include all model tests, and so polynomial expressions have been established covering this full range. The data has been made available as a database at Delft University of Technology. Students and engineers can obtain access by email request to J.G.denOuiden@tudelft.nl.

To recap, from Clement and Blount (Series 62) directly their observations were:

- Narrow transom improves performance in quartering seas and reduces tendency to broach but limits engine space aft.
- Possibly best to have constant deadrise aft of amidships to minimize drag.
- Slamming can be improved by raising chine at bow (higher deadrise)
- Deadrise as important as L/B for seagoing performance

Keuning and his colleagues at Delft University of Technology have completed a consistent set of model tests and analysis to enable resistance and trim evaluation

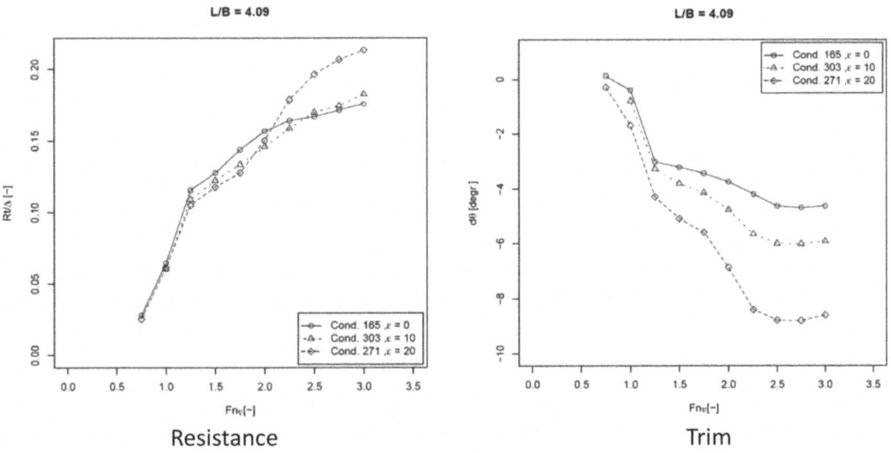

Fig. 4.34 Example resistance and trim graphs from the twist and rocker tests

including the effects of twist and rocker for a significant range of hull shapes and variations to a base design. The baseline plan form does include taper to the breadth toward the transom and a bow form more typical of utility craft rather than fast attack or racing hull forms, but this may be considered an attribute rather than out of date.

4.5 Analytical Equations Derived from Model Testing

Regression analysis has been used to develop polynomial equations for the prediction of resistance and trim of hard chine vessels for many of the model test series introduced in Sect. 4.3. This work has been focused on prediction for the speeds in the range up to $Fn_\nabla = 3.0$, where a hull is only partly supported by dynamic lift. The models available include:

- Series 62 and DSDS—developed by Delft University
- US Navy Academy—developed by the Academy
- Naples Series—developed by Naples University

Professor Radojcic of Belgrade University has also carried out extensive work analyzing several test series including Series 62 and 65, USCG and TUN series, and the Naples Series, documented in refs [24, 33, 34]. In addition, he has written a textbook gathering much of this material together [15].

In 1973, J A Mercier and D Savitsky published a regression analysis on a large number of model tests that included both round bilge and hard chine craft and a polynomial model for resistance at preplaning speeds (see Chap. 2, Sect. 2.3.1 and associated reference [27] of that chapter). This analysis was meant for preliminary assessment before hull characteristics were selected in detail. It may therefore be used now as a first pass, followed by the use of one or more of the other models discussed here to obtain a more detailed assessment of parameter variations, before model testing or/and CFD optimization.

We should mention that calm water performance is normally an initial focus while hull form optimization also requires attention to seakeeping. This may be considered a second round of the design spiral. Improving motions will likely need adjustments to some form parameters, so at least in the speed range up to about $Fn_\nabla = 3.5$ the consequences can be assessed by rerun of the appropriate polynomial model. Seakeeping is discussed in Chap. 5.

Issues to consider when using these polynomial models are the range of application, normally related to $L/\nabla^{1/3}$ and Fn_∇, and the parameters relevant to the polynomial model. We will give an overview here and refer readers to Appendix 1 for details of the models for practical use.

Figure 4.35 shows the ranges covered by the different model tests. The application areas focused on by Prof Radojcic and his team cover much of the parameter space

The Radojcic modeling covers most of the TUNS model space, the lower part of the USCG, and Series 62 at restricted slenderness range. His separate Series 62 model covers the whole slenderness range. The Hard chine hulls that are prismatic, and those with taper to the transom while retaining constant deadrise are thus solidly covered. Delft DSDS model covers a similar speed range and slenderness while also incorporating adjustment expressions for rocker and twist. Hulls with

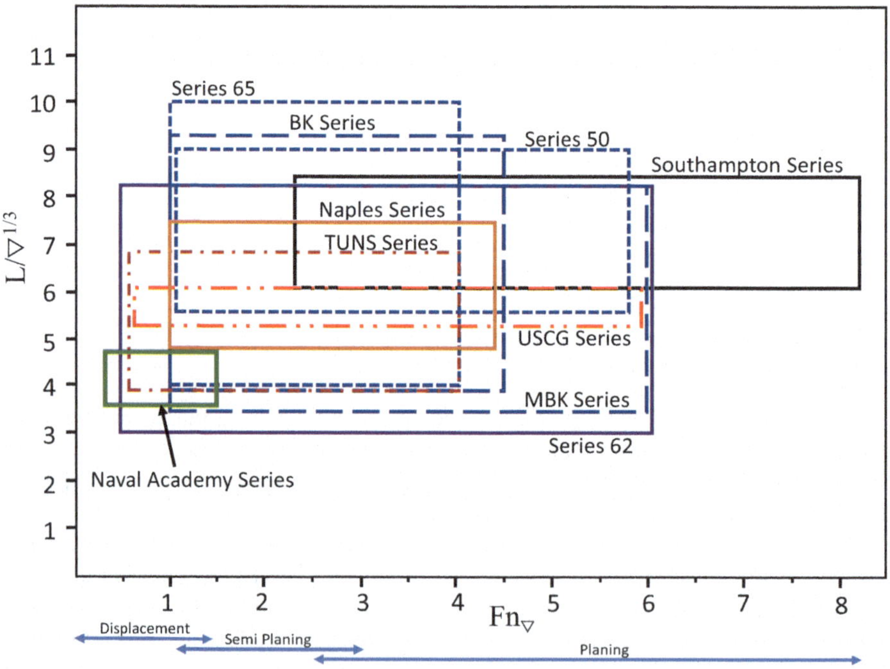

Fig. 4.35 Model tests programmes parameter coverage

twist and rocker have been studied by both Delft University and Naples University and so their models can be investigated for such designs. Finally, Prof Radojcic has produced a model for the Naples series hull form also.

In order to build a predictive equation by regression of model test data the key controlling (independent) parameters need to be assessed, and as far as possible interdependent parameters avoided.

Two key parameters are slenderness ($L/\nabla^{1/3}$) and planing surface aspect ratio (L/B). As speed increases from $Fn_\nabla = 0.5$ to 2.5 initially slenderness controls R/Δ. Above 2.5 R/Δ is primarily controlled by L/B, that is, planing surface aspect ratio, together with deadrise.

Above about 3.5 planing surface loading is important and so this with deadrise and L/B will control wetted surface area and so R/Δ. The speed range between 2.5 and 4.0 is where the relative hollow in the resistance curve will occur and R/Δ will be insensitive to slenderness or planing aspect ratio as a craft adjusts planing angle of attack with speed dependent on the longitudinal CG position.

Thus, L, B, ∇, and β, together with the position of CG are a starting point. An appropriate wetted surface area is also needed, to determine the friction component to add to the prediction of "residual" resistance, based on the friction resistance scaling with Reynolds Number, and the residual resistance scaling with Froude Number. This allows the equations to be developed from the model tests and projections for variable full-scale vessels made.

Radojcic uses $L = Lp$, $B = Bpx$, and $\beta = \beta_{Bpx}$, where Lp is the length over chines, Bpx is the maximum breadth at chines, and β_{Bpx} is the deadrise measured at Bpx. Due to the varied hull forms in different model test series, the dimensions adopted by different organizations also vary somewhat, so it is important that the dimensions are determined carefully for input to a model. We discuss and clarify this for each model in Appendix 1.

4.6 Model Test Issues for Planing Hulls

There are a number of issues to bear in mind concerning model testing and projection to full scale. There is no doubt that direct model testing is still the most credible method for determining drag of planing hulls. The scaling rules and drag component conversion method for calculating drag of planing hulls from model test results and scaling to full-scale craft is similar to that for conventional ships, that is, according to the Froude and Reynolds similarity, with pressure drag in terms of cubic, and friction in terms of square power relations. Here, we discuss some features related to the model tests and scaling of planing hulls that need special attention, as follows:

First, special attention must be paid on the laws of similarity of the dynamic trim for model and full-scale craft, which might be not so important on displacement ships since it doesn't change significantly between static and moving modes. The trim of a planing hull is rather different between the hull borne and planing mode, and this has significant influence on drag. Therefore, keeping the similarity of dynamic trim for both model and full-scale craft is an important issue.

Second is measurement of the wetted area, which varies with speed and thus dynamic trim, for calculating the friction drag. To accurately determine the wetted area, we have to consider *viscous flow effects* and so we will address this together with wetted area assessment.

Similarity

The similarity of value and position of acting forces defines the similarity of "dynamic trim," and such forces include gravity, hydrodynamic (and static pressure), and towing force, as follows:

1. Gravity similarity: Including similarity of gravity force and its location, for both longitudinal and vertical direction, is rather different from a displacement ship, for which only simulation in the longitudinal direction is needed. For a planing hull, the running trimming angle makes gravity not perpendicular to the baseline and change in CG height creates a different moment about the transom, as shown in Fig. 4.36.
2. In general, the hydrodynamic similarity of planing hulls can be obtained so long as geometric and dynamic trim similarities are available. However, attention must be paid to two issues: first, the model should not be too small; the model beam should not less than 30 cm, to minimize the effect of surface tension. Secondly, the model should be equipped with appendages as far as possible since these will influence the running attitude of the planing hull.
3. The model thrust line should be kept the same as the full-scale craft; however, this may be difficult, due to changes in trim angle and so thrust (towing) line with speed. Therefore, the direction of the towing line must be changed with speed, to make the angle of the towing line with the horizontal plane equal to the inclination angle of thrust plus trimming angle, as shown in Fig. 4.37. However, the dynamic trim angle is unknown before the test; therefore, some special towing equipment may be designed for the towing tank.

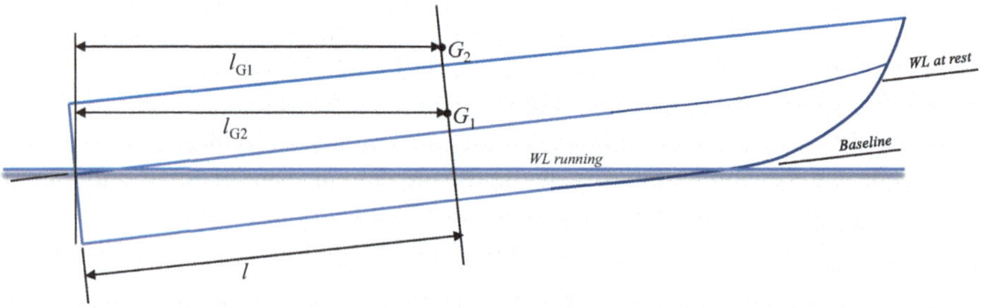

Fig. 4.36 Different moment due to different CG height of planing hull in operation

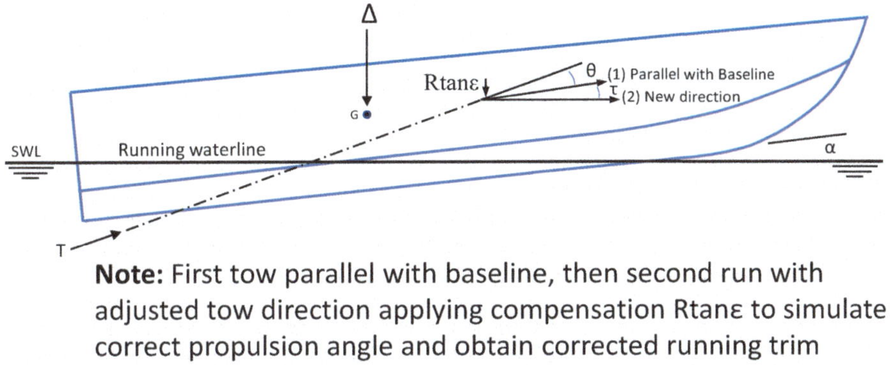

Fig. 4.37 Towing test method

In case of such special equipment not being available, the following method may be useful: selecting a point O on the towing line and carrying out the towing test with horizontal tow line. Using the measured trim angle τ and drag R, a thrust unloading weight R tan ε, where $\varepsilon = \theta + \tau$, θ, θ is angle between thrust line and base line of model, is calculated and applied on a second towing run, thus matching thrust line with the full-scale craft. If the trim angle changes significantly, additional iterations may be made to obtain the desired level of accuracy.

The model should be free to heave and pitch, and the trimming angle and wetted surface at every speed in tests must be measured. To measure the wetted surface by camera, it is better to make the model with transparent bottom made of plastic. However, errors may be introduced due to the presence of spray, so some coating material can be applied on the model surface, and the wetted area can be measured according to the washing off of the coating. It is also possible to paint a measurement grid on the bottom and take underwater photographs from the bottom. The measurement of wetted area still is a critical issue in model tests of planing hulls due to the forward flow from the stagnation line, and spray generation not scaling. One may expect therefore that model assessments might overestimate "spray wetted area" while wetting and so "spray drag" might also be higher than at full scale since full-scale spray impacting a full-scale craft will have a lower volumetric density.

Fixing the tow point at CG is still a generally applied method in model towing tests for planing hulls, based upon two assumptions: first, not considering the effect of running attitude, and second, the trimming is about the transverse axis at CG. The first assumption is obviously different from practice; however, considering the testing is at the initial design phase and the position of the thrust line still can't be determined precisely, and without appendages put on the model, the moment due to the appendages usually will compensate the moment caused by thrust. With respect to the second assumption, it can be considered as reasonable and accepted in general.

However, it must be noted that running attitude and drag of full-scale craft may be different from the model test. For this reason, some equipment for adjusting the trimming angle, such as stern wedge plate, stern flap, and stern interceptor may be mounted on planing hulls for maintaining and optimum trimming angle, and also for taking off smoothly and quickly.

Viscous Flow for Planing Hulls

The friction forces acting on a planing plate are shown in Fig. 3.3 in Chap. 3. Friction force F is parallel to the plate surface and its component along the direction of motion is friction drag, and with angle of attack α, thus $R_f = F \cos \alpha$.

The normal component of friction acting downwards partly reduces the hydrodynamic lift; however, it can be neglected due to its small value, at a small angle of attack α.

Strictly speaking, the drag due to viscous pressure is also present; however, it is small for high-speed planing hulls and can be included in residual drag. According to (4.3), m velocity on friction, the measurement of friction on a planing flat plate, for both large angle of attack ($\alpha = 12°$), small ($\alpha = 2 - 4°$) were carried out in ref [35], as shown in Figs. 4.38 and 4.39, where solid lines show the calculation results using Prandtl-Schlichting formula, and dotted lines show the calculated results considering variable velocity, and the dots are test results.

It can be shown that in the case of a small angle of attack ($\alpha = 2 - 4°$), the test results agree with Prandtl-Schlichting calculation results with the tests a little bit lower, suggesting the effect of nonuniform velocity is small. Residual drag $R_r = (P + P_0) \sin \alpha$, and the viscous pressure drag has been counted into pressure difference drag. Meanwhile $(P + P_0) \cos \alpha = \Delta$, the craft weight.

There is no difference between a planing hull and conventional displacement ship for calculation of the friction drag, that is, the calculation method using a flat plate in turbulent flow as the "Equivalent flat plate" to the actual hull surface, S, can be used for the friction calculation of planing hulls, that is,:

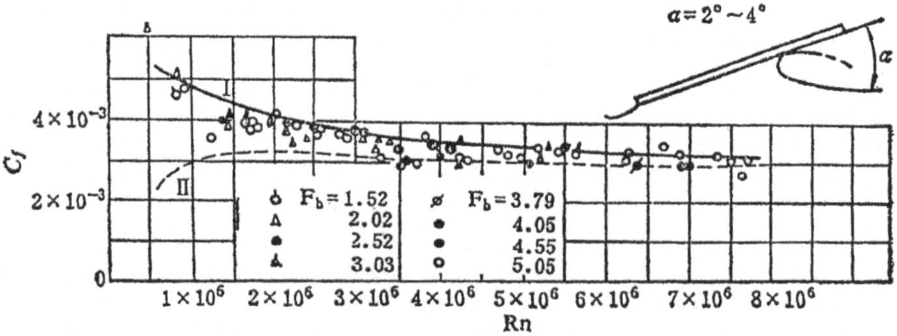

Fig. 4.38 Friction coefficients of planing flat plate with small angle of attack

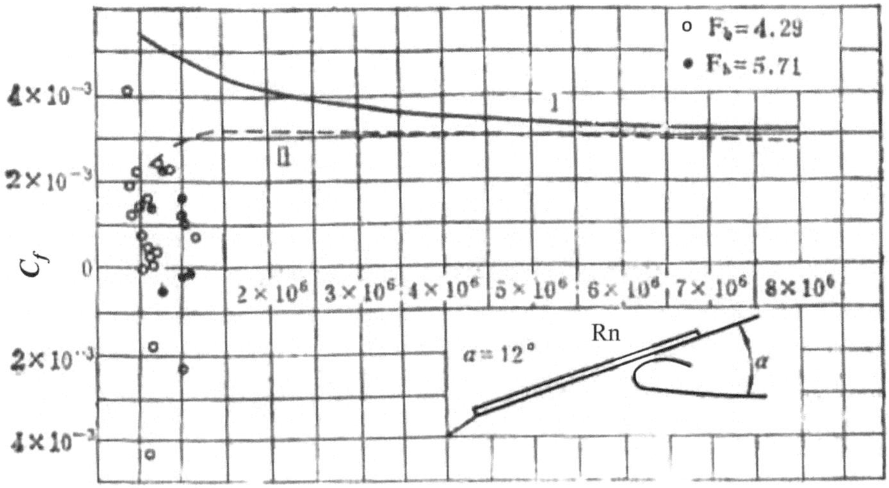

Fig. 4.39 Friction coefficients of planing flat plate with large angle of attack

$$R_f = C_f \frac{1}{2} \rho V_m^2 S \tag{4.23}$$

The friction drag coefficient $C_f = f(R_n)$ can be obtained according to the Prandtl-Schlichting formula or the ITTC formula issued in 1957.

With respect to the value V_m in (4.23), when calculating the friction drag of a flat plate by means of boundary layer theory, this can be considered as unchanged along the plate, since no pressure gradient is available there; however, for a plate with an angle of attack, the pressure along the plate varies so velocity V_m in (4.23) has to be taken as an average.

However, in the case of a large angle of attack ($\alpha = 12°$), as in Fig. 4.39, the test results are significantly lower than the calculations, and in some cases, even negative values appeared indicating the formation of a large area of reverse flow forward of the stagnation point, which moves downstream.

The calculation method using a friction coefficient equation for a flat plate in turbulent flow cannot be used for large angles of attack but is suitable to be used for small angles of attack. In general, for planing hulls will have an angle of attack between 2° and 6°.

There is also the issue of average flow velocity along the plate surface. For planing hulls with deadrise, Savitsky gave the following relationship between average speed V_m and incoming speed V [4]:

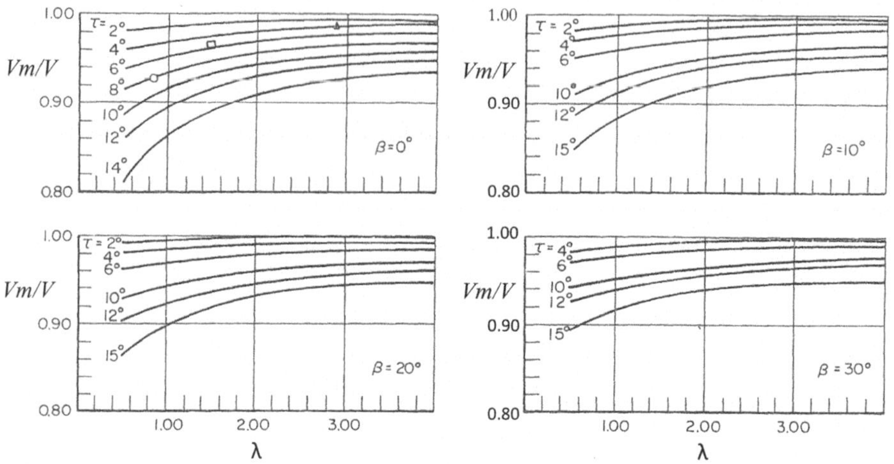

Fig. 4.40 Average flow speed under the bottom of planing surface. (Fig. 14 from Savitsky 1964—take new image)

$$\frac{V_m}{V} = \left[1 - \frac{0.012\lambda^{1/2}\alpha^{1.1} - 0.0065\beta\left(0.012\lambda^{1/2}\alpha^{1.1}\right)^{0.6}}{\lambda\cos\alpha}\right]^{1/2} \tag{4.24}$$

where α is the angle of attack and β is the deadrise angle. The equation can be presented in graphical form, as shown in Fig. 4.40 as presented in Ref [4]. The average velocity under the bottom of a planing flat plate (when $\beta = 0$) is given in the figure, and it can be seen that the average flow speed V_m is very close to the incoming flow speed V when angle of attack α is small, therefore the effect can be neglected. However, at a larger angle of attack, the effect should be considered, as shown in Fig. 4.39.

Wetted Area

The wetted area for calculating the friction drag of the planing surface with transverse deadrise should be the actual area, not the projected area on the horizontal plane. The true wetted area for trapezoidal planing surfaces is:

$$S_\beta = \frac{\lambda b^2}{\cos\beta} \tag{4.25}$$

The wetted length at the keel is different from that at the chine, and the average should be taken, as shown in Fig. 4.41, where O is the intersection of the keel and still water plane. C is the intersection of the still water plane with the chine.

The additional wetted region forward to the stagnation line approximately at OD is formed in front of line OC during planing, and this area should be included when

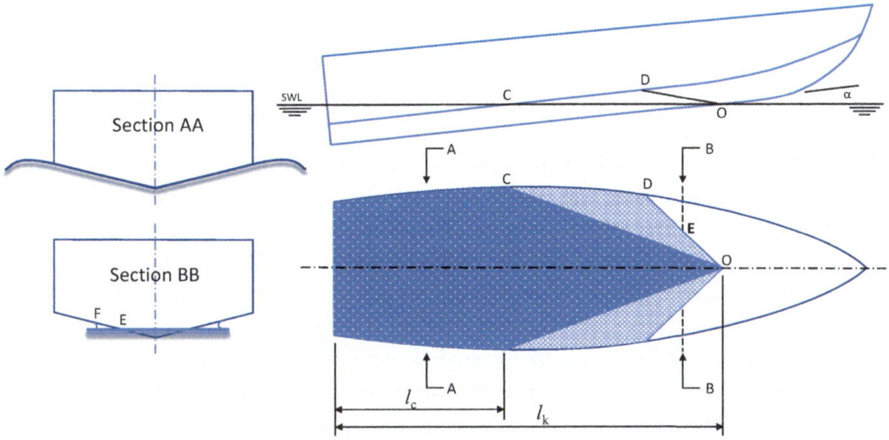

Fig. 4.41 Wetted area of planing surface with transverse deadrise angle

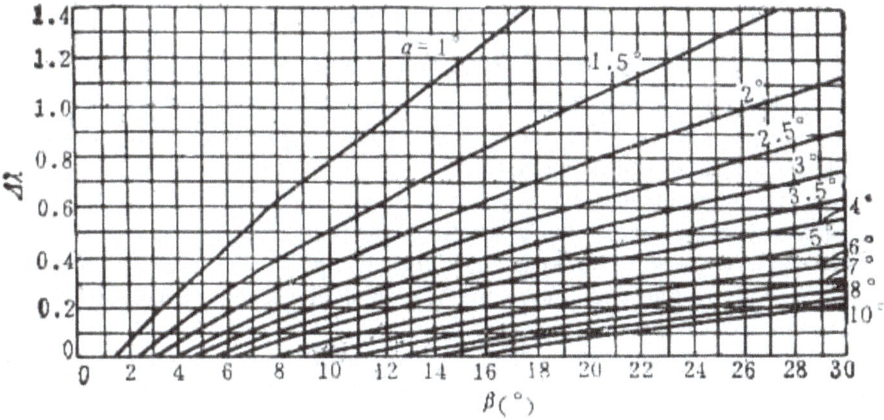

Fig. 4.42 Increment of length/beam ratio due to spray friction

calculating the wetted length-to-beam ratio and friction drag, which should be calculated equivalently into the additional wetted beam ratio $\delta\lambda$

Thus, $\lambda + \delta\lambda =$ (wetted area after OD)/(beam of craft). It may be noted that $\delta\lambda$ will change with deadrise and trim angle, as shown in Fig. 4.42 [4].

In determining the wetted area from model tests, the additional wetted length at the chine is the length between D and C and this cannot be measured, so $\delta\lambda$ cannot be determined from Fig. 4.42. However, P. Ward Brown [8] noted that in practice the wetted line at the leading edge is not a straight line but a curve protruding forward. When calculating the average wetted length, the additional value due to the protruding curve that should be included is 0.03 added to the average wetted length/beam ratio, as:

$$\lambda = \frac{1}{2}(\lambda_k + \lambda_c) + 0.03 \tag{4.26}$$

For the viscous flow condition on the bottom of a planing plate, the leading edge of the wetted area is not the intersection of the plate with the still water plane, but ahead at the spray root line OD, as shown in Fig. 4.41. For a hull with deadrise point O remains on the still water line, and as the position considered moves out toward the chine the spray root line moves forward. Therefore, taking the area after OC line for calculating lift is unsuitable, and for convenience, the test wetted area including spray root region, line OD is recommended in general for the calculation of lift and drag of the planing hulls since the curvature of this front line is small. However, it must be stressed that the issues mentioned above related to viscous flow still affect the calculations due to the uncertainty in exact dimensions and pressure in the region forward of the stagnation line.

4.7 Planing Hulls with Steps

4.7.1 Characteristics of Planing Hulls with Steps

To enhance planing efficiency one or more shallow steps can be made across the planing surface, as shown in Fig. 4.43.

For a planing hull with a step, fluid clears the craft bottom at the step making an air cavity and a water surface that reattaches to the bottom at a certain distance after the step and wets the hull to the stern edge. Thus, two planing surfaces are generated, on the forebody and aft body, respectively, with shorter planing length and higher aspect ratio that enhances the planing efficiency.

High planing efficiency means higher lift with lower drag, and if high lift can be obtained with a small angle of attack, it reduces pressure difference drag. Additionally, through higher lift efficiency, the step reduces the wetted surface, and therefore, the frictional drag is reduced. The main advantage of a stepped planing hull is its fine drag characteristics and lower variation of dynamic trim. Figure 4.44 shows the comparison of both drag and trimming curves of a planing hull with step (b) and conventional planing hull (a).

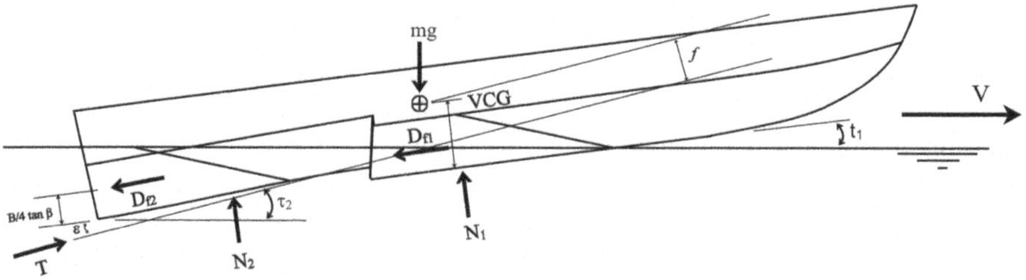

Fig. 4.43 Profile of planing hull with single step

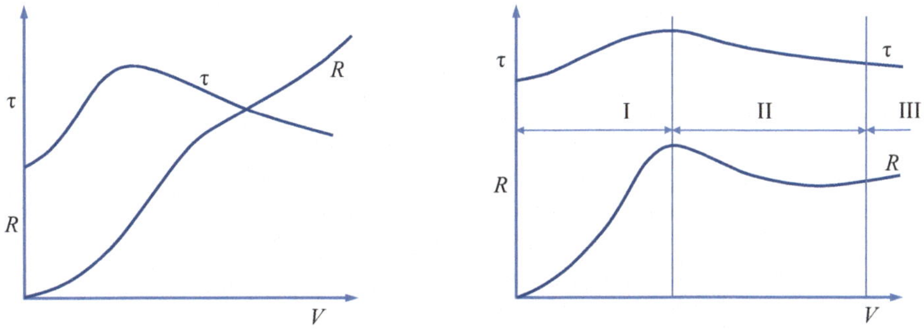

Fig. 4.44 Comparison of both drag and trim curves of planing hull with step and conventional planing hull without a step (**a**) on the left, and a stepped hull (**b**) on the right

From Fig. 4.44, one can see that at high speed the drag curve of a planing hull with a step is flatter than that of a conventional planing hull, and the trim angle is also lower. The characteristics of a stepped planing hull with speed can be divided into three stages:

First stage: In displacement mode to hump speed. Drag and trim increases with speed. The increase of drag is greater than conventional planing hulls due to turbulence behind the step, depending on the height of the step.

Second stage: From drag peak to drag trough. In this stage, the angle of attack is almost unchanged and so the pressure difference drag, and the lift coefficient, are unchanged. Since the lift (weight) is constant, and according to $L = C_L 0.5\rho V^2 S$, the wetted area should decrease with square power of increased speed according to (4.26), thus friction drag also remains unchanged if the effect of Reynolds number on the friction coefficient is neglected. So, in this stage, drag stays unchanged with speed, and the craft planes on the full width of two planing surfaces.

Third stage: Increasing speed after the drag trough: Since the wetted area decreases continuously with increase in speed, the planing surface length at chines may reduce until the leading edge intersects the step or/and transom and results in triangular planing areas commonly called a "chines-dry" situation. This will depend on the deadrise of the craft bottom and position and height of the step. A planing hull without spray strakes on the lower surface may experience unstable flow in this condition and should be designed to operate in 2nd stage with almost unchanged trim angle and drag, and planing on full beam of the craft. Spray strakes along the bottom surface can guide flow so as to maintain stable operation. The spray strakes perform a similar function to the chines at high speed.

4.7.2 Estimation of Drag of Stepped Planing Hulls

Drag estimation of the forward and aft planing surfaces would ideally be carried out separately. However, it is very difficult to do so due to the interaction of fluid deformation from the forward step on the aft planing surface that is difficult to determine independently.

At planing speed, the water flow should separate cleanly from the step. The turbulent reentrant flow behind the step at low speed will allow water vapor bubbles to be generated and eventually fill the space as true planing starts.

At the keel, there will be a static pressure head. Once clear of the step the water surface will rise toward static water level due to this pressure head unless the space behind the step is pressured. Normally the step extends to the chine and the pressure head at the chine is reduced close to atmospheric pressure. There is thus a gradient between chine and keel.

The water surface profile will follow a parabolic shape in the longitudinal direction. The same as happens behind the transom of a planing craft. In the case of a transom the flow not only rises to SWL but the energy available allows the flow to rise above it, forming another parabolic shape. At extreme speed, the energy may be so high that the flow breaks up forming what is referred to as a "rooster tail."

Behind a step underneath a stepped planing hull, the water surface will meet the rear planing surface some way back. At the keel line, the parabolic curve will be tightest while at the chine there will be little or no static pressure and so the curve will be almost straight.

Depending on the deadrise one can see that the front line of wetting for the rear planing surface will be a vee shape if the step height is constant across the beam and the step is also straight across under a hull with deadrise.

Following an early assessment of the phenomena, [36] Savitsky carried out testing and analysis with Morabito [37] to determine a relationship for the warped curved fluid surface behind a step. This work was in turn followed up by Svahn [38] who carried out a correlation exercise for drag and trim of a full-scale stepped planing vessel using these latest relationships to build an iterative model. The results were encouraging.

Yun and Bliault have used this work to build a similar model and extend this to air cavity hull forms [39].[2]

Similar to a single surface planing craft, at a given speed it is necessary to determine the wetted surface area to identify the drag, and the drag force will affect the trim and so the wetted areas. An iterative approach is necessary similar to that for a single planing surface hull but taking account of the two linked surfaces.

For a stepped hull it is useful to first make a simple estimate for the balance of vertical forces. This can then be used as a starting point for the assessment of drag and trim using the methods adopted in [37, 38]. Flow Chart 4.2 below shows the flow chart for this analysis. The parameters are as shown in Fig. 4.45.

It should be noted that the formulae have been derived empirically from model test observations rather than from pure analysis and so have a limited parameter range of applicability. Further, the "free surface" behind the step will not be clean as

[2]Full explanation of the model derived from Svahn's work [38] is given in Chapter 5 of [39] together with spreadsheets for analysis of stepped planing hulls.

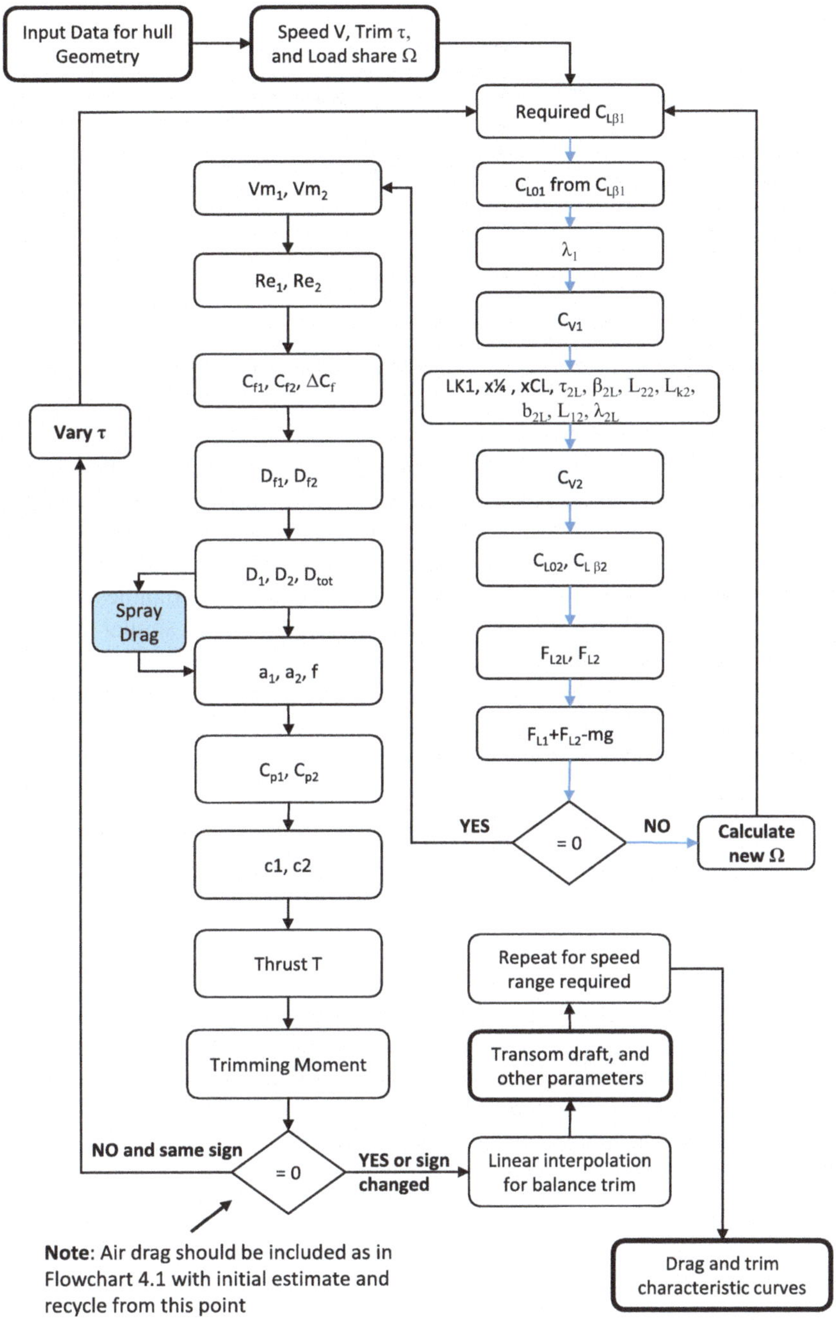

Flow Chart 4.2 Resistance and trim analysis of a stepped planing craft

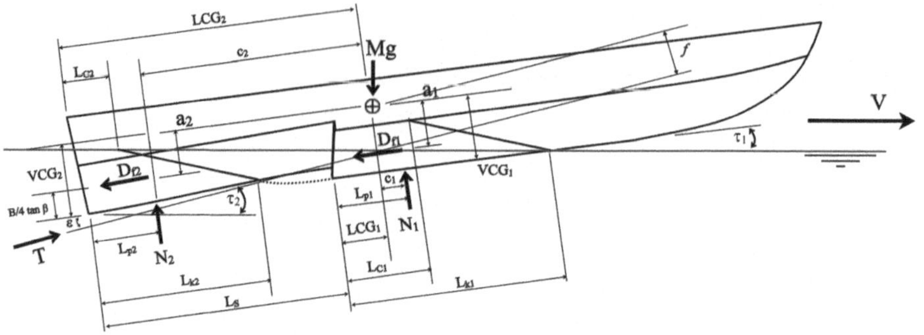

Fig. 4.45 Stepped planing craft geometry. (From [40], Chapter 5, Fig. 5.12)

there will be considerable vapor bubble content close to the surface. The reattached flow at the rear planing surface may have a slightly lower effective density and this in turn will affect the lift. While this effect will have been implicit in the observed results of [37], other geometries may display different characteristics.

For the purposes of our work here, we will assume the uncertainties created by water vapor and spray at the rear planing surface can be assumed similar to those for the forward planing surface.

To begin it is necessary to estimate a distribution of the weight supported by the two planing surfaces. The vertical force and moment balance for the stepped planing hull, see Fig. 4.45, can be written as follows:

$$\Delta = F_{L1} + F_{L2} \qquad \text{and,} \tag{4.27}$$

$$\Delta.l_G = F_{L1}.\left(l_{p1} + l_s\right) + F_{L2}.l_{p2} \qquad (=0 \text{ for equilibrium}) \tag{4.28}$$

where

Δ	displacement of the craft
F_{L1}, F_{L2}	vertical forces of forward and rear supporting surfaces (=$N_1 \cos \tau_1$, $N_2 \cos \tau_2$, vertical components of N_1, N_2 in Fig. 4.45)
l_{p1}, l_{p2}	distance between the center of action of forward and rear supporting surfaces from the step and transom, respectively
l_S	distance from transom to forward step
l_G	distance from transom to LCG

Typically, the forward planing surface supporting lift force is in the region 70–85% of displacement, and the rear supporting surface 15–30%, so for the first approximation, without knowing the centers of action the forward lift force can be estimated as:

$$F_{L1} = 0.75\ \ \Delta.1_G/1_S \tag{4.29}$$

Equation (4.29) can then be used as a first guess and (4.28) to determine an initial equilibrium using estimates for l_{p1} and l_{p2}. This is then input as a starting value for F_{L1} to the procedure to determine the overall equilibrium in Flow Chart 4.2 below.

The dynamic trim angle of a stepped craft can be estimated initially using the following equation (ignoring thrust components for the moment…):

$$\tau = \gamma + 2.5\ C_{FL1} \tag{4.30}$$

where:

$C_{FL1} = F_{L1}/(\rho \quad g \quad b_1^3)$	static load coefficient for the forward planing surface
b_1	beam of the forward planing surface
τ	trim angle of craft baseline at rest
γ	angle between baseline and line from keel at inner base of step to keel at transom

The angles of attack of forward and rear planing surfaces are $\alpha_1 = \tau + \gamma_{m1}$ and $\alpha_2 = \tau + \gamma_{m2}$, where γ_{m1} and γ_{m2} are the angles between the longitudinal mid-cross section with baselines of both forward and rear planing surface, respectively (Fig. 4.45).

It may be noted that flow from the first step will follow γ_{m1} It can be useful to minimize the necessary step height to have a positive value of γ_{m1} and make γ_{m1} and γ_{m2} the same. For steady flow and clean separation, it is important that the forward planing surface is trapezoidal for at least 0.5 B forward of the step.

With given loads Δ_1, Δ_2, and angle of attack of forward and rear planing surface, α_1, α_2, the wetted length/beam ratio λ_1, λ_2, and the lift centers l_{p1}, l_{p2} can be obtained by means of the calculation method for a single planing surface, see procedure and examples in [38, 39].

The calculation procedure can be using any method in Sect. 4.1 to obtain l_{p1}, l_{p2} individually then distribute the load once again for a second approximation. It should be noted that the two planing surfaces are linked geometrically and so equilibrium on the forward surface has to include the turning moment from the aft surface, thus the cycles have to achieve first vertical equilibrium and then moment equilibrium.

Here l_{p1} represents the distance between the lift center of forward planing surface and step and l_{p2} the distance between the lift center of rear planing surface and transom. Equations 4.27 and 4.28 can be used to evaluate equilibrium and can be repeated until the difference of obtained loads for the two next values converges to the required error. At equilibrium, the drag forces can then be determined.

This would be straightforward apart from the fact that the wetted lifting surface behind the step requires determination of the geometry of the cavity before the aft planing surface lift can be determined. This is addressed below, first some general observations regarding the arrangement of steps.

4.7.3 Arrangement of Step for Stepped Planing Hulls

In Eq. 4.28, above it can be seen that the position of the step relative to the center of gravity will affect the loading at the forward planing surface. Clearly, if LCG is aft of the step then more load will be applied to the aft surface. In the example geometry given, if l_G is at the step it will take 75% of the weight as intended. If CG is moved forward by 1/3 of craft length from the step all the load will be taken on the front planing surface.

Depending on the craft speed, it is possible to identify a suitable loading coefficient and thus planing area, in this case, to support 75% of the craft weight. Considering the intended hull overall geometry and knowing the forward planing area and so wetted length determined by the methods in Sect. 4.1 the position of the step can be determined measured aft from the leading edge of the planing surface, and so l_s. If this is less than the initial value used, the forward surface will take more than the assumed load and so either l_G has to be moved back or the step moved forward.

To determine the cavity geometry and lift provided by the aft planing surface it is first needed to select a height for the step and the geometry of the aft part of the hull. Typically, the aft keel line will form a straight line from the base of the step to the transom, the keel of which is in line with the craft baseline, Fig. 4.45. Similarly for the chines. Thus, the aft hull will have a slightly higher angle of attack than the forward hull unless the keel line forward also has an angle of attack relative to the baseline as in Fig. 4.45.

It is possible to introduce more than one step into a planing hull. If two or more are installed the flow across these aft surfaces will be very complex. While the aspect ratio is further increased, the turbulence and aeration will mean that lift may be rather lower and unsteady, making the craft itself unstable. While the forward planing surface will operate steadily the craft may be directionally unstable.

Analysis of multiple steps is very difficult, and such craft have generally been developed by building prototypes, testing, and modification – trial and error.

For a single-step craft, it is important also to correctly select the step arrangement, including the step height and craft length after the step. As discussed above such parameters are related to each other.

A planing craft will typically have its center of static buoyancy 10–20% aft of amidships and its CG in a similar position, this is because once planing the center of lift will be below this point. If we introduce a step, the hull is supported on two surfaces with a gap between them. While the aft surface may support just 30% of the weight its lever arm is significant (acting bow down) compared with the main lifting surface (acting bow up so long as CG is close to the step). In general, a step is located a little bit behind craft CG, and so the front planing surface will give bow-up moment.

The prerequisite for locating the step is that incoming fluid must be separated at the step and the length of air cavity must be shorter than craft length aft of the step to provide the length for the aft planing surface.

For the first prerequisite, as an initial estimate O M Faltinsen [40] suggested that it can be obtained by means of test results by L J Doctors [41] looking at the fluid separation condition for planing hulls with a square transom, and Froude number $Fn_D = V/(gD)^{0.5}$ with respect to draft at transom. This assumes that the pressure in the cavity is atmospheric.

Thus, the condition for fluid separation at the step is:

$$Fn_{Ds} > 2.5 \tag{4.31}$$

where

- V craft speed
- g gravity acceleration
- D_s draft at step considering the rise on planing and trim

For planing hulls, (4.33) can normally be satisfied since volumetric Froude number $Fn_\nabla \geq 3$.

With respect to the length of the air cavity after the step, Faltinsen suggested using an empirical equation provided by Savitsky [37], to obtain the profile of the free water surface at the center line of a planing craft, aft of the transom, as shown in Fig. 4.46, applying this to the step geometry. This is useful to make a first estimate of aft planing surface length.

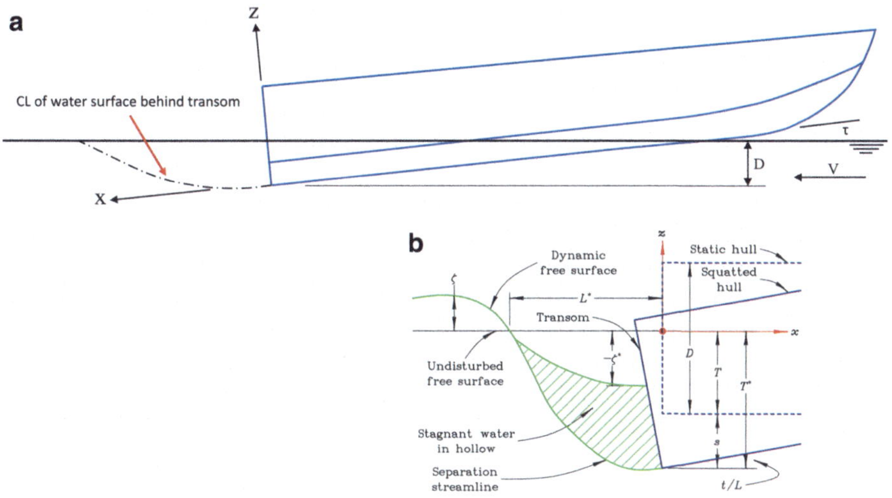

Fig. 4.46 (**a**) Longitudinal central profile of free water surface after the transom of planing hull at speed (after [40], Fig. 9.14). (**b**) Water surface geometry behind transom as it varies with speed. (From reference [41])

The equation describing the profile is:

$$\frac{Z}{B} = C_1 \left(\frac{X}{B}\right)^2 - C_2 \left(\frac{X}{B}\right)^{2.44} + C_3 \left(\frac{X}{B}\right) \quad (4.32)$$

where

$$C_1 = 0.02064 \left[\frac{\tau^{0.7}}{F_B^{0.6}}\right]^2 \quad (4.33)$$

$$C_2 = 0.00448 \left[\frac{\tau^{0.7}}{F_B^{0.6}}\right]^{2.44} \quad (4.34)$$

$$C_3 = 0.0108 \lambda_w \tau^{0.34} \quad (4.35)$$

where

B planing surface beam
τ trim angle, "deg"
F_B Beam Froude number
λ_w Average wetted surface ratio

The application range for this Savitsky empirical equation (4.34) is as follows:

$6 \text{ deg} < \tau < 14 \text{ deg}$,
$3 < F_B < 6$,
$10 \text{ deg} < \beta < 30 \text{ deg}$,
$0.34 (\tan \beta / \tan \tau) < \lambda_W < 3$, and
$0 < X/B < 6$

Figure 4.47 shows example results based on Equation (4.32) for the case where $\tau = 6°$, and $\lambda_W = 2$ prepared by Prof Faltinsen [40, Fig. 9.15]. In the figure, trim angle τ, and average wetted length/beam ratio λ_w are assumed as constant; however, actually they change with F_B. It can be seen from the figure that as Z/B increases X/B increases so that for Z/B = 0.03, X/B will be approximately 0.5. This is probably already longer than may be desirable and so Z/B may typically be set at 0.015 or 0.02B. Remember that this is the cavity length at the keel and the distance will be greater at the chine, depending on the deadrise and shape of the front step.

It may be noted that at the chine the step will create a triangular notch above the elevation of the flow that is separated at the chine forward of the step during planing with dry hull sides. This will allow the cavity to ventilate rather than build up pressure.

The wetted length of the two planing surfaces of a stepped craft is shorter compared with a conventional planing hull without a step. In design condition, the forward planing surface will be area CAOA′C′ shown in Fig. 4.48. The design challenge is to ensure that the wetted chine length CA is long enough so that at high speed the wetted surface front AOA′ does not move back to or beyond CC′ and

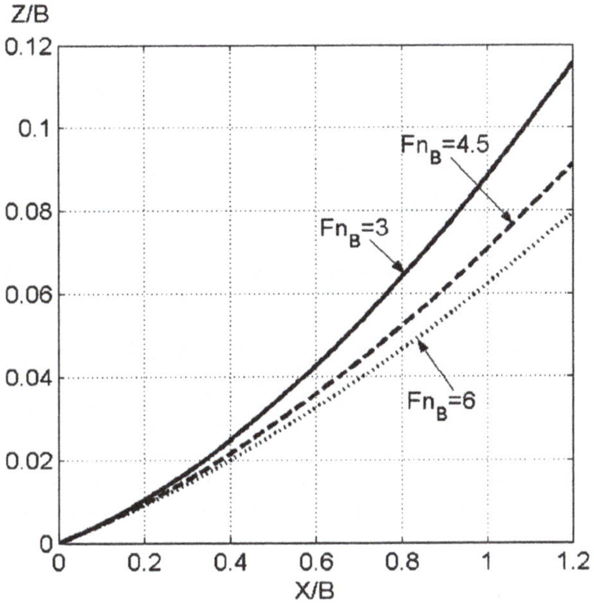

Fig. 4.47 Longitudinal central profile of free surface profile after step of planing hull with prismatic configuration and square stern, based on Savitsky's empirical equation

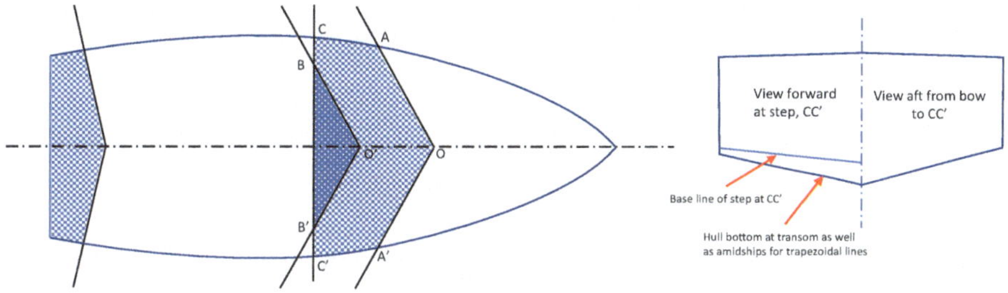

Fig. 4.48 Planing surface on bottom of stepped craft with transverse deadrise

cause unstable flow conditions. This is OK for craft with low deadrise (10–20°). For high-speed Deep Vee craft with a step, it will be necessary to install bottom strakes to stabilize the flow.

In order to avoid instability, the simplest method is to extend the forward planing surface sufficiently to avoid the triangular running operation (zero wetted chine length forward of the step). Another method is to change the plan profile of the step to be a "Reverse V profile," that is, CO′C′, as shown in Fig. 4.49a. This moves the leading edge of the rear planing surface to be closer to a straight line, so as to avoid the rear planing surface wetted area becoming triangular and maintain stable flow.

Recently, a so-called Reverse U plan profile of step has appeared on stepped craft, as shown in Fig. 4.49b, with step line CDD′C′, that is, keeping a straight line in the step center area, but having an inclined line at the outside edges, so as to enhance the planing efficiency compared to "Reverse V" type.

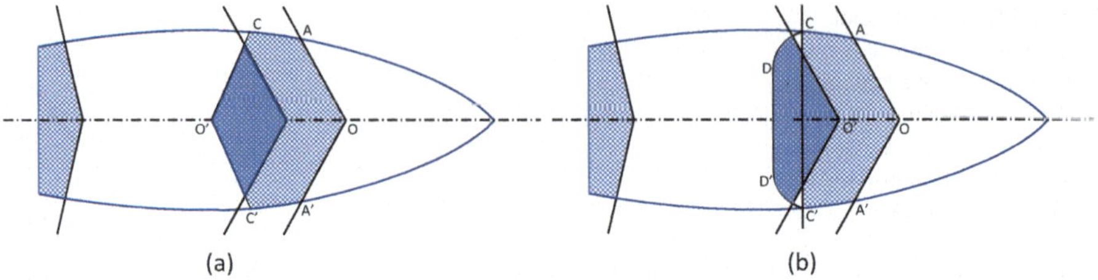

Fig. 4.49 Planing surface on bottom of stepped craft with "Reverse V" (**a**), "Reverse U" plane profile (**b**)

Having made the initial estimates and configuration choices for step position and form, and load sharing it is necessary to proceed through the iterative process depicted in Flowchart 4.2. The equations are detailed in Chapter 5 of [39][3] and [38]. It should be noted that the equations are empirically developed, so care is needed when setting up the craft geometry as the model may not converge to a result.

It may be noted that the test series at Southampton University (Sect. 4.4.8) included testing of a hull with one and two steps incorporated. This testing demonstrated a reduction in resistance compared to the same hull without a step. The same series appeared to indicate that the Savitsky empirical model was validated against the plane hull results. Further work is necessary to provide wide validation for the stepped hull empirical model discussed above.

4.8 Appendage and Air Resistance

Drag components discussed so far are for a bare hull, excluding appendage and air drag. Some references suggest that appendage drag can be estimated as for conventional ships, as a percentage of the total drag of bare hull in terms of the number of propellers, according to statistical data. However, the speed of a planing hull is higher, and the relative size of appendages is larger, so the appendage drag takes a higher percentage than that of conventional ships, in general, 15–30% of total drag. Therefore, the statistical method is unsuitable for high-speed planing hulls.

Since the main components of appendage drag are friction and form drag which are proportional to V^2, the total drag of appendages can also be taken as proportional to V^2. For displacement ships at lower speed, the ratio of appendage drag to bare hull frictional drag is unchanged through the speed range.

At high speeds with high percentage of wave-making drag that increases significantly with speed on a displacement form, the percent of appendage drag is rather different at various speeds.

[3]Reference [39] also presents spreadsheets for this procedure with instructions, available for download from Springer site.

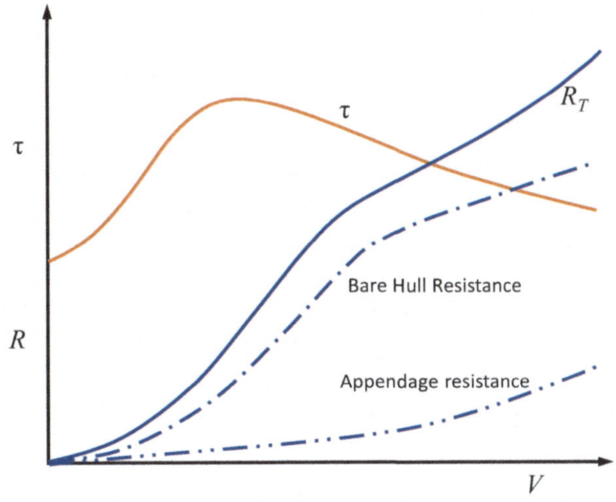

Fig. 4.50 Drag curves of appendage and bare planing hull

For a planing hull after take-off, the bare hull drag versus speed curve is relatively flat in a certain speed region since as speed increases the wetted area to support it decreases. Appendage drag therefore increases with speed more rapidly than that of the bare hull, contributing to a higher percent in total drag, as shown in Fig. 4.50.

Appendage drag can be estimated for planing craft either by means of model tests of craft with full appendages or by calculating drag components separately for individual appendages. If a "Percentage method" is used for estimating appendage drag, a correct prototype and speed have to be selected. Initial estimates will normally be by calculation and so we summarize these below.

4.8.1 Estimation of Appendage Drag Components

At the preliminary design stage, appendage drag can be calculated following semi-empirical equations provided by S F Hoerner in his book "Fluid Dynamic Drag" [42], as follows:

1. *Propulsion Shaft (or Quill Shaft)*

For a propulsion shaft without ventilation, the shaft drag (or quill shaft drag) can be calculated using the following equation:

$$R_{sh} = \frac{\rho}{2} V^2 l_s d_s \left(1.1 \sin^3 \varepsilon + \pi C_f\right) \tag{4.36}$$

where l_s Total length of shaft (or quill shaft), d_s diameter of shaft (or quill shaft), C_f friction coefficient in terms of Reynolds number, in which, incoming flow is assumed to be craft speed, and parallel to craft bottom. In this case Reynolds number with respect to shaft diameter $Rn_d = V.\ d_s/\eta$ ranges from 10^3 to 5.5×10^5

Calculations using this formulation have been correlated with experimental results. If the propulsion shaft is located in an extreme forward position or transiting through the air cavity of a stepped craft, the shaft may be ventilated, then the calculation can be carried out according to Chapter 10 of [42].

2. *Rudder and Bracket*

In general, the section profile of both a rudder and bracket is symmetric and streamlined, and its maximum thickness is close to centre chord. Experiments demonstrate that for such sections the drag depends mostly on plan profile, relative section thickness, and aspect ratio, and less on leading edge configuration and sweep angle. For a rudder and bracket without air ventilation, the drag can be calculated as follows:

Rudder Drag

$$R_R = \frac{\rho}{2} V^2 S_R 2C_f \left[1 + \frac{2t_R}{c_R} + 60\left(\frac{t_R}{c_R}\right)^4\right] \tag{4.37}$$

where:

S_R area of plane profile of rudder
t_R/c_R ratio of rudder max thickness of chord over chord length
C_f Schlichting or ITTC friction coefficient, when $Rn > 10^5$

Strut or Bracket Drag

$$R_{st} = \frac{\rho}{2} V^2 S_s 2C_f \left[1 + \frac{2t_{st}}{c_{st}} + 60\left(\frac{t_{st}}{c_{st}}\right)^4\right] \tag{4.38}$$

where:

S_s plan area of bracket
T_{st}/c_{st} maximum thickness of transverse section and as ratio of bracket chord
C_f Schlichting or ITTC friction coefficient, when $Rn > 10^5$.

For a rudder and bracket piercing the water surface, and generating spray and spray drag, then the spray drag can be calculated as:

$$\Delta R = 0.11 \frac{1}{2} \rho V^2 t_w^2 \tag{4.39}$$

where:

t_w max thickness of rudder or bracket at water surface.

For a high-speed craft with cavitation, then an air-ventilated rudder with wedge or elliptical profile may be used. Meanwhile, the rudder drag can be composed by friction, pressure difference, and spray drag, where friction drag can be calculated by the Schlichting equation in terms of Reynolds number with respect to chord length; coefficient of pressure drag by M. P. Tulin's equation [43], for the section with wedge and parabolic profile with zero cavitation number plus the correction value for cavitation number being not equal to zero.

While cavitation number is equal to zero ($\sigma = 0$), for a wedge form section the coefficient of pressure drag is as follows, [43]:

$$C_{RP}|_{\sigma=0} = \frac{2}{\pi}\left(\frac{t_r}{c_r}\right)^2 \tag{4.40}$$

For parabolic section:

$$C_{RP}|_{\sigma=0} = \frac{\pi}{8}\left(\frac{t_r}{c_r}\right)^2 \tag{4.41}$$

When cavitation number is not equal to zero ($\sigma > 0$), one can use Richard's approximate correction:

$$C_{RP}|_{\sigma>0} = C_{RP}|_{\sigma=0}(1+\sigma) \tag{4.42}$$

where cavitation number:

$$\sigma = \frac{gh_r}{V^2} \tag{4.43}$$

and h_r is wetted height of the rudder

In (4.43), cavitation number is calculated at the centre of wetted height $h_r/2$. Spray drag can be calculated by (4.40).

Total drag coefficient for an air ventilated rudder with wedge section is:

$$C_R = 2C_f + \frac{2}{\pi}\left(\frac{t_r}{c_r}\right)^2\left(1+\frac{gh_r}{V^2}\right) + \frac{0.11t_w^2}{h_r c_r} \tag{4.44}$$

For parabolic section it is:

$$C_R = 2C_f + \frac{\pi}{8}\left(\frac{t_r}{c_r}\right)^2\left(1+\frac{gh_r}{V^2}\right) + \frac{0.11t_w^2}{h_r c_r} \tag{4.45}$$

Equation (4.45) is suitable for aspect ratio, $h_r/c_r > 2.5$. When $h_r/c_r > 2.5$, the drag coefficient will be higher than that for (4.45) and the uprated value can be estimated as:

$$C_{R1} = C_R + \frac{\left(2.5 - \frac{h_r}{c_r}\right)^2}{150} \tag{4.46}$$

where C_R is from (4.45).

The total drag of rudder:

$$R_r = C_R \frac{1}{2} \rho V^2 h_r c_r \tag{4.47}$$

3. *Strut or Bracket Palm*

The strut (or bracket) is mounted on a craft hull generally by a palm, the palm drag can be calculated as:

$$R_P = 0.75 C_{RP} \sqrt[3]{\frac{h_p}{\delta}} b_w h_p \frac{\rho}{2} V^2 \tag{4.48}$$

where

h_{p}	thickness of palm protrusion on hull surface
b_W	width of leading edge of palm
$\delta = 0.016\ l_{\mathrm{xp}}$	thickness of boundary layer
l_{xp}	distance between stagnation line and palm
C_{RP}	drag coefficient, if palm is rectangular with round corner $C_{\mathrm{RP}} = 0.65$

The calculation shows the drag of a palm protruding from a hull surface is significant for a planing craft, and a designer must pay more attention and try to bury it into hull surface as deep as possible or use welded connection to a steel or aluminium hull.

4. *Interference Drag*

In case of many appendages arranged together, some hydrodynamic interference may occur, including the interference between individual appendages and appendages with hull. Investigations demonstrate the interferences between planing surface with rudder and bracket are most apparent, which can be estimated by:

$$\Delta R_1 = \frac{1}{2} \rho V^2 t^2 \left[0.75 \left(\frac{t}{c} \right) - \frac{0.0003}{\left(\frac{t}{c} \right)^2} \right] \tag{4.49}$$

where t/c is the section thickness to chord ratio of the rudder or bracket. Such calculation shows this is a small percentage of total appendage drag.

4.8.2 Air Drag

Air drag due to the portion of hull above the water surface can be estimated by:

$$R_a = C_a \frac{\rho_a}{2} (V \pm V_a)^2 S_x \tag{4.50}$$

where:

ρ_a air density
V_a wind speed, positive for heading wind, following negative
S_x projection area of hull transverse section
C_a air drag coefficient (0.5–0.7 depending on superstructure)

4.9 Influence of Appendages and Propellers on Running Trim

4.9.1 Influence of Appendages

Appendages will not only induce drag, but also change the running trim of a planing craft, due to trimming moments caused by both drag and lift acting on them (mainly from the propulsion shaft with inclination angle with incoming flow). It is not difficult to calculate the moment caused by drag of appendages. For appendages with frictional drag as the main drag component, the center of action can be estimated as the area center. For thick appendages, the drag is mostly due to pressure difference and then the drag center of action is located at maximum thickness of the section perpendicular to incoming flow.

Lift caused by an inclined shaft can be estimated by:

$$L_s = \frac{1}{2} \rho V^2 l_s d_s \left(1.1 \sin^2 \varepsilon \cos \varepsilon\right) \tag{4.51}$$

where

l_s total length of shaft
d_s shaft diameter

For incoming flow parallel to the craft bottom, the lift center can be put at the shaft CG. The running trim can be obtained using the equations for equilibrium of vertical force and longitudinal moment, caused by lift and moment arm.

It is possible the pressure field on the craft bottom may be affected due to the appendages, thus changing the lift acting on the planing surface. However, in general this effect is so small and in addition difficult to estimate, so that it can be ignored.

In order to estimate the influence of appendages on running attitude and performance of planing hulls, model tests with full appendages can be carried out in a towing tank. However, such methods still have not been generally recognized due to small size of model scale appendages introducing viscous scale effects. In addition, at preliminary design stage the propulsion system, which affects the running attitude is still not decided.

Modern planing vessels may well use Z-drive, water jet or surface piercing ventilated propellers for propulsion and in that case appendage evaluation can be left until the basic hull has been evaluated.

4.9.2 Influence of Propellers

Practice demonstrates the influence of propellers on running attitude is far more than appendages. For instance, the British planing fast patrol craft *Brave Borderer*[4] [44] installed originally three gas turbines driving three water propellers with three blades each in full cavitation, to achieve speeds up to 53 knots. However, in order to improve the vibration at stern, when the propellers were changed to five blades and still in full cavitation condition, the tests recorded the craft could only reach 48 knots due to lower running trim angle. Thus, it demonstrated the effects of propeller on the running attitude and performance of planing hulls. The causes of propeller influence on the performance of planing hulls can be outlined as follows:

- Trim angle caused by longitudinal moment of propeller thrust not passing through the CG of the craft.
- In general, the propellers are at angle to the incoming flow generating a "vertical force" along the propeller disc in addition to axial thrust, which influences the equilibrium of craft.
- The propeller local flow speed and pressure will alter the hydrodynamic forces on the planing surface.
- Additional resistance of rudder, behind and in the wake of the propeller, will be generated.

With respect to the "vertical force," acting perpendicularly on propeller disc located in the inclined incoming flow, it can be shown as in Fig. 4.51, that owing to the inclined flow an upward velocity component $V \sin \varepsilon$ is generated by the disc. For the blades located at up and down positions, that is, $\theta = 0$ and π, these velocity components are along the radial direction so as not to influence thrust and moment

[4] Brave Borderer class patrol boat by Vosper Ltd, Portsmouth, Great Britain, see summary at https://en.wikipedia.org/wiki/Brave-class_patrol_boat

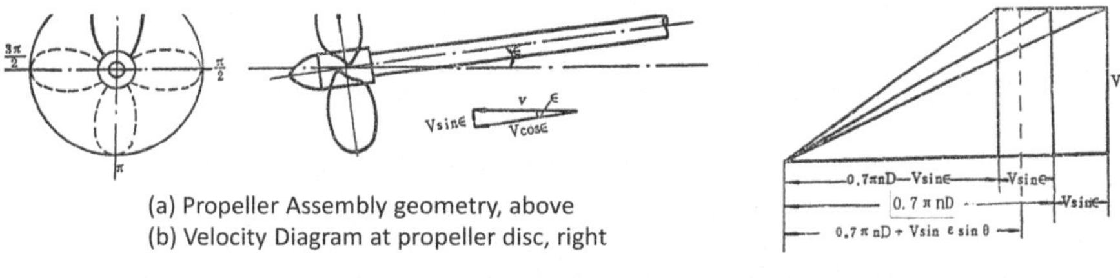

Fig. 4.51 Velocity diagram and so vertical force acting on propeller in inclined flow

on the blades; however, for a blade at $\theta = \pi/2$, the vertical velocity component aligns with the blade tangential velocity so as to increase original tangent velocity, and for a blade at $3\pi/2$, the tangential velocity decreases. Figure 4.51 shows the change of velocity triangle of blades at section of 0.7R with various θ.

At any rotational position of θ, the tangential velocity is $0.7\ \pi\ nD + V \sin \varepsilon \sin \theta$, and during one rotation of the blade the resultant of all tangential forces is a vertical force component and the torsion resultant caused by the tangential force is the propeller moment.

The value of vertical force for a noncavitating propeller can be calculated by means of vortex theory and other simplified calculation methods [45], that is, replacing the blade section at 0.7R to whole blade, and considering all forces concentrated on 0.7R, then using a quasi-steady analysis method, that is, taking tangential velocity at any instant as the velocity at a corresponding stable speed, $n(\theta)$. For various rotational positions of θ, the corresponding quasi-steady rotation speed $n(\theta)$ is different, as follows:

$$0.7\ \pi nD + V \sin \varepsilon \sin\ \theta = 0.7\ \pi Dn(\theta) \tag{4.52}$$

or:

$$\frac{n(\theta)}{n} = \frac{J}{J(\theta)} = 1 + \frac{J \sin \varepsilon \sin \theta}{0.7\pi} \tag{4.53}$$

where:

n	propeller rotation speed
$J = V/nD$	propeller speed of advance
$n(\theta)$	Quasi-steady speed
$J(\theta) = V/n(\theta)D$	Quasi-steady speed of advance

Since the propeller is under inclined incoming flow some errors may be present by using the above-mentioned method. An empirical correction coefficient C can be inserted into (4.51) as follows:

$$\frac{n(\theta)}{n} = \frac{J}{J(\theta)} = 1 + \frac{JC \sin \varepsilon \sin \theta}{0.7\pi} \tag{4.54}$$

Then thrust and moment coefficients $K_T(\theta)$, $K_Q(\theta)$, can be found from the performance curve of the propeller in axial flow, and the thrust and moment of every blade at various θ are:

$$\left.\begin{aligned} T_Z(\theta) &= \frac{1}{Z}\rho n^2(\theta)D^4K_T(\theta) \\ Q_Z(\theta) &= \frac{1}{Z}\rho n^2(\theta)D^5K_Q(\theta) \end{aligned}\right\} \tag{4.55}$$

Considering all forces concentrated at section of 0.7R, the tangential force can be obtained from the torsion moment as:

$$S_z(\theta) = \frac{1}{0.7RZ}\rho n^2(\theta)D^5K_Q(\theta) \tag{4.56}$$

The projection of this tangential force is the vertical force component

$$F_z(\theta) = \frac{1}{0.35Z}\rho n^2(\theta)D^4K_Q(\theta)\sin\theta \tag{4.57}$$

Since the position of blades at every instant is different for different θ the summation of force and moment acting on blades at different instants and θ are the resultant force and moment of the propeller. For a propeller with three blades, spacing should be $\Delta\theta = 2\pi/3$, and the resultant force and moment for three blades are:

$$T(\theta) = \sum_{i=0}^{2} T_z\left(\theta + i\frac{2\pi}{3}\right), \quad Q(\theta) = \sum_{i=0}^{2} Q_z\left(\theta + i\frac{2\pi}{3}\right) \quad \text{and}$$

$$F(\theta) = \sum_{i=0}^{2} F_z\left(\theta + i\frac{2\pi}{3}\right)$$

Thus, the revised values of thrust, moment, and vertical force in every rotation can be obtained.

The average values of force and moment of the propeller can be obtained, so long as to integrate the force and moment from 0 to 2π and divided by 2π. Since the integral value for each blade in each round does not relate to its initial position, the integral value for every blade is equal and the average vertical force is:

$$F = \frac{\rho n^2 D^4}{0.7\pi}\int_0^{2\pi}\left[1 + \frac{JC\sin\varepsilon\sin\theta}{0.7\pi}\right]^2 K_Q(\theta)\sin\theta d\theta \tag{4.58}$$

Corresponding thrust and moment are:

$$\left.\begin{aligned} T &= \frac{\rho n^2 D^4}{2\pi}\int_0^{2\pi}\left[1+\frac{JC\sin\varepsilon\sin\theta}{0.7\pi}\right]^2 K_T(\theta)d\theta \\ Q &= \frac{\rho n^2 D^5}{2\pi}\int_0^{2\pi}\left[1+\frac{JC\sin\varepsilon\sin\theta}{0.7\pi}\right]^2 K_Q(\theta)d\theta \end{aligned}\right\} \quad (4.59)$$

According to test results obtained by Gutsche, the empirical correction factor can be taken as 2.0, and gave good agreement in practice, except for the propellers with low load and large pitch. He also showed that for (4.58) and (4.59), the integral average value could be replaced by the average value of three at three different positions of θ, that is, $\pi/3$, π, $5\pi/3$, and obtained a satisfactory result.

The problem will be troublesome for a cavitating propeller since cavitation at different rotation positions of the propeller blade are different. The cavitation will be most serious at $\theta = \pi/2$, where the tangent velocity is maximum, but the cavitation will be lightest at $\theta = 3\pi/2$, so it is difficult to estimate the effect of cavitation at different positions of a blade.

Another important influence of a propeller on the planing surface is change of pressure field along the planing surface due to the presence of the propeller. For simplicity, the velocity and pressure fields caused by propellers can be investigated separately and combined to provide a nett effect. Such linear superposition is permitted in the case of a potential velocity field without considering the viscous effect. The velocity field around the propeller can be calculated using vortex theory, and we only need the surface velocity of flow due to the propeller.

According to vortex theory, the induced tangential velocity beyond the fluid tube of propeller should be equal to zero. Therefore, there is no tangential velocity increment on the planing surface due to the propeller.

The distribution of axial induced velocity beyond the fluid tube of propeller is shown in Fig. 4.52, which is asymmetric to the propeller disc with the axial induced velocity positive before the propeller disc and negative after the disc, and induced velocity decays rapidly with the distance from the fluid tube.

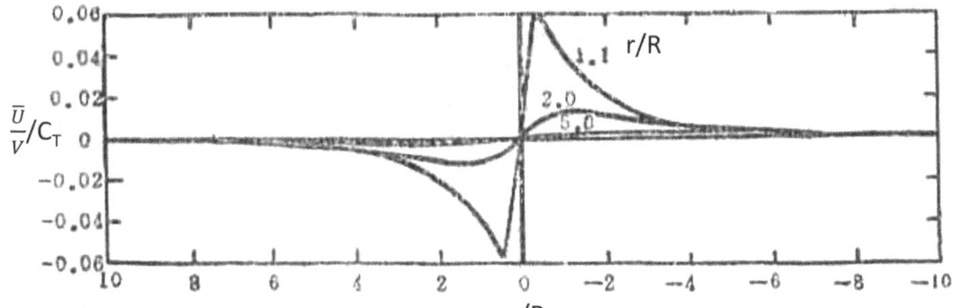

Fig. 4.52 Axial-induced velocity before and after the propeller disc

In Fig. 4.52, the abscissa is the relative longitudinal distance from propeller disc x/R, where R is propeller radius, and taking the disc as the origin the ordinate is relative axial induced velocity $\overline{u}/V$, divided by load coefficient C_T, here $\overline{u}$ is average axial velocity load coefficient induced by propeller incoming speed V, where $C_{\mathrm{T}} = T/[(0.5\rho V^2). (0.25\pi D^2)]$, where T thrust, D propeller diameter, and the distribution of axial induced velocity is function of load coefficient and advance speed of propeller disc.

Figure 4.52 shows the distribution of velocity field for the infinite fluid field. In the case of flow over the planing surface affecting the propeller inflow, in order to satisfy the boundary condition, the fluid potential field must be changed. For simplicity, the planing surface can be taken as a flat plate which is close to the practical condition of a planing hull, and in order to satisfy the flat plate boundary condition, that is, no normal velocity on the plane, a mirror image vortex should be installed with the plate as the mirror, thus the axial induced velocity on the plate must be two times that shown in Fig. 4.52.

Using the velocity distribution, the pressure distribution can be obtained by means of the Bernoulli equation. Since the axial induced velocity on the plate is two times of that in an infinite fluid field, we have:

$$p - p_0 = -\frac{\rho}{2} V^2 \frac{4\overline{u}}{V} \tag{4.60}$$

Integrating the pressure along the plate, the additional suction and pressure before and after the propeller disc caused by axial-induced velocity can be obtained. Since the induced velocities before and after the disc are in opposite directions:

$$\begin{aligned} N &= \int_{-y_1}^{y_2} dy \int_0^x (p - p_0) dx \\ &= -\int_{-y_1}^{y_2} dy \int_0^x \frac{\rho}{2} V^2 \frac{4\overline{u}}{V} dx \end{aligned}$$

Making x, y nondimensional with propeller diameter, R, we have

$$N = -2R^2 \frac{1}{2} \rho V^2 C_T \int_{-\frac{y_1}{R}}^{\frac{y_2}{R}} d\left(\frac{y}{R}\right) \int_0^{\frac{x}{R}} \frac{2\frac{\overline{u}}{V}}{C_T} d\left(\frac{x}{R}\right) \tag{4.61}$$

where y_1 and y_2 are the coordinates of the plate at left and right ends, and the integration from 0 to x/R can be carried out according to Fig. 4.52. Figure 4.53 shows the calculation results, where the abscissa is x/R, and the ordinate is the integration value. The resultant acting on the plate can be obtained by integrating once again along the y direction.

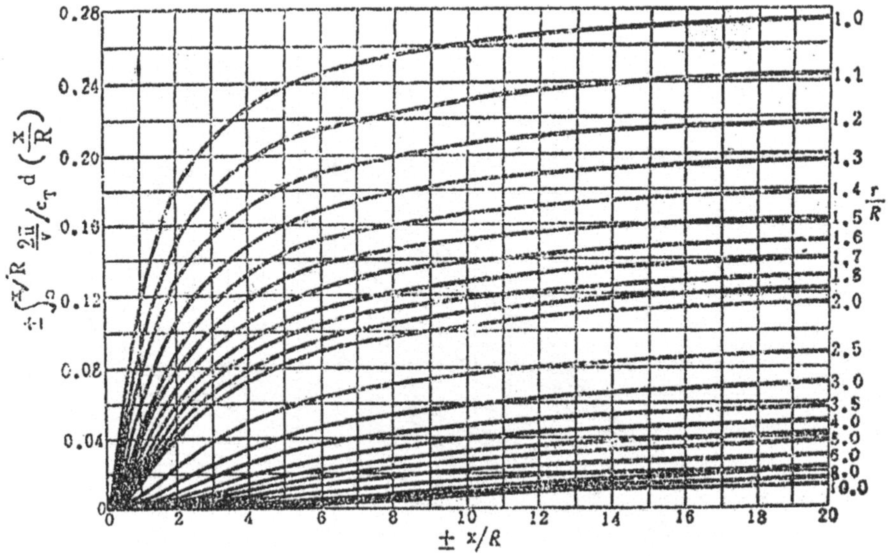

Fig. 4.53 Integral values of pressure coefficients at any point before and after propeller disc, at various *r*/*R*

Fig. 4.54 *y*/*R* diagram

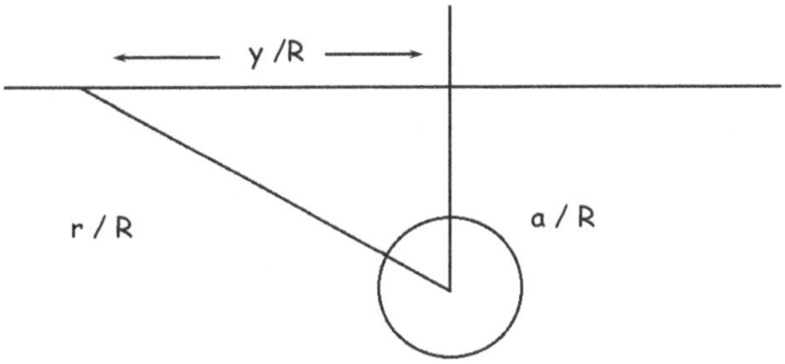

The relation between *y*/*R* and *r*/*R* is shown in Fig. 4.54 and the position of the resultant force can be obtained as:

$$\frac{\overline{x}}{R} = \frac{\int_{-\frac{y_1}{R}}^{\frac{y_2}{R}} d\left(\frac{y}{R}\right) \int_{0}^{\frac{x}{R}} \left(\frac{x}{R}\right) \frac{\frac{\overline{u}}{V}}{C_T} d\left(\frac{x}{R}\right)}{\int_{-\frac{y_1}{R}}^{\frac{y_2}{R}} d\left(\frac{y}{R}\right) \int_{0}^{\frac{x}{R}} \frac{\frac{\overline{u}}{V}}{C_T} d\left(\frac{x}{R}\right)} \tag{4.62}$$

where: *x*/*R* is the distance from propeller disc to pressure center at any relative radius *r*/*R*. In (4.62), the integral has been expressed as shown in Fig. 4.55.

Since propellers are arranged generally at the stern, and the suction before the propeller disc is greater than the push force behind the disc, the suction increases the

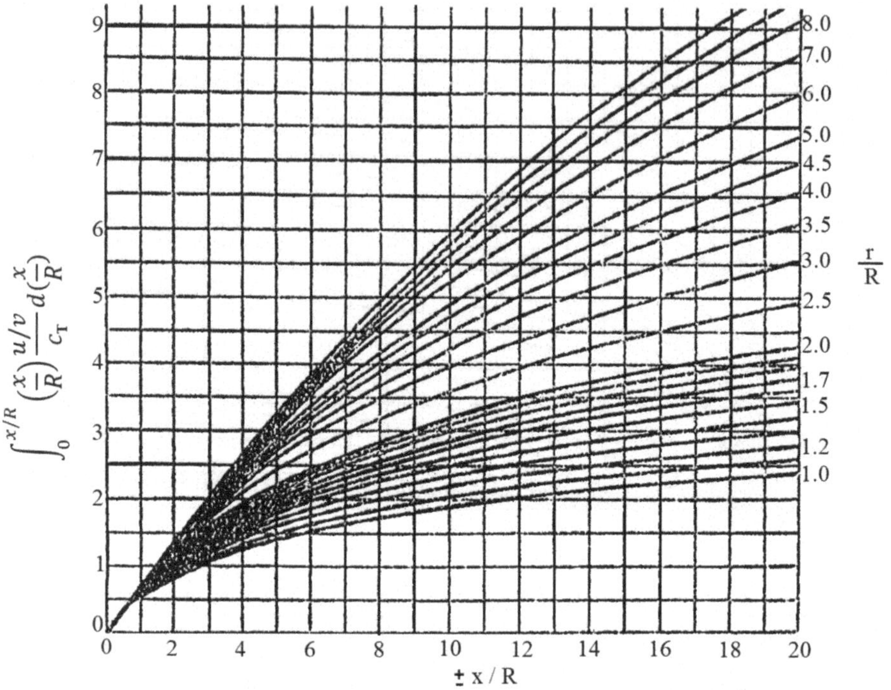

Fig. 4.55 Integral values of pressure moment coefficients along the X direction, at various r/R

load on the craft and also increases the trimming angle. For most craft with practical application, the craft CG is usually located ahead of the optimized position, so that the effects from the two sides mentioned above may counteract each other. Therefore, during initial estimation the influence of propellers on the planing surface pressure cannot be considered.

A rudder installed behind a propeller is in its wake and experiences an increase in drag from two actions. One is caused by axial induced speed of the propeller wake that generates flow speed into the rudder greater than the incoming flow speed, and the other is the distortion of the wake caused by induced tangential velocity which introduces an angle of attack to the rudder causing induced drag.

The rudder drag caused by axial induced velocity can be estimated using average induced wake velocity. The average axial-induced velocity of a whole wake is:

$$\overline{u} = \frac{\int_0^R u dr}{R}$$

Making it nondimensional relative to the incoming flow speed V and also considering the load factor C_T we have:

$$\frac{\overline{u}}{V} = C_T \int_0^{1.0} \frac{\frac{u}{V}}{C_T} d\left(\frac{r}{R}\right) \tag{4.63}$$

Then the increment of rudder drag, caused by axial induced velocity:

$$\begin{aligned} \Delta D_{r1} &= \frac{1}{2}\rho V^2 s_r \left(1 + \frac{\overline{u}}{V}\right)^2 C_{D_r} \\ &= \frac{1}{2}\rho V^2 s_r \left\{ \left[1 + C_T \int_0^{1.0} \frac{u/V}{C_T} d\left(\frac{r}{R}\right)\right]^2 C_{D_r} - C_{D_r} \right\} \end{aligned} \tag{4.64}$$

where:

s_r Rudder area
C_{Dr} Rudder drag coefficient, obtained from equations 4.44, 4.45, or 4.46

Regarding the integration in (4.64), Hough and Ordway, gave calculation results from several propellers as in Fig. 4.56, which can be applied in calculations [45].

In order to calculate the induced drag caused by the distortion of wake, first the distortion angle α has to be obtained, which in terms of axial induced velocity u and tangent induced velocity w, and w in terms of circulation distribution on the blade, as:

$$w = \frac{Z\Gamma(r)}{2\pi r} \tag{4.65}$$

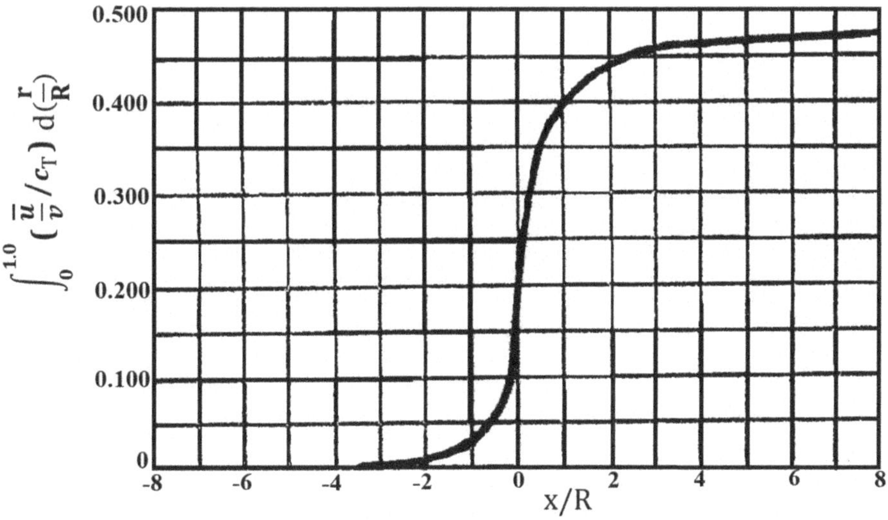

Fig. 4.56 Average axial induced velocity in propeller wake at different longitudinal position

where:

Z number of blades
$\Gamma(r)$ circulation distribution on blade along radius direction

Given $\Gamma(r)$, the induced tangent velocity, w at various radii can be obtained. For simplicity, taking the tangential induced velocity at r/R = 0.6 as the average value in the wake, and for several propellers, Hough and Ordway calculated the tangential induced velocity at r/R = 0.6 as:

$$\frac{\frac{w_{0.6}}{V}}{C_T} = 0.33\,J \tag{4.66}$$

where

V incoming flow speed
C_T load factor of propeller
$J = \frac{V}{nD}$ Speed of advance coefficient

Then the distortion angle at $r/R = 0.60$ is:

$$\tan\alpha = \frac{\frac{\overline{w}}{V}}{1 + \frac{\overline{u}}{V}} = \frac{0.33JC_T}{1 + C_T\int_0^{1.0} \frac{\frac{\overline{u}}{V}}{C_T}\, d\left(\frac{r}{R}\right)} \tag{4.67}$$

The integration in the equation can be obtained by use of Fig. 4.56.
The induced drag coefficient of rudder:

$$C_{D_i} = \frac{C_L^2}{\pi A}$$

The lift coefficient can be obtained approximately by two-dimensional thin wing results:

$$C_L = 2\pi\alpha$$

Then:

$$C_{D_i} = \frac{4\pi}{A}\alpha^2 \tag{4.68}$$

where A aspect ratio of rudder
Induced drag:

$$D_{r_i} = \frac{1}{2}\rho V^2\left[\left(1 + \frac{\overline{u}}{V}\right)^2 + \left(\frac{\overline{w}}{V}\right)^2\right] C_{D_i} s_r \tag{4.69}$$

The tangential average velocity in the above equation can be neglected, and by summation of drag increment caused by both induced drag and axial induced velocity, the total drag increment of rudder caused by propeller is:

$$\Delta D_\Gamma = \frac{1}{2}\rho V^2 s_r \left\{ \left[1 + C_T \int_0^{1.0} \frac{\frac{u}{V}}{C_T} d\left(\frac{r}{R}\right) \right]^2 (C_{D_r} + C_{D_i}) - C_{D_r} \right\} \tag{4.70}$$

Strictly speaking, the force caused by the propeller has to be included in the equilibrium equations. However, this is not generally employed considering practical difficulties and the additional force is just used for checking rudder design.

4.10 Resistance of Planing Craft in Shallow Water

For planing hulls operating at low speed in displacement mode, resistance in shallow water is the same as for conventional displacement ships, thus two speed ranges are relevant to be considered. When depth Froude number $Fn_h < 1$ that is, $V/(gh)^{0.5} < 1$, where h is water depth, this is a sub-critical speed. In this case the shallow water effect increases resistance, with the maximum increment at $Fn_h = 1$, the critical depth Froude number. When $Fn_h > 1$ the resistance augment gradually decreases with speed to a lower drag reaching that in deep water as speed increases above about $Fn_h = 1.5$. The augment reflects the effect of shallow water shortening hull-generated wavelengths, and so the speed at which critical wavelength is reached.

When a craft operates in planing mode, the main drag components are due to friction and forward spray, and the wave-making drag is small. The spray drag is from the flow forward of the stagnation line acting on the planing bottom and the effect of shallow water on the restricted flow constrained by the bottom will influence the spray drag component.

Using the simplified approach to planing, assuming small perturbation as well as high speed, and equating the planing surface to a thin wing, the boundary condition of a riverbed corresponds to an image vortex system with the bed as a mirror, as shown in Fig. 4.57. This reduces the flow speed at the lower surface of the wing increasing the hydrodynamic pressure and the planing efficiency. In order to maintain equilibrium, either the trim angle is reduced, or the hull will rise, the former decreasing the spray drag and the latter reducing the wetted area and the friction drag.

Planing hull drag in shallow water is therefore smaller than that in deep water. Figure 4.58 shows drag curves of model No. 1174 of "Series 62" made by A. Toro [46], where the shallow water drag and trim angle curves at three different water depths are shown. They show resistance and trim vs volume Froude number for relative depths h/b = 0.64, 1.024, and 1.706, where b is the hull breadth.

It can be seen from the figures that the drag peak appears at a lower speed for shallower water. At speeds up to a certain value drag in shallow water may be greater

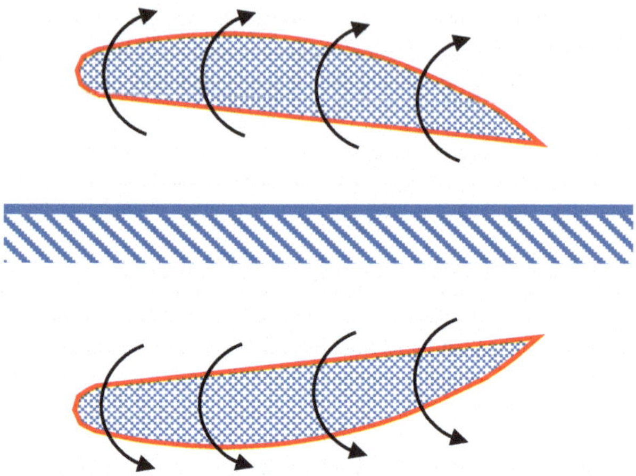

Fig. 4.57 Vortex system diagram

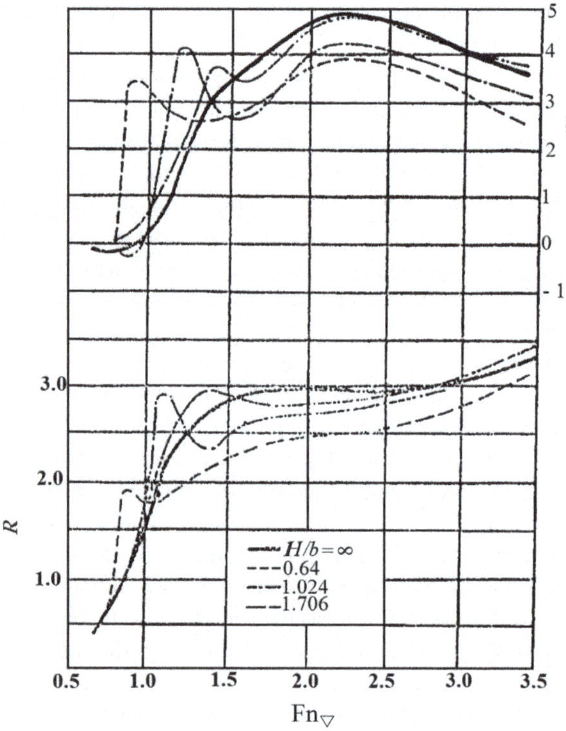

Fig. 4.58 Test results for the drag of model No. 1174 of 'Series 62' in shallow water, where R is model drag is in lbs

than that in deep water due to the effect of trim angle. Even though the hydrodynamic pressure and lift may be increased the increase of trim angle above the optimum causes an increase in drag, which can be seen in the figure.

According to the "Mirror image" model, the shallow water effect is not only due to the water depth but also relates to the craft profile and trim angle as the lift in water

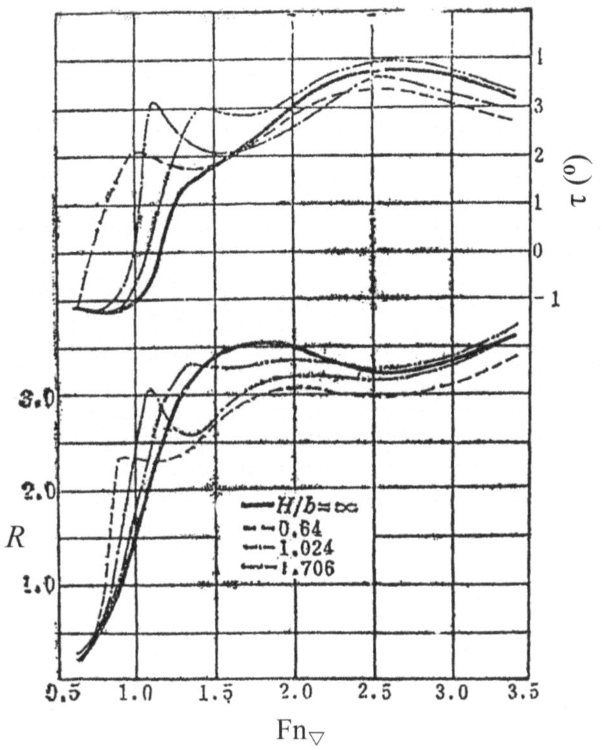

Fig. 4.59 Effect of CG moving forward of model No. 1174 "series 62" on the drag in shallow water

directly influences the vortex strength, which itself relates to the craft profile and trim angle. Therefore, for planing craft, the shallow water effect is different for different craft profile and trim angle.

Figure 4.59 shows the drag and trim curve of model No. 1174 of "series 62" in shallow water made by Toro after shifting the CG forward. Comparing with Fig. 4.58, it can be seen the shallow water effects change with different trimming angles.

Regarding the calculation of drag in shallow water, a theoretical method suitable for engineering application is still not available. For a simple two-dimensional planing flat plane, hydrodynamic pressure can be solved by means of jet theory with a bottom wall (seabed), according to the methods of M E Gorivin [47]. For a three-dimensional planing surface, it is more complicated.

The work by C. J. Reyling [35] is valuable for estimating the drag performance in shallow water. Using a semi-theoretical and semi-empirical method for calculating the flow, and via a large number of tests for planing surfaces with prismatic profile in different water depths, Reyling created correction equations for the influence of water depth. To simplify the problem, and based on the test results, he proposed that:

1. At fixed running attitude, shallow water influences the hydrodynamic pressure on the planing surface but not the static pressure.

2. When trim angle and wetted length are variable, it is not appropriate to use water depth as a parameter, but rather to use the distance between the transom and seabed as the parameter.

Thus, he obtained a correction equation referring to the effect of shallow water as follows:

Lift Coefficient

$$C_{Lb} = C_{LD\infty}\left(1 + \frac{0.21}{H'_z + \sin\alpha}\right) + C_{LS_\infty} \tag{4.71a}$$

where

$C_{LD\infty}$ dynamic lift coefficient in deep water
C_{LS_∞} static lift coefficient in shallow water
H'_z ratio of water depth at transom with beam of planing surface, $H'_z = H_z/b$

Moment Coefficient

$$C_{Mb} = C_{MD\infty}\left[1 + \frac{(0.124/\lambda) + 0.116}{H'_z + \sin\alpha}\right] + C_{MS\infty} \tag{4.71b}$$

where:

$C_{MD\infty}$ dynamic moment coefficient at deep water
$C_{MS\infty}$ static moment coefficient at deep water

Drag Coefficient

$$C_{Rb} = C_L \tan\alpha + \frac{C_f\lambda}{\cos\alpha\cos\beta} \tag{4.71c}$$

Here it is suggested to use the Schlichting equations for C_f.

Applying correction equations (4.71) to the semi-empirical and semi-theoretical method for calculating lift, drag, and trimming moments of a planing surface we can obtain the performance in shallow water. Basically, Equations (4.71) can be used in the calculation methods used in Sect. 4.1 above.

The calculation for shallow water is actually similar to that in deep water. At first, obtain the running attitude, that is, angle of attack α and wetted length/beam ratio λ, according to the equilibrium conditions, and then add the correction factors for shallow water from Eq. (4.69). The equilibrium can be obtained, as for the Shuford and Brown method:

$$C_b = \frac{\pi}{4}\sin(2\alpha)\cos\alpha\left[(1-\sin\beta)\frac{\lambda}{1+\lambda}+\frac{C_{DC}}{\pi}\lambda\sin(2\alpha)\cos\beta\right] \times\left(1+\frac{0.21}{H_z'+\sin\alpha}\right)+0.314\sin(2\alpha)\left(\frac{\lambda}{F_b}\right)^2 \tag{4.72}$$

$$C_M = \frac{\pi}{4}\lambda\sin(2\alpha)\left[\left(\frac{7}{8}\lambda-0.08\frac{\tan\beta}{\tan\alpha}\right)\frac{1-\sin\beta}{1+\lambda}+\frac{C_{DC}}{2\pi}\lambda\sin(2\alpha)\cos\beta\right] \times\left(1+\frac{(0.124/\lambda)+0.116}{H_z'+\sin\alpha}\right)+0.21\lambda\sin\alpha\left(\frac{\lambda}{F_b}\right)^2 \tag{4.73}$$

It is rather complicated to solve for α and λ from (4.72) and (4.73), particularly due to the uncertainty of the shallow water factor $H_z^{'}$, and so according to Reyling's definition:

$$H_z = H - l_k\sin\alpha \tag{4.74}$$

The geometric relation between them is as shown in Fig. 4.60. According to Brown's suggestion, the relationship between the wetted length at keel l_k and wetted length at the hard chine is:

$$\lambda_k - \lambda_c = (0.57+0.001\beta)\left(\frac{\tan\beta}{2}\tan\alpha-0.006\beta\right) \tag{4.75}$$

where:

$$\lambda_k = \frac{l_k}{b} \quad \text{and} \quad \lambda_c = \frac{l_c}{b}$$

Since

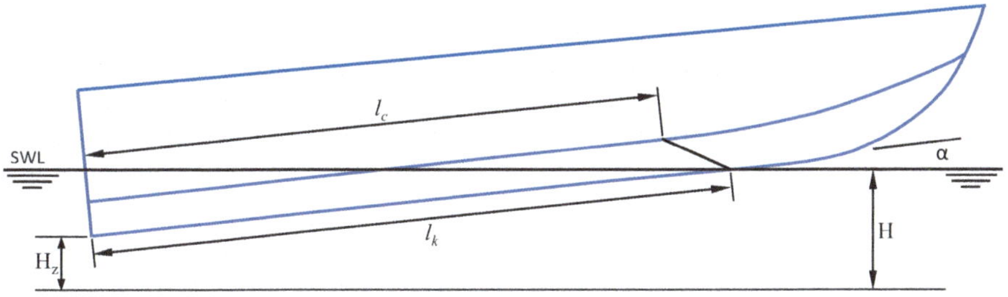

Fig. 4.60 Geometry figure

$$\lambda_k + \lambda_c = 2\lambda$$

Then

$$\lambda_k = \lambda + \frac{1}{2}(0.57 + 0.001\beta)\left(\frac{\tan\beta}{2}\tan\alpha - 0.006\beta\right) \tag{4.76}$$

Taking (4.74) in nondimensional form using craft beam b and using Eq. (4.76), we obtain:

$$H'_z = H' + \left[\lambda + \frac{1}{2}(0.57 + 0.001\beta)\left(\frac{\tan\beta}{2}\tan\alpha - 0.006\beta\right)\right]\sin\alpha \tag{4.77}$$

where $H' = H/b$

α and λ can be obtained by (4.77, 4.72, and 4.73) by numerical calculation. It is not difficult to calculate the drag after obtaining the running attitude.

Reyling showed that his shallow water correction equations (4.71) can be applied within the range of:

Speed	$1 < F_H < 5$
Wetted length beam ratio	$1 < \lambda < 4$
Deadrise angle	$\beta < 20°$
Trim angle	$4 < \alpha < 10°$
Relative water depth at transom	$Z' \leq 0.05$

The range is generally suitable for planing craft without steps, except where trim angle is larger than 10°. The trimming angle of planing craft is less than 10° except in extreme cases. Reyling also carried out tests of craft including $\alpha = 2°$ but got unsatisfactory results and so he excluded the craft with trimming angle below 4°.

It should be stressed that the equilibrium equation (4.77) for obtaining running attitude only reflects the relation between the "nominal wetted length" with running attitude, and which implies the wetted length being the distance from interaction of static water surface with keel to trailing edge, but not considering the effect of water surface elevation in front of the stagnation line on the change of wetted length, which is apparent at small trim angles. This is maybe one reason Reyling's tests at low trim gave "unsatisfactory" results, though the local flow turbulence clearly played a major role.

When considering the effect of local water surface elevation, Eq. (4.74) has to be adjusted; however, it is extremely difficult to make precise calculations, and many have tried to derive equations from experimental data for expressing such factors. After comparing some equations, we suggest using the equation, given earlier, developed by Russian professors P. A. Apuhetin and Y. I. Voitkonsky [48], as follows:

$$\frac{l_0}{l} = \frac{\lambda}{\lambda + 0.4} \tag{4.78}$$

where l_0 is "Nominal wetted length," and l is real wetted length considering the water surface elevation at the leading edge.

Equation (4.78) is simple and is based on the test results of planing flat plates. The shallow water drag of planing craft after correction of (4.77) by (4.78), agrees well with test results [49, 50] found by Z S Dong.

It must be stressed again that the speed range for suitable for the Reyling correction equation is wide with $1 < F_H < 5$. However, since the tests were carried out with the use of some measures to avoid side hull wetting, similar measures have to be taken to ensure the water separation at the bilge/chine of the hull, and the equation cannot be used for craft with lower speed where the side is wetted.

One may observe that investigation of shallow water effect on planing craft is still at an early stage compared with such issues for displacement ships, including the issues of operation of planing hulls before take-off, for which less data is available about the shallow water effects of craft with a hard chine, even though it may be considered basically as a displacement ship.

References

1. A.M. Robb, Theory of naval architecture, Chapter 32, in *Analysis of Trial Ship Records, 2 The Admiralty Constant* (Charles Griffin & Co Ltd, London, 1952), p. 580
2. Shinishi Niwa 丹羽诚, *High Speed Craft Technology*, Rudder, 1971, (in Japanese)
3. P.G. Alourdas, Planing Hull Resistance Calculation – The CAHI Method. Presentation made at SNAME Greek Section Meeting, 13th October 2016. The presentation is available online at: https://higherlogicdownload.s3.amazonaws.com/SNAME/a09ed13c-b8c0-4897-9e87-eb86f500359b/UploadedImages/2016-2017/Alourdas Complimentary Notes.pdf
4. D. Savitsky, Hydrodynamic Design of Planing Hulls. Marine Technol., 71–96 (1964)
5. C.L. Shuford Jr, A Theoretical and Experimental Study of a Planing Surface Including Effects of Cross Section and Plan Form, NACA TN 3939
6. E.P. Clement, J.D. Pope, Stepless and Stepped Planing Hull - Graphs for Performance Prediction and Design, DTMB Report 1490 January1961
7. E P Clement, 'Graphs for Predicting the Ideal High-Speed Resistance of Planing Catamarans', Int. Shipp. Prog. Vol.9 No.99 Nov.1967
8. P. Ward Brown, An Experimental and Theoretical Study of Planing Surface with Trim Flaps, Davidson Laboratory Report SIT-DL-71-1463 Apr.1971
9. D. Savitsky, P. Ward Brown, Procedures for hydrodynamic evaluation of planing hulls in smooth and rough water. Mar. Technol. **13** (October 1976). Published by Society of Naval Architects and Marine Engineers, New York, USA
10. D. Savitsky, M.F. DeLorme, R. Datla, Inclusion of whisker spray in performance prediction method for high-speed planing hulls. SNAME Mar. Technol. **44**(1), 35–56 (2007)
11. J.M. Altmeter, Resistance prediction of planing hulls: state of the art. Mar. Technol. **30**(4) (1993), 297–307, SNAME, New York, ISSN 0025-3316
12. A.B. Murray, The hydrodynamics of planing hulls. Trans Soc. Nav. Archit. Mar. Eng. **58**, 658–692 (1950)

13. M.M. Bun'kov, Study of the resistance of planing hulls, Boast and Yachts, No 2(48), 1974 in Russian, Naval Sea Systems Command Translation No 2148, 1987
14. K. Davidson, A. Saurez, Tests of 20 related models of V bottomed motorboats, EMB Series 50, David Taylor Model Basin Report 170, Bethesda Maryland USA, December 1948
15. D. Radojcic, *Reflections on Power Prediction Modelling of Conventional High-Speed Craft*, Springer Briefs in Applied Sciences and Technology (Springer, 2019), 11 pp. ISBN 978-3-319-94898-0
16. M. Morabito, Re-analysis of series 50 tests of V-bottom motor boats', January 2015, SNAME, New York, Transactions Vol 121, pp190-223 (Note that in Appendix 2. Resources at back of book is the complete compendium ref)
17. E.P. Clement, D.L. Blount, Resistance tests of a systematic series of planing hull forms. SNAME **71** (1963)
18. J B Hadler, E N Hubble, 'Prediction of the power performance of the Series 62 Planing Hull forms' Trans. SNAME 1971, 366 to 404.
19. J.A. Keuning, J. Gerritsma, Resistance test of a series of planing hull forms with 25 degrees deadrise angle. Int. Shipp. Prog. **29**(337) (1982)
20. W.X. Lu, A method for estimating the effective power of planing craft. China shipbuilding, 1984, No. 3, (in Chinese)
21. H.D. Holling, E.N. Hubble, Model Resistance Data of Series 65 Hull Forms Appliable to Hydrofoils and Planing Craft', NSRDC Report No 4121 May 1974
22. R.H. Compton, The resistance of a systematic series of semi-planing transom stern hulls. Mar. Technol. **23**(4) (1986), Published by SNAME
23. B.J. Metcalf, L. Faul, E. Bumiller, J. Slutsky, Resistance tests of a systematic series of US Coastguard planing hulls. Carderock Division, Naval Surface Warfare center, Bethesda, Report NSWCCD-50-2005/063 December 2005
24. D. Radojcic, An Approximate Method for calculation of resistance and trim of the planing hulls. Ship Science Report No 23, University of Southampton, December 1985
25. G. Delgado-Saldivar, Experimental Investigation of a New Series of Planing Hulls. M.Sc. Thesis, Technical University of Nova Scotia, Halifax, Nova Scotia, 1993
26. D.J. Taunton, D.A. Hudson, R.A. Shenoi, Characteristics of a series of high-speed hard chine planing hulls. Part I: Performance in calm water, Transactions of the Royal Institution of Naval Architects, Vol 152 Part B2 2011, ISSN 1740-0694. A systematic series programme at Southampton University
27. D.J. Taunton, D.A. Hudson, R.A. Shenoi, Characteristics of a series of high-speed hard chine planing hulls. Part II: Performance in waves, Transactions of the Royal Institution of Naval Architects, Vol 153 Part B1 2010, ISSN 1740-0694
28. D. Savitsky, M.F. DeLorme, R. Datla, Inclusion of whisker spray in performance prediction method for high-speed planing hulls. Mar. Technol. **44**(1), 35–56 (2007)
29. F. De Luca, C. Pensa, The Naples warped hard chine hulls systematic series. Ocean Eng. **139** (2017), 205–236. An extensive test series at University of Naples, Italy, including effect of transom interceptors for trim control. Open access article published by Elsevier under creative commons. It can be found at: www.dx.doi.org/10.1016/j.oceaneng.2017.04.038
30. J.A. Keuning, L. Hillege, The results of the Delft Systematic Deadrise Series. FAST 2017 Transactions Session B1, pp. 97–106, Data is available at www.dsds.tudelft.nl once you set up an account for access, the paper is via www.fast2017.com
31. J.A. Keuning, J. Gerritsma, Resistance Test Of A Series Of Planing Hull Forms With 25 Degrees Deadrise Angle. Int. Shipp. Prog. **29**(337) (1982)
32. J.A. Keuning, J. Gerritsma, P.F. Terwisga, Resistance tests of a series planing hull forms with 30 degrees deadrise angle and a calculation model based on this and similar series. Report No. 959, Delft University of Technology, 1993
33. D. Radojcic, A. Zgradic, M. Kalajdzic, A. Simic, Resistance Prediction for hard chine hulls in the pre-planing regime. Polish Maritime Res. **21**(2), 9–26 (2014). https://doi.org/10.2478/pomr-2014.0014

34. D. Radojcic, M. Kalajdzic, Resistance and trim modelling of the Naples hard chine systematic series. RINA Trans. **160**(Part B1), Int. J. Small Craft Technol. January–June 2018, pp B31–41, ISSN 0035-8967
35. C.J. Reyling, An Experimental Study of Planing Surface Operating on Shallow Water. Davidson Laboratory Report SIT-DL-76-1835 Sep.1976 Available at: https://apps.dtic.mil/sti/citations/ADA051604
36. D. Savitsky, Wake shapes behind planing hull forms, High Performance Marine Vehicles Conference, Shanghai 1988, CSNAME, VII, 1–15
37. D. Savitsky, M. Morabito, Surface Wave Contours Associated with the Forebody Wake of Stepped Planing Hulls. Davidson Laboratory Technical Memo #181, Stevens Institute of Technology, Hoboken, NJ USA, 2009
38. D. Svahn, Performance Prediction of hulls with transverse steps. Master's Thesis at Royal Institute of Technology, KTH, Centre for Naval Architecture, Stockholm, Sweden, 2009
39. G. Pavlov, L. Yun, S.-L. He, A. Bliault, *Air Lubricated and Air Cavity Ships, Development, Design and Application*. Springer Nature, April 2020, ISBN 978-1-0716-0423-6, 469 pp
40. O.M. Faltinsen, *Hydrodynamics of High-Speed Marine Vehicles*. Cambridge University Press, 2005, ISBN 978-0-521-84568-7, Chap 9.2.3, p355
41. L.J. Doctors, Hydrodynamics of the flow behind a transom stern. Proceedings of 29th Israel Conference on Mechanical Engineering, Haifa, Israel, Paper 20-1 replace with ICMRT05 paper
42. S.F. Hoerner, Fluid Dynamic Drag. Published by the Author, Hoerner Fluid Dynamics, P O Box 342, Brick Town New Jersey, 08723, USA, 1965. Also available via the American Institute of Aeronautics and Astronautics, New York, USA
43. V.L. Streeter (ed), *Cavitation Handbook of Fluid Dynamics*, chapter by M P Tulin, published by McGraw-Hill, New York, 1961
44. K.R. Suhrbier, On the influence of fully cavitating propellers on the interaction effects and dynamic stability of fast craft. Proceedings of FAST '95, Third International Conference on Fast Sea Transportation, Lubeck-Travemunde, Germany, September 25–27, 1995, vol 2, pp. 795–805
45. J.B. Hadler, The prediction of power performance on planing craft. Transactions SNAME, New York, Vol 74, 1966
46. A. Toro, Shallow water performance of a planing boat, University of Michigan Report No 019, Ann Arbor, April 1969
47. M.E. Gorivin, Theory of ideal fluid, Moscow, 1961 (in Russian)
48. Y.I. Voitkonsky, Water resistance of ships in motion, Shipbuilding, 1964, (in Russian)
49. Z.S. Dong, On the calculation of shallow water drag of planing craft. Journal, University of Naval Engineering of China, 1985, No. 3, (in Chinese)
50. Z.S. Dong, L. Wujiang, The Minimun Water Depth Calculation Without Shallow Water Effect on Resistance of Planing Hull. Third High Performance Marine Vessels Conference of China, Xiamen, China, 1992

Chapter 5
Planing Craft Stability, Maneuverability, and Seakeeping

In Chaps. 3 and 4, we have introduced the basic hydrodynamics of planing and evaluation of equilibrium of forces leading to drag and dynamic trim. We have also looked at the effect of hull geometry on this in the steady state. Now we turn to the dynamic response of a planing hull operating in a seaway.

Owing to the dynamic lift, a planing hull is lifted toward the sea surface from its displacement draft. This affects not only its resistance characteristics but also its stability, maneuverability, and seaworthiness, which will be described in succeeding sections of this chapter.

A planing craft is supported by dynamic forces that substantially change if the craft pitches, rolls, or yaws. If it rolls at speed, then due to the asymetry of the planing surfaces on the hull bottom, there will be both a roll righting moment together with an altered pitching moment. There may also be a yawing moment that will change the course. All of these coupled forces/moments will be nonlinear. It has been found by some studies that the basic pitch and heave response are quite close to linear for small motions, and the key challenge is therefore estimating the possible coupling of motions and identifying instability boundaries.

5.1 Transverse Stability of a Planing Hull

The stability of a planing hull at low speed operating in displacement mode is derived from its metacentric height in the same way as a conventional displacement ship. A planing craft has a larger static transverse metacentric height owing to its low length/beam ratio and so will be stiffer at slow speed.

Evaluation of static stability follows the same procedures as any buoyant vessel. National Regulatory bodies, Classification Societies, and the International Maritime Organization have set requirements for metacentric height and freeboard to guard against down-flooding and compartmentation to protect from flooding due to side or bottom damage. References to a selection of the Regulations and IMO

Z.-S. Dong et al., *High-Speed Monohull and Hydrofoil Craft*,
https://doi.org/10.1007/978-3-031-62760-6_5

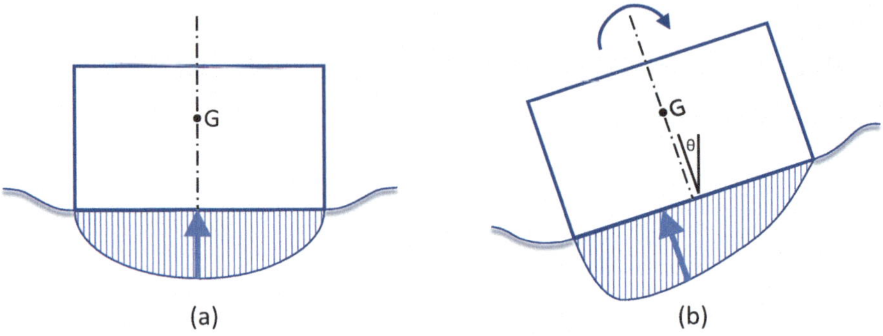

Fig. 5.1 Restoring moment of planing hull with flat bottom in heeling

documentation specific to high-speed craft plus a short commentary on their use are given in Appendix 2.

At high speed, the planing hull is supported primarily by the dynamic lift and, thus, it is the dynamic restoring moment that stabilizes a craft in heeling and pitching.

Calculation of dynamic stability is challenging due to the difficulty of calculating the quasistatic transverse and longitudinal shift of center of dynamic pressure of a craft heeling and trimming. However, dynamic transverse stability is generally larger than the static transverse stability for a craft with a flat bottom or "V" bottom with small deadrise.

Longitudinal stability, which is not a concern for displacement hulls, is a concern for planing hulls, especially for hulls with low wetted length/beam ratio. Longitudinal stability will be discussed later and here we first discuss transverse stability.

For the craft with a flat bottom, the dynamic restoring moment is caused by the shift of center of dynamic pressure to the heeling side as shown in Fig. 5.1.

When the craft is upright, the dynamic pressure on the hull bottom is symmetric as shown in Fig. 5.1a. When the craft heels with an angle θ, owing to the difference in draft port and starboard, the center of dynamic pressure shifts to the heeling side thus generating a restoring moment as shown in Fig. 5.1b. It is apparent that the restoring moment will increase both with heeling angle and with speed. This is why the dynamic heeling angle is smaller than the static heeling angle, so long as the restoring moment as the hull reaches planing is at least equal to the static buoyant restoring moment that it replaces.

For craft with deadrise, heeling corresponds to a change in effective deadrise angle on both sides of the keel. Figure 5.2a shows the pressure profile for a craft with deadrise when upright. There is a decrease in the effective deadrise angle on the heeling side, which increases the hydrodynamic pressure, and the opposite effect on the other side. This shifts the center of hydrodynamic pressure to the heeling side and provides a dynamic restoring moment. For craft with small deadrise angle, the restoring moment caused by hydrodynamic lift is positive, and the intersection of resultant force vector on each side is above the craft VCG as shown in Fig. 5.2b.

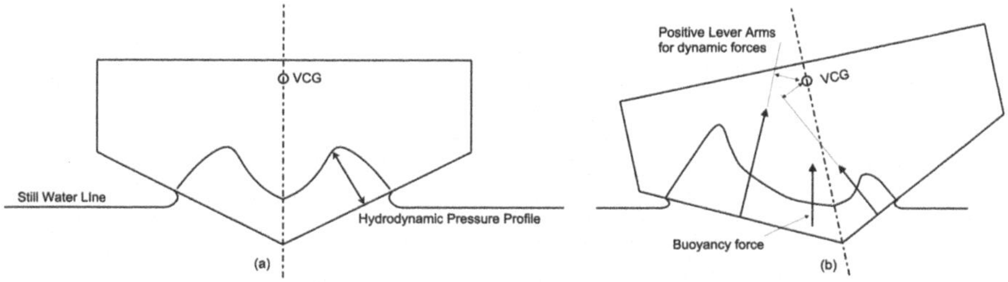

Fig. 5.2 Restoring moment of craft with deadrise. (**a**) Pressure profile for a craft with deadrise. (**b**) pressure profile with positive restoring moment when rolling

However, for hulls with larger deadrise angle and so deep V transverse sections, the restoring moment may be negative, particularly for craft with high VCG position. Therefore, high deadrise can be unfavorable for dynamic transverse stability.

The restoring moment generated by static buoyancy can be calculated without any difficulty so long as the running water line is known. The static transverse metacentric radius $r = I_X/\nabla$, and so the initial metacentric radius can be obtained from waterline area moment of inertia at the operational waterline I_X, and the displacement volume ∇ under the same waterline.

The operational waterline at speed can be obtained from the evaluation of equilibrium described in Sect. 4.1 of Chap. 4. Reference [1] introduced this method for calculating the quasi-static transverse metacentric height of a planing hull in planing mode. The calculation demonstrates that the transverse metacenter of a planing hull when planing is higher than in the static condition. This is because the change of operational waterline causes a reduction in underwater volume, which is greater than the reduction in waterplane area inertia. On the other hand, a reduction of quasistatic metacentric height can happen if a craft is running in "triangular" or chines dry configuration.

The transverse stability characteristic of a planing hull with speed can be shown by a restoring moment curve at initial heel angle. This can be obtained by applying an initial heeling angle θ_0, then measuring the change of heeling angle with various speeds and obtaining curves of $\theta = f(V)$, as shown in Fig. 5.3.[1]

At low speeds, the heeling angle does not change much or increases a little. During takeoff, the craft rises and waterplane area reduces, decreasing the restoring moment caused by buoyancy. The small dynamic restoring moment in this stage can compensate the loss of restoring moment by buoyancy in some cases. However, the dynamic restoring moment can also be negative with higher deadrise angle. On some craft, such decrease of stability during takeoff may not be obvious. After takeoff, the heeling angle decreases as speed increases and remains constant in the speed range corresponding to pure planing mode with negligible influence of buoyancy.

[1] Since when rolled the planing surface will be asymmetric, there will be a tendency to turn; thus, a rudder moment may be necessary to keep the vessel or model on a straight course. In waves, a snake-like course will be followed if the roll is maintained.

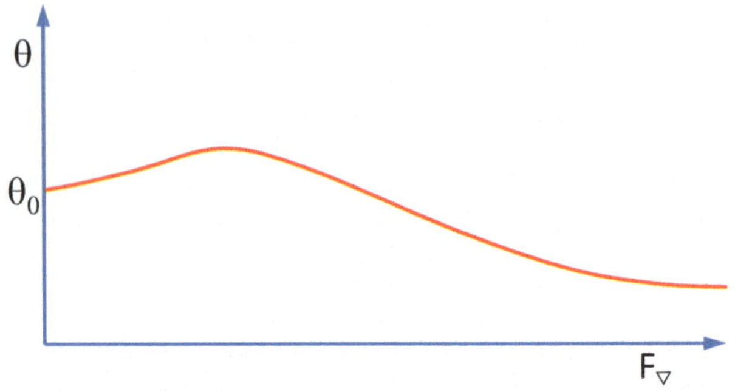

Fig. 5.3 Heeling angle versus speed curves

The transverse stability of high-speed craft can be estimated from the decrease of initial heeling angle when planing, as follows. The initial metacenter of a planing craft can be calculated according to the method indicated above to estimate the initial stability of the craft planing at different speeds. Assume the heeling moment acting on the steady craft is M_θ corresponding to an initial heeling angle θ_0, and heeling angle in planing mode is θ_c, then:

$$M_\theta = \Delta h_c \sin\theta_c \tag{5.1}$$

or

$$h_c = \frac{M_\theta}{\Delta \sin\theta_c} \tag{5.2}$$

Where h_c is the corresponding transverse metacenter height
Δ is the displacement of craft
Corresponding static transverse metacentric height:

$$h = \frac{M_\theta}{\Delta \sin\theta_0} \tag{5.3}$$

Of course, here, θ_0 and θ_c should be small values. If the heeling moments can be calculated for the planing craft at different speeds based on the pressure profile, then a dynamic metacentric height can be determined for use to develop a GZ curve for that speed.

From analysis of dynamic restoring moment, it may be shown that the main factors that influence the transverse stability are craft beam and transverse section of the bottom. The restoring moment increases with beam. Regarding bottom profile, the stability will be generally poor if the center of dynamic pressure on the bottom is concentrated near the keel and will be favorable when the center of dynamic pressure is near the bilge/chine.

Thus, apart from low transverse deadrise angle, adding a chine strake or bending the transverse sections downward at the chine is favorable to transverse stability. The CG position in the vertical direction does not influence the transverse stability much for craft with a flat bottom or with small deadrise; however, the influence of VCG is significant for deep "V" craft with a large deadrise angle.

Reference [2] suggests that for craft with variable deadrise angle from bow to stern, due to twist of the bottom profile making the flow rotate and thus generate local suction, this geometry can reduce transverse stability, particularly for craft that also have convex buttock lines on the bottom and for cases of bow down trim.

Some model tests have been carried out for transverse stability of a high-speed planing hull at a large heeling angle and demonstrated that the stability could be lost at large heel angles even if it was stable at small heel angles. Since the condition is rather complicated (multiple parameters with different tendencies), model tests with full appendages are recommended, and even to use a tank with variable atmospheric pressure for studying the effect of cavitation.

Transverse stability at large heeling angle and at high speed has not been studied widely as an important design criterion to date, and only the stability at large heeling angle of craft at zero or low speed, i.e., displacement operation, has been the focus. High-speed stability is therefore an important area for future research.

Faltinsen in reference [3] page 364, when discussing dynamic stability, makes the point that transverse stability has been shown to be dependent on trim angle when at high speed; thus, the roll restoring moment is a function of pitch. Ikeda and Katayama[2] have also shown that where the pitch period is half of the natural period in roll, that unsteady coupled motions can be induced.

At high speed, craft with lower L/B, trapezoidal aft lines, medium deadrise angle, and transverse sections with bilge and longitudinal anti-spray rails, etc., are recommended as favorable to transverse stability based on current knowledge.

5.2 Longitudinal Stability and Porpoising Motion of Planing Hulls

Longitudinal instability may occur on planing hulls. This is characterized by periodic pitching motion of craft running at steady speed in calm water, or strictly speaking coupled periodic pitching and heaving motion, the so-called "porpoising."

Porpoising motion cannot be "controlled" during operation. Rather it is necessary for a craft to be designed to operate outside the region where it would occur, so it is an important research topic. Owing to its complexity, experimental investigation has been the main approach for a long time and used as the main tool for judgment by designers for avoidance of porpoising. Although theoretical calculations have been carried out by means of short wing theory, this can only be applied to a hydroplane with high length to beam ratio but cannot be applied to a normal planing hull.

[2] Y Ikeda, T Katayama, 'Stability of High-Speed Craft', In: Contemporary ideas on Ship Stability, 1st Edition, pp401–409, Elsevier Science, Amsterdam, 2000, ISBN 978-008-054-715, at www.Elsevier.com

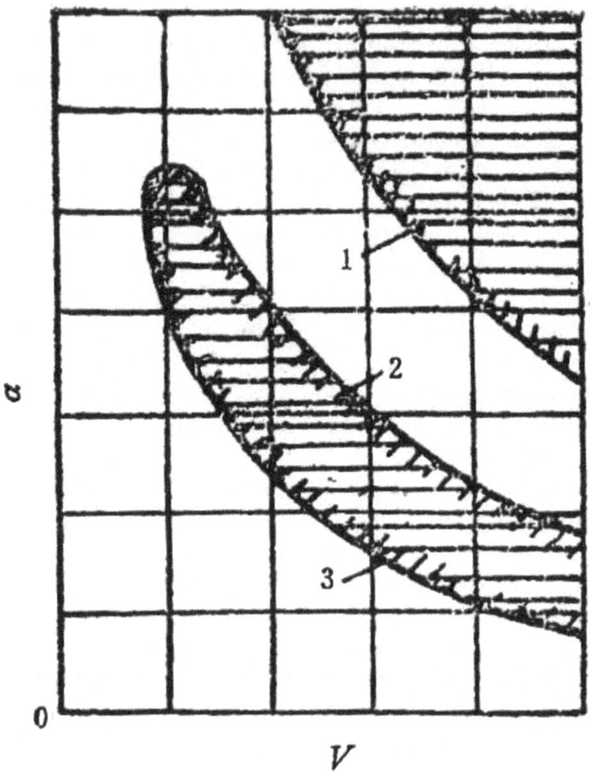

Fig. 5.4 Stable boundary limit for planing flat plate

5.2.1 *Experimental Investigation of Porpoising Motion*

In order to study porpoising motion, many model experimental investigations have been carried out in CAHI, the Central Aerodynamic and Hydrodynamic Research Institute of Russia, as well as by Day and Haag of the USA. CAHI compiled a set of graphs for judgment of porpoising motion, and separately Prof. D. Savitsky prepared a graph of the boundary limit for judging the porpoising motion of craft with a prismatic hull profile based on Day and Haag's work.

1. *CAHI method*

Experimental results from a planing plate showed that for a given loading and longitudinal moment of inertia the stability was found to be related to trim angle and speed. Thus, a longitudinal stability boundary limit for a typical flat plate can be obtained as shown in Fig. 5.4, with trim angle α and speed v as coordinates, and in Fig. 5.5 with wetted length/beam ratio λ and dynamic load coefficient C_b as coordinates.

In the figures, the hatched areas show the unstable regions, and it can be seen that there are three boundary lines.

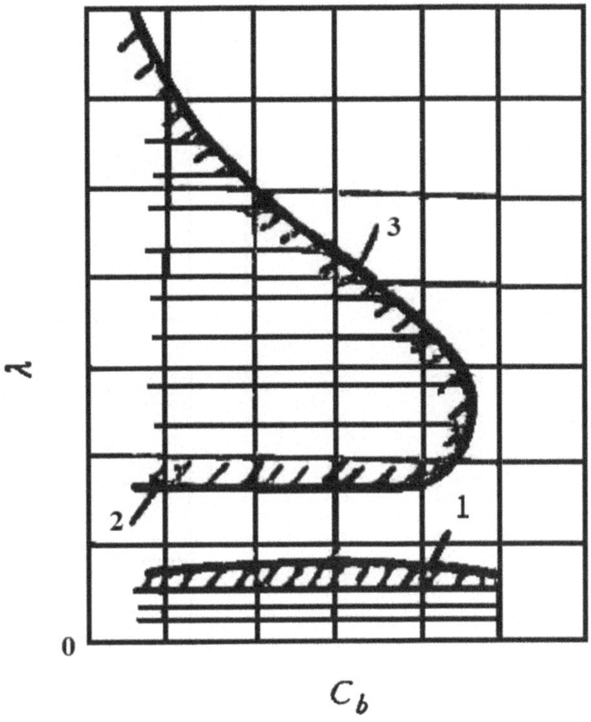

Fig. 5.5 Stable boundary limit, with λ and C_b as coordinates

The *first boundary limit* corresponds to large trim angle, on which wetted length/beam ratio λ is not related to C_b and $\lambda = 0.1$–0.3 for different parameters of plates. In this area, the features of unstable motions are vertical flutter at high frequency accompanied with slow change in trim angle.

The *second boundary limit* corresponds to small trim angle with constant λ, and the features of the unstable motions are pitching but without obvious heaving motion.

The *third boundary limit* corresponds to minimum trim angle, on which if speed increases (C_b decreases) wetted length/beam ratio increases and the unstable features are the coupled motions of pitching and heaving, i.e., the so-called porpoising motion.

Thus, the porpoising motion generally recognized as such (boundary 3) is not the only unstable motion. Nevertheless, for a planing hull, particularly for planing craft with no step, the higher trim angles for the first and medium trim of the second unstable boundary limits are not usually reached. The typical unstable condition is therefore porpoising motion, i.e., corresponding to the third unstable boundary limit.

It is possible for stepped hulls to reach the upper first unstable boundary limit causing vertical flutter with high frequency. However, the second unstable boundary limit can't be reached for a two supported surface configuration as long as the forward support surface is stable. For a hydroplane where larger angles of attack

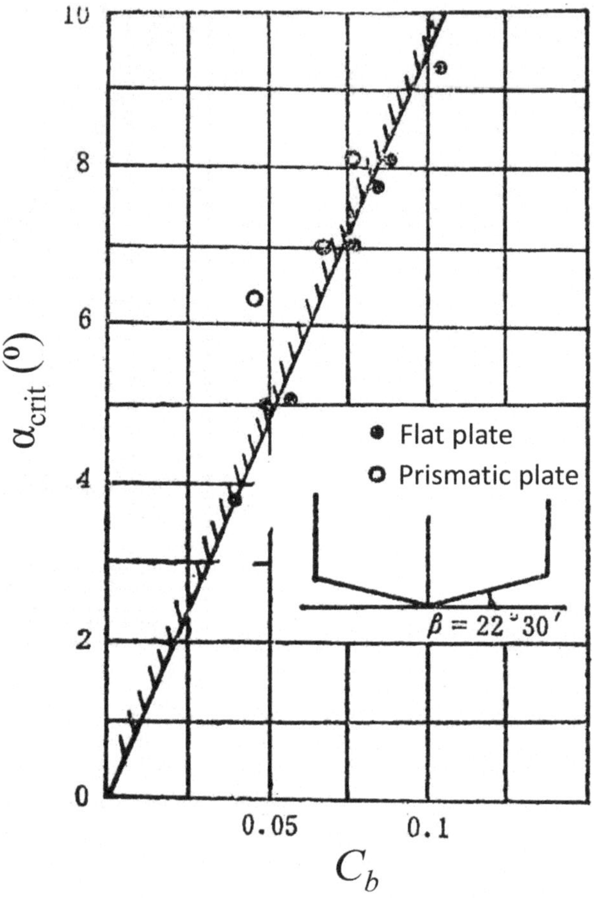

Fig. 5.6 Stability boundary limit for a flat plate and prismatic section. Black dots are plate. Open dots are prismatic section with 22 degrees deadrise

are present, the unstable phenomenon at the first and second boundary limit can possibly occur.

When considering the third unstable boundary limit, tests demonstrate that for a given flat plate, the ratio of critical angle of attack α_{cr} with dynamic load coefficient should be a constant value, i.e., $C_b/\alpha_{cr} = \text{const}$.

The stable boundary limit can be as shown in Fig. 5.6, with angle of attack α and dynamic load coefficient C_b as the coordinates.

Figure 5.6 shows the stability boundary for a given flat plate. For different plates, with different loading rates C_b/α_{cr}, there are different stability boundary limits. Tests demonstrate that C_b is a function of static load coefficient C_Δ, non-dimensional mass $\overline{m}$, and non-dimensional inertia $\overline{J}$:

$$C_b/\alpha_{cr} = C\left(C_\Delta,\ \overline{m, J}\right) \tag{5.4}$$

Where $\overline{m} = m/\left(\rho b^3/2\right)$, $\overline{J} = \frac{J}{\frac{1}{2}\rho b^5}$, and C is the dynamic load function.

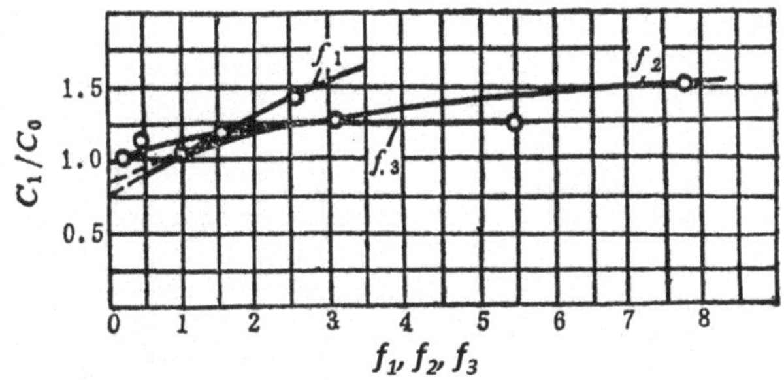

Fig. 5.7 The calculation graphics for third stability boundary limit

CAHI determined the relationship approximately of C_Δ, $\overline{m}$, and $\overline{J}$ with function C. They assumed three parameters influence function C separately, and through tests, the functional relationship as follows was determined:

$$C_1/C_0|_\Delta = f_1(C_{\Delta 1}/C_{\Delta 0}) \tag{5.5a}$$

$$C_1/C_0|_m = f_2(\overline{m}_1/\overline{m}_0)$$

$$C_1/C_0|_J = f_3(\overline{J_1}/\overline{J_0})$$

When $C_{\Delta 0} \to C_{\Delta 1}$, the ratio goes from $C_0 \to C_1$, and the same when $\overline{m}$ and $\overline{J}$ are changed. The curves of $f_1(C_{\Delta 1}/C_{\Delta 0})$, $f_2(\overline{m_1}/\overline{m_0})$, and $f_3(\overline{J_1}/\overline{J_0})$ are shown in Fig. 5.7. The corresponding parameters of the flat plate for these curves are $C_{\Delta 0} = 0.371$, $\overline{m_0} = 2.15$, and $\overline{J_0} = 8.23$. The corresponding ratio $C_0 = C_b/\alpha_{cr} = 1.13$.

When C_Δ, $\overline{m}$, and $\overline{J}$ are changed, f_1, f_2, and f_3 can be obtained from Fig. 5.7, meanwhile, this then corresponds to new stability boundary limits of $C_{\Delta 1}$, $\overline{m_1}$, and $\overline{J_1}$. Thus, the boundary can be determined using Fig. 5.7.

$$C_1\left(C_{\Delta 1}, \overline{m_1}, \overline{J_1}\right) = C_0\left(C_{\Delta 0}, \overline{m_0}, \overline{J_0}\right) f_1 f_2 f_3 \tag{5.5b}$$

For a craft with transverse deadrise and without a step, the CAHI tests demonstrated that for craft with small deadrise, the porpoising motion can be judged approximately from the stability boundary limit of a planing plate directly, so long as the deadrise angle is smaller than 15 deg. Figure 5.6 shows the test results of a model with prismatic cross section and deadrise $\beta = 22^0\ 30'$. One can see that the stability boundary limit is very close to that of a flat plate.

The requirements for the operational trim angle of a planing hull without a step to avoid porpoising is about 1 deg. less than the critical trim angle, so the critical trim angle as a function of $F_b = \frac{V}{\sqrt{gB}}$ has to be calculated by changing the curve $\alpha_{cr} - C_b$ to curve $\alpha_{cr} - F_b$. Then, plots of operational trim angle lower by about 1° from the critical curve can be drawn as shown in Fig. 5.8.

For a planing hull without a step, the porpoising motion is not a big concern in general, as long the hull planing surface profile is not too far away from a simple

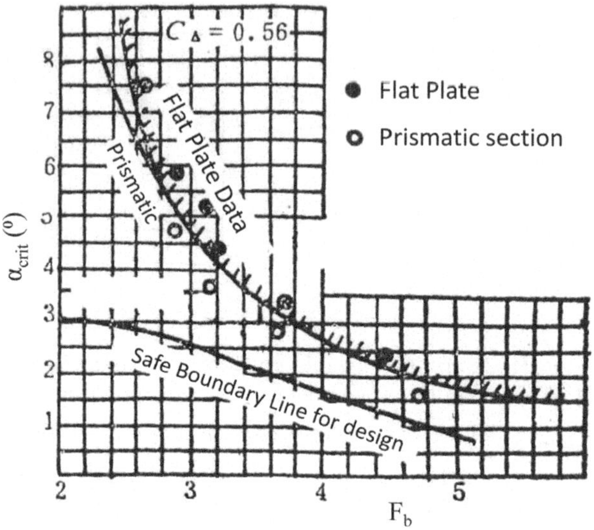

Fig. 5.8 The porpoising motion boundary limit of planing hulls

trapezoidal shape and the CG is not located too far aft. It is possible nevertheless for porpoising motion to occur for racing boats at very high speed.

For planing craft with a step, it is possible to have both fore and aft planing surfaces unstable. However, due to the twin planing surfaces and small range of operational trim angles, it is difficult for the first planing surface to reach porpoising instability.

Tests demonstrate that high frequency vertical flutter may occur when the forward planing surface is running in chines dry condition, generating water surface deformation at the step that makes the second planing surface exceed the first stability boundary limit. Such flutter also generates a large change in the angle of attack of the aft planing surface and can make the whole craft pitch heavily. In general, the porpoising motion will not occur so long as the forward planing surface does not reach the triangular wetted surface, chines-dry condition.

2. *D. Savitsky method*

Based on the results from tests carried out by Day and Haag,[3] Savitsky set up a porpoising motion boundary limit for the planing hull with prismatic profile [4] as shown in Fig. 5.9, where the abscissa is $(C_{L\beta}/2)^{0.5}$ and

$C_{L\beta}$ is the dynamic load coefficient $mg/(0.5\rho V^2 B^2)$
m is the craft mass
g is the acceleration due to gravity
V is the craft speed
B is the craft beam

[3] J P Day, R J Haag, "Planing Boat Porpoising," Thesis submitted to Webb Institute of naval Architecture, Glen Clove, N.Y., May 1952.

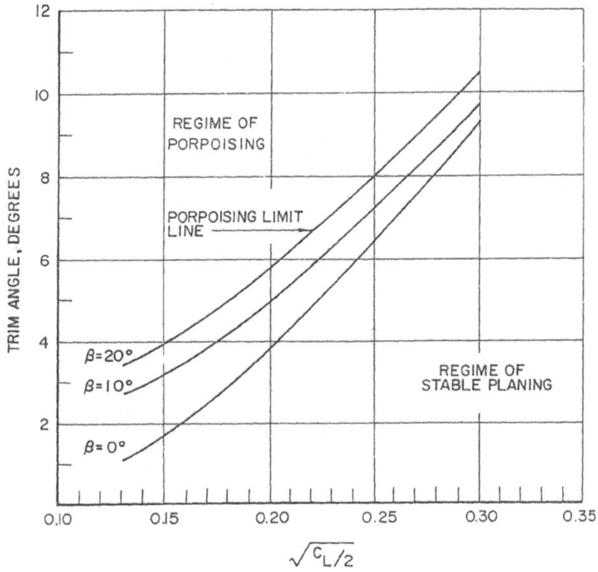

Fig. 5.9 Porpoising motion boundary limit for the planing hull with prismatic profile, suggested by Savitsky

Figure 5.9 shows the boundary limit of porpoising motion of craft with various deadrise angles β. When the running trim of the craft is located below the critical trim lines, no porpoising motion happens to the craft. From the figure, one can see that higher trim angle and lower dynamic load coefficient lead to porpoising motion, and decreasing the deadrise also can increase the probability of porpoising motion. So, the tendency for porpoising motion of a design can be judged by means of Fig. 5.9, and the safety margin of the design can also be estimated. Since the craft weight and speed, and so also the dynamic loading coefficient are givens in the calculation, the only parameter that can be varied is vessel trim. To avoid porpoising motion trim must be reduced below the curves in Fig. 5.9, with a safety margin of 1 degree.

Both CAHI and D. Savitsky methods are based upon test results of many model tests for judging the porpoising motion of planing hulls. We suggest using craft with smaller deadrise and transverse section and using low VCG for craft with larger deadrise and prismatic profile. Of course, designers can use both CAHI and Savitsky methods for comparison and choose the safe limit for a design.

The stability boundary limit also can be determined by model tests directly. For a given craft weight and speed, changing running trim angle by means of shifting CG forward and aft, the stability boundary limit point can be found at each speed. Plotting the points for various speeds, the stability boundary limit curves can be obtained. In addition to maintaining similarity of the craft profile and weight distribution, the longitudinal moment of inertia of the model has to be scaled properly to match with the prototype, just as in ordinary seakeeping model tests.

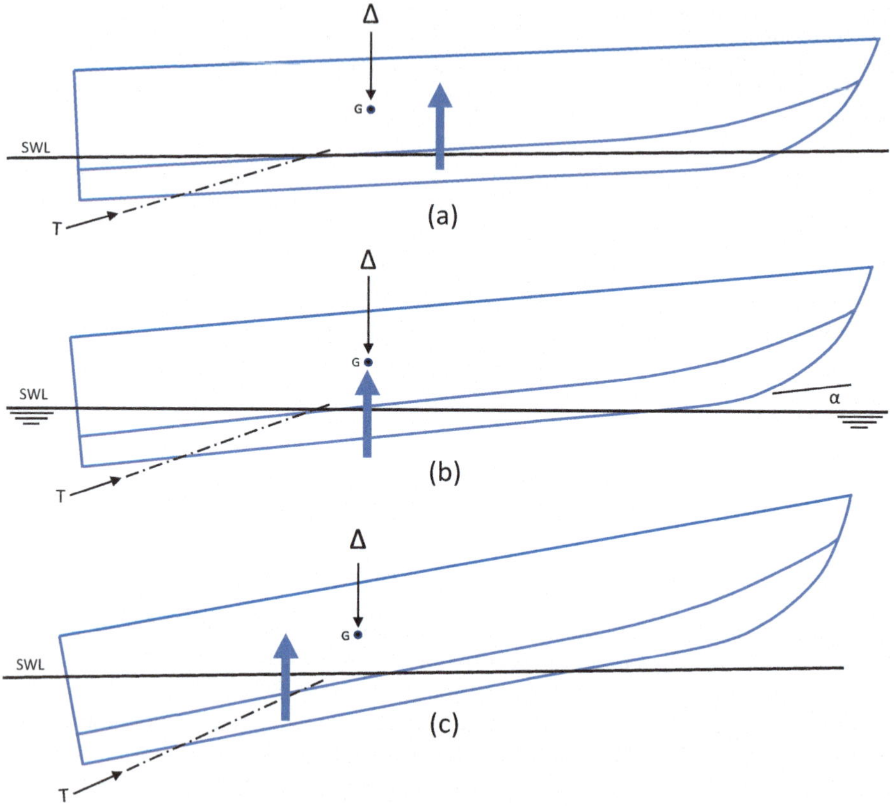

Fig. 5.10 Physical explanation of porpoising motion

5.2.1.1 Main Factors Causing Porpoising Motion

From both Figs. 5.6 and 5.9, the stability boundary limit for a planing plate and planing prismatic surface, respectively, one can see that larger trim angle and smaller dynamic load coefficient C_b lead to porpoising. For craft with given load and beam, increasing trim angle at a given speed decreases the wetted length, and decreasing coefficient C_b by increasing speed at given trim angle also decreases wetted length. Therefore, concentrating the load near stern will cause porpoising motion, which has been validated in practice. Essentially too low pitch inertia can lead to porpoising.

The porpoising motion can also be explained physically as due to the higher load per unit length at the stern. The variation of hydrodynamic load will be great, even though the variation of wetted length caused by external perturbation is little; thus, the craft cannot be kept in balance. If the external perturbation causes increase of wetted length, then the center of dynamic pressure moves forward, as shown in Fig. 5.10a, increasing the trim; meanwhile, dynamic lift moves backward for keeping the balance as shown in Fig. 5.10b. Due to the inertia effect, the shift of center of dynamic pressure does not stop moving until it is located aft of CG as shown in Fig. 5.10c, decreasing the wetted length and dynamic lift that reduces lower than the weight, thereby causing the heave and trim to reduce. Such coupled periodical heaving with pitching forms the porpoising motion.

Such physical explanation is not a complete explanation. However, excess shifting of CG and load relatively aft are the most important reasons causing the porpoising motion. Thus, shifting CG forward and installing stern plate, wedge, or interceptor plate for reducing trim angle can effectively avoid the porpoising motion. The fixed or adjustable stern interceptor plate is most effective for adjusting trim angle with speed, both for improving the porpoising motion and for keeping favorable running attitude during acceleration, powering performance, and seakeeping quality.

In addition, the craft moment of inertia also influences porpoising motion. Low moment of inertia, i.e., weight distribution close to CG, can improve the porpoising motion as shown in Fig. 5.10.

5.2.2 Theoretical Assessment of Longitudinal Stability of Planing Hulls

Analytical determination of dynamic stability is a complex subject, so we will introduce just some concepts and processing methods.

Since porpoising motion is an unstable motion, its assessment can be addressed through the investigation of motion stability and identification of the boundary to instability. W G A Perring and H. Glauert [5] were the first to theoretically predict the porpoising motion as an unstable motion in 1933. Then others including R. N. Lutowski, M Martin [6], and P. R. Payne [7] carried out work in this field.

Dynamic stability refers to a system's ability to restore to its original position when it is affected by external perturbation. The motion stability can be judged by following procedural steps:

- Setting up the differential equations of the system motion for both before and after external perturbation on the system.
- Obtaining the solutions for both cases.
- If the motion prediction after perturbation trends to the situation before the perturbation, then the solution can be considered as stable, otherwise, unstable. If the solutions of all motion differential equations are stable, then the system can be considered as stable.

If the system motion before perturbation can be deemed as relative stable situation, then the motion stability can be equated to stability at equilibrium, just as for investigation of static stability. Normally a system is considered stable if the static restoring moment for a range of roll or pitch angles is sufficient to return the vessel to upright equilibrium.

Here the external perturbation should be random, small, and instant, different from the studies of static stability where the process involves perturbation of only displacement but not of speed and acceleration. However, in this research, not only the displacement perturbation must be considered but also both speed and acceleration perturbations.

After taking the stability of motion system as the stability issue of system at equilibrium position, it is not necessary to investigate the motion differential equations before and after the perturbation, but only investigate the difference of two sets differential equations, i.e., judging the system stability via the solution of the perturbation differential equations. If after time t the solution of the perturbation differential equations trend to zero, then the system is stable, otherwise it is unstable.

We will first study the longitudinal stability, i.e., coupled heaving and pitching motions, and neglect the surging due to its small value. Then using linear theory, the perturbation forces (moments) are in linear relation with perturbation displacement and its derivatives; thus, the differential equations can be expressed as:

$$\left.\begin{array}{l}(Z_{\ddot{z}}-m)\ddot{Z}+Z_{\dot{z}}\dot{z}+Z_{z}z+Z_{\ddot{\theta}}\ddot{\theta}+Z_{\dot{\theta}}\dot{\theta}+Z_{\theta}\theta=0\\ M_{\ddot{z}}+M_{\dot{z}}\dot{z}+M_{z}z+\left(M_{\ddot{\theta}}-J_{y}\right)+M_{\dot{\theta}}\dot{\theta}+M_{\theta}\theta=0\end{array}\right\}\tag{5.6}$$

Where:

θ,$\dot{\theta}$,$\ddot{\theta}$ are the angular displacement, speed, and acceleration of pitching
m is the craft mass
J_y is the transverse moment of inertia of craft about CG (y axis)

The Z and M with subscripts represent the rate of change of vertical force and pitching moment caused by the perturbation expressed by subscripts. If $Z_{\ddot{\theta}}$ represents rate of change of vertical force caused by angular acceleration, $\ddot{\theta}=\frac{d^2\theta}{dt^2}$, then $Z_{\ddot{\theta}}\ddot{\theta}$ represents the force caused by angular acceleration. $M_{\dot{z}}$ represents the rate of change of pitching moment caused by heaving speed $\dot{z}=\frac{dz}{dt}$ and Z_z represents the rate of change of vertical force caused by heaving displacement, etc. All values with subscripts are the stability derivatives. The solution of the perturbation motion differential Eq. (5.6) can be obtained if the stability derivatives are available.

Figure 5.11 defines the coordinate system for investigating the porpoising motion. It must be noted that this coordinate system is different from the previous.

In order to judge dynamic stability, it is not necessary to obtain the solution of the motion equations directly, as it can be done by means of characteristic equations.

The general solution can be expressed as:

$$\left.\begin{array}{l}z=z_1e^{\sigma_1 t}+z_2e^{\sigma_2 t}+\cdots\\ \theta=\theta_1e^{\sigma_1 t}+\theta_2e^{\sigma_2 t}+\cdots\end{array}\right\}\tag{5.7}$$

Where $z_1, z_2 \ldots$ and $\theta_1, \theta_2 \ldots$ are the constants determined by initial condition $\sigma_1, \sigma_2 \ldots$ which are the roots for the corresponding characteristic equations.

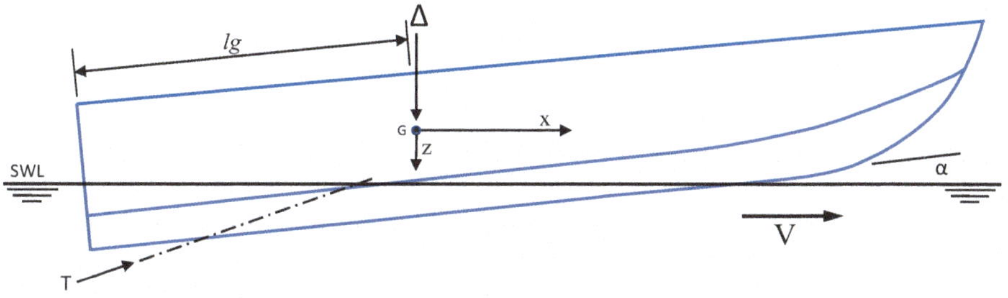

Fig. 5.11 Coordinate systems

Putting (5.7) into (5.6) and taking the non-zero solution, the following characteristic equation must be satisfied:

$$A\sigma^4 + B\sigma^3 + C\sigma^2 + D\sigma + E = 0 \tag{5.8}$$

The coefficients in (5.8) are:

$$\left.\begin{aligned} A &= (Z_{\ddot{z}} - m)\left(M_{\ddot{\theta}} - J_y\right) - M_{\ddot{z}} Z_{\ddot{\theta}} \\ B &= Z_{\dot{z}}\left(M_{\ddot{\theta}} - J_y\right) + (Z_{\ddot{z}} - m) M_{\dot{\theta}} - M_{\ddot{z}} Z_{\dot{\theta}} - M_{\dot{z}} Z_{\ddot{\theta}} \\ C &= Z_z\left(M_{\ddot{\theta}} - J_y\right) + Z_{\dot{z}} M_{\dot{\theta}} + (Z_{\ddot{z}} - m) M_{\theta} - M_{\ddot{z}} Z_{\theta} - M_z Z_{\ddot{\theta}} - M_{\dot{z}} Z_{\dot{\theta}} \\ D &= Z_z M_{\dot{\theta}} + Z_{\dot{z}} M_{\theta} - M_{\dot{z}} Z_{\theta} - M_z Z_{\dot{\theta}} \\ E &= Z_z M_{\theta} - M_z Z_{\theta} \end{aligned}\right\} \tag{5.9}$$

There are four roots σ_1, σ_2, σ_3, and σ_4 corresponding to the variables in (5.7). These roots may be real, imaginary, or conjugate complex values.

Taking one root as $\sigma_1 = \sigma_{1R} + i\sigma_{1I}$, the corresponding heaving response is:

$$z = z_1 \exp(\sigma_{1R} t)\,(\cos \sigma_{1I} t + i \sin \sigma_{1I} t) \tag{5.10}$$

This is an oscillatory response, and the oscillation will decay so long as the real part of characteristic root $\sigma_{1R} < 0$, so that when time $t \rightarrow \infty$, response $z \rightarrow 0$, and the response solution can be considered as stable. If $\sigma_{1\mathrm{R}} > 0$, the amplitude increases without limit as $t \rightarrow \infty$, and this condition is unstable. It is apparent that for a stable system, the real part of characteristic roots must be negative, or real part of max characteristic root $\sigma_{\mathrm{Rmax}} < 0$.

Thus, the system can be judged according to the real part of the characteristic roots by calculating the corresponding characteristic roots for the craft design with various parameters such as running attitude, hull profiles, load coefficients, CG positions, and speed and by plotting the boundary limit. Outside the limit, there may be porpoising motion. The real part of the maximum characteristic root $\sigma_{\mathrm{Rmx}} = 0$ is at a critical condition, so it can be considered as a boundary limit. The question then is how to obtain the roots of characteristic equations? The roots can be obtained from the stability derivatives in the perturbation differential equations.

5.2.2.1 The Determination of Stability Derivatives

This is a key problem for calculating the porpoising motion of a planing hull. Stability derivatives imply the change of forces and moments caused by displacement, speed, and acceleration and are referred to simply as displacement, speed, and acceleration derivatives. Since there is no theoretical method for precisely calculating the dynamic forces and moments acting on a planing hull, and thus the differential equations of motion, the following methods we present are approximate.

1. *The determination of displacement derivatives*

Displacement derivatives are also the static stability derivatives and include the change rates caused by heaving and pitching perturbation. These can be obtained directly by differentiating the hydrodynamic force and moment equations with respect to vertical displacement and pitching displacement, respectively. The semi-theoretical and semi-empirical methods described in Sect. 4.1 can be used to develop the force and moment equations. Using the Shuford-Brown formula [8, 9], we have non-dimensional lift coefficient:

$$-Z_s = \frac{\pi}{4}\sin 2\alpha\cos\alpha\left[(1-\sin\beta)\frac{\lambda}{1+\lambda}+\frac{C_{\mathrm{DC}}}{\pi}\lambda\sin 2\alpha\cos\beta+0.4\sec\alpha\left(\frac{\lambda}{F_{\mathrm{B}}}\right)^2\right] \tag{5.11}$$

Meanwhile the non-dimensional moment coefficient is:

$$M_s = \begin{array}{l}\frac{\pi}{4}\lambda\sin 2\alpha\left[\left(\frac{7}{8}\lambda-0.08\frac{\tan\beta}{\tan\alpha}\right)\frac{(1-\sin\beta)}{1+\lambda}+\frac{C_{\mathrm{DC}}}{2\pi}\lambda\sin 2\alpha\cos\beta\right.\\ \left.+\frac{0.4}{3}\sec\alpha\left(\frac{\lambda}{F_{\mathrm{B}}}\right)^2\right]\end{array} \tag{5.12}$$

Here, we use $-Z_s$ as the positive value for downward direction in this coordinate system.

Differentiating (5.11) and (5.12) with respect to vertical and pitching displacements, the displacement derivatives can be obtained. It must be noted that heaving perturbation only causes a change of wetted length, and no change of trim angle; however, pitching perturbation (assuming rotation about CG) causes both change of trim angle α and wetted length λ. Then we have the following partial differential relationships:

$$\left.\begin{aligned}\frac{\partial Z_s}{\partial z}&=\frac{\partial Z_s}{\partial \lambda}\frac{\partial \lambda}{\partial z}\\ \frac{\partial Z_\text{s}}{\partial \theta}&=\frac{\partial Z_\text{s}}{\partial \alpha}+\frac{\partial Z_\text{s}}{\partial \lambda}\frac{\partial \lambda}{\partial \alpha}\\ \frac{\partial M_\text{s}}{\partial z}&=\frac{\partial M_\text{s}}{\partial \lambda}\frac{\partial \lambda}{\partial z}\\ \frac{\partial M_\text{s}}{\partial \theta}&=\frac{\partial M_\text{s}}{\partial \alpha}+\frac{\partial M_\text{s}}{\partial \lambda}\frac{\partial \lambda}{\partial \alpha}\end{aligned}\right\} \tag{5.13}$$

Where $\frac{\partial \lambda}{\partial z}$ and $\frac{\partial \lambda}{\partial \alpha}$ can be obtained from of geometry of the hull and water surface.

The displacement and moment derivatives can be obtained for the other expressions from lift and moment characteristics.

2. *Determinations of speed and acceleration derivatives*

The speed and acceleration derivatives can be obtained from the hydrodynamic equations of steady-state motion, as the influence of perturbation speed and moment on hydrodynamic force can be determined in steady motion. The hydrodynamic force $\mathrm{d}F_\mathrm{D}$ at each strip ds can be obtained by means of virtual mass theory for a slender planing surface, i.e., the rate of change of momentum of a strip of water with respect to time, $m'\dot{\varsigma}\,\mathrm{d}s$, then:

$$\mathrm{d}F_\mathrm{D}=\frac{\mathrm{d}}{\mathrm{d}t}(m'\dot{\varsigma})\;\mathrm{d}s \tag{5.14}$$

Where $\dot{\varsigma}=\frac{\mathrm{d}\varsigma}{\mathrm{d}t}$ is the vertical speed of the section, and to differentiate from vertical displacement mentioned previously, here we use the coordinate system $O\xi\eta\zeta$.

Since ds changes with time, (5.14) can be written as:

$$\mathrm{d}F_\mathrm{D}=m'\frac{\mathrm{d}\dot{\varsigma}}{\mathrm{d}t}\;\mathrm{d}s+\dot{\varsigma}\frac{\mathrm{d}m'}{\mathrm{d}t}\;\mathrm{d}s \tag{5.15}$$

After integration, the total hydrodynamic force on a planing surface is:

$$F_\mathrm{D}=\int_0^l m'\frac{\mathrm{d}\dot{\varsigma}}{\mathrm{d}t}\mathrm{d}s+\int_0^l\dot{\varsigma}\frac{\mathrm{d}m'}{\mathrm{d}t}\mathrm{d}s \tag{5.16}$$

Equation (5.16) is suitable for any motion, especially for steady motion. Now we assume the craft is moving at steady speed along the X direction and superpose heaving and pitching small perturbations z, $\dot{z}$, $\ddot{z}$ and θ, $\dot{\theta}$, $\ddot{\theta}$. Expressing the integral of two terms on the right-hand side of (5.16) as a function of V, z, $\dot{z}$, $\ddot{z}$, θ, $\dot{\theta}$, $\ddot{\theta}$, we can obtain the relationship of vertical coordinate ς with θ via geometry and also the relation of $\dot{\varsigma}$ and $\frac{\mathrm{d}\dot{\varsigma}}{\mathrm{d}t}$ with perturbation values. The virtual mass m' and $\frac{\mathrm{d}m'}{\mathrm{d}t}$ also can be expressed as a function of perturbations. Substituting them into (5.16), the relationship of vertical hydrodynamic parameters with the perturbation values can be obtained.

Assuming the perturbation values are small, the values above second order can be neglected; thus, the linear relation of vertical forces with perturbation values and, consequently, the derivatives of perturbation values can be obtained easily, including the displacement derivatives, i.e., static derivatives.

In the paper of M. Martin [6], the static derivatives were obtained via the hydrodynamic force equations for the steady planing condition for more accurate predictions; however, the speed and acceleration derivatives were obtained by means of this above method.

Other methods for calculating the moment derivatives with respect to speed and acceleration are found in [5, 6].

According to stability theory, the motion stability can also be judged by the Routh-Hurwitz method [10], in which the motion can be deemed as stable as long as A, B, C, D, and E of the characteristic equation of motion (5.8) satisfy the following stability criteria:

$$\frac{B}{A} > 0,\ \frac{D}{A} > 0,\ \frac{E}{A} > 0,\ \frac{\mathrm{BCD} - \mathrm{AD}^2 - B^2E}{A^3} > 0 \tag{5.17}$$

After obtaining the derivatives, the characteristic coefficients of the equations can be obtained according to (5.9) and (5.17) and can be used to judge the motion stability.

It can be seen that in this calculation for predicting the porpoising motion, the assessment method is strictly theoretical; however, the prediction of derivatives is relatively approximate based on semi-empirical methods, mainly because the calculation of hydrodynamic forces and moments of a planing hull under perturbation is rather difficult.

At present, using the "virtual mass theory" for a slender planing surface may be a suitable method; however, a precise calculation method for virtual mass of sections still has to be improved, particularly for hard chine planing hull sections below the water surface.

The calculation method for hydrodynamic forces and moments under perturbation can nevertheless be extended to seakeeping assessment, as long as considering the wave as an external perturbation. This will be discussed in Sect. 5.4.

5.3 Maneuverability of a Planing Hull

For a planing hull at speed, owing to the great water pressure on the bottom and rather small underwater lateral projected area, the maneuverability of planing hull is rather different from the conventional ship. If propelled by conventional propellers, the underwater area of appendages is a large ratio of the total underwater lateral projected area of the craft and is very important for maneuverability. If using waterjet propulsion with almost no appendages, the maneuverability is mostly controlled by the waterjet and its auxiliary handling equipment. In case of a semi-submerged propeller (SSP), it is controlled by the immersed projected of the semi-submerged propellers.

5.3.1 Course Keeping Stability of Planing Hull

Without underwater appendages, no passive course keeping stability is available for planing hull with a flat bottom. For V profile craft with transverse deadrise, course keeping stability improves with increase of deadrise angle, and for craft using conventional propellers, the course stability may be fine due to the presence of lot of appendages at stern acting similar to a keel.

For craft with reducing deadrise toward the stern course stability may be poor if deadrise is as small as 1–2^0 at the stern. However, this can be improved using conventional propellers with appendages at the stern. Where waterjet and SSP are used, the course stability can be maintained by their control system installations or by adding vertical keels at stern bottom. For most planing hulls with deadrise between 10 and 20 degrees, the course keeping stability is acceptable due to sufficient restoring moment against yawing at high speed.

It must be noted that under the action of lateral wind, the ability to maintain a craft on a straight course without swaying is rather poor and the craft will exhibit swaying motion and heeling into the wind as shown in Fig. 5.12. This is due to increase in lift on the leeward side and decrease in lift on the windward side due to swaying motion.

Meanwhile, when using the rudder for maintaining course, heeling will be increased due to the action of rudder so the effects may cancel one another. In bow quartering wind, swaying will occur accompanied with yawing into the wind.

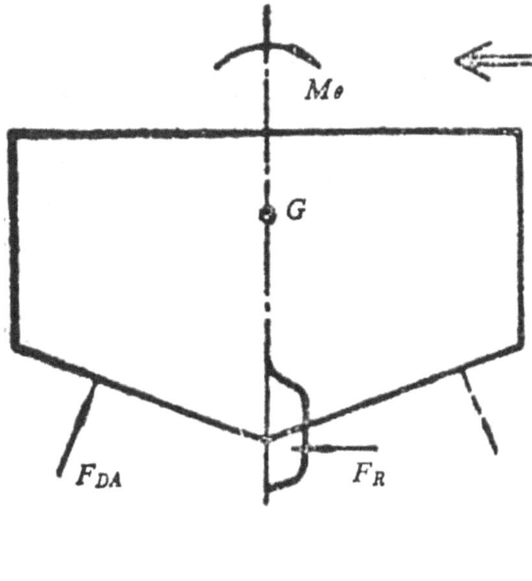

Fig. 5.12 Swaying and heeling due to lateral wind

Using rudder for maintaining course will significantly increase the heeling to headwind and may cause dangerous conditions particularly in waves where the impacts at the bow could be severe because of reduced deadrise angle on the heeling side. For craft with more appendages, this may be less serious due to the damping effect and restoring moment from the appendages. Adding stern skegs can decrease heeling in lateral winds as well.

5.3.2 *Turning Characteristic of a Planing Hull*

The turning diameter of a planing hull is rather large especially at high speed due to the small underwater lateral projected area of the hull and concentration of appendages at the stern. In order to improve turning characteristics, proper measures have to be taken with the stern appendages.

Owing to high speed and high inertia, swaying is possible during turning as shown in Fig. 5.13, which is more serious for craft with a flat bottom. The swaying significantly increases turning diameter (Fig. 5.14) and reduces maneuverability. Another turning characteristic of the planing hull is inward bank turning, which is different from displacement ships where inward bank turn only happens at initial instant of turning while banking outward afterward. For a planing hull, the inward bank turn is due to additional positive lift on the outside bottom and negative

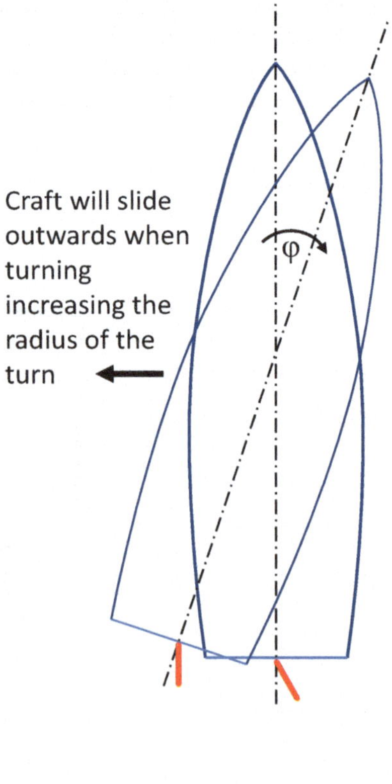

Fig. 5.13 Swaying phenomenon of planing hull

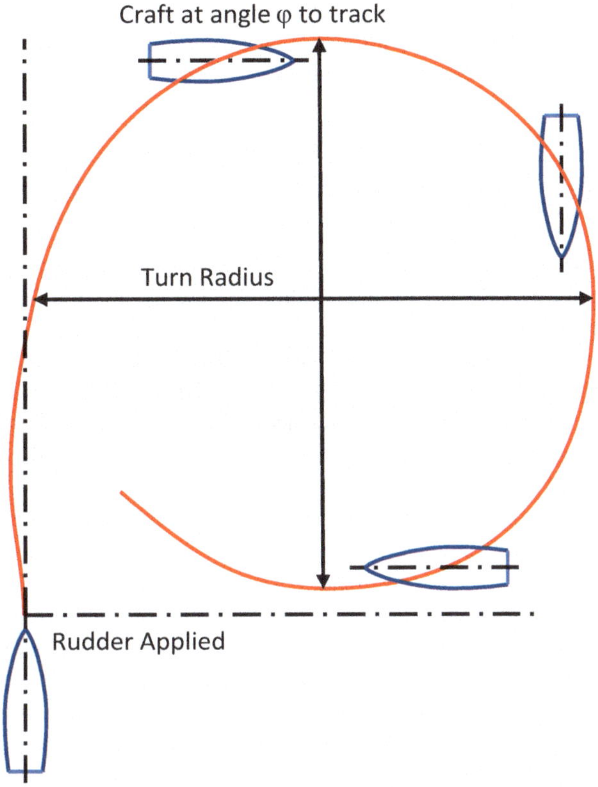

Fig. 5.14 Turning circle of planing hull

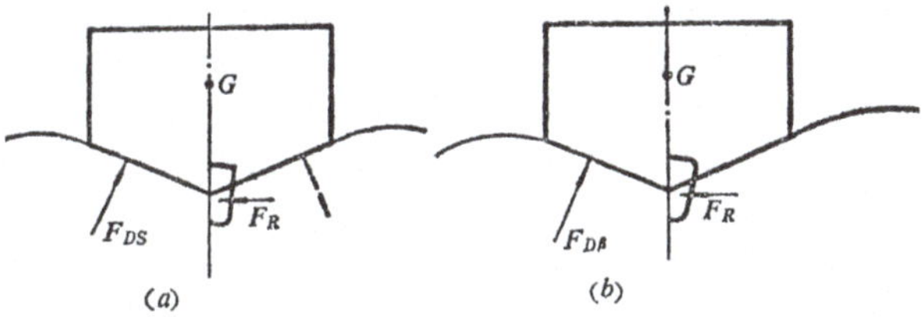

Fig. 5.15 Heeling moments acting on planing hull during turning: (**a**) swaying stage and (**b**) stable turning

additional lift on inside bottom caused by swaying not including the rudder effect as shown in Fig 5.15a. For a craft in a stable turn, there will be a swaying component (Fig. 5.14) in addition to the yaw into the turn caused by the rudder or propulsion devices. The additional hydrodynamic lift acts on the craft with deadrise so as to generate an inward banking moment and a banked turn as shown in Fig 5.15b. It must be noted that here we do not consider the appendages' drag which causes outward heeling, both in stable turning and in swaying stage. The heeling is caused by a combination of rudder force, hydrodynamic force on the bottom, and appendage drag as well. The inward banked turn with large inward heeling angle often happens on a planing hull due to high speed and small underwater lateral projected area that leads to hydrodynamic lift often larger than water damping force even equaling the transverse drag of appendages.

5.3.3 Features of Rudders on a Planing Hull

Due to high speed, small underwater lateral projected area and rudder placement directly aft of propellers, their efficiency is high and so only a small area is necessary.

Since planing hull trim varies with speed, the rudder efficiency also changes with speed and they have to be selected based on design speed. The rudder area of a planing hull is determined according to displacement and speed but not using a percentage of underwater lateral projected area to select dimensions as in displacement ships.

It must be noted that cavitation and air ventilation often occur on rudders of planing hulls. Rudder cavitation happens both from cavitation at rudders themselves and from propeller cavitation. In the case of cavitation being unavoidable, a supercavitating rudder with wedge section can be used to maintain stable cavitation and to avoid cavitation erosion and vibrations. The extension of cavitation from propellers can result in rudder erosion which happens on many craft. It seems there are no measures to solve such problems except for using materials that are most resistant to the erosion and changing the relative position between rudder and propeller.

In order to avoid suction of air into a rudder, the rudder is recommended to be installed under the hull bottom and not aft of the stern close to free water surface or piercing the water surface. For a rudder installed under the craft bottom with minimized gap to the hull surface, the aspect ratio effectively doubles and significantly decreases the free vortex at the top of the rudder. For this reason, the rudder chord at top often is lengthened forming a trapezoidal rudder as shown in Fig 5.16a. Thus, with the mirror image, a near optimum ellipse like circulation distribution is achieved. Regarding the peach shaped rudder (Fig 5.16b), it is designed to avoid floating debris attaching to the top of a trapezoidal rudder, and the profile of the peach like rudder is closer to semi-ellipse form.

Rudders on a planing hull are usually designed to be balanced or semi-balanced rudders.

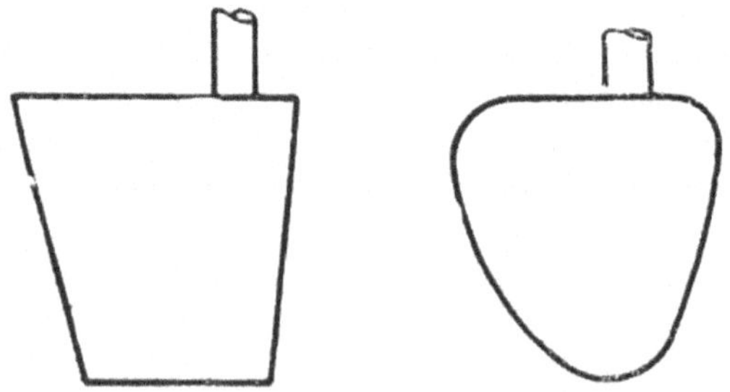

Fig. 5.16 Trapezoid and peach-like rudder

When using waterjet propulsion, the guide vanes aft of the waterjet has to be installed so as to provide the necessary forces and moments for turning, stopping, and reversing. In general, the installation is provided by the waterjet supplier rather than designed and manufactured by ship designers and builders.

For theoretical solution of maneuverability, and so sizing of rudders, the yawing and swaying perturbation differential equations have to be established and course stability has to be assessed by means of stability derivatives. For turning motion, the differential equations of motion in steady turning have to be formed, but calculation of the coefficients is quite difficult due to the asymmetric motion of the craft.

For conventional ships, the planar motion mechanism (PMM) can be used in a towing tank or maneuverability rotating tank to obtain the motion derivatives. However, for a high-speed planing craft, the maneuverability motion equations are not the coupled surging, swaying, and yawing motion equations, but coupled rolling, swaying, and yawing motion equations, due to large rolling present during turning motion of planing craft.

Thus, nonlinearity and viscous effects are important, and the hydrodynamic derivatives are difficult to obtain by experimental methods. Thus, the theoretical method for investigating the maneuverability of planing hull is still not well proven. Analysis of maneuverability and design of rudders is therefore generally based on practical experience.

5.4 Seakeeping Quality of Planing Hulls: Impact Forces

Seakeeping characteristics of planing hulls include impact forces, motions, accelerations, increase of resistance, and changes to maneuverability in a seaway. In Sect. 5.5, we will first discuss spray and slamming forces from waves, and then consider motion analysis and speed loss due to increased resistance.

Owing to high speed, small size, and high hydrodynamic pressure on the hull bottom, wave impact forces are significant, and this can limit the craft operation.

During the early development of seaplanes and flying boats in the 1920s and 1930s, the forces and spray during landing and takeoff were investigated through model tests, due to the importance of water drag during take-off and that of spray in affecting the vision of pilots and performance of air propellers.

Research on the response in waves, speed loss, and manoeuverability of planing craft was started later than the work for seaplanes and flying boats. This was due to the non-linearity of forces and responses and so difficulty of analysis compared to conventional ships where a linearised approach was possible. The research for planing craft began in the 1950's and 1960's.

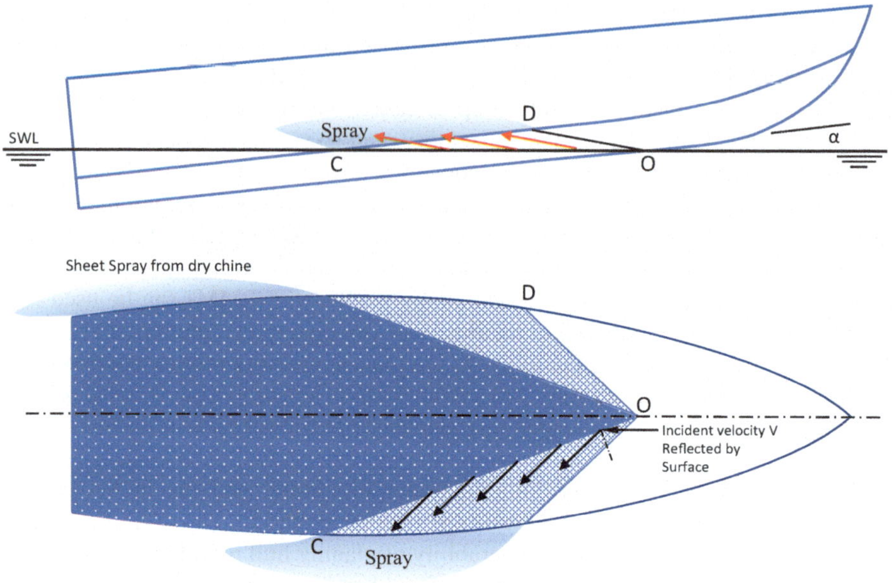

Fig. 5.17 Spray of craft planing in the calm water

5.4.1 Spray on Planing Hulls

The spray generated by planing hulls is due to hydrodynamic pressure acting on the bottom of hull and impacts its performance both on calm water and in waves. The spray of planing craft is more significant in waves.

There are two parts to spray from planing craft in calm water. One part is from the stagnation line of bottom OC, "forward spray," with thin water drops corresponding to the reflecting jet of a two-dimensional planing plate thrown out to the side due to the component of incoming water along the parallel stagnation line OC as shown in Fig. 5.17.

Another spray element, the so-called "sheet spray," is blown out from the hard chine starting at intersection point C, which starts as a continuous thin membrane like wide water spray. This sheet can form a conical water membrane with origin at point C. The height and width increase with speed and trim angle, and the top of the cone moves aft as point C moves aft. With further increase in speed, the trim angle starts to decrease, and the height and width of the spray decrease correspondingly, while being ejected from the dry length of the chine. Spray rails can depress and minimize this spray element.

Spray thrown forward has less influence on craft dynamics and can be neglected, so more attention has to be paid to the sheet spray thrown sideways and the "forward spray" or solid water comprising the upwelling that is thrown forward.

Tests demonstrate that the forward spray of craft in calm water is mainly related to the following factors:

1. *Dynamic load coefficient* C_B: this reflects the hydrodynamic pressure per unit area on the bottom and high C_B results in serious spray. C_B can also be related to the static load coefficient C_Δ and F_B, i.e., $C_B = 2C_\Delta / F_B{}^2$.

Figure 5.18 shows spray height versus static load coefficient C_Δ at given CG and speed. The measurement is carried out at a longitudinal position of one craft width ahead of the stern. The figure includes the effect of trim angle as well, since a change in C_Δ results in a change in trim angle τ for a given speed.

2. *Trim angle* τ*:* To consider the effect of trim angle separately, one may keep C_Δ and F_B unchanged, i.e., C_B unchanged, and change the trim angle via CG as shown in Fig. 5.19. It can be seen that the sheet spray increases with trim angle due to reduced wetted area and increased dynamic pressure.

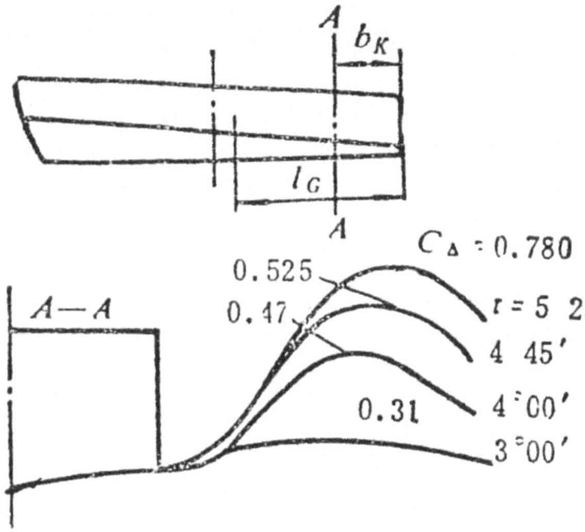

Fig. 5.18 C_Δ influence on spray height ($l_G/b = 2.5$, $F_b = 3.55$)

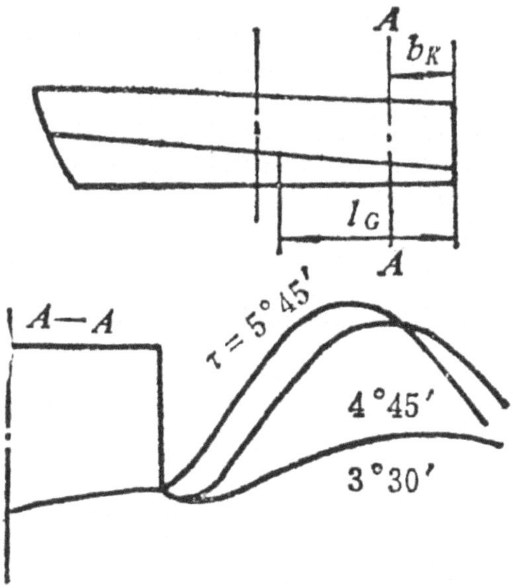

Fig. 5.19 Influence of trim angle on spray height ($C_\Delta = 0.625$, $F_b = 3.55$)

The dynamic pressure can be represented strictly by the lift coefficient and relationship between C_B and C_L where:

$$C_L = \Delta / \left(0.5\rho\ V^2 S\right), \text{and } C_L = C_B/\lambda \tag{5.18}$$

Thus, decrease of λ corresponds to an increase in load per unit area.

It may be seen from this that heavier loading and higher trim increase the height of the spray envelope emanating from the chine. This reflects the energy used and hence the induced drag. The objective therefore is to minimize ejected spray, and this minimizes drag.

3. *Craft profile*: The effect of bottom shape will be discussed in this section.

For a craft running in waves, there will be heavy spray, particularly at the bow, due to the impact of craft on the oncoming waves. The factors influencing the spray in calm water are the same in waves—load coefficient, trim angle, etc. However, owing to theoretical and experimental complications characterizing spray due to two-phase flow, it is difficult to establish the spray criteria. Nevertheless, some measures for improving the generation of spray in design are:

- Improving bow profile by avoiding large entrance angle of waterlines and steep buttock lines, so as to make smaller angle of attack at bow. Model tests suggest that entrance angle of 12 to 15 degrees is reasonable. Excessively small entrance angle may cause steeper buttock lines with flared shape on frames, both for bottom and sides.
- Improving amidships and aft shape, and applying twisting sections for depressing the main spray in calm water; flared sides forward can also improve deck wetness of craft in waves.
- Installing chine strakes can depress the spray, and will be more effective, if installed at the root of forward spray, i.e., starting a little forward of location C in Fig. 5.17. Chine strakes can be installed horizontally or inclined, but with a penalty of increased drag.
- Installing stern wedges or flaps may decrease the trim angle, the rooster tail stern wave, spray and also improve the working condition of propellers and rudders.
- Installing longitudinal anti-spray rails across the stagnation line (Fig. 5.20) can decrease "main" spray and also drag at high speed.

5.4.2 *Planing Hull Wave Slamming*

Hull impact is an important problem in the seakeeping quality of planing craft in waves. This happens generally in two ways:

- Direct impact by waves on craft bottom
- Waves generating vertical speed and acceleration which impact the hull bottom on water

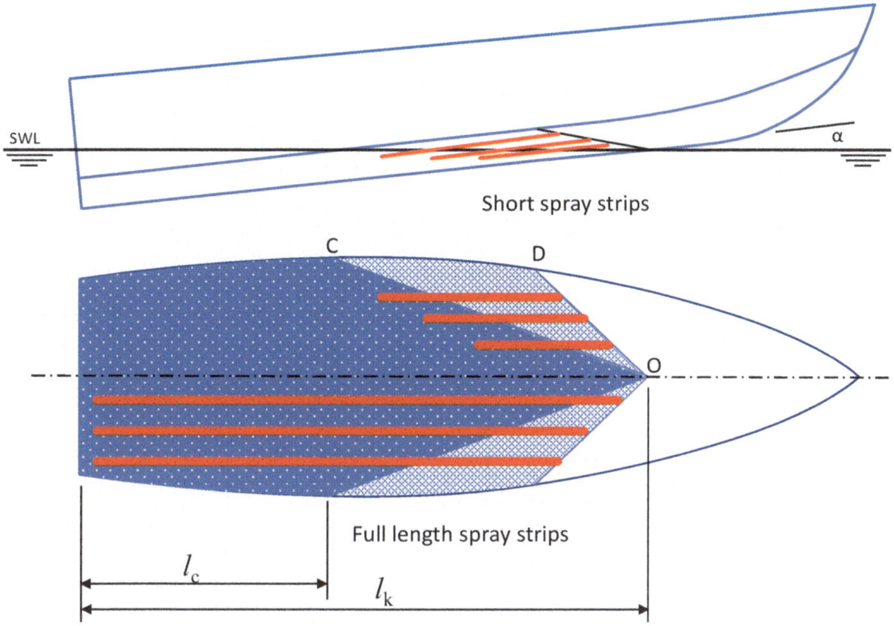

Fig. 5.20 Longitudinal anti-spray rail at bottom of craft

The impact loads may be as great as a factor of 2 to 10 times static pressure and will seriously influence normal operation of the craft. The impact force can be represented by an overload coefficient n, where the pressure $P \equiv (n + 1)\ \Delta$. In tests, the measurement used is the impact acceleration α, where $\alpha = (n + 1)g$. Thus, the overload coefficient can be obtained by measuring impact acceleration.

1. *Distribution of impact force along the craft length*

The early test results of full-scale craft released by Central Aero-Hydrodynamic Research Center of Russia, CAHI, demonstrated that the heaviest impact on a craft hull was at the bow, near #3 station (out of total 20 stations). Figures 5.21 and 5.22 show the test results of overload coefficients along craft length for craft without and with steps, respectively. The abscissa is the ratio of x, distance from transom, with craft beam B, and the ordinate is relative load coefficient. For craft without a step, the ratio of load coefficient at that station to that at CG is $(n + 1)/(n + 1)_G$, while for the stepped hull, the ratio is $(n + 1)/(n + 1)_S$, is measured relative to the step position.

From the figures, it can be seen that maximum impact load is at the bow for both craft and decreases gradually toward the stern. The overload of the plain hull is greater than that of stepped planing hull indicating poorer seakeeping quality of the plain hull compared to craft with a step. However, the results mentioned above were obtained for craft with certain size and waves. In practice, the distribution of impact force is relative to wavelength ratio and encounter frequency.

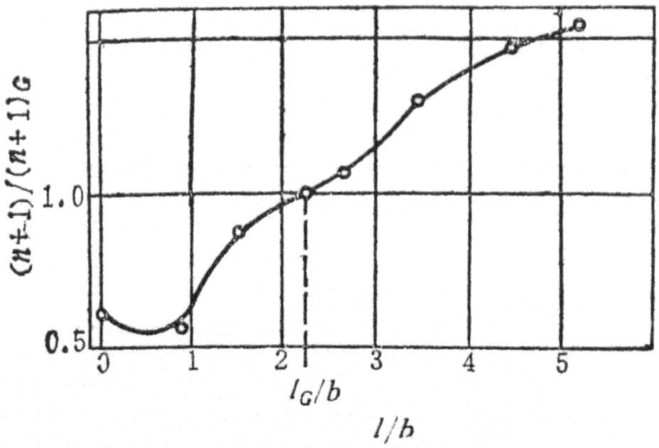

Fig. 5.21 Distribution of overload coefficient of craft without step along the craft length, $n + 1 = 3.4$ at CG ($C_{\Delta} = 0.56$, $F_B = 2.5$, Sea state 6). Note n is the acceleration, the plus 1 is to account for gravity. Normalized acceleration at CG

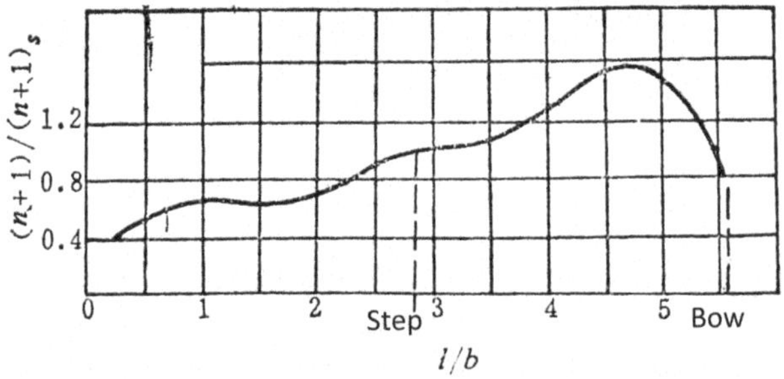

Fig. 5.22 Distribution of overload coefficient of craft with step along the craft length, $n + 1 = 5.4$ at step ($C_{\Delta} = 0.63$, $F_B = 2.85$, ss 5 ~ 6)

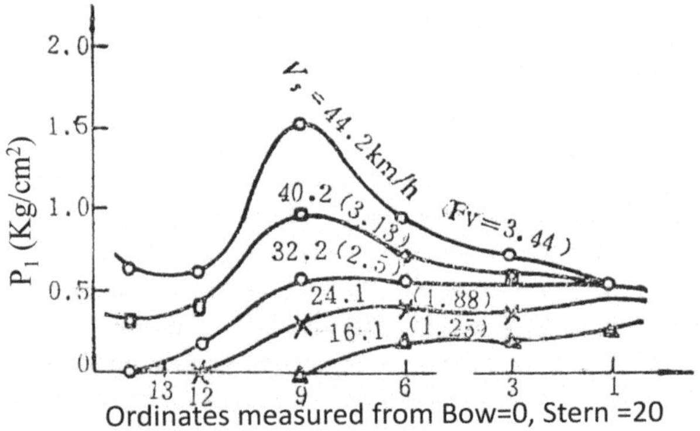

Fig. 5.23 Distribution of impact load along the craft length versus speed

Some model test results from Japan showed during lower speeds in certain waves that maximum impact force happened at the bow, but as speed increased, the maximum impact force shifted aft as shown in Fig. 5.23 [11]. The test craft had length 6 m, beam 2.44 m, draft 0.39 m, displacement 2.2 t, maximum speed 35knots, and deep *V* transverse section with hard chine. The test was in wave height 0.6 m, wavelength ratio $\lambda/L = 2.5$. The figure shows that as speed increases, the location of maximum impact force shifts aft. This is due to the boat taking off from the water

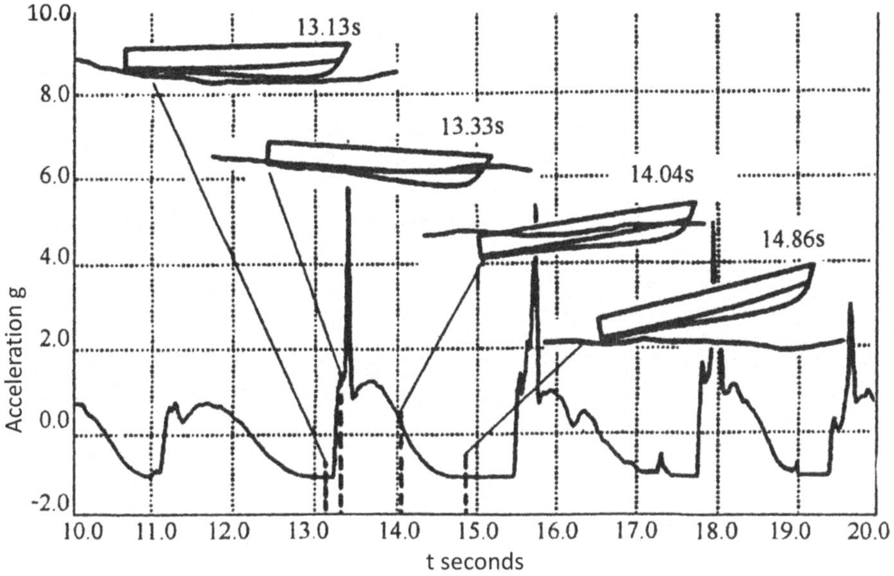

Fig. 5.24 Vertical acceleration time trace at CG of a patrol boat

surface and then dropping down into the water causing impacts with wave surface at high speeds. The contact area is random, so the position of maximum impact force is also random from bow to amidships and even toward the stern.

Figure 5.24 shows the vertical acceleration at CG of a patrol boat with length of 17.4 m running at speed 40 knots in SS5 waves, simulated by Lin et al., in 1955, where $Fn = 1.58$. The figure shows the relative position and orientation of the hull with waves at each step in the time history [12].

In Fig. 5.24, when $t = 13.13$ seconds, the craft takes off from water surface, with acceleration g downward. When $t = 13.33$ seconds, the craft drops down again, then meets with a big wave, and acceleration at 0.2 seconds time interval from that point increases to +1.3 g. After this, there is a very large acceleration close to 6 g. When $t = 14.86$ seconds, the craft takes off once again with acceleration 1 g downward.

In general, the maximum impact load close to the stern should be avoided, due to not only propulsion equipment being located there, but also if there is a flatter bottom there, it is easy to cause heavy impact. Of course, this is relative to wavelength ratio, speed, and running attitude.

Figure 5.25 [11] shows the test results of another larger craft with length 32 m, beam 8.5 m, draft 1.15 m, displacement 105 t, maximum speed 43knots, and a "w"-type transverse section profile. This craft did not take off due to its size. This plot also shows that maximum impact load moves aft with increase in speed.

This shows that the distribution of load is relative to speed, wave, and running attitude. In short, maximum impact force happens generally at the region from #3 to #10; therefore, the structural design of a hull in that region should be considered carefully.

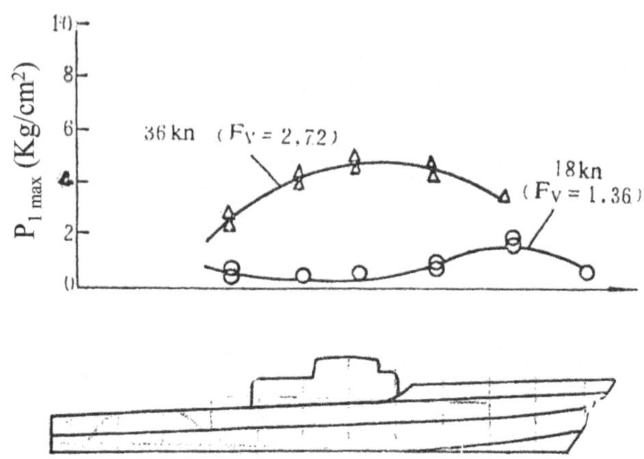

Fig. 5.25 Distribution of impact force on the bottom of bigger craft along the craft length

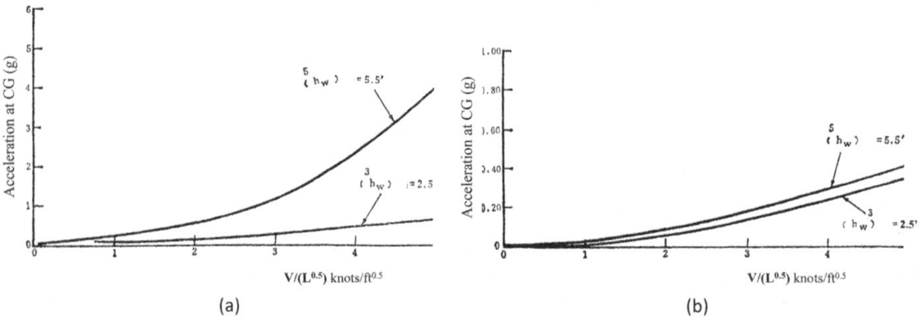

Fig. 5.26 Impact acceleration of planing hulls in irregular waves: (**a**) bow waves and (**b**) following waves

2. *The influence of sea sate and running trim on impact force*

The sea state in terms of wave height and the running trim of a planing hull in terms of speed and trim angle directly determine the impact force on a vessel.

The influence of wave height is the most obvious, and *impact acceleration* is in proportion to square power of wave height. Savitsky gave the vertical acceleration of planing hull at CG in irregular waves, as shown in Fig. 5.26 [4] for head sea waves. The abscissa is speed, with $V/(L^{0.5})$, where V is in knots and L is in feet. For $V/(L^{0.5}) > 2$, when wave height increases from $(h_w)_{average} = 2.5$ feet to 5.5 feet, impact acceleration increases up to five times, almost in proportion to square power of wave height. Figure 5.26b shows the vertical impact acceleration of craft in following waves, and it can be seen that the impact acceleration is reduced compared to head sea condition.

The influence of speed on *impact force* on the hull is in proportion to square power of speed. The running trim angle is also an important factor influencing the impact force because the vertical speed component is in proportion to the trim angle;

thus, it can be shown that the impact force at the bow section per unit length is in proportion to the square power of effective trim angle. So, reducing trim angle can significantly decrease the impact force on the bottom. Thus, using a stern flap, wedge plate or stern interceptor to optimize trim can minimize the running trim angle and so the impact force.

Savitsky and Brown gave the following empirical formula for average impact acceleration of planing hulls in waves [9]:

$$\overline{n}_{CG} = 0.0104\left(\frac{H_{!/3}}{b} + 0.084\right)\frac{\tau}{4}\left(\frac{5}{3} - \frac{\beta}{30}\right)\left(\frac{V_s}{\sqrt{L}}\right)^2\frac{L/b}{C_\Delta} \tag{5.19}$$

$$\overline{n}_{bow} = \overline{n}_{cg}\left[1 + \frac{3.8\left(\frac{L}{b} - 2.25\right)}{V_s/\sqrt{L}}\right] \tag{5.20}$$

Where $\overline{n}_{CG}$ is the average impact acceleration (in g) at CG

$\overline{n}_{bow}$ is the average impact acceleration at bow

b is the craft beam(feet)

$H_{1/3}$ is the significant wave height (feet)

τ is the trim angle of planing surface(deg)

β is the deadrise angle of bottom(deg)

L is the craft length(feet)

C_Δ is ∇/b^3, the static beam loading coefficient of the craft

Δ is the craft displacement(lbs)

∇ is the volumetric displacement of craft (ft^3)

V_s is the craft speed(knots)

V is the craft speed(ft/s)

The applied ranges of (5.19) and (5.20) are as follows:

$\Delta/(L/100)^3$ = 100 to 230, L/b = 3 to 5, τ (deg) = 3 to 7, β (deg) = 10 to 30, $H_{1/3}/b$ = 0.2 to 0.7, V_s / ($L^{0.5}$) =2 to 6, where V_s is craft speed in knots and $V_s/(L^{0.5})$ = 2 to 6 corresponding to $V/(L^{0.5})$ = 3.375 to 10.126 where V is in fps.

From the average acceleration calculated using (5.19) and (5.20), one can obtain the maximum average acceleration of 1/N values, as follows:

$$\overline{n}_{1/N} = \overline{n}(1 + \log_e N) \tag{5.21}$$

The significant acceleration (average highest 1/3) value and average value of highest 1/10 can be obtained by assuming exponential distribution as in (5.21).

Shinishi Niwa also gave an empirical formula for predicting the impact force on the bottom as follows [11]:

$$P = \frac{gV_s^2}{100} + (1 + n_1)\frac{g\Delta}{l_s \times b_{c\,max}} \tag{5.22}$$

Where P is the max impact pressure, kg/cm^2

V_s is the craft speed, knots

n_i is the overload coefficient at bow

g is the gravity acceleration

Δ is the craft weight, tonnes

L is the max length, m

b_{cmax} is the max width of hard chine, *m*

The formula for the overload coefficient at the bow is also in proportion to V^2 at high speed for $V/(L^{0.5}) > 2$

3. *Influence of craft profile on the impact force*

Generally speaking, the craft profile with good hydrodynamics in calm water (flatter bottom, lower deadrise) experiences higher impact force in waves.

(a) *Influence of deadrise on impact force*

The deadrise is a most important factor influencing the impact force. Figure 5.27 shows the tested values of vertical overload of prismatic sections with various

Fig. 5.27 $n \sim C_\Delta$ at different β

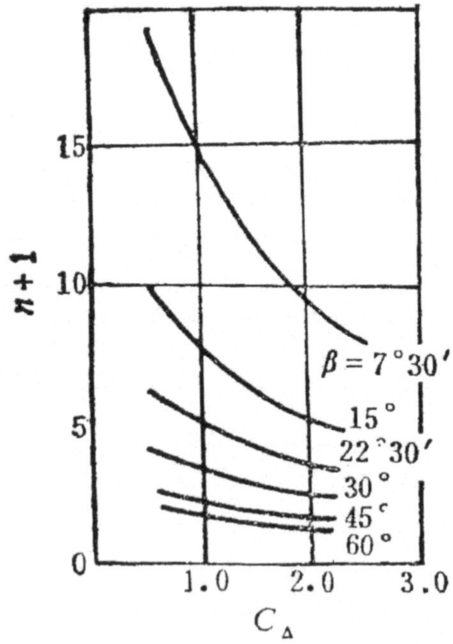

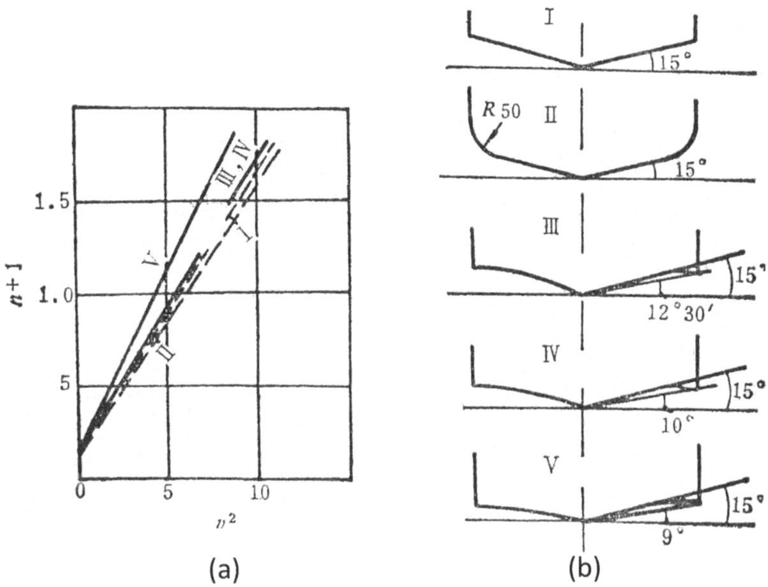

Fig. 5.28 Impact load of various hull cross sections with $\beta = 15°$, where $b = 300$ mm, $C_\Delta = 0.585$

deadrise values at different static load coefficient C_Δ. From the figure, it can be seen that as β increases, overload decreases.

(b) *Influence of transverse section shapes on impact force*

Figure 5.28 shows results from drop tests of five different prismatic sections with $\beta = 15^0$. Section I is straight deadrise, section II is round bilge, sections III, IV, and V are curved transversely with the inclination angle between the bilge-keel line and base line of $12^0 30'$, 10^0, and 9^0, respectively. The plots show the impact load coefficient versus square power of drop speed V_0.

It can be seen that impact force is minimum with round bilge and increases for curved sections, however, not significantly except for Section V. From the figure, it seems section IV is most suitable combining fine hydrodynamic properties with not too large impact load.

Vosper in Great Britain carried out seakeeping quality tests of four craft with different transverse sections in both regular and irregular waves, as shown in Fig. 5.29. Although the drag in calm water of A and B is lower than C and D, about 29% at high speed, the tests in waves demonstrate that the bow impact acceleration at 1/4 length is largest for C, minimum for D, and B a little better than A. However, if trim angle of A is decreased to 2 deg., then the bow impact acceleration of A was not only smaller than B but also better than D. Figure 5.30 shows the test results of model types B, C, D in irregular waves from [11].

From Fig. 5.30 it can be seen that the impact force acting on the craft with round bilge, Model D, is smallest while it had poor drag performance in calm water.

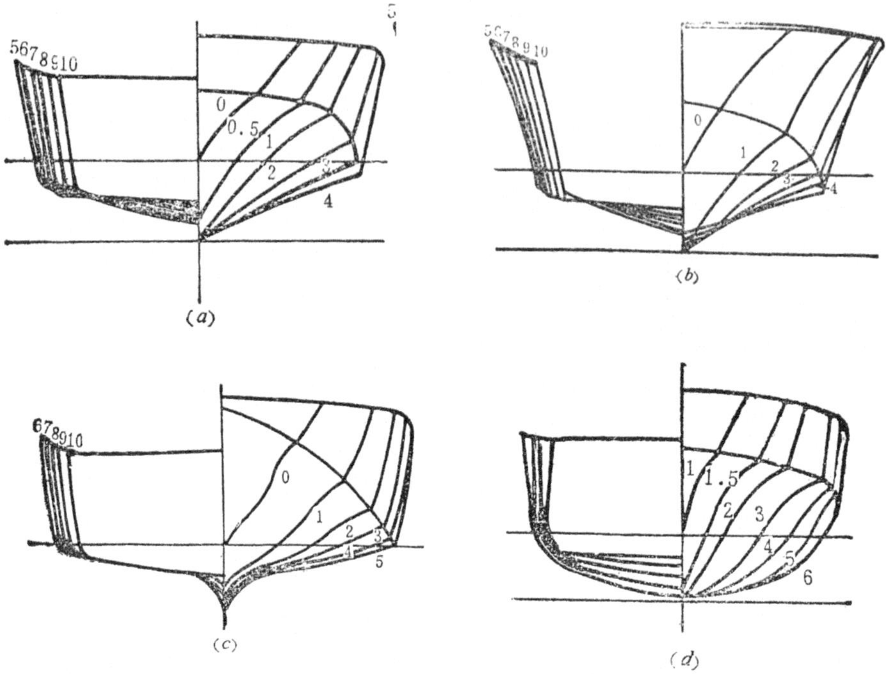

Fig. 5.29 Four models of Vosper Co, UK. (a) A type, (b) B, (c) C, (d) D. (From ref. [2] in ch 4)

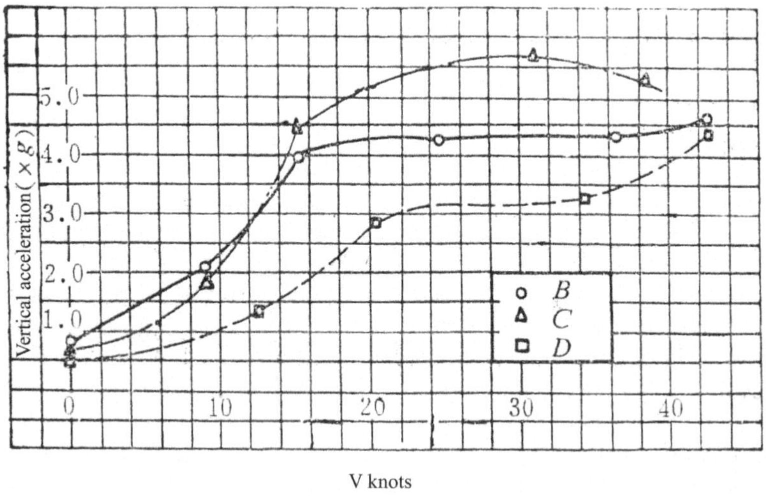

Fig. 5.30 Max acceleration versus average speed of planing hulls at bow in irregular heading waves

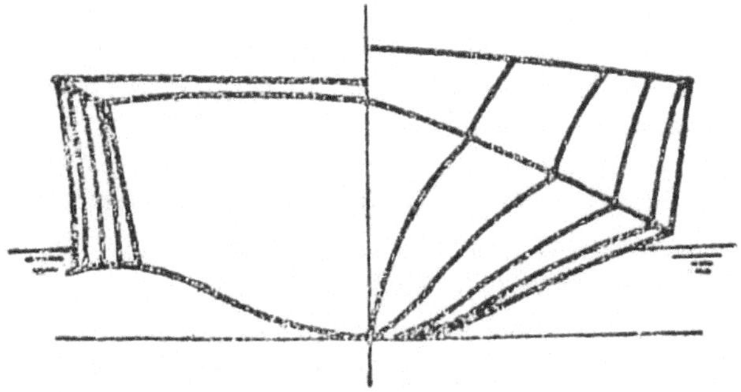

Fig. 5.31 Typical section of W craft

Meanwhile, the bow acceleration of craft C is large due to its small deadrise. The impact acceleration of Model B with deep V is similar to that of Model C at low speeds but decreasing at high speeds and almost matching Model D acceleration at speed 42 knots. This demonstrates that using deep V form on planing craft is appropriate.

Reference [11] also discusses the comparison of deep V with a 'W type' craft using the test results of two craft with same displacement, main dimensions, and CG position. The test results demonstrated that drag of the deep V type in calm water at low speed is better than the W type, but poorer than W type at high speed. The impact acceleration of the deep V is 1/2 to 1/3 of W type, that means V type craft can be operated in waves with 1.5 times the wave height that of W craft. The difference is due to the W craft configuration with flat and wide keel and hollow bilge as shown in Fig. 5.31, which maintains the hydrodynamic pressure at bottom favorable but increases the impact loads especially at the hollow bilge.

5.5 Seakeeping Quality of Planing Hulls: Motions and Speed Loss

5.5.1 Heaving, Pitching, and Speed Loss of Planing Craft in Waves

Heaving and pitching are the main parameters that characterize the motion of a planing hull in waves. A series of seakeeping tests of planing hull models with prismatic sections, in both regular and irregular waves, were carried out at the Davidson Laboratory of Stevens Institute of Technology, USA [13, 14]. The data provides results showing the influence of the main geometric parameters on motion response of planing hulls.

Three models with deadrise of 10^0, 20^0, 30^0 and prismatic section with a beam of 22.9 cm (9 inches) were used. Three length/beam ratios $l_{co} / b_c = 4, 5, 6$ of the model with $\beta = 20^0$ and one ratio $l_{co} / b_c = 5$ for the models with $\beta = 10^0$ and 30^0 were considered for testing. The displacement of the models corresponds to $C_\Delta = 0.608$

and 0.912, respectively. Longitudinal CG position is varied from 50–70% L from bow. Trim angles of 4^0 and 6^0 were set. Speed length ratios tested were $V_s/l_{co}^{0.5} = 2$, 4, and 6 measured in fps units. In the case of using international unit, m, the corresponding speed/length ratios are 3.63, 7.25, and 10.87.

In regular wave tests, six wavelength ratios were used, $\lambda/l_{co} = 1.0$, 1.5, 2.0, 3.0, 4.0, and 6.0. Main wave height is 25.4 mm (1″), whereas while investigating the linear relation between motion response and wave height, wave heights of 25.5.76.2 mm (1 to 3 inches) were used.

When the speed length ratio $V/L^{0.5} = 2$, corresponding to volumetric Froude number $F_{n\nabla} \approx 1.5$, the craft is still in displacement operation, so the test results are similar to displacement craft. Thus, the linear relationship for motion response with wave height causes increasing length/beam ratio to decrease motion response. When $V/L^{0.5} = 4$ or 6, corresponding to $F_\nabla = 3.0$ to 4.5 the craft runs in planing mode, and the wave response is rather different from a displacement ship, being nonlinear. The linear superposition method, applied usually for conventional displacement ships for predicting wave response from model test to irregular waves, cannot therefore be applied.

From the test results, key conclusions can be outlined as follows:

- When running in planing mode, $V_s/l_{co}^{0.5} = 4$ and 6, corresponding to $F_\nabla = 3.0$ to 4.5, no linear relationship between wave height and motion response is present, except in very long or short wavelengths, thus preventing the use of the spectral analysis method. However, the motion response is in proportion to $0.7 \sim 0.85$ power of wave height, a weak nonlinear relationship. Nevertheless, drag increment and impact acceleration are in "strong" nonlinear relationship with wave height.
- Increasing the static load coefficient C_Δ decreased the motion response and increased the drag in waves.
- Motion response, drag increment, and impact acceleration have all increased with trim angle, particularly with increasing speed. Therefore, craft with an adjustable transom plate for regulating the angle of attack are recommended for operation in waves. In emergency condition, shifting load forward is also suggested for reducing trim angle.
- At low speed, $V_s/l_{co}^{0.5} = 2$, the effect of β on the motion response, drag increment, and impact load is minimal. In planing mode, impact acceleration is affected by β and this is apparent. The effect of deadrise angle on the motion response and drag increment is not significant at $V_s/l_{co}^{0.5} = 4$; however, when $V_s/l_{co}^{0.5} = 6$, increasing deadrise reduced the motion responses significantly, including heaving and pitching, and correspondingly reduced the drag increment.
- When $V/L^{0.5} = 2$, similar to a displacement ship, increasing length/beam ratio can improve seakeeping quality. In planing mode, increasing length/beam ratio leads to increase in impact acceleration but not the drag increment. Regarding the motion response, this will be different for various wave lengths. For long waves, $\lambda/l_{co} > 3$, increase of length keeping beam unchanged causes larger motion response, and shorter wavelength, $\lambda/l_{co} < 3$, causes lower motion

response. Generally speaking, the influence of craft length on the motion response is less than a displacement craft.

The conclusions above have been validated in the tests of craft in irregular waves. From the test results above, G. Fridsma set up graphs and formulae for predicting the motion responses of planing hull in waves, i.e., heaving, pitching, drag increment, and impact acceleration that are detailed in the references [13, 14].

Based on this work, Savitsky gave a typical example for showing heaving and pitching versus length/beam ratio of a planing hull running in regular seas, as shown in Fig. 5.32 [4]. In the figure, the abscissa is speed/length ratio (units of knots and feet). When $V/L^{0.5} < 2$ a craft runs in displacement mode and the motion in waves is large; however, for a craft running in planing mode, $V/L^{0.5} > 2$, the craft may be skipping from peak to another peak, the motion will be relatively less, and heaving to wave height ratio increases with speed, but pitching to wave height ratio decreases, or stays the same. The figure shows the effect of wavelength ratio λ/L_m, (L_w/LWL), showing that motion response increases with wavelength to craft length ratio, due to the reduction of the skipping phenomenon, so as to increase the heaving and pitching response.

Sometimes two peaks of motion response versus wavelength ratio of planing hulls running at high speed in regular waves have appeared in test results as shown in Fig. 5.33, which is due to skipping of craft between two peaks of waves close to each

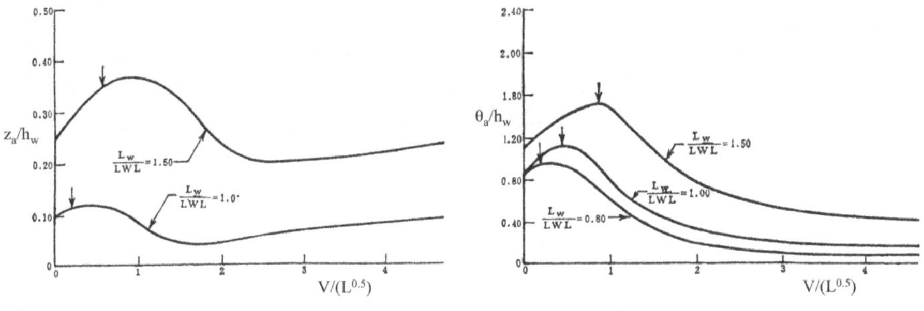

Fig. 5.32 Heaving and pitching motion of planing craft in heading regular waves

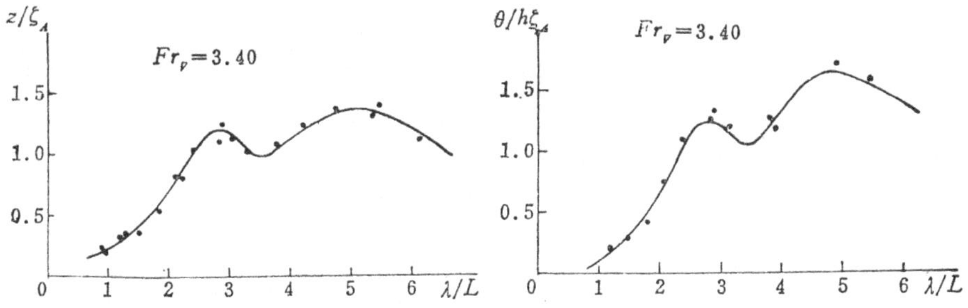

Fig. 5.33 Heaving and pitching motion of planing hulls in regular heading waves with two peaks

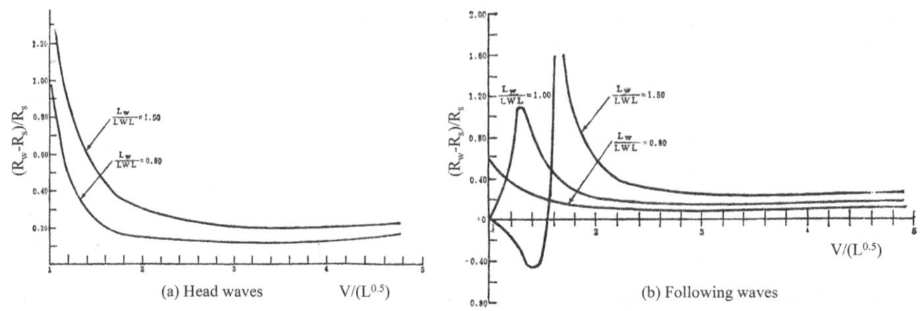

Fig. 5.34 drag increment of planing hull in regular waves: (**a**) head seas and (**b**) following seas

other that creates the dip. The figure shows heave and pitch amplitude for a speed of $Fn_\nabla = 3.4$ for varying relative wavelength to craft length λ/L.

A typical example of drag increment for planing hulls in waves is as shown in Fig. 5.34 [4], where (a) shows drag increment versus speed/length ratio in head waves, and (b) shows the drag increment in following waves. Here the ordinate is $(R_w - R_s)/R_s$ where R_w is the total resistance measured in waves and R_s is the total resistance in smooth water.

It can be seen that in head waves, the drag increases inversely with speed/length ratio and is almost unchanged after a certain speed. Since drag increases with speed in calm water, the absolute value of the drag increment is increased. The figure also shows the effect of wavelength ratio. The drag increment increases with wavelength ratio due to the increase of motion response. In following waves, a peak of drag increment in the low-speed region is present. At high speed, the craft is also skipping between the wave peaks and the drag increment trends flat.

Savitsky and Brown [9] also gave the empirical equations for predicting the drag increment of planing craft in waves as follows:

$$\left.\begin{array}{ll} V_s/\sqrt{L}=2 & \dfrac{R_{Aw}}{2b^3} = 6b \times 10^{-6}\left(\dfrac{H_{1/3}}{b}+0.5\right)\dfrac{\left(\dfrac{L}{b}\right)^3}{C_\Delta} + 0.0043(\tau - 4) \\ V_s/\sqrt{L}=4 & \dfrac{R_{Aw}}{\Delta} = \dfrac{0.3H_{1/3}/b}{1+2H_{1/3}/b}\left(1.76 - \dfrac{x}{6} - 2\tau_{am}^2\beta\right) \\ V_s/\sqrt{L}=6 & \dfrac{R_{Aw}}{2b^3} = \dfrac{0.158H_{1/3}/b}{1+\left(H_{1/3}/b\right)\left[0.12\beta - 21C_\Delta(5.6 - L/b) + 7.5(6 - L/b)\right]} \end{array}\right\} \quad (5.23)$$

Where R_{Aw} is the drag increment in waves, lbs

λ is the water density, lb./ft^3

Other notations are the same as in (5.19) and (5.20). The application range is also as in (5.19) and (5.20).

Since the nonlinear components of drag increment and vertical acceleration are greater at high speed, the prediction of impact acceleration and drag increment of planing hulls running at high speed is less precise.

5.5.2 *Broaching of Planing Hulls in Following Waves*

Similar to displacement ships, broaching can occur to planing hulls when running in following waves. "broaching" is a motion phenomenon of losing course stability and handling, as well as becoming broadsided to the waves when planing craft operate in following waves. Consequently, high heeling angles may occur and even overturning transversely when the craft is broaching to beam-on to the waves.

In view of these serious consequences, a shipbuilder will pay great attention and interest in investigating the phenomena. However, due to the complex nature of the phenomenon, only the reasons causing broaching based on observations and experimental investigations are presented here.

- For a craft running in following seas, when the bow is located at a wave hollow and stern at peak, the bow may be dropping down and stern may be lifted up from water surface as shown in Fig. 5.35. Meanwhile, the center of underwater lateral projected area of the craft is shifting forward, and the craft loses course stability. The wave may apply a great turning moment on the craft, especially in stern quartering waves, thus causing heeling of the craft. Such a situation is not possible in head seas due to short time duration with higher encounter frequency.
- In view of the circular motion of water molecules in waves, the bow located at a wave trough will experience downward wave–induced velocity that leads to decreased angle of attack of incoming water. However, the stern is in the wave crest with upward wave–induced velocity and increased angle of attack of incoming water. Consequently, such actions will tend to make the craft pitch down, possibly leading to plough-in of the craft.
- The wave-induced velocities also decrease the incoming velocity to rudders decreasing their effectiveness and making it difficult to maintain course.
- Craft heeling also leads to turning moment of craft in following seas.

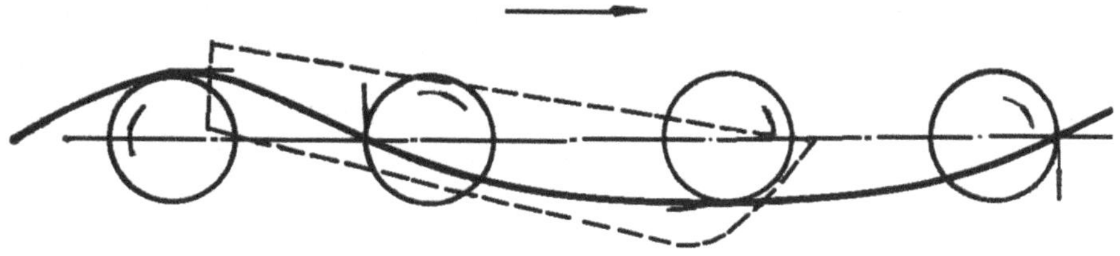

Fig. 5.35 Effects of following seas and trajectory circular motion of waves

From the above explanation, the loss of control is related to wave length ratio, as well as wave/craft speed ratio.

When wavelength is much smaller than craft length, i.e., where $\lambda/L << 1$, the craft positions over several waves. If wavelength is significantly greater than craft length $\lambda/L >> 1$, then the craft will follow the wave perfectly, and all of the situations mentioned above will not cause plough-in, stern takeoff, or losing control of the craft. Only in the case of $\lambda/L = 2$, or close to such condition, as shown in Fig. 5.35, the loss of control tends to happen. Many tests have verified that the phenomenon happens at $1 \leq \lambda \leq 3$.

In case of great difference between craft speed V with wave speed C, i.e., $V/C >> 1$ or $V/C << 1$, such phenomenon also will not happen, as encounter frequency is high for the former, similar to head sea waves. Only when craft speed is close to wave speed, will broaching be possible as shown in Fig. 5.35.

Figure 5.36 shows test results of control loss of a craft, with length 11.4 m, maximum beam 2.70 m, draft 0.55 m, displacement 5.6 t, speed 10knots, and having deep V transverse section with hard chine. During the tests, wave course angle $x = 0^0$, i.e., pure following waves, while wavelength/height ratio $\lambda/H_w = 15$. The ordinate is λ/L, abscissa is Fn. Curves for $V=C$ are plotted in the figure. From the figure, it can be seen that loss of control occurred at $1 < <\lambda \leq 3$ and where $V \approx C$.

Since sea waves usually are gravity waves, their wavelength and speed have the following relationship:

$$C = \sqrt{g\lambda/2\pi} \tag{5.24}$$

The range of craft speed for losing control (broaching) down wave is when both following conditions of $1 \leq \lambda \leq 3$ and $V \approx C$ are met, thus using (5.24) we find that $0.4 \leq \text{Fn} \leq 0.7$.

If we extend the craft and wave speed ratio to $V/C = 0.5$ to 1.5, then the corresponding range of craft speed for broaching will be $0.2 \leq \text{Fn} \leq 1.04$.

So, the craft speed running in following waves at which the craft may broach is in displacement operation and transit modes.

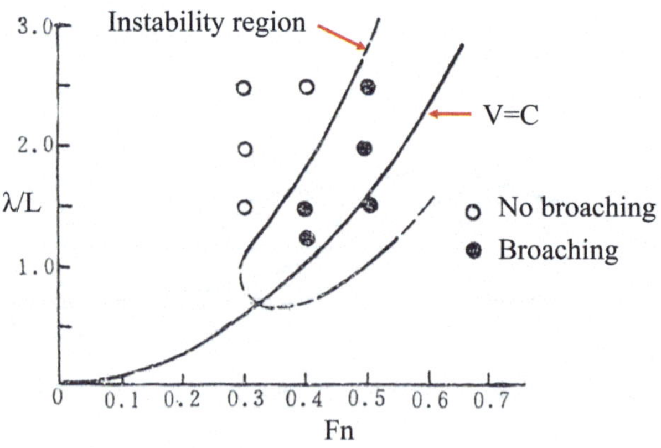

Fig. 5.36 Test results for speed range of a craft losing handling

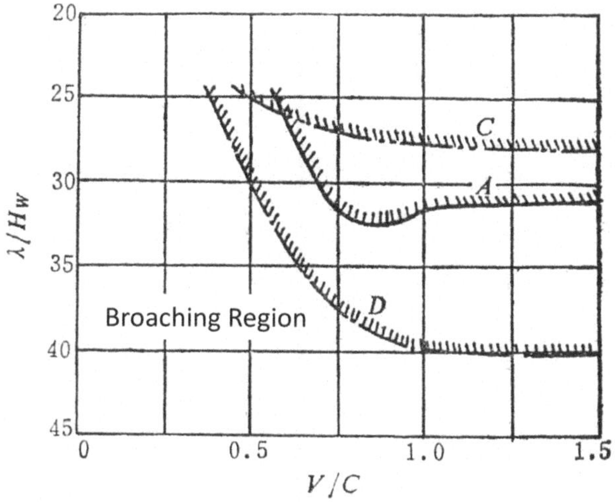

Fig. 5.37 Test results for handling loss by three models

Aside from this, losing control relates to wave characteristics. For small wavelength to height ratio λ/H_w, i.e., short and steep waves, the broaching range is larger due to heavier pitching response and thus serious plough-in tendency of the craft. However, for large λ/H_w, i.e., low long waves (swell), the broaching may never happen at any λ/l_{co} and V/C.

The craft profile also influences the broaching phenomenon. Vosper Company carried out tests of three models referred to above as A, C, and D (Fig. 5.29), at various speeds and wave parameters and identified boundary limits as shown in Fig. 5.37. It can be seen from the figure that the speed range for broaching of type D with round bilge is the largest, followed by type A with deep V hull, and C is the smallest, due to its sharp stern bottom keel which increases the underwater lateral projected area. In addition, the outward curved lines at the bow provide higher bow up restoring moment during plough in.

Thus, the characteristics for craft profile to improve the handling in following seas can be outlined as:

- Using sharp stern keel or skegs for increasing underwater lateral projected area and/or increasing the deadrise at stern.
- Decreasing the stern hydrodynamic lift (by increasing deadrise, decreasing stern beam, avoiding inner hollow of stern buttock lines, etc.) for reducing plough-in and stern lifting phenomena.
- Increasing the fullness of bow to increase the restoring moment after plough-in and decreasing the lateral projected area at bow. However, this may cause larger impact loads in head seas. Savitsky recommended a compromise bow external fuller section, as shown in Fig. 5.38. J.Stoltz and J.G.Koelbel [15] recommended a craft profile with double hard chine bow section for increasing bow up moment, and stern section with deadrise continued to the transom as shown in Fig. 5.39.
- Increasing rudder area, particularly the rudder in the wake of propeller, for improving the handling.

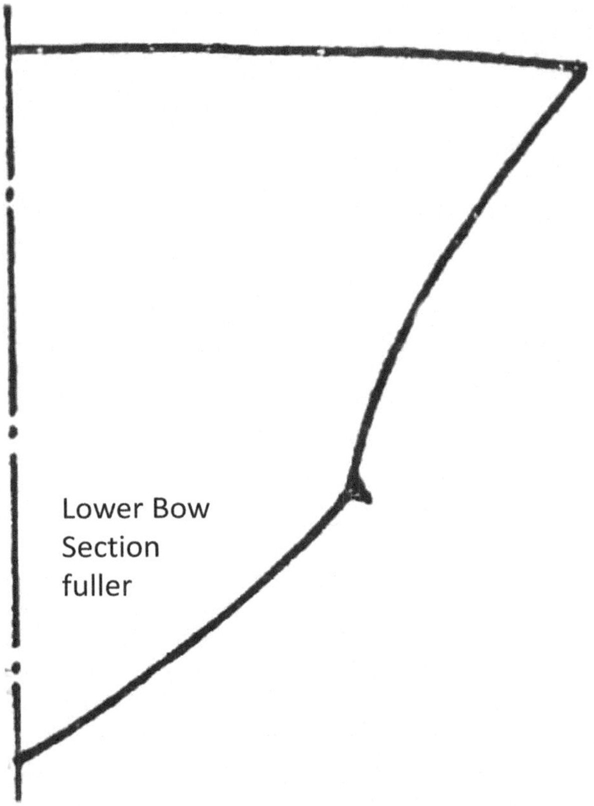

Fig. 5.38 The external bow section recommended by D. Savitsky

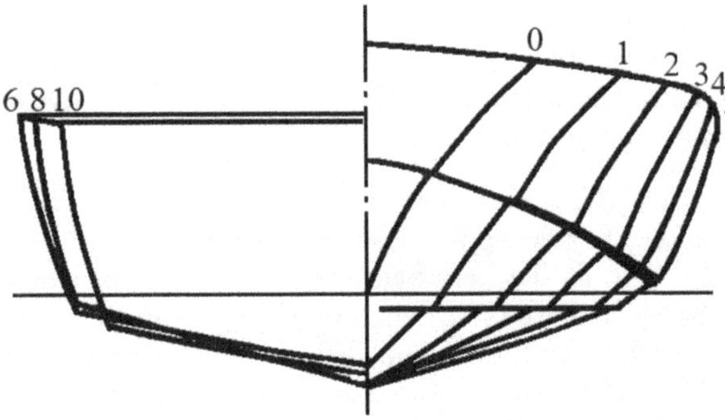

Fig. 5.39 Double hard chine bow section and stern section with large deadrise

5.6 Prediction of Seakeeping Quality of Planing Hulls

Japanese scientists Masatoshi Bessho, Masaaki Yoji, and Masahiko Komatsu investigated the prediction of seakeeping quality of planing craft by means of strip theory, as used for conventional displacement ships, in [16], and compared the results with

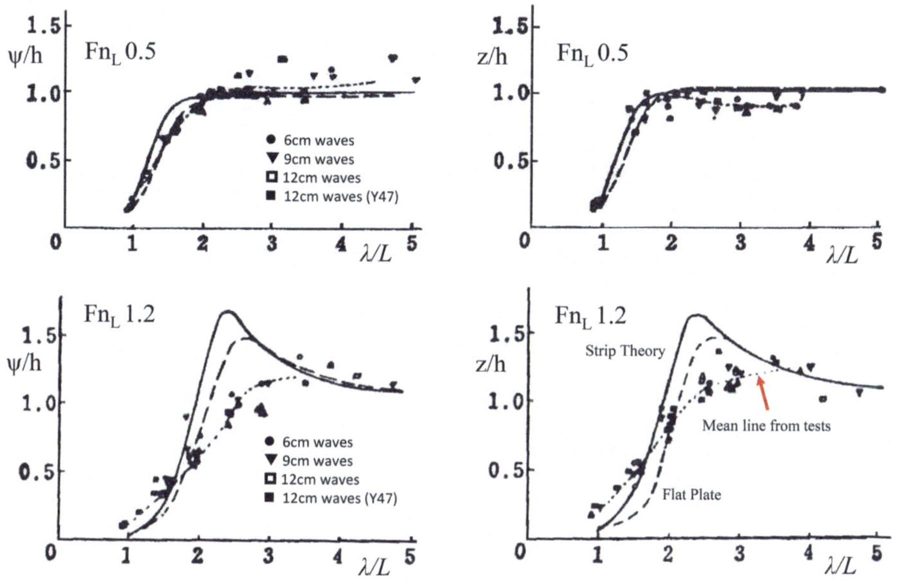

Fig. 5.40 Comparison of calculation values of pitch and heave motions of planing hull by means of strip method with test results $Fr = 0.5$ and 1.2

the testing of a torpedo boat of the ministry of defense. It was found that the comparison is satisfactory at low speed, i.e., displacement operation; however, when planing, the difference between them is rather great.

Figure 5.40 shows the comparison of the calculated heaving and pitching values with test results of the model in head regular waves.

Figure 5.40 shows the test results of heaving and pitching at two speeds, $Fn_L = 0.5$ and 1.2. Data for model wave heights of 6, 9, and 12 cm, for a model of the Y47 torpedo boat are shown by different shaped dots. The theoretical calculations by means of strip method are shown by solid lines in the figures, the added mass and damping results calculated using a flat plate are shown by dashed lines, and average test results are shown by dotted lines.

From the comparison, one can see that in displacement mode at $Fn_L = 0.5$, the calculation agrees well with test results. However, in planing mode at $Fn_L = 1.2$, the difference between them is great. In the region of small wavelength ratio λ/L, i.e., with high encounter frequency, the theoretical values are lower, but they are higher at lower encounter frequency. At the critical wavelength ratio around $\lambda/L = 2$, the model tests do not show the amplified peak suggested by strip theory at the higher speed. At very great wave to craft length ratio, i.e., very low encounter frequency, the two values are close.

Thus, the conventional strip method cannot be applied to accurately predict motions of planing hulls in waves as the dynamic lift on the bottom in waves is rather different from the forces acting on a displacement ship. In addition, the nonlinear effects on a high-speed planing hull clearly cannot be represented by the strip method which is based upon the linear theory.

At present, the model experimental method in irregular waves for predicting the seakeeping quality of a planing hull is the most credible method; however, it is very expensive. Designers usually expect to use simple methods for predicting seakeeping quality at the initial design stage to select the best possible design. They are provided with two options:

1. Prediction, using plots set up from experimental series results
2. Theoretical computer calculation

Each of these will now be introduced briefly.

5.6.1 Prediction from Model Series Experimental Results

Owing to the strong nonlinear motion of planing hulls in waves, using model tests to predict the seakeeping quality of planing hulls is still the preferred method. The procedure is basically as same as for conventional ships. Since the impact acceleration is a key topic for planing hulls, usually towing tests of high-speed planing hulls are carried out in head sea waves.

In tests, the heaving and pitching motion at bow and CG, and the drag increment should be measured. Owing to the bottom slamming of high-speed planing hulls, a lot of high-frequency noise accompanies the measured data, and post-processing using filters forms an important part of the work.

Due to nonlinearity, using spectral analysis from the test results of craft in regular waves to predict the seakeeping quality of full-scale craft in irregular waves may not be accurate and can introduce significant errors. Therefore, the model experiments have to be carried out in irregular head sea waves of correct significant height to predict the seakeeping quality of the full-scale craft directly. However, a comparison with the tests in regular waves can also be carried out. It is found that the agreement for heave and pitch motion response may be reasonable, but large differences are observed of the impact acceleration and the drag increment, particularly where skipping motion occurs in waves.

According to our experience, the predicted motion responses by model tests of planing hulls, particularly impact acceleration, may be greater than measured data of full-scale craft, possibly due to the nonlinearity causing higher damping at full scale and so reducing the motion response, particularly in the resonant region. Therefore, the model predictions should be rather conservative and safe.

G. Fridsma compiled test results of planing hulls with prismatic sections in irregular waves in the Davidson Laboratory of Stevens Institute of Technology, USA, and presented them in graphical form for predicting the seakeeping quality of planing hulls [13, 14]. The application range is listed in Table 5.1. The prediction includes the average value of the 1/10 highest values, $Z_{1/10}$ and $\theta_{1/10}$, the drag

Table 5.1 Application range from Fridsma

Parameter	Static load coefficient C_Δ	Length beam ratio *L/B*	$\frac{C_\Delta}{(L/B)}$	Trim angle τ	Dead rise angle β	Relative significant wave height $H_{1/3}/B$	Speed length ratio $V_s/\sqrt{L}$
Range	$0.3 \sim 0.9$	$3 \sim 6$	$0.06 \sim 0.18$	$3 \sim 7^0$	$10^0 \sim 30^0$	≤ 0.8	≤ 6

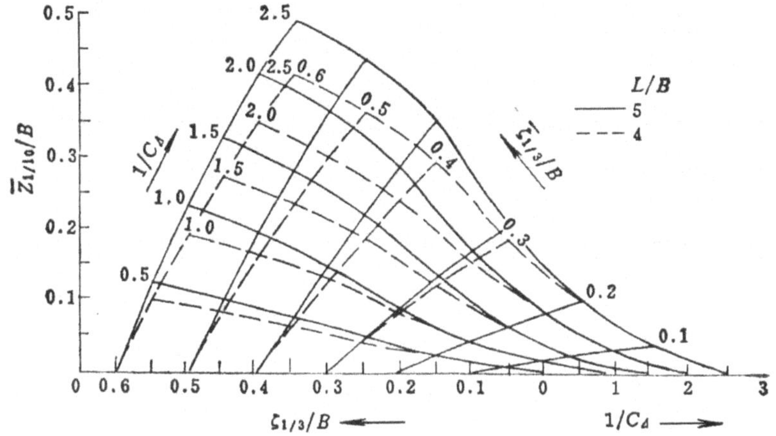

Fig. 5.41 Heaving amplitude at $Vs/\ L^{0.5} = 4$, $\tau = 4^0$

increment in waves, and the average values of vertical acceleration at *L*/10 from the bow, and at CG.

(a) *Calculation of motion amplitude*

There are a total of five carpet plots, corresponding to heaving and pitching amplitude respectively at speed length ratio $Vs/\ L^{0.5} = 2$, 4, and 6. Figure 5.41 shows an example for heaving amplitude at $Vs/\ L^{0.5} = 4$, while Fig. 5.42 shows pitching amplitude at $Vs/\ L^{0.5} = 4$, where speed V is knots, Loa *L* is in feet, and beam *B* is in feet. Note that wave height $H_{1/3}$ is represented by $\varsigma_{1/3}$ in the diagrams.

The plots all correspond to trim angle $\tau = 4^0$ and deadrise $\beta = 20^0$. When $\tau \neq 4^0$, the values need to be adjusted using correction equations as below.

When $\tau \neq 4^0$ and $\beta \neq 4^0$, the correction equation for heaving amplitude is:

$$H_{1/10}/B = \left[h_{1/10}/B\right]_{\text{graph}} \times\ F \times\ G \tag{5.25}$$

Correction equation for pitching amplitude is:

$$\theta_{1/10} = \left(\theta_{1/10}\right)_{\text{graph}} \times F \times G \tag{5.26}$$

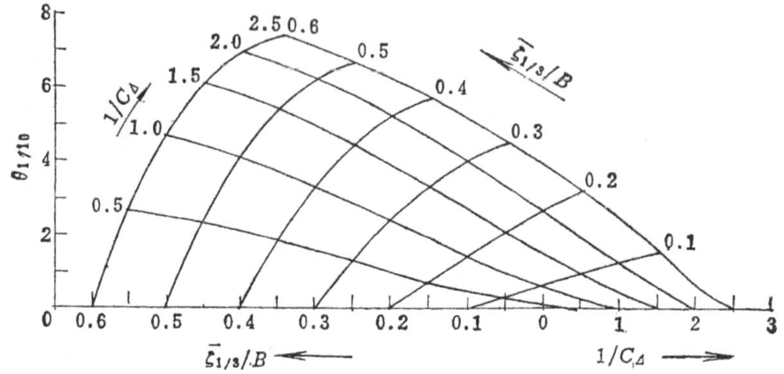

Fig. 5.42 Pitching amplitude at $Vs/L^{0.5} = 4$, $\tau = 4^0$

Correction value F at trim angle τ is:

$$F = 1 + \frac{V/\sqrt{L}}{24}(\tau - 4) \tag{5.27}$$

Correction value G at deadrise β when $Vs/L^{0.5} > 4$

$$G = 0.56 + 0.11\frac{V}{\sqrt{L}} + 0.11\left(\frac{\beta}{10}\right)^2\left(1 - \frac{V/\sqrt{L}}{4}\right) \tag{5.28}$$

When $Vs/L^{0.5} \leq 4$ $G = 1$

If B and β change longitudinally, it is suggested that the average value at 80% length from the stern be taken.

(b) *Calculation of drag increment*

The plots for predicting the drag increment in waves are shown in Figs. 5.43 and 5.44, obtained at standard conditions of $C_\Delta = 0.6$ and $L/B = 5$.

First, from Fig. 5.43, at given trim angle τ, deadrise β, and wave height/beam ratio $H_{1/3}/B$, find maximum drag increment $(R_{AW}/\gamma B^3)_{max}$ and corresponding $(V/L^{0.5})_{max}$. If sea state is different from the graphs, carry out interpolation, and then according to $(V/L^{0.5})$ and $(V/L^{0.5})_{max}$, obtain $(V_s/V_s)_{max}$. From Fig. 5.44 using $(V_s/V_{s)max}$ and $H_{1/3}/B$, find $R_{AW}/(R_{AW})_{max}$. Finally, the corresponding drag increment is:

$$\frac{R_{Aw}}{\gamma B^3} = \frac{R_{Aw}}{(R_{Aw})_{max}}\left(\frac{R_{Aw}}{\gamma B^3}\right)_{max} \times E \tag{5.29}$$

Where E is the correction factor for C_Δ and L/B and it is different for different $V/L^{0.5}$, as shown in Table 5.2.

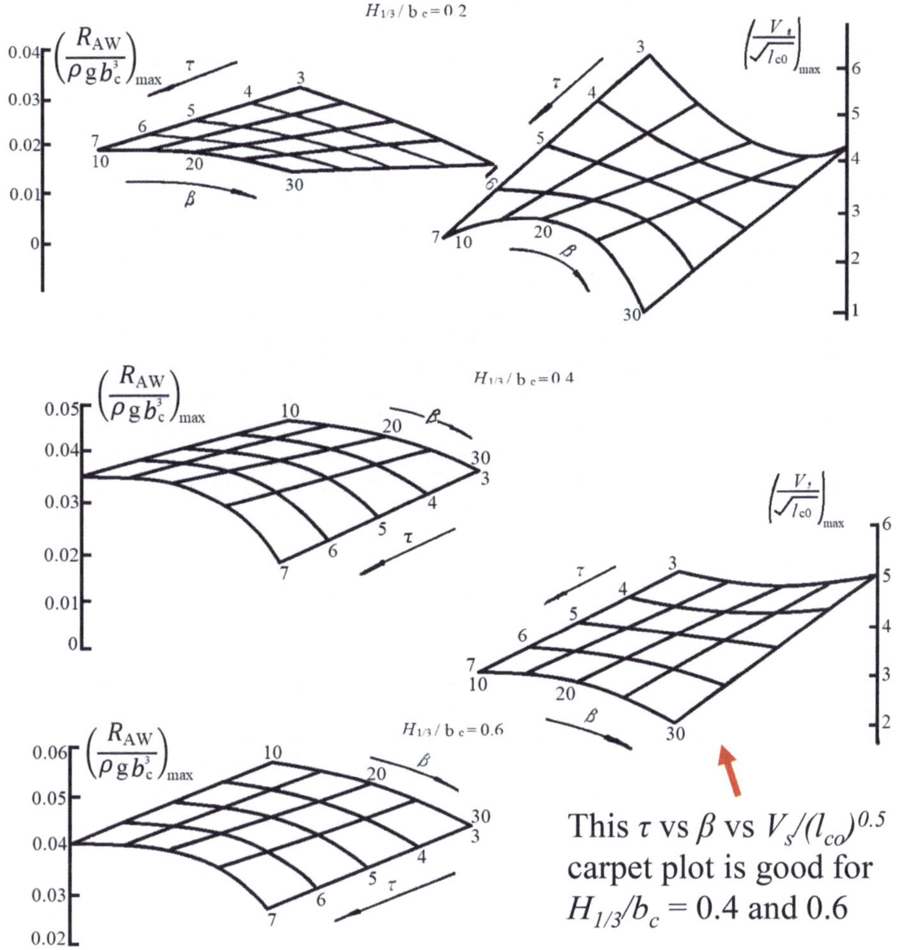

Fig. 5.43 Max additional drag and its corresponding speed $C_\Delta = 0.6$, $L/B = 5$

Fig. 5.44 Generalized additional drag curve

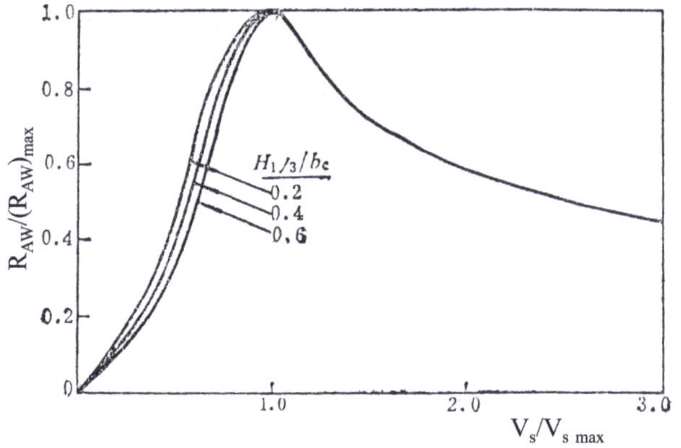

Table 5.2 Correction factor E for C_Δ and L/B

$V/\sqrt{L}$	E
2	$1+\left\{\left[(L/B)^2/25\right]-1\right\}/\left\{1+0.895\left[(H_{1/3}/B)-0.6\right]\right\}$
4	$1+10H_{1/3}/B\left[\frac{C_\Delta}{L/B}-0.12\right]$
6	$1+2H_{1/3}/B[0.9(C_\Delta-0.6)-0.7(C_\Delta-0.6)^2]$

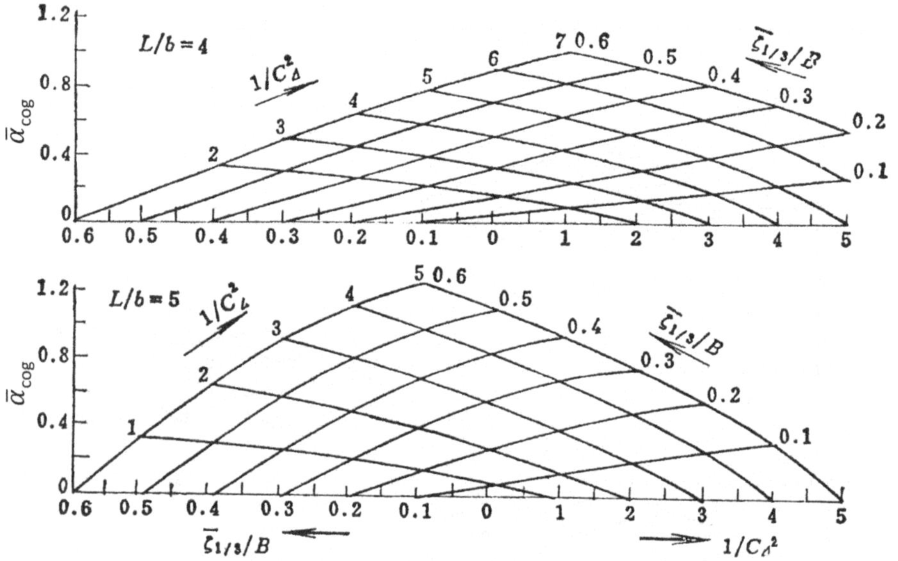

Fig. 5.45 Average acceleration at CG at $V/\sqrt{L}=4$ ($\tau=4°, \beta=20°$)

(c) *Calculation of impact acceleration*

There are six carpet plots for impact acceleration, corresponding to average acceleration at bow and CG respectively, at three speeds, $V/L^{0.5}=2$, 4, and 6. Each graph has two sub-graphs for $L/B=4$ and 5, respectively. Figure 5.45 shows the average acceleration at CG at $V/L^{0.5}=4$, and Fig. 5.46 shows average acceleration at bow (10% length from bow end) at $V/L^{0.5}=4$. Please note that acceleration is here denoted by α.

First, according to $V/L^{0.5}$ and L/B, find the corresponding graph and obtain the average acceleration α_0 from given $1/C_\Delta^2$ and $H_{1/3}/B$. Interpolation has to be used for intermediate values of speed/length ratio and length/beam ratio. Correct the acceleration value for different trim angle τ and deadrise angle β from the correction equation:

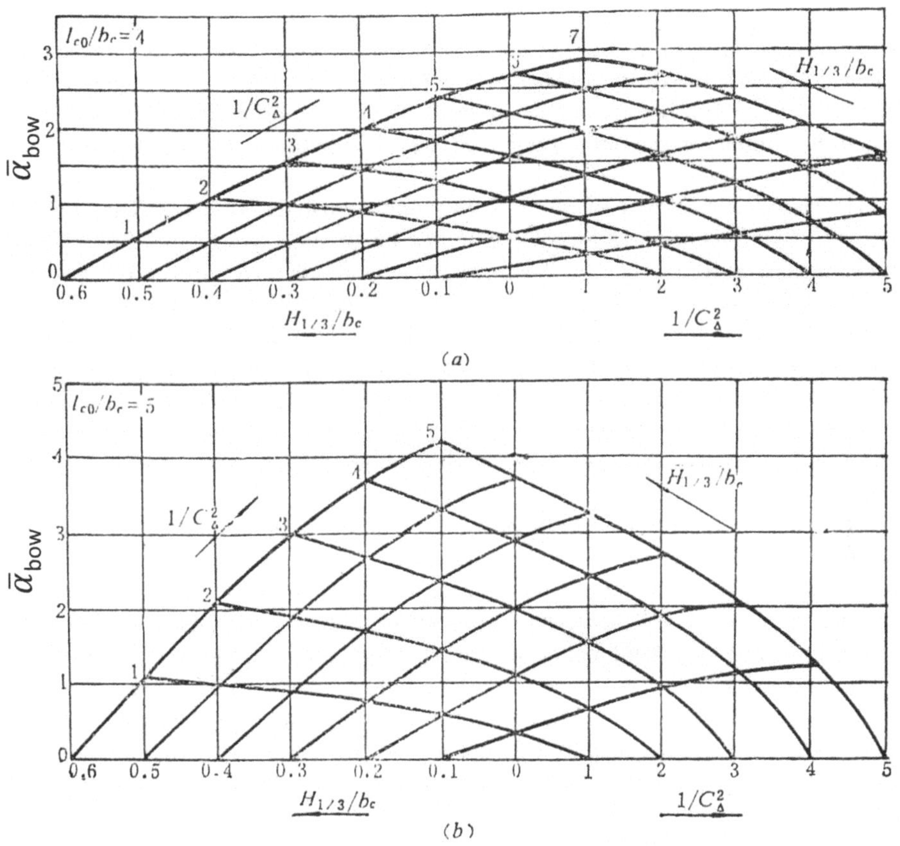

Fig. 5.46 Average acceleration at bow at $V/\sqrt{L} = 4$ ($\tau = 4°, \beta = 20°$)

$$a = a_0\left[\frac{\tau}{4°}\left(\frac{5}{3} - \frac{\beta}{30°}\right)\right] \tag{5.30}$$

Where α_0 is the acceleration value from the graph.

Later, E. E. Zarnik and C. R. Turner extended the same model tests with higher $L_{co}/B_c = 7$ and deadrise angles of 10, 20, and 30 degrees, and with $L_{co}/B_c = 9$ and deadrise angle of 10 degrees. The test results were compiled and presented in graphical form as an addition to the Fridsma plots applied to planing hulls with high length beam ratio, as given in [15].

At the University of Southampton in Great Britain, a seakeeping test series was carried out by D J Taunton, D A Hudson, and R A Shenoi, [16], continuing from the calm water program described in Chap. 4, Sect. 4.4.8.

A total of 41 tests were carried out on the four models, A, B, C, D, together with the single step C1 and double step C2 models. Tests were carried out predominantly at a single $H_{1/3}$ and radius of gyration Gyy in pitch while speed was varied and separately the sea spectra Tz. The spectrum used was based on a one-year data set taken in the Solent, which is between Southampton and the Isle of Wight, with a limited fetch appropriate for smaller coastal vessels generating a limited bandwidth spectrum similar to the joint north sea wave amplitude probability (Jonswap) spectrum.

The testing was carried out at GKN Westland Aerospace No 3 Test Tank in East Cowes, IoW. This has a length of 198 m and maximum carriage speed of 15 m/s. Wave making is from a flap-type wavemaker able to create irregular wave spectra.

The results from the test runs were prepared as probability distributions using the methodology developed by Fridsma [14] and Zarnic [17]. It was noted that Fridsma used an exponential distribution for vertical accelerations, and a Rayleigh or Cartwright and Longuet-Higgins distribution for heave and pitch motions to fit test data and allow projections. This study found that the Gamma distribution was a better fit for the accelerations than the exponential distribution (which is a specific case of the Gamma distribution).

The data sought from the testing is the peaks of the motion or acceleration traces. In case of regular waves, the average of a repeated series can be taken and individual peaks need to be obtained and correlated for both the waves and the motions. Taunton et al. used a peak detection algorithm to detect the peaks in their data logger time traces, maxima being of interest for pitch and heave, while minima or maximum deceleration for the vertical accelerations. The parallel wave data traces were analyzed for the related peaks and troughs. This data was then gathered into histograms, and further using the appropriate probability density function, an expected value was determined for the motion or acceleration for comparison purposes.

By fitting their model test data to the Longuet-Higgins probability distribution for pitch and heave and the Gamma distribution for accelerations, they generated mean, RMS, and bandwidth data that could be used to predict any level of occurrence using the response distribution for the specific hull and sea state. This is particularly useful for assessing exposure of an operating crew to accelerations. The authors prepare and discuss example results based on their model C2 at a full-scale length of 15 m and at a design speed of 65 knots in a 0.5 m sea state. Following the approach of ISO standard 2631, they show that the allowable 8-hour exposure according to European Council Directive 2002/44/EC is extremely short, which one may expect for such a high-speed vessel. Mitigating factors that would be in place for an actual vessel would be sprung/damped seats for the crew, which would not be "working" in the sense that the EC Directive is proposed. Nevertheless, such results do highlight that for high-speed planing craft, both the outfit and the operational limits need careful attention.

One important conclusion from this work was that there was very little difference in acceleration response of the stepped hulls C1 and C2 from the parent hull without steps C. Thus, while a stepped hull can improve drag, and can lower trim variation, it is not a solution to reducing vertical acceleration. An alternative geometry with a deep forefoot and hard chine aft can deliver improvements, as we discuss in Chap. 6.

The question remains, "How may we predict seakeeping quality of a hull using analytical methods?" We discuss this aspect below.

5.6.2 Theoretical Calculation for Predicting the Seakeeping Quality of Planing Hulls

Linear strip theory can successfully predict the seakeeping quality of a displacement ship, including the motion response in waves. However, for planing hulls, studies by Masatoshi Bessho of Japan [17] showed that significant errors may be introduced on acceleration and drag increment due to hydrodynamic forces at high speeds. In addition, the nonlinear relation between the motion response and wave height also makes it difficult to obtain the solutions for motion response in irregular waves by means of spectral analysis using linear superposition. Therefore, some special calculation methods for prediction of seakeeping quality for planing hulls are required.

(a) *M. Martin's linear method* [18]

Based upon the theoretical calculation in Sect. 5.2 for predicting the boundary limit of porpoising motion, Martin obtained the motion response of planing hulls in regular waves by means of a linear method. This method is also based on strip theory; however, the hydrodynamic lift is added in the linear differential equations for coupled heaving and pitching as shown in Eq. (5.6). The difference between them is adding wave force to the equations.

The stability derivatives in the equations are unchanged due to the added wave forces, i.e., Shuford's hydrodynamic force equations in steady planing mode to obtain the displacement derivatives, and variation of momentum caused by virtual mass to obtain the speed and acceleration derivatives can be used.

Regarding the wave force, Martin uses strip theory taking the strip as a two-dimensional body to set up the variation of momentum virtual mass under the waves and integrates it along the craft length to obtain the hydrodynamic forces under the wave surface and then applies a three-dimensional correction.

Finally, subtracting the hydrodynamic force of the body in steady motion in calm water gives the wave force.

The wave force includes a hydrodynamic increment caused by wave inclination angle, forces induced by wave-induced particle orbital velocities and accelerations, a buoyancy increment caused by the waves, and forces due to variation of wetted length by the wave. There are five moments corresponding to the forces.

Adding the wave forces (and moments) into the motion differential Eq. (5.6), and solving them, the heaving and pitching response amplitude and frequency in waves can be obtained. The solutions of the equations are harmonic due to the wave forces being harmonic functions.

Comparison of Martin's theoretical calculation with test results shows good agreement at $V/L^{0.5} = 4$, but large differences at $V/L^{0.5} = 6$, especially in the resonance region, where the theoretical results are significantly larger than the test results; possibly due to the nonlinear terms, these cannot be neglected. The agreement is not good at $V/L^{0.5} = 2$ as well, as the wetted part of craft above the hard chine is not considered.

(b) *E. E. Zarnic's nonlinear method* [19]

Based on the virtual mass theory of a slender planing surface (Chap. 3, Sect. 3.4), Zarnic derived the hydrodynamic forces of a planing surface in waves without neglecting the second-order terms of the wave forces as Martin did. He assumed the following:

- Inclination angle was a small value due to wavelength greater than craft length.
- Horizontal component of wave orbital motion was a small value compared with craft speed.
- Coupled motion of surge is negligible and so only coupled heaving and pitching motion are considered.
- The deformation of wave profile due to the craft is negligible.

Using the virtual mass theory of a slender planing surface, Zarnic obtained the force acting on a transverse strip per unit length on a planing surface as follows:

$$\frac{\mathrm{d}p}{\mathrm{d}x} = -\left[\frac{D}{Dt}(m'v) + \frac{1}{2}C_{\mathrm{DC}}\rho b v^2\right] \tag{5.31}$$

Where m' is the virtual mass on section

ρ is the water density

v is the vertical velocity

C_{Dc} is the drag coefficient for the transverse flow

b is the section width

The first term on the right of (5.31) is vertical lift caused by the variation of virtual mass:

$$\begin{aligned} \frac{D}{Dt}(m'v) &= v\frac{\mathrm{d}m'}{\mathrm{d}t} + m'\frac{\mathrm{d}v}{\mathrm{d}t} \\ &= v\frac{\mathrm{d}m'}{\mathrm{d}x}\frac{\mathrm{d}x}{\mathrm{d}t} + m'\frac{\mathrm{d}v}{\mathrm{d}t} \\ &= v_\infty v\frac{\mathrm{d}m'}{\mathrm{d}x} + m'\frac{\mathrm{d}v}{\mathrm{d}t} \end{aligned} \tag{5.32}$$

Where $\frac{\mathrm{d}x}{\mathrm{d}t} = V_\infty \cos \tau_\infty \approx V_\infty$

In Zarnic's hydrodynamic expressions for the planing surface, he retains the nonlinear terms of some interactions, among them the second term on right of (5.31) which is a second-order term of vertical speed component v, i.e., transverse flow drag. From (3.50), (see also Fig. 3.7), the relationship of virtual mass with angle of attack τ is to the power of 1/2, thus:

$$m' = \frac{1}{2}\pi\rho\left(\frac{b}{2}\right)^2\left(1 + \frac{1}{2}\sqrt{\frac{2x\tau}{b}}\right)$$

So, the normal force in (3.51) with the nonlinear term of τ is proportional to $\tau^{3/2}$:

$$P_{m'} = \frac{1}{2}\pi\rho\left(\frac{b}{2}\right)^2 V_{\infty}^2\tau\left(1 + \frac{1}{2}\sqrt{\frac{2x\tau}{b}}\right)$$

Zarnic obtained nonlinear coupled heaving and pitching motion equations and calculated the heaving and pitching response of the prismatic planing model, the model testing of which was carried out by Fridsma, by means of a numerical calculation method.

The comparison of the calculation with test results demonstrates that at $V_s/l_{co}^{0.5} = 4$ and 6, the agreement is good and better than Martin's for motion responses. The agreement of acceleration response at high wavelength ratios is also fine; however there is some difference between them at small wavelength to craft length ratio $\lambda/l_{co} < 2$ particularly at high-speed $V_s/l_{co}^{0.5} = 6$ on bow acceleration.

Since the relationship between wave response and wave height is nonlinear, strictly speaking spectral analysis cannot be applied to predict the wave response of craft in irregular waves by means of Martin's linear or Zarnic's nonlinear analysis.

Zarnic used his nonlinear equations, and time history method to calculate the motion response of craft in irregular waves [20], and in order to save computational time, he superposed ten harmonic waves with random phases to form the irregular waves. The Pierson-Moskowitz wave spectrum with fully developed sea state was applied as the calculated wave spectrum in Fridsma's model tests with prismatic sections and $l_{co}/b_c = 5$, $\beta = 20^0$, and $V_s/l_{co} = 6$ for several sea states, as well as deadrise β of 10, 20 and 30 degrees, and at $V_s/l_{co}^{0.5} = 4$ with one sea state. The test results show that the calculations of heaving and pitching are close to test results below the middle sea state; however, the calculations of vertical acceleration response are only ½ of test results.

Based on the Zarnic's nonlinear equations, Dai, Song, and Li of Harbin Shipbuilding Engineering University, China, carried out response analysis of craft in irregular waves [21]. They used the spectral analysis method to calculate the motion spectrum, including heaving, pitching, and vertical acceleration responses in irregular waves. They also used the time history method to collate the calculation results. The comparison demonstrated that there are no large differences between the two methods. They also compared their results with Fridsma's model test results and demonstrated rather good agreement at middle speed and sea state; however, they also obtained large differences at high sea states similar to Zarnic's except that the vertical acceleration seemed to agree with model test results better than Zarnic's predictions.

It can be concluded from this that linearized prediction can be applied for pitch and heave motion of a planing craft. The approach of Taunton et al. [16] can provide reliable results. The challenge remains to accurately predict roll motion, owing to asymmetric hydrodynamic forces acting on planing hulls, craft geometry with curved transverse section, and difficulties with calculation of virtual mass. This needs further investigation.

5.7 Influence of Planing Hull Geometry on Seaworthiness

Concluding the discussion of seaworthiness performance of planing hulls, the influence of hull geometry on performance may be outlined as follows.

5.7.1 Geometric Parameters of Craft Configuration

The geometry of a planing hull is characterized by the hard chine and its characteristic parameters, as only the portion below the hard chine contacts with the water surface in the planing mode.

1. *Beam (maximum beam or beam at midsection)*

The hard chine beam is a most important parameter. Large beam corresponds to high aspect ratio and increases the planing efficiency. If the beam is fixed, the CG can be shifted for maintaining a favorable trim angle. If the CG position is fixed, then a favorable beam is available corresponding to a favorable trim angle, so during design, beam has to be considered along with CG position.

For given displacement and design speed, increase in beam means a decrease in C_{Δ} or C_{B}, which increases motion responses and impact acceleration and decreases spray, and all these should be considered together. Beam is also limited by general arrangement and structure weight.

In order to have a high planing efficiency, the maximum beam position is usually before the midsection, between 1/4 and 1/2 length of the craft, and aft of the baseline C_{B} of the craft in displacement mode, so as to provide positive trim angle favorable for takeoff. The determination of hard chine beam should be considered for meeting the requirements of static stability. In the primary design stage, the following empirical formula for prediction of initial stability metacenter height h_0 can be used:

$$h_0 = K_1 \frac{B^2}{T} + K_2 T - Z_g \tag{5.33}$$

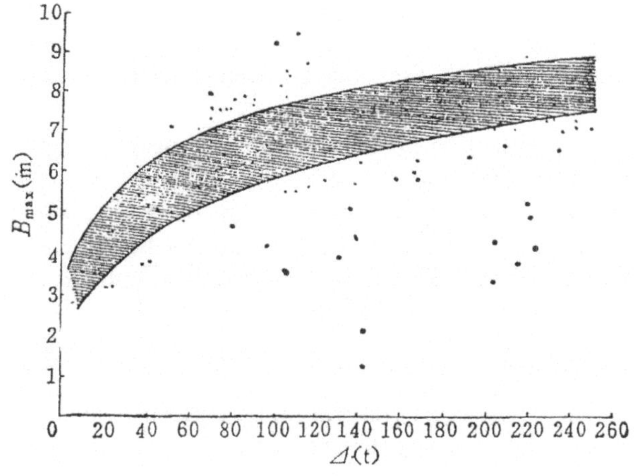

Fig. 5.47 Statistical relationship between the max deck beam and displacement of planing hulls

Where, K_1 is the coefficient related to water plane configuration and block coefficient $K_1 = 0.12\text{–}0.14$

B is the average width of hard chine and is the average value of maximum width at hard chine and transom width

T is the average draft and can take for estimation $T = 0.3\Delta^{1/3}$

K_2 is the coefficient related to transverse sections; for hard chine, 0.66–0.7

Z_g is the CG height above keel

Δ is the craft displacement

Deck beam is generally a little bit larger than hard chine beam due to flare at the forward part of the craft for deflection of spray, with a ratio of deck beam to hard chine beam of B_c/B_D between 0.75 and 0.95, generally within 0.85 to 0.95. Statistics of maximum deck beam versus displacement gathered in China are shown in Fig. 5.47 [21], which can be used to determine the hard chine beam.

2. *Stern width*

In order to enhance planing efficiency, the planing area should concentrate close to the forward stagnation line, with small stern beam. However, this is limited by general arrangement and requirement of seakeeping quality because decrease of stern beam increases trim angle and is unfavorable to motion response, impact force, and spray in waves. In general, the trim angle of craft at maximum and cruising speed is a little bit lower than optimum angle, and this also limits stern beam.

In addition, the beam at the hard chine has to be considered for arranging the propulsion installation. When using conventional propulsion, enough space between propellers at stern has to be considered, while when using waterjet propulsion, the size for arranging the water inlet has to be accounted for so as to decrease the effective area of planing surface and hydrodynamic forces. When using a semi-submersible propeller (SSP), aside from considering stern beam, the geometry of the transom has to be arranged for installing the SSP.

In general, the ratio of hard chine beam at midsection B_{cmax} to that at the stern B_T will be around 0.85 to 1.0 for modern craft, while earlier in development, craft had smaller transoms for propeller arrangements and so the ratio was between 0.65 and 0.85.

It may also be noted that the work of many researchers has been based on a prismatic section for the planing surface, which is consistent also with minimizing deviation to flow caused by the hull geometry. A reduced stern width may be useful for reducing drag at low speed and so this might be part of the design consideration. It certainly was for the designs of the mid-twentieth century. The modern approach is use of a prismatic section aft.

3. *Craft length*

The influence of craft length on performance is small, while the length of hard chine influences drag directly only at low speed, and it influences the drag at high-speed planing mode only via the influence of CG position. Increased length may increase a little bit of impact force. In long waves, $\lambda/l_{co} > 3$, motion response will increase, and in short waves, it decreases. The length must be considered for the arrangement of propulsion installation. For conventional propellers, it is not helpful to take excessively short length which results in larger inclination angle of shafts, while for waterjet propulsion, enough space for the waterjet inlet is necessary. For SSP, no obvious requirement is available, though this type of propulsion is generally associated with racing craft that have high *L*/*B* and deep *V* form. In general, length/beam ratio is L_c/B_c between 4 and 6. Fig. 5.48 shows statistical data of length versus displacement curve of craft for reference.

4. *Draft and depth*

The draft of a planing hull influences the static buoyancy and low speed operation but has no direct influence on the planing performance, so it is normally determined after the beam, length, and displacement.

The determination of craft depth derives from the control of deck wetness of craft in waves and overall strength of craft. Excessively low depth is unfavorable to the operation of craft in waves, and overall strength of craft; however, excessively high

Fig. 5.48 Statistical data for length displacement relationship of planing craft

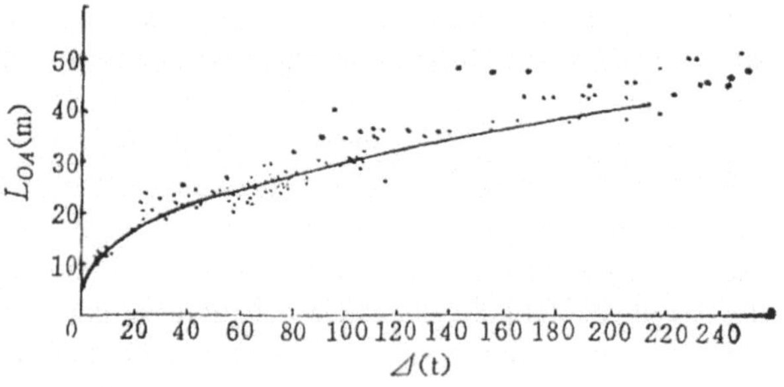

depth may increase hull weight and raise the CG. Figures 5.49 and 5.50 show statistical data of depth with length and beam respectively [21].

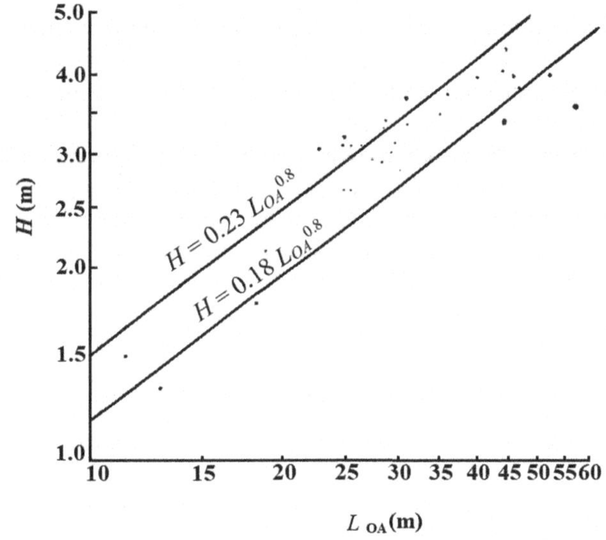

Fig. 5.49 Statistical data for depth versus length of craft

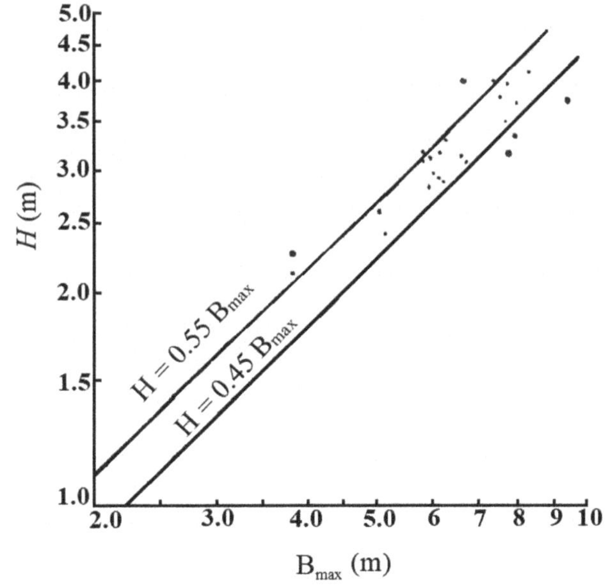

Fig. 5.50 Statistical data for craft depth versus deck beam

5. *Longitudinal CG position*

Longitudinal CG position significantly influences the craft performance via trim angle. For drag, shifting the CG aft corresponds to larger beam enhancing planing efficiency and decreasing the drag. However, the drag variation during the takeoff process has to be considered as larger initial trim may cause increasing drag during takeoff and so high hump drag. Shifting CG excessively forward causes small trim angle, larger drag at design speed but a reduction in hump drag. So, it is important to consider these factors in selecting the best longitudinal CG location.

Aside from drag considerations, significant aft CG position may cause larger trim angle, initiating porpoising motion and causing unfavorable impact forces in waves. On the other hand, smaller trim angle is unfavorable to deck wetness in head sea waves and for broaching motion of craft in following waves.

In general, the CG on a craft with conventional propulsion installation is usually a little bit further forward due to the general arrangement compared to waterjet, stern drive, outboard, or semi submerged transom mounted propellers (SSP). However, in case of using waterjet and SSP, the CG can be more easily adjusted to meet the powering performance and seakeeping quality. This is also an important reason for modern planing craft to use stern drive, waterjet, and SSP.

6. *Area load coefficient* $A_p/\nabla^{2/3}$

The area load coefficient $A_p/\nabla^{2/3}$ represents loading on the planing surface similar to the static load coefficient $\Delta/\gamma B^3$. However, it more clearly represents the load per unit area, just as displacement length coefficient, $\Delta/(L/100)^3$, on conventional displacement ships.

The area load coefficient $A_p/\nabla^{2/3}$ significantly influences planing performance and drag. However it should be considered along with CG position (and so trim) for optimising performance.

Figure 5.51 shows drag versus different area load coefficients from test results of a planing craft model. CG is aft of the center of planing area A_p by 8%, i.e., $(X_A - X_g)/L = 8\%$, when $A_p/\nabla^{2/3} = 5.5$, 7.0, and 8.5 [21].

Figure 5.52 shows some results from model tests of the same 45.4 tonne displacement planing hull concept with the CG moved between 0 and 12% aft of the center of lift when $A_p/\nabla^{2/3} = 7$.

7. *Section Form*

1. *Transverse sections*

The profile of the transverse section significantly influences the drag, wave impact force, as well as transverse stability, and designers have to consider a proper deadrise for meeting performance requirements and transverse stability. For planing hulls operating in coastal seas with not very large waves, the deadrise angle at midsection can be taken as $\beta_{ms} = 15^0$–17^0. For craft running in high waves, in order to enhance seakeeping quality, $\beta_{ms} = 20^0$–25^0, while for inland water planing hulls $\beta_{ms} = 10^0$–12^0.

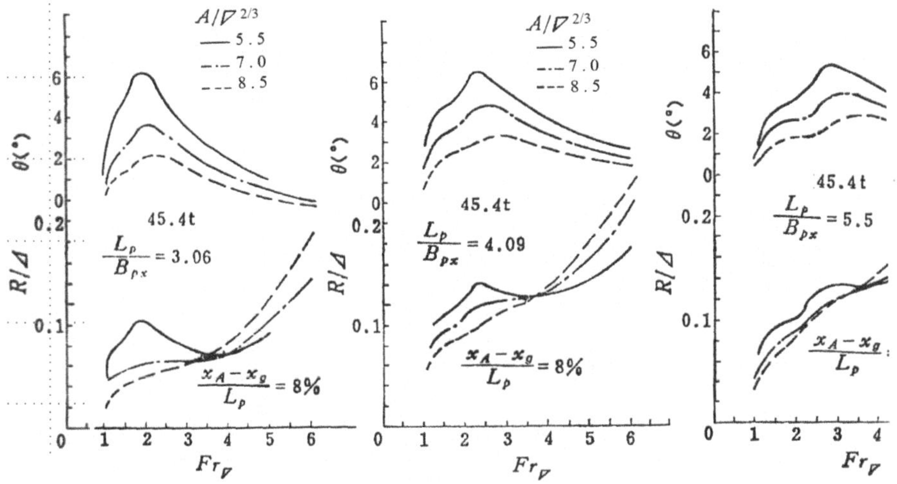

Fig. 5.51 Resistance of planing craft at various area load coefficients

Fig. 5.52 Drag for varying CG

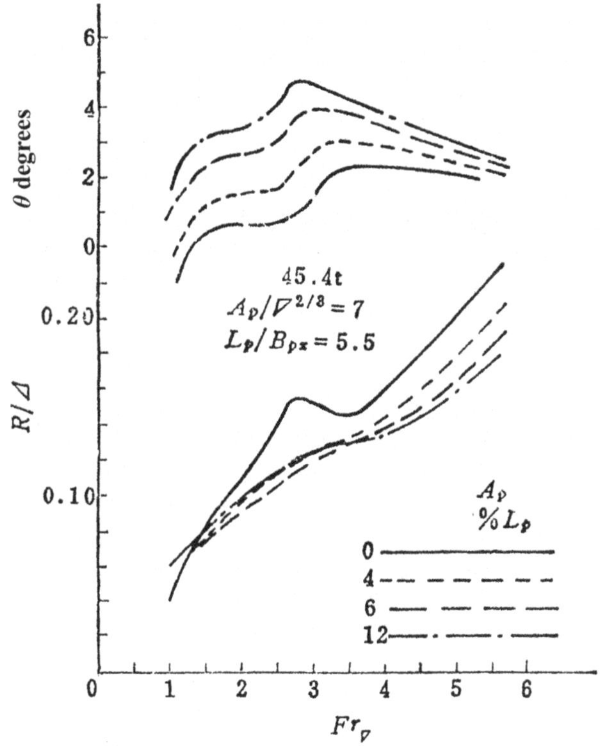

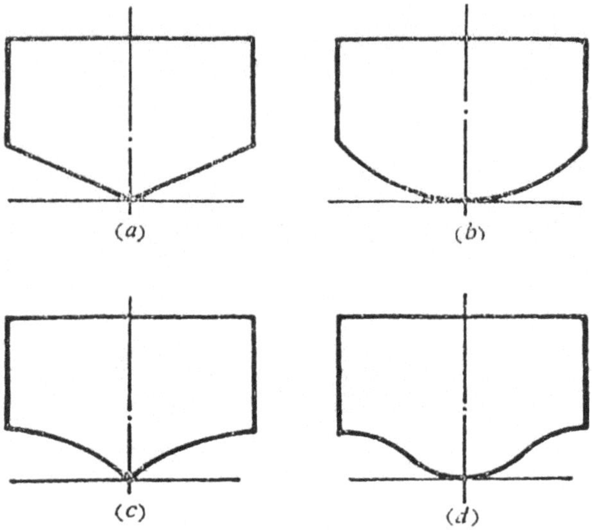

Fig. 5.53 Various transverse profiles

Some profiles for transverse sections are shown in Fig. 5.53, such as straight line deadrise, protruding outward deadrise, inward hollow deadrise (transverse curve), and wave shape.

Designing a transverse section with inward hollow deadrise can meet requirements for both drag and wave impact force; however, excess inner hollow may cause increase of local impact force. The outward curve section can reduce wave impact force and, however, leads to large spray and unfavorable planing efficiency. The wave shape type is a compromise between high planing efficiency of the inward hollow shape and lower impact force of the outward curve shape. An ideal seems to be difficult to reach for all such profiles. For the bilge bending downward to form an inverted "w" shape, its planing efficiency is high, however, with larger impact force than deep V section. The side can be flared in order to reduce spray and deck wetness.

To reduce impact forces at the bow, the deadrise angle may be increased gradually from amidships to bow; at $l_{co}/4$ from stem, use $\beta_{1/4} \geq 20$ to 30 degrees and at #3 from stem $\beta_{\#3} < 30$ to 35 degrees; however, an excessively sharp bow is unfavorable in terms of broaching. A double hard chine as shown in Fig. 5.39, or an outward curved type for bow transverse section as shown in Fig. 5.38, may also be used. Due to small impact loads at the stern, the transverse deadrise angle can be decreased gradually from amidships to stern for improving the planing efficiency. Stern deadrise angle can be reduced to as little as 0 to 2 degrees. Such decrease should be smooth, with variation of β as shown in Fig. 5.54b, d for best fit to propellers. However, this may lead to vortices under the bottom if the angle change is too great and give unfavorable course keeping stability. The deadrise may also be kept unchanged for easier fabrication of the hull as in (a).

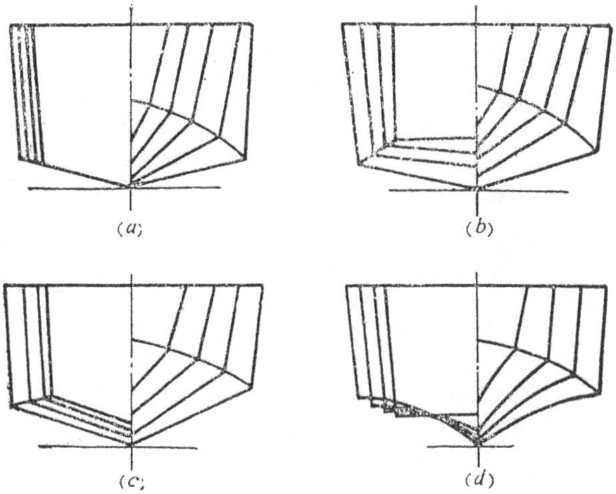

Fig. 5.54 β changed longitudinally

2. *Longitudinal sections*

In order to easily arrange the propellers at stern, the keel line may be raised up a little toward the stern. The lowest position of keel may be between stations #5 and #6. Having a full stem shape may give larger forward buoyancy favorable to takeoff and improving broaching resistance in following waves. It is suggested that longitudinal buttocks should be straight at the stern and avoid curving outward. Having a little hollow may enhance the planing efficiency and decrease hump drag; however, it is unfavorable for broaching and plough-in in following waves. The intersection angles between the longitudinal buttocks and waterlines directly influence trim angle at low speed and take-off speed; however, excessively large angle may cause large spray, and it seems that staying less than 6° is optimal.

5.8 Special Measures for Improving Planing Hull Performance

In order to improve performance of planing hulls, some special auxiliary equipment may be installed on the craft as follows.

1. *Transom flap*

This is a small fixed or hinged planing surface added aft of the transom with a little inclination angle downward, creating a planing surface with camber for enhancing the planing efficiency, adjusting trim angle, as well as improving the working condition of propellers and rudders. The most reasonable measure is installing an adjustable flap at the transom, which not only meets the requirements mentioned above but also reasonably adjusts the trim angle in waves.

2. *Transom wedge plate*

The function of transom wedge plate is similar to the transom flap for improving the planing efficiency. Double wedge plates are also suitable, i.e., adding another wedge plate at mid of the craft, corresponding to two planing surfaces with camber profile for improving planing efficiency but not reducing the seakeeping quality as in stepped planing craft, possibly due to small height of wedge plate with smaller cavity than that of a stepped craft.

3. *Anti-spray rail(bilge plate)*

Anti-spray rails depress the spray and increase local pressure at the chines. J. G. Koelbel has referred to test results showing that the suppression of spray is better using a bilge spray rail with small declination angle of 15 degrees rather than a large angle of 45 degrees.

4. *Transom interceptor plate*

A vertical plate, called an interceptor, can be put mounted at the lower edge of the transom. This can be moved down below the transom line to raise the local pressure and perform the same function as the transom flap or wedge plate. In addition, due to its placement directly under the bottom and normal to the flow, the regulation of trim should be more effective. It has been widely applied due its advantage of easy installation and control and provides nice running attitude in different operational regimes [22]

5. *Longitudinal anti-spray rail*

Spray rails have been introduced in Sect. 5.4. Typically, three rails are used on the planing surface. The most important is to have these located across the leading-edge boundary of the planing wetted surface. Deep V craft tend to extend the rails right aft. As discussed earlier, spray rails can be important also for maneuvering stability at speed.

6. *Alternative hull geometries*

In recent years, a lot of researchers and ship designers have been involved in the investigation of novel types of craft and made great progress in developing planing catamarans [23], sea-knife craft, and wave-piercing craft, which have significantly improved seaworthiness of planing craft.

References

1. J. Iwai, Stability of hard chine Boats at High Speeds (2). Ship Sci. **35**(1) (1982). (in Japanese)
2. P. Du Cane, *High Speed Small Craft*, 4th edn. (1973)
3. O.M. Faltinsen, *Hydrodynamics of High-Speed Marine Vehicles* (Cambridge University Press, UK, 2005)

4. D. Savitsky, On the seakeeping of planing hulls. SNAME Marine Technol. **5**(2), 164–174 (1968)
5. W.G.A Perring, H. Glauert, Stability on the water of a seaplane in the planing condition. Aeronautical Research Council, Technical Reports, Vol. 42 (September 1933). Hard copy recorded by SLUB Dresden, no electronic
6. M. Martin, Theoretical prediction of porpoising instability of high-speed planing boats. SNAME JSR **22**(1) (1978)
7. P.R. Payne, Coupled pitch and heave porpoising instability in hydrodynamics planing. J. Hydronaut. **8**(2) (1974)
8. C.L. Shuford Jnr, *A Theoretical and Experimental Study of Planing Surface Including Effects of Cross Section and Plan*, NACA TN 3939
9. D. Savitsky, P. Ward Brown, Procedures for hydrodynamic evaluation of planing hulls in smooth and rough water. SNAME Mar. Technol. **13**(4) (1976)
10. R.C. Dorf, R.H. Bishop, *Modern Control Systems* (California, Addison Wesley Longman. Inc., Monlo Park, 1998)
11. S. Niwa, *High Speed Craft Technology*, vol in Japanese (Rudder, 1971)
12. Z.S. Dong, *High Speed Crafts Dynamics* (Press of Central China University of Technology, China, 1991)., (in Chinese)
13. G. Fridsma, *A Systematic Study of the Rough-Water Performance of Planing Boats, Part 1, Regular Waves* (Davidson Laboratory, Stevens Institute of Technology, Report SIT-DL-69-1275, November 1969)
14. G. Fridsma, *A Systematic Study of the Rough-Water Performance of Planing Boats, Part 2, Irregular Waves* (Davidson Laboratory, Stevens Institute of Technology, Report SIT-DL-71-1495, March 1971)
15. M. Bessho, M. Yoji, M. Komatsu, Study on the movement of high-speed boats in the waves, in *Proceedings of the Japan Shipbuilding Association*, No. 35 (1974), p. 6
16. D.J. Taunton, D.A. Hudson, R.A. Shenoi, Characteristics of a series of high-speed hard chine planing hulls—Part II: Performance in waves. Int. J. Small Craft Technol., Trans RINA **153**(Part B1), B1–B23 (2011)
17. E.E. Zarnic, C.R. Turner, *Rough Water Performance of High Length to Beam Ratio Planing Boats*, DTNSRDC/SPD-0973-01,1981.12
18. M. Martin, Theoretical prediction of motions of high-speed planing boats in waves. JSR **22**(3), 140–169 (1978)
19. E E Zarnic, *A Nonlinear Mathematical Model of Motions of a Planing Boat in Regular Waves*, DTNSRDC-78/032
20. E.E. Zarnic, *A Nonlinear Mathematical Model of Motions of a Planing Boat in Irregular Waves*, DTNSRDC/-SPD-0867-01
21. Y.S. Dai, J.Z. Song, W.Y. Li, On the motion of high-speed craft in waves. J. Harbin Univ. Technol., Harbin, China (March 1981). (in Chinese)
22. F. De Luca, C. Pensa, Experimental investigation on conventional and unconventional Interceptors. Int. J. Small Craft Technol., Trans RINA **154**(Part B2), 65–72, ISSN 0035.8967, (2012)
23. A. Bliault, L. Yun, H.Z. Rong, in *High Speed Catamarans and Multihulls—Technology, Performance and Applications*, Chapter 8 (Springer Nature, 2019), 746 pp, ISBN 978-1-4939-7889-2

Chapter 6
Deep Vee Semi-planing Craft

6.1 Basic Concepts and Features

The Deep Vee semi-planing craft is a specific geometry with a large deadrise angle and hard chine section profile used for improved performance in a sea state.

Planing hulls generally have deadrise combined with hard chine cross section, the so-called trapezoidal V-type profile extended aft. A number also has deadrise angle in excess of 20 degrees and up to 25 degrees all the way to the stern. Such craft are referred to as "Deep Vee" craft.

We have discussed the performance of planing craft with deadrise, including the deeper vee form in Chaps. 4 and 5. As particularly discussed in Chap. 5, planing craft operating in an open seaway need a higher deadrise for motion performance and low added drag. By definition, these high-speed craft ride on the surface rather than cut through it. The lateral profile will display an overhung bow to provide supporting buoyancy as the craft rides a wave. While a sharper vee shape will lessen the vertical forces through a wave peak, if the shape of the bow were changed to slice through the wave, it would have less ability to recover in a wave trough and so would give a much wetter ride in a seaway.

If we consider speeds down toward the semi-planing regime, wave-piercing can also be considered to improve seakeeping, but how to minimize green water on deck? In this chapter, we will discuss design possibilities for Deep Vee semi-planing craft with the vee and chine form using high deadrise as an alternative to taking the rounded hull forms of Chap. 2 and going to extreme L/B for seakeeping. We will also introduce bow forms for motion damping that follow similar logic to wave-piercing fast catamarans [1].

The Deep Vee geometry for these vessels was an evolution from the hard chine planing hull. For example, the fast patrol craft SAR33[1] weighing 170 t with the

[1] https://en.sg.gov.tr/sar-33-class-cg-boats

Z.-S. Dong et al., *High-Speed Monohull and Hydrofoil Craft*,
https://doi.org/10.1007/978-3-031-62760-6_6

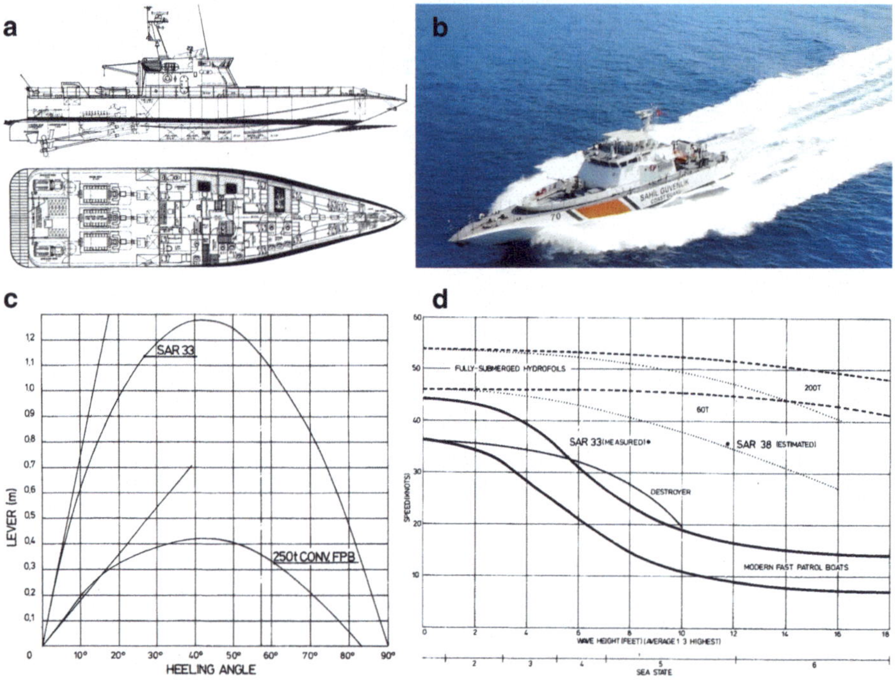

Fig. 6.1 SAR33 GA. (**a**) General arrangement. (**b**) Photo. (**c**) Stability curve. (**d**)Seakeeping comparison—ability to maintain speed

speed of $Fn_L = 1.22$, $Fn_\nabla = 2.7$, close to critical speed for planing, which was developed in the 1970s.

Designers started developing fast interceptor vessels with larger displacements that resulted in specific semi-displacement craft that extended the speed range further down into the transition mode, for example, the NHP48 design at $Fn_L = 0.88$, $Fn_\nabla = 2.33$ [2].

With the development of powerful and light high-speed diesel engines, enhancing craft speed became easier, but developments in improving seakeeping quality have been slower. The strong requirements for seakeeping performance, especially for craft equipped with guided missile weapons, lead to the development of Deep Vee craft for the same service speed range as the semi-planing craft, discussed in Chap. 2.

The Deep Vee form for this duty continues to develop and improve in performance. During the early stages of development, it was not so much different from a hard chine planing craft, except for higher deadrise, as the general arrangement for the SAR33 shown in Fig. 6.1a demonstrates. Figure 6.1b shows the craft in service, while Figure 6.1c shows a comparison of the stability curve of this craft against an equivalent round bilge design, showing the much higher righting lever. Figure 6.1d shows a comparison of seakeeping performance for different vessel types where the SAR33 has a clear improvement in performance compared to low deadrise FPB craft.

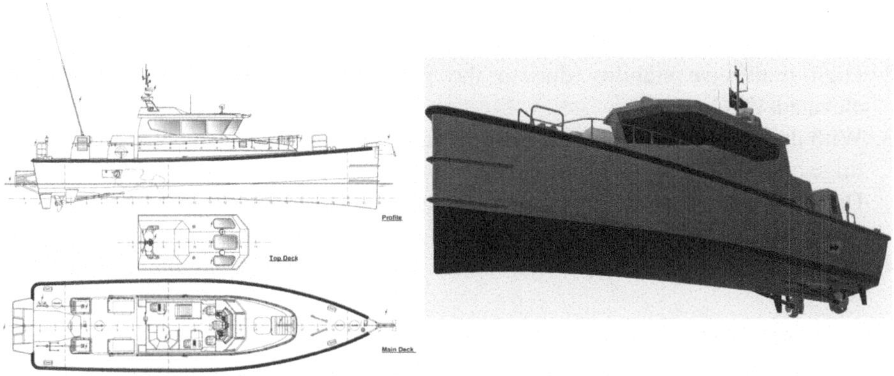

Fig. 6.2 Damen Stan 2205 Patrol Craft, general arrangement, and 3D model viewed from underside forward of the bow to show underwater shape, fins, and propulsion

While in medium (coastal SS5) weather, the long-overhung bow gave good seakeeping, concepts with a more vertical stem line were proposed, perhaps even with a bulb form below the surface, to give enhanced seakeeping in heavier weather during long-duration patrols.

The Geneva Hydrodynamic Institute of Switzerland has itemized the following characteristics in addition to a large deadrise angle of $\beta \geq 20°$ [3] for a Deep Vee craft intended as a small fast warship operating in the transitional speed range. Figure 6.2 shows a typical hull with these characteristics, the Damen Stan Patrol 2205:

- Special bow lines with the keel extending below the baseline
- Deadrise angle aft of midship section kept constant or increased slightly until close to stern
- Longitudinal keel line rising from bow to midship, but not after midship, as other craft
- Two stern bilge stabilizer fins

While the features of these craft are mainly for improvement of seakeeping quality, additionally, they provide a good compromise to meet the requirements for good powering, stability, maneuverability, and hull structure weight. Different profiles and other features have been trialed in other Deep Vee projects. We will discuss several of these in this chapter.

Advantages

1. High-quality seakeeping is a significant advantage of Deep Vee craft. Compared to round bilge craft, the motion response of Deep Vee craft in waves, including pitching, heaving, and vertical accelerations, is reduced. Particularly the slamming, green water shipment, and drag increment in waves are reduced significantly, so allowing higher speed in waves compared to round bilge craft. The

rolling response of Deep Vee craft is close to the round bilge craft fitted with bilge keels.
2. High transverse stability due to the generally wide hull beam, even with elevated VCG.
3. With deep and wide cross sections, the maneuverability is also better than round bilge craft, with a small turning radius and inward heel angle during turning. The Deep Vee section at the stern can maintain good course keeping ability and avoid broaching.
4. A wide beam creates a large hold volume and deck area, a favorable arrangement.

Main Disadvantages

1. Deep Vee craft drag in calm water is larger than that of round bilge craft, due to its lower length/beam ratio. The drag penalty can be up to 10% for initial designs; however, after improvements, it can be reduced to under 5% for final designs. At Froude number above $Fn_L = 0.8$, the drag is similar to that of the round bilge craft.
2. The weight of Deep Vee craft is a little higher (1–4%) than that of round bilge craft due to its wide beam, heavier deck, and hull structure weight.
3. From the project design of frigate-size ships, Kehoe et al. [4] found that:
   (a) For a vessel with structure weighing 1000 t, the structural weight of Deep Vee craft was close to that of the round bilge craft.
   (b) For a 2400 t structural weight frigate, the structural weight increase of Deep Vee craft is about 30 t, which is 1.5% higher than for the round bilge craft.
   (c) For a ship of 4500 t, the structural weight increase is about 115 t, which is 2.5% higher than the round bilge craft as increment of empty ship weight
4. The Deep Vee ship is not favorable to the arrangement of conventional propulsion equipment due to narrow bottom of hull and deep draft at stern, which leads to raising the power plant resulting in propeller shafts with larger inclination angles and possibly extending below the baseline.

6.2 Drag of Deep Vee Semi-planing Craft in Calm Water

In general, the drag in calm water is greater than that of well-designed round bilge craft with the same displacement, mainly due to a lower length/beam ratio and larger wetted area. A low length/beam ratio leads to a larger entrance angle, increasing the wave-making drag, and a larger wetted area leads to higher friction drag.

Typical drag of initial designs of Deep Vee craft in calm water was larger than round bilge craft by 15–20%. E H. Serter [2] presented the test results of model bare hull drag in calm water of the Deep Vee fast interceptor NHP48 displacing 400 t and

Table 6.1 Increment of drag of NHP48 vs PR72 in calm water

Speed scale ratio	20 knots drag increment	30 knots drag increment	40 knots drag increment
5.7:1	>22%	>23%	>18%

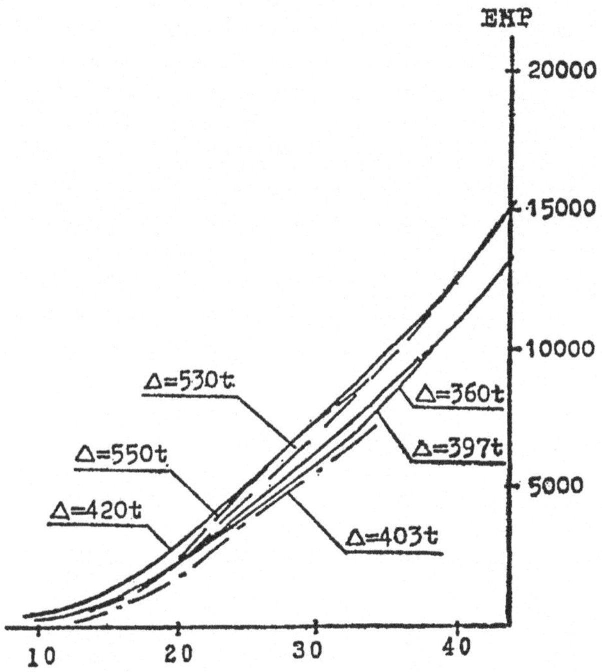

Fig. 6.3 Effective power of Deep Vee NHP48 and round bilge craft PR72. (From Serter [2])

a round bilge corvette PR72[2] with 560 t displacement and found that the drag of the Deep Vee craft at equivalent displacement was increased by ~20% compared with the round bilge craft, as shown in Table 6.1. The maximum speed is reduced by 1.5–2.0 knots due to the increase in drag.

Serter noted that the comparison above was for the bare hull. Appendages such as bilge keel and even anti-roll controllable fins have to be mounted on the round bilge craft for improving the rolling performance, which will add drag to the craft in calm water by 2–3% for bilge keels and 3% for anti-roll fins. However, due to the chines in its design, it is not necessary for Deep V craft to mount additional bilge keels or anti-rolling fins and so the real increase in drag of these craft compared to round bilge is only about 15%.

Additionally, while having a large deck area and hold volume, the size and displacement of Deep Vee craft will be smaller than the round bilge craft for the same outfit equipment. For NPH48, the decrease of displacement is about 20~30 t, so the actual drag increment in calm water is even smaller. Figure 6.3 shows the

[2]https://en.wikipedia.org/wiki/PR-72P-class_corvette

comparison of the effective power of NHP48 weighing 360 t with PR72 weighing 397 t and 403 t, respectively, and NHP48 weighing 420 t with PR72 weighing 530 t and 550 t, respectively. One can see that the drag differences are small with this adjustment in displacement.

From the comparison, one can see that the drag of the Deep Vee craft at low speed is larger; however, the drag gradually matches that of the round bilge craft as speed increases. This is due to the planing configuration of a Deep Vee hull, which develops hydrodynamic pressure on the hull bottom, lifting the craft gradually with increasing speed so as to decrease the wetted surface area and drag.

Many investigations have been carried out to improve the speed performance of Deep Vee craft in calm water, studying the effects of size, parameters of hull configuration, deadrise angle, longitudinal position of C_B, stern plate, etc., as discussed in Chap. 4. These developments have improved the speed performance and constrain the drag increment within 5–10% compared with round bilge craft. It should be noted however that the objective of the Deep Vee semi-planing design is to achieve the best compromise between minimizing drag and delivery of the best possible motion performance in a seaway.

To summarize, the effect of hull configuration of Deep Vee semi-planing craft on the drag in calm water is as follows:

6.2.1 Effect of Length Beam Ratio L/B

Deep Vee craft drag in calm water increases inversely with L/B, as shown in Fig. 6.4, obtained from series tests on Deep Vee models carried out by Hu Dong Shipyard Group, China [5]. The results show that the residuary drag of models per unit displacement decreases significantly for increase in L/B from 5 to 9, especially at high speed. The wetted surface area of the hull increases with L/B increasing the friction drag; however, the increase is less than the reduction in residual drag. Meanwhile, the structural weight increases with L/B, so an optimum L/B considering these factors has to be selected.

6.2.2 Effect of Deadrise Angle β

From Fig. 6.4, one can see in the case of low L/B, the residual drag of the model in calm water with $\beta = 25°$ is lower than that with $\beta = 15°$ but are same above $L/B = 7$. The wetted surface area will be larger, and the holds will have less volume for higher deadrise at a given displacement. Thus, to maintain the same effective hold volume, the drag of the craft with larger deadrise is greater, and this will be more apparent for high-length beam ratios.

Fig. 6.4 Effect of L/B and β to drag

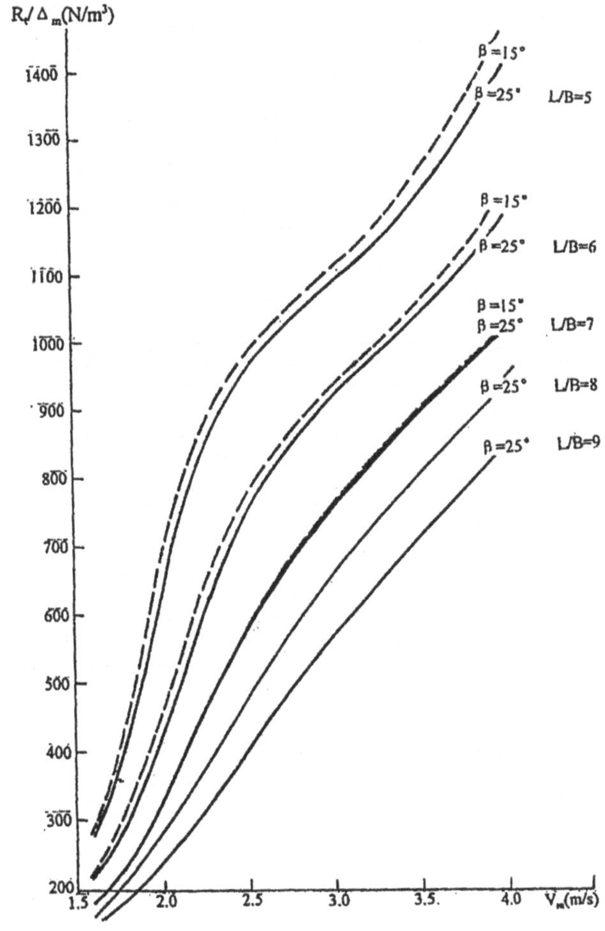

6.2.3 *Effect of Longitudinal Position of* C_B (X_C)

The longitudinal position of C_B influences the running trim. Fig. 6.5 shows the change of longitudinal position of C_B on total bare hull drag of a Deep Vee model with L/B = 8, $\beta = 25°$ and longitudinal position of C_B at −3.48% [5]. Tests were carried out with CB moved back and forward by 5%L from the design value of −3.48%L from amidships. The change of C_B is reflected in the change of running attitude as well. From the figure, one can see that at low speed, optimum X_c is forward of amidships and it moves aft with increase in speed.

When the craft is running, the trimming angle changes gradually from bow-down to bow-up due to the hydrodynamic moment from wavemaking increasing with speed up to the critical speed at approx. $Fn_L = 0.75$. Figure 6.6 shows the change of running trim angle with Froude Number for Deep Vee model with $L/B = 8$, and $\beta = 25°$ at different slendernes and values of $L/\nabla^{1/3}$.

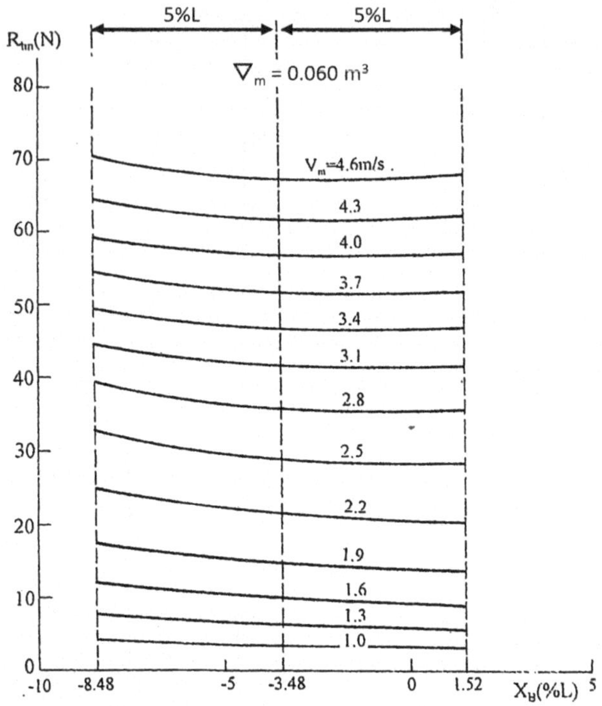

Fig. 6.5 Effect of longitudinal position of CB to drag

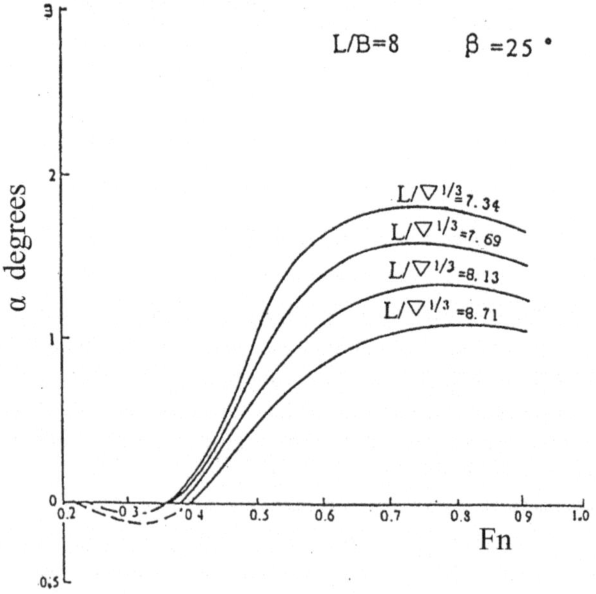

Fig. 6.6 Running trimming angle versus speed

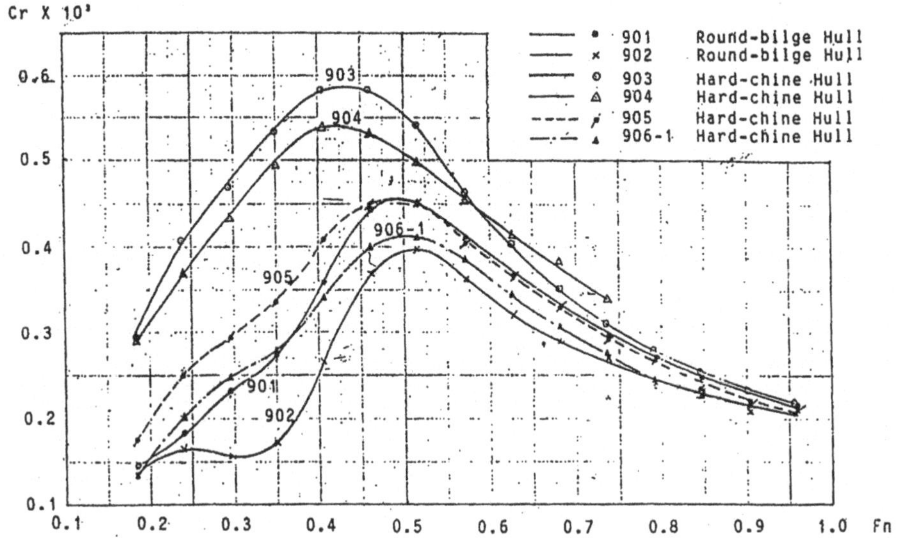

Fig. 6.7 Residual drag coefficient of Deep Vee and round bilge craft

It can be seen that the craft is trimming bow down only at low speed, $F_n \leq 0.4$ while at $F_n \geq 0.4$, the trim angle changes to a positive value, reaching a maximum at $Fn_L = 0.75$ and the trimming angle is small at $1 - 2°$for a slender craft. From the point of view of drag in calm water, trim angle of about 1° is favorable.

Summing up, it is possible to reduce the drag of Deep Vee craft in calm water. Increasing the length/beam ratio is one effective measure, particularly for high-speed designs. Reducing deadrise angle properly is also an effective measure, however, with reduced seakeeping performance.

Presently, the deadrise angle is commonly selected at $20 - 25°$ with some researchers suggesting keeping a large deadrise angle at amidships and decreasing it gradually toward the stern so as to decrease drag in calm water without a penalty on seakeeping (a warped geometry). However, the deterioration of course keeping quality has to be considered as well, particularly in following waves. Using a stern plate, stern wedge or interceptors also provides good results, especially for craft with a low length/beam ratio running at trim angles in excess of 1.5 ° . The trim angle can be reduced to about 1° so as to improve not only drag on calm water but also seakeeping quality.

The application of the equipment mentioned above should be determined by specific requirements for the performance and considering the change of structural weight for such equipment. At present, for Deep Vee craft at $0.5 < F_n < 0.7$, the drag in calm water is greater than well-designed round bilge craft by about 5~10 % , and at speeds of $F_n > 0.8$, it will be the same as or better than that of round bilge craft.

Figure 6.7 presents the test results of residual drag of a series of Deep Vee models developed and carried out by Hudong Shipyard and CSSRC, China, where 901 and 902 represent round bilge craft and 903 and 904 represent the initial designs of Deep

Table 6.2 Drag results of Deep Vee craft in calm water(model test results with round bilge model 902 as reference)

V	901	902	903	904	905	906.1	906.2	Fn
1.3	101.2	100	148.7	140.0	121.6	112.6	122.9	0.240
1.6	112.3	100	165.1	163.0	132.8	123.9	126.5	0.295
1.9	116.8	100	173.5	171.3	137.8	126.0	126.8	0.350
2.2	113.7	100	156.1	153.4	129.7	118.1	116.9	0.406
2.5	107.9	100	134.4	129.4	116.4	109.0	107.9	0.461
2.8	107.0	100	124.1	122.5	114.8	109.0	107.7	0.516
3.1	105.2	100	117.6	120.1	111.8	108.4	107.3	0.571
3.4	104.7	100	114.4	120.0	111.8	107.4	106.2	0.627
3.7	104.0	100	111.9	119.5	111.0	105.9	105.7	0.682
4.0	102.6	100	109.3	119.2	108.5	104.9	103.9	0.737
4.3	101.4	100	108.0	–	107.4	104.1	102.4	0.793
4.6	99.7	100	106.1	–	105.2	–	–	0.848
4.9	98.3	100	104.8	–	105.0	–	–	0.903
5.2	97.5	100	102.6	–	104.7	–	–	0.959

Vee craft. After gradual improvements, such as 906.1, the residual drag coefficient is better than that of 901 and even matches with the well-designed round bilge craft 902. Table 6.2 provides the drag data in calm water [6], which shows the increment of drag of model 906.1 at $F_n \geq 0.7$ is less than 5% and the increment of drag even smaller at 3–4% for the improved model 906.2.

Additionally, when using conventional propellers on Deep Vee craft, the influence of the thrust reduction factor has to be considered. Remote controlled self-propulsion model tests demonstrate that when increasing L/B of Deep Vee craft, the thrust reduction coefficient t increases at low speed while keeping the wake coefficient w constant, so as to decrease hull efficiency. Also, t decreases and hull efficiency increases as speed is increased, resulting in good performance at high L/B of Deep Vee craft.

The influence of β at low L/B shows that the wake fraction w remains unchanged; however, the thrust reduction coefficient t will be higher at smaller β than that at larger β, leading to low hull efficiency, and t will be increased with speed. Therefore, transverse sections with small β ($<15°$) are not favorable to the propulsion efficiency of these craft, particularly at high speed.

When using water jet propulsion, the thrust reduction factor disappears. Therefore, it is reasonable to use water jet propulsion on Deep Vee craft at high speed, both from the point of view of arrangement of power plant and propulsive efficiency.

6.3 Seakeeping Quality of Deep Vee Semi-planing Craft

The fine seakeeping quality is a main advantage of these craft, and a lot of test results demonstrate that the motion response in waves is reduced significantly compared to round bilge craft with the same displacement, particularly the impact loads in head sea waves. The smaller wave drag increment leads to operational speeds greater than that of round bilge craft in waves, in spite of the calm water speeds lower than round bilge craft. In addition, according to experienced captains, the speed loss of craft in waves is mostly because of slamming and the need to reduce speed. For safety, captains are obliged to throttle down the speed after slamming events to a craft in waves begin. However, due to reduction in severity of impacts on Deep Vee craft, they can be kept at high speed in rough seas.

6.3.1 Heaving, Pitching, and Vertical Acceleration of Deep Vee Craft in Waves

Comparing the Deep Vee craft with round bilge craft with same displacement, the decrease in heaving, pitching, and vertical accelerations of Deep Vee craft in waves is about 10~20 % , due to larger damping coefficients of the transverse section and lower C_B compared with round bilge craft. Table 6.3 shows a comparison of motion response from model test results of Deep Vee craft with lowered forefoot and a typical round bilge craft obtained at CSSC in China [5]. The body plans are shown in Fig. 6.8.

Table 6.3 Motion response of Deep Vee craft and round bilge craft in waves $\Delta = 519.4\text{t}$, $H_{1/3} = 2.4\text{m}$, $T_1 = 6\text{s}$

	Speed (knots)	Round bilge craft 902	Deep Vee craft 904	Reduction value %
Heaving (m)	17.6	0.74	0.6	−19
	35.3	0.99	0.89	−10
Pitching (deg)	17.6	3.4	3.1	−9
	35.3	2.3	2.0	−13
Bow acceleration	17.6	0.65 g	0.58 g	−11
	35.3	0.70 g	0.58 g	−17
Acceleration amidships	17.6	0.18 g	0.15 g	−17
	35.3	0.35 g	0.28 g	−20
Stern acceleration	17.6	0.34 g	0.32 g	−6
	35.3	0.51 g	0.44 g	−14

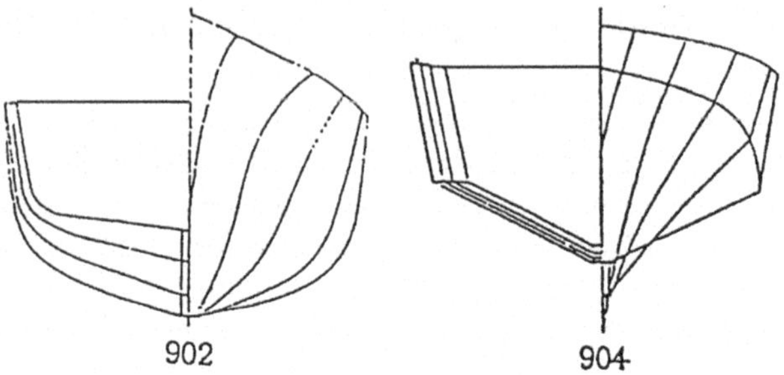

Fig. 6.8 Body plan of round bilge craft 902 and Deep Vee craft 904

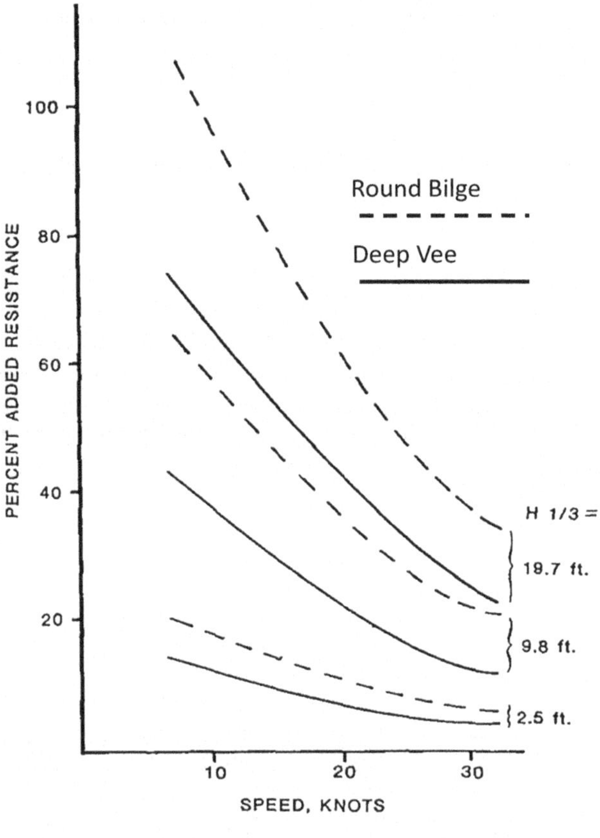

Fig. 6.9 Drag increments of 2400 t Deep Vee ship and round bilge ship in waves. Dash is round bilge. Line is Vee

6.3.2 Drag Increment of Craft in Waves

The drag increment of Deep Vee craft is far smaller than round bilge craft in waves, which has been validated by a number of model test results. Figure 6.9 shows the test results of drag increment in waves (in % compared with the drag in calm water) of a

420 ft. Deep Vee frigate weighing 4500 t, and an equivalent round bilge craft, at wave heights of 2.5, 10, and 20 ft. (SS 3, 5, and 6), given by Kehoe, Brower, and Serter [4]. This shows the clear advantage of the Deep Vee form.

Table 6.4 presents a comparison of the drag increments of a Chinese series of Deep Vee craft models, 1735#, with round bilge craft, at different speeds of 15~30 knots, and sea states 4, 5, 6, where $\Delta R0$ is the drag increment of round bilge craft in waves, and $\Delta R1$ the drag increment of Deep Vee craft in waves.

The drag increment of the Deep Vee craft is smaller than round bilge in waves by up to 25~30 % , so in spite of drag in calm water being larger than round bilge craft, Deep Vee craft can achieve higher speeds than round bilge craft in a sea state.

6.3.3 Slamming in Waves

The slamming of Deep Vee craft in waves is significantly lower than round bilge craft. A comparison of slamming probability of Kehoe's 4500 t Deep Vee ship with a round bilge ship in head sea waves is shown in Fig. 6.10 [4]. This is a critical seakeeping quality in favor of the Deep Vee ship and leads to no voluntary speed reduction in waves, even up to Sea State 7. The slamming frequency of the 4500 t Deep Vee ship at a speed of 30 knots is only about 12 per hour, far lower than the US Navy criterion of 20 per hour.

Figure 6.11 shows the maximum speed of the 4500 t frigate with Deep Vee and round bilge configuration in various sea states according to drag increments and slamming of the ships, carried out by Kehoe et al. [4]. From the figure, it can be seen that the speed of the Deep Vee ship does not decrease much with increased sea state, whereas the speed of the round bilge ship is reduced significantly. In addition, green water on deck, propeller emergence out of the water surface, etc. will also be a concern for round bilge hulls. To sum up, the seakeeping quality of the Deep Vee is much better than round bilge ships.

Figure 6.12 shows the slamming frequency of a 1700 t frigate with Deep Vee hull and a round bilge type in different sea states, from model test results carried out by Hu Dong Shipyard and CSSRC of China [5], where 1735 and 1736 are Deep Vee ships and the comparison ship is of round bilge form. From the figure, it can be seen that the slamming frequencies of a Deep Vee ship are decreased significantly compared with round bilge ship at sea states of $H_{1/3} = 4.545$m, particularly at high speed.

Table 6.4 Comparison of drag increments of Deep Vee craft and round bilge craft in waves

	Sea state 4 $H_{1/3} = 2.0\text{m}T_1 = 6.835$			Sea state 6 $H_{1/3} = 4.545\text{m}, T_1 = 8.525$			Sea state 7 $H_{1/3} = 4.545\text{m}, T_1 = 8.525$		
	Drag increment(KN)		%	Drag increment(KN)		%	Drag increment(KN)		%
Speed knots	Round bilge ΔR_0	Deep Vee, ΔR_1	$\left(1 - {}^{\Delta R_0}/_{\Delta R_1}\right)\%$	Round bilge ΔR_0	Deep Vee ΔR_1	$\left(1 - {}^{\Delta R_0}/_{\Delta R_1}\right)\%$	Round bilge ΔR_0	Deep Vee ΔR_1	$\left(1 - {}^{\Delta R_0}/_{\Delta R_1}\right)\%$
15	29.40	24.56	−29	120.84	105.97	−25	170.40	153.40	−23
20	29.98	25.96	−33	126.42	115.81	−29	178.28	168.75	−27
25	33.60	27.40	−29	145.01	119.81	−29	202.74	170.82	−28
30	41.42	32.68	−26	179.92	153.89	−20	256.55	225.55	−18
Average %			−29	Average %		−26	Average %		−24

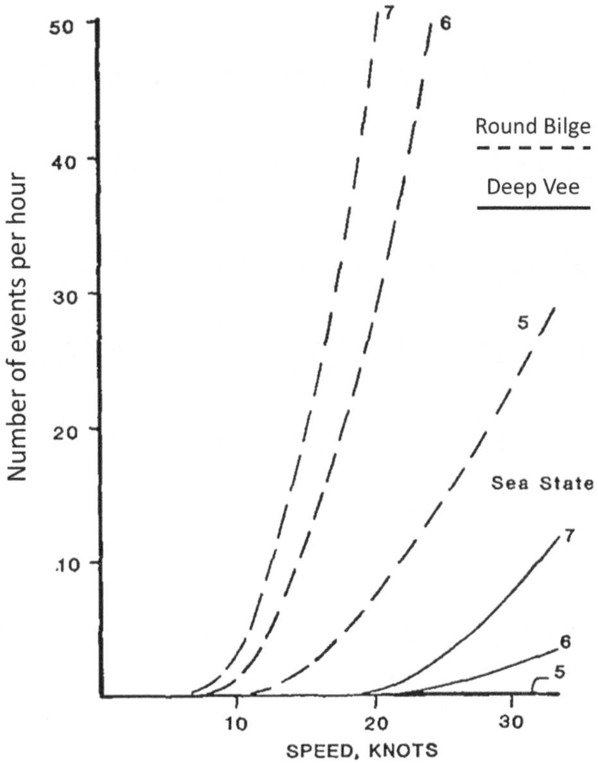

Fig. 6.10 Slamming of Deep "Vee" 2400t vessel compared with round bilge. (Ref. fig. 4 in Kehoe alt ref. 72)

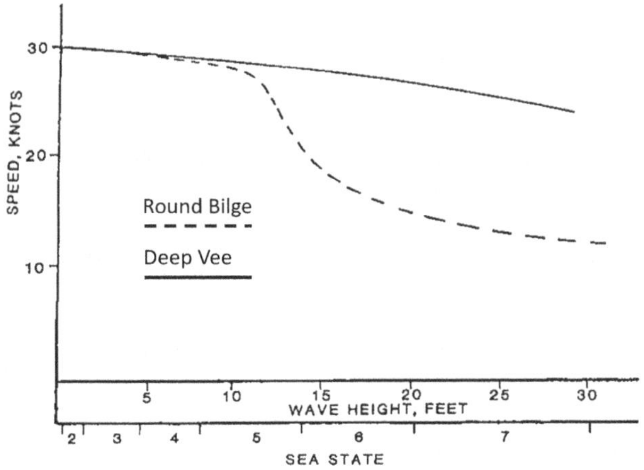

Fig. 6.11 Maximum speed for 2400t frigate with Deep Vee and round bilge against sea states. (Fig. 7 in Kehoe al ref. 72)

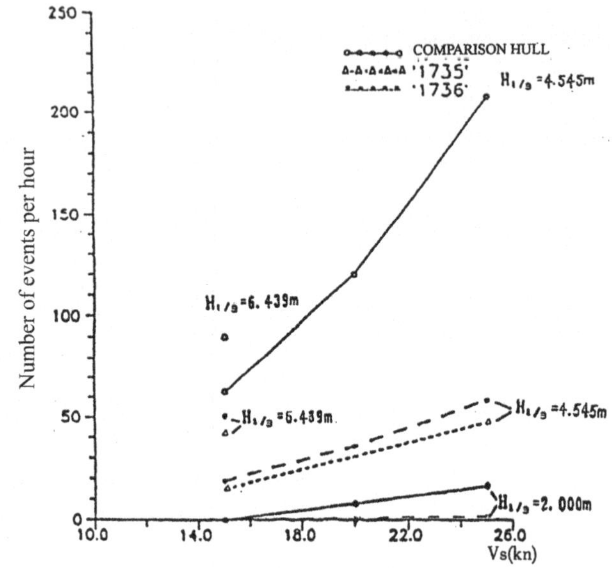

Fig. 6.12 Comparison of slamming frequency of 1700 t frigate with Deep Vee and round bilge

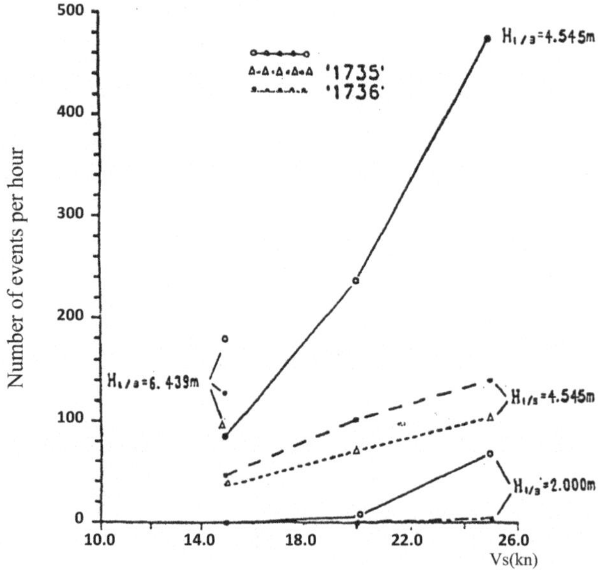

Fig. 6.13 Frequency of wetted deck for Deep Vee and round bilge in various sea states

6.3.4 *Green Water and Surface Emergence of Propellers*

Figure 6.13 shows the frequency of green water on deck for both Deep Vee and round bilge craft in different sea states, which again shows the obvious advantages of Deep Vee craft. Figure 6.14 shows the frequency of propeller emergence, where 1/3 part of the propeller is out of water, in different sea states, which also shows the obvious advantage of the Deep Vee design.

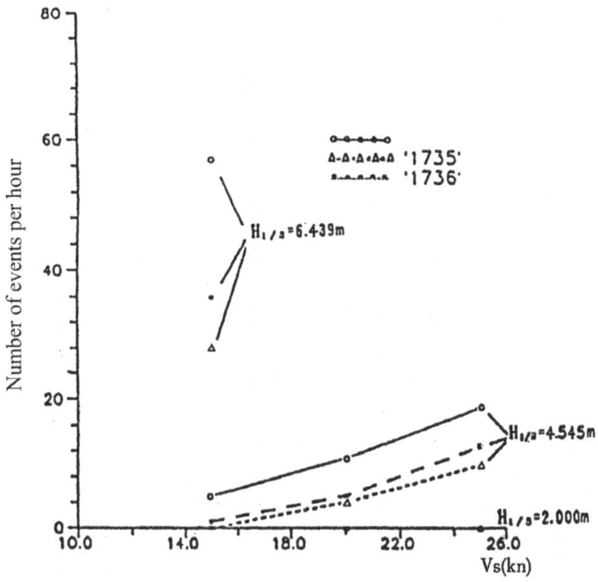

Fig. 6.14 Emergence of propeller at different sea states for Deep Vee and round bilge

6.3.5 Roll and Rolling Acceleration in Waves

In general, the rolling response of Deep Vee craft without a bilge keel is similar to that of a round bilge craft with bilge keels. Figure 6.15 [3] shows the rolling angle and transverse accelerations at 5 m over the water line of both 1000 t Deep Vee and round bilge ships at $V_s = 26$ knots and significant wave height $H_{1/3} = 3.75$m with different wave directions. Bilge keels and stern skegs are mounted on the round bilge ship, while stern skegs and short stern bilge keels are also mounted on the Deep Vee ship. It can be seen that the values of both ships are very close.

In spite of the high CG of Deep Vee vessels, they have better transverse stability as the metacenter is higher, giving shorter inherent rolling periods than round bilge craft.

It is easier to mount anti-rolling bilge strakes on round bilge craft for improving the rolling performance. However, on Deep Vee craft, it must be mounted on the sides above the chine and is not as convenient as on the round bilge craft, unless they are treated as chine rails for faster craft with the same section. Automatic controllable stern and side interceptors can be mounted on Deep Vee craft to improve the rolling performance, so there are useful design options.

Many tests demonstrate that the rolling performance of Deep Vee craft at high speed is better than round bilge craft due to an increase in roll damping from hydrodynamic pressure, so there is less requirement for additional outfit than on round bilge craft.

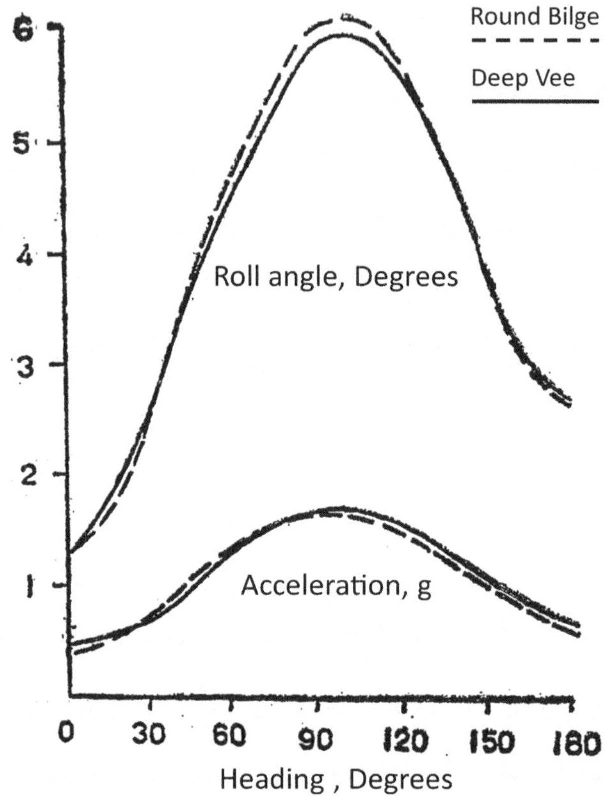

Fig. 6.15 Rolling and transverse acceleration in sea state for 1000t Deep Vee and round bilge

6.3.6 Influence of Size and Profile on the Seakeeping Quality of Deep Vee Craft

6.3.6.1 Influence of Length/Beam Ratio L/B

High length/beam ratio is favorable to the seakeeping quality. Towing tests of China's Series of Deep Vee models were conducted in regular head sea waves with L/B between 5 and 9. Using the linear superposition method, the single amplitude motion responses in irregular waves are predicted and shown in Table 6.5.

From the table, one can see that the motion responses and drag increment decrease with L/B. Thus, higher L/B is preferred for seakeeping quality. However, this leads to an increase in hull structural weight and power, as well as fuel consumption, and so these should be considered.

Table 6.5 Significant values of single amplitude of motion response per unit wave height ($\Delta = 620\text{t}, T_1 = 6\text{s}$)

	$\beta = 25°$						$\beta = 15°$				
	L/B Vs knots	5	6	7	8	9	5	6	7	8	9
Heave	18.2	0.44	0.40	0.35	0.32	0.27	0.45	0.41	0.36	0.32	0.28
	22.7	0.48	0.44	0.39	0.35	0.31	0.49	0.45	0.40	0.36	0.32
	27.2	0.51	0.46	0.42	0.38	0.34	0.52	0.47	0.43	0.39	0.35
Pitch (°/m)	18.2	1.67	1.52	1.38	1.25	1.11	1.73	1.56	1.43	1.30	1.17
	22.7	1.62	1.47	1.33	1.22	1.10	1.65	1.51	1.38	1.25	1.14
	27.2	1.54	1.40	1.27	1.15	1.04	1.60	1.44	1.30	1.17	1.05
Bow acceleration	18.2	2.55	2.41	2.24	2.08	1.92	2.53	2.41	2.26	2.12	1.98
	22.7	2.85	2.73	2.58	2.43	2.30	2.89	2.76	2.61	2.48	2.36
	27.2	3.05	2.93	2.80	2.66	2.58	3.21	3.09	2.94	2.80	2.65
Stern acceleration	18.2	1.45	1.34	1.24	1.15	1.05	1.48	1.38	1.29	1.19	1.08
	22.7	1.55	1.48	1.39	1.30	1.22	1.62	1.53	1.44	1.34	1.25
	27.2	1.80	1.68	1.56	1.44	1.33	1.98	1.85	1.73	1.59	1.47
Average drag increment kN/m^2	18.2	7.43	6.58	5.85	5.15	4.57	7.78	6.88	6.08	5.48	4.93
	22.7	8.58	7.58	6.75	5.99	5.36	8.81	7.92	7.09	6.34	5.63
	27.2	9.63	8.67	7.77	6.90	6.09	9.70	8.68	7.79	7.04	6.32

6.3.6.2 Influence of Deadrise β

Table 6.5 presents the motion response of Deep Vee craft with deadrise $\beta = 25°$ and 15° with the same L/B. It shows that small β reduces the seakeeping quality with larger motion responses of pitching, vertical acceleration, and drag increment, but with less influence on the heaving response.

Small β is favored for low drag in calm water. Varying β from amidship to stern is used on some projects of Deep Vee craft for improving drag in calm water; however, it reduces the seakeeping quality. In addition, small β at the stern will lead to reduction in course keeping stability, particularly for Deep Vee craft with bow keel line lowered, and attention has to be paid to broaching in following seas as well. Small β at stern will also lead to size increase of the stern skegs to provide necessary directional stability.

6.4 Maneuverability of Deep Vee Craft

Compared with round bilge craft, the center of underwater lateral projected area of Deep Vee craft is further aft due to the fact that the draft of round bilge craft at stern is usually shallower than the Deep Vee craft in spite of the longitudinal keel line of the latter rising up from bow to stern gradually. Consequently, the course stability of Deep Vee craft is better than round bilge craft.

The turning performance of Deep Vee craft is similar to round bilge craft, but due to the damping force (moment) of Deep Vee craft in turning being greater, this leads to a greater speed loss for Deep Vee craft than round bilge craft during turning. In addition, the wider and shorter hull of the Deep Vee craft with the same displacement leads to smaller turning radius of Deep Vee craft.

Another feature of Deep Vee craft maneuverability is a small outward heeling angle during turning, and sometimes an inward heeling angle due to the hydrodynamic pressure acting on the bottom of the craft, particularly during high-speed turning. Since the hydrodynamic pressure acting on the bottom decreases the outward heeling moment, the Deep Vee craft are able to turn at high speed with full rudder angle without an excessively large heeling angle outward, as a round bilge craft sometimes experiences.

Deep Vee craft are able to maintain good course keeping stability and maneuverability simultaneously, which for round bilge craft is a greater challenge.

Additionally, the stopping distance and time are shorter for Deep Vee craft than round bilge craft due to deeper stern draft and larger drag during emergency stopping.

Table 6.6 shows a comparison of tests of maneuverability; zigzag motion with a rudder angle of 10 °, stopping at speed ahead (no reverse thrust), stopping with reversed thrust, and course keeping ability with reversed thrust on one of Chinese Deep Vee series 1730 and three round bilge craft [6].

The 1730 craft has the following characteristics $L/B = 8$, $\beta = 25$ °, Deep Vee form. The craft A, B, and C are round bilge vessels of different characteristics. T' is the nondimensional rudder response index (or stability index), while T_L' is the nondimensional yawing stagnation, ψ_{ov} is the yawing angle lag, and P is the maneuverability coefficient.

From Table 6.6, one can see that for a Deep Vee vessel with smaller rudder area, the relative turning diameter $\overline{d}_{min} = d/L$ is equal to that of round bilge craft or a little bit better. Smaller rudder response index (stability index) T' of the Deep Vee demonstrates fine course keeping stability The stopping time and distance at forward and reverse power of Deep Vee ship are smaller than round bilge vessels. The maximum heeling angle of the Deep Vee craft is smaller than round bilge vessel and with inward heeling angle ϕ_{max} of the former in a turn. All these conclusively demonstrate the advantages of the Deep Vee form for maneuverability.

6.5 Stability of Deep Vee Craft

With a wider beam and shorter length, the transverse stability of a Deep Vee is generally better than the round bilge craft, even with a higher CG of the former. In general, the metacentric height of round bilge craft is 0.8–1.2 m and 1.4–1.8 m for Deep Vee craft.

The stability of Deep Vee craft in waves is also better than round bilge craft. The stability loss of Deep Vee craft running in following waves with the same speed as

Table 6.6 Maneuverability comparison of Deep Vee and round bilge craft

Ship name	Coefficient of rudder area	Speed V_0 m/s	Fn	$\bar{d}_{min}$ (35° rudder angle)	10 ° /1°Zigzag tests				Stopping tests				Reversed navigation		Heeling angle, ϕ_{max} (deg)
									Inertia		Reversed speed				
					T'	T_L'	ψ_{ov} deg	P	I_2'	t_z'	1.99	3.27	Straight navigation	Change course	
1730	0.02	1.98	0.301	3.85	0.381	0.395	1.73	0.487	6.63	17.55	–	–	OK	OK	
		2.40	0.365	4.22	0.370	0.382	1.77	0.485	–	–	–	–	–	–	
		3.25	0.494	4.70	–	–	–	–	–	–	–	–	–	–	−2.73
Round bilge A	0.025	1.60	0.281	4.25	–	–	–	–	–	–	–	–	OK	Not OK	–
		2.30	0.404	4.70	1.090	1.000	3.20	0.314	–	–	–	–	–	–	–
Round bilge B	0.0266	1.95	0.295	3.95	0.843	0.460	3.20	0.424	–	–	–	–	OK	Not OK	–
		3.30	0.499	4.60	0.820	0.290	3.90	0.449	–	–	–	–	–	–	10.2
Round bilge C	0.0276	2.00	0.303	4.00	0.680	0.550	2.10	–	7.39	26.20	2.31	4.45	OK	OK	–
		3.00	0.455	5.25	0.650	0.470	1.90	–	–	–	–	–	–	–	–

waves is smaller than round bilge craft due to its wider stern beam and deeper stern draft. These have been validated by a number of analyses.

Good transverse stability and course keeping stability of the Deep Vee craft in waves, particularly in following waves, leads to a small probability of broaching compared to round bilge craft and is one of the important advantages of the design.

6.6 Semi-submerged Bow for Deep Vee Craft

On a typical Deep Vee semi-planing craft, the longitudinal keel usually rises up from bow to stern gradually and sometimes the bow extends below the baseline, which is favorable to improved motion response in waves. Such lowering of the bow profile does not influence the course keeping ability of craft on calm water; however, when in a seaway, particularly in case of bow locating at wave peak and stern at trough, some problems of course keeping may occur for which stern skegs may need to be mounted. Mostly one skeg is mounted along the keel, not only for course stability but also to increase rolling damping and improve roll performance.

A stern skeg is also able to prevent the interference of propellers between the port and starboard side, thus improving the working condition of the propellers. Since bilge keels are difficult to be mounted on the Deep Vee craft, short stern bilge skegs are suggested to be mounted on the Deep Vee craft for increasing the roll damping.

In addition, stern flaps, wedge, or interceptor can also be installed on Deep Vee craft to adjust running attitude and maintain optimum trim angle. Since the rolling performance of Deep Vee craft is the only motion that is worse compared to the round bilge craft, controllable interceptors at the stern can be a useful measure for improving the rolling performance of the craft which cannot easily be realized on round bilge craft due to its stern profile unless this is adjusted to the almost square section as that tested by VTT (ref [23] in Chap. 4). Anti-rolling fins can also be mounted on the bottom of Deep Vee craft, however, with a penalty on drag.

A configuration with a semi-submerged bow (SSB) is a development project of Deep Vee semi-planing craft with the bow lowered down as shown in Fig. 6.16. The idea is to lower the buoyancy so as to make small change of buoyancy under the interference of waves and increase bow vertical damping and added mass, and consequently improve the seakeeping quality.

The SSB may be a body of revolution or a body with elliptical section. It may extend to under the keel or attach to the keel. The SSB displaced volume may be 2~10% of craft displacement. Larger volume and deeper SSB are favorable to seakeeping quality but will increase the drag.

Different from a conventional bulbous bow, the SSB sits lower with its main function to improve seakeeping, so it is difficult to create favorable interference with the surface waves generated by the hull. Figure 6.17 presents test results of vertical acceleration at the bow of a craft with different positions and sizes of SSB [7] at a speed of 30 knots in wave height of 2 m. For the SSB with 6% of volume and attached to the bow centered on the keel, the bow vertical acceleration is decreased

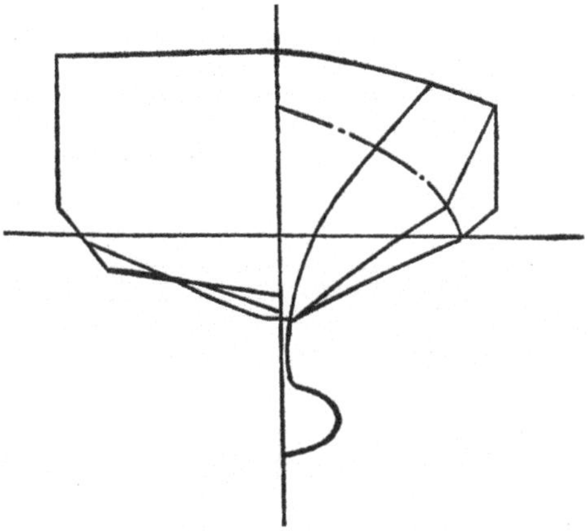

Fig. 6.16 Body plan with SSB

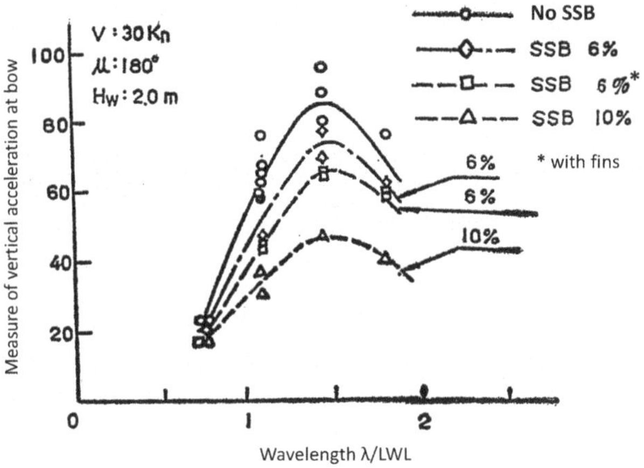

Fig. 6.17 Vertical acceleration of craft with different SSB in regular waves

significantly. It is higher when the SSB is lowered further down and volume increased to 10% when the vertical bow acceleration is reduced up to 50%. However, it is impractical to use such a big SSB for reasons such as draft and connection structural strength, together with the increase in drag in calm water.

Using an SSB with fins to reduce its size and improve the efficiency was investigated, as shown in Fig. 6.18 [8], where the SSB volume was only 2% of ship displacement with $l/b = 10$, $l/t = 19$, where l, b, and t are the length, width, and thickness of the SSB, respectively.

Model tests show no increase in drag while pitching and vertical acceleration reduced by about 35~45 % . However, no improvement was noticed for heaving motion. Therefore, using SSB with fins can decrease its size significantly while improving the seakeeping quality.

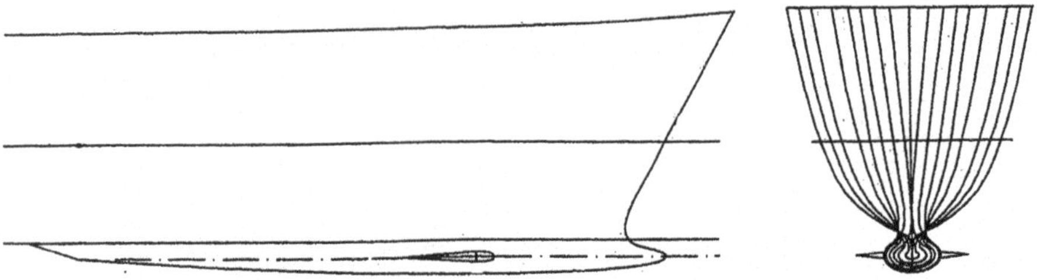

Fig. 6.18 Arrangement of SSB with fins

6.7 Development of the Axe Bow

The Axe Bow is a recently developed design based on the semi-planing Deep Vee hull form with the aim of improving the seakeeping quality of high-speed vessels such as fast crew boats. It was developed by J. A. Keuning of the Ship Hydromechanics Department, Delft University of Technology, and researchers of Damen Shipyards in Holland.

Based upon the experience of crews operating in the North Sea, the operators felt that vertical acceleration was the most important criterion concerning seakeeping quality, particularly speed loss of a craft in waves, as we have mentioned in previous chapters. Keuning selected the criterion for judging the seakeeping quality as the maximum value of vertical acceleration rather than the significant or average value, which are the standards for seakeeping quality for displacement ships. It is clear that this suggestion is rather different from that used for judging the seakeeping quality previously; however, it is supported by most navigators and crew of fast vessels.

J. A. Keuning [9], quoted studies as concluding:

> Generally spoken all crews imposed a voluntary speed reduction at roughly the same conditions. It also showed clearly that the real measure for imposing a voluntary speed reduction was not the prevailing magnitude of the significant amplitude of the motions or vertical accelerations at that time, but the occurrence of the high peaks in particular in the vertical acceleration. The occurrence of 'one big skim' provoked a speed reduction by the crew just to 'prevent it from happening again'. In fact such a reaction is more or less in line with a well-known more general aspect of human behavior, namely that most people are inclined to react to 'extremes' rather than to 'averages'.

During initial research, the concept was referred to as the "Enlarged Ship Concept" (ESC), that is, keeping the beam unchanged and extending the ship length by 25~50 % . The vertical acceleration of a ship in waves decreases significantly when a ship is extended [10].

For small craft, keeping the beam unchanged and extending length by 25~50% will not lead to significant structural weight increase, since the thickness of the hull structure plates is usually not determined by hull strength but by the corrosion and rigidity requirements of the small ship. Thus, the decrease of drag can compensate the increase of the engine's power, caused by the increased displacement of the ship.

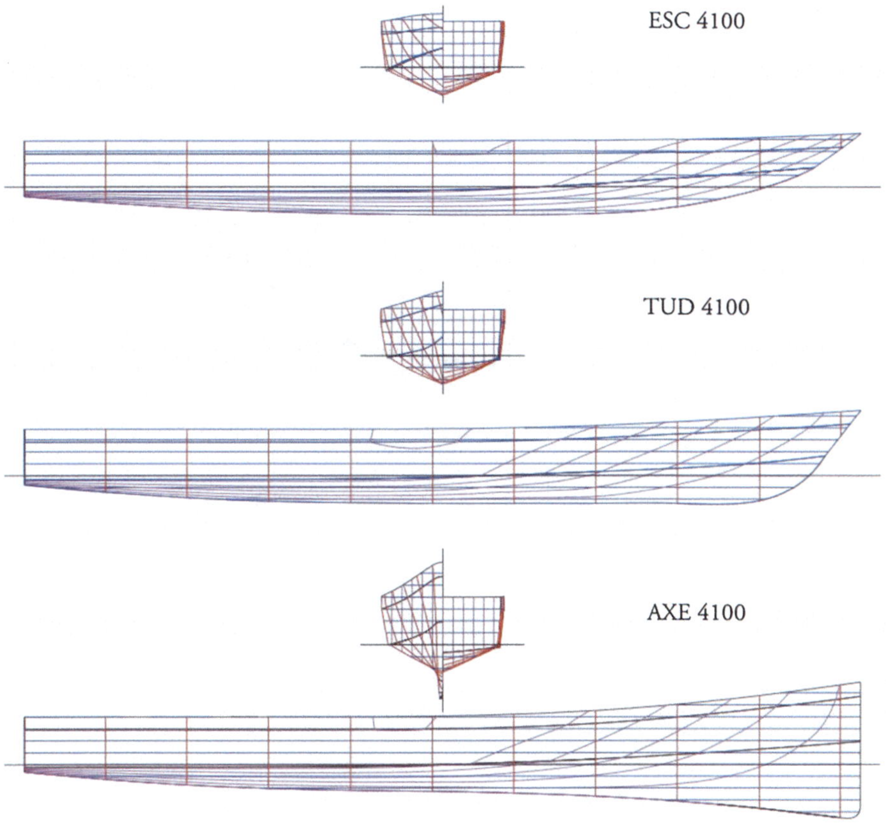

Fig. 6.19 Lines of EPC, revised EPC at bow (TUD), and Axe. (From Keuning et al. [12] Fig. 2)

So, for small ships, extending the length while keeping the beam unchanged will not increase the displacement of the ship significantly but improve seakeeping quality significantly. However, for large ships, the issues are rather different, and it has to be studied properly to determine the suitable length/beam ratio of the ship.

As a second step, they made the bow lines fuller to decrease the vertical acceleration [11]. Based upon such experience, Keuning et al. presented the concept of the Axe Bow profile [12], as shown in Fig. 6.19, where a comparison of the lines is shown of extension type ESC 4100, full bow type TUD 4000, and the Axe bow. Figure 6.20 shows the probability distribution of vertical accelerations of the three types of ships in Sea State 5.

The Axe Bow vessel is characterized by a higher length beam ratio, about 20~50% higher than the Deep Vee ships. The second feature is its axe profile bow, which perhaps is the main difference from the Deep Vee ships. The stem is perpendicular to the water line with bow keel descending forwards, flared lines above the design water line, and with same profile of water lines both above and

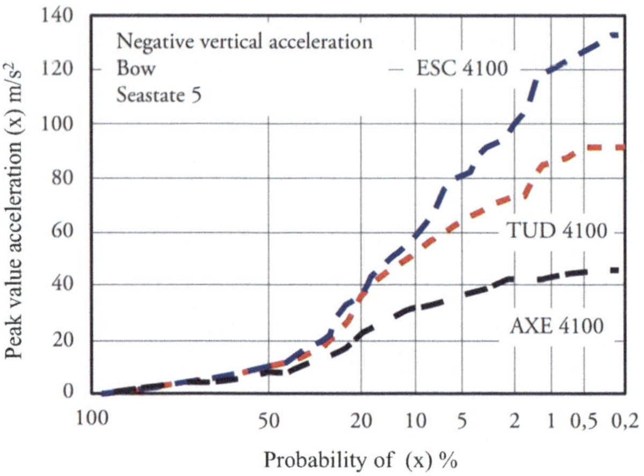

Fig. 6.20 Probability distribution of bow vertical accelerations of three ship types. (From Keuning et al. ref. [12] Fig. 3)

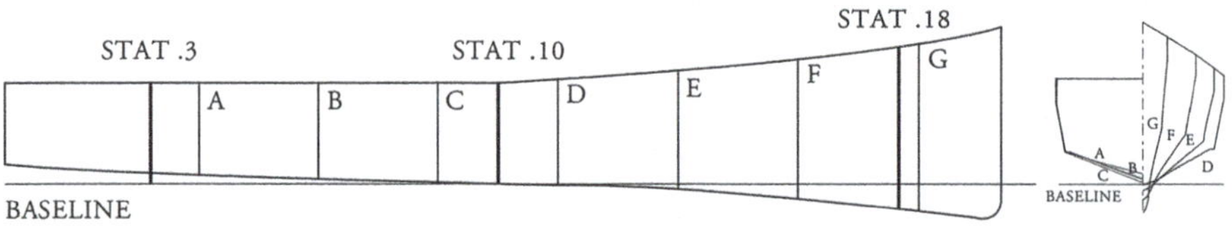

Fig. 6.21 Lines of Axe ship. (From Keuning et al. ref. [12] Fig. 12)

Fig. 6.22 Profiles of Axe ships—Damen FCS 7011 and 4008

under the design waterline at the bow. This is different from conventional ships with a constricted body plan under the water line and flare above the water line.

This profile keeps the change of buoyancy and its center to a minimum during heaving, minimizing interaction with pitch motion. Figures 6.21 and 6.22 show the profile lines and a view of a constructed craft.

The variation of buoyancy at the bow in waves is reduced so as to decrease the bow vertical acceleration of the Axe Bow ship, however, the amplitude of heave and pitching will be increased due to lower damping.

In order to improve the amplitude of heave and pitch motion anti-spray rails have to be mounted at the bow over/at the waterline as shown in Fig. 6.22 on FCS 4008. Having reduced pitching motions freeboard at the bow has to be increased along with incorporating flare at the bow to decrease green water shipment.

The powering performance and seakeeping quality analysis of an Axe Bow ship has been carried out in the Naval University of Engineering of China and comparison with a Deep Vee ship weighing 200 t has also been carried out at the university [13, 14]. The model test results show that when $F_n \geq 0.7$, the '"Axe Bow" with an extended length of 20% demonstrates the advantage of drag but has a disadvantage in the low-speed range. The results also showed significant improvement of seakeeping quality not only in reducing the maximum value of vertical acceleration but also in reducing heaving and pitching motions.

To sum up, the Axe is a novel type of craft developed based on the Deep Vee craft and has been applied to medium size craft. However, other performances such as stability, maneuverability, broaching in following waves, overall seakeeping quality, and the sizing concerning the Axe Bow craft have to be investigated further.

References

1. A. Bliault, L. Yun, H.Z. Rong, *High Speed Catamarans and Multihulls—Technology, Performance and Applications*, Chapter 8. Springer Nature (2019), 746 pp, ISBN 978-1-4939-7889-2
2. E.H. Serter, FAC Hull forms, a comparative study of two different concepts. Int. Def. Rev. **14**, 917–924 (1981)
3. J.W. Kehoe, K.S. Brower, E.H. Serter, Deep-Vee hulls improved seakeeping for small fast warships. Int. Def. Rev. **19**(11) (1986)
4. J.W. Kehoe, K.S. Brower, E.H. Serter, The deep Vee hull form—Improves seakeeping and combat system performance. Nav. Eng. J. **99**, 39–54 (1987)
5. Technical Report, Research and development of Craft type "Deep Vee", Hu Dong shipyard Group, China Ship Scientific Research Center (CSSRC), Jan 2002 (in Chinese)
6. Z. Chenxia, et al., Research of Deep-Vee Hull Forms (I), in *HPV'92 China Proceedings*, pp 33–39, November 12–15, 1992, Shanghai
7. K. Kihara, S. Maiba, Development of a new type of high-speed boat for a comfortable ride during navigation. Mitsubishi Heavy Industries Technical Report, Vol. 25(6), November 1988 (in Japanese)
8. L. Jide, Z. Heng, T. Minggi, Z. Xiaodong, The test research on high seakeeping performance displacement hybrid Monohull ship. Shipbuilding of China, Vol. 48(3), Sept 2007 (in Chinese)
9. J.A. Keuning, *"Grinding the bow" or "How to improve the operability of fast monohulls"*. Delft University of Technology, Ship Hydromechanics Laboratory Report No. 1501-P, 2006, Publication: International Shipbuilding Progress, Vol. 53(4), 2006, ISSN 0020-868X
10. J.A. Keuning, J. Pinkster, Further Design and Seakeeping Investigations into the Enlarged Ship Concept, in *FAST Conference Proceedings*, (Sydney, 1997)
11. J.A. Keuning, S. Toxopeus, J. Pinkster, The Effect of Bow Shape on the Seakeeping Performance of a Fast Monohull, in *FAST Conference Proceedings Southampton*, (September 2001)
12. J.A. Keuning, J.L. Gelling, Recent developments in the design of Fast Ships. Ship Sci. Technol. **5**(9), 57–68 (July 2011)—Cartagena, Colombia, 2011, also report No 1704-P at TU Delft, 2010 There are also HIPER 2006 and HIPER 2008 papers by Keuning on this subject

13. S. Wei, Q. Zheng, X.P. Gao, Z.S. Dong, Model Experimental Investigation on Drag Performance, in *Proceedings of International High Performance Marine Vessels (HPMV) Conference*, April 14–17, 2011, Shanghai.
14. Q.W. Wu, Q. Zheng, X.P. Gao, Z.S. Dong, Nice Seakeeping Quality of Craft with Axe Bow, in *Proceedings of International High Performance Marine Vessels (HPMV) Conference*, April 19–22, 2012, Shanghai

Part II
Hydrofoil Craft

Chapter 7
Introduction to Hydrofoil Craft

7.1 Introduction

Since resistance of a high-speed vessel increases rapidly with speed, the design idea for a planing hull was to reduce the hull surface area exposed to the water surface. However, a planing hull is still not completely clear from the water surface. Early in the twentieth century, a number of inventors had the idea to use lifting foils so that the hull was completely clear of the water in normal operation. In this case, the foils provided lift to balance the vessel weight, a little like the wings of an aircraft, but much smaller since water is 820 times denser than air.

$$L = 0.5\, \rho_w\, V^2\, S\, n$$

where L represent lift, ρ_w density of water (air), V craft speed, S area of hydrofoil (airfoil), and n is the number of individual foils immersed in the water. Since a foil shape will generate a large lift compared with its drag [1, 2], the vessel drag will be reduced. By lifting the hull out of the water, the buoyancy forces are replaced by the dynamic lift forces, and so inertial loads applied to a hull by waves will be minimized and the vessel ride will improve significantly. In addition, there will be no slamming forces on a hydrofoil within its design envelope.

Figure 7.1 shows a hydrofoil craft, USS "PGH-2" (*Tucumcari*), that had its lifting foils completely submerged, running at high speed with the hull clear of the water surface sufficient to avoid waves impacting the hull.

The location of the immersed foils relative to the water surface influences the lift that can be generated. As the foil approaches the surface, the flow over the top surface is reduced to zero, and in the limit, the lower foil surface acts simply as a planing surface. So long as the foil is immersed by greater than about a chord length, it will be able to operate as a "normal" lifting foil with almost no degradation of lift due to the free surface. To an extent, this property gives a self-stabilizing effect for hydrofoils in heave. To complete this effect without using control surfaces like

Z.-S. Dong et al., *High-Speed Monohull and Hydrofoil Craft*,
https://doi.org/10.1007/978-3-031-62760-6_7

Fig. 7.1 PGH-2 *Tucumcari*

elevators on an airplane wing, it is necessary to give the foils a dihedral and design their area so that the outer part of the foil breaks the water surface in normal operation. We explain this further below.

Thus, below a relative submerged depth larger than 1, that is, $\overline{h} = h/c > 1$, where $\overline{h}$ denotes the relative submerged depth of hydrofoil, h is the submerged depth, and c foil chord, the influence of submerged depth on dynamic lift from the foil is close to zero. If the foils were horizontal, there would be no change in lift force if the foil rises (the vessel heaves) or if the foil banks (the vessel rolls). Only if the vessel were to pitch would the higher angle of attack generate higher lift force. Thus, a hydrofoil with fully submerged foils needs to have control surfaces installed and has to be equipped with a control system to adjust the control surfaces automatically to maintain steady foil immersion and the vessel attitude in pitch and roll.

Such a hydrofoil vessel has the advantage of good seaworthiness, due to lower disturbance by surface waves on the hydrofoils. The penalty is high construction and maintenance cost of the craft due to the complexity of the hydrofoils and control.

When a hydrofoil operates with $h/c < 0.5$, the influence of the free surface increases, as noted above. In order to minimize the vessel draft, hydrofoils designed for river or lake operation typically are designed with $h/c = 0.2 - 0.3$ and are called shallow submersed hydrofoils. Such vessels have foils with significant dihedral and surface area that is above the water surface at service speed. In this way, if the vessel pitches or rolls, the foils will automatically submerse or emerge, providing stabilizing forces in pitch, roll, and heave. These vessels do not need control surfaces or control systems as the fully submerged foil craft do.

In terms of stability, there are thus two types of hydrofoil, that is, craft with and without passive stability. Inherent stability can be achieved by using the physical phenomenon of the water surface effect, so in general, passive stability is available for hydrofoil vessels with shallow submersible foils. This is not available on a fully submerged hydrofoil. For fully submerged hydrofoils, an automatic control system for adjusting the incidence angle of foils or control surfaces has to be installed.

7.2 Transverse Arrangement of Foils

There are a number of possible geometric arrangements for foils attached under a hull by vertical or canted supports. These are shown as cross-section diagrams in Fig. 7.2 below. The different features are as follows:

Some characteristics of the foil configurations shown in Fig. 7.2 are as follows:

(a) Fully submersed flat foils, as shown in Fig. 7.2a, have some stability in heave and roll (see Fig. 7.3a) due to their proximity to the surface. While not optimal, it is possible for such a craft to be operable. The stabilization is low.
(b) Surface piercing tapered foils, Fig. 7.2b, have features that are between geometries (a) and (c). For shallow water operation, the flat part of the foils provides efficient lift and the outer inclined part of the foils gives improved stability.
(c) Transverse stability for surface piercing V-type foils, as shown in Fig. 7.2c, is provided not only due to surface effect but also due to a higher rate of change of wetted surface of V foils so as to improve the restoring moment when heeling. Such craft have good transverse stability, as shown in Fig. 7.3b. This figure shows the effect of foil giving positive transverse stability to the flattened foil. However, this type of foil also has lower hydrodynamic efficiency due to the surface piercing of the outer foils. The configuration is cost-effective compared with the fully submersed foil system, however, since it has passive stability and does not need an active control system. The configuration has therefore become popular for coastal and river ferry craft in Russia and the Mediterranean Sea.
(d) The "separated" configurations, a shown in Fig. 7.2d, f, have to be sized to provide the necessary lift force and so are suited to smaller or more lightly loaded craft. By having canted lifting surfaces (d), they provide necessary

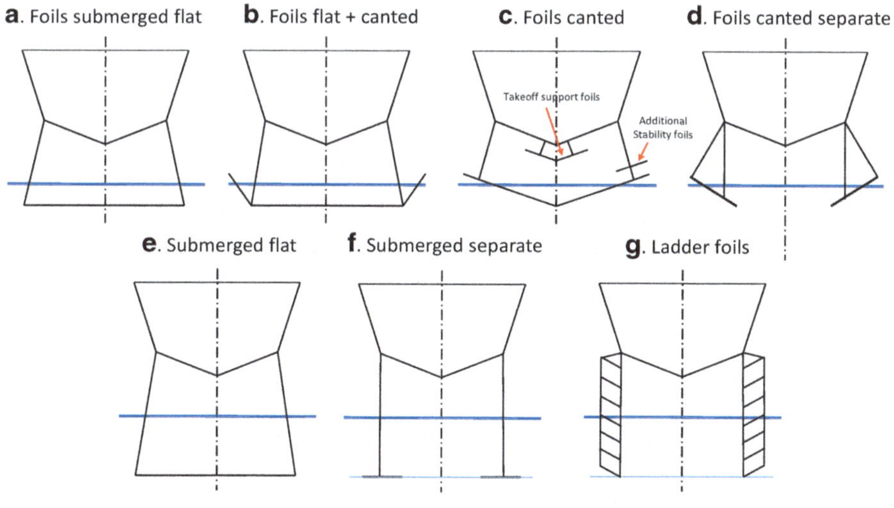

Fig. 7.2 Foil cross-section diagrams and annotation

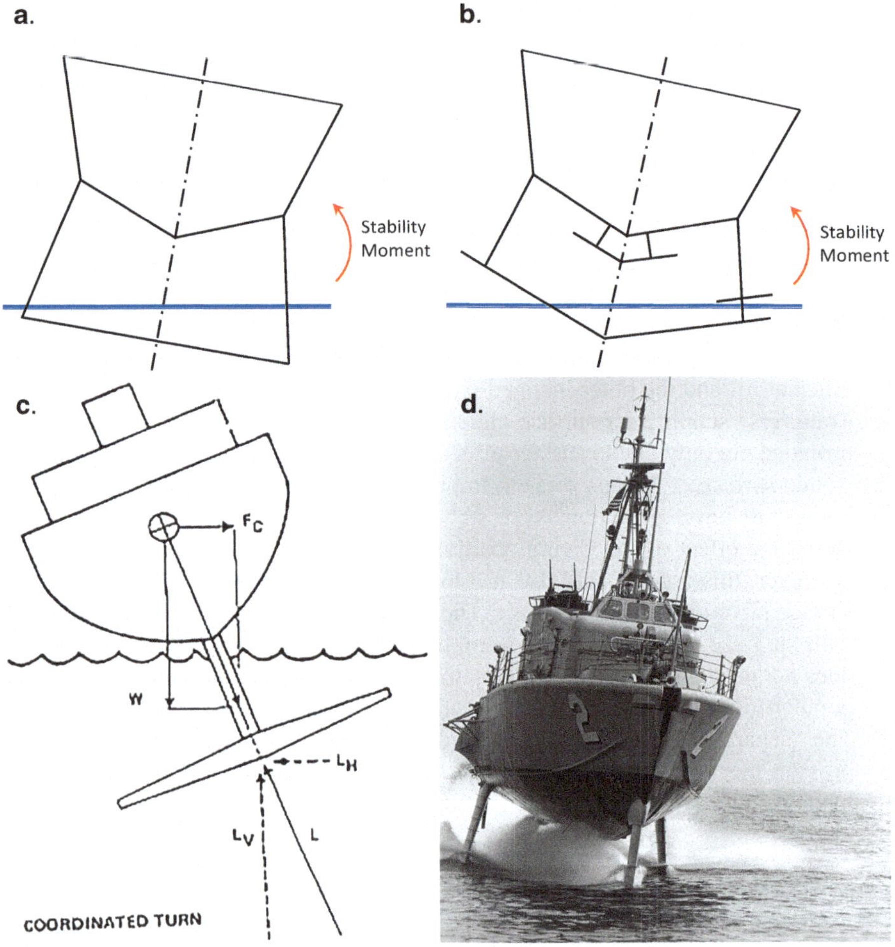

Fig. 7.3 Forces and moments acting on hydrofoils when rolling and turning. (**a**) and (**b**), rolling moments. (**c**) Forces on a submerged hydrofoil in a turn. (**d**) Tucumcari during a coordinated turn

stability moments, and being separated they have the opportunity to be rotatable out of the water when the craft is at rest, so minimizing the draft at the quayside. Retractable foils do require a more complex structural arrangement.

(e) Deep and fully submersible foils with no passive stability, as shown in Fig. 7.2e, f, provide more efficient lift and less wave perturbation. In addition, the stably coordinated turn, or banked turn, with a small turning diameter can be performed by differential use of elevator controls. The advantage of the banked turn is a small turning diameter, and lower sideways g force on the crew, giving an improved feeling of safety and stability. Figure 7.3c shows the forces acting

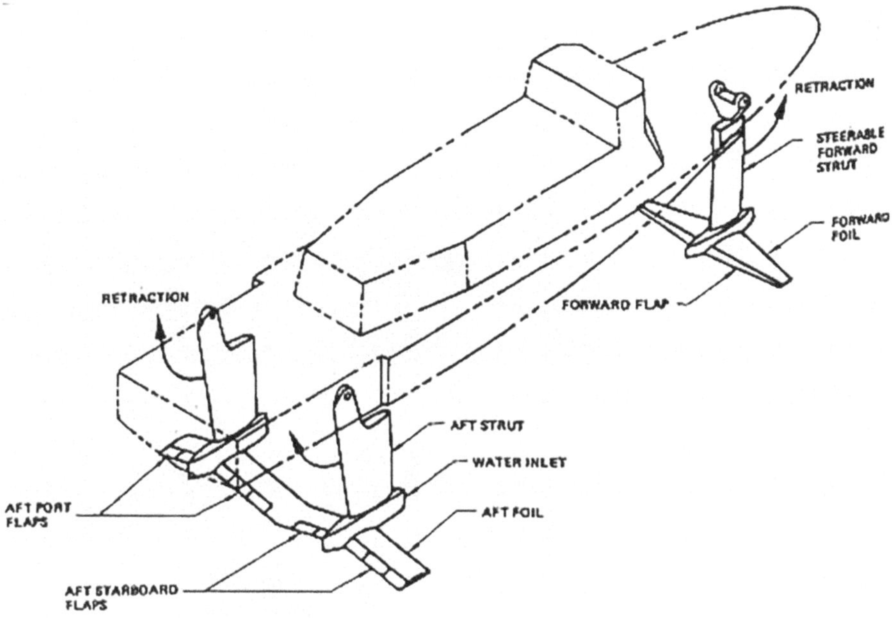

Fig. 7.4 PHM foil arrangement

on the foils and craft with the coordinated or banked turn, and Fig. 7.3d shows the US Navy PHM in a banked turn.

Because of the deep submergence of the foils, the draft of this configuration is also deep, so the foils have to be retractable to reduce draft during docking and operate at slow speed in shallow water. The mechanism for this is as follows:

- The bow foil is retractable by rotating longitudinally, as shown in Fig. 7.4, that shows the foil arrangement of the Patrol Hydrofoil Missile (PHM) that succeeded the prototype PHG-2. The penalty of the system is increased hull weight and craft construction cost; however, for docking and maintenance, such a system is necessary, and so the system has been installed on both military and commercial ferry craft.
- PGH-2 stern foils are retracted by rotation transversely, as shown in Fig. 7.5, which shows the foil system retracted. The advantage of the split rear foil is lower influence of downwash flow from the bow foil. Interaction between the foils reduces the hydrodynamic performance, both for lift and dynamic response.

(f) A ladder-type foil system, as shown in Figs. 7.2g and 7.6, Bell HD-4 Hydrodome [3] was an early configuration pioneered by Bell and others in the early 1900s when they were experimenting with hydrofoil boats. The idea was to provide a compact arrangement that allowed foils to rise out of the water as

Fig. 7.5 *Tucumcari* with retracted foils

Fig. 7.6 Early ladder foil craft—Bell Hydrodome HD-4

speed increased and so lower the immersed area. This provided a passive stabilization and allowed the hull to ride high out of the water. Russian shallow water hydrofoils using a configuration based on Fig. 7.2b have adopted additional "biplane" foils mainly to assist takeoff, see Fig. 7.7a, b.

Key to Kolkhida General Arrangement

1 waste oil collection tank; 2 contaminated water tank; 3 sewage water tank; 4 fuel tank; 5 foil borne waterline; 6 floating waterline; 7 hull baseline; 8 hydraulic station; 9 station for fuel oil filling; firefighting; and waste water scavenge; 10 condition monitoring; 11 control post; 12/13 rear passenger saloon; 14 luggage; 15 stern promenade deck; 16 auxiliary power unit; 17 exhaust duct; 18 air intake duct; 19 main passenger saloon (91 pax); 20 emergency equipment; 21 air conditioning (there is one unit in each passenger cabin); 22 bow passenger saloon; 23 vessel Centre line; 24 toilet.

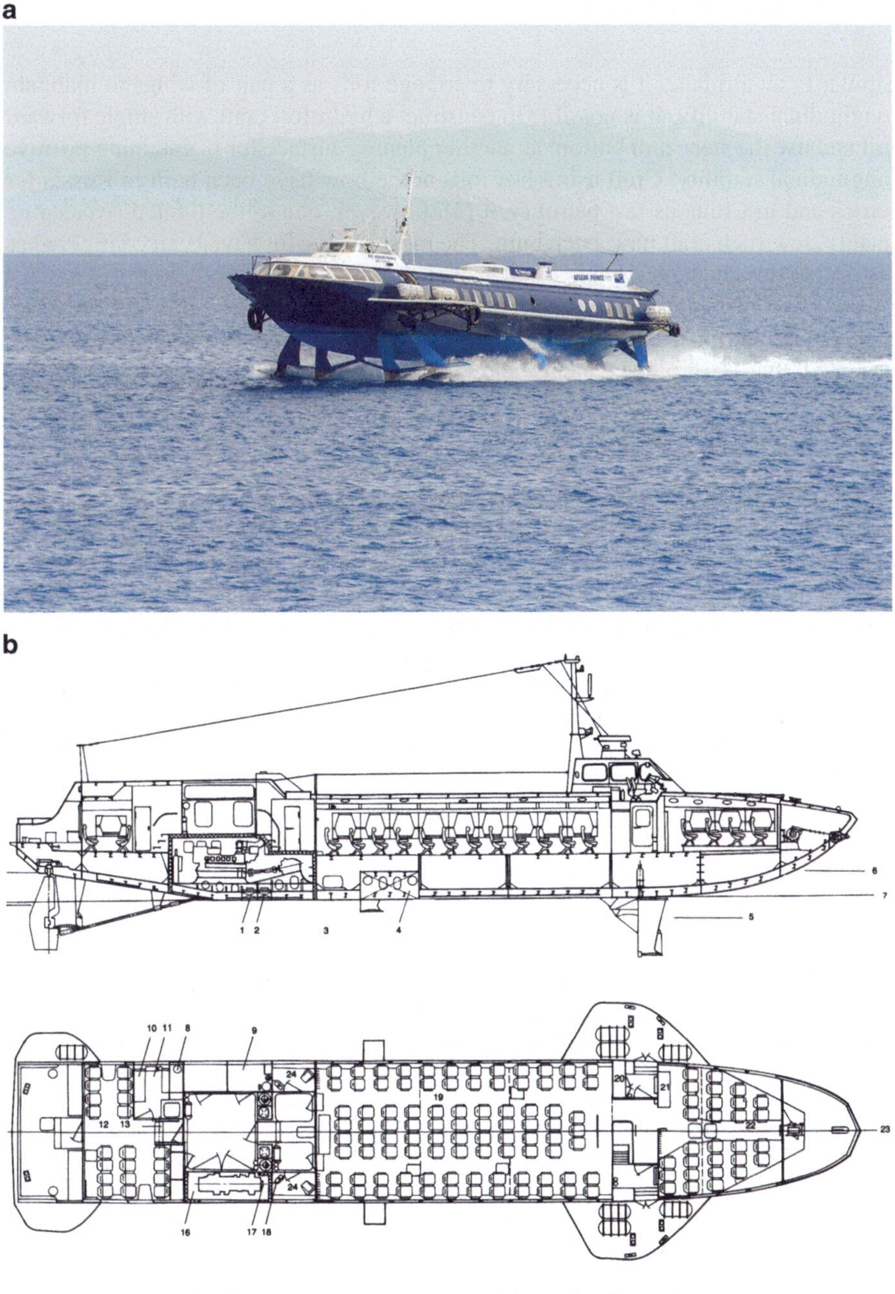

Fig. 7.7 (**a**) Photo of Aegean Prince. A Kolkhida craft in service from Rhodes to Kos with sister Flying Poseidon. (**b**) Kolkhida general arrangement

7.3 Plan Arrangement of Foils

Similar to an airplane, it is necessary to arrange foils as a pair of wings to maintain longitudinal stability. It is possible to construct a hydrofoil craft with single forward foil and use the stern hull bottom as another planing surface for maintaining positive longitudinal stability. Craft using just foils at the bow have been built in Russia for ferries and in China as fast patrol craft [3]; however, due to the limited seakeeping quality, few such craft have been built. The most successful have been configured as fast ferries for sheltered waters (see Chap. 11).

Based on different requirements for the combination of powering performance, seakeeping, and maneuverability, three types of hydrofoil plan arrangement can be designed, that is, airplane, canard, and tandem configuration, as shown in Fig. 7.8 [4, 5].

1. *Airplane Configuration*

The airplane hydrofoil configuration supports between 65% and 95% of craft weight on the forward foil(s), and the rest is supported by the rear foil, as shown in Fig. 7.8 (top figure). The front foil is placed at a position $0<x/L_f< 0.35$, where x is the distance between craft CG forward to the forward foil and L_f is the distance between the two foils. Since the load on the forward foil is greater, the configuration will be similar to an aircraft with a main wing and a small tail stabilizer.

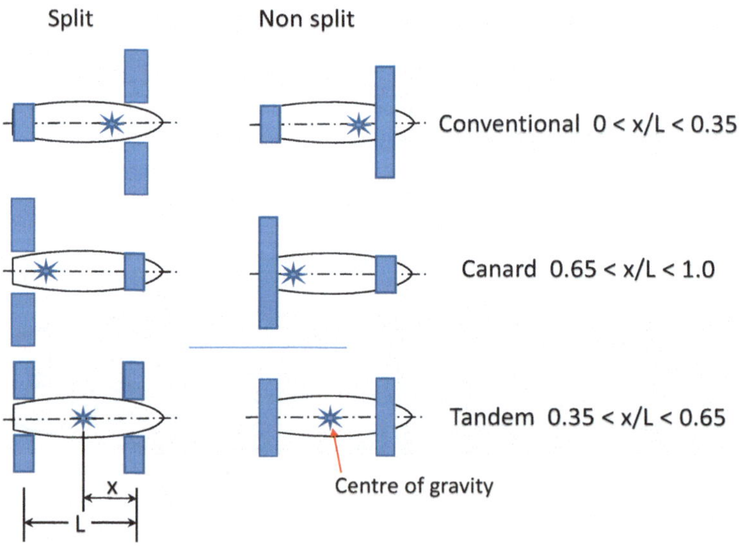

Fig. 7.8 Foil plan configurations

The configuration with an airplane foil arrangement suits a vessel where more weight is located forward, with a hull body that is fuller forward of amidships and the center of buoyancy forward, so as to fit with the CG position. This configuration is easier to take off onto the foils and has good transverse stability. It also has less tendency to plough in (bow pitch down) when running in waves. However, due to the wave perturbation on the large front foil, the penalty of this configuration is lower seaworthiness, and in addition, the hull borne resistance also increases.

There are two subtypes for this arrangement, that is, split and nonsplit type of main foil, as shown in Fig. 7.8.

2. *Canard Configuration*

In contrast to the airplane configuration, between 65% to 95% of craft weight is supported by the rear foil, on the canard configuration, that is, $0.65<x/L<1.0$, and only a little weight is supported by the forward foil. This configuration suits a craft with a full rear body and lean forebody, where both CG and CB are located aft of amidships.

The transverse stability is fine, just as for the airplane configuration, and apart from this, the seakeeping quality is fine due to the lower wave perturbation on the smaller forward foil, so avoiding main foil ventilation and craft plough-in. In addition, with most parts of the craft mass concentrated aft, there is less tendency to dangerous "Broaching" of the craft when running in following seas.

With the forward foil located far from the CG, there is a large arm for generating longitudinal moment to prevent plough-in, and improve maneuverability, while using smaller forces. This latter makes an automated control system more efficient.

This foil configuration is used on craft with high seaworthiness requirements; however, the penalty is longer take-off time from planing to foiling.

The configuration is suitable for high-speed craft requiring powerful power plant, and (generally) water jet propulsion, due to its space requirement in the after part of a craft, and this is why the US naval hydrofoils PGH and PHM have this type of foil arrangement and automatic ride control system (RCS).

3. *Tandem Arrangement*

This type of foil arrangement is in between the airplane and canard arrangement, that is, the load of both forward and rear foil is almost the same, and $0.35<x/L<0.65$ spread about amidships. The configuration has a softer roll stability as the foils will have a lower width (span) than the canard or airplane configuration main foil configurations.

The forward foil may cause more inflow disturbance to the aft foil and reduce the effectiveness of this foil so the configuration suits high L/B craft. When the forward foil runs in the water, the water surface will be deformed and form a hollow behind it due to the downwash, the width of which is equal to the foil span at the foil, which constricts gradually in terms of the distance after the foil, and finally, the two edges of the hollow span will be closed to a point. Since, in general, the distance between the forward and rear foil of a hydrofoil is less than the span of hollow, the influence of the forward foil on the rear foil is unavoidable.

The influence of forward foil on the rear is to decrease the effective draft of the rear foil and create downwash flow incoming to the rear foil. This will decrease its lift and may generate cavitation of rear foil.

Due to such interference, the hydrodynamic performance of hydrofoils with tandem foil configuration will be less efficient than that with both airplane and canard foil configurations. This encouraged the split foil arrangements, as shown in Fig. 7.8, to reduce such influence. While the interference is significant, for shallow draft hydrofoils, such as those designed for Russian rivers, the Black Sea, and the Mediterranean Sea, it was convenient to use a very high L/B for the hull design, and this helped to improve overall efficiency while maintaining a "passive" hydrofoil system design. Fig. 7.6 shows a Kolkhida class craft with tandem design foils, the forward foils being larger than the stern foils to take more of the weight, and having partial biplane outer foils to assist takeoff and roll stability.

7.4 Hydrofoil Development

Development of hydrofoils by inventors in the late nineteenth and early twentieth centuries went in parallel with powered flight, and in a number of cases, inventors experimenting with flight also used foils as a means of taking off from water [6].

In the USA, a hydrofoil patent was registered in the late 1880s, and the Meachan brothers successfully operated the world's first 14-feet hydrofoil on the lake by Chicago, Illinois in 1894, before the Wright brothers successfully tested their first powered airplane in 1905.

In the early part of the twentieth century, inventors experimented with the different transverse and plan arrangement of foils described in Sects. 7.2 and 7.3 above, mainly focusing on shallow immersion foils, ladder arrangements, etc., as at this time the technology for automated controls did not exist, so the foil structures had to be "fixed."

Enrico Forlanini began experiments in 1898 using the ladder foil arrangement with four ladders, two each at bow and stern. In 1906, he built a full-size craft and tested it on Lake Maggiore in Italy. Initially, he used flat ladder steps that gave unsteady rising and descending of the hull, so Vee-shaped foils were later experimented with to smooth this out. He used a 45 kW motor and achieved a speed of 38 knots.

Several other pioneers also experimented with hydrofoil craft, including A Crocco and O Ricanoldi in 1907 and Alexander Graham Bell, who visited Forlanini in 1910 and then began his own experiments with air propulsion, which he called "Hydrodome." He achieved a speed of 70.86 mph (61.5 knots) in September 1919 with his HD-4 (Fig. 7.6) using two 260 kW aircraft engines, a world record speed that stood for 10 years.

In 1927, Baron Hans von Schertel [3] began his own experimental work in Germany to improve the takeoff and landing performance of flying boats, an aircraft type that was gaining popularity for passenger flight at that time. The Baron built seven different prototype boats to test foil configurations up until 1936. Initially, he concentrated on submerged foils, but finding it difficult to control steady immersion depth, he moved on to surface piercing geometries. The last test craft, with a V-form front foil and rectangular-shaped rear foils, was successful enough to perform a demonstration run on the Rhine between Mainz and Cologne.

In the Second World War, a huge development program for hydrofoils was undergone in Germany, including the construction of six 20-tonne surface piercing type hydrofoils, intended for operation along the Norwegian coastline for surveillance, and another three 60-tonne hydrofoils, aimed at supporting the North African campaigns. However, they were destroyed by the Alliance Air Force in an air attack, while one craft was on test and another three under construction.

This work formed the starting point for the development of the Supramar hydrofoil company in Switzerland at the beginning of the 1950s. At this point, the concept was being developed in a number of different countries somewhat independently influenced by the environment their designs needed to adapt to, whether relatively calm rivers and lakes and coastal areas of Europe or the rougher offshore conditions of the east coast of the USA and Canada. We will begin with Russia and Europe and then look at USA and Canada, with, finally, a note on other countries.

Russia

Research and development on hydrofoils [3, 7–10] started in the early 1950s at Sormovo Shipyard and at the Hydrofoil Central Design Bureau in the Former USSR, under the leadership of Dr R.Y. Alexeyev in Gorki City (now Nizhny Novgorod). Dr Alexeev, leading his research team, developed hydrofoil craft targeted at the extensive Volga river system and linked lakes. The concepts were efficient for passenger transport and a comfortable alternative in areas where road or rail transport was not highly developed at the time.

The operational requirements suited the shallow draft hydrofoil and allowed speeds as high as 50 knots. At this speed, the foils and propulsion propellers met the cavitation barrier limiting further development. As an alternative concept, he focused further research on wing-in-ground craft in the 1960s, aimed at aircraft-like speeds while staying close to the water surface [9]. In the meantime, the design and production of passenger hydrofoils became a successful industrial endeavor.

By 1957, the Design Bureau and Sormovo shipyard had developed passenger hydrofoils with shallow draft for river and lake operation and successfully put them into service for high-speed transportation, with speeds in the range of 60–90 km/h. Construction of the various types of hydrofoils with shallow submersible foil systems reached more than 300 vessels, such as "Raketa," "Meteor," etc. for the domestic market and for export.

During the 1960s and 1970s, the Former USSR developed almost one new design of hydrofoil every year, totaling 19 types of hydrofoil for inland rivers, lakes, and coastal service. Most of these were shallow submersible foil designs with passive stability. See our summary data tables at the end of this chapter for sample details and [10–12] for more extensive details.

Fig. 7.9 Meteor

Meanwhile, the Alexeyev Design Bureau also developed a seagoing hydrofoil with a deep Vee surface piercing foil system and passive stability (Fig. 7.2c). In order to improve the seakeeping, an augmented stability system with air ventilation in the foil upper surface was installed on these designs. The ventilation reduced the variation of foil loading as waves passed softening the ride. This passive system was cost-effective and maintained passive stability; however, the ride performance was not as smooth as the fully submersed hydrofoil system with active controls developed in the USA.

A classic design from this period for inland waters is the "Meteor" shown in Fig. 7.9. with leading particulars as follows:

Loa	34.5 m
Boa	6.0 m
Draft, foil borne	1.2 m
Hull borne	2.4 m
Displacement (full loaded)	54.3 t
Main engines	2 × M50 12c diesels, 816 kW each
Hull material	Riveted aluminum alloy
Passengers	116
Full speed	35 knots
Cruising speed	32 knots

The craft has surface piercing hydrofoils close to the hull underside, mid-mounted engines, and weight almost equally distributed between forward and stern foils. The foil configuration includes additional small-span foils attached to the front foil support struts that enter the water when the main foils reach one chord depth. The additional lift then brings the main foils back into shallow immersion operation so ensuring the vessel draft stays shallow.

The first Meteor was launched in 1960, succeeding the earlier Raketa. The smaller Raketa was launched in 1957 with 58 seats. Over 300 craft were built and operated for almost half a century in Russian inland waterways. There were about 100 craft built to the Meteor design and its very similar predecessor, the Raketa, and these had great success in the Black Sea, Adriatic, and Aegean Sea as fast ferries as well as the Russian river systems.

Figure 7.7a, b shows the Russian Kolkhida hydrofoil designed with a little deeper hydrofoil draft for rougher offshore conditions. The overall foil configuration followed the same logic, with subsidiary foils for takeoff and foil draft passive control, and ventilated upper surface foils for seakeeping. Their key data are as follows:

Loa	34.5 m
Boa	5.8 m
Draft, foil borne	1.9 m
Hull borne	3.5 m
Displacement (full loaded)	72 t
Main engines	2 × MTU 12V96 TC82, 960 kw each
Hull material	Riveted aluminum alloy
Crew	6
Passengers	120
Full speed	36 knots
Cruising speed	34 knots

The largest passenger hydrofoil developed in Russia is the "Cyclone," which has seats for 250 passengers on two deck levels, see Fig. 7.10. This vessel was built at the Feodosia shipyard on the Crimean coast and went into operation on the Black Sea in 1987 after trials in 1986.

The craft key data are as follows:

Loa	44.2 m
Boa	12.5 m
Hull width	7.3 m
Draft, foil borne	2.4 m
Hull borne	4.3 m
Displacement (full loaded)	137 t
Dry weight	100.8 t
Main engines	1 × M37 5150 kW gas turbine
	1 × 735 kW diesel for slow running

Fig. 7.10 250-passenger Cyclone

Propulsion	1 × propeller
Foils	Surface piercing
Hull material	Aluminum alloy
Crew	9
Passengers	250 on two deck levels
Full speed	44 knots
Cruising speed	42 knots

Russia has also developed a number of hydrofoils for military applications, including the "Osa," "Turya," and "Matka" classes of missile and torpedo hydrofoil craft, which had a single bow foil configuration all based on the same hull. This was a means to increase the speed in operation of planing fast patrol craft and improve seaworthiness without the additional complication of a propulsion shaft system at a greater angle to link with an aft foil. A Turya class vessel is shown at speed in Fig. 7.11a, while Fig. 7.11b shows a "Matka" class hydrofoil with surface piercing forward foil.

Key data for the Matka class are as follows:

Full displacement	260 t
Loa	39.3 m
Beam, for hull	7.6 m
Foil span	11.3 m
Draft on foils	3 m
Speed	40 knots

a

b

Fig. 7.11 (**a**) Turya class vessel at speed. (**b**) Matka

Fig. 7.12 Mukha

Main engines	Three M504 diesel, 3700 kW ea
Outfit	Guide missiles

The biggest hydrofoil in Russia is "Mukha" (Sokol class), shown in Fig. 7.12 [13]. This large and powerful military craft is aimed at offshore conditions and has fixed surface piercing foils at bow, amidships, and stern. Key data for this craft are as follows:

Displacement	470 t
Loa	49.97 m
Foil span	21.13 m
Hull beam	8.59 m
Draft (hull)	4.00 m
Draft (foil)	7.26 m
(on foils)	2.16 m
Main engines	CODOG,
	2 × M10D Gas turbines 15,000kW ea
	2 × diesels 1630 kW total
Speed	50 knots

Fig. 7.13 PT 20 Frechia del sole

Italy, Switzerland

Based upon the Schertel-Sachsenberg Hydrofoil Syndicate and its license, Supramar Ltd of Switzerland was founded in 1952, [1, 7, 10]. The company prototype design was the hydrofoil PT-10. This prototype started the world's first hydrofoil commercial route for passenger transportation, operating in Lake Maggiore, Italy, in 1953. From the PT-10, the company developed PT-20, as shown in Fig. 7.13 [3, 14], with surface piercing, nonsplit, tandem arrangement. The foils have a V configuration, as shown in Fig. 7.13, to give a foil system with passive stability.

From that point, Supramar developed a series of hydrofoil passenger craft of increasing size. The foil configuration was simply scaled with each larger craft from PT 20 to 50 and 150. Figure 7.14a shows the front foil configuration of the PT 150, while Fig. 7.14b shows the PT50 at speed. All of these craft were powered by diesel engines, with long and inclined shafts for power transmission to propeller supports at the rear foils. This was a similar arrangement to that used by the Russian craft but with a sharper angle for the shafts due to the higher hull flying height for Supramar craft. The PT 50 was a very popular craft that was also built under license by other shipyards.

To improve seakeeping quality, the larger PT-150 had an air ventilation system installed. The system was rather simple and simply feeding atmospheric air down through a row of holes in the foil to ventilate the submerged foil upper surface to passively control the lift in a seaway.

Fig. 7.14 (a) PT 50 front foil system. (b) PT 50 at speed

Since the method, for adjusting the foil forces, and moments acting on the craft, was due to reducing the lift of the foils, the penalty of this system was decreased hydrodynamic efficiency of the foil system.

Rodriquez Cantieri Navali Spa, Italy, built a number of the Supramar designs under licence from 1956 and from the early 1960s began to develop its own designs based on the Supramar patents. They produced a series of hydrofoils for coastal operation, the RHS 70, 110, 140, 150, 160, and 200, and became the world's largest hydrofoil producer outside Russia in the 1980s and 1990s. Key data for the RHS 150 and 200 are given in Table 7.1 at the end of this chapter. Following the approach in Supramar itself, the design series are essentially similar and scaled in dimensions. A significant number have been built, most for operation in the Mediterranean around the coasts of Italy and Greece, as well as in the waters around Hong Kong from the mid-1960s. Rodriquez RHS 150 and RHS 200 are shown in Fig. 7.15 below.

In Norway, the Westermoen shipyard in Mandal took a licence with Supramar and began construction in the early 1960s, initially with the PT 50 and subsequently with the PT 75 and 150. In 1967, Westermoen carried out trials with a small craft on stabilization using foil air ventilation and subsequently incorporated this into the craft that they built. Their hydrofoils have been operated in Norway and Scandinavia in the 1970s and 1980s.

More recently, Rodriquez has developed a 243-seat passenger hydrofoil with a fully submersed foil system in 2009 [3], partly funded by a European Union R&D program, as shown in Fig. 7.16. Two craft of this type, designated FSH-38 by Rodriquez were built. The FSH-38 has length 37.25 m, beam 8 m, draft 4.15 m, and can reach 45 knots powered by two MTU16V4000M70 diesel engines, each of 2320 kW.

The first craft has mid-mounted engines driving two long canted shafts for simplified power transmission compared to the more complex arrangements employed by designers in the United States and Canada. The second craft had a Z drive rather than the canted propeller shafts, while the engine bay was maintained centrally, and the Z drive struts were ahead of the aft main foil. The foil plan is a canard configuration with a smaller central bow strut and foil and a main lifting foil halfway between amidships and the stern. Trials were carried out in 2011 to identify the relative advantages of the two configurations [15]. The Z drive version was projected to give a higher speed. The target was 45 knots, while the first craft achieved 42.5 knots. Takeoff for both craft is about 27 knots, depending on the environment.

USA

After the early experiments of Alexander Graham Bell and his associate Casey Baldwin in the early 1920s, the focus of Aeromarine research turned to seaplanes and flying boats—aircraft that could take off from water—and so the attention was on stepped planing hulls rather than foils.

In the late 1950s the US Marine Administration (MARAD) sponsored work looking at hydrofoils as a means of transporting people and freight along coastlines at higher than 50 knots. The first craft built was the Grumman "Denison" [3], with forward surface piercing Vee foils, and a submerged stern foil and Z-drive powered by a gas turbine.

The success of this craft led Grumman to design a 64-tonne passenger ferry called the "Dolphin" for 116 passengers. With its fully submerged foils fore and aft, it was able to reach up to 48 knots in up to 3m seas, an amazing achievement. The vessel

Fig. 7.15 (**a**) Rodriquez RHS-150. (**b**) Rodriquez RHS 200

Fig. 7.16 FSH 38

was operated in the Canary Islands, Bahamas, and Virgin Islands, each for a period, but each service was not able to prove reliable for various reasons. Essentially, the craft was ahead of its time and propulsion, and control systems needed maturing to achieve reliability.

Later in the 1960s, hydrofoil research and development in the USA focused on the craft for military application, with both high speed and seaworthiness. This led to hydrofoil designs of the fully submersed type with active control systems. The research was started based on the US Navy specified missions of Patrol Craft Hydrofoil (PCH), and Patrol Gunboat Hydrofoil (PGH) [3, 16]. The program began with experimental craft.

Boeing was awarded a contract in 1960 to build the PCH-1 *High Point* with fully submersed foils, Fig. 7.17. The craft had a canard split foil layout, and foil retraction system vertically through watertight spaces in the hull. Key data for this craft were as follows:

Loa	35.3 m
Hull Beam	9.53 m
Draft foil retracted	2.00 m
Draft foil extended	6.04 m
Draft on foils	2.60 m
Full loaded displacement	112 t
Manning	18

Fig. 7.17 PCH-1 *High Point* at speed and (a) launching a Harpoon missile

The craft was tested for offshore anti-submarine operations, as a small, high-speed sonar platform, and equipped with anti-submarine torpedoes. It was delivered in 1963.

Variable experience by USN during the trials program led the navy to sponsor Boeing to build a small experimental catamaran hydrofoil, the "FRESH-1," which was used to investigate foil cavitation and automated control systems for fully submerged foils. This work gave Boeing the basis for their later developments for the Patrol Hydrofoil program. PCH-1 was transferred to the David Taylor Model

Fig. 7.18 AGEH-1 *Plainview*

Basin in 1965 for use as a full-scale test bed and was operated in this mode until 1984.

Meanwhile, the US Navy also tendered a proposal for an Auxiliary General Experimental Hydrofoil, christened *Plainview* (AGEH-1) in 1961. Grumman were the designers, while eventually, in 1964, construction was awarded to Lockheed Shipbuilding. While considerably larger than *Dolphin*, it retained the foil layout (Fig. 7.18). Key data are as below:

Loa	64.6 m
Hull beam	12.2 m (21.58 m max over foils)
Draft, foils retracted	1.9 m
Foils extended	7.6 m
On foils	3.0
Full loaded displacement	328.0 t
Manning	25

AGEH-1 had a long service record after initial teething troubles mirroring that of PCH-1. *Plainview* eventually took part in a wide range of operational trials, including torpedo launch, missile launch, RPV operations, and various classified tasks. It remained in service until 1978.

Fig. 7.19 PGH-1 Flagstaff

Using experience gained from the extensive trials and operations with PCH-1 and AGEH-1, the US Navy developed two competitive designs for a Patrol Gunboat Hydrofoil, the *Flagstaff* (PGH-1), Fig. 7.19, and *Tucumcari* (PGH-2), Fig. 7.5. Both vessels had deep submersed foils and automated control systems. PGH-1 had an airplane plan split foil system with flap control and water propeller propulsion. PGH-2 had a canard, split foil system with flap control and water jet propulsion. Both were delivered in 1966. The leading particulars are:

	PGH-1	PGH-2
Loa	22.2 m	22.7 m
Beam hull	6.5 m	5.9 m
Over foils	11.3 m	10.76 m
Draft foil retracted	1.2 m	1.5 m
Foil extended	4.26 m	4.33 m
Full loaded displacement	72 t	57.4 t
Manning	13	13

The PGH design was targeted as a patrol boat for supporting amphibious warfare, including coastal surveillance, fast personal transportation, and deception and decoy of hostile forces. At the end of the 1960s, experience from operational and technical trials with the two PGH craft gave the basis for specification of a patrol hydrofoil intended as a common platform for NATO forces, and so potential sales to several nations as well as the USA.

The design was designated Patrol Hydrofoil, Missile (PHM). The PHM aimed to cope with the increasing military risk from hostile guided missile naval vessels for the NATO partners. It was based upon the successful development of *Tucumcari* and developed at first in Europe by a joint USN and Italian Navy team.

In 1972, the USA together with Italy signed a contract for co-design and construction of the PHM-1, approximately twice the size of *Tucumcari*. The specification was for a displacement of 230 t, speed 50 knots, equipped with Harpoon guided missiles, and a 76 mm OTO Melara gun forward of the wheelhouse. The foil system adopted was a canard, nonsplit, fully submersed design, and foil automatic control with flap using flaps. The forward foil was retractable by forward rotation, and the rear foils retracted by aft rotation. The first craft of this series was *Pegasus* (PHM-1); Fig. 7.20 shows the PHM fleet at speed.

The big change for the PHM design was the propulsion, a waterjet system developed by Aerojet General, which had its inlet integrated with the aft foil and skegs and the exhaust in the main hull stern. Boeing Marine Systems commissioned the first of class in 1977, with the remaining five vessels following its trials and acceptance of the class in 1981 and 1982. Italy also built seven vessels to this class, and later, in 1991, Sumitomo in Japan built a further two craft for coastal patrol duties

Fig. 7.20 PHM fleet at speed

The leading particulars of this craft are:

Loa	40.5 m
Beam	8.6 m
Overall aft foil span	14.5 m
Draft, foil retracted	1.9 m
Foil extended	7.1 m
Full loaded displacement	241.3 t
Foil borne propulsion	one LM2500 gas turbine
Hull borne propulsion	2 MTU, MB8V331TC81 diesels
Propulsors	2 Aerojet water jets
Hull material	Welded 5456 aluminum
Foil and struts material	Welded 17-4PH stainless steel
Complement	23 officers and enlisted personnel

The arrangement of the power plant is shown in Fig. 7.21

In addition to developing the naval hydrofoil, Boeing also developed a hydrofoil design for passenger ferry application. Using the experience from the PHM class, Boeing Marine Systems constructed the "Jetfoil,", with a fully submersed foil system, and water jet propulsion using the same approach of pitot type inlet at the lower end of the stern central strut feeding water to two jet pumps mounted in the stern of the hull, as shown in Fig. 7.22. This was a simplification of the earlier PHM arrangement. Leading particulars of the Jetfoil 929 are as follows:

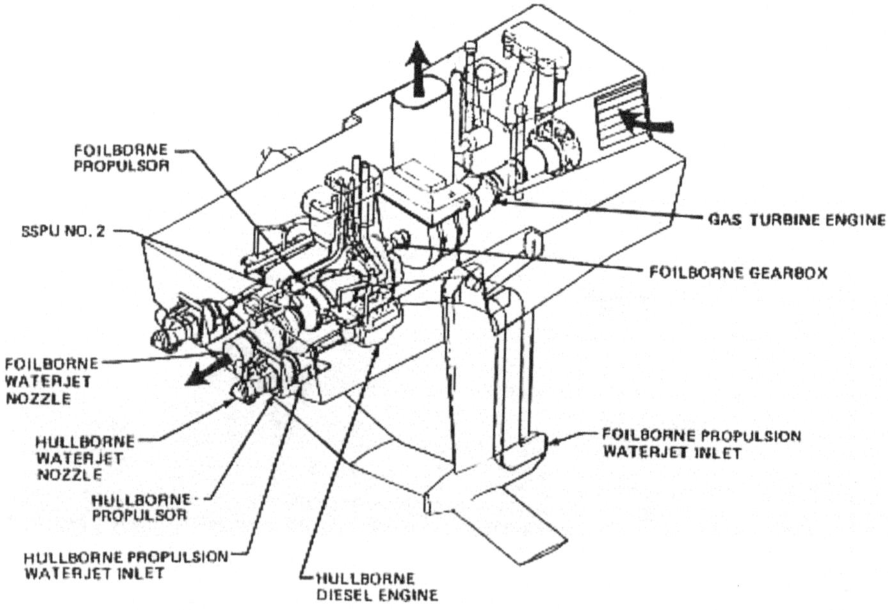

Fig. 7.21 PHM power plant

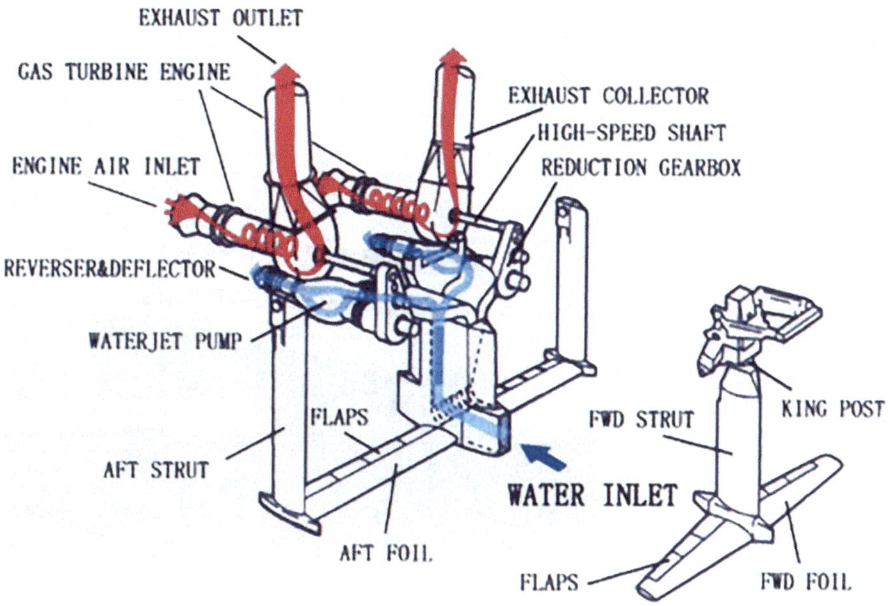

Fig. 7.22 Jetfoil Power plant layout

Loa	27.4 m
Beam	9.5 m
Displacement	110 tonnes
Draft, foil retracted	2.0 m
Foil extended	5.0 m
Hull clearance on foils	2.5 m
Passengers	250 up to 350
Crew	8
Speed in service/max	50 knots, 43 in SS 4/5)
Foil borne propulsion	2 × Allison 501-K20A gas turbine
	2 × Rocketdyne water jets
Hull material	Welded 5456 aluminum
Foil and strut material	Welded 17-4PH stainless steel

Reference [17] gives a summary of the craft development and experience in operation from sea trials, while [18] gives a detailed account of the trials of craft 011 and the hydrodynamic performance leading to a number of design improvements for future craft in the series. Key aspects studied during development of Jetfoil were design improvements to limit cavitation on the foils, and limits to flap operation before cavitation was induced so reducing their effect. The Jetfoil initially had commercial success with Far East Hydrofoils in Hong Kong servicing the Hong Kong to Macau service in competition with catamarans and Surface Effect Ships;

Fig. 7.23 Jetfoil 929 at speed

Fig. 7.24 Chinese Hu Chwan bow hydrofoil patrol boat

this later extended to craft built under licence in Japan, craft for Indonesia, and across the English Channel from Belgium to England. A total of 27 craft were built in the USA to 929-100 and 929-115 specification between 1974 and 1986, while another 11 were built to 929-117 specification by Kawasaki between 1989 and 1991 [10, 1992-3 edition] (Fig. 7.23).

Other Countries

China started hydrofoil research and development in 1958 and constructed a number of naval single foil type torpedo boats in the early 1960s, as shown in Fig. 7.24 [3, 19]; however, due to the poor seakeeping quality in Chinese coastal operations, the construction of further craft was suspended.

Fig. 7.25 Chinese Chang Jiang Hydrofoil (Russian Raketa design built under licence)

Development for civil passenger hydrofoils operating in inland waters also started in 1958 with the type "Chang Jiang hydrofoil," as shown in Fig. 7.25, operated on the Chang Jiang (Yangtze) River between the Wu Han and Jiu Jiang City.

Research continued with designs for experimental craft and jet turbine propulsion at the China Ship Science Research Center (CSSRC), including the type "101" test hydrofoil with a semi-automatic foil control system and the hydrofoil type "7211" with fully submerged foils. Unfortunately, these designs did not lead to the construction and implementation of craft in service. It was at the beginning of the 1990s that experience from hydrofoils, such as the Jetfoils operating on services from Hong Kong to Macao, gave impetus to a renewed interest in building a Chinese passenger hydrofoil for offshore service.

Since 1993, in partnership with other marine technology units and by arrangements with Boeing, CSSRC designed and constructed two "PS-30" hydrofoils with fully submerged foil systems and water jet propulsion for the "Hong Kong Far East Hydrofoil Navigation Co" and named "Arctic Star" and "South Pole Star," as shown in Fig. 7.26.

Leading particulars of the PS-30 are:

Loa	29.1 m
Boa	9.2 m
Hull beam	8.6 m
Power plant	2 × Allison-501 KF Gas turbines, 2318 kw each
Max speed	93.0 km/h
Max operating sea state	Sea state 4
Passengers	294

Fig. 7.26 PS-30 Far East Hydrofoil

Fig. 7.27 Chinese river hydrofoil ferry *Mountain Ge Lu*

Meanwhile, CSSRC also developed a series of inland river hydrofoils using a shallow submersible surface piercing foil system, for example, the *Mountain Ge Lu* with 80 seats, as shown in Fig. 7.27, and *Yuan Zhou* with 110 seats.

A significant market for high-speed water transportation opened up in China during the 1980s and 1990s, leading many Chinese shipping companies to purchase hydrofoils from the international market. This included Boeing Jetfoils on the Hong Kong to Macau route and Russian hydrofoils that were operated along extensive stretches of the Yangtze River above Shanghai and the Heilong Jiang (Amur) River in North East China.

In *Japan*, Kawasaki Engineering Co. purchased a licence to build US Boeing's "Jetfoil" hydrofoil and constructed several such craft, operated on Japanese coastal passenger routes. See (Boeing 929 - Wikipedia).

Shipbuilders in Japan and Norway also took licences to the Supramar surface piercing hydrofoil technology, building craft in series for fast passenger services in both countries and operated on a number of coastal passenger ferry routes reducing journey times (Westermoen Hydrofoil - Wikipedia).

Earlier, in the 1960s, in *Canada*, the Canadian Navy designed and constructed a prototype hydrofoil for the anti-submarine and patrol mission, targeted at a service speed of 60 knots in a seaway, long range, and low-speed performance for long-duration surveillance.

The FHE-400 was conceived prior to the advent of fully submerged foil systems and so had a canard plan layout and main foil system just aft of amidships supporting 90% of the craft weight. The foils were a deep vee shape in transverse section. Propulsion was via z drives to podded propellers at the base of the two main struts and two further propellers higher up the outboard foil struts for hull borne operation. The whole system was fixed.

The outline arrangement is shown in Fig. 7.28 and the craft in operation in Fig. 7.29. The main upper and lower foils both had flaps installed to assist ride height adjustment and coordinating turns, while instead of aft rudders the Bras D'Or forward diamond-shaped foil system was rotatable to enable turning.

The *Bras d'Or*, with a displacement of 260 tonnes was the largest such craft in the world at that time. It was built by the De Havilland Aircraft Company. It was powered by a Pratt and Whitney gas turbine of 22,500 shp and a gearbox system to distribute power to the lower propellers. A Paxman Ventura diesel of 2000 shp mounted in an aft engine room also has a gearbox and split drive train down to the two upper propellers for hull borne operation. The hull was a high L/B slender design following practice for high-speed frigates. In operation, it achieved 50 knots in 3.5 m seas and close to 60 knots in calm conditions, so meeting the original design brief. Hull clearance on foils to calm water was 3.2 m while when floating the draft to the base of foils was 7.1 m.

While a technical success in operation, Bras D'Or was complex and highly powered, and its 7 m floating draft restricted in its potential deployment. The program was therefore closed out once the vessel had completed its experimental and demonstration operations.

7.5 Development Opportunities for Hydrofoils

The reader may note that hydrofoil development has been enabled by increasing available power from diesel and gas turbine engines. Through the twentieth century, there has been increasing experience transferring this power through long canted shafts to the propellers. Shaft whirling for such configurations is a significant problem to get right, in addition to the propeller being canted rather than horizontal. Successful design experience with this enabled greater hull clearance and so seakeeping ability. Then in the second half of the century advances in gear trains

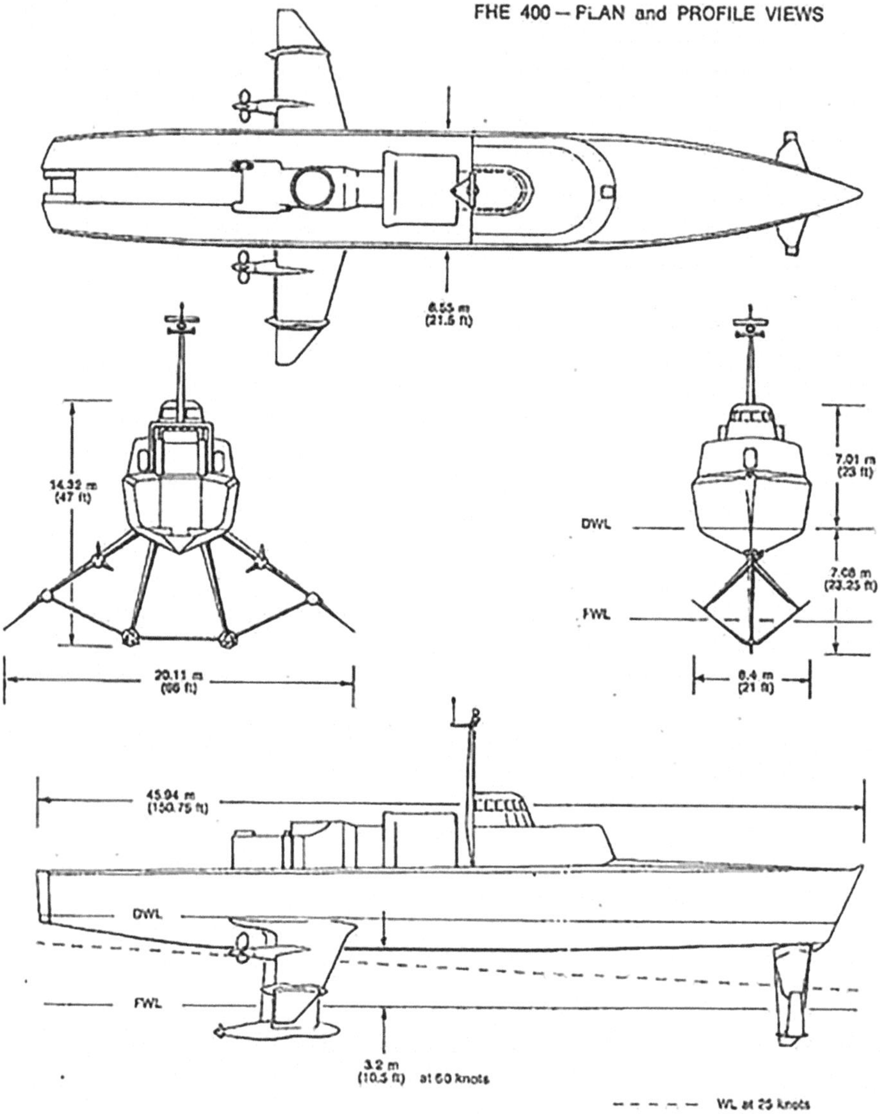

Fig. 7.28 GA of FHE-400

enabled Z drives to podded propellers. Then later, after propellers were meeting their limits due to cavitation, water jets were introduced and steadily improved with craft such as the Boeing Jetfoil. Waterjets can be designed for high efficiency at high vessel speeds.

Fig. 7.29 FHE-400 *Bras d'Or*

Rather like aircraft, hydrofoils suit a mission where the payload is relatively low, so passengers rather than freight, or in military terms a high-speed interdiction task rather than a long-duration patrol.

Cost and Speed

Although high speed and fine seaworthiness are available simultaneously on the fully submersed foil system, such designs have high construction and maintenance costs. In addition, such craft lack passive stability. Much work was carried out on the control systems for PHM and Jetfoil so that there was control systems backup, and that 'crash landing' was practical in a similar way to tests for aircraft ditching.

In the case of the PHM, the craft proved highly operable, but endurance was a problem both in duration and in seakeeping performance when loitering compared with semidisplacement patrol craft, so after a number of years' operation, the US navy suspended the program after the six PHM commissioned in the 1980s.

Technical Challenges

The concept that matured with the PHM and Jetfoil based on fully submerged foils and water jet propulsion solved many problems experienced with earlier propeller-driven designs. Fifty-knot operating speed became practical in a significant seaway. This introduced a significant new problem of cavitation, both for the water jet system inlet and pump design, and for the foils themselves. It was this same challenge that led Dr. Alexeyev in Russia to transfer his research focus from hydrofoils to WIG after the 1970s. Meanwhile, his design bureau has continued to refine surface piercing hydrofoil technology for craft in the speed range of 30–45 knots for rivers, lakes, and more benign seas. Likewise, the designers at Supramar and Rodriquez.

Currently, then, a designer may follow either route—surface piercing or fully submerged hydrofoil design—to achieve practical and efficient craft for commercial operation in inland or coastal conditions.

Ability to Scale Up

While foil lift increases in relation to its area, and so the square of a scale dimension, the mass of a craft will tend to increase with the cube of this same scaling. This effectively limits foil borne operation to lightly loaded and generally small vessels. The designers referred to in this chapter have had success with craft up to about 250 tonnes displacement and passenger capacity between 250 and 30. At least to date, this has been found to be a practical limit in terms of size.

We have seen early on in the chapter that folding systems for the foils have been successfully developed to allow docking for maintenance. Normally, though, for passenger terminal operations, the vessel configuration has to be fixed. For the surface piercing craft, a small adjustment to terminal quayside arrangements is all that is required. For submerged foil craft, there need to be deeper approach channels and water depth at the quay for convenient operation. In locations such as Japan and Norway, this is not a problem, nor is it in Hong Kong. Many large maritime cities nevertheless have restricted draft at their quaysides, and this has limited expansion of the commercial hydrofoil market. A designer has to take terminal and route water depths carefully into account.

Potential

The lift/drag ratio of a hydrofoil at perhaps 18 compared to 2–3 for a boat hull is a very attractive prospect for a designer. Given mission requirements within the kind of dimensions introduced in this chapter, it is possible to design a vessel that "flies over the sea" having a smooth passenger experience that is hardly attainable by other marine vessels. The small practical dimensions do nevertheless limit endurance or range.

The current challenges for transitioning from diesel or gas turbine to fuel cell and electric propulsion currently close this design envelope further, but this is likely to change within a decade (say after 2030). The one exception to this in terms of endurance is the large foiling competition yachts. Since 2000, much research has gone into these foil systems, and the technology is now mature enough for standard designs to be specified for the America's Cup and the Vendee Globe races.

7.6 Summary Data for Sample Hydrofoils (Tables 7.1, 7.2, 7.3, 7.4, and 7.5)

Table 7.1 Key data for early commercial Russian and Supramar PT craft

Name	Raketa	Meteor	Kolkhida	PT 20	PT 50	RHS 110	PT 75
Country	Russia	Russia	Russia	Switzerland	Switzerland	Switzerland	Switzerland
Builder	Krasnoye Sormovo, Gorki	Krasnoye Sormovo, Gorki	Poti, Georgia	Rodriquez, Messina, Italy	Rodriquez, Messina, Italy	Rodriquez, Messina, Italy	Vosper Thornycroft, Paulsgrove, UK
Delivery Date (first)	1957	1960	1981	1956	1957	1973	1975
Passengers Crew	64 na	116 5	120 6	68 2	118 2	110 2	160 na
Displacement, t Lightweight, t	27 20	54.3	72 56	33 26	68.5 48	54	80 60.3
Loa, m	27	34.5	34.5	20.6	27.9	25.6	30
Hull beam Boa, m	5 6	6 9.5	5.8 10.3	5.06 7.14	6.11 10.65	5.95 9.2	5.8 11.6
Draught, m							
Hull borne Foil borne	1.8 1.1	2.4 1.2	3.5 1.9	2.69 1.14	3.5 1.42	2.3 1.25	4.0 1.96
Machinery	1 x M401A	2 x M50-12	2 x MTU12V396 TC82	1 x MB820Db	2 x MB820 Db	2 x MB12V493 TY71	2 x MB12V652 SB7
Power, kW	1 x 668	2 x 816	2 x 965	1 x 1000	2 x 1000	2 x 1000	2 x 1224
Speed, knots							
Maximum Operating	33.5 33.5	35 32 in SS3	36 34 in SS3	38 34	37 34	40 37	39 36
Range, nm	262	324	200	243	290	260	180
Seakeeping On foils Off foils	River/lake conditions	SS3 SS4	SS5 SS6	SS3 SS4	SS3 SS4 coastal	SS3 SS4 coastal	SS4 SS6 coastal

(continued)

Table 7.1 (continued)

Name	Raketa	Meteor	Kolkhida	PT 20	PT 50	RHS 110	PT 75
Hull type	Monohull	Monohull	Monohull	Monohull	Monohull	Monohull	Monohull
Structure	Aluminum	Aluminum	Aluminum	Peraluman Alu/Mg	Aluminum and Steel	Riveted Aluminum	Aluminum
Propulsion	1 x propeller	2 x propeller	2 x propeller	1 x propeller	2 x propeller	2 x propeller	2 x propeller
Foil type	Surface piercing	Surface piercing	Surface piercing	Surface piercing	Surface piercing	Surface piercing	Surface piercing
Controls	Auxiliary lift foils	Auxiliary lift foils	Auxiliary lift foils, main foil flaps		Flaps	Flaps front and rear	Air stabilization
Fr_L	1.05	0.89	0.95	1.23	1.06	1.20	1.08

Notes: Fr_L is based on hull LOA at operational speed. This is just in the planing region. The foils on the other hand lift these craft out of the water at 30–50% of this speed

Table 7.2 Leading particulars for later commercial hydrofoils

Name	Tayfun	RHS 150	RHS 200	Foilmaster	North Star	Mountain Ge-Lu
Country	Russia	Italy	Italy	Italy	China	China
Delivery Date (first)	1969	1980	1981	1994	1994	1996
Builder	Almaz Shipyard	Rodriquez	Rodriquez	Rodriquez	n/a	n/a
Passengers	250	150	210 - 238	242	294	82
Displacement, t	135	65.5	130	112	118	32
Loa, m	42.55	28.7	35.8	31.2	29.1	22.96
Hull beam Boa, m	8.3 n/a	5.6 11	7.6 14.5	6.78 13.3	8.6 9.2	4.8 7.6
Draught, m						
Hull borne Foil borne	4.6 2.0	3.1 1.4	4.55 2.05	3.89 1.45	4.5 n/a	2.05 1.1
Machinery	2 x MTU 16V396TB84,	2 x MTU 12V331TC82,	2 x MTU 16V652TB71,	2 x MTU 16V396TE74	Allison-501KF, gas turbine	KTA19-M2
Power kW	2 × 1900	2 × 1052	2 × 1914	2 × 1550	2 × 2865 (max 3180)	2 × 461
Speed, knots						
Full Operating	40 37	35 32.5	37 35	37 35	50 43	38 37
Range, nm	295	130	200	150	100	216
Seakeeping	SS4	SS4	SS4	SS4	SS4	SS4
Hull type	Monohull	Monohull	Monohull	Monohull	Monohull	Monohull
Structure	Al Alloy	Al Alloy	Al Alloy	Al Alloy	Al Alloy	Al Alloy
Propulsion	Propeller	Propeller FPP	Propeller FPP	Water jet	Water jet	Propeller FPP
Foil type	Surface piercing plus automatic control	Surface piercing	Surface piercing	Full submerged plus automatic control	Full submerged plus automatic control	Surface piercing
Foil load Fore/rear	n/a	60/40	60/40	n/a	30/70	52/49
Fr_L	1.01	1.07	1.02	1.09	1.52	1.3

Table 7.3 Example commercial hydrofoils

Name	Jetfoil 929	Westfoil	MEC-1	Rainbow SS-400	Hayate TSL-F	Foilcat	FSH 38
Country	USA	USA	Italy	Japan	Japan	Norway	Italy
Builder	Boeing	Westport Shipyard	Rodriquez	Mitsubishi, Shimonoseki	Kawasaki, Kobe	Fjellstrand	Rodriquez
Delivery Date (first)	1980–1990	1991	1992	1993	1994	1995	2010 2 prototypes
Passengers + crew	450 + 8	149	146	341 + 4	14 crew	403	245
Displacement, t	110	89.4	n/a	75 (35 dwt)	n/a	n/a	140
Loa, m	27.4	25.6	25.0	33.24	17.1	35	37.25
Hull beam Boa, m	9.5 11	7.6 8.0	6.7 8.4	12.0 13.2	6.2 6.36	12 12	8.0 8.0
Draught, m							
Hull borne Foil borne	5.0 2.0	4.75 1.83	2.8 0.95	4.5 2.1	3.1 1.6	4.7 2.55	4.3
Machinery	2 x Allison 501K20A GT	4 DDC 12V-92TA	2 MWM TBD604 BV8	4 MHIS 16R-MTK	Single Gas Turbine	2 GE-LM500 GT	2 MTU 16V4000M70
Power, kW	2 × 2450	4 × 625	2 × 840	4 × 2100	1 × 2820	2 × 4026	2 × 2320
Speed, knots							
Maximum Operating	59 50	50 42	38 34	45 40	41 41	52 45	47 45
Range, nm	405	730	200	n/a	n/a	300	430
Seakeeping	SS5 (3.65m max)	SS 4	SS4	SS4	SS5	SS4	SS6
Hull type	Monohull	Monohull	Monohull	Catamaran	Monohull	Catamaran	Monohull

Structure	Al Alloy	Al Alloy	Al Alloy	Al Alloy	Al Alloy	Al Alloy	Al Alloy
Propulsion	2 x Waterjets Rocketdyne 20	2 x Ducted Air propellers	Hydraulic driven Rex-roth Propeller pod	2 x Water jet	2 x Hamilton Water jet	2 x Water jet	2 x FPP or 1 x contra-rotating podded unit
Foil type	Fully submerged plus automatic control	Fully submerged plus automatic control	Surface piercing plus control	Fully submerged plus automatic control	Fully submerged plus submerged buoyancy	Fully submerged plus automatic control	Fully submerged plus automatic control
Fr_L	1.86	1.63	1.24	1.28	1.63	1.43	1.26

Table 7.4 Example military hydrofoils

Type	PCH-1 High Point	AGEH-1 Plainview	FHE-400 HMCS Bras D'Or	PGH-1 Flagstaff	PGH-2 Tucumcari	PHM-1 Pegasus	Sarancha (Almaz Project 1240 Uragan)
Country	USA	USA	Canada	USA	USA	USA	Former USSR
Builder	Boeing	Grumman	DeHavilland	Grumman	Boeing	Boeing	Almaz Shipyard
Mission	Antisubmarine Warfare ASW	ASW	ASW	Gunboat	Gunboat	Guided Missile	Guided Missile
Delivery Year	1963	1965	1968	1968	1968	1975	1977
Crew	18	25	25	13	13	24	40
Displacement, t	108–131	328	236	72	58	235	280–330
Loa, m	35.3	64.6	49.95	22.2	22.7	40.5	50.6
Beam, m	9.53	12.3	6.5	6.5	5.9	8.9	11
Draft, m							3.5 foils retr.
Hull borne Foil borne	6.04 2.6	7.6 3.0	7.16 2.3	4.26 1.2	4.33 1.5	7.1 2.7	7.3 2.8
Max speed, Cruise speed, kt	50 max 30-40 cruise	50	63	40	50	50	58 52
Machinery Foil borne	2 × RR Proteus 1273	2 × GE LM1500 GT	GT Gas turbine	RR Tyne Gas turbine	RR Proteus Gas turbine	GT Gas turbine	2 × GTD-M10 Gas turbines
kW	2 × 3154	2 × 10,760	16,325	2486	2450	15,000	2 × 11,130
Hull Borne	GM 12V71	GM12V71	Diesel	2 × 6M6V	Diesel	Diesel	2 × M-401
kW	390	2 × 390	1090	150	90	2 × 440	2 × 668
Propulsion	Z drive 5 b 34" prop	2 × Z drive 5 b 40" prop	Z Drive SC props	Z drive SC prop	Water Jet	Water Jet	Z Drive SC Prop × 4

Range, n mile	n/a	500	500	800	800	600	700
Seakeeping	SS 4 - 5	SS 5	SS 4-5	SS 4	SS 4	SS 5	SS 5
Main Weapon	2 × Mk32 Torpedoes	6 × anti Submarine Torpedo	4 x Anti-sub Torpedo	152 mm M551 howitzer rapid fire gun	Artillery 4 × 40 mm	Artillery 1 × 76 mm 2 × 4 Guide missile "Harpoon"	4 × GM SS-N-9 2 × GM Anti-air, SA-N-4
Foil features	Fully submerged, retractable, automatic control	Full sub-merged auto-control retractable foils	Surface piercing foil	Fully submerged, retractable automatic control	Full submerged automatic control retractable foils	Full sub-merged automatic control retractable Foils	Surface piercing fwd foil fully submerged rear retractable foils
Foil Loading	68% aft 30% fwd	10% aft 90%fwd	90% main 10% fwd	30% aft 70% fwd	70% aft 30% fwd	68.2% aft 31.8% fwd	30% aft 70% fwd
Fr_L	1.38	1.02	1.53	1.39	1.72	1.29	1.42

Table 7.5 Example key data for some Russian military hydrofoils

Name	Turya	Matka.	Babochka	Mukha	Muravey
Project	Project 206M Shtorm	Project 206MR	Project 1141	Project 1145.1 Sokol class	Project 133 Antares
Construction	Zelenodolsk and Yaroslavl Yards	Izhora shipyard, St Petersburg	Gorkiy Shipyard Zelenodolsk	Feodosiya Yard Crimea	Feodosiya Yard Crimea
Mission	Torpedo boat	Missile boat	ASW Corvette	ASW Corvette	Patrol boat
Year of building	1972	1978	1977	1987	1983 - 1989
Crew	30	33	36	39	30
Displacement, t Standard/Full	220 250	225 260	349 465	364 470	180 230
Length, m	39.6	39.6	50.2	49.97	38.6
Beam m					
Hull maximum Over foils	7.6 12.3	7.6 12.3	9.76 20	8.5 na	7.6 na
Draft, with foil Draft, hull	3.8 3.0	3.8 3.0	6.9 2.17	7.26 2.16	4.4 1.9
Max speed, kts	40	40	52	40	40
Main engines	3x M504, diesels 3700 kw ea	3x M-504 diesel 3700 kw ea	3x M10 Gas turbines 13,350 kw ea 2 x Diesels 1630 kw total	2x M10D Gas turbines 15,000kw ea 2 x diesels 1630 kw total	2x M70FRU Gas turbines 8150 kw ea
Range, Miles@kts	600/35 (foil born)	600/35 (foil born)	na	750/35	na
	1450/14 (hull born)	1500/14 (hull born)	2000/8	na	900/12 (hull born)
Foils	Bow foils	Bow foils	Surface piercing bow, fully submerged rear	Surface piercing bow, fully submerged rear	Surface piercing bow, fully submerged rear
Propulsion	3 x propellers	3 x propellers	3 x Z drive double propellers 2 Water Jets	2 x Z drive double propellers	2 x Z drive double propellers

References

1. S.F. Hoerner, *Fluid Dynamic Drag*. Published by the Author, Hoerner Fluid Dynamics, P.O. Box 342, Brick Town, New Jersey, USA 08723, 1965
2. I.H. Abbot, A.E. Von Doenhoff, *Theory of Wing Sections* (Dover Publications, New York, 1958), Library of Congress Catalog 60-1601, 693 pp
3. L. Yun, A. Bliault, *High performance Marine Vessels* (Springer, New York, 2012), chapter 5, pp. 163–171
4. D. Savitsky, Planing craft, chapter IV. Naval Eng. J. **97**(2) (1985), USA
5. Z.S. Dong, Dynamics for High Speed Crafts (Naval Engineering University, China, March 1985) (in Chinese)
6. H.F. King, 'Aeromarine Origins' (Putnam, Great Russell Street, London, 1966), 93 pp
7. L.E. Zhao et al., *Theory and Design of High Performance Marine Vessels* (National Defense Industrial Press, China, Jan 2009) (in Chinese)
8. B.Q. Li et al., *Ocean Going High Performance Marine Vessels in 21st Century* (National Defense Industrial Press, China, July 2001) (in Chinese)
9. T.S. Guan et al., *Basic knowledge for Naval High Speed craft* (National Defense Industrial Press, China, July 1992) (in Chinese)
10. Jane's High Speed Marine Craft and Air Cushion Vehicles, 1967–1993, Jane's Publishing Company, London, UK, ISBN 0 7106-0851-9
11. Jane's Fast Sea Transportation Annual Publication 1994–2012, Jane's Publishing Company, London, UK, ISBN-13:978-0710628893
12. R. McLeavy, Jane's Pocket Book 21, Hovercraft and Hydrofoils, 1980, ISBN 0 354 01212 6, 251 pp
13. Jane's High Speed Marine Craft and Air Cushion Vehicles, 1988 edition, Jane's Publishing Company, London, ISBN 0 7106-0851-9
14. P. Du Cane, *High Speed Small Craft*, 3rd & 4th edition, 1964, 1972, Temple Press Books, 469 pp
15. R. Rossi, G. Biancuzzo, Development of the new Rodriquez 38m FSH Fully Submerged Hydrofoil, 9th Conference High Speed Marine Vessels, Naples 25–27 May 2011
16. G.R. Lamb, High Speed Naval Vessel Technology Development Plan, NSWC, CD-20-TR-2003/69, May 2003, MD, (from Google network)
17. V.H. Salisbury, Jetfoil in Operation, AIAA/SNAME Advanced Marine Vehicles Conference, October 2–4, 1979, Baltimore, Maryland, USA, published by the American Institute of Aeronautics and Astronautics, 1979, pp 63–71
18. A.E. Noreen, P.R. Gill, W.M. Felfel, Foilborne hydrodynamic performance of Jetfoil, AIAA/SNAME Advanced Marine Vehicles Conference, October 2–4, 1979, Baltimore, Maryland, USA, published by the American Institute of Aeronautics and Astronautics, 1979, pp 12–21
19. L. Dong et al., A summary on development of hydrofoil craft, Proceedings of HPMV conference, April 8–11, 2010, Shanghai, China

Chapter 8
Introduction to Hydrofoil Wing Theory

The hydrodynamic properties of hydrofoil lifting surfaces on a hydrofoil craft directly influence its performance. We have described a wide range of hydrofoil geometries and vessel configurations in Chap. 7. The hydrodynamic properties of the various hydrofoil configurations vary significantly, and in addition, the foil section profiles have specific characteristics.

In principle, there are no fundamental differences between the foils on a hydrofoil and the airfoils on an airplane, so much of the theory and methods used in wing section theory [1, 2], can be extended to hydrofoils. The main differences between them are:

- Hydrofoils usually operate near the water free surface, and are significantly influenced by the free water surface boundary. There is no equivalent influence on an airplane wing that is operated in the unlimited air medium.
- Hydrofoils operating above a certain high speed will experience cavitation due to suction pressure on their upper surface going below local water vapor pressure. There is no such physical phenomenon for airplane wings.

In this chapter, we introduce wing section theory for fully immersed foil sections. The "Theory of a thin section wing" considered initially in two dimensions is principally discussed, as the thickness of foils generally used on hydrofoil craft are small. In the Chap. 9, we will look at the effects of free surface proximity, turbulent flow, and cavitation. For a more complete treatment of hydrodynamic potential theory for airfoils, see [1], and for hydrofoils see [3, 4] pp 178–220. For hydrodynamic potential theory, see [5].

Z.-S. Dong et al., *High-Speed Monohull and Hydrofoil Craft*,
https://doi.org/10.1007/978-3-031-62760-6_8

8.1 Theory of Two-Dimensional Thin Wing

8.1.1 Basic Assumptions

We make three key assumptions for the theoretical presentation in this chapter.

First: we assume the fluid is non-viscous, i.e., having ideal potential flow. The key to this is considering kinetic and inertial forces without the influence of viscosity, with the viscous force component being added separately (superposition is used). Viscous forces are further considered in Chap. 9.

Second: we use a two-dimensional geometry where the wingspan is unlimited.

Third: we assume the foil section fulfills the three requirements for thin wing sections, whereby the maximum cross section thickness is small, the camber line curvature is small, and the angle of attack toward oncoming flow is also small.

These requirements meet a criterion for small perturbation of the flow field by the wing, i.e., the perturbation speed is significantly smaller than the oncoming flow speed. This then allows us to neglect the second order of perturbation speed in analytical formulations. The concept of a "thin wing" is equivalent to one with small perturbation of flow. In addition, since the hydrofoil speed is usually high, the inertia force is often significantly greater than gravity, so gravity can be neglected in the Lagrange integral equation[1] if this is used to obtain the pressure field. However, when using the solution for wing lift by means of Zhoukovsky's theory[2] such an assumption may not be applicable so in that case, we do not neglect gravity force.

8.1.2 The Boundary Condition and Its Solution

The best-known method for processing the problem of the flow around a wing section is that using the boundary condition to solve the dynamic equations. Regarding the potential flow, we use the boundary condition to solve the continuity equation, i.e., the Laplace equation[3]:

$$\nabla^2 \phi = 0 \tag{8.1}$$

where ϕ is the velocity potential. The above equation can be solved uniquely so long as the boundary condition is specified. For a two-dimensional wing moving in an unlimited medium at speed V_∞, the boundary conditions are that the flow speed at

[1]For Lagrangian Mechanics explanation, see Lagrangian mechanics—Wikipedia

[2]The Kutta-Zhoukovsky (Joukowski) theorem, see Kutta–Joukowski theorem—Wikipedia

[3]Laplace's equation, see Laplace's equation—Wikipedia

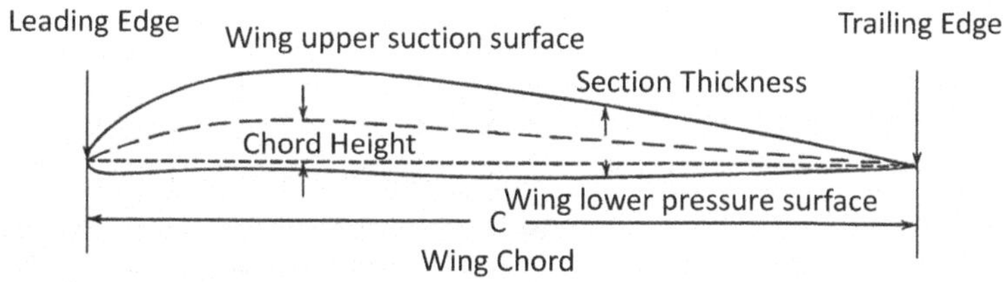

Fig. 8.1 Geometrical profile and nomenclature of wing section

Fig. 8.2 Coordinate system of two-dimensional thin wing

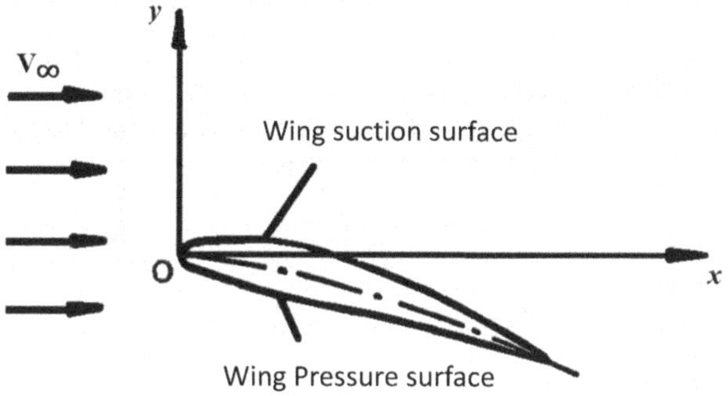

infinite far distance is equal to zero, and no flow penetrates the wing surface. For the purpose of analysis, we consider the wing is fixed, and flow with uniform speed V_∞ oncoming to the wing.

In order to discuss the boundary conditions, including the actual wing surfaces, we need to introduce the nomenclature for the geometrical profile of the wing, Fig. 8.1.

The front edge of a wing is called the "leading edge." The back edge is called the "trailing edge." The straight line connecting the leading and trailing edges is the "chord." Its length is the "chord length." The wing's upper surface is the "wing back" or "suction surface," while the lower surface is called the "pressure surface." The perpendicular distance between both surfaces of the wing is the "wing thickness." The line connecting the midpoints of the various wing thicknesses is the "camber line," and the distance between the "camber line" and "chord line" is called the "camber."

The coordinate system for a wing in uniform oncoming flow can be established as shown in Fig. 8.2, with the origin of the axes located at the leading edge and X axis parallel to oncoming flow.

The equation for the wing surface can be expressed as:

$$y = \eta(x)$$

According to the condition with flow not penetrating the wing surface (wing section rather than source/camber line), the boundary condition for the body surface is:

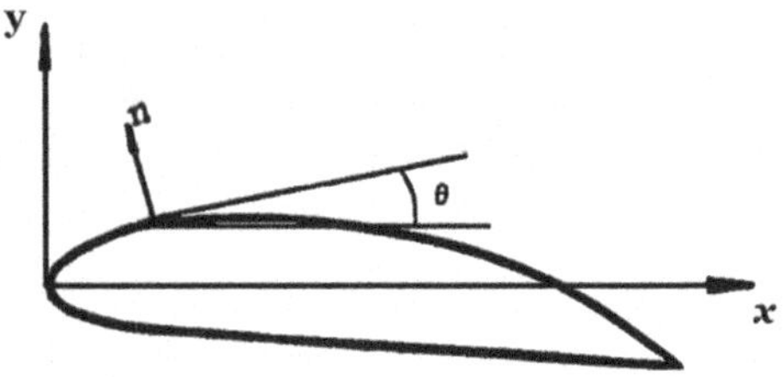

Fig. 8.3 The normal direction of any point on wing surface

$$V_n = \frac{\partial \phi}{\partial n} = V_\infty \cos(n, x) \tag{8.2}$$

Where n is the normal direction of any point on the wing surface, and the angle between the tangent to the surface and X axis at that point is θ, see Fig. 8.3.

Then:

$$\cos\ (n, x) = \text{-}\sin\theta \,\text{at position}\ x$$

For a thin wing, θ is small, and so:

$$\sin\theta \approx \tan\theta = \frac{d\eta(x)}{dx} \quad \text{and} \quad V_n \approx V_x$$

Putting this into (8.2) we obtain approximately

$$V_n = V_y = V_\infty \frac{d\eta(x)}{dx} \tag{8.3}$$

Defining the ordinates of the upper and lower surfaces of the wing as $\eta_u(x)$ and $\eta_i(x)$ respectively, then the boundary condition can be expressed as:

$$\left.\begin{aligned} V_{yu} &= V_\infty \frac{d\eta_u(x)}{dx} \\ V_{yi} &= V_\infty \frac{d\eta_i(x)}{dx} \end{aligned}\right\} \tag{8.4}$$

The upper and lower surface ordinates of the wing can also be written as:

$$\left.\begin{aligned} \eta_u &= \frac{1}{2}(\eta_u + \eta_l) + \frac{1}{2}(\eta_u - \eta_l) \\ \eta_l &= \frac{1}{2}(\eta_u + \eta_l) - \frac{1}{2}(\eta_u - \eta_l) \end{aligned}\right\} \tag{8.5}$$

where

$\frac{1}{2}(\eta_u + \eta_l)$ represents the coordinates of the camber line

$\frac{1}{2}(\eta_u - \eta_l)$ represents the coordinates of upper wing surface measured from the camber line

$-\frac{1}{2}(\eta_u - \eta_l)$ represents the coordinates of the lower surface measured from the camber line.

The boundary condition (8.4) can also be written as two parts, i.e., camber part and thickness part respectively. For the camber part, it can be divided into two parts, the angle of attack, and camber part respectively, in terms of coordinates from baseline X axis:

$$\eta_c = -\alpha x + \eta_b = \frac{1}{2}(\eta_u + \eta_l) \tag{8.6}$$

where

α angle of attack of chord, i.e., angle between chord and oncoming flow V_∞
η_b the coordinate of camber line from chord line, thus from (8.6):

$$\eta_b = \alpha x + \frac{1}{2}(\eta_u + \eta_l) \tag{8.7}$$

These linear components are illustrated in Fig. 8.4. The boundary condition for a thin wing can also be divided to three parts as in the right-hand side of (8.8) below:

$$V_{yu} = \left.\frac{\partial \varphi}{\partial y}\right|_{y=+0} = V_\infty \frac{\partial \eta_u}{\partial x} = V_\infty \frac{\partial}{\partial x}\left[\frac{1}{2}(\eta_u + \eta_i) + \frac{1}{2}(\eta_u - \eta_i)\right]$$

$$= -V_\infty \alpha + V_\infty \frac{\partial \eta_b}{\partial x} + V_\infty \frac{\partial \eta_t}{\partial x} \tag{8.8a}$$

$$V_{yi} = \left.\frac{\partial \varphi}{\partial y}\right|_{y=-0} = -V_\infty \alpha + V_\infty \frac{\partial \eta_b}{\partial x} - V_\infty \frac{\partial \eta_t}{\partial x} \tag{8.8b}$$

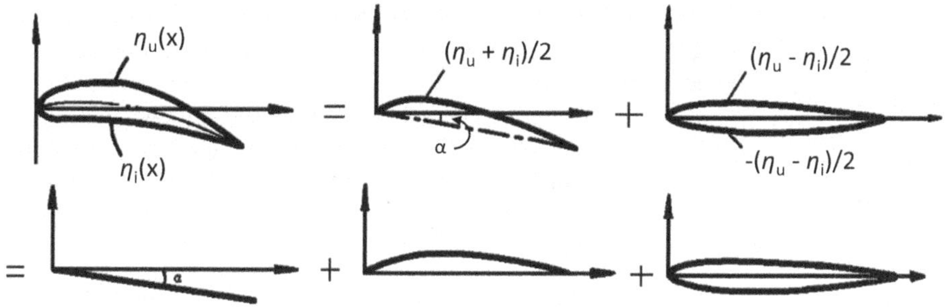

Fig. 8.4 Separation of boundary conditions for a thin wing

We assume for the moment that the boundary condition for upper and lower surfaces is approximately zero at the upper (so $y = +0$) and lower sides (where $y = -0$) of the X axis, without significant error for the thin wing. Then the basic equation $\nabla^2\phi = 0$ can be divided in three parts, and use $\varphi = \varphi_1 + \varphi_2 + \varphi_3$. The components may be described as follows. *First,* φ_1 represents the perturbation velocity potential induced by the flat plate without thickness and with an angle of attack α.

Second, φ_2 represents the perturbation velocity potential induced by a "wing" with no thickness and zero angle of attack, and only with camber.

Finally φ_3 represents the perturbation velocity potential induced by a symmetric wing with a certain thickness profile.

The total perturbation velocity potential of the thin wing is the superposition of the three parts. Thus, the three parts of the boundary condition can generate three components of velocity potential:

$$\begin{aligned} &\nabla^2\phi_1 = 0, \text{and meet boundary condition } \left.\frac{\partial \phi}{\partial y}\right|_{y=\pm 0} = -V_\infty \alpha \\ &\nabla^2\phi_2 = 0, \text{and meet boundary condition } \left.\frac{\partial \phi}{\partial y}\right|_{y=\pm 0} = V_\infty \frac{\partial \eta_b}{\partial x} \\ &\nabla^2\phi_3 = 0, \text{and meet boundary condition } \left.\frac{\partial \phi}{\partial y}\right|_{y=\pm 0} = \pm V_\infty \frac{\partial \eta_t}{\partial x} \end{aligned} \tag{8.8c}$$

Combining φ_1, φ_2, and φ_3, we obtain $\nabla^2\phi = \nabla^2(\phi_1 + \phi_2 + \phi_3) = 0$ by linear superposition and also meet the original boundary condition for an (infinitely) thin wing.

8.1.3 Vortex Model of Two-Dimensional Thin Wing

When studying the lift of a thin wing in an ideal fluid, a symmetric wing section does not itself generate lift, thus, it is only necessary to calculate the superposition of two parts, i.e., that induced by angle of attack and camber. The total lift should be equal to the lift of the superposed two parts.

Since an ideal fluid is used for the investigations, with a two-dimensional wing, there will be no drag of the two-dimensional wing. The thin thickness wing can be neglected for simplification, and then only two parts are required in (8.8a), and only two components, i.e., angle of attack and camber are linked to the boundary condition, as in Eq. 8.9 below.

It must be noted that the camber coordinate, η_b, is a Y axis distance measured from the chord line to the camber line, which is different from the "camber" given in general references for wing sections. Generally, the camber of a wing is a distance measured perpendicularly from the chord line, with a difference of $(1 - \eta_b \cos \alpha)$ due to angle of attack α. However, since the angle of attack for the thin wing we are considering is very small, the difference can be neglected. Therefore, the camber herewith can be taken as the camber coordinate presented in normal reference data for a wing plus the perpendicular distance, x_α, caused by the angle of attack. The boundary conditions are thus:

$$\left.\begin{aligned} \left.\frac{\partial\phi}{\partial y}\right|_{y=+0} &= -V_\infty\alpha + V_\infty\frac{\partial\eta_b}{\partial x} \\ \left.\frac{\partial\phi}{\partial y}\right|_{y=-0} &= -V_\infty\alpha + V_\infty\frac{\partial\eta_b}{\partial x} \end{aligned}\right\} \tag{8.9}$$

From (8.9), it can be seen that the normal velocity at both the upper and lower surface is equal, as:

$$\left.\frac{\partial\phi}{\partial y}\right|_{y=+0} = \left.\frac{\partial\phi}{\partial y}\right|_{y=-0}$$

Or:

$$\frac{\partial\phi(x,y)}{\partial y} = \frac{\partial\phi(x,-y)}{\partial y} \tag{8.10}$$

Integrating (8.10) along the y direction, then:

$$\phi(x,y) = -\phi(x,-y) \quad (\text{when}, y = +/-0) \tag{8.11}$$

And

$$\frac{\partial\phi(x,y)}{\partial \mathrm{x}} = -\frac{\partial\phi(x,-y)}{\partial x} \quad (\text{when}, y = +/-0) \tag{8.12}$$

The perturbation velocity in Eq. (8.12) at upper and lower surface of the wing with camber along the X direction is asymmetric.

In order to respect the surface condition of (8.10) and (8.12), vortices must be distributed over the wing surface. A simplified assumption for a thin wing along the X axis $0 < x < c$, is that $y = 0$, where c is chord length, may be used. Therefore, the vortex $\gamma(\xi)$ is applied along the X axis from $0 < x < c$, with $y = 0$, as shown in

Fig. 8.5 Vortex model for a thin wing

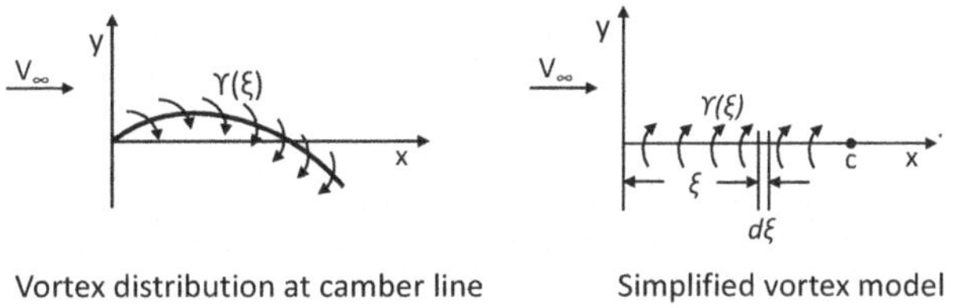

Fig. 8.5 right hand diagram. The flow velocity field is continuous, except at the upper and lower surface of the wing.

Regarding the two edges, i.e., $x = 0$ and $y = 0$ for the leading edge, and $x = c$, $y = 0$ for the trailing edge, the velocity should be continuous before and after these points, thus at the trailing edge:

$$\left.\frac{\partial \phi}{\partial y}\right|_{x=c} = \left.\frac{\partial \phi}{\partial x}\right|_{x=c} = 0$$

The vortex strength here is: $\gamma(c) = 0$.

This is the so-called Kutta-Zhoukovsky[4] condition in ordinary wing theory.

The situation at the leading edge is rather more complicated. Since an angle of attack is present, the leading edge itself is not a stagnation point. Thus, the velocity of flow around a sharp leading edge is theoretically infinite, and so $\gamma(0) \neq 0$ and creates suction on the leading edge.

8.1.4 Solution of Flow Around the Two-Dimensional Thin Wing

According to the vortex model described above, the flow field is a uniform oncoming flow field plus the perturbation velocity field induced by the vortex layer distributed at $0 < x \leq c$, $y = 0$, with strength of $\gamma(\xi)$. So the whole flow field can be determined, and the vortex strength $\gamma(\xi)$ can be obtained in accordance with the boundary condition of the body.

According to the Biot-Savart equation,[5] the induced velocity on a unit length of wing surface, $d\xi$ induced by the vortex with circulation of $\gamma(\xi)\,d\xi$ is:

$$dv = \frac{\gamma(\xi)\,d\xi}{2\pi R}$$

Where R is a distance from any point in the field to the unit length, $d\xi$, and the induced velocity at the point with coordinate of x is:

$$dv_y = \frac{\gamma(\xi)\,d\xi}{2\pi(\xi - x)} \tag{8.13}$$

The total induced velocity at that point generated by the whole vortex layer on the wing surface is:

[4]See Kutta–Joukowski theorem—Wikipedia

[5]The Biot-Savart Relation (aerodynamic vortex circulation) at Biot–Savart law—Wikipedia

$$v_y = \frac{1}{2\pi}\int_0^c \frac{\gamma(\xi)}{\xi - x}d\xi \tag{8.14}$$

According to the boundary conditions we have from (8.9):

$$v_y = \frac{\partial \varphi}{\partial y}\bigg|_{y=\pm 0} = -V_\infty \alpha + V_\infty \frac{d\eta_b}{dx} \tag{8.9a}$$

Making (8.14) equal to (8.9a) then we have:

$$\frac{1}{2\pi}\int_0^c \frac{\gamma(\xi)}{\xi - x}d\xi = -V_\infty \alpha + V_\infty \frac{d\eta_b}{dx} \tag{8.15}$$

The velocity field $\gamma(\xi)$, can be solved by means of solving integral Eq. (8.15), with a given angle of attack α and wing camber profile η_b. In Eq. (8.15) when $x = \xi$, the function to be integrated tends to infinity. In order to simplify, converting the variable ξ as follows:

$$\xi = \frac{c}{2}(1 - \cos\theta) \tag{8.16}$$

And

$$d\xi = \frac{c}{2}\sin\theta d\theta \tag{8.17}$$

Then

$$x = \frac{c}{2}(1 - \cos\beta) \tag{8.18}$$

Figure 8.6 gives the geometric relation of ξ with θ, and x with β. Putting equations from (8.16) to (8.18) into (8.15), we then have:

$$\frac{1}{2\pi}\int_0^x \frac{\gamma(\theta)\sin\theta}{\cos\theta - \cos\beta}d\theta = V_\infty\left(\alpha - \frac{d\eta_b}{dx}\right) \tag{8.19}$$

Transforming $\gamma(\theta)$ into a trigonometric function we can obtain:

$$\gamma(\theta) = 2V_\infty\left[A_0\cot\frac{\theta}{2} + \sum\nolimits_{n=1}^{\infty} A_n \sin n\theta\right] \tag{8.20}$$

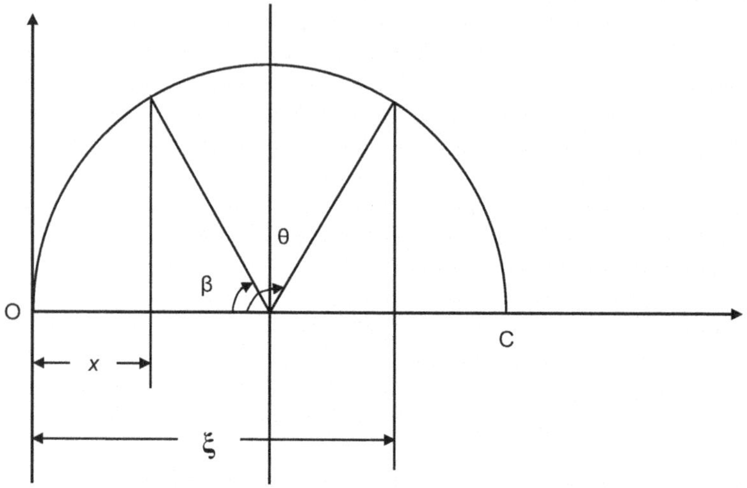

Fig. 8.6 Geometric relations relating to circulation around a wing section

Inserting this into (8.19), after simplification we derive:

$$\frac{d\eta_b}{dx} = \alpha - A_0 + \sum_{n=1}^{\infty} A_n \cos n\beta \tag{8.21}$$

Equation (8.21) gives the relation between the boundary condition and vortex strength of the wing surface. So, from this equation, the distribution of vortices on the wing surface can be obtained in terms of the wing boundary condition.

The coefficients A_0 and A_n can be given in the terms of integration of $\frac{d\eta_b}{dx}$, the rate of change of the surface boundary generating the perturbation, thus:

$$\left.\begin{aligned} \alpha - A_0 &= \frac{1}{\pi}\int_0^{\pi} \frac{d\eta_b}{dx} d\beta \\ A_n &= \frac{1}{\pi}\int_0^{\pi} \frac{d\eta_b}{dx} \cos n\beta d\beta \end{aligned}\right\} \tag{8.22}$$

This can then be expressed by the various coefficients in a Fourier Series,[6] in which only cosine terms are non-zero, and all of sine terms are equal to zero.

From (8.22), one can see various coefficients in the expression for the vortex layer (8.20) A_0 and A_n can be obtained at a given thin wing angle of attack α and the camber line distribution to obtain the vortex strength $\gamma(\theta)$. Following this, the velocity and pressure distribution of the flow field around the section can be obtained

[6] See Fourier series—Wikipedia, also Fast Fourier transform—Wikipedia for evaluation

in terms of Kutta-Zhoukovsky theory. From this the lift force and moment acting on the wing can be obtained.

Here we have to remember that to obtain the pressure distribution on the wing surface, it is needed to add the influence of thickness on the pressure and then superpose this on the flow around the "thin" wing section as discussed under Sect. 8.1.2 above.

Thus, one first takes the distribution of sources and sinks on the symmetric axis instead of the wing profile as in Fig. 8.5. Secondly the source and sink strength can be solved in terms of the boundary condition of a symmetric wing, and thus obtain the distribution of velocity and pressure, and superpose it to the velocity and pressure field induced by the vortex flow from the camber line and angle of attack. Once the velocity and pressure distribution are determined one can determine forces and moments on a wing.

8.1.5 Forces and Moments Acting on a Thin Wing[7]

According to the Kutta-Zhoukovsky theory, at the unit section $d\xi$ of a thin wing in oncoming flow velocity V_∞, with vortex circulation $d\Gamma = \gamma(\xi)d\xi$, and the lift acting on the wing is:

$$dL = \rho V_\infty d\Gamma = \rho V_\infty \gamma(\xi) d\xi$$

The total lift acting on the thin wing is:

$$\begin{aligned} L &= \rho V_\infty \int_0^c \gamma(\xi) d\xi \\ &= \rho V_\infty \int_0^\pi 2V_\infty \left[A_0 \cot\frac{\theta}{2} + \sum_{n=1}^{\infty} A_n \sin n\theta \right] \frac{c}{2} \sin\theta d\theta \end{aligned}$$

After deriving the trigonometric function of this equation, the total lift can be expressed as:

$$L = \rho V_\infty^2 c\pi \left(A_0 + \frac{A_1}{2} \right)$$

The corresponding lift coefficient is then:

[7] See also Airfoil—Wikipedia

$$C_L = \frac{L}{\frac{1}{2}\rho V_\infty^2 c} = 2\pi\left(A_0 + \frac{A_1}{2}\right) \tag{8.23}$$

From (8.23), one can see that lift relates to the coefficients of two items in the vortex expression 8.21.

Similarly with the moment coefficient:

$$C_M = \frac{M}{\frac{1}{2}\rho V_\infty^2 c^2} = -\frac{\pi}{2}\left(A_0 + A_1 - \frac{1}{2}A_2\right) \tag{8.24}$$

The moment coefficients only relate to the coefficients of first three items of the vortex expression. Putting the coefficients A_0 and A_n of (8.22) and wing surface boundary condition $\frac{d\eta_b}{dx}$ into (8.23) and (8.24), we then obtain:

$$C_L = 2\pi\left[\alpha + \frac{1}{\pi}\int_0^\pi \frac{d\eta_b}{dx}(\cos\beta - 1)d\beta\right] = 2\pi(\alpha - \alpha_0) \tag{8.25}$$

$$C_M = 2\left[\mu_0 + \frac{\pi}{4}\alpha_0\right] - \frac{1}{4}C_L \tag{8.26}$$

where

$$\mu_0 = \int_0^\pi \frac{d\eta_b}{dx}(1 - \cos 2\beta)d\beta \tag{8.27}$$

and

$$\alpha_0 = -\frac{1}{\pi}\int_0^\pi \frac{d\eta_b}{dx}(\cos\beta - 1)d\beta \tag{8.28}$$

The term α_0 is the angle of attack at zero lift, thus when angle of attack is α_0, the wing lift is equal to zero.

Using (8.25) and (8.26), for a given camber distribution of a thin wing $\eta_b(x)$ and angle of attack α, the lift and moment coefficients can be obtained, as in the following examples.

(1) Flat Plate Wing.

For flat plate wing, we have $\eta_b(x) = 0$, i.e., no camber available so $\frac{d\eta_b}{dx} = 0$. Then:

$$\left.\begin{aligned} C_L &= 2\pi\alpha \\ C_m &= -\frac{\pi}{2}\alpha \end{aligned}\right\}. \tag{8.29}$$

Then one can see the center of lift is located at ¼ of chord length from the leading edge, because:

$$C_m = -C_L \cdot \frac{x_L}{c} \quad \text{and so} \quad x_L = \frac{1}{4}c$$

Where x_L is the center of the lift, located from the leading edge of the wing as shown in Fig. 8.7 below.

(2) Wing with Camber Line Profile.

Assume the expression for camber of a cross-section profile wing is as follows:

$$\eta_b = f\left[1 - 4\left(\frac{x}{c} - \frac{1}{2}\right)^2\right] \tag{8.30}$$

Where:

f max camber
c chord length

The origin of this camber curve is at the leading edge, as in Fig. 8.8.

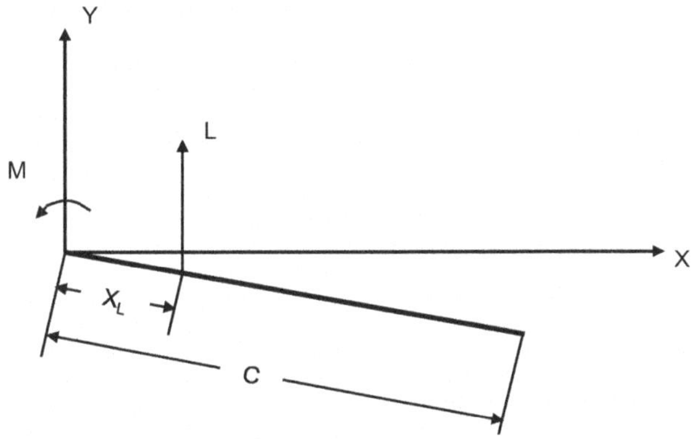

Fig. 8.7 Nomenclature for plate wing

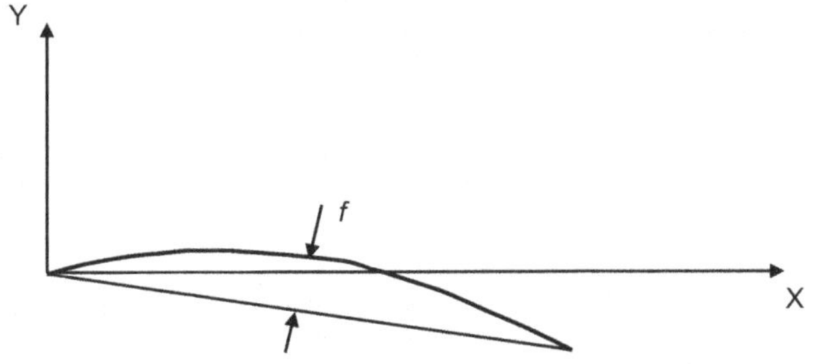

Fig. 8.8 Wing with camber line profile

From (8.30), we then derive:

$$\frac{d\eta_b}{dx} = -8f\left(\frac{x}{c} - \frac{1}{2}\right) \cdot \frac{1}{c} = -8\bar{f}\left(\frac{x}{c} - \frac{1}{2}\right)$$

Where: $\bar{f} = \frac{f}{c}$, the relative camber taken as proportion of chord c.
From (8.18), we have:

$$\frac{\mathrm{x}}{\mathrm{c}} = \frac{1}{2}(1 - \cos\beta) \text{ and } \frac{d\eta_b}{dx} = 4\bar{f}\cos\beta$$

Then the various coefficients to determine C_L and C_M from (8.23) and (8.24) are:

$$A_0 = \alpha - \frac{1}{\pi}\int_0^{\pi} 4\bar{f}\cos\beta d\beta = \alpha$$

$$A_1 = \frac{2}{\pi}\int_0^{\pi} 4\bar{f}\cos^2\beta d\beta = \frac{8\bar{f}}{\pi}\left[\frac{\beta}{2} + \frac{1}{2}\sin\beta\cos\beta\right]_0^{\pi} = 4\bar{f}$$

$$A_2 = \frac{8\bar{f}}{\pi}\int_0^{\pi} \cos\beta\cos 2\beta d\beta = 0$$

So, the lift and moment expressed as coefficients are:

$$\left.\begin{aligned} C_L &= 2\pi(\alpha + 2\bar{f}) \\ C_m &= -\frac{\pi}{2}(\alpha + 4\bar{f}) \end{aligned}\right\} \tag{8.31}$$

From the (8.31), one can see the angle of attack at zero lift of the camber line profile thin wing (in radians) is:

$$\alpha_0 = -2\bar{f}$$

The zero lift definition is shown in Fig. 8.9, which can be extended to any thin wing with x-wise symmetric camber with respect to the chord center point, with reasonable accuracy.

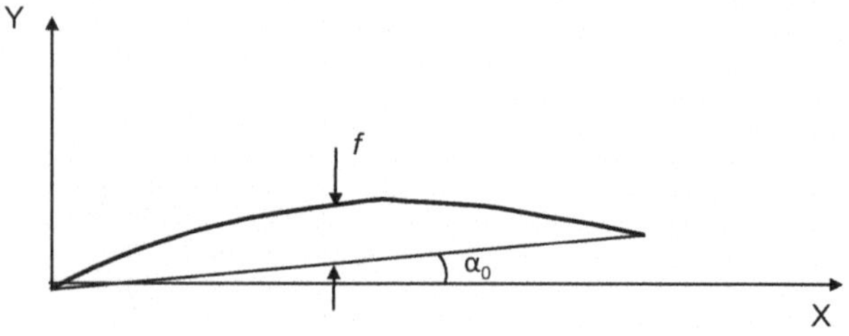

Fig. 8.9 Camber line wing—zero lift angle diagram

The center of lift of the camber line profile wing is located at:

$$\frac{X_L}{C} = -\frac{C_m}{C_L} = \frac{\frac{\pi}{2}(\alpha + 4\bar{f})}{2\pi(\alpha + 2\bar{f})} = \frac{1}{4}\frac{(\alpha + 4\bar{f})}{(\alpha + 2\bar{f})} \tag{8.32}$$

When $\bar{f} = 0$, i.e., flat plate wing, the acting point of resultant force is located at ¼c from leading edge. When $\bar{f} \neq 0$ but $\alpha = 0$, the acting point of the lift resultant is located at ½c from the leading edge.

The lift can be divided into two parts: the lift induced by angle of attack located at ¼c, and that induced by camber and acting at ½c. Thus, the resultant induced by both acts at a point between ¼c and ½c. As angle of attack is increased the acting point of the resultant trends to ¼c, which can be applied for any thin wing with symmetric camber, with respect to the chord center point.

Finally, the issues of the lift direction and suction at the leading edge acting on the thin wing need to be considered. Taking the flat plate wing as an example, one can see from the force acting on a planing surface as introduced in previous chapters, the pressure acting on the underside of the flat plate is perpendicular to it. This is similar to the flat wing, where the lift should be perpendicular to the plate surface, but not oncoming flow, and causes the drag of the plate relative to the oncoming flow direction. However, in non-viscous flow, no drag should be created on the wing.

Actually, for the plate wing, the leading edge is not the forward stagnation point, and for a wing with an angle of attack, generally the forward stagnation point is located at the front part of the lower surface and the flow is around the leading edge from the lower wing surface. Meanwhile, the leading edge is an inflexion point and the distance between the stagnation point and the leading edge with "reverse flow" causes a suction forward at the leading edge, the value of which can be obtained by means of circulation theory.[8]

The value of this suction is exactly equal to the drag, and compensates each other. However, on the planing plate, the suction at leading edge is not available, due to the flow only exiting on lower surface, and not around the leading edge, therefore, drag exists on a "planing" foil even in ideal flow. For a thin wing with camber, the situation is similar. The existence of suction at the leading edge causes the lift to be perpendicular to oncoming flow.

8.1.6 Simplified Lifting Line Model

When analyzing flow circulation issues, a simpler model may be applied instead of the vortex layer model distributed along the camber of the wing described above. Instead, one can consider the vortex concentrated at a certain point on the wing

[8] See also Stagnation point—Wikipedia, and Stagnation point flow—Wikipedia

section. Using a point vortex instead of a distribution across the wing area is the so-called the lifting line model.

Regarding using a line with vortex points (lifting line) span-wise to replace the vortex layer (lifting surface), principally the obtained lift and moment for both methods should be the same, so that the total circulation is the same, and the lifting line should be located at the center of lift along the chord.

By taking one vortex line to replace a whole vortex layer to represent the lifting line model, there is only one unknown value, i.e., the total vortex, and thus only one point on the wing surface strictly meets the surface boundary condition.

For the flat plate wing, it is not difficult to prove that when the lifting line is located at ¼c from the leading edge, where c is chord length, the calculated lift and moment are equal to that determined from the vortex layer distribution method. This is so long as the point ¾c from leading edge meets the boundary condition of the wing surface.

It is the case that for the camber line profile wing, when the lifting line is located at ¼c and meeting the boundary layer of wing surface at ¾c the obtained total lift is also equal to that determined by the vortex layer distribution method. However, the calculated moment is rather different. So, one can calculate the lift according to the method mentioned above for simplification, with some differences to the moment.

Regarding velocity and pressure distribution on the wing surface, the difference is more significant. Therefore, the simplified lifting line theory is used only for calculating the total lift acting on the wing. That can at least simplify the issues related to the complicated circulation flow problems without significant error.

8.2 Influence of Limited Span, the Three-Dimensional Thin Wing

8.2.1 Flow Situation and Lifting Line Model of Wing with Limited Span

In case of a three-dimensional wing with limited span, the flow will not be two dimensional, but three dimensional, with strong influence of the wing tip on the flow.

Owing to the camber or angle of attack of the wing, a negative pressure exists on the upper surface, and positive pressure on lower surface of the wing. This causes circular flow from lower to upper at the tips of the wing, and influences across the whole wingspan, generating flow along the span, which directs toward the central line of symmetry of the wing at its upper surface and away from it at lower surface, as shown in Fig. 8.10.

Such span-wise flow is very apparent at the wing tips, and weakens gradually toward the centerline. It increases inversely with span length. In general, the span/chord ratio affects the strength of the span-wise flow in proportion to:

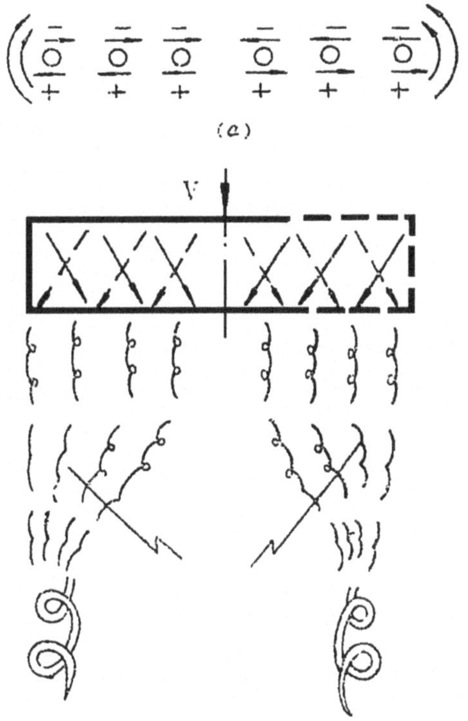

Fig. 8.10 Flow nearby the surface of limited span wing

$$A = \frac{l^2}{S}. \tag{8.33}$$

where:

l span length
S plan projected area of wing plane

For a rectangular wing $S = l \times c$, where c is chord length, and so $A = l/c$.

For a limited span wing, when describing the vortex model, the vortex will not terminate at the span tips according to Milne Thomson's theory,[9] but continue downstream with the oncoming flow, to form a free vortex at the tips, and extend to the far distance, rather than the vortex layer distributing on the wing surface. In addition, anywhere along the wingspan, the flow speed difference span-wise on both upper and lower wing surface creates free vortices anywhere along the span at the trailing edge, extending downstream. Closer to the wing tip, increasing strength of free vortices exists, and so the free vortices are formed into two streams, extending to infinite distance, as shown in Fig. 8.10.

[9] See L. M. Milne-Thomson—Wikipedia and his books

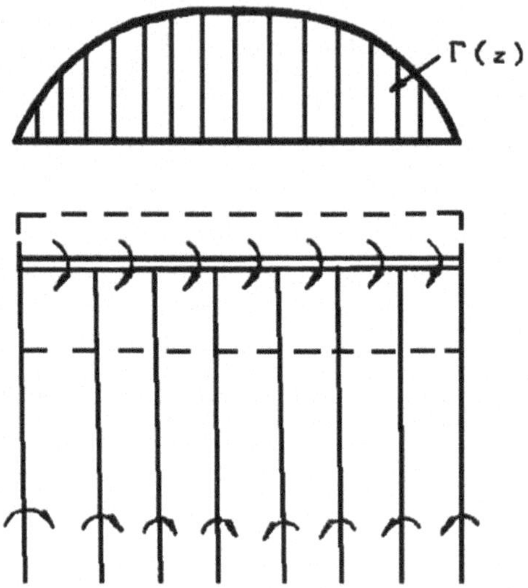

Fig. 8.11 Lifting line "Π" model

When the wing span is longer, the lifting line model can be used to describe the flow, i.e., assuming the attached vortex is concentrated at the ¼c point, and the attached vortices along the whole span are decreasing relative to the free vortices, and heading downstream, thus, the circulation of the attached vortices along the span is the variable $\Gamma(z)$..

The circulation difference of an attached vortex at a certain point is $d\Gamma(z)$, i.e., the circulation strength of a free vortex at z. This is the lifting line "Π" model of a limited span wing, as shown in Fig. 8.11. Then, we take the influence of the wing tip as the influence of the total free vortex flow, and the flow at every section of the wing can still be assumed as the two-dimensional flow to calculate the forces acting on the wing.

8.2.2 Downwash Caused by Free Vortices

According to the "Π" model of the lifting line of a limited span wing, the influence of the tip is converse to the influence of induced velocity by the free vortices on the wing surface, which is referred to as downwash velocity, because the direction of such velocity is downward in case of positive lift being generated.

We can calculate the downwash velocity at the attached vortices, induced by free vortices. First, we locate the coordinate origin at the intersection of attached vortices with the central span line. We take the x axis as the oncoming flow direction, y axis as the upward direction perpendicular to the oncoming flow, and z axis along the span direction, as shown in Fig. 8.12

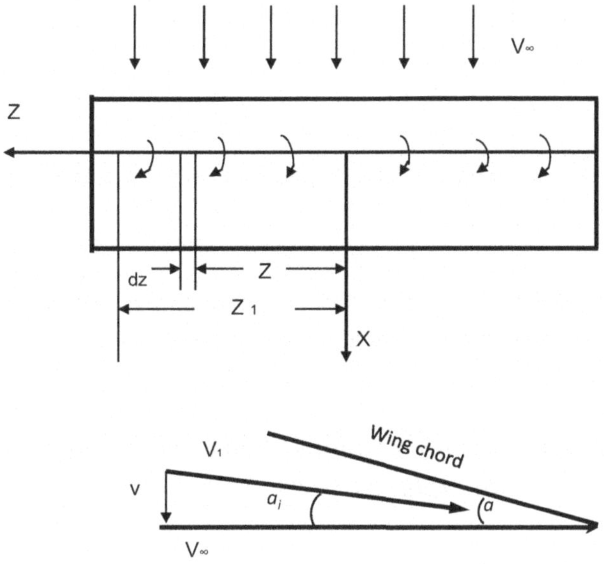

Fig. 8.12 Downwash caused by free vortices

Since the free vortices are semi-unlimited long vortices, the induced velocity at z, caused by free vortices at any point z_1 is:

$$dv = \frac{d\Gamma}{4\pi(z_1 - z)}$$

Where $d\Gamma$ free vortex strength at z_1.

The induced velocity at z, caused by the free vortices is:

$$v(z) = \int_{\frac{-l}{2}}^{\frac{l}{2}} \frac{\frac{d\Gamma}{dz_1}}{4\pi(z_1 - z)} dz_1 \tag{8.34}$$

While $z_1 \rightarrow z$ the integral function trends to infinity; however, it can be proved that when $z_1 = z$ the point of divergence can be avoided, and in principle the integration is convergent, thus the main integration of (8.34) can be achieved. Then the downwash velocity v can be obtained so long as the circulation distribution on the attached vortex of the wing Γ(z) is given. The resultant of downwash velocity v with oncoming velocity V_∞ changes the direction of oncoming flow, decreasing the angle of attack by the downwash angle, with value of:

$$\alpha_i = \text{arc} \quad \tan \frac{v}{V_\infty} \tag{8.35}$$

The change to oncoming velocity, caused by the downwash velocity is smaller than the downwash angle, and can be neglected.

The lift will be decreased due to the downwash flow (reduced effective angle of attack), and the lift coefficient will be changed to:

$$C_L = 2\pi(\alpha_c - \alpha_i)$$

In addition, reduced drag occurs and so reduced drag coefficient:

$$C_{Di} = C_L \cdot \alpha_i$$

Where α_c is the effective angle of attack and is equal to the geometric angle of attack minus angle of attack with zero lift, i.e., $\alpha_c = \alpha - \alpha_0$.

8.2.3 *Lift on a Limited Span Wing and Equation for Induced Drag*

The lift acting on a limited span wing and its induced drag can be obtained, so long as the distribution of attached vortices along the span is available. Research demonstrates that for a wing with elliptical plan profile, the circulation along the span is also elliptical in profile, i.e., where the wing plan configuration is:

$$\frac{c^2}{c_0^2} + \frac{z^2}{\left(\frac{l}{2}\right)^2} = 1$$

The circulation distribution on the wing is:

$$\frac{\Gamma^2}{\Gamma_0^2} + \frac{z^2}{\left(\frac{l}{2}\right)^2} = 1$$

where:

c chord length at any point of the wing,
c_0 max chord length at the center of span
l span
Γ circulation at any point of the wing
Γ_0 circulation at max span.

It can be demonstrated that the downwash velocity is equal at any section, irrespective of ordinate z, and obtaining the calculation result (here neglecting the complicated derivation), as:

$$v = -\frac{\Gamma_0}{2l}. \tag{8.36}$$

The corresponding downwash angle is

$$\alpha_i = -\frac{v}{V_\infty} = \frac{\Gamma_0}{2lV_\infty}. \tag{8.37}$$

Lift coefficient is then

$$C_L = 2\pi(\alpha_c - \alpha_i) = 2\pi\left(\alpha_c - \frac{\Gamma_0}{2lV_\infty}\right) \tag{8.38}$$

According to the Kutta-Zhoukovsky Theory, the lift coefficient can be written as:

$$C_L = \frac{\int_{\frac{-l}{2}}^{\frac{l}{2}} \rho V_\infty \Gamma dz}{\frac{1}{2}\rho V_\infty^2 S} = \frac{\pi \Gamma_0 l}{2V_\infty S}$$

Where S is the horizontal plane projected area of the wing. Thus, (8.37) may be rewritten as

$$\alpha_i = \frac{\Gamma_0}{2lV_\infty} = \frac{\pi l \Gamma_0}{2V_\infty S} \cdot \frac{S}{\pi l^2} = \frac{C_L}{\pi A} \tag{8.39}$$

And so

$$C_L = \frac{2\pi}{1 + \frac{2}{A}}(\alpha - \alpha_0) \tag{8.40}$$

The downwash angle changes the direction of the lift resultant and forms the induced drag in horizontal plane. The induced drag coefficient is

$$C_{Di} = C_L \cdot \alpha_i = C_L^2/\pi A. \tag{8.41}$$

Regarding a wing without elliptical plan profile, this can be solved upon the basis of the ellipse profile, and carrying out some corrections. The downwash angle and induced drag can be expressed as follows:

$$\alpha_i = \frac{C_L}{\pi A}(1 + \tau) \tag{8.42}$$

$$C_{Di} = \frac{C_L^2}{\pi A}(1 + \delta) \tag{8.43}$$

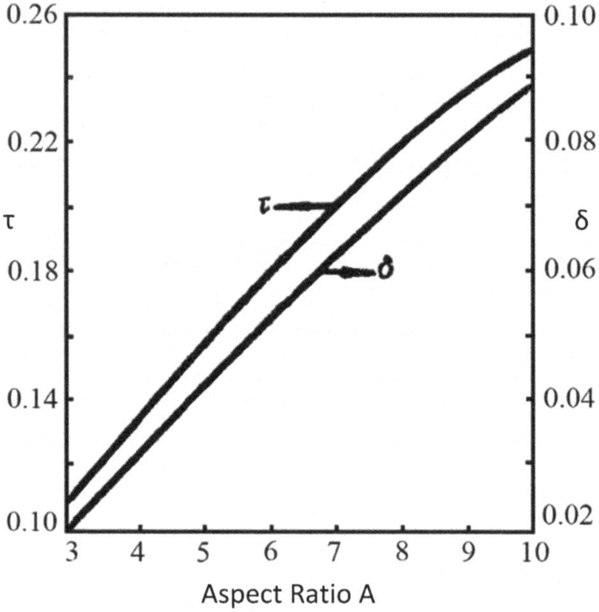

Fig. 8.13 Correction coefficients for rectangular wing

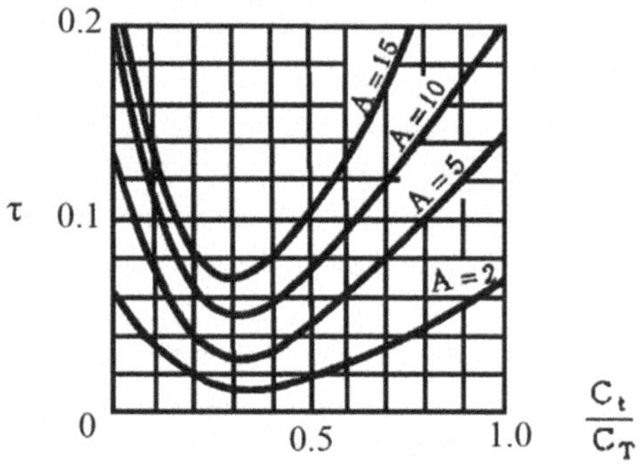

Fig. 8.14 Value of τ for the trapezium profile wing, as shown in Fig. 8.15

Correspondent to (8.42), the lift coefficient:

$$C_L = \frac{2\pi}{1 + \frac{2}{A}(1+\tau)}(\alpha - \alpha_0) \tag{8.44}$$

For a wing with rectangular plan profile, τ. and δ are a function of aspect ratio A, as shown in Fig. 8.13. For a wing with a trapezium plan profile as shown in Fig. 8.15, the τ. and δ are shown in Figs. 8.14 and 8.15, where τ. is shown in Fig. 8.14 directly, and δ value obtained by the following equation:

$$\delta = \delta' A$$

Where $\delta^{'}$ can be obtained from Fig. 8.15. The values τ and $\delta^{'}$ are a function of C_t/C_T, where C_t is chord at tip and C_T is max chord length.

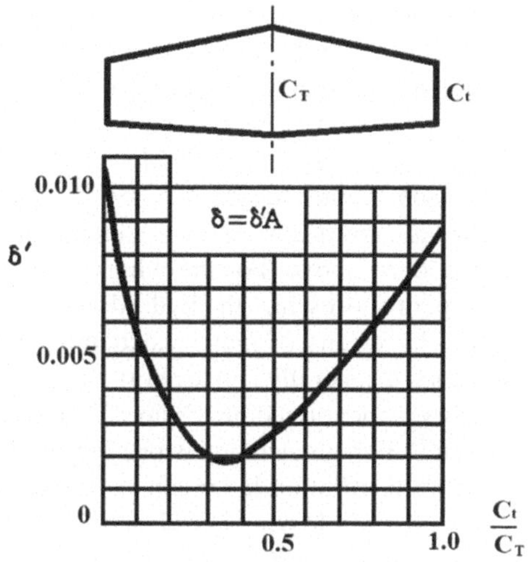

Fig. 8.15 Value of δ' for trapezium profile wing

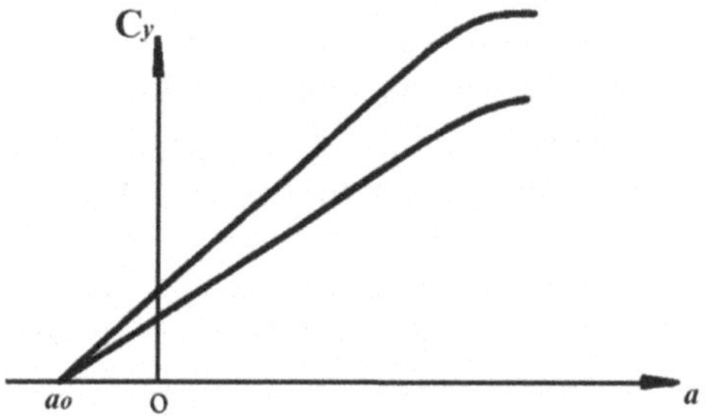

Fig. 8.16 Lift coefficient curves

8.2.4 Explanations

1. For a wing with a *small angle of attack*, i.e., in the range where the lift coefficient slope is linear, the slope for a limited span wing is:

$$\frac{dC_L}{d\alpha} = \frac{2\pi}{1 + \frac{2}{A}(1 + \tau)} < 2\pi \tag{8.45}$$

This is as shown in Fig. 8.16. The lift coefficient slope decreases with aspect ratio.

In the case where $\frac{dC_L}{d\alpha}$ decreases, the correspondent wing efficiency also decreases. When the aspect ratio is constant, the efficiency for the wing with elliptical plane shape is highest, due to the downwash velocity being equal at any point across the span, so generating minimum induced drag.

2. The calculation of influence of *limited wing span* by means of the lifting line model can only be applied to a thin wing with high aspect ratio, as it assumes the influence of span attributing to the downwash is induced by free vortices, and taking the flow at every section as two dimensional, but neglecting the influence of transverse flow. The whole calculation is limited to linear theory. In general, the aspect ratio must be $A \geq 4 - 5$ or at least $A \geq 3$. When $A < 3$, the transverse flow is significant, and will create errors when using the lifting line theory mentioned above. For the wings with $1 \leq A \leq 3$ the so-called low aspect ratio wing, or short wing, the hydrodynamic characteristics can be calculated successfully by means of many calculation models, and its correspondent processing methods, generally called short wing theory, one of which is the "transverse drag theory," mentioned in Chap. 2, and also the Weissinger method, mentioned there.[10] For greater precision, the model of a two-dimensional thin wing with vortex distribution layer can be applied, so as to distribute the attached vortices along the whole wing chord, and free vortex at every point of wing surface, the so called "lifting surface theory" model. It is more complicated, however, for the short wing with lower aspect ratio, especially at $A \approx 1$ one is obliged to use this method. For a wing with extremely short span $A << 1$, a so called super short wing, can be calculated by Jones' slender wing theory, introduced in Chap. 2, Sect. 2.4, here the transverse flow becomes the principal flow, and the longitudinal flow is small, and can be neglected.
3. For a *high aspect ratio wing*, lifting line theory can be further simplified, i.e., assuming the free vortices only reduce at the tips, and keeping the attached vortex strength across the whole span constant, thus, no free vortex reduction, and forming a simplified lifting line model with so called "single Π vortex" model, as shown in Fig. 8.17. Since the vortex strength here is constant, the boundary condition can be kept, i.e., at point P, at ¾ Chord, of span centerline. Other points will not exactly meet the boundary condition. In spite of this, the method is useful for an estimate.
4. Limited span influences the wing efficiency, and the tips influence principally the flow from the positive pressure zone to the negative pressure zone. In order to reduce such effect, tip plates (sometimes called fences) can be mounted at the wing ends, for preventing such transverse circular flow, as shown in Fig. 8.18. In the case of wing with tip fences, the transverse flow, and downwash flow, will be decreased, and the downwash angle can be estimated approximately with the following equation, via a coefficient K_e.

$$\Delta\alpha = \frac{C_L}{\pi A} K_e \tag{8.46}$$

[10] See NASA report TN 149 Dec 1947 for practical discussion at 19930082544.pdf (nasa.gov) See also NASA Report 929 here https://ntrs.nasa.gov/citations/19930092031

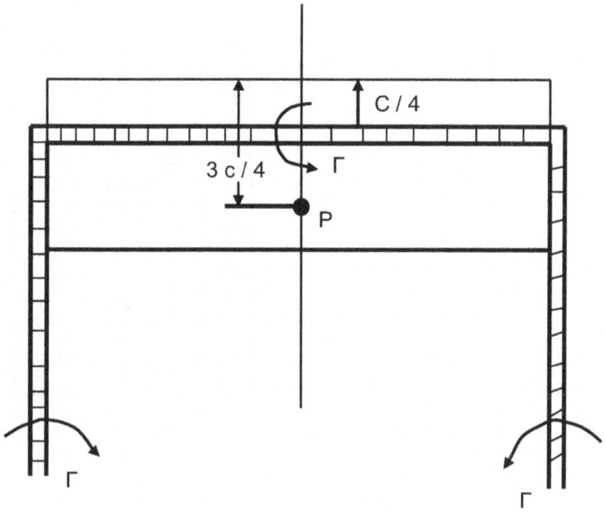

Fig. 8.17 Simplified "Π" vortex model

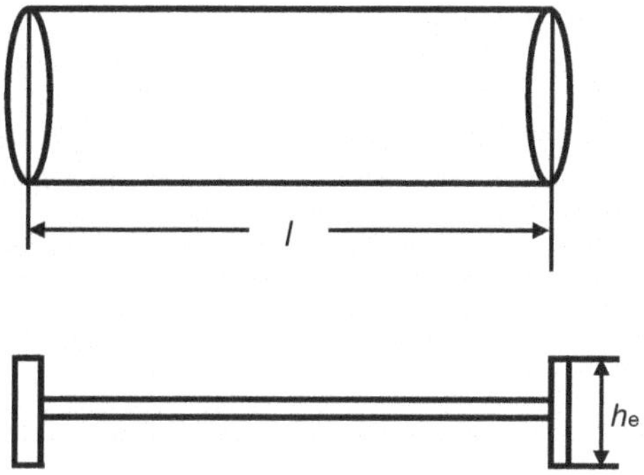

Fig. 8.18 Wing tip fences

Where K_e influence coefficient of tip, expressed as:

$$K_e = \frac{1}{1 + 2\frac{h_e}{l}} \tag{8.47}$$

Where

h_e tip height
l span length

For convenience, (8.47) can be expressed by a curve, as shown in Fig. 8.19.

Fig. 8.19 $K_e = f(h_e/l)$ curve

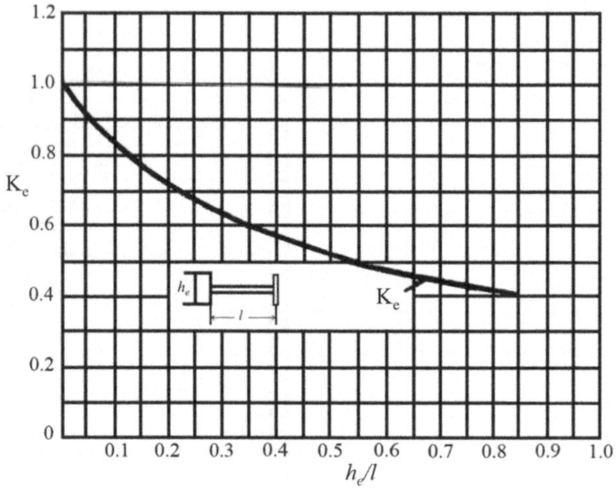

8.3 Effect of Wing Configuration on Hydrodynamic Performance

The wing geometry discussed above relates to the horizontal plane, and dihedral (upward or downward) of the wing. Several different configurations have been introduced in Chap. 7, and here we introduce the influence of sweepback and dihedral of a hydrofoil wing on its hydrodynamic performance.

8.3.1 Calculation of Hydrodynamic Performance of Swept-Back Wing

For an airplane wing, generally a swept back wing is only applied to aircraft with sub or supersonic speed, however for hydrofoil craft sweep back can defer cavitation on the hydrofoil surfaces.

1. Weissinger Method for Calculating Hydrodynamics of a Swept Back Foil

The lifting line method, applied to the limited span wing with rectangular or trapezoidal plan view, but without sweep back, due to that the downwash velocity at point z of the wing by means of lifting line theory gives the following relation:

$$C_L = \int_{-\frac{l}{2}}^{\frac{l}{2}} \frac{\frac{d\Gamma}{dz_1}}{4\pi(z_1 - z)} dz_1 \tag{8.34}$$

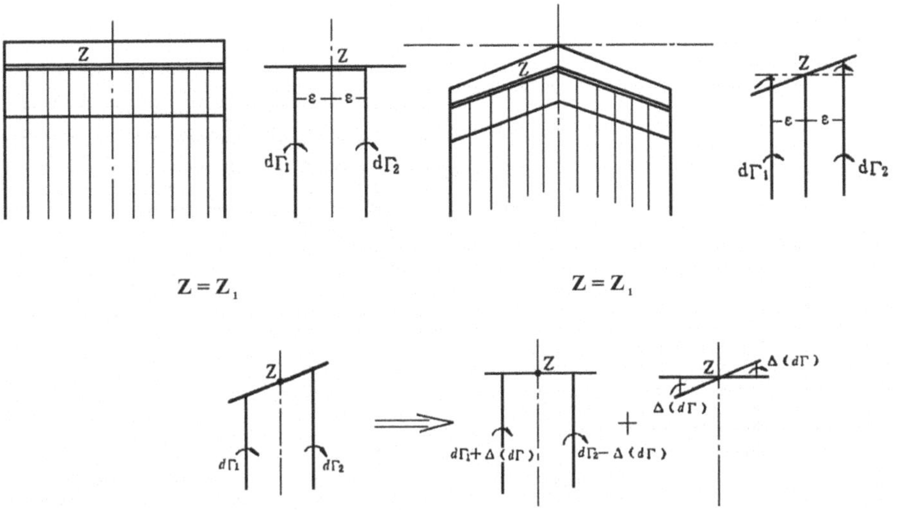

Fig. 8.20 Free vortex series on swept back wing

The point of divergence for this relation is at $z_1 = z$ where the integral value trends to infinity. However, for a flat and straight plan the point of divergence can be avoided, and the peak value of the integral restrained. Nevertheless, for a swept back wing, the integral peak value is un-restrained. Here we cannot prove the constraint analytically, but need to interpret it physically.

Let us consider a flat and straight wing taking z_1 as a single point of z with a distance of ε from the centerline. Then the direction of velocity induced by the two free vortices, at two points located a distance ε from z, is inverse, as shown in Fig. 8.20.

We assume the strength of the two free vortices is $d\Gamma_1$ and $d\Gamma_2$ respectively, when $\varepsilon \to 0$, $d\Gamma_1 = d\Gamma_2$, thus the induced velocity generated by the two free vortices is canceled out. This means that at $z_1 = z$, the induced velocity induced by a free vortex does not equal to infinity (trend to zero), i.e., may cancel the point of divergence mathematically.

However, for the swept back wing the situation is rather different. In this case, considering the field close to z at distance ε when $\varepsilon \to 0$, the velocity induced by $d\Gamma_1$ and $d\Gamma_2$ cannot be cancelled. Dividing $d\Gamma_1$ and $d\Gamma_2$ into two parts corresponding to "plan and straight part" plus "value difference," and when $\varepsilon \to 0$, the induced velocity of plan straight part is canceled but the induced velocity induced by "value difference" should be superposed, due to the inverse direction of the induced velocity. This value trends to infinity when $\varepsilon \to 0$, i.e., in mathematics, when $z_1 = z$, this point in the integral function of (8.34) is a point of divergence, and integration is indeterminate.

Weissinger presented a method similar to lifting line theory, i.e., took the attached vortex concentrated at the ¼ chord, and meeting the boundary condition of wing

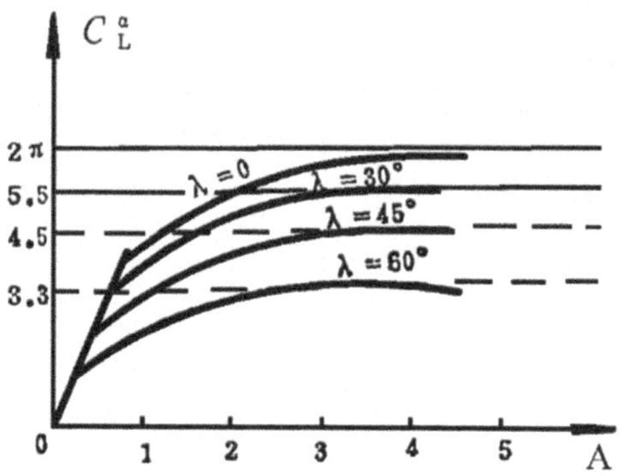

Fig. 8.21 Lift coefficient slope of swept back wing versus aspect ratio

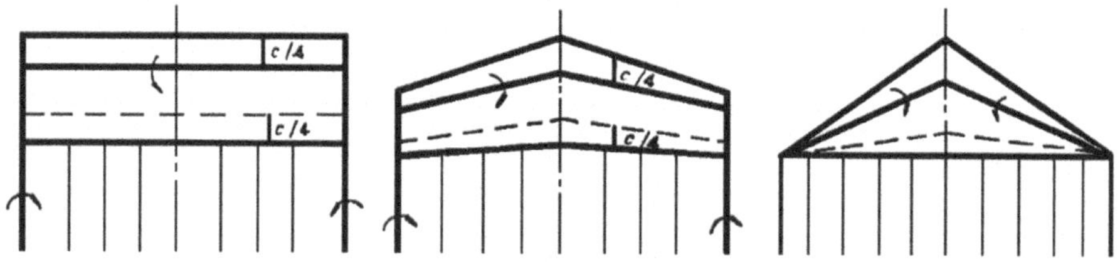

Fig. 8.22 Vortex lines distribution of wings with typical plan profiles

surface at ¾ chord, so as to avoid the peculiar integration of intersection of vortex lines.

Regarding different sweep back angle Λ, the lift coefficient curve slope, in terms of Weissinger method is shown in Fig. 8.21, where the sweep back angles are $\Lambda = 0$, 30, 45 and 60 degrees respectively. The ordinate represents the lift coefficient slope, $C_L^\alpha = \frac{dC_L}{d\alpha}$ while the abscissa represents the aspect ratio $A = l^2/S$. When $A \to \infty$ the corresponding values C_L^α are equal to 2π, 5.5, 4.5 and 3.3 respectively, thus, it can be seen that C_L^α apparently decreases inversely with the sweep back angle. It is easy to understand physically, since the oncoming flow to the swept back wing can be divided to two parts, i.e., perpendicular and parallel to the span, and only the velocity component perpendicular to the span, $V_\infty \cos\Lambda$, can generate the lift, while the velocity component parallel to the span, $V_\infty \sin\Lambda$, does not generate any lift.

The calculation result for limited span by means of the Weissinger method agrees with practical results, and also it can be applied to lower aspect ratio wings where $1.5 \leq A \leq 3$, where the general lifting line theory cannot be accurately applied. In addition, it is simpler compared with the lifting surface theory, and so is an effective method to calculate the hydrodynamic characteristics of a foil with low aspect ratio.

When using Weissinger method to calculate the wing with varying chord along the span, one can still use the attached vortex located at ¼-chord length, and meet the boundary condition at ¾-chord length. The distribution of vortices for three typical plan profile wings, and the locations meeting the boundary condition are shown in Fig. 8.22.

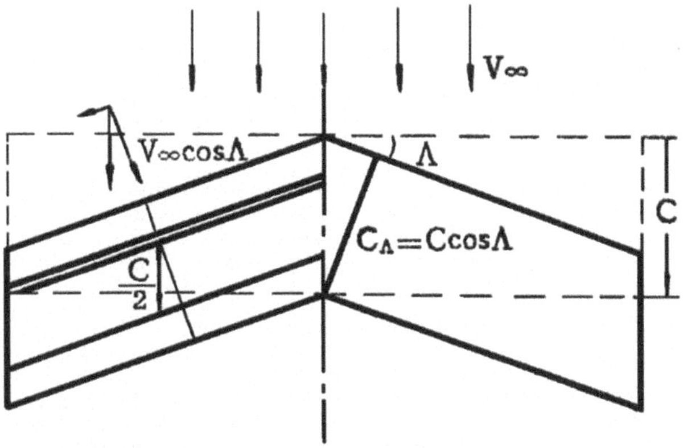

Fig. 8.23 Vortex geometry for swept back wing

2. Approximate Estimation of Sweep Back Effect.

The Weissinger method is actually too complicated for engineering practice, so a simple approximate estimation may be useful. Comparing the wing with sweep back angle Λ, with a rectangular wing, with same plan area, chord length, camber, and projected span length, the differences of hydrodynamic characteristics are as follows:

(a) Only the velocity component, perpendicular to the span provides lift, so the lift should be proportional to $V^2_\infty \cos^2 \Lambda$ while the lift of a rectangular wing is proportional to V^2_∞.
(b) It can be proved in terms of geometry that as the direction of oncoming flow changes from V_∞ to $V_\infty \cos \Lambda$, the geometrical angle of attack will be increased by $1/\cos \Lambda$ ratio to the angle of attack of a rectangular wing. Zero lift angle of attack also increases with the same multiple. According to the resultant of these two terms mentioned above, the lift of a swept back wing should be a multiple $\cos \Lambda$ of the rectangular wing.
(c) According to Weissinger's vortex model for a swept back wing the distance to ¾-chord is shorter than rectangular wing by the multiple $\cos \Lambda$, as shown in Fig. 8.23. However, for the same wing configuration the boundary condition for both swept back and rectangular wings is the same, i.e., the induced velocity in the normal direction and at that point is the same. Since the distance is shortened by $\cos \Lambda$ the attached vortex should be decreased by $\cos \Lambda$. As the length of the attached vortex of the swept back wing is greater than that of rectangular plan by $1/\cos \Lambda$ the overall strength of attached vortex is the same for both, but the local vortex strength of swept back wing is smaller than rectangular wing by the factor $\cos \Lambda$.

Integrating the effects of the three terms mentioned above, the lift coefficient of the swept back wing may be written as (refer also back to 8.40 and 8.44):

$$C_L = 2\pi\left(\alpha_e - \frac{C_L}{\pi A}\cos \Lambda\right)\cos \Lambda$$

Where $\alpha_e = \alpha - \alpha_0$ is the effective angle of attack, i.e., the sum of the nominal geometric angle of attack, (angle of attack of span centerline section) plus zero lift angle of attack.

Transforming the above equation, we obtain:

$$C_L = \frac{2\pi}{\frac{1}{\cos\Lambda} + \frac{2}{A}\cos\Lambda}\alpha_e \tag{8.48}$$

The slope of the lift coefficient:

$$C_L^\alpha = \frac{dC_L}{d\alpha} = \frac{2\pi A\cos\Lambda}{A + 2\cos^2\Lambda} \tag{8.49}$$

Induced drag coefficient:

$$C_{Di} = \frac{C_L^2}{\pi A}\cos\Lambda \tag{8.50}$$

Using (8.49) and calculating the two-dimensional swept back wing with unlimited span length, and $\Lambda = 30\,°$, 45° and 60 °, then:

$$C_L^\alpha = 2\pi\cos 30\,° \approx 5.45$$

$$C_L^\alpha = 2\pi\cos 45\,° \approx 4.45$$

$$C_L^\alpha = 2\pi\cos 60\,° \approx 3.14$$

Which is close to the calculation by the Weissinger method, as shown in Fig. 8.21. Another approximation can be expressed as:

$$C_L^\alpha = \frac{2\pi A}{2 + \sqrt{4 + \left(\frac{A}{\cos\Lambda}\right)^2}} \tag{8.51}$$

Or in a more simplified expression as:

$$C_L^\alpha = \frac{2\pi}{1 + \frac{2}{A}}(1.1\cos\Lambda) \tag{8.52}$$

In some practical engineering calculations, for greater simplification one can take the projected area perpendicular to the oncoming flow of swept back wing as the calculation area, keeping the other factors unchanged. Then the lift and induced drag can be determined using:

$$L = C_L \frac{1}{2} \rho V_\infty^2 S \cos \Lambda \tag{8.53}$$

$$D_i = C_{D_i} \frac{1}{2} \rho V_\infty^2 S \cos \Lambda \tag{8.54}$$

where:

C_L and C_{D_i} are the lift and induced drag coefficients of a straight flat wing with corresponding sections, and
V_∞ real oncoming flow velocity
S real projected area of swept back wing.

In general, a hydrofoil sweep will be less than 30 degrees and so such an estimate should be realistic.

8.3.2 Effect of Wing Dihedral

In hydrofoil craft, where wings with dihedral are used, the normal dihedral (V geometry) wing is applied for obtaining fine transverse stability for surface piercing foils, and anhedral (inverse V) wing for avoiding the emersion of foil tips from the water surface during heeling of fully submerged foils. The dihedral angle is expressed by β, the dihedral is $\beta > 0$ and anhedral is $\beta < 0$, as shown in Fig. 8.24.

For a wing with dihedral, lifting line or lifting surface theory can be applied for hydrodynamic calculation, and there are no differences between the dihedral and flat plane wings, just a rather complicated calculation due to the dihedral. The details

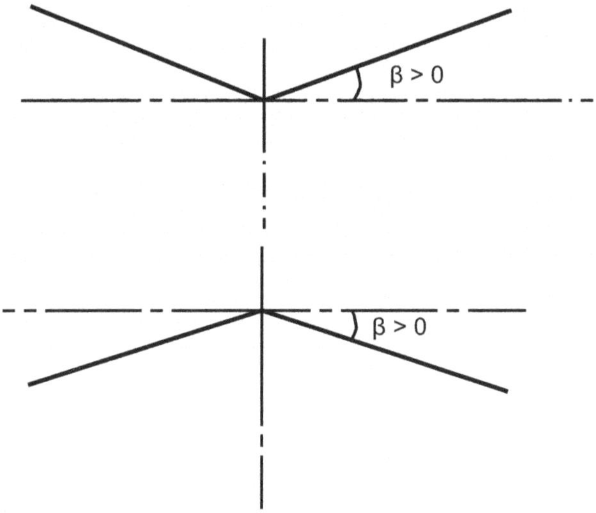

Fig. 8.24 Hydrofoil "dihedral"

will not be discussed here but we would like to introduce an approximate estimation for a dihedral wing, for use in engineering practice.

The influence of dihedral concerns two factors as follows:

1. the lift of a dihedral wing acts perpendicular to the wing surface; thus, the real lift is the component of lift on the vertical plane, i.e. times $\cos\beta$ to the total lift. The relation is to the wing area, i.e., taking the projected area on horizontal plane as the calculation area for the lift.
2. "dihedral" changes the geometric angle of attack of the wing relevant to oncoming flow, i.e., the real angle of attack is equal to the nominal angle of attack multiplied by $\cos\beta$:

$$\alpha_c = \alpha\, cos\ \beta$$

Where:

α_c effective angle of attack, i.e., calculated angle of attack
α the geometric angle of attack at the center of wing span

Then the lift coefficient of a two-dimensional wing with dihedral angle β is:

$$C_{LB} = 2\pi(\alpha\, cos\, \beta - \alpha_0) \tag{8.55}$$

where:

α_0 zero lift angle of attack, relevant to the characteristics of wing sections, but irrelevant to the dihedral angle.

For three-dimensional wings, the calculation for the velocity induced by the free vortices is rather complicated, as the velocity induced by free vortices on each side of the wing surface is different. The downwash velocity of a wing with dihedral is that of a flat wing multiplied by $\cos\beta$. The lift coefficient can be calculated approximately as $C_{L\beta} \approx C_L \cos\beta$, and the downwash angle as:

$$\Delta\alpha = \frac{C_{L\beta}}{\pi A} \tag{8.56}$$

Since the downwash angle is relative to the geometric angle of attack, and it must be reduced by downwash angle of attack, the lift coefficient of a three-dimensional wing with aspect ratio A, and dihedral angle β is:

$$C_{L\beta} = 2\pi\left[\left(\alpha - \frac{C_{L\beta}}{\pi A}\right) cos\ \beta - \alpha_0\right] \tag{8.57}$$

which can be rewritten as:

$$C_{L\beta} = \frac{2\pi}{\frac{1}{\cos\beta} + \frac{2}{A}} \left(\alpha - \frac{\alpha_0}{\cos\beta} \right) \tag{8.58}$$

Equation (8.58) does not consider the influence of plan profile, so the equation may be rewritten as follows, considering the correction of plan profile:

$$C_{L\beta} = \frac{2\pi}{\frac{1}{\cos\beta} + \frac{2}{A}(1+\tau)} \left(\alpha - \frac{\alpha_0}{\cos\beta} \right) \tag{8.59}$$

Correspondent lift:

$$\mathrm{L} = C_{L\beta} \frac{\rho}{2} V_\infty^2 S\cos\beta \tag{8.60}$$

Coefficient of induced drag:

$$\mathrm{C}_{Di} = \frac{C_{L\beta}^2}{\pi A}(1+\delta) \tag{8.61}$$

Induced drag:

$$D_i = C_{Di} \frac{\rho}{2} V_\infty^2 S \tag{8.62}$$

where

τ see Figs. 8.13, 8.14 and 8.15
δ see Figs. 8.13, 8.14 and 8.15
S Area of hydrofoil wing in its plane.

8.4 Experimental Reference Data for Wing Sections

The presentation concerning the characteristics of wing sections and plans above assumes the wings are in ideal flow, and does not consider the influence of viscosity of the flow of a real fluid. When including viscosity, the drag, including friction and form drag, the so-called profile drag, and the lift all need to be addressed. Therefore, in order to obtain precisely the hydrodynamic characteristics, the best method may be model experiments in a wind tunnel, or taking the publicly available experimental results of wing sections in wind tunnel.

Thanks to the earlier and wider development of the aviation industry, far more than the hydrofoil, the data from reference test series of aircraft wings can be used as a starting point for hydrofoils such as [1, 2]. There are also some data directly available for hydrofoil forms as we present below.

However, the difference between airfoils and hydrofoils is "Cavitation," which may occur on hydrofoils at high speed. Thus when using the reference data of well proven aircraft wing sections, in addition to the nice hydrodynamic properties, i.e., high lift drag ratio, measures for avoiding cavitation have to be considered carefully. (see next chapter in detail). An example is the uniform distribution of negative pressure on back surface of wing avoiding a high suction peak.

We give below some data as reference material, taken from publicly available materials.

8.4.1 *Some Test References for Airfoils Frequently Used for Hydrofoils*

8.4.1.1 NACA Series Test Results on Aircraft Wings, USA

NACA [1] presented series test results of wing sections, including a four-digit series, five-digit series, NACA1 series, NACA6 series, NACA7 series, and other special series of geometrical forms.

The NACA four-digit series uses four digits to represent wing profile parameters, e.g., NACA2415. The first digit presents the percentage of maximum relative camber, and in this example $\overline{f}=2\%$. The second digit presents the percent of chord that maximum camber located from the leading edge as a percentage, e.g., in this item is 40%. The last two digits present the percent of maximum relative thickness, and in this item $\overline{t}=15\%$.

In NACA five-digit series, e.g., NACA23012, the first digit still presents relative camber, e.g., $\overline{f}=2\%$. second and third digits present the relative distance of maximum camber location from the leading edge, and times two, in this example, is 15% and last two digits present maximum relative thickness, $\overline{t}=12\%$. The best of this series is NACA23012, which has been applied widely.

NACA6 is also represented by five digits, e.g., NACA63-206. The first digit represents 6 series, second digit represents the percentage of distance of the minimum pressure point located from leading edge, in terms of percent of thickness at zero angle of attack; in this example, 30%. The digits after the small bar represent the designed lift coefficient times ten, e.g., $C_L=0.2$ in this example. The last two digits represent the percentage of relative thickness, e.g., $\overline{t}=6\%$. Sometimes a subscript is available after the second digit, e.g., NACA65_3 — 218. The subscript denotes the range of permissible positive/negative change on lift coefficient, and times ten, and at this change range, the pressure slope at upper and lower surface is located at good range. In this example, the permissible change range of lift coefficient, ±30%.

Sometimes, the letter A is inserted instead of short bar, e.g., NACA64A210, denotes at 0.8 chord length to trailing edge of upper and lower wing surface is a straight line. In addition, a subscript α added after the digit, e.g.,

NACA65$_1$ — 212 $\alpha = 0.6$, denotes the load is distributed uniformly from the leading edge to 0.6c, and after 0.6c to trailing edge, the pressure drops down to zero in a straight line. In case of no subscript α denotes the load from leading to trailing edge is uniform. The hydrodynamic property of the 6 series is rather smooth with minimum pressure point moving afterward and is favorable to deferring cavitation, and so application to the hydrofoil craft.

As an example of properties, Table 8.1 gives the offsets of NACA64-208 wing section profile, and Fig. 8.25a, b shows the correspondent lift, moment, and drag coefficient. Figure 8.25a shows the lift coefficient curve, $C_L = f(\alpha)$, and moment coefficient curve $C_m = f(\alpha)$. about the point at ¼ c, where abscissa represents angle of attack α, in degrees, Fig. 8.25b shows drag coefficient $C_D = f(C_L)$, and the moment coefficient $C_{mo} = f(C_L)$ about the hydrodynamic center (a, c). The tests were carried out for three Reynolds number $R_n = 3 \times 10^6$, 6×10^6, and 9×10^6 for

Table 8.1 Properties of the NACA 64-208 wing section

NACA64-208			
Upper surface		Lower surface	
$\frac{x}{c}\%$	$\frac{y_u}{c}\%$	$\frac{x}{c}\%$	$\frac{y_l}{c}\%$
0	0	0	0
0.445	0.706	0.555	−0.606
0.688	0.862	0.812	−0.722
1.180	1.110	1.320	−0.896
2.421	1.549	2.579	−1.177
4.912	2.189	5.088	−1.557
7.410	2.681	7.590	−1.833
9.909	3.089	10.091	−2.055
14.915	3.741	15.085	−2.395
19.924	4.232	20.076	−2.640
24.935	4.598	25.065	−2.808
29.948	4.856	30.052	−2.912
34.961	5.009	35.039	−2.949
39.974	5.063	40.026	−2.921
44.988	4.978	45.012	−2.788
50.000	4.787	50.000	−2.581
55.011	4.506	54.989	−2.316
60.020	4.152	59.980	−2.010
65.027	3.733	64.973	−1.673
70.031	3.263	69.969	−1.319
75.032	2.749	74.968	−0.959
80.031	2.200	79.969	−0.628
85.027	1.634	84.973	−0.028
90.019	1.069	89.981	−0.033
95.010	0.522	94.990	0.110
100.00	0	100.00	0

Leading edge (L.E) radius:0.455% chord length
Slope of radius through L.E: 0.084

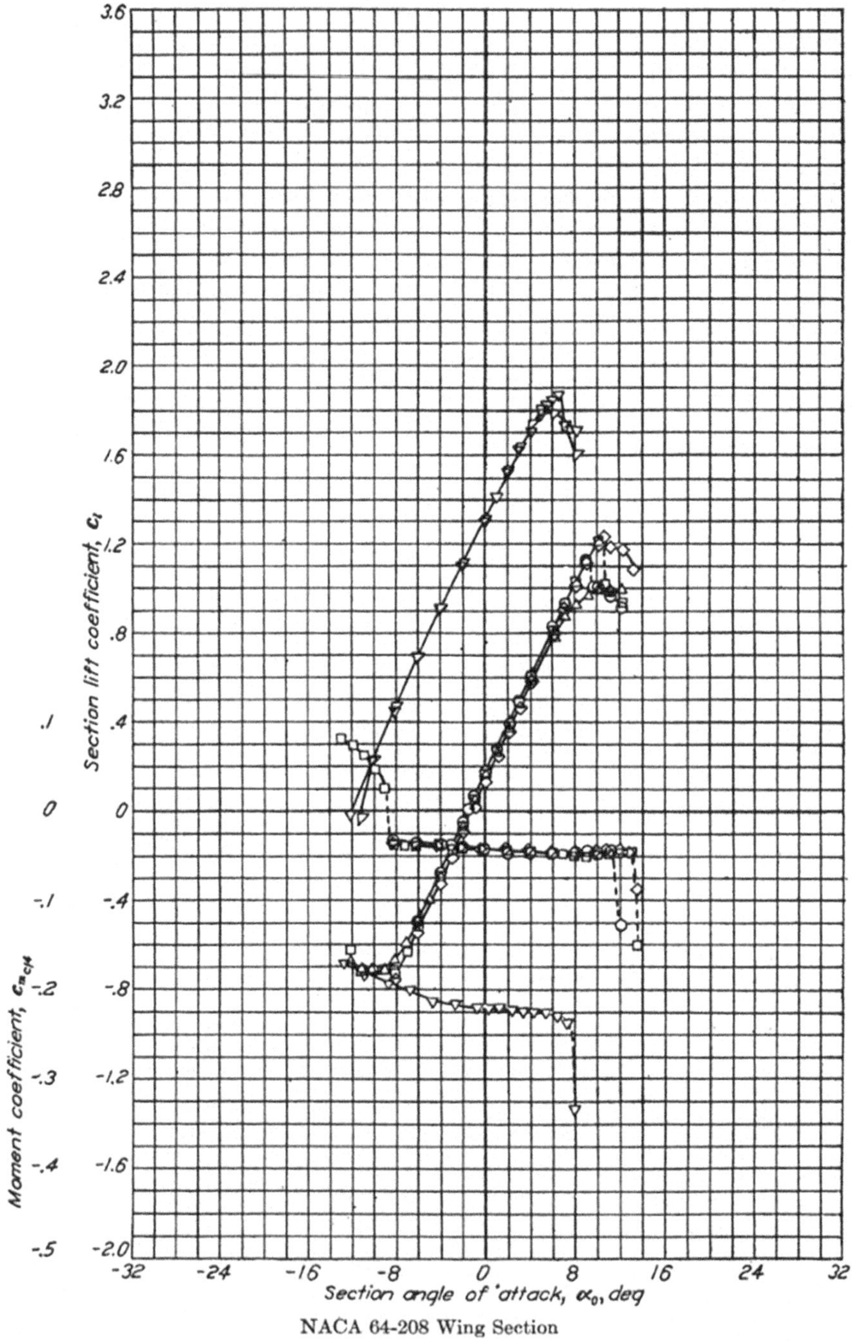

Fig. 8.25a NACA64-208 wing section characteristics lift

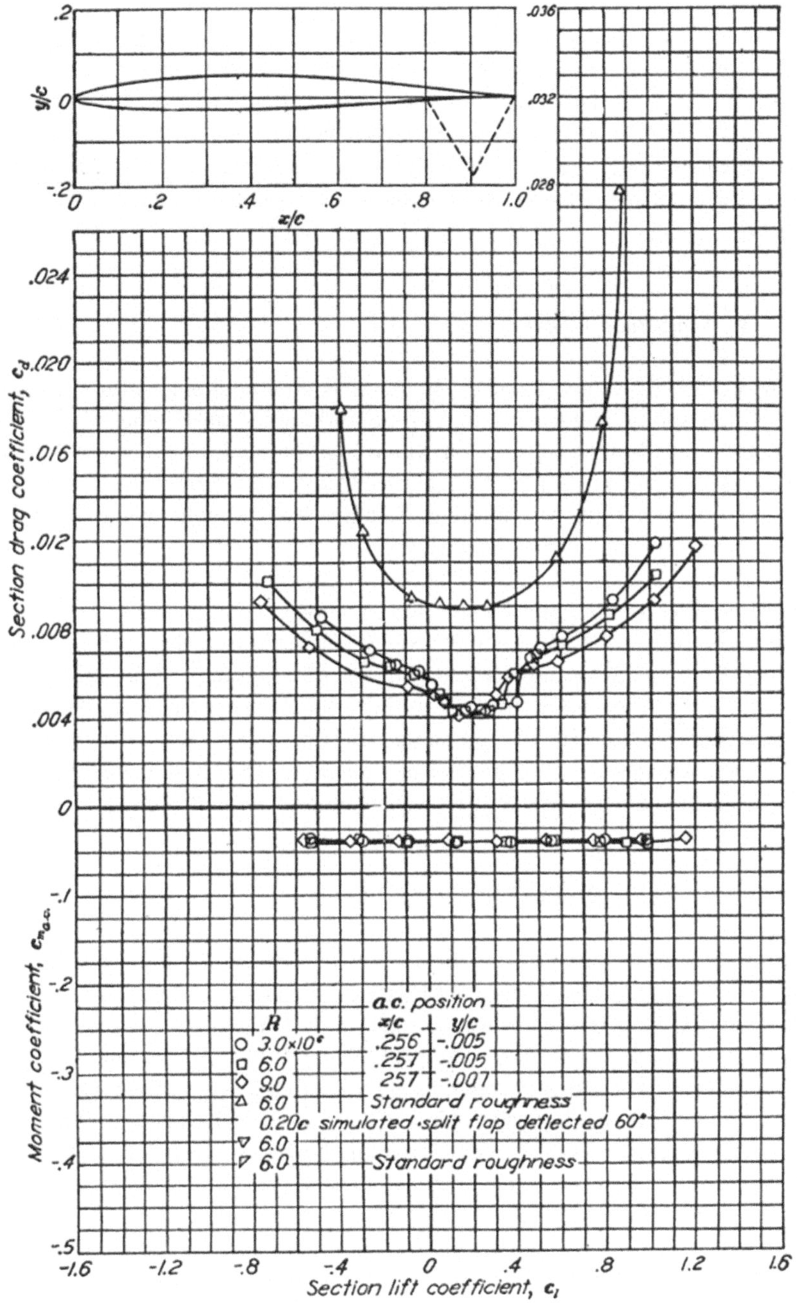

NACA 64-208 Wing Section (*Continued*)

Fig. 8.25b NACA64-208 wing section characteristics drag

smooth wing surface. In addition, tests at $R_n = 6 \times 10^6$ with wing surface standard roughness were also carried out. Standard roughness implies the diameter of small spherical particles located nearby the leading edge at small range is 0.28 mm, when chord length, $c = 61$ cm. Since the influence of roughness on the hydrofoil is significantly greater than that on the air wing, it is suggested, that the test results have to be taken, under the standard roughness. Figure 8.25a, b also shows the test results of the wing with a flap with inclination angle of 60° located at 0.2c from the trailing edge.

8.4.1.2 E Wing Profile Series

For a hydrofoil, resistance against cavitation is an important factor. In order to improve such ability, R Eppler and Y T Shen designed an E wing profile series, and YS wing profile series, [6–9], on which high resistance against cavitation has been proven. Among them, the E836 and E838 are symmetric wing profiles. Their offsets are shown in Table 8.2, and Figs. 8.26 and 8.27 show the wing sections, and their correspondent cavitation characteristics.

The thickness ratio of E836, $\bar{t} = 12.66\%$, E838 $\bar{t} = 18.42\%$.

E874 and YS920 wing profiles are two asymmetric section wing profiles, E874 with thickness ratio $\bar{t} = 7.91\%$. while YS-920 has thickness ratio $\bar{t} = 9\%$. Table 8.3 gives the offsets of E874 and YS-920 wing profile. Figure 8.28 presents the wing profile and cavitation characteristics of E874 wing profile; Fig. 8.29 shows the cavitation characteristics of YS-920 and a comparison with two NACA wing profiles. One can see that the cavitation characteristic of YS-920 is better than NACA wing profile. Cavitation characteristics will be introduced in Chap. 9.

8.4.1.3 Other Wing Profile Series [10]

Apart from the above wing profile series, there are some profile series applied for the hydrofoil craft such as UK RAF(or RAE) series, the German Gottingen profile, and the Russian ЧАГИ or ВИГМ series (CHAGI or WIGM).

Figure 8.30 shows the RAF-6 wing profile, and Table 8.4 presents its offsets. It can be seen that the features of the wing section are a lower surface which is a flat and straight line with leading edge radius of $R/y_{max} = 0.1$, radius of trailing edge $R/y_{max} = 0.08$, where y_{max} is the profile maximum thickness. The center of section area $x_g = 0.37c$ and $y_g = 0.41y_{max}$. The section area is $0.74\,cy_{max}$ and the inertia moment about the coordinate origin is $J_0 = 0.0472{y_{max}}^3$.

Table 8.5 above gives the offsets of symmetric section profile of RAE101. Table 8.6 presents the offsets of Gottingen wing profiles, where 443 is a symmetric wing profile while 490, 623 and Munk6 are asymmetric wing profiles. When the trailing profile is hollow, such as Gottingen 490 and 623, the chord coordinates can be given as offsets from a base line as shown in Fig. 8.31.

Table 8.2 Offsets of E836 and E838 wing profiles

E836, 12.66%				E838, 18.42%			
x/c/%	y_u/c/%	x/c/%	y_l/c/%	x/c/%	y_u/c/%	x/c/%	y_l/c/%
0	0	0	0	0	0	0	0
0.307	0.560	0.307	−0.560	0.322	0.757	0.322	−0.757
1.176	1.218	1.176	−1.218	1.225	1.631	1.225	−1.631
2.539	1.903	2.539	−1.903	2.626	2.549	2.626	−2.549
4.376	2.584	4.376	−2.584	4.498	3.480	4.498	−3.480
6.669	3.241	6.669	−3.241	6.816	4.400	6.816	−4.400
9.397	3.858	9.397	−3.858	9.550	5.289	9.550	−5.289
12.535	4.423	12.535	−4.423	12.667	6.131	12.667	−6.131
16.049	4.929	16.049	−4.929	16.131	6.905	16.131	−6.905
19.906	5.365	19.906	−5.365	19.902	7.595	19.902	−7.595
24.064	5.727	24.064	−5.727	23.938	8.185	23.938	−8.185
28.480	6.009	28.480	−6.009	28.195	8.659	28.195	−8.659
33.106	6.205	33.106	−6.205	32.624	8.999	32.624	−8.999
37.891	6.310	37.891	−6.310	37.177	9.184	37.177	−9.184
42.784	6.320	42.784	−6.320	41.818	9.182	41.818	−9.182
47.730	6.223	47.730	−6.223	46.524	8.979	46.524	−8.979
52.684	6.006	52.684	−6.006	51.277	8.570	51.277	−8.570
57.615	5.662	57.615	−5.662	56.059	7.962	56.059	−7.962
62.490	5.200	62.490	−5.200	60.856	7.173	60.856	−7.173
67.277	4.633	67.277	−4.633	65.663	6.246	65.663	−6.246
71.945	3.983	71.945	−3.983	70.444	5.256	70.444	−5.256
76.465	3.286	76.465	−3.286	75.130	4.270	75.130	−4.270
80.789	2.599	80.789	−2.599	79.642	3.336	79.642	−3.336
84.843	1.966	84.843	−1.966	83.896	2.494	83.896	−2.494
88.551	1.413	88.551	−1.413	87.806	1.770	87.806	−1.770
91.838	0.956	91.838	−0.956	91.289	1.179	91.289	−1.179
94.634	0.593	94.634	−0.593	94.263	0.719	94.263	−0.719
96.891	0.312	96.891	−0.312	96.671	0.372	96.671	−0.372
98.576	0.118	98.576	−0.118	98.474	0.138	98.474	−0.138
99.636	0.022	99.636	−0.022	99.609	0.026	99.609	−0.026
100.000	0	100.000	0	100.000	0	100.000	0
$C_{mo} = 0.0000$				$C_{mo} = 0.0000$			

8.4.2 *Correction to Lift Coefficient and Approximate Estimation of Viscous Drag*

In order to estimate the hydrodynamic characteristics of a hydrofoil before formally selecting the wing profile, an approximate method is often used.

Many analyses demonstrate that use of Eq. (8.52) for the thin wing lift coefficients, in terms of wing theory mentioned above, give results a little bit

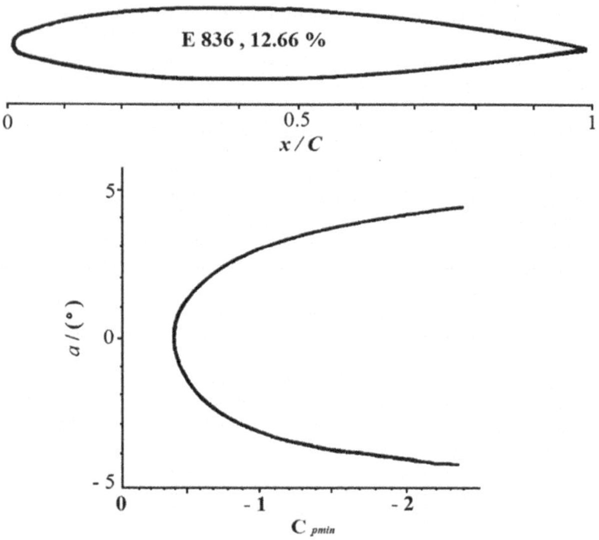

Fig. 8.26 E836 wing profile and its cavitation characteristics

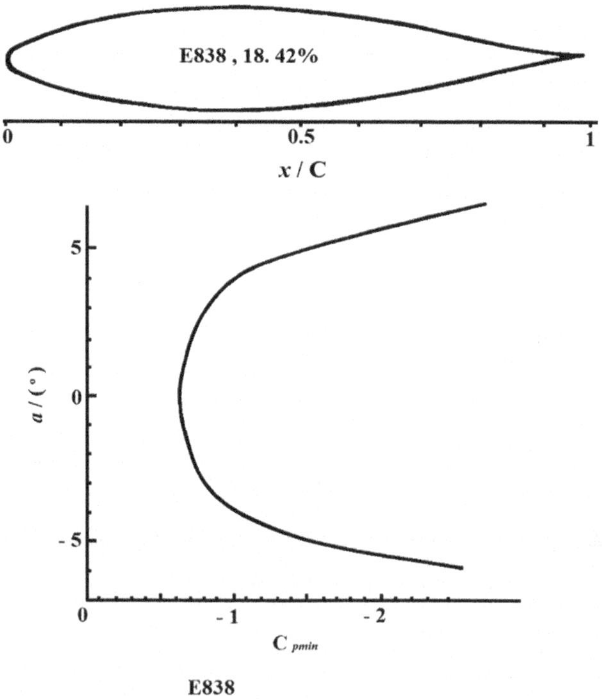

Fig. 8.27 E838 wing profile and its cavitation characteristics

Table 8.3 Offsets of E874 and YS-920 wing profile

E874, 7.91%				YS-920, 9%			
x/c/%	$y_u/c/\%$	*x/c/%*	$y_l/c/\%$	*x/c/%*	$y_u/c/\%$	*x/c/%*	$y_l/c/\%$
0.019	0.097	0.023	−0.102	0.009	0.064	0.040	−0.129
0.180	0.311	0.191	−0.315	0.152	0.274	0.240	−0.344
0.476	0.550	0.490	−0.547	0.429	0.514	0.570	−0.576
0.900	0.803	0.916	−0.789	0.836	0.768	1.030	−0.812
1.450	1.069	1.465	−1.038	1.374	1.031	1.620	−1.054
2.121	1.342	2.131	−1.288	2.039	1.304	2.335	−1.298
2.910	1.620	2.913	−1.535	2.829	1.583	3.172	−1.543
3.814	1.901	3.807	−1.777	3.740	1.866	4.126	−1.785
4.831	2.182	4.811	−2.009	4.769	2.152	5.195	−2.023
5.958	2.460	5.920	−2.226	5.914	2.437	6.375	−2.253
7.191	2.733	7.133	−2.425	7.171	2.722	7.662	−2.475
8.527	2.999	8.447	−2.596	8.537	3.003	9.053	−2.686
9.962	3.257	9.870	−2.732	10.008	3.280	10.544	−2.883
11.493	3.504	11.409	−2.842	11.580	3.550	12.129	−8.065
13.114	3.738	13.058	−2.929	13.250	3.814	13.905	−3.228
14.822	3.958	14.814	−2.996	15.012	4.069	15.567	−3.371
16.613	4.161	16.672	−3.045	16.863	4.313	17.408	−3.490
18.480	4.345	18.627	−3.076	18.797	4.548	19.325	−3.580
20.420	4.508	20.675	−3.091	20.809	4.770	21.310	−3.632
22.426	4.647	22.809	−3.090	22.894	4.979	23.375	−3.631
24.494	4.760	25.025	−3.076	25.046	5.175	25.535	−3.590
26.618	4.839	27.316	−3.049	27.260	5.356	27.782	−3.523
28.799	4.877	29.676	−3.010	29.531	5.521	30.109	−3.435
31.044	4.873	32.098	−2.961	31.851	5.670	32.509	−3.330
33.354	4.836	34.576	−2.902	34.215	5.803	34.973	−3.211
35.723	4.773	37.102	−2.834	36.617	5.918	37.494	−3.081
38.145	4.867	39.670	−2.759	39.050	6.014	40.063	−2.943
40.612	4.581	42.272	−2.678	41.508	6.092	42.673	−2.798
43.118	4.458	44.901	−2.592	43.985	6.152	45.315	−2.648
45.655	4.320	47.548	−2.500	46.474	6.191	47.980	−2.496
48.217	4.169	50.206	−2.406	48.968	6.211	50.661	−2.342
50.795	4.008	52.867	−2.308	51.461	6.211	53.348	−2.188
53.382	3.838	55.524	−2.208	53.946	6.191	56.034	−2.035
55.971	3.662	59.168	−2.108	56.416	6.150	58.709	−1.885
58.553	3.481	60.792	−2.006	58.866	6.089	61.364	−1.738
61.120	3.296	63.387	−1.905	61.288	6.006	63.993	−1.595
63.664	3.110	65.947	−1.804	63.677	5.903	66.585	−1.457
66.178	2.924	68.462	−1.704	66.025	5.778	69.133	−1.326
68.654	2.739	70.926	−1.606	68.327	5.631	71.628	−1.200
71.082	2.556	73.331	−1.509	70.575	5.462	74.063	−1.081
73.457	2.376	75.669	−1.415	72.765	5.271	76.429	−0.970

(continued)

Table 8.3 (continued)

E874, 7.91%				YS-920, 9%			
$x/c/\%$	$y_u/c/\%$	$x/c/\%$	$y_l/c/\%$	$x/c/\%$	$y_u/c/\%$	$x/c/\%$	$y_l/c/\%$
75.770	2.200	77.934	−1.323	74.890	5.054	78.719	−0.866
78.014	2.028	80.119	−1.233	76.943	4.812	80.925	−0.769
80.182	1.863	82.217	−1.145	78.915	4.532	83.039	−0.680
82.265	1.703	84.222	−1.060	80.826	4.197	85.056	−0.599
84.259	1.549	86.127	−0.977	82.699	3.831	86.968	−0.525
86.155	1.402	87.927	−0.897	84.522	3.459	88.768	−0.457
87.948	1.261	89.616	−0.818	86.285	3.087	90.450	−0.396
89.633	1.127	91.189	−0.740	87.976	2.725	92.009	−0.339
91.202	0.999	92.642	−0.663	89.587	2.377	93.438	−0.287
92.652	0.876	93.969	−0.587	91.107	2.048	94.733	−0.236
93.978	0.759	95.166	−0.509	92.527	1.739	95.891	−0.182
95.174	0.645	96.231	−0.428	93.838	1.452	96.917	−0.124
96.238	0.532	97.163	−0.338	95.031	1.184	97.811	−0.069
97.169	0.415	97.972	−0.240	96.101	0.932	98.568	−0.026
97.976	0.295	98.661	−0.147	97.053	0.692	99.178	0.001
98.663	0.183	99.223	−0.072	97.890	0.474	99.629	0.009
99.225	0.093	99.646	−0.025	98.607	0.290	99.906	0.005
99.696	0.034	99.910	−0.004	99.194	0.149	100.00	−0.000
99.910	0.006	100.000	−0.000	99.633	0.057		
100.000	0			99.907	0.012		
				100.000	0		
$C_{mo} = -0.0147$							

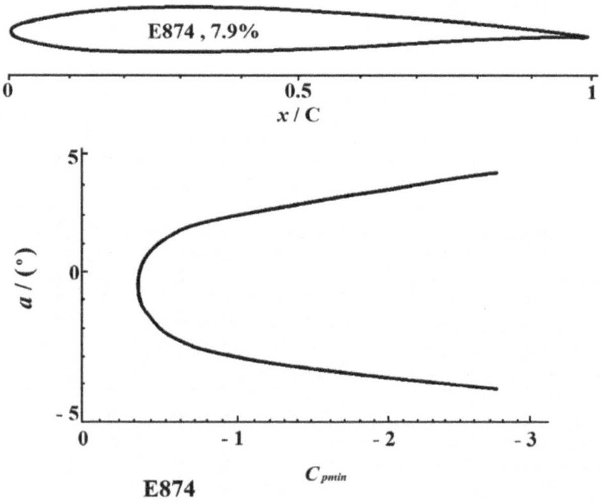

Fig. 8.28 E874 wing profile and its cavitation characteristics

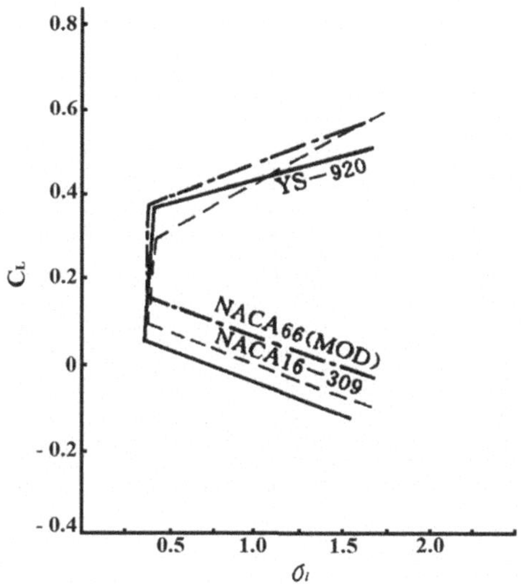

Fig. 8.29 YS-920 cavitation characteristics compare with NACA 16 and 66

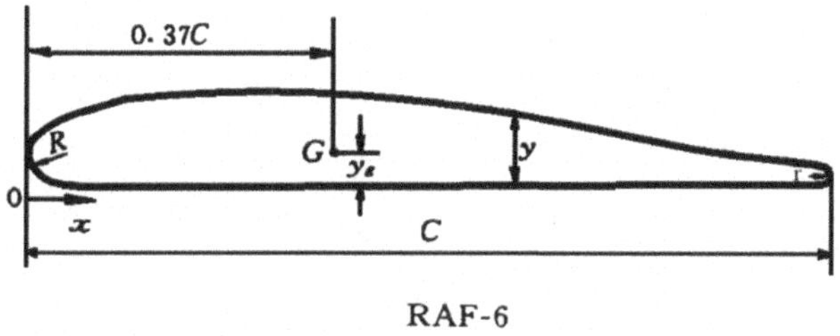

Fig. 8.30 RAF-6 wing profile

Table 8.4 offsets of RAF-6 airfoil section

x/c	0.00	0.025	0.05	0.10	0.20	0.30	0.40	0.50	0.60	0.70	0.80	0.90	1.00
y/y_{max}	0.1	0.41	0.59	0.73	0.95	0.998	0.99	0.95	0.87	0.74	0.56	0.35	0.08

Table 8.5 Offset of symmetric profile RAE101

x/*c*	100 *y*/*c*	*x*/*c*	100 *y*/*c*	*x*/*c*	100 *y*/*c*	*x*/*c*	100 *y*/*c*
0.0	0.0	0.07	3.0959	0.35	4.9531	0.75	2.2357
0.005	0.8705	0.08	3.2822	0.40	4.8005	0.80	1.7886
0.01	1.2265	0.09	3.4519	0.45	4.5657	0.85	1.3414
0.02	1.7215	0.10	3.6073	0.50	4.2670	0.90	0.8943
0.03	2.0924	0.15	4.2202	0.55	3.9183	0.95	0.4471
0.04	2.3974	0.20	4.6304	0.60	3.5310	0.98	0.1711
0.05	2.6593	0.25	4.8851	0.65	3.1154	1.0	0.0
0.06	2.8898	0.30	4.9969	0.70	2.6807		

Table 8.6 Gottingen wing profile

490			623		Munk6		443	
$x/c/\%$	$y/c/\%$	$y_l/c/\%$	$y_u/c/\%$	$y_l/c/\%$	$y_u/c/\%$	$y_l/c/\%$	$y_u/c/\%$	$y_l/c/\%$
0	2.0	2.0	3.25	3.25	0.00	0.00	0.00	0.00
1.25	3.60	0.85	5.45	1.95	1.98	−1.76	0.60	−0.60
2.50	4.60	0.50	6.45	1.50	2.81	−2.20	0.85	−0.85
5.0	5.95	0.15	7.90	0.90	4.03	−2.73	1.15	−1.15
7.5	7.00	0.00	9.05	0.35	4.94	−3.03	1.45	−1.45
10	7.70	0.00	9.90	0.20	5.71	−3.24	1.60	−1.60
15	8.65	0.20	10.95	0.10	6.82	−3.47	1.90	−1.90
20	9.20	0.40	11.55	0.05	7.55	−3.62	2.15	−2.15
30	9.60	0.95	12.00	0.00	8.22	−3.70	2.50	−2.50
40	9.05	0.80	11.70	0.00	8.05	−3.90	2.50	−2.50
50	8.55	0.80	10.65	0.00	7.26	−3.94	2.35	−2.35
60	7.45	0.60	9.15	0.00	6.03	−3.82	2.05	−2.05
70	6.05	0.40	7.35	0.00	4.58	−3.48	1.60	−1.60
80	4.40	0.15	5.15	0.00	3.06	−2.83	1.15	−1.15
90	2.50	0.00	2.80	0.00	1.55	−1.77	0.65	−0.65
95	1.45	0.05	1.60	0.00	0.88	−1.08	0.30	−0.30
100	0.15	0.15	0.30	0.00	0.00	0.00	0.00	0.00

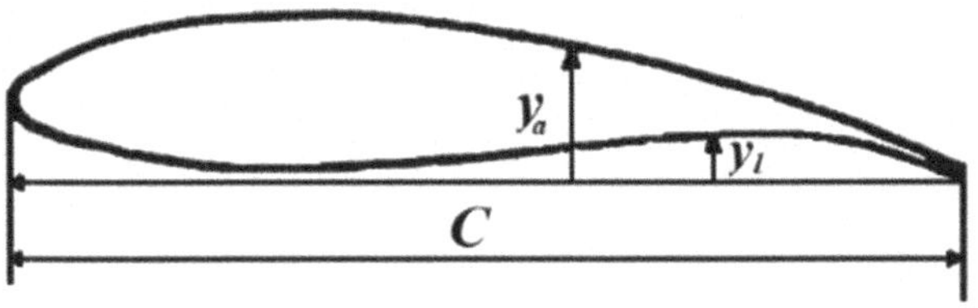

Fig. 8.31 Coordinates of wing profile chord

different from the test results, thus, a correction equation for lift coefficient can be used as:

$$C_L = 2\pi\eta(\alpha - \alpha_0) \tag{8.63}$$

Where: correction coefficient $\eta = 0.8$ to 0.9.

The difference, between the tests and theoretical calculation, is due to the influence of viscosity and wing thickness. Viscosity creates a boundary layer on the wing surface, so as to change the pressure distribution on the surface, and so affects the lift. According to linear theory, wing thickness in ideal flow does not influence lift, however in viscous flow physical thickness generates a different pressure distribution, influencing directly the thickness of the boundary layer, and the boundary layer influences inversely on the pressure distribution. Tests demonstrate that the slope of the lift coefficient increases inversely with the section and boundary layer thickness. Both of these two influences are introduced as viscous correction factors.

The correction coefficient η changes with different wing section forms, however for thin wings the range is small. For limited span wing, the definition of lift coefficient is:

$$C_{\mathrm{L}} = \frac{2\pi\eta}{1 + \frac{2\eta}{A}(1+\tau)}(\alpha - \alpha_0) \tag{8.64}$$

Where τ can be obtained from Figs. 8.13, 8.14.

Approximate Estimation of the Viscous Drag

If there is no suitable test reference for the wing section, the viscous drag of the wing can be estimated with the following equation:

$$C_{\mathrm{DV}} = C_f + C_\phi = 2C_f\left(1 + 10\bar{t}^2\right)\left[1 + (\Delta C_L)^2\right] \tag{8.65}$$

where:

C_f friction drag coefficient
C_ϕ form drag coefficient
$\bar{t}$ relative thickness, i.e., ratio of max thickness with chord length
ΔC_L the difference of lift coefficient at specified wing, C_L with the lift coefficient at minimum drag C_{LO}.

The number 2, at the right side of above equation, is to account for the upper and lower two surfaces of the wing, while the wing projected area S is used in the calculation of form drag coefficient.

The friction drag coefficient C_f can be calculated in terms of the Prandtl equation[11] and ITTC equation,[12] normally using a roughness allowance of $\Delta C_f = 0.4 \times 10^{-3}$ for air wings and $\Delta C_f = 0.8 \times 10^{-3}$ for hydrofoils. Due to considering the high sensitivity of roughness on hydrofoils, calculation of C_f also can be adopted from test results of NACA wings, using standard roughness:

$$C_f = 0.08 R_n^{-\frac{1}{8}} \tag{8.66}$$

where:

$$R_n = \frac{V_\infty c}{\nu}.$$

Reynolds number with respect to wing chord;
No roughness allowance should be added in (8.66).

[11] For Prandtl's boundary layer theory see Boundary layer—Wikipedia

[12] ITTC friction drag procedures see 75-02-02-01.pdf (ittc.info) and 75-02-02-02.pdf (ittc.info)

In some references, the following equation for calculating the friction drag coefficient is used allowing for different wing surface roughness:

$$C_f = \left[1.89 + 1.62\log\left(\frac{C}{K_S}\right)\right]^{-2.5} \tag{8.67}$$

Where: K_S is equivalent sand particle diameter, which can be taken according to following table. In (8.67) above, log indicates log, i.e., $\log_{10}$.

Surface condition	Corresponding sand particle diameter, $K_S \times 10^5$ mm
Mirror surface	0.0–0.02
Polishing metal or wood	0.02–0.04
Natural metal plate	0.04–0.08
Optimum spray painting	0.08–0.12
Ordinary spray painting	0.12–0.40
Spray painting at batch production	0.40–4.06

For practical calculation, a comparison between the calculation results in terms of (8.67) and Prandtl's and ITTC equations can be used and take the larger value for the calculated friction drag.

In addition, the following equation also can be recommended for calculating the viscous drag coefficient:

$$C_{DV} = 2C_f(1 + 1.2\bar{t}) + 0.11(\Delta C_L) \tag{8.68}$$

According to the [4], the following equation can be used, if the maximum thickness is located at 40 % $\sim$ 50% chord length:

$$C_{DV} = 2C_f\left[(1 + 1.2\bar{t}) + 60\bar{t}^4\right] \tag{8.69}$$

Here the second item represents the friction drag increment due to the wing thickness, causing increase of average flow speed compared to a correspondent flat plate, and the third item is drag due to induced pressure difference.

In case of small hydrofoil size or low craft speed, so that $R_n < 3 \times 10^5$ with respect to chord length, the Blasius[13] equation can be used to obtain the friction drag coefficient, for laminar boundary flow of the hydrofoils:

$$C_f = \frac{1.328}{\sqrt{R_n}}$$

When $3 \times 10^5 < R_n < 10^7$ the following equation can be used for calculating the friction drag coefficient:

[13] Blasius equation see Blasius boundary layer—Wikipedia

$$C_f = \frac{0.455}{(\ln R_n)^{2.58}} - \frac{1700}{R_n}$$

Where ln R_n is $\log_e$ of R_n.

References

1. I.H. Abbott, A.E. Von Doenhoff, *Theory of Wing Sections* (Dover Publications, New York, 1959)
2. S.F. Hoerner, H.V. Borst, *Fluid-Dynamic Lift*, 2nd edn. (Published by Mrs Liselotte A Hoerner, Chapter VIII Hydrodynamic Lift, 1985)
3. J.E. Kerwin, *Lecture Notes Hydrofoils and Propellers* (from MIT Mechanical Engineering Course, Cambridge, MA, March 2001) Pdf available at https://ocw.mit.edu/courses/mechanical-engineering/2-23-hydrofoils-and-propellers-spring-2007/course-notes/
4. O.M. Faltinsen, *Hydrodynamics of High Speed Marine Vehicles* (Cambridge University Press, 2005., ISBN 978-0-521l-84568-7), p. 454
5. Sir Horace Lamb, 'Hydrodynamics', Cambridge University Press, various editions from 1879 to 1943, and currently available from Dover Publications as a paperback reprint of 1932 edition, 738 pp, ISBN 9780486602561
6. R. Eppler, *Airfoil Design and Data* (Springer-Verlag, Berlin, 1990)
7. R. Eppler, Y.T. Shen, Wing Sections for Hydrofoils—Part 1, Symmetrical Profiles. SNAME J. Ship Res. **23**(3), 209–217 (1979)
8. Y.T. Shen, R. Eppler, Wing sections for hydrofoils—part II, nonsymmetrical profiles. SNAME Journal of Ship Research **25**(3), 191–200 (1981)
9. Y.T. Shen, Wing sections for hydrofoils—part III, experimental verifications. SNAME Journal of Ship Research **29**(1), 39–50 (1985)
10. Xian Fu Wang, 'Marine Air-Hydrofoils Theory', National Defense Industry Press, China, 10. 1998. (in Chinese)

Chapter 9
Hydrofoil Wing Theory Influence of Free Surface and Cavitation

The principal differences between a hydrofoil and an airfoil are the viscous fluid versus gaseous media, influence of the water free surface, and the existence of cavitation for the hydrofoil. Each of these is an important element for investigation of hydrofoil performance. We will cover them each in turn in this chapter, building on the foil theory discussed in Chap. 8.

9.1 Influence of the Free Surface

When a hydrofoil is moving close to the free surface, induced wave making can be observed. The wave making has a relationship to the Froude number with respect to the depth of hydrofoil, $Fn_h = V/(\mathrm{gh})^{0.5}$, where h is the submerged depth of the hydrofoil. When $Fn_h < 1$ wave making is small. When F_{nh} is close to 1 wave making is more apparent, with spray being caused by breaking of the wave, as shown in Fig 9.1a.

When $Fn_h > 1$ spray disappears, and a flatter wave profile appears downstream of the hydrofoil. This further flattens with increase of Fn_h. When the foil span is limited, the water surface at the tip of the foil span is protruded slightly, and a concave closure with protruded edge is formed after the hydrofoil, as shown in Fig 9.1b.

When measuring the lift acting on a hydrofoil, it is found that the lift coefficient stays constant when the Froude number with respect to foil chord $Fn_c = V/(\mathrm{gc})^{0.5} > 4\sim5$, and does not change further with changing Fn_c. This means the gravity effect can be neglected in such conditions. A typical curve for the lift on a hydrofoil versus its depth and angle of attack can be obtained as shown in Fig. 9.2. One can see the lift coefficient slope and absolute value of angle of attack at zero lift α_0 increase with hydrofoil relative depth *h*/*c*.

A typical pressure distribution curve on the surfaces of a hydrofoil is shown in Fig. 9.3. It can be seen that the absolute value of negative pressure on upper surface

Z.-S. Dong et al., *High-Speed Monohull and Hydrofoil Craft*,
https://doi.org/10.1007/978-3-031-62760-6_9

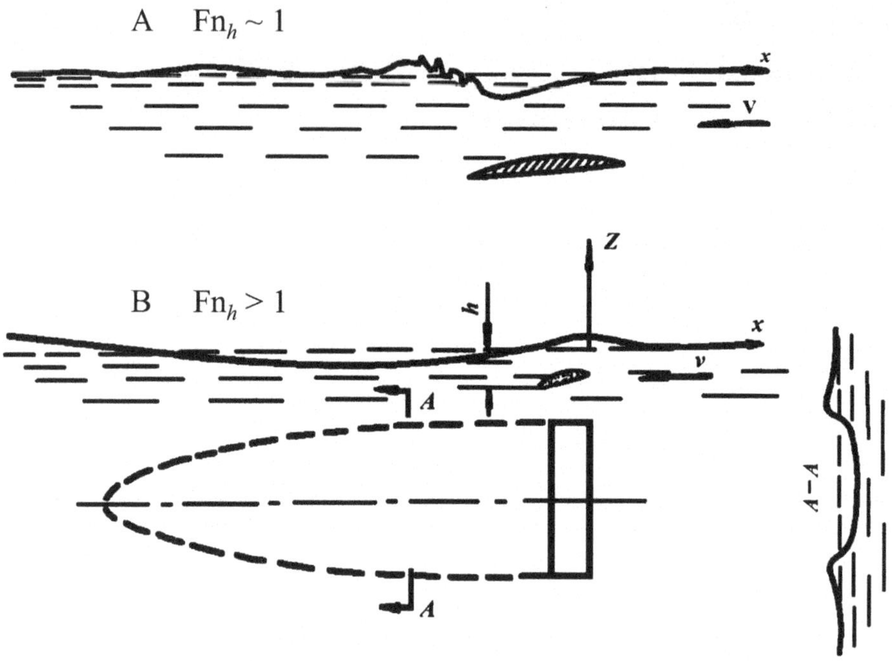

Fig. 9.1 Hydrofoil wave making close to free surface

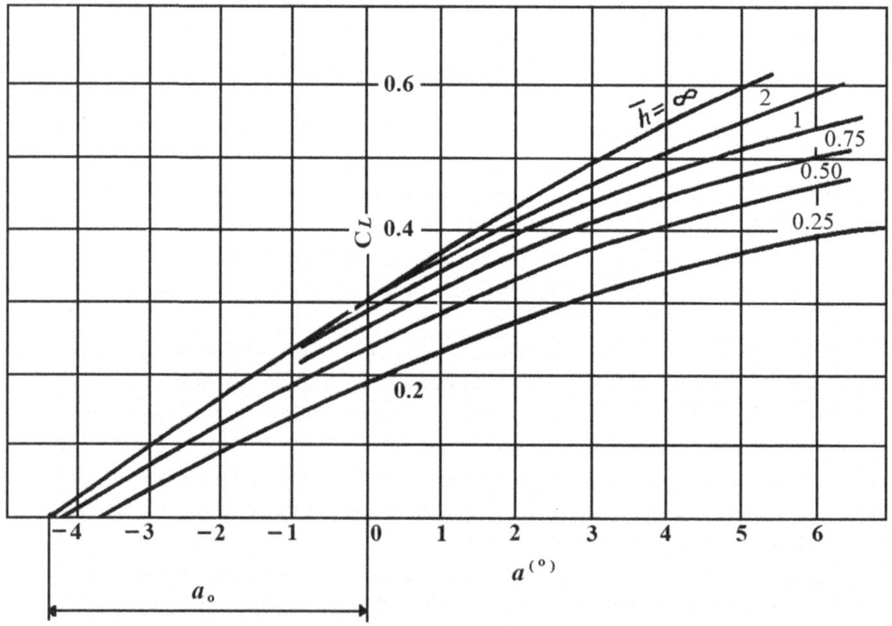

Fig. 9.2 Lift versus submerged depth of foil at high F_n

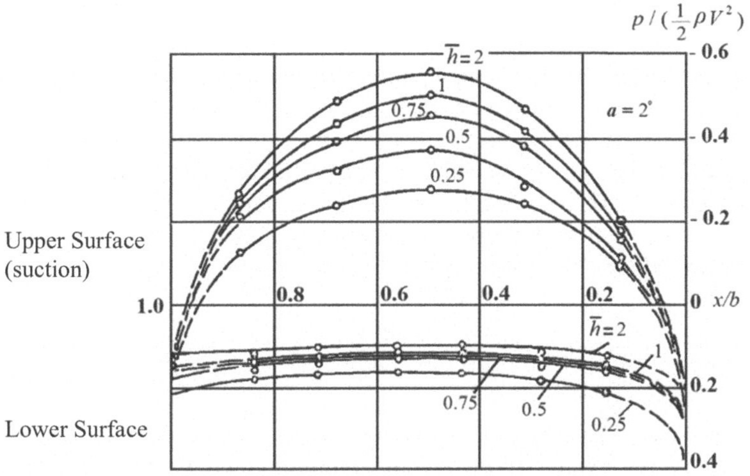

Fig. 9.3 Pressure distribution at the lower/upper surface of a hydrofoil at high F_n. (The foil has a camber profile and cross section with 7.5% thickness)

is decreased significantly, and positive pressure on the underside increased, with decrease of h/c, thus, the total lift is decreased.

The suction pressure on the upper surface cannot be less than atmospheric pressure. Once suction pressure reduces to water vapor pressure for the depth of immersion of the foil, induced water vapor will create cavitation. While at deeper draft the suction may not reach local vapor pressure, the effect will be apparent with shallow draft, and at the extreme condition of foil depth equal to zero, the upper surface pressure is simply atmospheric and then the hydrofoil is in a planing condition.

From the point of view of solution of hydrodynamic equations accounting for the existence of a free surface, one more boundary condition—the free surface boundary condition, has to be satisfied, while the continuity equation and dynamics equation are unchanged. Where a hydrofoil is located close to the free surface the condition for the flow around the hydrofoil in equation 9.1a should be satisfied.

$$\frac{\partial \varphi}{\partial \mathrm{t}} + \frac{1}{2}\left(V_x^2 + V_y^2 + V_z^2\right) + g\eta_S(x, z) = 0 \tag{9.1a}$$

Where the axis system is established on the hydrofoil moving at constant linear velocity, the free surface boundary condition then becomes:

$$\frac{\partial \phi}{\partial \mathrm{x}} + \frac{1}{2}\left[\left(\frac{\partial \phi}{\partial x}\right)^2 + \left(\frac{\partial \phi}{\partial y}\right)^2 + \left(\frac{\partial \phi}{\partial z}\right)^2\right] + g\eta_S(x, z) = 0 \tag{9.1b}$$

Where $\frac{\partial \phi}{\partial x}, \frac{\partial \phi}{\partial y}, \frac{\partial \phi}{\partial z}$, are disturbance velocities in directions of x, y, and z, respectively.

In normal conditions, the free surface profile $\eta_s(x, z)$ is not given, and it is difficult to solve the basic equation with the additional boundary conditions.

We therefore use simplified assumptions for a high-speed foil, i.e., neglecting the gravity term, assuming small perturbation, neglecting the secondary perturbation term and considering the free surface wave being a wave with small amplitude. The free surface boundary condition can then be replaced by the condition to be satisfied when axis plane XOZ is on the horizontal plane, with $y = 0$ and considering the hydrofoil is in constant and uniform incoming flow. Then the free surface condition can be simplified to:

$$v_x\Big|_{y=0} = \frac{\partial \phi}{\partial x}\Big|_{y=0} = 0 \tag{9.1c}$$

Not considering the perturbation velocity in the moving direction, on the wave free surface has been validated by tests of a thin foil moving at high speed under the free surface. The tested perturbation speed on the water surface along the motion direction is nearly equal to zero. This could be verified by a simple example—putting a slice of wood on a calm water surface and for hydrofoil moving at high speed through the water surface. One would see the wood piece moving only in horizontal direction, but no motion in vertical direction.

This forms a starting point for our exploration of the behavior of submerged foils using laminar flow and linear theory, similar to our exposition in Chap. 8. Once real fluid characteristics are considered, the behavior of the water surface is affected as shown in Fig. 9.1. We will introduce approaches to this later in the chapter.

9.2 Mirror Image Vortex Model on Free Surface and Its Use

The simplified free surface boundary condition (9.1c) means that for the hydrofoil moving nearby the free water surface, the free surface can be replaced by a mirror vortex, i.e., the free surface condition can be modeled by a mirror image around the inflexion point. So, for a point vortex with strength of Υ on the hydrofoil under the water surface, then the same point vortex, at same direction at the point of symmetry with respect to the free surface, should be set up as shown in Fig. 9.4, where p denotes a point vortex. For a series of vortices on the hydrofoil, the vortex series should be set up at the line of symmetry. It is not difficult to realize such vortex distribution can satisfy the simplified free surface boundary condition—there is no perturbation velocity in the horizontal plane, but only in the vertical plane.

Meanwhile, for sources distributed on the foil surface (i.e., following the thickness profile of the foil), an equivalent distribution should be set up about the axis of symmetry to maintain the free surface boundary condition.

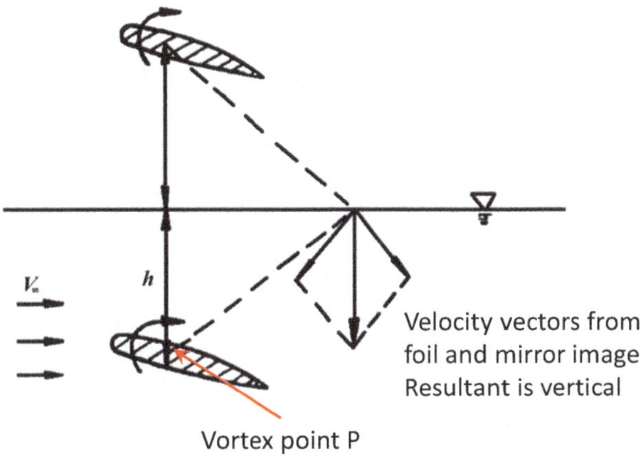

Fig. 9.4 Mirror image vortex of hydrofoil under the free surface

For a three-dimensional thin foil, the influence of perturbation velocity in the X direction will not exist so long as the free vortex series is parallel to the horizontal plane, and so the mirror image vortex series method can still be used.

Use of the mirror image vortex to replace the free surface can also explain physically why the free surface can decrease the hydrofoil lift. This is due to the velocity induced by the reflected vortex nearby the hydrofoil being reverse direction to the incoming flow, and such additional perturbation velocity decreases the flow velocity nearby the upper and lower surfaces of the hydrofoil.

The decrease of flow velocity at upper foil surface leads to decrease of the absolute value of dynamic pressure on the upper surface, and the decrease of flow velocity on lower foil surface leads to the decrease of positive pressure of lower surface. Due to the upper surface being close to the reflected vortex, the velocity is significantly reduced, so reducing the dynamic pressure and lift force.

For a hydrofoil with a large angle of attack, the corresponding vortex circulation is also increased, so increasing the reverse perturbation velocity field, and decreasing the lift of hydrofoil. This is why the lift coefficient slope, $dC_L/d\alpha$, decreases.

The deformation of the water surface nearby the hydrofoil, as shown in Fig. 9.1 will affect the absolute value of angle of attack at zero lift close to free surface. The water surface deflection bends the incoming flow to the hydrofoil, so effectively decreasing the hydrofoil camber, and decreasing the absolute value of angle of attack at zero lift. The water surface deformation magnitude is relative to the submerged depth and the hydrofoil section profile, including its camber and thickness.

9.2.1 Using a Mirror Image Vortex Model to Calculate Free Surface Effect

In order to simplify the calculation, here we introduce the simplified lifting line model, i.e., "Π" model, using a "Π" vortex to replace the hydrofoil, and assume the

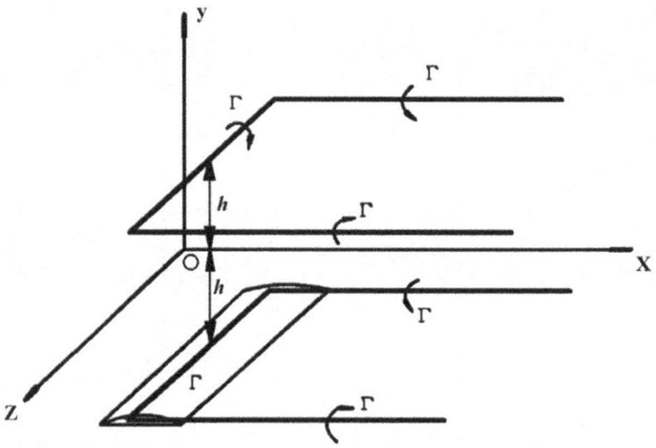

Fig. 9.5 Simplified mirror vortex model of free water surface

free vortex is concentrated at the tip of the span, and a corresponding image vortex, also a mirror "Π" vortex, as shown in Fig. 9.5. The coordinate system XOZ is formed on the horizontal plane, where the Y axis is upward positive.

We assume the strength of the vortex attached to the hydrofoil is constant, and equal to Γ, the submerged depth of hydrofoil h, chord c, span l, and mirror image vortex at h over the water surface, and in accordance with the lifting line theory. Further the attached vortex is located at ¼c, and satisfies the boundary condition of foil surface at ¾c, to obtain the vortex circulation Γ. Thanks to the simplified lifting line model, Γ being constant, satisfying the boundary condition only on centerline of span at ¾c, the induced velocity can be calculated as follows.

9.2.2 *Velocity Induced by Attached Vortex*

The Biot-Savart Theory[1] for defining the induced velocity by a limited length vortex is used to calculate the induced velocity, with the following relationship:

$$v_A = \frac{\Gamma}{4\pi r}(\cos\theta_1 - \cos\theta_2) \tag{9.2}$$

Where Γ is the circulation of this vortex section, r the vertical distance of point A to vortex section, and θ_1 and θ_2 are angles between the lines from the tips of the vortex section to point A with vortex section vector respectively, as shown in Fig. 9.6.

[1]Biot–Savart law—Wikipedia.

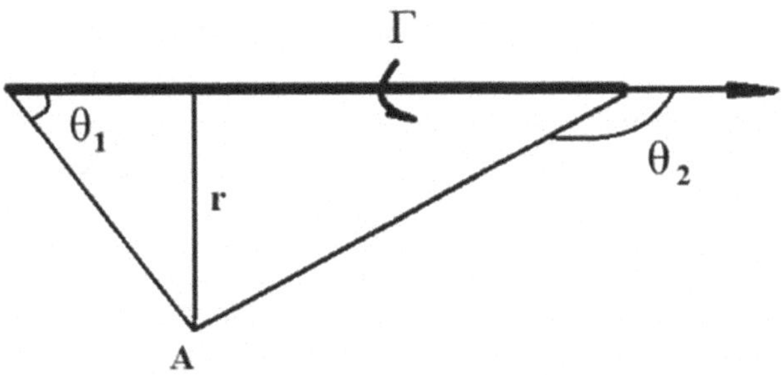

Fig. 9.6 velocity induced by attached vortex

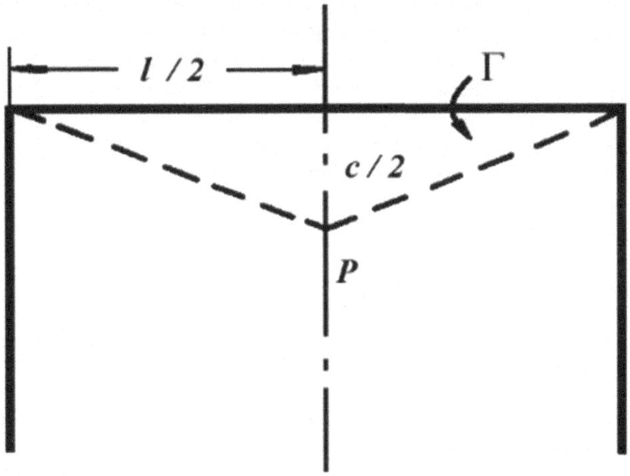

Fig. 9.7 Velocity induced by attached vortex

In the case of the right side of the vortex extending to infinity and forming a semi-infinite vortex, then we take $\theta_2 = \pi$, $\cos\theta_2 = -1$ and if the left side also extends to infinity $\cos\theta_1 = 1$, then $v_A = \Gamma/(2\pi r)$.

For a hydrofoil, putting the corresponding values into (9.2) then:

$$
\begin{aligned}
v_1 &= \frac{\Gamma}{4\pi \frac{c}{2}}\left[\frac{\frac{l}{2}}{\sqrt{\left(\frac{l}{2}\right)^2+\left(\frac{c}{2}\right)^2}} - \frac{-\frac{l}{2}}{\sqrt{\left(\frac{l}{2}\right)^2+\left(\frac{c}{2}\right)^2}}\right] \\
&= \frac{\Gamma}{\pi c}\frac{l}{\sqrt{l^2+c^2}}
\end{aligned}
\tag{9.3}
$$

The direction of induced velocity is vertically downward is as shown in Fig. 9.7.

9.2.3 Velocity Induced by Mirror Image Vortex

Following the approach above for the attached vortex system, the imaginary induced velocity for the mirror image vortex system will be:

$$v_2 = \frac{\Gamma}{4\pi\sqrt{(2h)^2 + \left(\frac{c}{2}\right)^2}}\left[\frac{\frac{l}{2}}{\sqrt{(2h)^2 + \left(\frac{c}{2}\right)^2 + \left(\frac{l}{2}\right)^2}} - \frac{-\frac{l}{2}}{\sqrt{(2h)^2 + \left(\frac{c}{2}\right)^2 + \left(\frac{l}{2}\right)^2}}\right]$$

The relationship between the various items is shown in Fig. 9.8. The direction of induced velocity AP line is downward, and its vertical component is:

$$\begin{aligned} v_2' &= \frac{\Gamma}{4\pi\sqrt{(2h)^2 + \left(\frac{c}{2}\right)^2}} \frac{l}{\sqrt{(2h)^2 + \left(\frac{c}{2}\right)^2 + \left(\frac{l}{2}\right)^2}} \cdot \frac{\frac{c}{2}}{\sqrt{(2h)^2 + \left(\frac{c}{2}\right)^2}} \\ &= \frac{\Gamma}{\pi} \frac{cl}{\left(16h^2 + c^2\right)\sqrt{16h^2 + l^2 + c^2}} \end{aligned} \tag{9.4}$$

Considering next the free vortices we obtain as follows.

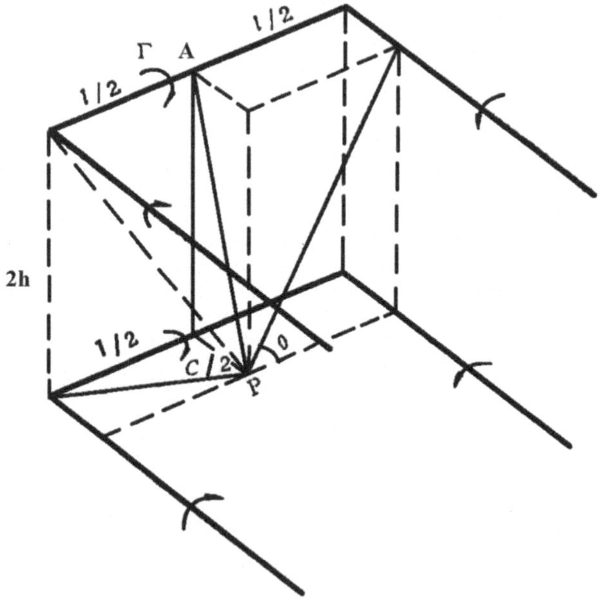

Fig. 9.8 Relationship between various vortices

9.2.4 Induced Velocity from Free Vortex

$$v_3 = \frac{2\Gamma}{4\pi\frac{l}{2}}\left[1+\frac{\frac{c}{2}}{\sqrt{\left(\frac{l}{2}\right)^2+\left(\frac{c}{2}\right)^2}}\right] = \frac{\Gamma}{\pi l}\left[1+\frac{c}{\sqrt{l^2+c^2}}\right] \tag{9.5}$$

The direction of the induced velocity from the free vortex is vertical down.

9.2.5 Induced Velocity from Mirror Image Free Vortex

$$v_4 = \frac{2\Gamma}{4\pi\sqrt{(2h)^2+\left(\frac{l}{2}\right)^2}}\left[\frac{\frac{c}{2}}{\sqrt{(2h)^2+\left(\frac{c}{2}\right)^2+\left(\frac{l}{2}\right)^2}}+1\right]$$

The vertical component is $v_4 \cos\theta$, see Fig. 9.8.
The vertical component is thus:

$$\begin{aligned} v_4' &= \frac{2\Gamma}{4\pi\sqrt{(2h)^2+\left(\frac{l}{2}\right)^2}}\left[\frac{\frac{c}{2}}{\sqrt{(2h)^2+\left(\frac{c}{2}\right)^2+\left(\frac{l}{2}\right)^2}}+1\right]\cdot\frac{\frac{l}{2}}{\sqrt{(2h)^2+\left(\frac{l}{2}\right)^2}} \\ &= \frac{\Gamma l}{\pi(16h^2+l^2)}\left[\frac{c}{\sqrt{16h^2+c^2+l^2}}+1\right] \end{aligned} \tag{9.6}$$

Adding the four induced velocities and relating to the boundary condition, we have:

$$v_1 + v_2' + v_3 + v_4' = -V_\infty\left(\alpha - \left.\frac{d\eta}{dx}\right|_P\right) \tag{9.7}$$

Where $\left.\frac{d\eta}{dx}\right|_P$ is the slope of foil surface at point P.

The corresponding camber equation can be expressed by $y = \eta(x)$. Since only lift is to be discussed here, the foil thickness can be neglected. Putting (9.3) and (9.6) into (9.7) we obtain the following relation:

$$\begin{aligned}&\frac{\Gamma}{\pi}\left[\frac{1}{cl}\sqrt{l^2+c^2}+\frac{1}{l}\right]+\frac{\Gamma}{\pi}\frac{cl}{\left(16h^2+c^2\right)\sqrt{16h^2+l^2+c^2}}\\&+\frac{\Gamma cl}{\pi\left(16h^2+l^2\right)}\left[\frac{1}{\sqrt{16h^2+l^2+c^2}}+\frac{1}{c}\right]=-V_\infty\left(\alpha-\left.\frac{d\eta}{dx}\right|_P\right)\end{aligned} \tag{9.8}$$

Here the hydrofoil profile has chord c, span l, and foil surface slope at point P is $\left.\frac{d\eta}{dx}\right|_P$. The submerged depth is h, angle of attack α, and incoming flow speed V_∞.

Then from (9.8) we obtain the vortex circulation per unit length Γ and lift and induced drag. Since we wish to obtain the influence of the free water surface, the comparison of vortex strength for with/without free water surface will be carried out next.

First, in case of no free water surface, with the hydrofoil at the same working condition, the induced velocity consists of only two parts, v_l and v_3, thus the boundary conditions are:

$$\begin{aligned}v_1+v_3&=\frac{\Gamma_0}{\pi}\left[\frac{l}{c\sqrt{l^2+c^2}}+\frac{1}{l}+\frac{c}{l\sqrt{l^2+c^2}}\right]\\&=\frac{\Gamma_0}{\pi}\left[\frac{1}{cl}\sqrt{l^2+c^2}+\frac{1}{l}\right]=-V_\infty\left(\alpha-\left.\frac{d\eta}{dx}\right|_P\right)\end{aligned} \tag{9.9}$$

The vortex strength Γ_0 in the case of no free water surface can be solved using (9.9). From and (9.8) and (9.9) the ratio of vortex strength between with and without the free water surface can be obtained as:

$$\begin{aligned}\frac{\Gamma}{\Gamma_0}&=\left\{\frac{\left(\frac{1}{cl}\sqrt{l^2+c^2}+\frac{1}{l}\right)+\frac{cl}{\left(16h^2+l^2\right)}\left[\frac{1}{\sqrt{16h^2+l^2+c^2}}+\frac{1}{c}\right]+\frac{1}{\left(16h^2+c^2\right)\sqrt{16h^2+l^2+c^2}}}{\frac{1}{cl}\sqrt{l^2+c^2}+\frac{1}{l}}\right\}^{-1}\\&=\left\{1+\frac{\frac{cl}{\sqrt{16h^2+l^2+c^2}}\left[\frac{1}{\left(16h^2+l^2\right)}+\frac{1}{\left(16h^2+c^2\right)}\right]+\frac{l}{16h^2+l^2}}{\frac{1}{cl}\sqrt{l^2+c^2}+\frac{1}{l}}\right\}^{-1}\end{aligned} \tag{9.10}$$

Changing the above equation to be non-dimensional by putting:

$$A=\frac{l}{c},\quad \bar{h}=\frac{h}{c},\quad \text{and}\quad \frac{\Gamma}{\Gamma_0}=\frac{C_L}{C_{L0}}$$

Where

C_L Lift coefficient for free water surface present.
C_{L0} Lift coefficient for no free water surface.

Then:

$$\frac{C_L}{C_{L0}} = \left\{ 1 + \frac{\dfrac{A^2}{\sqrt{16\bar{h}^2 + A^2 + 1}}\left(\dfrac{1}{16\bar{h}^2 + A^2} + \dfrac{1}{16\bar{h}^2 + 1}\right) + \dfrac{A^2}{16\bar{h}^2 + A^2}}{\sqrt{A^2 + 1} + 1} \right\}^{-1} \tag{9.11}$$

Equation (9.11) represents the effect of the free water surface on the hydrofoil lift coefficient, and one can see that when $\bar{h}$ decreases, C_L/C_{L0} decreases as well.

In the limit as $\bar{h} \to \infty$, i.e., the hydrofoil submerged depth is infinity, then from (9.11) we can see that $C_L/C_{L0} = 1$ and as $\bar{h} \to 0$, i.e., hydrofoil moving on the water free surface, then (9.11) becomes:

$$\frac{C_L}{C_{L0}} = \left\{ 1 + \frac{\dfrac{A^2}{\sqrt{A^2+1}}\left(\dfrac{1}{A^2} + 1\right) + 1}{\sqrt{A^2+1}+1} \right\}^{-1} = \left\{ 1 + \frac{\dfrac{1}{\sqrt{A^2+1}} + \dfrac{A^2}{\sqrt{A^2+1^2}} + 1}{\sqrt{A^2+1}+1} \right\}^{-1}$$

$$= \left\{ 1 + \frac{\dfrac{1 + A^2 + \sqrt{A^2+1}}{\sqrt{A^2+1}}}{\sqrt{A^2+1}+1} \right\}^{-1} = \left\{ 1 + \frac{\sqrt{A^2+1}+1}{\sqrt{A^2+1}+1} \right\}^{-1} = \frac{1}{2}$$

This last is the thin foil analog of the planing plate. When $\bar{h} = 0$, only the lower surface touches the water and the hydrofoil can be taken as a planing plate, demonstrating that the mirror image vortex model is reasonable.

When $A \to \infty$, i.e., the two-dimensional foil condition, the Eq. (9.11) should be changed a little bit, as in (9.12) below.

$$\frac{C_L}{C_{L0}} = \left\{ 1 + \frac{\dfrac{A}{\sqrt{16\bar{h}^2 + A^2 + 1}}\left(\dfrac{1}{16\bar{h}^2 + A^2} + \dfrac{1}{16\bar{h}^2 + 1}\right) + \dfrac{A^2}{16\bar{h}^2 + A^2}}{\dfrac{\sqrt{A^2 + 1} + 1}{A}} \right\}^{-1}$$

$$= \left\{ 1 + \frac{\dfrac{1}{\sqrt{\dfrac{16\bar{h}^2}{A^2} + 1 + \dfrac{1}{A^2}}}\left(\dfrac{1}{16\bar{h}^2 + A^2} + \dfrac{1}{16\bar{h}^2 + 1}\right) + \dfrac{1}{\dfrac{16\bar{h}^2}{A^2} + A}}{\sqrt{1 + \dfrac{1}{A^2} + \dfrac{1}{A}}} \right\}^{-1} \tag{9.12}$$

$$= \left\{ 1 + \frac{1}{16\bar{\mathrm{h}}^2 + 1} \right\}^{-1} = \left\{ \frac{16\bar{\mathrm{h}}^2 + 1 + 1}{16\bar{\mathrm{h}}^2 + 1} \right\}^{-1} = \frac{16\bar{\mathrm{h}}^2 + 1}{16\bar{\mathrm{h}}^2 + 2}$$

With respect to the effect of the free water surface to the induced drag, according to the lifting line theory, the effect of limited span can be realized by introducing the downwash velocity induced by the free vortex. The downwash velocity can be calculated from the free vortex at the position of the attached vortex, i.e., ¼c.

When a free water surface is present, downwash velocity is equal to the sum of downwash velocity induced by both free vortex and reflected image free vortex at the position of the attached vortex. Comparing the calculation of downwash angle and induced drag, in terms of above method compared with test results, it has been found that the calculation gave smaller results than the tests. So, summing up, when $h \to \infty$, $C_L/C_{LO} = 1$ and when h $\to 0$ $C_L/C_{LO} = ½$.

K.L.Wadlin and C.L.Shuford [1] analyzed a hydrofoil with influence from the free water surface, and using a method similar to the Weissinger method as well as considering the influence of a free surface on the additional downwash velocity at ¾c rather than at ¼c. They then calculated the induced drag and found the calculation agreed with the test results.

It is suggested to use an additional downwash velocity at ¾c to calculate additional downwash angle, and calculate the hydrofoil induced drag from the free water surface in two parts, with the drag induced by free vortex of hydrofoil itself, as:

$$C_{\text{Dio}} = \frac{C_L^2}{\pi A}(1+\delta) \tag{9.13}$$

The second part is the additional induced drag from the free surface. The downwash velocity at ¾c, is not only induced by the reflected image free vortex v_4' but also by reflected image attached vortex v_2' so the attached downwash angle is:

$$\Delta\alpha' = -\frac{v_2' + v_4'}{V_\infty}$$

Putting (9.4) and (9.6) into the above equation, then:

$$\Delta\alpha' = -\frac{\Gamma}{V_\infty \pi l}K \tag{9.14}$$

Where:

$$\begin{aligned} K &= \frac{cl^2}{16h^2 + l^2}\left(\frac{1}{\sqrt{16h^2 + l^2 + c^2}} + \frac{1}{c}\right) + \frac{cl^2}{\left(16h^2 + c^2\right)\sqrt{16h^2 + l^2 + c^2}} \\ &= \frac{A^2}{16\bar{h}^2 + A^2}\left(\frac{1}{\sqrt{16\bar{h}^2 + A^2 + 1}} + 1\right) + \frac{A^2}{\left(16\bar{h}^2 + 1\right)\sqrt{16\bar{h}^2 + A^2 + 1}} \end{aligned} \tag{9.15}$$

According to Zhoukovsky's theory:

$$L = \rho V_\infty \Gamma l, \quad \text{and } C_L = \frac{L}{0.5\rho V^2 cl} = \frac{\rho V_\infty \Gamma l}{0.5\rho V_\infty{}^2 cl} = \frac{2\Gamma}{V_\infty c}$$

Thus: $\frac{\Gamma}{V_\infty} = \frac{c}{2} C_L$

Putting this into (9.14), and adding a correction factor for the plan profile of the hydrofoil, we then obtain:

$$\Delta \alpha' = \frac{C_L}{\pi A}(1+\tau)\frac{K}{2} \tag{9.16}$$

The corresponding induced drag coefficient is:

$$\Delta C_{Di} = \frac{C_L^2}{2\pi A} K(1+\delta) \tag{9.17}$$

The total downwash angle is:

$$\Delta \alpha_T = \frac{C_L}{\pi A}(1+\tau) + \frac{C_L}{2\pi A} K(1+\tau) = \frac{C_L}{\pi A}(1+\tau)\left[1+\frac{K}{2}\right] \tag{9.18}$$

The total induced drag coefficient is:

$$C_{Di} = \frac{C_L^2}{\pi A}(1+\delta)\left[1+\frac{K}{2}\right] \tag{9.19}$$

The coefficient *K*/2 is an influence factor of the free surface on induced drag. From Eq. (9.15), one can see when $\overline{h} \to \infty$, $K = 0$, so there is no additional induced drag, while when $\overline{h} \to 0$, $K = \sqrt{1+A^2} + 1$, and when $\overline{h}$ moves between $0 \to \infty$, *K* is changed between $0 \to \left[\sqrt{1+A^2}+1\right]$.

According to (9.11), one can calculate $\frac{C_L}{C_{L0}} = f\left(A, \overline{h}\right)$, as shown in Fig. 9.9; and according to (9.17), calculate the additional induced drag coefficient, in terms of free water surface effect, i.e., $\Delta C_{Di} = f\left(\overline{h}, C_L\right)$, which was presented in a paper "Study of Hydrofoil Sea craft", provided by Grumman Co, US, where (9.17) was revised slightly as follows:

$$\Delta C_{Di} = \frac{C_L^2}{2\pi A} K(1+\delta) = \frac{C_L^2}{8\pi}(1+\delta)\left(4\frac{K}{A}\right) = \frac{C_L^2}{8\pi}(1+\delta)K_1$$

Curves for $K_1 = \left(\overline{h}, A\right)$ were developed as shown in Fig. 9.10.

From Figs. 9.9 and 9.10, one can see the relationship between the submerged depth $\overline{h}$, the lift coefficient, and the increment of drag coefficient is non-linear.

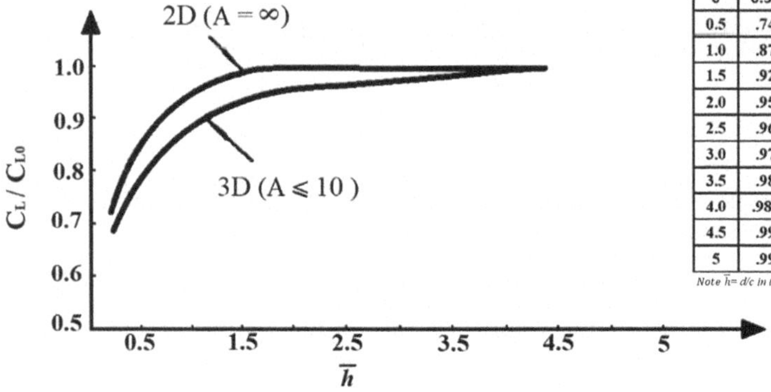

d/c	A=4	A=6	A=10	A=8
0	0.50	0.50	0.50	0.50
0.5	.74	.77	.78	.82
1.0	.87	.87	.88	.94
1.5	.92	.92	.92	.97
2.0	.95	.94	.93	.975
2.5	.96	.96	.95	.98
3.0	.97	.97	.96	.985
3.5	.98	.975	.965	.99
4.0	.989	.98	.970	.992
4.5	.99	.985	.975	.994
5	.99	.99	.98	.995

Note $\bar{h}$= d/c in table

Fig. 9.9 Effect of free water surface to lift coefficient

Fig. 9.10 Additional drag coefficient K_I induced by free water surface

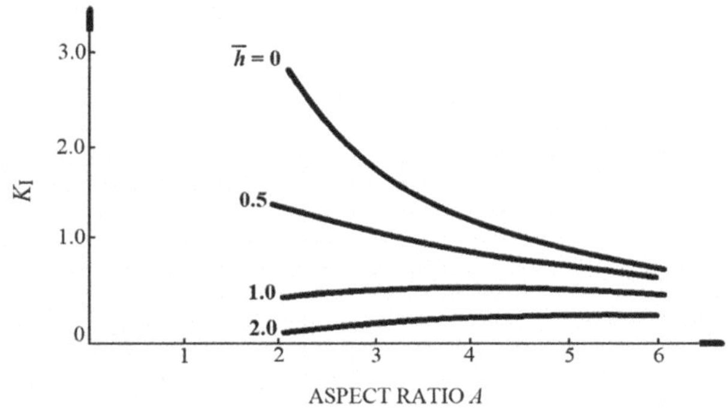

The influence increases with the decrease of $\bar{h}$, and is significant when $\bar{h} < 1$. When $\bar{h} > 3.0 \sim 4.0$ the influence of the free water surface is minimal.

Wadlin and Shuford [1] carried out a comparison between the calculation results by this method and test results, and found that when $\bar{h} > 0.5$ the lift coefficient agreed well between calculation and tests, and so did the drag coefficient, but not quite as good correlation as lift coefficient. When $\bar{h} < 0.5$ the error increased.

This lifting line mirror image vortex method for calculating the influence of the free water surface is a simplified theoretical method. It is based upon the assumption of high speed and small perturbation and so it is only suitable for thin foils at high F_n. The calculation should be correct so long as the hydrofoil is thin, with $F_n > 4 \sim 5$, however tests demonstrate that when $\bar{h} < 0.5$ the theory is not suitable even though the two conditions mentioned above are satisfied.

This is due to the model not reflecting correctly local deformation of the water surface nearby the hydrofoil. The water surface nearby the hydrofoil and incoming flow are deflected toward the hydrofoil, decreasing the zero lift angle of attack. This effect is strongest at shallow submerged depth, and neglecting this will cause the error in hydrodynamic properties. This is a disadvantage of this method.

It is difficult to find a simplifying method similar to the above using small perturbation simplification when considering the influence of gravity and low speed. When considering the water surface motion while using the mirror image vortex model, the vortex strength on the hydrofoil and on the mirror image vortex should be different due to the free surface deformation.

It has been suggested to deem the strength of both attached and mirror image vortex as the same, then add a velocity potential of wave motion on the free surface, and to integrate the differential equation with aid of numerical calculation. Tetsuo Nishiyama (西山哲男) in Japan carried out such calculations and the results demonstrated that there was no great influence of gravity on the lift of a hydrofoil moving at high speed, and so the accuracy of calculation can be improved only at low F_n. Nevertheless, it didn't significantly improve the calculation precision for shallow foil depth, and is only suitable for great depth ($\bar{h} > 0.5$) the same as the simplified high speed small perturbation method. This is due to the lifting line model not responding correctly to the water deformation nearby the foil.

Regarding wave making drag in terms of the existence of the free water surface, this can be calculated using the lifting line model at high F_n, neglecting the high order items as follows:

$$C_W = \frac{C_{L_\infty}^2}{2F_n^2}\left(1 - \frac{2\pi}{F_n^2}\right) \tag{9.20}$$

Where:

C_{L_∞} lift coefficient at infinite submerged depth
$F_n = \frac{V}{\sqrt{gc}}$ Froude number with respect to chord

Or using:

$$C_W = \frac{C_{L_h}^2}{2F_n^2} e^{-\frac{2}{F_{n_h}^2}} \tag{9.21}$$

Where:

C_{L_h} Lift coefficient at shallow submerged depth
F_n Froude number with respect to chord as in (9.20)
$Fn_h = \frac{V}{\sqrt{gh}}$ Froude number with respect to nominal submerged depth

From the discussion above one can see that it is very difficult to calculate the influence due to water surface deformation nearby the foil by means of a lifting line model. However, the lifting surface theory is too complicated for a purely analytical approach. Some empirical methods are therefore applied at the initial stages of design as we will discuss below, and CFD together with model testing is used for detailed analysis at a later stage.

9.3 Estimation of Free Water Surface Effect

Test results of a hydrofoil moving under the free water surface, as shown in Fig. 9.2, and calculation of the free water surface influence using the mirror image vortex model on the lift acting on the hydrofoil are as follows:

1. Free water surface deformation drives decrease of lift coefficient slope $dC_L/d\alpha$, due to the decrease of lift calculated by the mirrored perturbation velocity field.
2. Mirror image free vortex creates increased downwash, leading to decrease of lift coefficient and increase of drag coefficient
3. Owing to the deflected flow caused by the local deformation of water surface nearby the hydrofoil leading edge, the zero lift angle of attack reduces

The approximate estimation of the effect of the free water surface deformation is based upon adjustment from calculation of hydrofoil at infinite submerged depth to deal with the effects of the three points mentioned above. The equations for lift coefficient and drag coefficient have been given in Chap. 8, as:

$$C_L = \frac{2\pi\eta}{1+\frac{2\eta}{A}(1+\tau)K_e}(\alpha-\alpha_0) \tag{9.22}$$

$$C_{D_i} = \frac{C_L^2}{\pi A}(1+\delta)K_e \tag{9.23}$$

Where:

η Coefficient for effect of viscosity (0.8–0.9)
τ, δ plan profile modification coefficient, given by Figs. 8.14 or 8.15
K_e foil tip coefficient, in terms of (Fig. 8.19)

Equation (9.22) requires a factor for increased downwash angle η as included in Eq. (8.64). In order to consider the effects of free surface mentioned above in (9.22) and (9.23), then (9.22) can be rewritten in its original form as:

$$\begin{aligned} C_L &= 2\pi\eta(\alpha-\alpha_0-\Delta\alpha) \\ &= 2\pi\eta\left[\alpha-\alpha_0-\frac{C_L}{\pi A}(1+\tau)K_e\right] \end{aligned}$$

Assuming the change of lift coefficient slope caused by the free water surface is Υ, then the total downwash angle in free water surface is $\Delta\alpha_\gamma$, as in (9.18). The change of zero lift angle of attack can be expressed by $\Delta\alpha_0$. Then the hydrofoil lift coefficient close to the free water surface is:

$$C_L = \frac{2\pi\eta\gamma}{1+\frac{2\eta\gamma}{A}(1+\tau+\sigma)K_e}[\alpha-(\alpha_0+\Delta\alpha_0)] \tag{9.24}$$

Where: $\sigma = \tau K_e/2$

The corresponding induced drag coefficient is:

$$C_{D_i} = \frac{C_L^2}{\pi A}(1 + \delta + \sigma)K_e \tag{9.25}$$

The residual issues are how to define the correction coefficients Υ, and σ, the change of zero lift angle of attack $\Delta\alpha_0$, and the approximate estimation method is as follows.

Regarding to the tip influence coefficient K_e, the equation mentioned previously (8.47) is formed for the condition of no free water surface present, however when present the effect of tips at upper surface and lower surface of the foil are different, which can be seen clearly from the action of the mirror image vortex.

The direction of transverse flow at the lower surface of a foil is induced by the mirror image free vortex, however, it will be reversed at upper surface, as shown in Fig. 9.11.

So, for transverse flow, the effect of the upper surface tip is greater than the lower surface tip, so the different effect of both upper and lower tip has to be considered. The following empirical equations are recommended for calculating the corresponding height of tips:

$$\left(\frac{h_e}{l}\right)E = \frac{h_l}{l}\left(\frac{2h_e - h_l}{2h_e}\right) + \frac{h_u}{l}\left(\frac{2(h_e + h_u)}{2h_e + h_u}\right) \tag{9.26}$$

K_e can be obtained from (8.47) so long as taking $\left(\frac{h_e}{l}\right)_E$ to replace $\frac{h_e}{l}$, so that:

$$K_e = \frac{1}{1 + 2\left(\frac{h_e}{l}\right)_E} \tag{9.27}$$

The meaning of items in (9.26) is shown in Fig. 9.12.

Essentially the mirror image vortex model is considering the hydrofoil with free water surface correspondingly as a low foil of a biplane (with symmetrical airfoils),

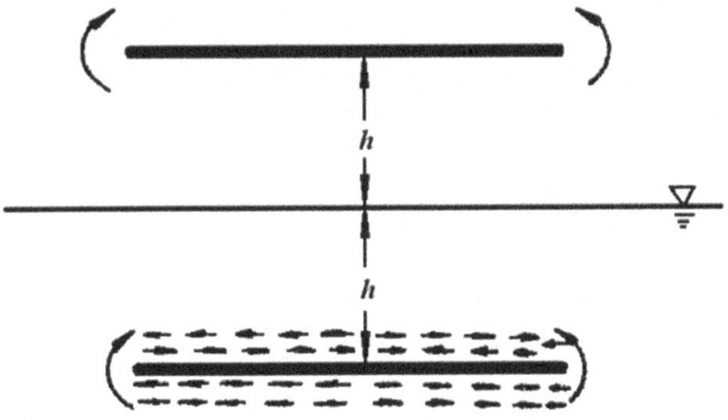

Fig. 9.11 Transverse flow velocity on foil and induced velocity on foil surface induced by its mirror image free vortex

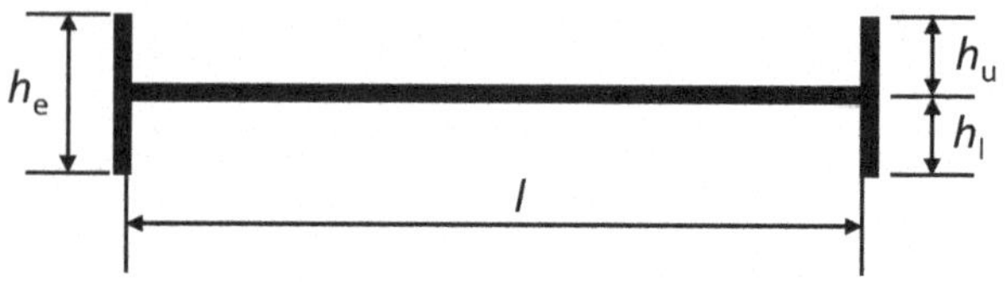

Fig. 9.12 Nomenclature for calculation of K_e

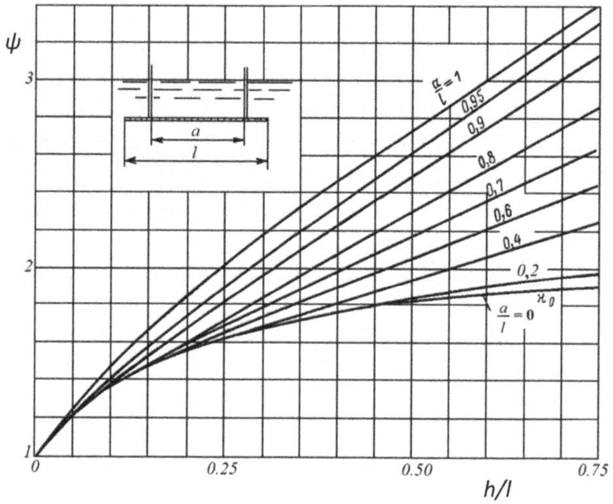

Fig. 9.13 Downwash correction factor $\psi = f\left(\frac{h}{l}, \frac{a}{l}\right)$ for hydrofoil with struts

so, the hydrofoil performance under the free water surface can be defined by aerodynamic test results of a biplane.

According to test results [2] the factor Υ of lift coefficient slope can be calculated as:

$$\gamma = 1 - (0.5 + \bar{t})\exp[-2(\overline{h})^{0.6} \tag{9.28}$$

Where:

$\bar{t} = \frac{t}{c}$ Relative thickness of hydrofoil;
$\overline{h} = \frac{h}{c}$ Relative submerged depth of hydrofoil

The correction factor for downwash angle can be defined by the following empirical equation:

$$\sigma = \frac{2}{\psi} - 1 \tag{9.29}$$

Where $\psi = f\,(h/l,\ a/l)$ as defined by Fig. 9.13 that plots ψ of hydrofoil with twin struts in various submerged depths. Here the submerged depth and space of strut is

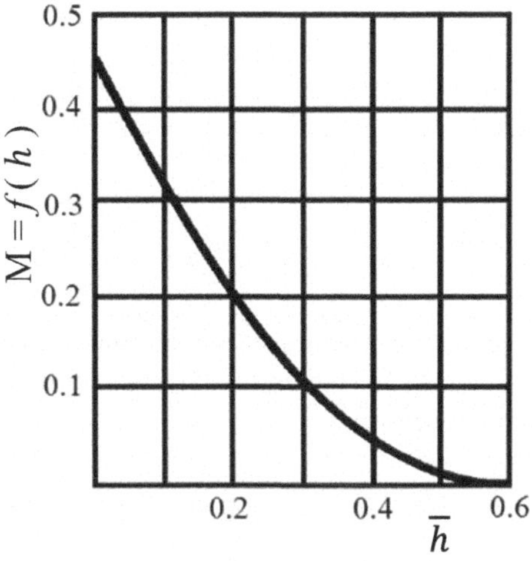

Fig. 9.14 Coefficient $m = f(\overline{h})$

relative to span, when $a = 0$, denotes no strut, and $h \to \infty, \psi \to 2, \sigma \to 0$, denotes no free water surface influence.

From Fig. 9.13, one can see for a strut present at a certain depth, for example that $\psi = 2$ for $a/l = 1$, and $h/l = 0.25$. Then the increase of downwash caused by the free surface compensates for the decrease of that by the strut, and for no free water surface present, when $\psi > 2$, the advantage of strut is greater than disadvantage of free water surface.

Change of zero lift angle of attack can be determined by the empirical equation:

$$\Delta a_0 = m\bar{t} \tag{9.30}$$

Coefficient $m = f(\overline{h})$ is a function of relative submerged depth $\overline{h}$ as shown in Fig. 9.14. The zero lift angle varies in terms of relative depth and relative thickness $\bar{t} = t/c$ since water surface deformation is affected by these parameters.

According to biplane theory, the coefficient of viscosity can be taken as $\eta = 0.88$, so, the lift coefficient accounting for the influence of free water surface is:

$$C_{Lh} = \frac{1.76\pi\gamma}{1 + \frac{1.76\gamma}{A}\left(\tau + \frac{2}{\partial}\right)K_e}[a - (a_0 + m\bar{t})] \tag{9.31}$$

The corresponding induced drag coefficient:

$$C_{D_i} = \frac{C_{L_h}^2}{\pi A}\left(\delta + \frac{2}{\partial}\right)K_e \tag{9.32}$$

For a thin foil the zero lift angle of attack in (9.31) can be taken as $\alpha_0 = -2\bar{f}$.

Using a circular disc test to define the influence coefficient of a free water surface S.G. Chugenov(С.Д.Чудинов) [3] considered the influence of the free water surface on the virtual mass of a circular disc moving under the water surface, and derived the free water surface influence coefficient, as well as influence coefficient of lift coefficient slope to be:

$$\gamma = 1 - \frac{1}{2}\exp\left[-2.37(\bar{h})^{2/3}\right] \tag{9.33}$$

For the correction factor of downwash, he used a different expression, and took $1 + \frac{K}{2} = \xi$ in the induced drag coefficient Eq. (9.18) then downwash can be represented as:

$$\Delta\alpha_{\mathrm{T}} = \frac{C_L}{\pi A}(1+\tau)\zeta \tag{9.34}$$

According to the circular disc test, the influence coefficient ζ due to the free water surface can be expressed as:

$$\zeta = 1 + \exp\left[-5.35(\bar{\mathrm{h}}/_{\mathrm{A}})^{2/3}\right] \tag{9.35}$$

The change of zero lift angle due to free surface is function of relative thickness $\bar{t}$ and coefficient Υ and can be expressed by:

$$\Delta\alpha_0 = \eta\bar{t}\left(\frac{1}{\gamma} - 1\right) \tag{9.36}$$

The corresponding lift coefficient is:

$$C_{L_h} = \frac{2\pi\eta\gamma}{1 + \frac{2\eta\gamma\zeta}{A}(1+\tau)K_e}[\alpha - (\alpha_0 + \Delta\alpha_0)] \tag{9.37}$$

In Chugenov's expression, the coefficient of viscosity can be taken as $\eta = 0.85$. The zero-lift angle of a thin foil is then $\alpha_0 = -\eta 2\bar{f}$ and the induced drag coefficient will be:

$$C_{D_i} = \frac{C_{L_h}^2}{\pi A}(1+\delta)\zeta K_e \tag{9.38}$$

The graphs showing Υ, calculated by (9.33) and ζ by (9.35), as a function of average immersion are as shown in Figs. 9.15 and 9.16.

It should be noted that the method above is for a fully submerged hydrofoil.

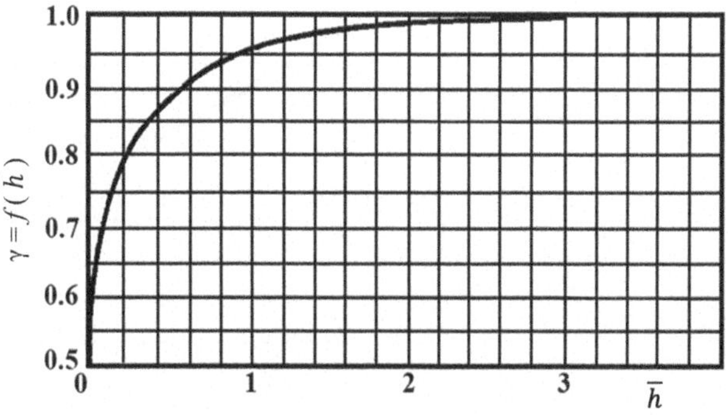

Fig. 9.15 $\gamma = f(\bar{h})$ from 9.33

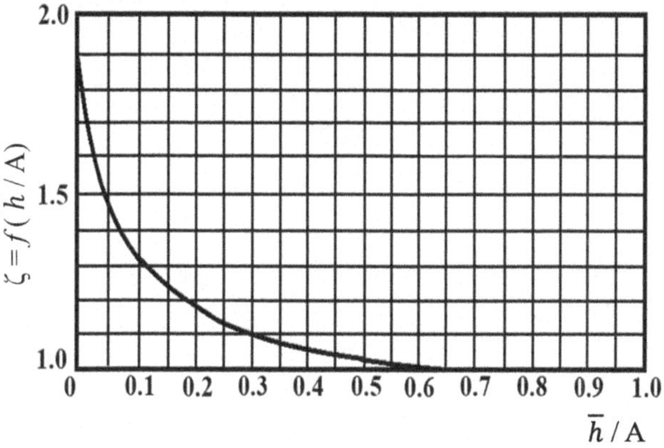

Fig. 9.16 $\zeta = f\left(\frac{\bar{h}}{A}\right)$ from 9.35

9.3.1 Estimation of Free Water Surface Influence for a Surface Piercing Hydrofoil

For a surface piercing hydrofoil without air ventilation on its foil back, the free water surface influence can be calculated by considering the equivalent flat foil, i.e., taking the average submerged depth of the surface piercing foil as the corresponding flat foil depth, and modifying the angle of attack as follows.

9.3.2 Average Submerged Depth of V Profile Foil

$$h_{cp} = \frac{1}{2} h_{\max}$$

For circular arc outer curve profile of surface piercing foil

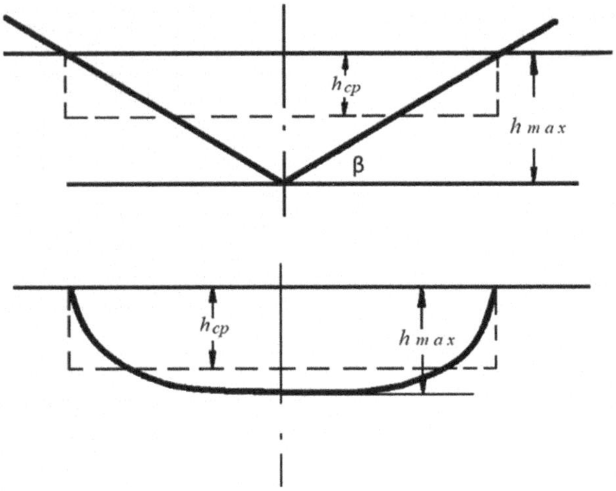

Fig. 9.17 average submerged depth of the surface piercing foil

$$h_{cp} = \frac{2}{3} h_{\max}$$

The average submerged depth of a surface piercing hydrofoil is shown in Fig. 9.17.

9.3.3 Correction to Angle of Attack

In Chap. 8, it was mentioned that for a foil with upward dihedral, the actual angle of attack would be smaller than the nominal angle of attack, being modified by cos β, thus:

$$\alpha = \alpha_A \cos \beta$$

Where

α_A angle of attack at center of foil, i.e., nominal angle of attack
β upward dihedral angle

The zero lift angle of attack would be increased by 1/ cosβ

That is $\alpha_0 = -\frac{2\bar{f}}{\cos \beta}$, where the section relative camber is $\bar{f} = \frac{f}{c}$

Considering this correction, the lift and drag coefficients can be calculated in relation to a corresponding flat foil, and so the lift coefficient can be expressed as:

$$C_{L_{hV}} = \frac{2\pi\eta\gamma}{\frac{1}{\cos \beta} + \frac{2\eta\gamma}{A}\left(\frac{2}{\partial} + \tau\right)K_e}\left[\alpha + \left(\frac{2\bar{f}}{\cos \beta} - \Delta\alpha_0\right)\right] \tag{9.39}$$

Where $\eta = 0.88$

Or using:

$$C_{L_{hV}} = \frac{2\pi\eta\gamma}{\frac{1}{\cos\beta} + \frac{2\eta\gamma\zeta}{A}(1+\tau)K_e}\left[\alpha + \left(\frac{2\overline{f}}{\cos\beta} - \Delta\alpha_0\right)\right] \tag{9.40}$$

Where $\eta = 0.85$

Induced drag can be obtained by (9.32) or (9.38) as well. Apart from the above equations, other more approximate equations can be used, as follows:

$$\frac{C_{Lh}}{C_{L\infty}} = 1 - 0.5\exp\left(-2.5\overline{h}\right) \tag{9.41}$$

Or

$$\frac{C_{L_h}}{C_{L_\infty}} = 1 - 0.5(10)^{-\overline{h}} \tag{9.42}$$

Or

$$\frac{C_{L_h}}{C_{L_\infty}} = K\left[1 - e^{-3\left(\overline{h}+0.1\right)}\right] \tag{9.43}$$

Where $K = 0.98 \sim 1.0$ for flat surface foil, and using the lower value for small aspect ratio foils. $K = 0.94 \sim 0.98$ for a V profile foil. For V profile foil, another special approximate equation is available as follows:

$$\frac{C_{L_h}}{C_{L_\infty}} = 1 - 0.422\ \ \exp\ \ \left(-1.454\overline{h_{cp}}\right) \tag{9.44}$$

Where $\overline{h_{cp}}$ average relative submerged depth.

Other than induced drag, there are wave making and viscous drags to consider. Wave making drag can be calculated using relations (9.19) or (9.20) and viscous drag can be obtained by test results for relevant foil sections, or equations in Sect. 8.4 used.

9.3.4 Hydrofoil Performance from Model Tests Under Free Water Surface

Hydrofoil model tests can be performed in a towing tank or circulating water tunnel to measure the characteristics of hydrofoils at various angles of attack and submerged depth. Since there are many hydrofoil sections and geometries, the working envelope is large, and a comprehensive series of test results are not available.

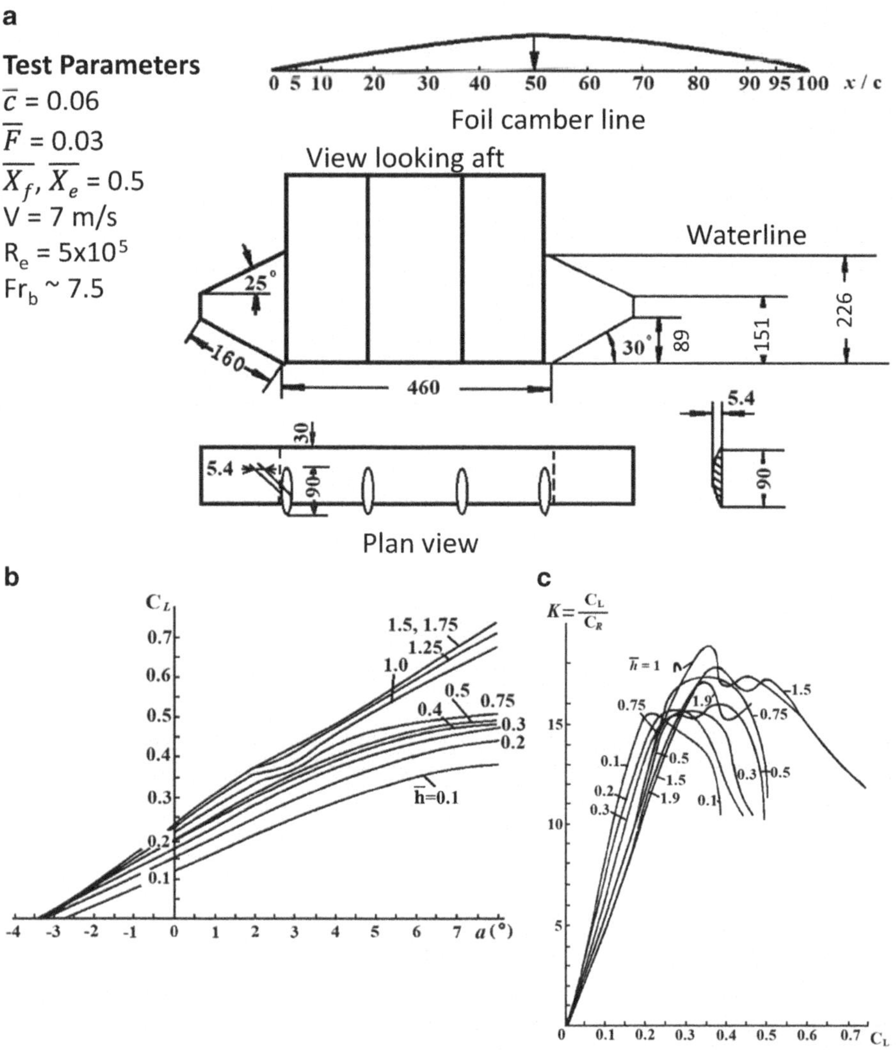

Fig. 9.18 Geometry and test results for trapezoid hydrofoil. (**a**) geometry diagrams, (**b**) $C_L = f(\alpha)$, (**c**) $K = f(C_L)$

Individual hydrofoil model test results for specific types of hydrofoil can be compiled and plotted against $C_L = f(\alpha)$ and $K = C_L/C_D = f(C_L)$ curves at various submerged depth $\bar{h}$ for design, including the definition of area and design angle of attack. Examples are shown in Figs. 9.18 and 9.19 for hydrofoils with trapezium and rectangular profiles.

The figures are taken from reference [4]. Such material can be used for estimating the hydrodynamic properties of hydrofoils with similar overall geometry. In the case

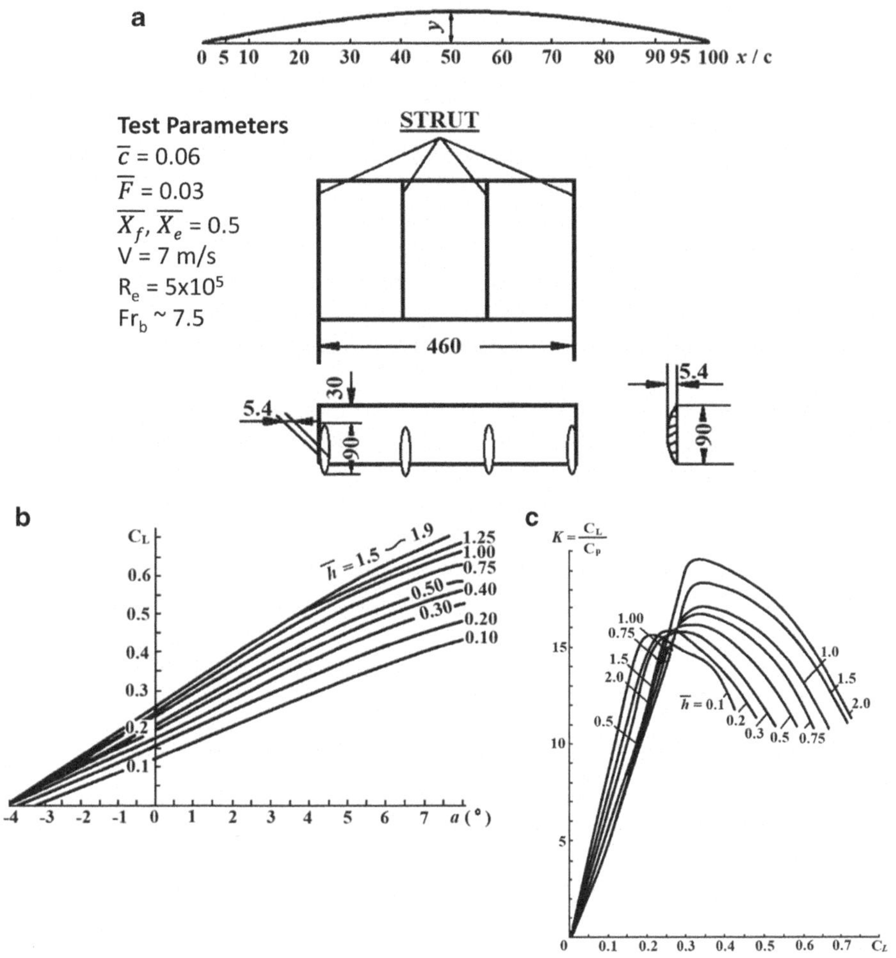

Fig. 9.19 Geometry and test results for rectangular hydrofoil. (**a**) geometry diagrams, (**b**) $C_L = f(\alpha)$, (**c**) $K = f(C_L)$

of different aspect ratio, and number of struts, the change of lift can be estimated approximately by the above methods.

The effect of different aspect ratio, and number of struts, and the change to drag can be estimated approximately in terms of the number of struts.

Figure 9.18 shows the variation in the lift coefficient against angle of attack for foil immersion between $\bar{h} = 0.1$ up to 1.75. Fig. 9.18c shows the variation of K with C_L for the same range of $\bar{h}$.

One can see that the highest lift to drag ratio is in the range of C_L from 0.3 to 0.6, which equates to an angle of attack between 3° and 4°. Similar results are shown for the geometry in Fig. 9.19.

9.4 Hydrofoil Cavitation and Its Avoidance

For a hydrofoil moving at sufficiently high speed, the pressure at the points on the foil back drop to the saturated vapor pressure of water. Molecules of water start to vaporize and form cavitation bubbles. When cavitation occurs at a foil surface, the flow is not continuous any more. A part of the hydrofoil contacts with water flow, and other parts contact with cavitation vapor. In consequence, the flow will be more complicated than the flow around a fully wetted hydrofoil.

Due to the flow around the hydrofoil being discontinuous, the hydrodynamic force acting on hydrofoil is different from fully wetted flow. Cavitation will reduce the hydrodynamic properties by reducing lift and increasing drag, thus it should be avoided as far as possible.

However, at high enough speed it cannot be avoided and then the hydrodynamic analysis should be carried out accounting for cavitation, and try to improve the flow, i.e., to design a hydrofoil section profile with stable cavitation over the back surface. For this reason, we will discuss methods for deferring or avoiding the cavitation, the rationale of cavitation and its effects, and evaluation of cavitation on performance. We will look at hydrodynamic analysis of a hydrofoil after cavitation appears, and hydrofoil profiles for super cavitation. The subject is complex, and so in this book we provide an overview.

9.4.1 *The Formation of Cavitation and Its Effects*

As a body moves in water the pressure at its surface changes, with the zone of low-pressure (accelerated velocity) dropping to saturated vapor pressure. Water starts to vaporize, and produce cavitation. Such low-pressure zones will appear at a hydrofoil back with camber and at angle of attack. They also occur on symmetric foils with certain thickness and zero angle of attack, such as struts, rudders, and other profiles such as guided missiles with cylindrical bodies, torpedoes, or submarines, bridge posts, dams, etc.

Tests demonstrate that saturated vapor pressure is not the unique condition for the generation of the cavitation. In purified water with no impurities or dissolved air gas molecules then cavitation will not occur, even at a pressure below to saturated vapor pressure. Air molecules or impurities are a necessary condition for cavitation. Water normally contains a lot of air and impurities and so in engineering one takes saturated vapor pressure as the initial pressure for cavitation inception.

The vapor pressure varies with water temperature, and so the inception of cavitation varies. Table 9.1 gives water vapor pressure versus temperature.

Table 9.1 Vapor pressure versus water temperature

Water temperature t°C	5	10	15	20	30	40	50	60	100
Vapor pressure p_d, kg/m^2	80	125	174	238	433	752	1258	2031	10,330

Since the minimum pressure is located at a certain point or local area of flow around the body, the cavitation will occur at certain points or local areas, and change with incoming flow condition. At increased speed or angle of attack, where the area of saturated vapor pressure extends increasingly, the corresponding range of cavitation expands gradually, to the whole of one or both sides of the body, or the whole foil back for a hydrofoil. The first stage of cavitation is forming the initial cavitation bubbles at the foil back and extending to full cavitation on the foil back a speed increases. Once the cavity covers the back surface completely the second cavitation stage starts, referred to as 'super cavitation'.

During the *first stage of cavitation*, i.e., local cavitation, the bubbles may not stay at a local position and will be pushed downstream by the flow vorticity. The pressure downstream is greater than saturated vapor pressure, so as to close the cavity.

In the first stage of cavitation, formation and collapse of the cavitation happens continuously and unstably. The collapse of cavitation bubbles will produce a great impacting force acting on the body surface, so as to cause material "Erosion" and the appearance of surface pitting, which can seriously damage the structure. Such unstable formation and collapse of cavitation, also creates pressure vibration, and vibration noise.

When a hydrofoil is at this cavitation stage, due to the renewed application of pressure at the foil back as bubbles collapse there is no obvious influence of cavitation on the total lift of the hydrofoil, apart from some rapid variation due to turbulence.

In the *second stage of cavitation*, cavitation wholly covers the foil back, which experiences saturated vapor pressure constantly, even at increasing speed, so in this cavitation stage, foil lift will be smaller than in the no cavitation case, but with increased drag, so decreasing hydrofoil efficiency. Due to the cavitation covering the foil back, collapse of the cavitation will occur aft of the hydrofoil, thus no erosion is caused on the hydrofoil surface.

In general, at its operating condition, a hydrofoil should be designed to work with no cavitation, the so-called sub-cavitating hydrofoil on hydrofoil craft. If the speed of hydrofoil craft is high enough, and cavitation is impossible to be avoided, then hydrofoil should be designed to operate in the second stage of cavitation, i.e., working at super cavitation.

It is best to avoid the first stage of cavitation, due to cavitation erosion, cavitation noise, and high frequency vibrations, as well as the unstable pressure, causing the poor operating condition of the hydrofoil craft, with poor hydrofoil life, and with significantly increased drag. The challenge for design is that a hydrofoil has to accelerate from rest to the operating speed and thus pass through the speed region where bubble and partial cavitation will occur. This has led to development of foil cross section geometries with elements to trip full surface cavitation at low speed.

9.4.2 Cavitation Inception of Hydrofoil

The criterion for cavitation inception is that the lowest pressure on the foil back reaches saturated vapor pressure at the working temperature of the hydrofoil, $p_{\min} \le p_{\mathrm{d}}$.

Presented as a non-dimensional pressure coefficient, we have:

$$\xi = \frac{p_0 - p}{\frac{1}{2}\rho V^2}$$

The maximum pressure drop coefficient at lowest pressure point is:

$$\xi_{\max} = \frac{p_0 - p_{\min}}{\frac{1}{2}\rho V^2}$$

Where:

p_0 static pressure on hydrofoil section
$p_{\min}$ lowest pressure on hydrofoil section
p_{d} vapor pressure on hydrofoil section
V hydrofoil speed

Defining cavitation number as $\sigma = \frac{p_0 - p_d}{\frac{1}{2}\rho V^2}$, the condition for cavitation occurring is:

$$\xi_{\max} \ge \frac{p_0 - p_d}{\frac{1}{2}\rho V^2} = \sigma \tag{9.45}$$

Thus, cavitation will appear at the foil back, so long as the maximum pressure drop coefficient is greater than or equal to cavitation number.

From the definition of cavitation number σ it can be seen that σ varies primarily in terms of the operating condition of the hydrofoil—submerged depth, water temperature, and speed. The foil section profile and angle of attack will have a secondary effect if the foil is assumed thin and with low angle of attack, and may be ignored on initial investigation.

The cavitation number can be defined so long as the operating condition is defined. The maximum pressure drop coefficient $\xi_{\max}$ is a characteristic parameter of a foil section, which is a function of foil section profile and angle of attack. For a given hydrofoil section, the maximum pressure drop coefficient $\xi_{\max}$, at different angles of attack can be obtained by test results and lifting surface theory, and compared with σ, for judging cavitation occurrence or not on the hydrofoil.

In general, the cavitation curve for a given hydrofoil can be determined and allow assessment of the safe working zone of the foil without cavitation. The cavitation curve is for a given hydrofoil, for maximum pressure drop equal to cavitation number $\xi_{\max} = \sigma$. The curve of foil angle of attack corresponding with this critical

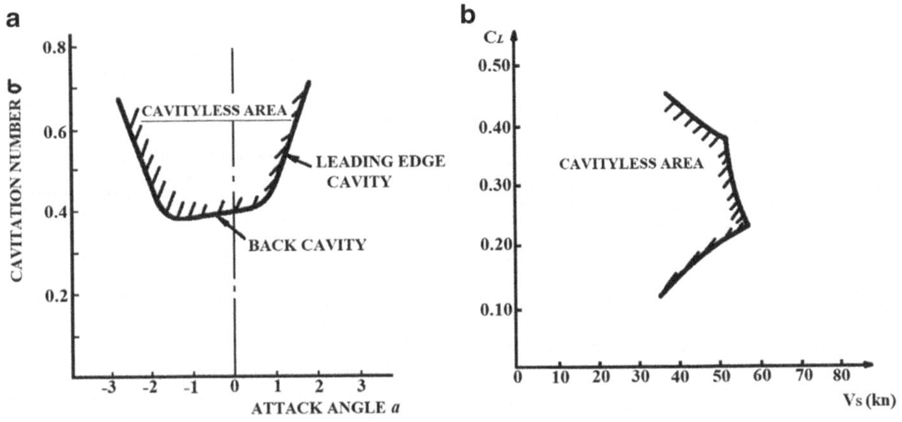

Fig. 9.20 Cavitation curves (**a**) vs angle of attack (**b**) vs lift coefficient and speed

cavitation number, is shown in Fig. 9.20a. Cavitation will not occur when the working condition is located within the boundary line, and vice versa. The cavitation boundary line also can be expressed in relation to lift coefficient and hydrofoil speed as parameters, as shown in Fig. 9.20b, since for a given hydrofoil and foil submerged depth the lift coefficient is related to angle of attack, and cavitation number corresponding to speed.

Approximate calculation equations for judging the appearance of cavitation can also be used. The pressure drop on a hydrofoil surface can be taken as the superposition of a part caused by symmetric thickness with no lift available and a part with lift, where the lift part includes the effect of camber and angle of attack. The pressure drop caused by foil symmetric thickness can be expressed approximately as:

$$\xi_{1\max} \approx 2.1\frac{t}{2x} \tag{9.46}$$

Where

- t Max thickness of hydrofoil
- x distance of maximum thickness location to leading edge of foil

The maximum pressure drop caused by lift can be expressed approximately as:

$$\xi_{2\max} \approx KC_L \tag{9.47}$$

Where

- C_L lift coefficient of hydrofoil;
- K parameter relevant to the position of maximum thickness. When located at 50% chord $K = 0.7$, at 30% chord $K = 0.85$.

So the total pressure drop coefficient is:

$$\xi_{\max} = 2.1\frac{t}{2x} + KC_L \tag{9.48}$$

When water temperature is 15 °C, $p_d = 174$ kg/m^2 and neglecting the static water head of hydrofoil takes atmosphere pressure $p_O = 10{,}330$ kg/m^2 the relationship between the cavitation number and hydrofoil speed is:

$$V = \sqrt{\frac{p_0 - p_d}{\frac{1}{2}\rho\sigma}} \approx \frac{14.0}{\sqrt{\sigma}}\,\mathrm{m/\,sec} \tag{9.49}$$

The condition for no cavitation is determined by putting $\xi_{\max} \leq \sigma$ into above equation, and taking $\xi_{\max}$ from (9.48) then the critical speed for avoiding cavitation is:

$$V_{cr} \leq \frac{14}{\sqrt{2.1\frac{t}{2x} + KC_L}}\,\mathrm{m/\,sec} \tag{9.50}$$

And

$$V_{cr} \leq \frac{27}{\sqrt{2.1\frac{t}{2x} + KC_L}}\ \text{knots} \tag{9.51}$$

9.4.3 *Avoiding Hydrofoil Cavitation: The Features of Sub-Cavitating Hydrofoils*

Knowing the causes for inception of cavitation as above, one can see the means for avoiding the cavitation are by enhancing the cavitation number σ and by decreasing the maximum pressure drop coefficient $\xi_{\max}$.

Cavitation number is mostly controlled by hydrofoil speed; however, that is usually pre-defined by the mission of the hydrofoil craft. Generally, for speed $V_s \leq 50$ knots, cavitation may be avoided, however if $V_s > 50$ knots, it is difficult to avoid serious cavitation, and the design approach to use supercavitating foils should be taken.

The decrease of maximum pressure drop coefficient $\xi_{\max}$ should not influence the decrease of total lift coefficient C_L as the decrease of C_L will cause the hydrofoil area to be increased, and so hydrofoil efficiency.

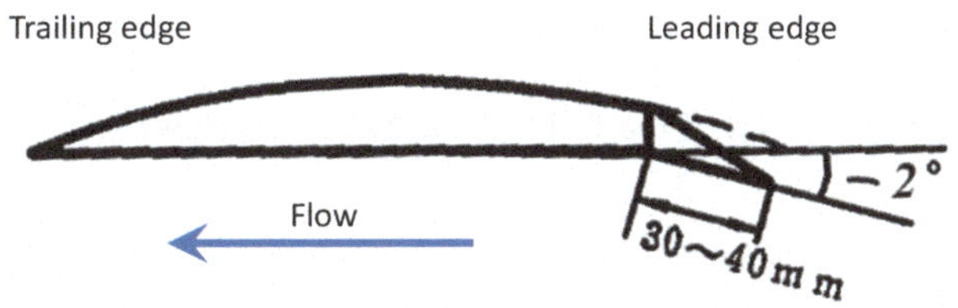

Fig. 9.21 Camber section with a small downward section at leading edge

Generally designers use means for decreasing ξ_{max} such as:

1. Selecting a hydrofoil section with uniform distribution of negative pressure on the foil back to decrease the maximum pressure drop at the same C_L. Carefully designed hydrofoil sections can give uniform negative pressure distribution. For example, the NACA-16 series profile, E series profile, and YS hydrofoil series.

   The leading edge section of the YS series has rather uniform negative pressure distribution. The lower surface is a straight line and upper surface is circular curve profile, with circular leading edge with small curvature radius. This geometry can be simply fabricated, so it is used as hydrofoil at critical speed.
2. The foil section thickness should be as small as possible, since this only generates a pressure drop and does not provide lift. Small thickness can enable uniform distribution of the negative pressure zone. According to operational experience with modern hydrofoils, for craft with $V_s \leq 50$ kts, the relative thickness of hydrofoil is typically about $\overline{t} = 6–8\%$.
3. Making the angle of attack close or equal to the condition of "shock-free entry." This implies no reverse flow from foil lower surface to upper surface, i.e., no flow around the leading edge. This is to avoid suction pressure at the leading edge and stagnation point toward the underside. For the leading edge, the angle of attack for "shock-free entry" is the zero lift angle of attack; however, this small angle of attack is impractical for the requirement of lift. In order to solve this problem, one may use a "crescent" section, or add a small downward section at the leading edge of the camber line, as shown in Fig. 9.21, which can maintain incoming flow with "shock free entry" at larger angle of attack. Experience from Russia showed that a section with a width of about 30–40 mm, and angle of about 2°, gave a good result. A Russian hydrofoil used this as an effective means to avoid cavitation on a foil with chord length of c = 930 mm, and small section located at a distance of 30 mm from leading edge, with downward angle of 2°. In order to avoid a great negative pressure at the leading edge, for using leading edge curved or crescent section profile, the leading edge should not be a sharp profile, and rather have a small curvature radius.

9.4.4 *Hydrofoil with Swept Plan Form*

A swept planform hydrofoil can generate lower incoming flow speed with the same section thickness, so as to decrease the maximum pressure drop, with the aim to defer hydrofoil cavitation.

In case of being unable to avoid cavitation after taking the steps described above, one can design the hydrofoil to operate with full cavitation over the upper surface, to avoid erosion of the foil from bubble or partial cavitation.

9.4.5 Hydrofoil Air Ventilation

From the point of view of hydrodynamics, air ventilation has a similar effect to cavitation. As the pressure on the foil upper surface is usually lower than atmospheric pressure, if the hydrofoil pierces the water surface as a surface piercing hydrofoil, or the strut of a hydrofoil pierces the water surface, air has an easy path to penetrate to the foil upper surface. In addition, air can penetrate to the upper surface via the free vortex at leading edge. All of this is called air ventilation. For air ventilation, atmospheric pressure acts on upper surface, and reduces the hydrodynamic performance.

Similar to cavitation, which destroys the continuity of flow, an air cavity adds one more boundary condition in the flow basic equations to be satisfied. This is the geometry of the cavity surface rather than the foil upper surface. However, *natural* air ventilation is unstable. It has been found in model tests that periodical pitching and heaving will occur due to *natural* air ventilation. Therefore, generally it is avoided on hydrofoil craft. To prevent ingress of air "fences" can be installed on struts and foils.

An alternative, in order to improve the maneuverability or sea-keeping of hydrofoil craft, an *artificial* air ventilation can be used. On hydrofoil craft with supercavitating hydrofoils, in order to operate stably during take-off, *artificial* air ventilation also can be applied, to maintain stability of the air cavity and craft motion. To do this ventilation passages can be bored through the foils from outer tips down to the lower area and vent holes arranged along the upper surface some way back from the leading edge.

Definition of a clear boundary for application of hydrofoil air ventilation seems not to be found so far, however study has shown that separation of the flow boundary layer is the first condition for air ventilation. The depth under the free water surface has direct influence, and minimizing the gradient of negative pressure on the upper foil surface, which may lead to flow boundary separation close to the surface, may be a favorite to avoid air ventilation. Alternatively increasing the submerged depth of the hydrofoil to increase the static pressure at the hydrofoil will defer the air ventilation. However, it is still difficult to establish accurate quantitative relationships to "design" the air ventilation analytically.

Generally, the means for avoiding the air ventilation are:

1. Decreasing the negative pressure gradient on the foil upper surface, so as to achieve a flat pressure distribution on foil, and small pressure drop on back surface of foil. This is similar to the means for avoiding the cavitation, on one hand, and on the other hand, is resisting air ventilation,

2. Installing air prevention fences is recommended on wave piercing hydrofoil craft, located on the upper surface of the foil piercing the water surface, and nearby the water surface, along the chord direction. For fully submerged hydrofoil craft, the fences can be positioned on the both sides of struts, piercing the water surface, and nearby the water surface. Practice demonstrates that such air preventing fences are effective for preventing air ventilation.

Regarding artificial air ventilation, the general approach is to duct lightly pressurized air into the lifting foils and to holes on the foil upper surface, for delivery to the space above the foil. The pressure may be equal to atmosphere pressure, which requires no power, or greater than atmosphere pressure, which needs additional power. The calculation of the hydrodynamic characteristics of such artificial air ventilation is similar to that for cavitation discussed above.

9.5 Hydrodynamic Characteristics of Cavitating Hydrofoils

For a high-speed craft, it is impossible to avoid cavitation, so the hydrodynamic characteristics of a cavitating foil need to be defined. At lower speed, the cavitation is only located over local areas of the back surface and not over the whole foil. This cavitation does not influence the steady hydrofoil lift and drag, causing just unsteady effects. We are concerned here with the hydrodynamic characteristics of cavitation at the second stage, i.e., cavitation is over the whole back surface.

9.5.1 Cavitation Flow and Its Simplified Model

The flow around the cavitation zone of a hydrofoil is very complex. There is a lot of vapor/air mixture, the density of which is significantly smaller than water, filling the cavitation zone. There is no sharp boundary surface between the cavitation cavity and main flow, i.e., between the liquid and gas, but a boundary area filled with large amount of small water drops and vapor.

At the downstream edge of cavitation, the flow is turbulent. The pressure beyond the cavitation cavity is greater than that in the cavity and so there is a back flow at the end of cavitation. This zone is filled with significantly chaotic and unstable air bubbles and vortices, which will be small and collapse downstream. Further downstream vortices will diffuse and flow stability is recovered.

M.P. Tulin presented a cavitation flow sketch [5], as shown in Fig. 9.22. The zone before the cavitation inception is called the near cavitation zone, and the flow zone after cavitation is called the far tail flow zone.

For such complex flow, it is very difficult to process even with the aid of mainframe computers, so a simplified method has to be set up in engineering practice. For research focused on forces acting on the flow around the foil body

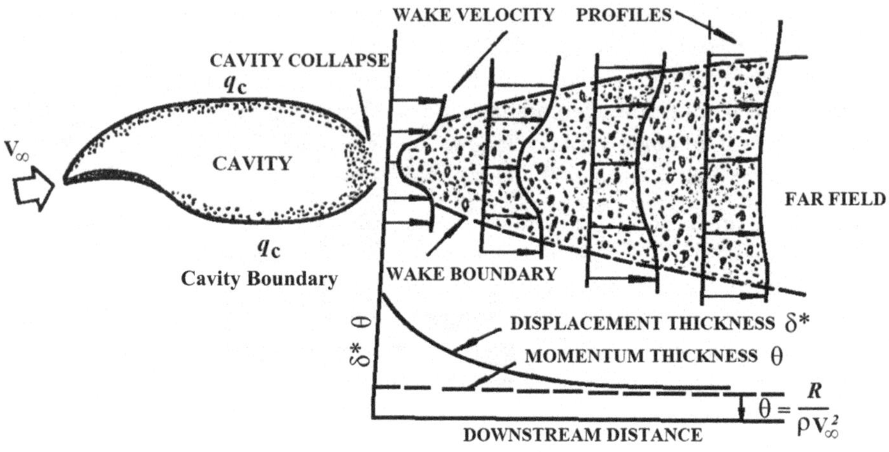

Fig. 9.22 Cavitation flow sketch

Fig. 9.23 Simplified flow field of fully cavitating hydrofoil

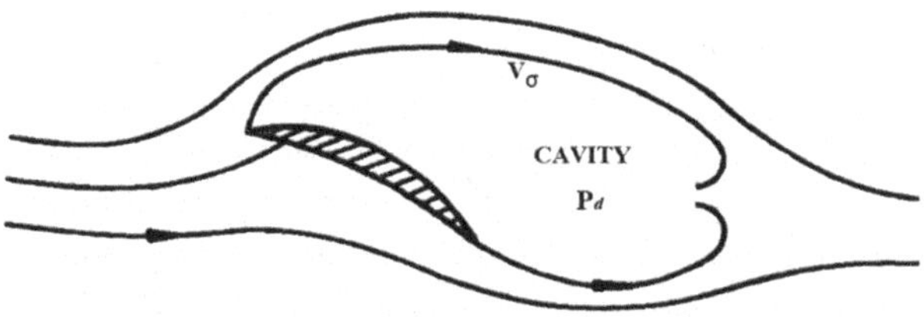

and noise during the collapse of cavitation, the investigation can be concentrated at the near cavitation zone, and neglecting the far tail flow zone, to form a simplified cavitation flow field as shown in Fig. 9.23.

In this simplified sketch, it is assumed that the cavitation boundary surface is one of vapor and water, the inner space is a cavitation cavity where pressure is constant, and equal to a vapor pressure p_σ of the saturated water at given temperature, and the potential flow is outside of this boundary. Flow viscosity is only considered in the boundary layer of flow around the foil, i.e., the pressure surface.

The residual issue is how to process the cavitation downstream edge, i.e., the simplified flow at the end of the cavity. The flow does not seal the tail end of cavitation cavity, as potential flow exists beyond the cavitation cavity, so a stagnation point at the end of the cavitation cavity has to exist, which will be in contradiction with the assumption of that the pressure at the cavitation boundary should be equal to saturated vapor pressure. In order to solve this issue, researchers have presented many models for two-dimensional steady cavitation flow, as shown in Fig. 9.24, and summarized below.

(a) *Image model*, which was presented by Riabouchinsky, for analyzing the cavitation flow around a vertical flat plate, and then developed for any angle of attack of hydrofoil. It uses a mirror image hydrofoil, taking the center of cavitation

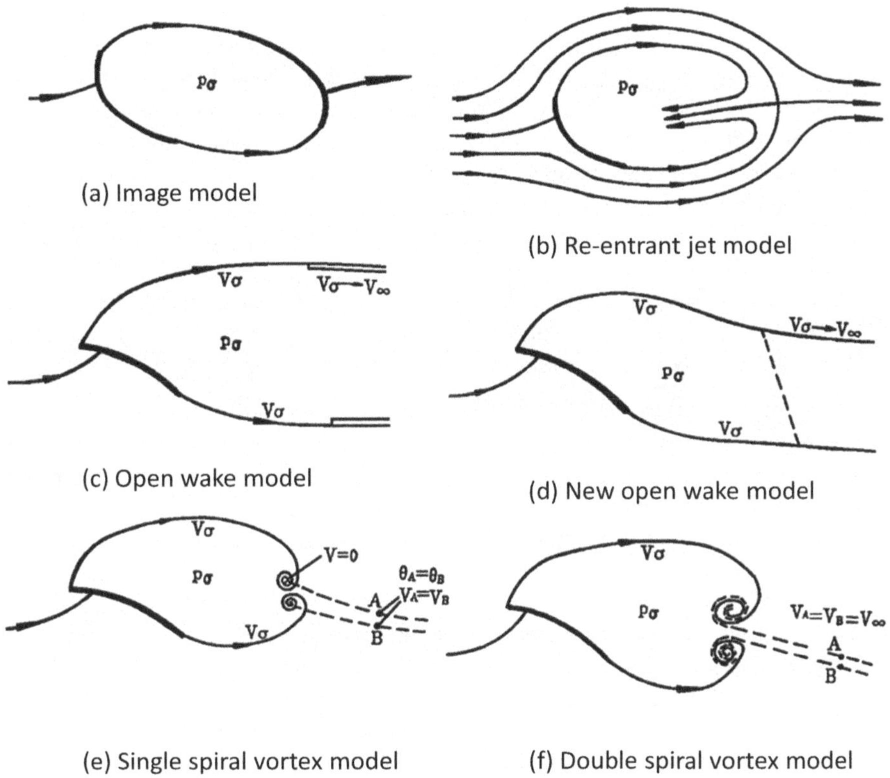

Fig. 9.24 various models for two dimensional cavitation flows

cavity as the point of symmetry, at the front end edge of cavitation, to downstream end of cavitation, as shown in Fig. 9.24a.

(b) *Re-entrant jet model*: Efros (Эфрос) presented a re-entrant jet model, as shown in Fig. 9.24b, where stagnation exists at the front end of cavitation, and a back jet coming into the end of cavity so that a point of stagnation occurs and the rear closure.
(c) *Open tail flow model*, as shown in Fig. 9.24c: dividing tail flow into two parts. In the tail flow nearby the body, pressure is maintained at p_σ, and the flow speed at the upper flow boundary is $V_\sigma =$ constant. Two parallel lines corresponding to two flat plates extending to infinite far distance end the cavitation. The direction of velocity on the parallel line stays constant, and is the same as the incoming flow direction. The value decreases from V_σ to $V_{\square}$, and pressure increases from p_σ to p_0.
(d) *New open tail flow model*, presented by Y.T.Wu in 1962, as shown in Fig. 9.24d. This was to improve the open tail flow model. It keeps the near tail zone the same as the open model, but taking two flow lines instead of two parallel lines in the far tail flow zone. The velocity direction and value gradually change to that of

the incoming flow at infinity downstream. At the end of cavitation, i.e., the initial points of two parallel lines, the direction and values of speed are the same.

(e) *Single screw vortex model*: Using a pair of screw vortices to end the cavitation cavity, which may also consider single and twin-screw vortex models. In the former, a flow line is led from both screw centers to extend to infinity, and the screw vortex generates the speed from V_σ at the center to V = 0, but the speed on the flow line induced from the center, changes from 0 to V□ at infinity, and the velocity potential on both flow lines at the corresponding position is equal, as shown in Fig. 9.24e.

(f) *Twin screw vortex model*: The flow speed on the cavitation boundary from V_σ via the vortex decreases to $V = 0$ at the vortex center, to V□ at infinity, so the speed is equal to V□ on both flow lines, as shown in Fig. 9.24f.

All of these models simplify the real model of cavitation tail flow, so it is impossible to obtain a solution the same as that in real conditions. It is clear that if the cavitation number is small enough and the cavitation length is great enough, the influence of cavitation end edge on flow around body is small. In this case neglecting the chaotic flow at the cavitation closure will not have a significant effect, however, for cavitation length close to body length, such neglect will lead to errors.

From the comparison of several models, it seems the back flow model is closest to the real condition of cavitation closure. This model must include a second surface, which leads to a great difficulty mathematically where the cavitation cavity length is equal to the body length. Modeling with a mirror image cannot be applied, so the cavitation cavity should be very long. In addition, it will lead to complicated numerical elliptical integration mathematically. In contrast, the open model is rather easy to be processed mathematically.

Y.T. Wu solved the problem of the cavitation flow around any foil profile. However, in this model a tail flow with same order of cavitation width exists at the cavitation end, which seems rather different from the real condition. In addition, the open model removes the inflexion point, which is also unreasonable.

The width of the far tail flow on the screw vortex model is rather small. This is called a semi-closed model. An inflexion point is also avoided at the cavitation end due to existence of the vortex, this seems rather reasonable physically and is also not too complicated mathematically, so such a model is more widely used.

The problem of solving cavitation flow is similar to ordinary hydrodynamic problems, and it can be solved by means of analytical expressions, i.e., using theoretical expressions to define a numerical solution. Since the cavitation flow is rather complicated, significant simplification has to be applied for solution with the analytical expression method.

There are two methods for solving the cavitation flow. One of these is the method presented by Kirchoff,[2] using free flow line theory to process the problem of flow around the body. Although a cavitation boundary profile is not given in the physical

[2]Kirchhoff equations—Wikipedia, note also H Lamb, Hydrodynamics, p 501.

plane, it can be defined in the complex potential plane, so long as taking the cavitation boundary flow line as the zero flow line. Then it is the real axis corresponding to the complex potential plane. Meanwhile, the cavitation flow line profile in the complex potential plane also can be defined. At constant velocity, it is a circular curve corresponding to the complex velocity plane. Thus, the transform function between the complex potential plane and complex velocity plane can be obtained by means of conformal transformation and the complex potential.

The Kirchoff problem is concerned with the separation of flow around the body, with the tail flow extended to infinity, and Kirchoff only obtained the solution in the vertical plane. Later Rayleigh[3] solved the separated flow from body problem on planes with any angle of attack. Levi-Civita[4] developed this to an arbitrary curved body. However, all of them focused on tail flow with infinite length, corresponding to cavitation number $\sigma = 0$ [6].

Y.T. Wu solved the problem with limited length tail flow at $\sigma > 0$ by means of the open model. Even though it is rather complicated, the basic idea is the same [7–9], and this free flow theory is used in cavitation flow problems, as it maintains the non-linearity of boundary conditions, so it is referred to as a nonlinear theory of cavitation flow.

Another method is a simplified linear method, similar to wing theory, using the high-speed small perturbation simplification, and neglecting all second order perturbations. The foil surface and cavitation boundary conditions on the real axis are satisfied approximately, to simplify the problem. For $\sigma = 0$ M. P. Tulin applied a thin wing analog method, and wisely transformed the cavitation flow to the flow around a thin wing, then used the thin wing solution to solve the hydrofoil cavitation problem [10].

For $\sigma > 0$ J. A. Geurst transformed it to a mixed boundary problem in a complex plane function to solve the problem [11]. Shi-Tang Dong took a closed and open model separately to solve the overall cavitation and local cavitation flow around the body with linear solutions [12, 13].

The problem can be simplified by means of a linearized method, but it is less clear in conception than free flow theory with velocity graphic expression, so Fuo-Kung Tang presented a perturbation velocity graphic method using linear theory [14], and tried to concentrate the advantages of both to solve the problem clearly and simply. He created a new approach to linearized theory based on this.

The cavitation theory has been limited to a two-dimensional problem in engineering practice, so we introduce the hydrodynamic calculation of two-dimensional cavitation flow as follows.

[3] Lord Rayleigh Cavitation modelling—Wikipedia.

[4] Levi-Civita Tullio Levi-Civita—Wikipedia.

9.5.2 Hydrodynamics of Two Dimensional Fully Cavitating Hydrofoil (at σ = 0)

$\sigma = 0$ denotes that pressure in a cavitation cavity is equal to incoming flow pressure, with cavitation length extending to infinity, as in the Kirchoff model, so the outcome of separated flow from the body can be applied directly. When $\boldsymbol{\sigma} = \mathbf{0}$ and the end of cavitation goes to infinity, the tail flow model is not so important and is convenient in the mathematical processing.

9.5.3 Flat Foil at σ = 0

Kirchoff's solution for separated flow around a vertical flat plate (Fig. 9.25) is:

$$C_p = \frac{2\pi}{4+\pi} \approx 0.88 \tag{9.52}$$

Kirhoff was using this to calculate the drag of a flat plate and found significant error compared with test results, due to assuming tail flow being a dead flow zone, where the speed of the free flow located on the boundary of the dead flow zone is changed suddenly, which was rather different from the practical flow.

However, in cavitating flow, the free flow line model is appropriate, as there is no flow in the cavitation cavity, so the speed at the cavitation boundary is actually changed suddenly. So, this research result is usable, and close to practice, so long as at positive small $\boldsymbol{\sigma}$ but not when equal to zero.

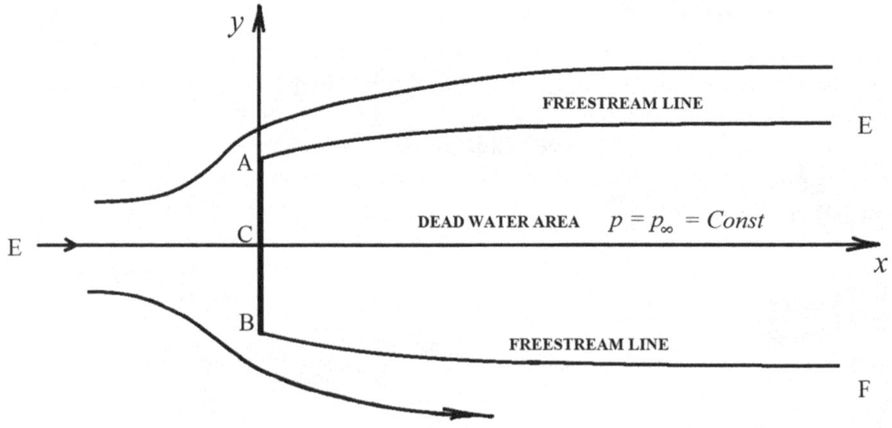

Fig. 9.25 Kirchhoff's flow model around a vertical flat plate

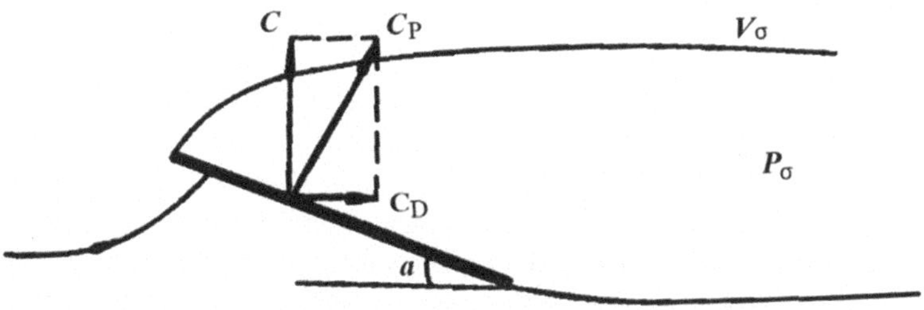

Fig. 9.26 separated flow around the inclined flat plate from Rayleigh

Later, Rayleigh obtained the pressure coefficient of separated flow around a flat plate at any angle of attack, i.e., at $\boldsymbol{\sigma} = \mathbf{0}$ see Fig. 9.26. The pressure coefficient of a flat plate foil in full cavitation:

$$C_p = \frac{2\pi \sin \alpha}{4 + \pi \sin \alpha} \tag{9.53}$$

Corresponding lift coefficient:

$$C_L = \frac{2\pi \sin \alpha}{4 + \pi \sin \alpha} \cos \alpha \tag{9.54}$$

Corresponding drag coefficient:

$$C_D = \frac{2\pi \sin^2 \alpha}{4 + \pi \sin \alpha} \tag{9.55}$$

The detailed solutions for Kirchoff and Rayleigh's method can be obtained in hydrodynamics textbooks [6].

There is no linearized simplification in free flow theory, so it can be applied for any angle of attack. For small angles of attack α, the Eqs. (9.54) and (9.55) can be simplified as:

$$C_L = \frac{\pi}{2} \alpha \tag{9.56}$$

$$C_D = \frac{\pi}{2} \alpha^2 \tag{9.57}$$

It is interesting to compare the calculation result for a fully cavitating flat plate foil with two-dimensional flow fully around the flat plate foil and planing flat plate, as follows:

	Lift coefficient C_L	Drag coefficient C_D
Flow fully around the flat plate	$2\pi\alpha$	0
Planing flat plate	$\pi\alpha$	$\pi\alpha^2$
Fully cavitating flat plate	$\pi\alpha/2$	$\pi\alpha^2/2$

The lift coefficient slope of a planing flat plate is half that of a fully immersed flat plate, and the lift coefficient of the flat plate with full cavitation is half of that of a planing flat plate, and similarly for the drag coefficient. Due to the suction force available at the leading edge of a flat plate with fully immersed flow, the form drag is equal to zero. Such drag on a planing flat plate is referred to as spray drag and called cavitation drag at a cavitating foil. Both of these are pressure difference drag components.

Levi-Civita solved the problem of the separated flow around any section profile by means of free flowline theory.[5] Since flow is nonlinear around the body of a profile section, it is too complicated to apply the Levi-Civita solution for normal engineering practice, and the simplified linear theory will be more favorable.

9.5.4 Linearized Theoretical Solution for Camber Foil at $\sigma = 0$

For a thin foil with $\boldsymbol{\sigma} = \mathbf{0}$, Tulin used a linearized thin foil analog method, and achieved a simplified outcome [10], with lift and drag coefficients for a flat plate foil with $\boldsymbol{\sigma} = \mathbf{0}$, as in (9.56) and (9.57), the same as the results of theoretical calculation by free flow line theory with small angle of attack α.

For a segment angle of a circular curve of 2γ, with angle of attack α of circular curve foil, as shown in Fig. 9.27, Tulin derived the following:

$$\left.\begin{aligned} C_L &= \frac{\pi}{2}\left(\alpha + \frac{7}{8}\gamma\right) \\ C_D &= \frac{\pi}{2}\left(\alpha + \frac{\gamma}{4}\right)^2 \end{aligned}\right\} \tag{9.58}$$

Where

α geometric angle of attack, measured from chord line
γ half of center angle of circular curve

Using relative camber $\bar{f} = f/c$ then: (where f here is max height from chord line)

[5] See (PDF) On the lift and drag of cavitating profiles and the maximum lift and drag (researchgate.net).

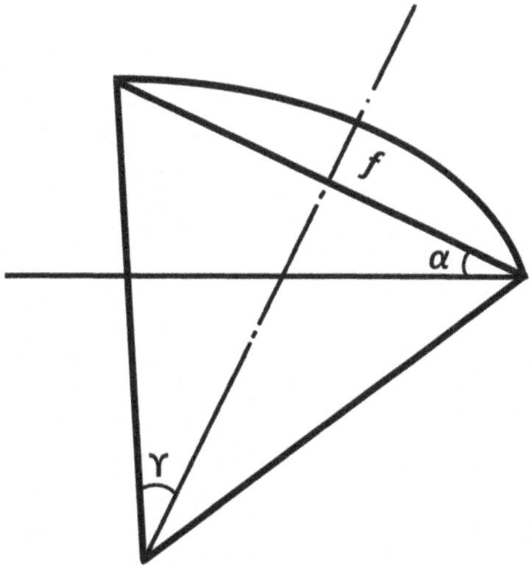

Fig. 9.27 circular curve foil diagram

$$\left.\begin{aligned} C_L &= \frac{\pi}{2}\left(\alpha + \frac{7}{2}\bar{f}\right) \\ C_D &= \frac{\pi}{2}\left(\alpha + \bar{f}\right)^2 \end{aligned}\right\} \quad (9.59)$$

From (9.58) or (9.59), it can be seen that the camber foil with full cavitation is similar to that with full surrounding flow. The lift can be divided into parts related to angle of attack and camber respectively, and the latter can be presented as the zero angle of attack for the thin foil.

To distinguish between the two for a cavitating foil, the latter can be described as "Camber equivalent to angle of attack with same lift," which is not equal to zero lift angle of attack. For different camber, the expression of camber equivalent to angle of attack with same lift result is different as well. The value $7/2\bar{f}$ in (9.59) is only for a circular curve foil. The difference of the cavitating foil from flow fully around the foil is drag on the cavitating foil, even when not considering the viscosity, and also "camber equivalent to angle of attack with same drag" is available, the value of which is different from "camber equivalent to angle of attack with same drag" for the fully immersed foil.

Due to the cavitation drag on a cavitating foil, increasing the lift by means of increasing angle of attack and camber is rather different, as shown in (9.58) or (9.59) as follows. When increasing camber, the increase of lift is significantly greater than that of drag. When increasing angle of attack, the increase of lift and drag is almost the same. This can be demonstrated quantitatively by the comparison of the lift/drag ratio of a flat plate foil with angle of attack, and a foil with circular curve camber and without angle of attack, as follows:

For flat plate foil:

$$\frac{C_L}{C_D} = \frac{\frac{\pi}{2}\alpha}{\frac{\pi}{2}\alpha^2} = \frac{\pi}{2}\frac{1}{C_L} \tag{9.60}$$

For circular curve foil with zero angle of attack:

$$\frac{C_L}{C_D} = \frac{\frac{\pi}{2}\left(\alpha + \frac{7}{2}\bar{f}\right)}{\frac{\pi}{2}\left(\alpha + \bar{f}\right)^2} = \frac{49}{4}\frac{\pi}{2}\frac{1}{C_L} \tag{9.61}$$

In order to compare the efficiency at same lift, taking lift/drag ratio as a function of C_L, one can see that at the same lift value, the increase of efficiency by means of camber is up to 12 times greater than that of angle of attack. Of course this is an extreme condition, and it is impossible practically to form full cavitation at zero angle of attack, so a certain angle of attack is necessary. Meanwhile the lift/drag ratio is:

$$\frac{C_L}{C_D} = \frac{\alpha + \frac{7}{2}\bar{f}}{\left(\alpha + \bar{f}\right)^2} = \left(\frac{1 + \frac{7}{2}\frac{\bar{f}}{\alpha}}{1 + \frac{\bar{f}}{\alpha}}\right)^2 \frac{\pi}{2}\frac{1}{C_L} \tag{9.62}$$

This means the efficiency of a camber foil is higher than a flat plate foil by $\left(\frac{1+\frac{7}{2}\frac{\bar{f}}{\alpha}}{1+\frac{\bar{f}}{\alpha}}\right)^2$ times, which is greater than 1 and decreases as angle of attack is increased.

For other camber foils, it is similar, thus in fully cavitating flow, the efficiency of a flat plate foil is least. Since the back surface of a fully cavitating foil is located in a cavitation cavity, the profile of the foil back does not influence the hydrodynamics.

This conclusion for the flat plate foil implies that all foils where the lower surface (pressure surface) is a flat plane will have low efficiency in cavitating flow, including the rounded leading-edge foil, which is considered as the better profile in sub-cavitating flow.

Due to this it is best to take the angle of attack of a fully cavitating hydrofoil as small as possible, to start cavitation at the leading edge and keep the foil back located in a cavity.

The angle of attack may be a little greater than "shock free entry angle," in terms of sharpness of the leading edge and other factors. Thus, for the cavitating foil, the "shock free entry angle" is an important parameter. For a circular curve foil, research demonstrates "shock free entry angle" is about $\Upsilon/4$, where Υ is half of segment angle of the circular curve of the camber line.

When "shock free entry angle" is achieved the lift/drag ratio is:

$$\frac{C_L}{C_D} = \frac{\frac{\nu}{4} + \frac{7}{8}\nu}{\left(\frac{\nu}{4} + \frac{\nu}{4}\right)^2} = \frac{81}{16}\frac{\pi}{2}\frac{1}{C_L} \tag{9.63}$$

This is almost five times that of a flat plate foil, though practically it may be less than this since angle of attack is a little bit greater than this mentioned above.

Based on a linearized assumption Tulin obtained hydrodynamic coefficients of a camber foil at zero cavitation number, however, only suitable for a large angle of attack. For a camber foil with a large angle of attack, the Rayleigh calculation results can be developed and applied approximately with aid of "camber equivalent to angle of attack with same result." For a flat plate foil with large angle of attack, Rayleigh's lift coefficient is:

$$C_L = \frac{2\pi \sin \alpha}{4 + \pi \sin \alpha} \cos \alpha$$

Changing angle of attack α to angle of attack α plus angle of attack with respect to camber α_c, equivalent to the same lift, we obtain:

$$C_L = \frac{2\pi \sin(\alpha + \alpha_c)}{4 + \pi \sin(\alpha + \alpha_c)} \cos \alpha_n \tag{9.64}$$

Where α_n is the angle between the total pressure vector and the vertical, as shown in Fig. 9.28. In general, the direction of the total pressure vector is almost perpendicular to the chord line, and so we can take $\alpha_n \approx \alpha$. The drag coefficient can also be dealt with by the same method, but it must be noted, that "camber equivalent to angle of attack with same drag" is different from "camber equivalent to angle of attack with same lift."

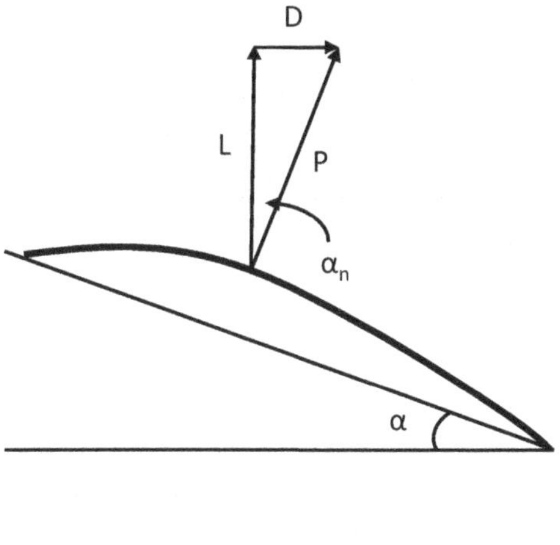

Fig. 9.28 definition diagram for α_n

Thus:

$$C_L = \frac{2\pi \sin(\alpha + \alpha_c)}{4 + \pi \sin(\alpha + \alpha_c)} \cos\alpha \tag{9.65}$$

$$C_D = \frac{2\pi \sin(\alpha + \alpha_{c1})}{4 + \pi \sin(\alpha + \alpha_{c1})} \sin\alpha \tag{9.66}$$

It may be noted that for a circular curve foil, $\alpha_c = \frac{7}{2}\overline{f}$, $\alpha_{c1} = \overline{f}$.

9.5.5 Two-Dimensional Fully Cavitating Foil at σ > 0

For a two-dimensional fully cavitating foil at $\sigma > 0$, a linearized method can be used; however, the calculation is still complex, so here we introduce a simple approximate estimation method as follows.

9.5.6 Betz's Zero Approximation Method

For $\sigma > 0$ cavitation length is limited. We take a small unit on the foil face surface as ds on which is acting pressure p, the pressure in the cavitation cavity is p_σ, the cavitation boundary speed is V_σ, and incoming flow speed is V_∞ as shown in Fig. 9.29.

The pressure coefficient on whole foil:

$$\begin{aligned} C_P(\sigma) &= \frac{\int_0^c (p - p_\sigma)ds}{\frac{1}{2}\rho V_\infty^2 c} \\ &= \frac{\int_0^c (p - p_0)ds + \int_0^c (p_0 - p_\sigma)ds}{\frac{1}{2}\rho V_\infty^2 c} \\ &= \frac{\int_0^c (p - p_0)ds}{\frac{1}{2}\rho V_\infty^2 c} + \frac{p_0 - p_\sigma}{\frac{1}{2}\rho V_\infty^2} \end{aligned} \tag{9.67a}$$

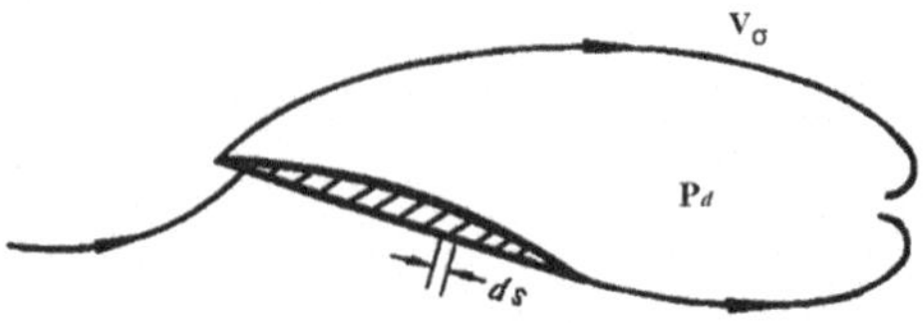

Fig. 9.29 Diagram for definition of Betz's method

Where the first item in (9.67a) corresponds to the foil pressure coefficient at zero cavitation number, and the second item corresponds to the definition of cavitation as follows:

$$C_p(\sigma) = C_p(0) + \sigma \tag{9.67b}$$

This is Betz's result with zero approximation, i.e., at $\sigma > 0$ the pressure coefficient is equal to the pressure coefficient at $\sigma = 0$ plus cavitation number.

This method is very simple to apply for engineering practice since the calculation result for $\sigma = 0$ is easily to be obtained. However it is rather rough, only being usable for very small cavitation number because it assumes $\frac{\int_0^c (p - p_0) ds}{\frac{1}{2}\rho V_\infty^2 c}$ is the pressure coefficient $C_p(0)$ at $\sigma = 0$. It seems correct in form; however, the pressure here p should be the pressure at $\sigma > 0$, but not at $\sigma = 0$, and for real $C_p(0)$, the p in integration has to be taken as that when $\sigma > 0$, i.e., $p(\sigma) = p(0)$, and this is very approximate.

9.5.7 Richardt's Zero Approximation

Richardt determined the pressure coefficient at $\sigma > 0$ by another approximate equation. According to Bernoulli's equation, the pressure relationship can be transformed to a velocity relationship as:

$$p - p_\sigma = \frac{1}{2}\rho\left(V_\sigma^2 - V^2\right), \text{and so: } \frac{p - p_\sigma}{\frac{1}{2}\rho V_\infty^2} = \frac{V_\sigma^8}{V_\infty^2}\left(1 - \frac{V^2}{V_\sigma^2}\right)$$

Where p and V are pressure and velocity at any point on the foil surface. If we put this into a pressure coefficient expression, then:

$$\begin{aligned} C_p(\sigma) &= \frac{\int_0^c (p - p_0) ds}{\frac{1}{2}\rho V_\infty^2 c} = \frac{V_\sigma^2}{V_\infty^2 c}\int_0^c \left(1 - \frac{V^2}{V_\sigma^2}\right) ds \\ &= \frac{1 + \sigma}{c}\int_0^c \left(1 - \frac{V^2}{V_\sigma^2}\right) ds \end{aligned}$$

Using the relationship $V_\sigma^2 / V_\infty^2 = 1 + \sigma$ in the above equation, an equation can be derived, taking Bernoulli's equation at two points, one on the cavitation flow line and one infinitely far forward. Richardt assumed $V / V_\infty = f(s)$ is function of foil profile, then

$$C_p(\sigma) = (1 + \sigma)\frac{1}{c}\int_0^c \left[1 - f^2(s)\right] ds = (1 + \sigma)F \tag{9.68}$$

Where

$F = \frac{1}{c}\int_0^c \left[1 - f^2(s)\right] ds$ is a function of the foil profile.

When $\sigma = 0$ above equation changes to $C_p\ (0) = F$, so that

$$C_p\ () = (1 +)\ C_p\ (0) \tag{9.69}$$

Since the above assumes V/V_∞ is only the function of foil profile, it is not precise. Actually, the speed on the foil surface is not only a function of foil profile, but also of cavitation number, and so $V/V_\infty = f(s, \sigma)$, i.e., $f(s, \sigma)$ also is a function of foil profile and cavitation number. When putting $\sigma = 0$ into (9.68) and defining F, we obtain the solution of $F(s, 0)$ at zero cavitation number but not at $\sigma > 0$ of the solution of $F(s, \sigma)$ so it is rather approximate in (9.69). This is nevertheless suitable for small σ values. In general, the result of Richardt's solution is better than Betz's.

By the way, why we call the approximation mentioned above as zero approximation? This is due to when σ is a small value, the function of σ at the area close to $\sigma = 0$ can be developed as a power series.

In Richardt's equation, F can be developed to the following relationship:

$$F = F(0) + \alpha_1\sigma + \alpha_2\sigma^2 + \alpha_3\sigma^3 + \cdots\cdots$$

The first item $F(0)$ is the value at $\sigma = 0$. At present, we only take the first item, i.e., the zero power of σ so call this a zero-order approximation. For Betz's approximation, the surface pressure p also can be developed at close to $\sigma = 0$, and we also take a zero approximation.

9.5.8 Numerical Calculation of Cavitating Hydrofoil at $\sigma > 0$

After numerical calculation in terms of cavitation theory, the approximate expressions for hydrodynamic coefficients at $\sigma > 0$ can be obtained with more precision [15]. Equation (9.70) is the hydrodynamic coefficient equation developed by means of comparing the numerical calculation results with the result at $\sigma = 0$:

$$\left.\begin{aligned} C_L(\sigma) &\approx C_L(0)(1+\sigma)\left[1 + \frac{9\pi}{32}\frac{(1-0.5\sigma)}{(1+\sigma)}\frac{\sigma^2}{C_D(0)} \times \left(1 - \frac{\frac{\pi}{2}\alpha_c}{C_L(0)}\right)\right] \\ C_D(\sigma) &\approx C_D(0)(1+\sigma)\left[1 + \frac{9\pi}{32}\frac{(1-0.5\sigma)}{(1+\sigma)}\frac{\sigma^2}{C_D(0)} \times \left(1 - \frac{\alpha_c}{4\sqrt{C_D(0)}}\right)\right] \end{aligned}\right\} \tag{9.70}$$

Where

$C_L\ (0)$, $C_D\ (0)$ are lift and drag coefficients at $\sigma = 0$ respectively

α_c camber equivalent to the angle of attack with same lift

Where the second order item of σ is reserved, one can see (9.70) is better than zero order approximation, but still neglects the higher order of σ. When $\sigma^2/C_D(0) \ll 1$, the Richardt's zero order approximation can be obtained so long as the second item is included in the bracket. Thus, the lift and drag coefficients at $\sigma > 0$ can be obtained based on the calculation results at $\sigma = 0$. For a circular curve foil, the following approximate equation can be used:

$$\left.\begin{aligned} C_L(\sigma) &\approx \frac{\pi}{2}(1+\sigma)\left(\alpha+\frac{7\nu}{8}\right)+\sigma^2(1+\sigma)\frac{3\pi}{32}\times\frac{1}{\alpha+\frac{\nu}{4}} \\ C_D(\sigma) &\approx \frac{\pi}{2}(1+\sigma)\left(\alpha+\frac{\nu}{4}\right)^2+\sigma^2(1+\sigma)\frac{3\pi}{32}\times\left(\frac{\alpha+\frac{7\nu}{24}}{\alpha+\frac{\nu}{4}}\right)^2 \end{aligned}\right\} \tag{9.71}$$

Where

ν half of circular center angle

The first term in both Eqs. 9.70 and 9.71 is Richardt's zero order approximation, here the second and third order items are added, but neglecting higher order of σ.

For a two-dimensional cavitating foil, it is not difficult to carry out the numerical calculation for the hydrodynamic coefficient of various flow models; however, such calculation only can be applied for comparison qualitatively in engineering practice, as it is a simplification. It is therefore suggested to use the approximate method for engineering practice.

9.6 Influence of Free Water Surface and Correction for Limited Span

The influence of a free water surface on the pressure distribution around a fully cavitating hydrofoil and the numerical calculation for a fully cavitating hydrofoil with limited span have been studied based on the simplified flow analytical model for the cavitation boundary. However, the calculation is complicated and has poor credibility, so here we introduce some approximate estimation methods based on simplified processing from Tulin's work [16].

1. *Influence of Free Water Surface on Cavitating Foil*

For a fully cavitating foil with zero cavitation number ($\sigma = 0$) under the free water surface, Tulin's linearized thin foil analog method can be applied, using a mirror image to represent the influence of the free water surface to carry out the calculation. V. E. Johnson developed the approximate calculation equation as follows:

$$\left.\begin{aligned} C_{\mathrm{L}}\left(0,\ \frac{h}{c}\right) &= \frac{\pi}{2} f_{\alpha}\left[\alpha + \frac{f_K}{f_{\alpha}}\frac{2}{\pi} C_{Lf}\right] \\ &= \frac{\pi}{2} f_{\alpha}\alpha + f_K C_{Lf} \\ C_{\mathrm{D}}\left(0,\ \frac{h}{c}\right) &= \frac{C_{\mathrm{D}}(0)}{C_{\mathrm{L}}(0)} C_{\mathrm{L}}\left(0,\ \frac{h}{c}\right) \end{aligned}\right\} \tag{9.72}$$

Where

$C_L\left(0,\ \frac{h}{c}\right)$ and $C_D\left(0,\ \frac{h}{c}\right)$ are the lift and drag coefficient for $\sigma = 0$, at relative submerged depth h/c.

$f_{\alpha} = \left\{1 - \frac{1}{2}\left[\left(1 + \sqrt{\frac{h}{c}}\right)\left(1 + 2\sqrt{\frac{h}{c}} + \frac{5}{4}\frac{h}{c}\right)^{-1}\right]\right\}^{-1}$ is the lift coefficient correction due to the angle of attack, where

$C_{Lf} = \frac{\pi}{2}\alpha_c$ lift coefficient due to the camber;
$f_K = (1 - \beta) + f_{\alpha}\beta$ lift correction factor due to the camber;

Where

α_c camber equivalent to angle of attack with same lift
β parameter due to camber,

For a circular curve foil, $\alpha_c = \frac{7}{8}$, $\gamma = \frac{7}{2}\bar{f}$ (Υ is half of center angle of the circular curve segment, where $\bar{f}$ is relative camber). When there is no precise value available for β, the approximate value can be taken as $\beta = 0.3$.

For $\sigma > 0$ values can be calculated by (9.72) at $\sigma = 0$, then carry out the correction of cavitation number in terms of method outlined in Sect. 3.

2. *Influence of Limited Span on Cavitating Foil*

Johnson took Tulin's analog method for a thin foil and carried out the correction for limited span of a fully cavitating foil, at $\sigma = 0$, using the concept of downwash correction of the flow fully around the foil. If there is free water surface to consider, the calculation form is as same as for a thin foil:

$$C_L(0) = \frac{1}{1 + \frac{2}{A}}\frac{\pi}{2}(\alpha + \alpha_c) \tag{9.73}$$

Where $\frac{\pi}{2}(\alpha + \alpha_c)$ is lift coefficient of a two-dimensional cavitating foil, so one can see the correction for limited span is similar to that in flow fully around the foil.

When the free water surface is considered, C.C. Hsu presented an approximate equation as follows [15]:

$$
\left.\begin{aligned}
C_{\mathrm{L}}\left(0,\ \frac{h}{c}\right) &= \frac{A}{f_a+1+A}C_{L,2D}\left(0,\ \frac{h}{c}\right)\\
C_{\mathrm{D}}\left(0,\ \frac{h}{c}\right) &= \frac{C_D}{C_L}C_L
\end{aligned}\right\} \tag{9.74}
$$

Where

A aspect ratio of hydrofoil;
$C_{L,2D}\left(0,\ \frac{h}{c}\right)$ lift coefficient of two-dimensional hydrofoil at $\sigma = 0$
$\frac{h}{c}$ relative depth of submergence
$\frac{C_D}{C_L} = \left(\alpha+\alpha_i+\frac{A_1}{2}\right)^2\left(\alpha+\alpha_i+C_{Lf}\right)^{-1}$
$\alpha_i = \frac{C_L}{\pi A}$ downwash angle
f_2 and C_{Lf} have the same meaning as in (9.72)
A_1 Coefficient in the expression for the thin foil vortex with same result as shown in (8.20)

Equations (9.73) and (9.74) are only suitable for foils with high aspect ratio. For low aspect ratio where ($A \le 1$) at zero cavitation number, the approximate equation is:

$$
\left.\begin{aligned}
C_{\mathrm{L}}\left(0,\ \frac{h}{c}\right) &= \frac{A}{1.5f_a+1}C_{L2D}\left(0,\ \frac{h}{c}\right)\\
C_{\mathrm{D}}\left(0,\ \frac{h}{c}\right) &= \left(\alpha+\frac{A_1}{2}\right)^2\left(\alpha+C_{Lf}\right)^{-1}C_L
\end{aligned}\right\} \tag{9.75}
$$

The definition of items in the equation are the same as that before.

For the short foil with medium span ($1 < A < 3$), there are still no suitable equations so far. C.C. Hsu has suggested adding a modification coefficient in (9.74) as:

$$
\left.\begin{aligned}
C_L\left(0,\ \frac{h}{c}\right) &= \frac{A}{(f_a+1+A)\alpha_c}C_{L2D}\left(0,\ \frac{h}{c}\right)\\
C_D\left(0,\ \frac{h}{c}\right) &= \frac{C_D}{C_L}C_L
\end{aligned}\right\} \tag{9.76}
$$

Where

$$
\frac{C_D}{C_L} = \left(\alpha+\alpha_c\alpha_i+\frac{A_1}{2}\right)^2\left(\alpha+\alpha_c\alpha_i+C_{Lf}\right)^{-1}
$$

Correction coefficients α_c and α_i are functions of aspect ratio, A, defined by experience, which C.C. Hsu gave as:

$$\alpha_c = \begin{cases} 1 + \dfrac{1}{5A} \\ 1 + \dfrac{1}{7A} \\ 1 + \dfrac{1}{9A} \end{cases} \text{When } A \geq 1 \text{ and } \frac{h}{c} = \begin{cases} \infty \\ 3 \\ 2 \end{cases}$$

$$\alpha_i = 1 - A^{-3} \text{ When } A \geq 1$$

The expressions mentioned above are for $\sigma = 0$. For $\sigma > 0$, we still can use (9.70) so long as adding one additional correction coefficient due to aspect ratio, as α_i or α_d:

$$C_L\left(\sigma, \frac{h}{c}\right) = C_L\left(0, \frac{h}{c}\right)(1+\sigma)\left[1 - \frac{9\pi(1-0.5\sigma)}{32\ (1+\sigma)} \times \frac{\sigma^2}{C_D\left(0, \frac{h}{c}\right)}\left(1 - \frac{\frac{\pi}{2}\alpha_c}{C_L\left(0, \frac{h}{c}\right)}\right)\right]\alpha_i$$

$$C_D\left(\sigma, \frac{h}{c}\right) = C_D\left(0, \frac{h}{c}\right)(1+\sigma)\left[1 + \frac{9\pi(1-0.5\sigma)}{32\ (1+\sigma)} \times \frac{\sigma^2}{C_D\left(0, \frac{h}{c}\right)}\left(1 - \frac{\alpha_c}{4\sqrt{C_D\left(0, \frac{h}{c}\right)}}\right)\right]\alpha_d \tag{9.77}$$

Where the correction coefficients are $\alpha_l = \alpha_d = 1 + \sigma\left[4AC_L\left(0, \frac{h}{c}\right)\right]^{-1}$

9.7 Section Profiles for Super-Cavitating Foils

We call the section profile suitable for a fully cavitating mode a super-cavitating foil profile. When discussing the hydrodynamic performance of a fully cavitating hydrofoil, there are two points of initial understanding for super-cavitating foil to consider:

1. The lower surface profile is a critical factor in defining the performance.
2. The hydrodynamic performance of a foil with flat lower surface and straight form is worse than that of a camber form.

The residual issues are what form of camber is better for the super-cavitating foil and, other than the lower surface, what characteristics are suitable.

9.7.1 Foil Profile with Low Drag at Zero Cavitation Number

From Tulin's method for solving the issue of the thin foil with the same result as the fully cavitating foil at $\sigma = 0$, it is noticed that the hydrofoil lift coefficient is equal to the moment coefficient about the leading edge of the thin foil with the same result. Meanwhile, the drag coefficient is in second order ratio to the lift coefficient of a thin foil, with the same result.

So, for obtaining a high lift/drag ratio super-cavitating hydrofoil, the moment coefficient at a certain lift coefficient of the equivalent thin foil with same result is high, and so it is advisable that the lift of a thin foil with same result should be distributed closer as to the trailing edge. A foil profile with low drag at $\sigma = 0$ can be obtained according to such principle.

When the camber form $\eta = \eta_0(\xi)$ is given, the vortex strength can be obtained according to the theory of a two-dimensional thin foil, see (8.20) reproduced below.

$$\gamma(\theta) = 2V_\infty \left[A_0 \cot\frac{\theta}{2} + \sum_{n=1}^{\infty} A_n \sin n\theta \right] \tag{8.20}$$

Where the various coefficients can be obtained according to (8.22)

$$\begin{cases} A_0 = \alpha - \dfrac{1}{\pi}\displaystyle\int_0^\pi \frac{d\eta_0}{d\xi} d\beta \\ A_n = \dfrac{2}{\pi}\displaystyle\int_0^\pi \frac{d\eta_0}{d\xi} \cos n\beta d\beta \end{cases} \tag{8.22}$$

The lift and moment coefficients of the equivalent thin foil with the same result can be obtained using the first three terms A_0, A_1, and A_2:

$$\widetilde{C}_m = \frac{\pi}{2}\left(A_0 + A_1 - \frac{1}{2}A_2\right)$$

Please note the moment is positive in counter clockwise direction.

$$\widetilde{C}_L = 2\pi\left(A_0 + \frac{A_1}{2}\right)$$

We can now determine the maximum lift/drag ratio, i.e., based upon the condition of maximum $\frac{\widetilde{C}_m}{\frac{1}{8\pi}\widetilde{C}_L^2}$ of the equivalent thin foil (or thin foil with same result), to find its foil profile $\eta = \eta_0(\xi)$, then transform it to z plane, and obtain the foil profile of a super cavitating hydrofoil, $y = y_0(x)$.

To simplify the expression, we assume $A_0 = 0$, which means the angle of attack is $\alpha = \frac{1}{\pi}\int_0^\pi \frac{d\eta_0}{d\xi} d\beta$, and the rotational axis ξ is parallel to incoming flow; and setting $\alpha = \frac{1}{\pi}\int_0^\pi \frac{d\eta_0}{d\xi} d\beta = 0$ one then gets an expression for lift/drag ratio as follows:

$$\frac{C_L}{C_D}=\frac{\widetilde{C}_m}{\frac{1}{8\pi}\widetilde{C}_L^2}=\frac{\frac{\pi}{2}\left(A_1-\frac{1}{2}A_2\right)}{\frac{1}{8\pi}(\pi A_1)^2}=\frac{4\left(A_1-\frac{1}{2}A_2\right)}{A_1^2}\times\frac{\frac{\pi}{2}\left(A_1-\frac{A_2}{2}\right)}{C_L}=4\left(1-\frac{1}{2}\frac{A_2}{A_1}\right)\frac{\pi}{2C_L} \tag{9.78}$$

Now we need to find the maximum value of C_L/C_D at a certain C_L, i.e., the minimum value of A_2/A_1, with the restriction of no cavitation on the lower surface of the foil. This corresponds to the pressure on the lower surface to be greater than p_0 at any point, or the perturbation velocity at any point is smaller than zero or the circulation distribution of the equivalent thin foil at any point is positive, as follows:

$$\gamma(\theta)=2V_\infty\sum_{n=1}^{\infty}A_n\sin\ n\theta\geq 0\ \ (0\leq\theta\leq\pi) \tag{9.79}$$

So, the issue is finding the minimum value of A_2/A_1 based upon the condition (9.79) then obtaining the corresponding foil profile. Tulin [10] took the first two terms of (9.79) and neglected other terms, thus:

$$\gamma(\theta)=2V_\infty(A_1\sin\theta+A_2\sin 2\theta)\geq 0\ \ (0\leq\theta\leq\pi)$$

And $1+2\frac{A_2}{A_1}\cos\theta\geq 0\ (0\leq\theta\leq\pi)$

The minimum value of A_2/A_1 was -½ and corresponding lift/drag ratio of the optimum foil was:

$$\frac{C_L}{C_D}=4\left(1+\frac{1}{4}\right)^2\frac{\pi}{2C_L}=\frac{25}{4}\frac{\pi}{2C_L}$$

Being 25/4 times the flat plane, thus its foil profile:

$$\frac{y}{c}=\frac{A_1}{2}\left[\frac{x}{c}+\frac{8}{3}\left(\frac{x}{c}\right)^{\frac{3}{2}}-4\left(\frac{x}{c}\right)^2\right] \tag{9.80}$$

A_1 could be related to the lift coefficient, and he took the design lift coefficient C_{Ld} to replace A_1, and under the condition:

$$C_{Ld}=\widetilde{C}_M=\frac{\pi}{2}\left(A_1-\frac{A_2}{2}\right)=\frac{\pi}{2}A_1\left(1-\frac{1}{2}\frac{A_2}{A_1}\right)$$
$$=\frac{\pi}{2}\frac{5}{4}A_1=\frac{5\pi}{8}A_1$$

Putting this into the expression for foil profile (9.80) then:

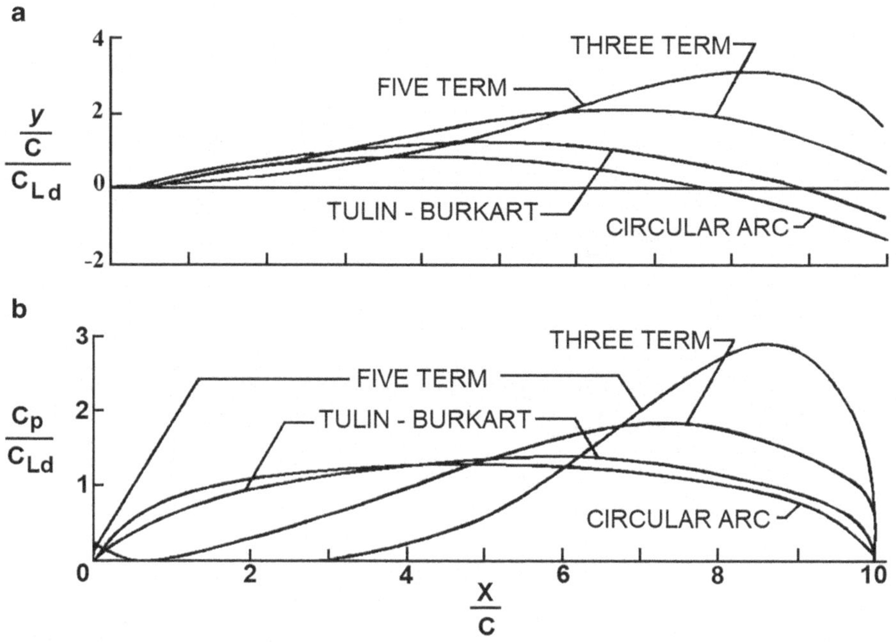

Fig. 9.30 Various super-cavitating foil profiles at $\sigma = 0$

$$\frac{y}{c} = \frac{4C_{Ld}}{5\pi}\left[\frac{x}{c} + \frac{8}{3}\left(\frac{x}{c}\right)^{\frac{3}{2}} - 4\left(\frac{x}{c}\right)^{2}\right] \tag{9.81}$$

Equations (9.80) or (9.81) present the foil profile called generally "Tulin-Burkhart two term foil," or "T-B" foil profile, where x and y are the ordinate and abscissa of foil camber, c chord length, and C_{Ld} design lift coefficient for the hydrofoil. It should be noticed that the x axis is not along the chord direction, but along the direction of

$$\alpha = \frac{1}{\pi}\int_0^{\pi} \frac{d\eta}{d\xi} d\beta = 0$$

so when $x = c$, $y \neq 0$, as shown in Fig. 9.30.

When calculating the circular curve foil at same incoming flow, then the lift/drag ratio can be obtained as:

$$\frac{C_L}{C_D} = \frac{81}{16}\frac{\pi}{2}\frac{1}{C_L}$$

This is five times the flat plane L/D, and a little bit worse compared to the two-term foil. Later, V. E. Johnson [16] took three and five terms in the vortex strength distribution series respectively and generated three and five terms foils. The foil profiles and lift ratios were as follows:

Three terms foil:

$$\frac{y}{c}=\frac{2C_{Ld}}{15\pi}\left[5\left(\frac{x}{c}\right)-20\left(\frac{x}{c}\right)^{\frac{3}{2}}+80\left(\frac{x}{c}\right)^{2}-64\left(\frac{x}{c}\right)^{\frac{3}{2}}\right]$$

$$\frac{C_L}{C_D}=9\left(\frac{\pi}{2C_L}\right)$$

Five terms foil:

$$\frac{y}{c}=\frac{2C_{Ld}}{525\pi}\left[210\left(\frac{x}{c}\right)-2240\left(\frac{x}{c}\right)^{\frac{3}{2}}+12600\left(\frac{x}{c}\right)^{2}-30912\left(\frac{x}{c}\right)^{\frac{5}{2}}+35840\left(\frac{x}{c}\right)^{3}-15360\left(\frac{x}{c}\right)^{\frac{7}{2}}\right]$$

$$\frac{C_L}{C_D}=\frac{100}{9}\left(\frac{\pi}{2C_L}\right)$$

The foil profiles are shown in Fig. 9.30. one can see that if more terms are taken in $\gamma(\theta)$, the lift/drag ratio is increased, and the corresponding camber of the profile moves rearward. This agrees with the result mentioned above that moving the center of lift of an equivalent thin foil rearward could improve the hydrodynamic performance of a cavitating hydrofoil.

When using a super-cavitating foil profile, the following points have to be noted.

1. The lift/drag ratio here mentioned is implying the value at the fixed coordinate condition of "shock free entry"; however, in practice, the angle of attack according to the solved condition only avoids cavitation on the lower foil surface, but can't ensure a full cavity on the back surface of the foil. Actually, to keep cavitation starting from the foil leading edge, a positive angle of attack $\alpha > 0$ has to be kept and then the lift/drag ratio of various profiles will be lower than that mentioned above. This can be estimated as:

   (a) Two terms foil: $\dfrac{C_L}{C_D}=\dfrac{\alpha+\frac{2}{\pi}C_{Ld}}{\left(\alpha+\frac{4}{5\pi}C_{Ld}\right)^2}$

   (b) Three terms foil: $\dfrac{C_L}{C_D}=\dfrac{\alpha+\frac{2}{\pi}C_{Ld}}{\left(\alpha+\frac{2}{3\pi}C_{Ld}\right)^2}$

   (c) Five terms foil: $\dfrac{C_L}{C_D}=\dfrac{\alpha+\frac{2}{\pi}C_{Ld}}{\left(\alpha+\frac{3}{5\pi}C_{Ld}\right)^2}$

2. The result above is at zero cavitation number $\sigma=0$, while when $\sigma>0$, it will be rather different. However, in practice, a super cavitating foil designed based on $\sigma=0$, is also better when $\sigma>0$ so long as the cavitation number is not so large, so in engineering practice, the foil profile at $\sigma=0$ can be used directly.

3. For a foil profile with low drag, one can see that the more terms used (5 rather than 3) implies higher camber, but it is not implying the form of face surface of foil. When adding foil thickness, the camber of the back surface will be larger, so potentially causing intersection of the foil back surface with the cavitation boundary, and devastating the stability of the cavitation cavity. It has been found that a foil profile with more than five terms cannot be applied in practice due to this issue. Supercavitating hydrofoils with a circular camber line and T_B profile are suitable, and both these have been applied on the Chinese supercavitating test hydrofoil type CXH-6, and on the bow foil of the Canadian anti-submarine hydrofoil "Bras D"or (FHE400) respectively.

9.7.2 Foil Profile of Super-Cavitating Hydrofoil

The foil back surface of a super-cavitating hydrofoil is positioned in a fully cavitating cavity. The pressure in the cavity is constant and equal to cavitation pressure.

The hydrodynamic performance of a super cavitating foil is related to the profile of foil face, and profile of foil back surface has no influence in this operational phase. The foil back profile, including dimensional error, surface smoothness, etc., has almost no influence on its hydrodynamic performance. However, there are other important requirements for the back surface profile of a super-cavitating hydrofoil that should be satisfied.

First, the strength of the foil section demands a certain thickness that should be satisfied. In addition, the foil back surface profile should not intersect with cavitation flow line, so the maximum thickness should be located at the trailing edge. Second, in order to generate cavitation starting from the leading edge, it should be sharp to generate the cavitation. These requirements lead to the wedge profile foil section as shown in Fig. 9.31.

In practice, a too sharp leading edge is easy to be broken, so in order to keep the foil back surface filled with full cavitation, some means for making agitated flow should be made for inducing the cavitation, as shown in Fig. 9.31, [17].

Additional measures that can be taken are:

1. An agitation thread mounted after leading edge at 5% length of chord.
2. Agitation wedge mounted after leading edge, as a step, for separating the flow and exciting the cavitation.

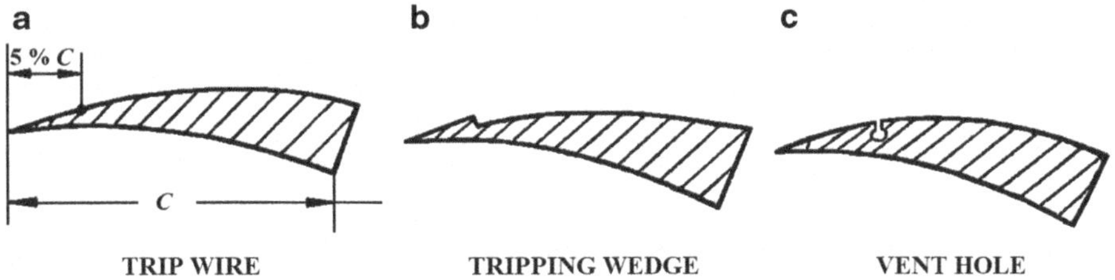

Fig. 9.31 Section profile of super-cavitating hydrofoil and its agitating means (**a**) agitating thread; (**b**) agitating wedge; (**c**) air ventilation hole

3. Making holes at back surface of foil for air ventilation, and inducing the cavitation. The generation of cavitation is related directly to the molecular weight of the water, and on some foil profiles the air ventilation makes the foil simply ventilate the whole foil back surface.

At present, the application of super-cavitating hydrofoils applied on hydrofoil craft is not wide, due to the technical challenges for acceleration from slow to high speed while maintaining stable flow. In addition to problems with efficiency and partial cavity instability from local cavitation before full cavitation and its control, the erosion of foil upper surfaces is a challenge. However, full ventilation has been applied widely in the design and engineering practice of semi-submerged propellers with good results. For hydrofoil craft, the transition and associated stability are the key issues, rather than operation at high speed when full cavitation can be maintained.

References

1. K.L. Wadlin, C.L. Shuford, A Theoretical and Experimental Investigation of the Lift and Drag Characteristics of Hydrofoils at Subcritical and Supercritical Speed, NACA TR 1732, 1955 available at UNT public digital library: A Theoretical and Experimental Investigation of the Lift and Drag Characteristics of Hydrofoils at Subcritical and Supercritical Speeds – UNT Digital Library
2. Y.I. Voitkonsky, *Water Resistance of Moving Ships, Leningrad*, 1964 (in Russian)
3. S.D. Chujino, *On the Lift of Hydrofoil with Limited Span*, Selected papers on hydrofoil craft, 1964 (in Chinese)
4. T.-S. Kuan, *Fundamental Conception for the Design of Military Craft* (National Defense Industry Press, 1992) (in Chinese)
5. M.P. Tulin, Supercavitating flow – small perturbation theory. J. Ship Res. SNAME **8**, 16 (1964)
6. N.I. Kochin, *Theoretical Hydrodynamics* (High Education Press, China, 1952) (in Chinese)
7. T.Y. Wu, A free streamline theory for two-dimensional cavitating hydrofoils. J. Math. Phys. **35**, 236 (1956)
8. T.Y. Wu, A wake model for free-streamline flow theory, Part I. J. Fluid Mech. **13**, part 2 (1962)
9. T.Y. Wu, A wake model for free-streamline flow theory part II. J. Fluid Mech. **15**, Part 1 (1964)
10. M.P. Tulin, M.P. Burkhart, *Linearized Theory for Flow About Lifting Foils at Zero Cavitation*, DTMB Report C-638, (1955)
11. J.A. Geurst, Linearized theory for partially Cavitating-hydrofoils. Int. Shipbuild. Prog. **6**(60), 369–384 (1959)
12. S.-T. Dong, Linearized theory on two-dimensional super-cavitating hydrofoil. China Shipbuilding **49**, part 1 (1963) (in Chinese)
13. S.-T. Dong, Linearized theory on super-cavitating & partial cavitation hydrofoil. China Shipbuilding **54**(part 1) (1964) (in Chinese)
14. F.-K. Tang, Velocity-graphical method for linearized cavitation hydrodynamics, in *First National Conference on Hydrodynamics, China*, (1963) (in Chinese)
15. C.C. Hsu, Some remarks on the Progress of cavity flow studies. J. Fluid Eng. ASME Trans. Ser. **1** (1975)
16. V.E. Johnson, The influence of depth of submersion, aspect ratio, and thickness of a supercavitating hydrofoil operating at zero cavitation number, in *2nd Symposium on Naval Hydrodynamics*, (1958)
17. B.A. Kolezaiv, et al, *Design Handbook for Dynamic Supported Craft*, Leningrad, Shipbuilding, (in Russian)

Chapter 10
Hydrofoil Craft Initial Design and Performance

In Chaps 8 and 9, we have discussed first the physics of foils in a perfect fluid, and then in a real fluid. Our approach has been from classic hydrodynamic and aerodynamic theory, together with some comparisons to physical testing. In Chap. 9, we have also discussed the consequences of cavitation for high-speed hydrofoils. In this chapter, we will use this as background to an approach for setting the dimensions and configuration of hydrofoils on a hull and then discuss the performance of hydrofoil craft.

We have introduced hydrofoil configurations in Chap. 7, so we start from the definition of foil arrangements and dimensioning, and then follow with the introduction of performance aspects such as speed performance, stability, maneuverability, and seakeeping. Finally, we touch on control theory and its use for dynamic stabilization.

10.1 The Arrangement, Type, and Definition of Hydrofoil Size

10.1.1 Arrangement

As we introduced in Chap. 7, there are three main foil arrangements for hydrofoil craft, the Airplane, Canard and Tandem arrangements, see Fig. 7.8. Hydrofoils may have their lifting surfaces completely submerged or piercing the surface, and have a range of different transverse geometries, as shown in Fig. 7.2. A fully submerged foil may be designed to operate at a particular submergence depending on the seakeeping requirement—the wave height limit for operations.

To provide passive rather than actively controlled roll and heave stability foils may be designed with dihedral, with variable chord, and to be surface piercing, as also shown in Fig. 7.2.

Z.-S. Dong et al., *High-Speed Monohull and Hydrofoil Craft*,
https://doi.org/10.1007/978-3-031-62760-6_10

10.1.2 Geometry and Sizing

Depending to the requirements for lift to be generated by the foils, the geometry can be configured to keep the craft fully supported at the intended clearance from the waterline. The geometry can be set up to provide additional lift as wave peaks pass; from an increased wetted area, or active control surfaces to vary camber and so lift force.

Hydrofoil craft have a wide range of speeds in foil borne operation; however, foil size has to be initially defined for its design operational speed. For most hydrofoil craft, the takeoff speed is between 50% and 60% of design speed and the dynamic pressure head due to speed is lower than that at design speed by 2/3~3/4, so the lift should be compensated by other means, such as flaps or auxiliary hydrofoils, as shown in Fig. 7.2.

After selecting the arrangement and type of hydrofoils and the requirement of lift, the foil geometric size and profile can be defined. The hydrofoil lift can be expressed as:

$$L = C_L \frac{1}{2} \rho V^2 S$$

After defining the design speed, the lift is in terms of projected area S and lift coefficient C_L, and C_L is related to geometric characteristics of the foil, including aspect ratio, section profile, plan profile, and angle of attack. According to Eq. (8.64) in Chap. 8, the lift coefficient of a foil with limited span is:

$$C_L = \frac{2\pi\eta}{1 + \frac{2\eta}{A}(1+\tau)}(\alpha - \alpha_o) \tag{8.64}$$

where:

η correction factor for viscosity, normally η is 0.8 ~ 0.9
$A = S/c^2$ aspect ratio of hydrofoil
c chord length of hydrofoil,
α geometric angle of attack
α_0 zero lift angle of attack of hydrofoil section
τ correction for foil plan profile, where $\tau = 0$ for an elliptical form.

For hydrofoils with "sweep" or "dihedral" downward, the value of correction factor τ should be in accordance with Chap. 8 (Figs. 8.13 and 8.14).

Since there are many factors influencing the lift coefficient, the definition of geometric size and form of a foil to be used has to be defined in a stepwise manner from an initial configuration and dimensions through refinement to a final configuration. Some parameters can be determined through engineering experience taken from existing hydrofoil craft and others through prototype testing. The design process then follows the classic spiral of specification, analysis and testing with feedback, and refinement until the desired performance parameters are met.

10.1.3 Initial Definition of Hydrofoil Area

Hydrofoil area can be defined from the load per unit hydrofoil area. Very light loading will lead to too large foil area, and friction drag; however, too high loading will lead to high pressure drag, and lead to cavitation and air ventilation. Reference [1] gave statistical data of loading per unit area, Δ/S, from a large number of Russian hydrofoil craft with non-automatic control foil systems, and hydrofoil craft of other countries, where Δ craft weight and S foil area. This data can be used to assist in design. According to this data a regression equation can be obtained as follows:

$$\Delta/_S = 10\left\{1 - \exp\left[-\left(\frac{V}{90}\right)^3\right]\right\} \tag{10.1}$$

where speed V is in km/h.

Based on statistical data from existing hydrofoil craft with a speed range of $50 \sim 90$ km/h, an alternative regression equation can used for craft in the speed range of 50–90 km/h as follows:

$$\Delta/_S = A\rho V^2 \tag{10.2}$$

where coefficient A is in terms of speed unit:

$A = 0.1 \times 10^{-3}$ when V is in m/s.
$A = 7.7 \times 10^{-6}$when V in km/h.
$A = 0.26 \times 10^{-4}$ when V in knots.

It should be noted that (10.2) applies only for $V_S = (50 \sim 90)$ km/h..

Figure 10.1 shows a generalized relationship between foil loading and both takeoff and minimum service speeds taken from designs in the USA ([2], Fig. 67) that may give a useful start point for designers setting initial dimensions for hydrofoil craft beginning with hydrofoil area.

10.1.4 Hydrofoil Dimensions and Plan Profile

After the definition of hydrofoil total plan area, the area and longitudinal position of forward and aft foils can be defined in terms of the load distribution to each foil so that the size and plan profile of each foil can be defined according to the area required.

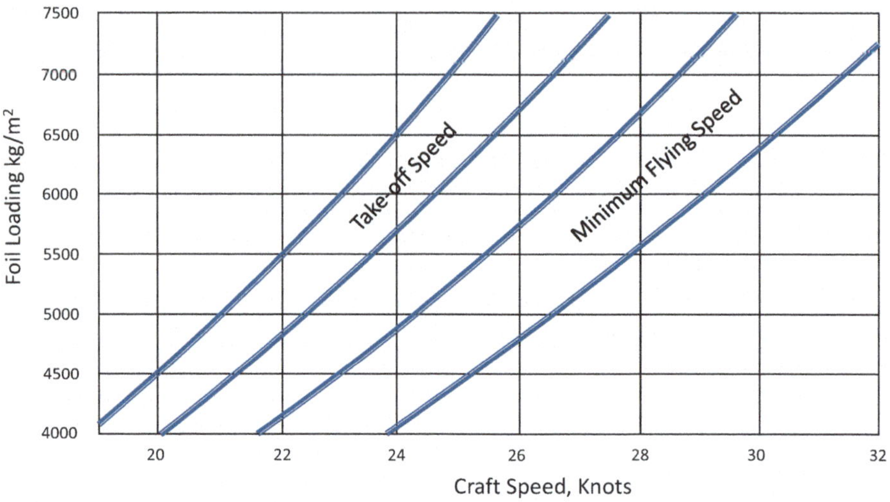

Fig. 10.1 Relation between hydrofoil unit loading and craft takeoff and service speed

The distribution of load and its longitudinal position directly influences the external loads applied to the craft hull structure.

Regarding drag, the wave interference between the forward and aft foils; and the downwash of the forward foil on the aft foil has to be considered, especially for the tandem foil arrangement. For the tandem arrangement of foils, [1] suggests the space between both forward and aft foils to optimize wave interference is:

$$V\sqrt{\frac{l_f}{g}} < L_d < 3V\sqrt{\frac{l_f}{g}} \tag{10.3}$$

where

V (m/s) is the craft speed.
l_f (m) span of forward foil
l_d (m) space between forward and aft foils
g (m/s^2) gravity acceleration (9.81 m/s^2).

For airplane and canard arrangements, the wave making interference should be lower than a tandem arrangement. The downwash of the forward foil influencing the aft foil from the tandem arrangement of foils is greater than the other two layouts. The airplane arrangement is better since the aft foil is much smaller so the effect is less. The aft foil of the canard arrangement is also less subject to downwash from the forward foil. The Canard and airplane configurations allow the "main" lifting foil to have a higher aspect ratio giving it higher efficiency. The distribution of load will be discussed in some detail in relation to sea-keeping quality in Sect. 10.3.

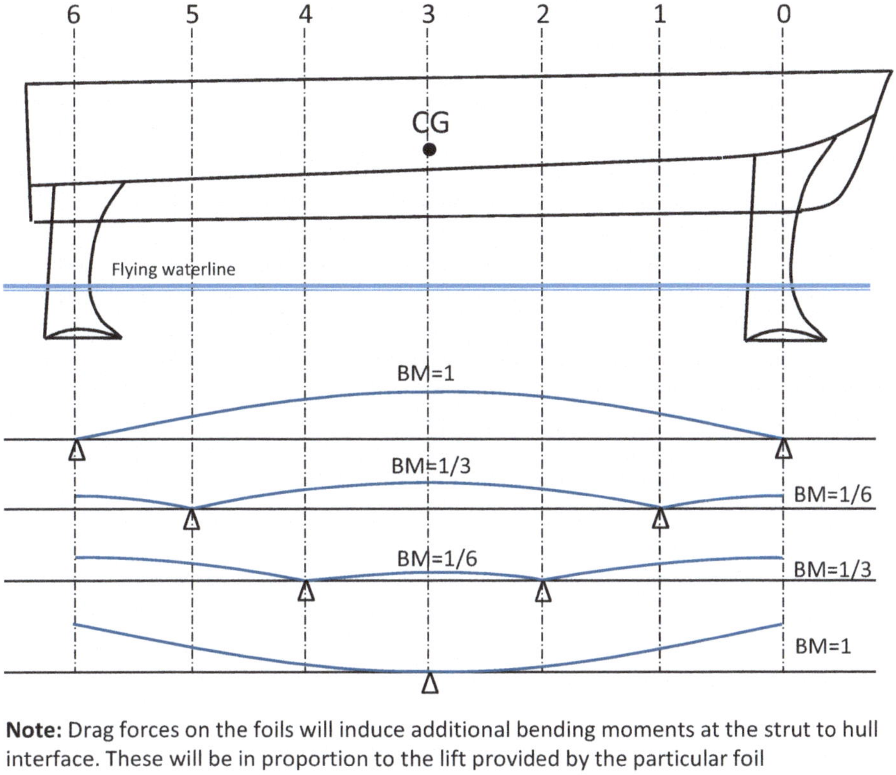

Fig. 10.2 Example relationship of vertical bending moment of hydrofoil craft operating on calm water with changing position of foils

The load distribution between the foils also strongly influences the external loading on the hull. Since a hydrofoil inputs loads to the hull structure at fixed points, bending moments and stresses can be determined treating the hull as a simply supported beam.

Figure 10.2 shows the relation between the bending moment and position of both foils of hydrofoil craft in calm water. Here it is assumed for demonstration purposes that the weight distribution of the craft is uniform and so does not influence analysis directly. In foil borne operation, the forward and aft foil supports the whole craft weight Δ, with CG at point G, and the figure shows the distribution of bending moment where forward and aft foils are located at ordinates 0, 6; 1, 5; 2, 4; and 3, 3, respectively.

From the figure, one can see that when both foils are located at the bow and stern as 0, 6 or concentrated at ordinate 3 the bending moment acting on the craft is maximum, and located at the CG or bow/stern. In this case, the load distribution is considered equal for the tandem arrangement; nevertheless, for the same tandem arrangement with foils located at ordinates 1, 5 and 2, 4, the maximum bending moment may be decreased.

The minimum bending moment is with the foil arrangement at ¼ craft length if the weight distribution is uniform. In general, the main engines and transmission mean that CG is towards the stern, and so a position of the foils at stations 5 and 1 or 2 may minimize maximum bending moment. For the canard arrangement with the foils at 1, 5 or 1, 4, the bending moment may be minimized.

The bending moment distribution for the airplane arrangement may correspond to the canard arrangement in reverse, with the main foil at ordinate 1 or 2 and the rear stabilizing foil between 5 and 6. Figure 10.2 is intended to give an impression, to demonstrate that the bending moment acting on the craft is important to the positioning of both foils related to craft CG operating foil borne and in turn influences the hull structure weight and craft displacement.

After the definition of the area distribution between the forward and aft foils, the most important issue is selecting the aspect ratio of foils. This is an important parameter influencing the lift/drag ratio, which a designer is seeking the best possible value; however, the influence of the forward foil tip vortex on the aft foil and interference between both foils due to downwash must be considered carefully.

Reference [1] suggests the aspect ratio of the main foil (A) can be selected in relation to craft speed using the relation as follows:

$$A = \frac{300}{V_S}(1 \pm 0.4) \tag{10.4}$$

where V_S craft speed (knots).

So for a 40 knot craft an aspect ratio between 5 and 10 may be suitable while at 50 knots a range between 3.6 and 8.4 is suggested with start points at 7.5 and 6 for the main lifting foil. After the definition of foil area and aspect ratio, it is not difficult to define the chord length; however, an initial check of foil strength related to foil size has to be carried out. The foil dimensions can then be finalized based on aspect ratio A, related to design size, mass, hull structure loading, and interference between both foils.

Considering deferring cavitation on high-speed hydrofoil craft, the foil plan form may use a sweep. This increases the chord length of the foil section in the direction of incoming flow, to decrease the relative thickness of the foil section at the same thickness of foil, and decrease the negative pressure peak.

In general, a sweep between 30° and 60° will defer cavitation. Figure 10.3 shows the permitted profile relative thickness to avoid cavitation on a swept foil against speed [1]. Here the so-called foil relative section thickness profile is related to the requirement for foil structural strength, that is, the ratio of maximum thickness of foil with chord length perpendicular to the span.

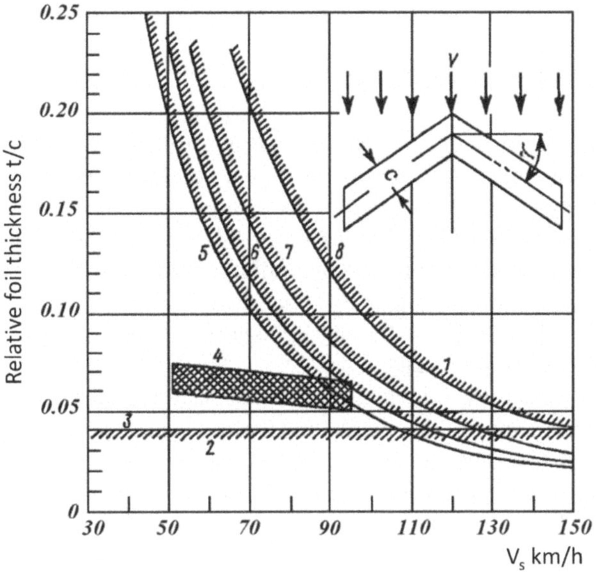

Fig. 10.3 Relationship of relative thickness of foil and sweep angle with speed for avoiding cavitation

Key to diagram is as follows:

1 Speed limit to avoid supercavitation

2 t/c lower limit to avoid supercavitation

3 Safe lower design limit to t/c

4 Safe design area for t/c adopted by Russian hydrofoils

5, 6, 7, 8 Speed limits for foils with sweep 0, 15, 30, 45 degrees

10.1.5 Foil Section Profile

The foil section thickness is defined according to the foil strength, and the section profile is defined in terms of foil efficiency (lift/drag ratio) and cavitation. For higher speed craft where $V_s \geq 50$ knots deferring cavitation is a key parameter, and suggests using a foil section with uniform distribution of negative pressure, such as NACA-6 series, E foil series, and YS foil series, see Sect. 8.4.

In order to simply configure the foil section, a circular bowed back profile is suggested, and using a "bending downward" of the camber line about 30–40 mm at the foil leading edge. This will induce flow at the sharp leading edge to give "shock-free entrance of flow," see Fig. 9.21. In addition, using a small radius circular leading edge connected to the bowed back section profile has the same aim.

For hydrofoil craft at $V_s > 60$ knots cavitation is impossible to avoid and is used positively by applying a super-cavitating foil profile, see Figs. 9.30 and 9.31.

While super-cavitation will not lower efficiency at high speed, there is lower efficiency at low speed, and in addition, partial cavitation and erosion will occur at transit speeds. Cavitation vibration and noise will occur due to the unsteady cavitation, as well as problems concerned with the cavitation erosion caused on other bodies close to the hydrofoils, which prevents the practical application of the super-cavitating hydrofoil on the hydrofoil craft at speed higher than 60 knots so far.

An effective tool, that is, "air ventilation," at the leading edge of the super-cavitating hydrofoil is available to avoid cavitation erosion due to partial unsteady cavitation at transit speeds. This has been applied on the bow hydrofoil of Canadian hydrofoil craft "Bras D'or," on which double steps are also installed on the back of the forward hydrofoil.

10.1.6 Hydrofoil Takeoff Aids and Struts

10.1.6.1 Takeoff Aids

A hydrofoil is initially designed according to the lift requirement at design speed. However, during transit speed and acceleration, hydrofoil lift is far lower than the weight of craft due to the lower speed. The takeoff time will be long with just foils designed for the foil borne speed. There is also a drag hump at transition speed in the drag speed curve for a monohull hydrofoil. The main engines may therefore have to run at maximum power, which can be harmful. In order to speed up the takeoff, auxiliary foil installation as takeoff aid may be necessary unless active flaps or elevators are installed.

Generally, it is suggested that the take-off speed should be about 50% and 60% of design speed. Meanwhile, the hydrodynamic pressure at the hydrofoil is about 25% to 36% of that at design speed, so the lift coefficient or foil area should be enhanced to give the required lift for take-off.

For hydrofoil craft with a controllable foil system, a similar result can be reached by using foil flaps and angling them for take-off aid and not adding any auxiliary installation.

For hydrofoil craft with an air-ventilated hydrofoil system, the lift coefficient can be enhanced by means of changing the air ventilation value, including designing it to be used close to foiling elevation and when at service speed.

For hydrofoils with shallow submerged and fixed hydrofoil systems, increase in lift coefficient can be realized by changing the submerged depth and trim angle. However, it is difficult to enhance the lift coefficient by a factor of 3–4, so one is obliged to increase the foil area, that is, install "takeoff aid foils" above the normal foil waterline. These only provide the lift during takeoff and do not contact the water surface or provide lift after take-off to avoid increase of drag at design speed.

Takeoff aids may be a separate installation, mounted at the bow or connecting with the main foil and in front of the main foil for providing additional lift, raising the bow and increasing the angle of attack of the main foils. The takeoff aids can also be mounted with a high installation angle of attack, greater than the main foil by 2° to decrease their area.

Cavitation of both main foil and take-off aids foils should be avoided during takeoff. A simple rectangular plan foil is generally applied for such separate takeoff aid foils.

Takeoff aid foils connected with the main foil are usually mounted at the sides of the main foil, to form a surface piercing side foil higher on the support strut. This will provide a larger foil area during operation at low speed with deeper draft, and foil area will be decreased gradually with increase of speed, to minimize drag, as shown in Fig. 10.4. The additional surface piercing side foil will also improve transverse stability.

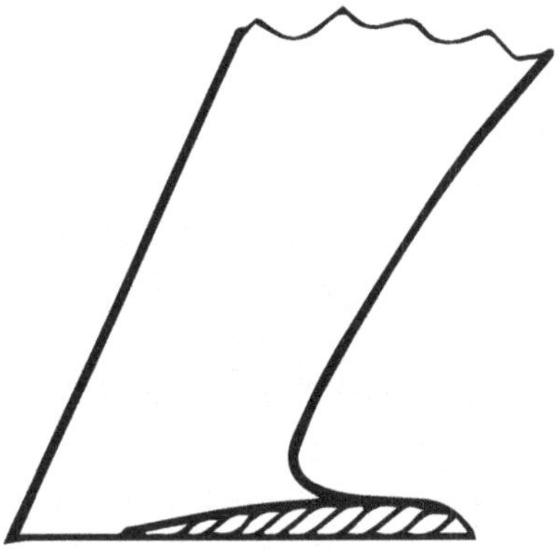

Fig. 10.4 Takeoff aid additional foils

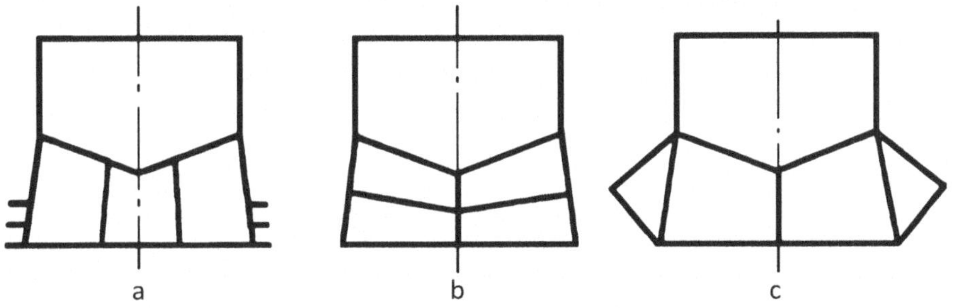

Fig. 10.5 Lateral profile of strut

Sometimes separate takeoff aid foils as well as takeoff aids direct on the main foils such as moveable flaps are installed together to obtain faster takeoff at lower engine power.

10.1.6.2 Struts

A strut is a necessary part for connecting the hydrofoils and hull but creates drag, so should be the smallest size to meet the necessary strength and structural stiffness requirements. Struts should be of symmetric section in the direction of incoming flow, as well as having a sharp leading edge and trailing edge. The strut leading edge should be set back from the foil leading edge, as shown in Fig. 10.5. For low-speed hydrofoil craft, a round leading edge of strut cross section may be used. For higher-speed craft, a slender leading edge may be used, and for craft with very high speed, especially for super-cavitating foils, the straight wedge section is better, similar to the foils shown in Fig. 9.30. The strut trailing edge breaking the water surface generates suction and ventilation behind the strut for reducing drag.

To minimize the effect of ventilation being drawn down from the water surface over struts or surface piercing foils, fences are often installed, as illustrated in

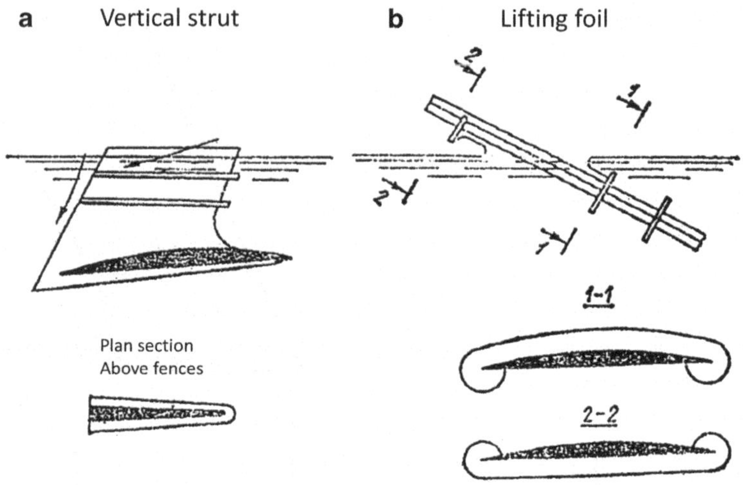

Fig. 10.6 Installation of struts and interceptor fences on the strut of surface piercing hydrofoils

Fig. 10.6. Fences that will normally be submerged after takeoff may just cover the upper surface. Fences on a canted foil lower surface can also help minimize upward lateral flow just above the water surface to maximize efficiency. At the connection of a strut with a hydrofoil, care should be taken so that the strut does not destroy the suction on the foil back as far as possible and to avoid air suction and cavitation. Strut leading edge interface should be 30–40% of chord length aft of the leading edge, and the strut leading edge should be inclined at least for the first part above the foil to defer cavitation, and not snag floating debris in the water. The trailing edge for high-speed cavitating hydrofoils should be extended aft of the foil trailing edge to avoid the strut straight blunt trailing edge being located at the suction surface, as shown Fig. 10.6a. In order to avoid air sucking down to the upper surface of the foil along the strut, an air interceptor fence should be mounted on the strut, and also on surface piercing foils near the water surface, as shown in Fig. 10.6b, which shows the mounting and sketch of the interceptor fences for a surface piercing foil.

10.2 Hydrofoil Drag

Hydrofoil craft drag at high speed is usually lower than that at low-speed hull borne drag, so the profile of the Drag/Speed curve hump drag has a characteristic "hump" shape (see Fig. 11.3). The drag peak at the 'hump' occurs as the hydrofoil goes through take-off. Before the region of hump speed, the hydrofoil craft is in hull borne mode and the drag components are similar to a displacement ship, but with additional drag of foils and struts.

In the hump speed region, the drag is maximum, and after hump speed the hydrofoil lift takes the craft above the water surface gradually, with change of running attitude and decrease of drag. On the drag curve, one can see the drag decreases inversely with speed; the trough point of the drag curve being once the takeoff phase is complete and the vessel is at design elevation above the water surface. As the speed is further increased the foil drag and lift both increase and the foil submerged depth may also decrease, so strut drag also decreases, and a flat drag section appears on the drag curve, after which drag increases with speed following the normal square law relation.

Hydrofoil craft drag analysis is often focused on the drag of a craft in foil borne mode, even though hump drag is important for hydrofoil craft, and strongly affects the performance during takeoff. However, just as for displacement ships, it is difficult to calculate accurately and theoretically the drag during the take-off, and it can only be reliably obtained by model testing. The test method for both displacement ships and hydrofoil craft is the same, so here we will refer readers to traditional Naval Architecture texts.

When looking at the hump speed range, this is where in order to reduce the drag hump, take-off aids are necessary for a fixed foil system, while the foil system of fully submerged hydrofoil craft can use controllable flaps to create higher camber at low speed.

It must be noted that conversion of model drag to full-scale craft drag is according to the cube of linear scale ratio (λ^3). For hydrofoils, due to the small foil size of scaled models, towing tank testing as for a displacement craft is difficult. It is normal therefore to test foils in a flowing water basin and scale these results for design. Large-scale free running models can be used for performance evaluation, or "small-scale" prototypes, as was the case for the US development programs.

10.2.1 Determination of Craft Running Attitude in Foil Borne Operation

When supported by the hydrofoils rather than buoyancy, almost the total drag is generated from friction drag on the foils and struts and the rearwards lift vector, both of which are dynamic loads proportional to V^2. The drag varies with running attitude, so the desired running attitude needs to be defined. Generally, this can be specified by accounting for the position of the craft CG and load distribution of both forward and aft foils.

The required foil lift and distribution will influence the foils' design submerged depth and angle of attack to meet force and moment equilibrium for the craft. The foil drag and strut drag can be obtained so long as the submerged depth and angle of attack are specified as a start point.

For hydrofoil craft at speed with sub-cavitating foils the typical range of submerged depth and angle of attack is as follows: shallow submerged depth foils $0.2 \leq \overline{h} \leq 0.5$, deep submerged depth foil $\overline{h} > 1.0 \sim 1.5$. The foil angle of attack at design speed is normally $\alpha \approx 1 \sim 2°$.

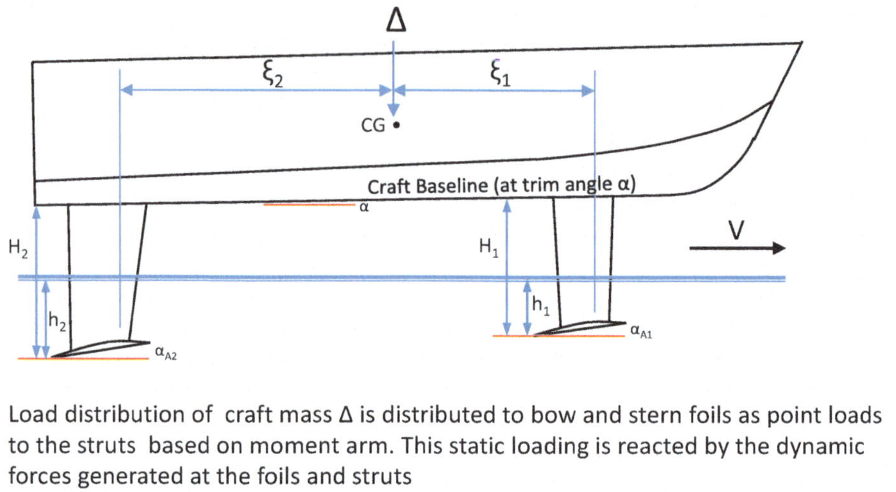

Fig. 10.7 Running attitude of hydrofoil craft at foil borne operation

The running attitude should be defined first via vertical force equilibrium, then the drag components and trimming moments for rotational equilibrium be determined. Due to the interaction between the parameters, an iterative approach has to be followed to determine equilibrium, similar to the approach used to determine equilibrium for planing craft. For a hydrofoil, the foil lift varies with submerged depth, so after the initial estimate, depending on the result the foil depth might be raised or lowered a little to reduce or increase total lift. Alternatively, the angle of attack or the foil areas can be adjusted so as to achieve vertical equilibrium. A similar approach can be taken to equilibrium in pitch, either by adjusting vessel trim (effectively adjusting foil angle of attack) or by adjusting the distribution of foil plan areas. Once equilibrium is available, drag can be evaluated.

Figure 10.7 shows the running attitude of a hydrofoil craft. In the figure, the following parameters are indicated. Craft displacement Δ, with load distribution and the distance between bow and stern foils and craft CG as ξ_1 and ξ_2. The perpendicular distance of bow and stern foil from baseline are H_1 and H_2, and installation angle of both foils (intersection angle with base line) are α_{A1} and α_{A2}. We also define the craft speed V, running attitude craft trim angle α, submerged depth of bow foil h_1 and stern foil h_2..

Equilibrium can be initially estimated using the force and moment equilibrium equations as follows:

$$\Delta = L_1 + L_2 = C_{L1}\frac{1}{2}\rho V^2 S_1 + C_{L2}\frac{1}{2}\rho V^2 S_2 \tag{10.5}$$

$$C_{L1}\frac{1}{2}\rho V^2 S_1 \xi_1 = C_{L2}\frac{1}{2}\rho V^2 S_2 \xi_2 \tag{10.6}$$

where:

S_1, S_2 projected area of bow and stern foils
L_1, L_2 lift of bow and stern foil respectively
C_{L1}, C_{L2} lift coefficient of bow and stern foils.

For the first approximation, the drag and thrust can be assumed acting at the same horizontal level, with no trimming moment. This does need to be considered once initial assessments have been made, particularly if the propeller thrust line will induce a significant pitching moment.

The lift coefficient of both foils can be obtained according to the equations mentioned above, then the equilibrium position can be obtained via the lift coefficient equations, including the effect of downwash, see Eq. (8.35):

$$C_{L1} = \frac{2\eta\pi\gamma_1}{1 + \frac{2\eta\gamma_1\xi_1}{A_1}(1+\tau_1)K_{e1}}[\alpha_1 - (\alpha_{01} + \Delta\alpha_{01})]$$

$$C_{L2} = \frac{2\eta\pi\gamma_2}{1 + \frac{2\eta\gamma_2\xi_2}{A_2}(1+\tau_2)K_{e2}}[\alpha_2 - (\alpha_{02} + \Delta\alpha_{02})]$$

Here the real angle of attack of both bow and stern foils are $\alpha_1 = \alpha + \alpha_{A1}$ and $\alpha_2 = \alpha + \alpha_{A2}$ respectively. Since the installation angles α_{A1} and α_{A2} are given and only α is an unknown value, the water surface influence correction coefficient in the equations γ, and $\Delta\alpha_0$ are functions of the bow and stern foil submerged depth h_1 and $h_{2,}$ and the relation of h_1 and h_2 is:

$$h_1 = h_2 + (H_1\text{-}H_2) + (\xi_1 + \xi_2)\alpha$$

Here H_1 and H_2 are the foil depths from the craft baseline to account for different design submerged depths of the bow and stern foils, as shown in Fig. 10.7.

The only two unknown values are α and h_2 (or h_1) in the two lift coefficient equations, and it can be solved by means of an iterative method with the aid of the computer. This applies where the hydrofoil geometry is considered fixed for the determination of equilibrium. Particularly if the vertical or trim equilibrium is close, small adjustments to foil plan and angle of attack might be carried out at this stage.

Finally, there are two points that have to be explained. The calculated running attitude is a first approximation, without considering the influence of moment caused by thrust and drag, and lift acting on the appendages. If a more precise calculation is necessary, these factors should be included, particularly the drag, and moment generated by appendages, or large shaft bosses due to some power transmission systems. In addition, appendages might also generate lift when running with a trim angle, which can be calculated using slender body theory, as has been discussed in Chap. 3. Readers can also see this in Chap. 19 of [3].

10.2.2 Interaction Between Bow and Stern Foils

Hydrodynamic interference between the bow and stern foils is not included in the first approximation, and in general, the stern foil is located in the wake of the bow foil. In the bow foil wake a downwash is created, which affects the aft foil by lowering the water-free surface. When considering this influence, the following equations can be used for estimation.

(a) Lowering of water surface aft of bow foil

$$\frac{\Delta h}{C_2} = -C_{L1}\left[\exp\left(-\frac{1}{F_n^{2q}}\right) + \frac{0.6}{A_1}\left(F_n^{2q} - 2\right)\right]\sin\frac{x}{c_1 F_n^{2q}} \tag{10.7}$$

where:

Δh lowering value of water surface
C_1, C_2 chord length of bow and stern foils
C_{L1} lift coefficient of bow foil
$F_n = \frac{V}{\sqrt{gc_2}}$ Froude number with respect to stern foil chord
$q = \exp\left(\frac{-0.73}{A_1^{0.2}}\right)$
A_1 aspect ratio of bow foil
x space between bow and stern foil,

(b) Downwash angle at stern foil induced by bow foil

$$\Delta\alpha_o = -\frac{3}{4}\frac{C_{L1}}{F_n^{2q}}\left[\exp\left(\frac{-1}{F_n^{2q}}\right) + \frac{0.6}{A_1}\left(F_n^{2q} - 2\right)\right] COS\frac{x}{c_2 F_n^{2q}} \tag{10.8}$$

The nomenclature in (10.8) is the same as in (10.7).

During the calculation for the stern foil, the submerged depth should be reduced by Δh, and the angle of attack for the stern foil will be effectively decreased by the downwash angle induced by bow foil $\Delta\alpha_0$. An alternative way to look at this is that to achieve the lift calculated without the downwash effect, the foil will need to be lowered and angle of attack increased to compensate.

In the case of the stern foil located outside the bow foil wake, then the influence mentioned above can be neglected. In general, the influence of bow foil on the aft foil is significant in the airplane and tandem foil arrangement, unless the split type of bow foil is applied, to move the stern foil out of the concave region. In this case, the stern foil is located between the downwash of bow foils.

For the canard arrangement, the situation is rather different due to the width of the bow foil being smaller than the stern foil. The water surface trough is small in height and width. In addition, only the mid part of the stern foil is located in the downwash and the two sides are located in free flow. The non-uniform downwash field will promote cavitation earlier. It is suggested that the stern foil angle of attack may be varied spanwise to compensate the nonuniform downwash; however, the result will change if craft speed changes and so also the downwash field. If a split type of stern foil is used in this case, the main part or whole part of the stern foil will be located in incoming free flow, and such influence minimized.

10.2.3 Estimation of Drag

10.2.3.1 Hydrofoil Drag

The estimation of foil drag has been discussed in Sects. 8.4, 9.2, and 9.3. Here we summarize it as follows:

(a) Viscous drag

This includes friction and viscous pressure drag, which can be obtained by wind tunnel test, or Eq. (8.65):

$$\mathrm{C_{DV}} = C_f + C_\Phi = 2C_f\left(1 + 10\overline{t^2}\right)\left[1 + (\Delta C_L)^2\right] \tag{8.65}$$

where:

C_{f} friction coefficient, using Prandtl or Schoenherr equations
C_O coefficient for viscous-pressure drag (form drag)
$\bar{t}$ relative thickness of foil section
ΔC_L the difference between the foil in operation C_L and minimum C_L.

where C_{f} includes roughness allowance, for hydrofoil craft $\Delta C_f = 0.8 \times 10^{-3}$, or using the test results directly of NACA with standard wing roughness, and use:

$$C_f = 0.8 \times {R_n}^{1/8} \tag{8.66}$$

where R_n = V × C/γ, the Reynolds number with respect to chord. One can also use (8.67) for calculating friction coefficient C_f. Viscous drag can be obtained by approximate equation, as (8.68):

$$C_{DV} = 2C_f[(1 + 1.2\bar{t}) + 0.11(\Delta C_L)] \tag{8.68}$$

where ΔC_L is as in (8.65).

Reference [4] suggests taking the following equation for drag coefficient if the maximum thickness is located at 40 % ~ 50% of the chord:

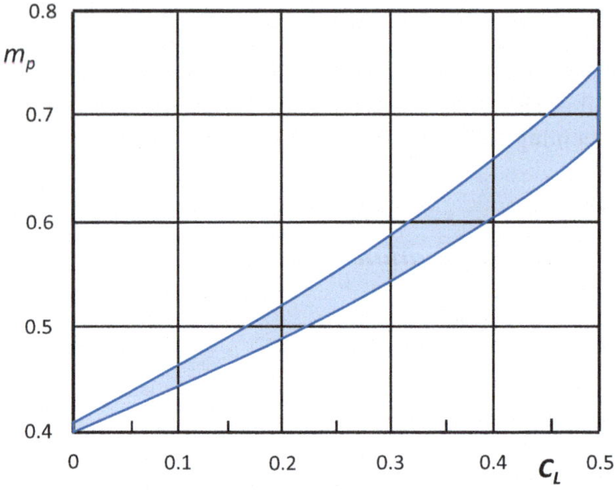

Fig. 10.8 $m_p = f(C_L)$, where the blue region shows the range of m_p which may vary up to 10% at any given C_L

$$C_{DV} = 2C_f\left[(1 + 1.2\bar{t}) + 60\bar{t}^4\right] \quad (8.69)$$

Reference [1] proposes the effect of a free surface on the viscous drag of shallow submerged hydrofoil craft should be considered, and proposes the following equation:

$$C_{DV} = 2C_f\left\{1 + \left[(m_p + 0.5)\frac{\varphi(\bar{h})}{k_\varphi} - 0.5\right]C_L\right\} \quad (10.9)$$

where $m_p = f(C_L)$ and can be taken from Fig. 10.8, and φ and k_φ below are correction factors for free surface effect on lift coefficient:

$$\phi(\bar{h}) = 1 - \exp\left[-2(\bar{h})^{0.6}\right] \quad (10.10)$$

$$k_\varphi = 1 - (0.5 + \bar{c})\exp\left[-2(\bar{h})^{0.6}\right] \quad (10.11)$$

(b) Induced drag

Using (9.32):

$$C_{D_i} = \frac{C_{L_h}^2}{\pi A}\left(\delta + \frac{2}{\partial}\right)K_e \quad (9.32)$$

where:

C_{Lh} lift coefficient in terms of free water surface effect
A aspect ratio of foil.
δ correction factor for hydrofoil plan profile influence on induced drag, as shown in Figs. 8.13 and 8.15
K_e correction factor of foil tips, as shown in Fig. 8.19
∂ correction factor for foil strut, as shown in Fig. 9.13.

or using (9.38):

$$C_{Di} = \frac{C_{Lh}^2}{\pi A}(1+\delta)\varsigma K_e \tag{9.38}$$

where

$$\varsigma = 1 + \exp\left[-5.35\left(\bar{h}/A\right)^{2/3}\right] \tag{9.33}$$

This is the influence coefficient for free water surface on downwash angle. Other nomenclature is as same as in (9.30).

(c) Wave-making drag

Using (9.20) or (9.21) for estimation:

$$C_w = \frac{C_{L\infty}^2}{2F_n^2}\left(1 - \frac{2\pi}{F_n^2}\right) \tag{9.20}$$

$$C_w = \frac{C_{Lh}^2}{2F_n^2}\exp\left(-\frac{2}{F_{nh}^2}\right) \tag{9.21}$$

where:

$C_{L\infty}$ lift coefficient for infinite depth of foil
C_{Lh} lift coefficient for shallow submerged depth
F_n Froude number with respect to chord
$F_{nh} = V/(gh)^{0.5}$ submerged depth Froude number

(d) Local protrusion drag on foils

In the case of foil with small protrusions, the drag coefficient can be calculated as:

$$\Delta C_D = K\sqrt[3]{h/x} \tag{10.12}$$

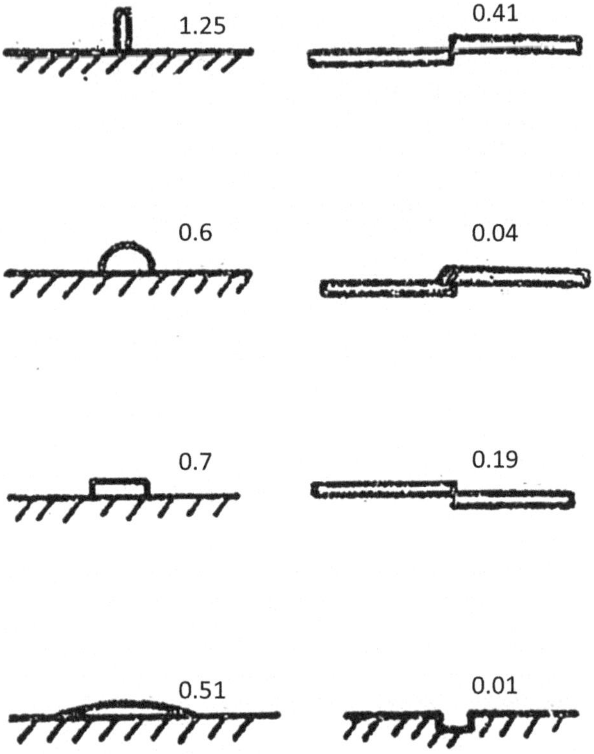

Fig. 10.9 K value for local protrusion drag—welds and surface imperfections

where

h/x the ratio of local protrusion height with distance from the leading edge when $h/x < 0.02$,

K shape coefficient of protruding body, from Figs. 10.9 and 10.10.

For a supercavitating hydrofoil the calculation method can be taken from Sect. 9.5 of Chap. 9.

When one looks at the potential delta to C_D from these imperfections or protruding bodies, it can be seen that the consequences are serious; hence, great care is necessary to avoid these drag generators on a hydrofoil lifting surface, or indeed on a strut.

10.2.3.2 Strut Drag

A foil strut generally creates drag but not lift and so should be as small as possible. Struts actually have to be sized to transmit the lift from foils, torque due to maneuvering, and loads from a propeller transmission arrangement. Struts are also dimensioned for the submerged depth of foils, including seakeeping quality

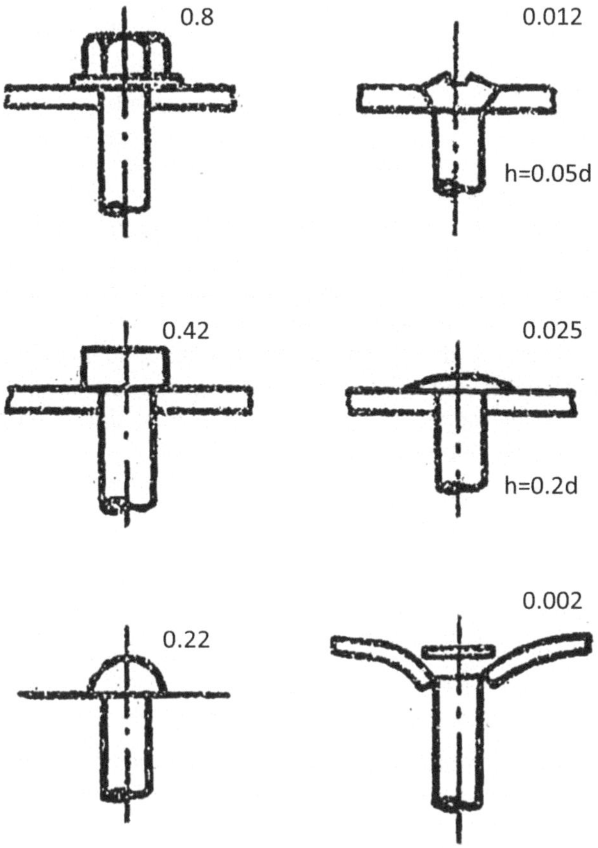

Fig. 10.10 K value for local protrusion drag—bolts and rivets

requirements, etc. Strut drag includes viscous, spray drag, and interference drag between foils, which can be obtained by following empirical equations as follows:

(a) Viscous drag

This can be determined by reference to wing section test data, and in the case of insufficient reference data, the equation recommended by NACA (see (8.65)) can be used as follows:

$$C_{DV} = C_f + C_\phi = 2C_f\left(1 + 10\overline{t^2}\right)\left[1 + (\Delta C_L)^2\right] \tag{8.65}$$

where ΔC_L is the difference between the hydrofoil lift coefficient at operation mode and at maximum drag. Since the struts are symmetric generally, at zero angle of attack, so $\Delta C_L = 0$ and (8.65) changes to:

$$C_{DS} = 2C_f\left(1 + 10\bar{t}^2\right) \tag{10.13}$$

where: $C_f = 0.08R_n^{-1/8}$, $R_n = \frac{V_c}{v}$, the foil chord Reynolds number, which also can be calculated by (8.69).

(b) Spray drag

Due to a strut piercing through the water surface strut spray is created as the hydrofoil craft is at high speed. The Froude number with respect to the strut length is far greater than 4 ~ 5, and generally up to 8~10, so gravity can be neglected so as to not consider the wave-making drag.

The spray drag is generated due to the breaking of the wave in front of the strut at high speed, and blown out to the sides and forward into the atmosphere to form the spray and spray drag, so it is an inertial drag, and can be calculated with the following empirical equation:

$$D_s = C_{st}\frac{1}{2}\rho V^2 t_s c_s \tag{10.14}$$

where

t_s strut thickness
c_s strut chord length
C_{st} spray drag coefficient, can be taken from Table 10.1.

Reference [4] suggests that for various sections with relative thickness $\bar{t} < 0.20$, the value of (C_{st}, c_s) can be reduced to a constant and spray drag expressed as:

$$D_s = 0.24\frac{1}{2}\rho V^2 t_s^2 \tag{10.15}$$

(c) Interference drag between strut and foil

The strut lower edge connects with the hydrofoil, and they interact with each other. The strut interferes with the flow on foil back and changes the pressure distribution. In return, the foil also changes the pressure distribution of flow around the strut. Simply speaking, the forward stagnation at the strut generates a pressure increase of the foil at its forward part, and creates additional drag, which is rather difficult to calculate, and so estimation may be carried out by referring to the following test results

$$D_s = C_{Ds} 0.5\rho V^2 .t.t_s \tag{10.16}$$

where:

t hydrofoil thickness,
t_s strut thickness

Table 10.1 Automatic control systems in several hydrofoil craft of US

		"High point" PCH-1					
		original	Amended	AGEH-1	PGH-1	PGH-2	Jetfoil-1
	Pitching	Vertical Gyro, (VG)	VG	VG	VG	VG	VG
	Rolling	VG	VG	VG	VG	VG	VG
	Attitude	Ultrasound (US)	US/radar	US	Radar	US	US
	Vertical acceleration	Servo-accelerator (SA)	SA	SA	SA	SA	SA
	Roll angular velocity	Angular velocity gyro (AVG)	Roll angular electronic line differential (RAELD)	AVG	AVG	Roll differential	Roll differential
	Pitch angular velocity	AVG	Pitch AELD	AVG	AVG	Pitch AELD	Pitch AELD
	Yawing angular velocity	AVG	AVG	AVG	AVG	AVG + lateral accelerator	AVG + lateral accelerator
Computer		Simulated electronic panel (SEP)	SLP	SEP	SEP	Simulated electronic metal screened panel (SEMSP)	SEMSP
Actuator system		3000 lb/ in^2 hydraulic pump,	3000 lb/ in^2 hydraulic pump	3000 lb/ in^2 hydraulic pump, 100% margin system, tandem Actuators	3000 lb/ in^2 hydraulic pump,	3000 lb/ in^2 hydraulic pump	3000 lb/ in^2 hydraulic pump
Force generator	Foil lift	Tail flap	Tail flap (TF)	Whole rotating foil (WRF)	WRF	TF	TF
	Rudder	Strut flap +rudder	Whole rotating fore strut	Whole rotating aft strut(WRRS)	WRRS	Whole rotating fore strut (WRFS)	WRFS
Hydrofoil arrangement		Canard	Canard	Airplane	Airplane	Canard	Canard

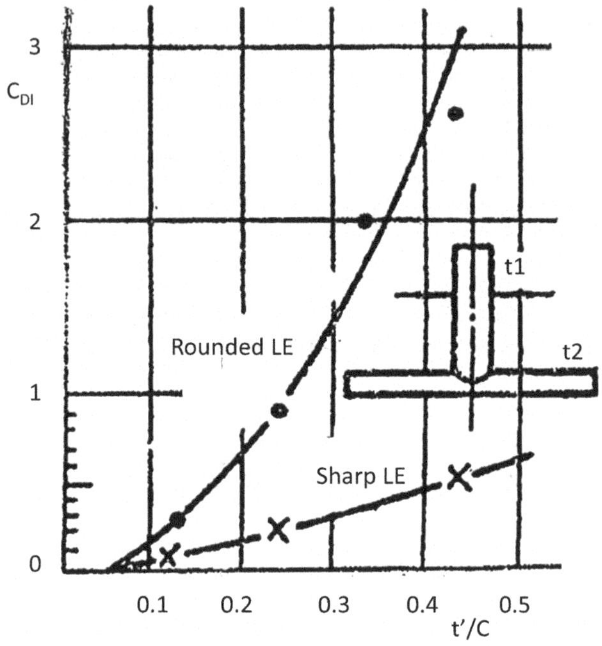

Fig. 10.11 Interference drag coefficient of strut and foil interface

C_{DI} interference drag coefficient defined by Fig. 10.11, which presents the drag for two conditions, that is, struts with round leading edge, and struts without sharp leading edge. The value t' is the average between t and t_s. Thus, the interference drag relates to the frontal area of the joint. A sharp leading edge will lead to lower interference drag than a blunter rounded leading edge.

Reference [5] gives test results of various hydrofoils with struts, in which the lift and drag coefficients can be obtained, and used for design where the desired profile is similar to the tested version. Examples of two of these configurations tested are presented in Figs 9.18 and 9.19.

When using super-cavitating or air-ventilated hydrofoils, the correspondent strut with cavitation (air ventilation) section should be used, with a symmetric form of wedge or sharp leading edge section profiles. When using a symmetric wedge section, the pressure difference drag coefficient, at zero cavitation number, can be taken from the following approximate equation, and in linear proportion to the half wedge angle ε [4]:

$$C_{DO} = \frac{4}{\pi}\varepsilon \tag{10.17}$$

The drag coefficient at nonzero cavitation number can be obtained by the following approximate equation:

$$C_{D\sigma} = C_{Do}(1+\sigma) = \frac{4}{\pi}\varepsilon(1+\sigma) \tag{10.18}$$

In the case of using a sharp leading edge section, with cavitation number σ, the pressure difference coefficient is:

$$C_{D\sigma} = \frac{\pi}{8}\overline{t^2}(1+\sigma) \tag{10-18a}$$

The calculation for other drags, friction, and spray drag, is as same as that for noncavitating strut.

10.2.3.3 Appendage Drag

As the drag of a hydrofoil and its strut are small, the percentage of appendage drag is relatively great. Similar to planing craft, it is not suitable to estimate the appendage drag as a percentage of bare hull drag. Rather, it should be calculated separately.

Some appendages are similar to those on a planing hull, so it need not be repeated here. However, the propulsion transmission casing is a special appendage in hydrofoil craft, and this drag can be a great proportion of total appendage drag, due to the significant volume for accommodating the bearings and propeller shaft power transmission. The friction and form drag can be estimated with the following empirical equation:

$$C_{DP} = C_f\left[1 + \frac{1}{2}\frac{d}{l} + \frac{1}{6}\left(\frac{d}{l}\right)^4\right] \tag{10.19}$$

The drag of the corresponding transmission package:

$$D_P = C_{DP}\frac{1}{2}\rho V^2 S_P \tag{10.20}$$

where:

$C_f$friction coefficient at the correspondent Reynolds number with respect to the transmission package length
$d$maximum diameter of transmission package
$l$length of transmission package
$S_P$wetted surface area, can be estimated as:

$$S_P = S_o \times 3\frac{l}{d}$$

S_o front projection area of the transmission pack.

In addition, [4] gives the following empirical equation based on planing boat design to derive C_{DP}:

$$C_{DP} = C_f \left[1 + 1.5 \left(\frac{d}{l} \right)^{3/2} + 7 \left(\frac{d}{l} \right)^3 \right] \tag{10.21}$$

10.2.3.4 Air Drag

This can be estimated by:

$$R_\alpha = C_\alpha \times \frac{1}{2} \rho_\alpha V_\alpha^2 S \tag{10.22}$$

where:

ρ_α air density, 0.125 kg sec /m^4
V_α incoming air speed, which is craft speed plus wind speed
S frontal projection area of hull above water surface
C_α air drag coefficient, take as 0.4~0.6;.

When estimating drag from model test data and the above approximate estimation, the total power of hydrofoil craft can be estimated by a coefficient method known as the K_η method, discussed in Chap. 3, Sect. 3.3.2, with typical data shown in Fig. 3.13.

For hydrofoil craft with single bow hydrofoil, the drag of bow foil and aft planing surface can be calculated separately, using the force balance equation, similar to the approach for hydrofoil craft as above.

10.3 Stability of Hydrofoil Craft

The stability of hydrofoil craft in slow speed hull borne operation is the same as a displacement ship, with the restoring force (moment) provided by movement of the center of hull underwater displacement to provide a righting moment acting about the VCG. In foil borne operation a hydrofoil hull is above the water surface and stability has to be provided dynamically by the hydrofoils. Another difference from a displacement ship is that heaving, or vertical stability, is different for hydrofoil craft as that also has to be provided by dynamic force generation at the foils. The foils are acting in a similar manner to the wings and tail stabilizer of an airplane.

10.3.1 Vertical Stability of Hydrofoil Craft

The vertical stability of a displacement ship is provided automatically by the change of buoyancy with draft. Vertical stability for a hydrofoil craft in foil borne operation mode is provided by changes to the foil lift. This is rather different in various types of foil arrangement. The mechanisms for surface piercing and fully submerged hydrofoils are as follows:

(1) Shallow submerged hydrofoil craft

Vertical stability is provided by the effect of the shallow submerged foil depth. The lift acting on the foil increases with submerged depth, so with increase in submerged depth the additional foil lift provides restoring force, and vice versa.

Two points have to be noted. First, the submerged depth should be "shallow" enough. It is well known from the previous section that for relative submerged depth at $\overline{h} = h/c \geq 1.0 \sim 1.5$, lift can be changed with submerged depth, nevertheless, at this depth the restoring force is low, so in order to maintain vertical stability the relative submerged depth should be $\overline{h} \leq 0.5$.

Secondly, even though a well-designed shallow submerged depth hydrofoil craft maintains vertical stability in foil borne mode there may be issues during the transition from hull born mode. During the takeoff from hull mode to foil borne mode, the craft rises and the foils may transit from deep submerged to shallow submerged.

Meanwhile, the vertical stability provided by buoyancy *decreases* gradually, and that provided by foil lift *increases* gradually. A certain condition may be reached where the increase of vertical stability provided by foil lift does not compensate the decrease of buoyancy. In this case, the craft may not be able to safely transit to hydro-foiling mode.

For shallow submerged depth hydrofoil craft, the checking of stability should not only be carried out for foil borne mode but also during takeoff. This issue also relates to both longitudinal and transverse stability. In general, some additional means should be mounted on the shallow submerged hydrofoil craft for keeping positive stability during transition, such as surface piercing side foils, or mid foils between the hull bottom and main foil.

(2) Surface piercing foil

The V-type foil is a regular type of surface piercing foil to achieve changing the lift with the aid of changing the submerged foil area, and creating positive vertical stability. Of course, the surface piercing foil has to be operated close to the water surface, with a shallow submergence effect. In general, the vertical stability of surface piercing foils is better than a shallow submerged flat foil, with good stability during takeoff, due to the large range of submerged depth, the deeper submerged depth of a V foil can provide higher restoring force during the take-off. In addition, variation of foil chord can be used to tune the variation of lift force with foil submergence.

(3) Fully submerged foil with control surfaces

Lift can be controlled on this type, for providing restoring force and vertical stability by active use of control surfaces. For deep submerged hydrofoil craft, the lift controls keep positive stability by change of angle of attack of whole foil or flap, or by air ventilation on the foil back.

The power consumption for rotation of a whole foil is greatest and so only suitable for small hydrofoil craft. The adjustable flap is the most reasonable type from the point of view of power consumption and efficiency. A foil submerged in seawater, has a loading that is significantly larger than a wing in air. A foil is also sensitive to external perturbation due to waves, and other problems such as erosion and this applies to control surfaces. The control of hydrofoil flap is more complicated than the control of a wing flap. In particular, the safety consequences of "mal-operation" of a hydrofoil flap may be serious. In relation to vertical stability, the flap positions required to maintain stability during takeoff are important to define. The alternative of using air ventilation on a foil back is rather simple, however, with lower hydrodynamic efficiency than a flap.

Fully submerged hydrofoils can maintain positive vertical stability, and transverse and longitudinal stability through active operation of the control surfaces. Furthermore, these active controls can decrease the perturbation from waves, and decrease motion response of craft significantly increasing seaworthiness.

The three types of vertical stability controls mentioned above generate the necessary restoring force of foils under the external vertical perturbation. However, it must be noted that a single foil with positive vertical stability does not maintain positive vertical stability for a whole hydrofoil craft, as there are two foils on the craft, with a certain space between them.

If a hydrofoil craft descends due to a vertical perturbation, and a larger restoring force is generated by the stern foil than the bow foil then a bow-down pitching will occur, so decreasing the angle of attack of foils and lift. The craft will continue to descend. Vertical stability therefore connects strongly with longitudinal stability.

When analyzing vertical stability a designer should check that as a hydrofoil heaves, any out-of-balance forces creating a pitching moment are minimized. This will minimize any coupling effect between pitch and heave.

10.3.2 Longitudinal Stability of Hydrofoil Craft

The longitudinal stability of hydrofoil craft is provided by lift adjustment of both bow and stern foils. Similar to an airplane it cannot be provided by a single foil. For the deep submerged hydrofoil with active controls, lift can be controlled according to requirements so the necessary longitudinal stability can be guaranteed by reasonable design of automatic control system.

For fixed hydrofoils, an arrangement of foils is required that maintains positive longitudinal stability when an external perturbation is applied. Typically a wave

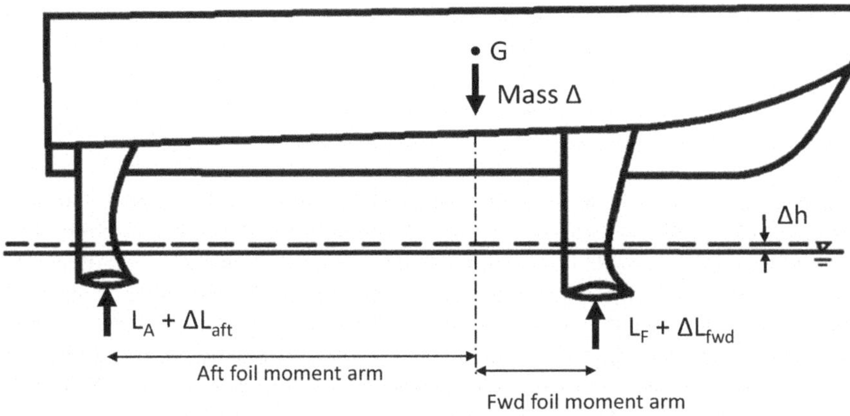

Fig. 10.12 Longitudinal balance of foil loading

passing the craft will progressively apply differential forces to the bow and stern foils, so generating pitching moment.

Basic Requirement for the Arrangement of Both Forward and Aft Hydrofoils Here the stability we discussed above is implying the quasistatic stability of hydrofoil craft in foil borne mode, and only considering the displacement of motion, but not motion velocity and acceleration, that is, the motion is "quasi-static."

In the case of hydrofoil craft with the same load for both forward and aft foils, the equilibrium requirement is met, and with $L_F = L_A, \overline{h}_E = \overline{h}_A \ldots\ldots$, as shown in Fig. 10.12.

When the craft experiences a vertical external perturbation downward, that is, $+\Delta h$ the response for both forward and aft foils are the same. Thus, lift increment $\Delta L_{fwd} = \Delta L_{aft}$, so the craft moves up in heave, and returns to the original position with positive vertical stability. The variation with heave is equal, i.e. $\frac{dL}{dh}\big|_F = \frac{dL}{dh}\big|_A$, which can be called as the "foil stiffness," just as the stiffness of a spring.

If the stiffness of the bow foil is greater than the stern, that is, $\frac{dL}{dh}\big|_F > \frac{dL}{dh}\big|_A$, then the lift increment of the bow foil is greater than the stern foil, $\Delta L_{fore} > \Delta L_{aft}$, and makes the bow foil rise faster than the stern foil, trimming the bow up. This increases the bow foil angle of attack and increases bow up trimming again. . ., meanwhile, due to the bow pitching up, the draft of the bow foil is smaller than the aft foil, so decreasing the bow foil lift, and restoring to the original craft elevation.

If the bow foil stiffness is smaller than the stern foil, $\frac{dL}{dh}\big|_F < \frac{dL}{dh}\big|_A$, then for craft pitching down, the stern foil lift is greater than that of the bow foil, and the bow trims down decreasing the angle of attack and lift of both foils. The craft may not return to the original equilibrium position, dependent on the relation of lift change in terms of angle of attack and heaving displacement, and may cause the craft to plough into the water.

So from the point of view of static longitudinal stability, the stiffness of bow foil should be greater than stern foil, that is,

$$\left.\frac{dL}{dh}\right|_F > \left.\frac{dL}{dh}\right|_{\mathrm{A}} \tag{10.23}$$

Regarding pitching, for bow pitching up, if we assume the center of rotation is between the two foils the craft motion will be as follows. Upwards pitching will cause submerged depth of the bow foil to decrease with an increased angle of attack. Meanwhile, the stern foil will increase both submerged depth and angle of attack.

The stern foil movement causes a restoring moment to reduce pitch. However, the moment acting on the bow foil to pitch down will be a balance between the lift increment from the increment of angle of attack and lift decrement in terms of decrement of submerged depth. From the point of view of keeping positive longitudinal stability, it is needed that the lift decrement by submerged depth is greater than the lift increment by increment of angle of attack, that is, with larger $\frac{dL}{dh}$ and smaller $\frac{dL}{d\alpha}$, and such requirement also for the pitching down case. $\frac{dL}{d\alpha}$ is generally called "foil damping."

"Stiffness" and "Damping" for Various Hydrofoils

From the above discussion, one can see that from the point of view of stability, it is helpful if the forward foil has high stiffness with low damping. In terms of profile and dimensions, if lift can be expressed by $L = C_L \frac{\rho}{2} V^2 S$, then stiffness:

$$\frac{dL}{dh} = \frac{\rho}{2} V^2 \frac{d}{dh} \ (C_L S) \ = \frac{\rho}{2} V^2 \left[\frac{dC_L}{dh} \cdot S + \frac{dS}{dh} \cdot C_L \right] \tag{10.24}$$

Since $\frac{\rho}{2} V^2$ is constant with submerged depth, and for a fully submerged foil, $\frac{dS}{dh} = 0$, so only the first item changes in the bracket for the fully submerged foil.

For a surface piercing foil, the two items are not zero. $\frac{dC_L}{dh}$ represents the effect of shallow submerged depth. When h decreases, $\frac{dC_L}{dh}$ increases. Finally the lifting surface change can be calculated using the projected area of a foil with upward dihedral:

$$\frac{dS}{dh} = \frac{2c}{\tan \beta}$$

where:

c chord length of foil,
β dihedral upward angle.

So the stiffness of a shallow submerged foil increases inversely with submerged depth and angle of dihedral. The stiffness of a surface piercing foil is relative to the submerged depth and area of foil. It is greater than a flat shallow submerged foil.

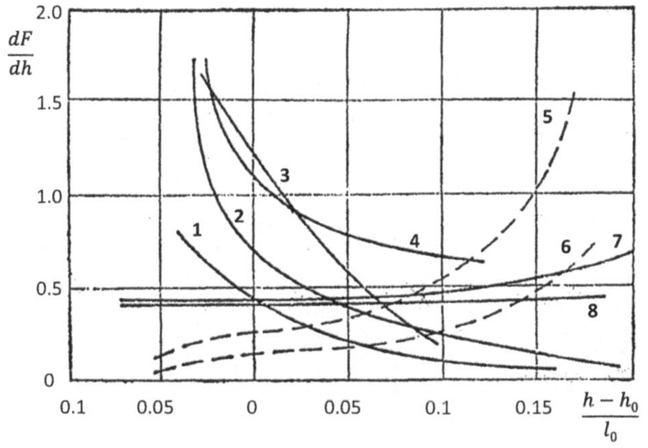

Fig. 10.13 Relationship of stiffness with submerged depth
Key to diagram 1 Flat fully submerged foil $\left(\overline{h}_0 = 0.5,\ \alpha = 1\,°\right)$ 2 Low-submerged flat foil 3 Flat foil with inclined stabilizers and deeply submerged elements 4 Flat foil with inclined stabilizers 5 Optimal curve for the bow foil 6 Optimal curve for the stern foil 7 Deep-submerged foil of the Schertel-Saxenberg scheme (PT-50) 8 V-shaped foil $\left(\beta = 26.5\,°,\ \overline{h}_0 = 0.5\right)$

Reference [1] gives the stiffness curves related to the submerged depth of various foils, as shown in Fig. 10.13 below. This shows the foil stiffness dF/dh ($= dL/dh$ in (10.24 above) is related to relative submerged depth $(h - h_O)/l_O$, where h and h_O are actual submerged depth and design submerged depth in foil borne operation respectively and l_O is the foil span in foil borne operation. One can see from the figure that stiffness is great at shallow submerged depth, and reduces with increase of submerged depth.

For a craft operating in small waves, it can respond sensitively to the small waves, and following the waves. However, in larger waves, the response is relatively less sensitive.

The stiffness of a flat shallow submerged foil is shown in curves 1 and 2. A hydrofoil with side stability foils is shown in lines 3 and 4. For a V-type surface piercing foil, due to the foil chord increasing gradually upward, the stiffness increases steadily with submerged depth (curve 7 and 8).

The ideal stiffness of a fixed foil relative to submerged depth is shown in curves 5 and 6 of Fig. 10.13, that is, stiffness increases with submerged depth, and it is more important at greater submerged depth to maintain enough restoring force in high waves. For a stern foil, the stiffness should be smaller than the bow foil, while following the requirements mentioned above, for to maintain longitudinal stability, as shown in curve 6 in Fig. 10.13.

In order to improve the longitudinal stability of fixed foil hydrofoil craft, the stiffness of both bow and stern foils has to be designed carefully, such as increasing the angle of attack of takeoff aids and increasing the chord of surface piercing foils.

Additionally, it is possible to mount a middle foil between hull and main foil or increase the space between the bow and stern foils.

For fully submerged automatic control hydrofoil craft, the stiffness can be specified in terms of the rate of change of foil lift. Thus, foil damping can be expressed as:

$$\frac{dL}{d\alpha} = \frac{dC_L}{d\alpha}\frac{1}{2}\rho V^2 S \tag{10.25}$$

The rate of change of lift coefficient with angle of attack, $\frac{dC_L}{d\alpha}$, is controlled by foil section profile and also foil type. For foils with upward dihedral the effective angle of attack decreases by cos β, so the damping will also be decreased by cos β compared with flat and fully submerged foils with the same section profile and same effective area.

Longitudinal Stability for Various Foil Arrangements
The characteristics and advantages/disadvantages related to longitudinal stability for various arrangements of foils can be compared in terms of their stiffness and damping. We consider the main arrangements in turn as follows:

(a) Airplane Arrangement

For bow and stern foils with the same section profile and type, the stiffness of the bow foil will be greater due to the larger bow foil area.

This system and variations with foil areas close to the tandem foil arrangement were applied on the shallow fully submerged foil system of early Russian hydrofoil craft (Alekseyev's system). This kept the stiffness of bow foil greater than stern foil, avoiding pitch-down tendency in waves.

In later designs, a ladder-type foil was used at both sides on the forward foil to further improve its stiffness and both longitudinal and transverse stability. Surface piercing foils have also been used on the sides of the forward foil struts, for improving stability.

For the airplane foil arrangement with surface piercing side foils, the load on the forward foil should be not too great, as a larger forward foil size may create too great stiffness and so heavier motion in waves.

The three ladder foils applied on Alexander Graham Bell's hydrofoil craft in the airplane arrangement, as shown in Fig. 7.7, is an example using separated bow foils rather than a single bow foil.

(b) Tandem Arrangement

In general, the load on the forward foil is designed to be somewhat greater than at the aft foil. The Von Schertel-Sachsenburg hydrofoil system used this for early designs, with a surface piercing foil arrangement. The load at the forward foil is about 53% of the total for give higher stiffness. Later Supramar also applied this arrangement, and amended the stern foil to be a fully submerged shallow foil, to reduce its stiffness, and increase the damping. This system has well-balanced

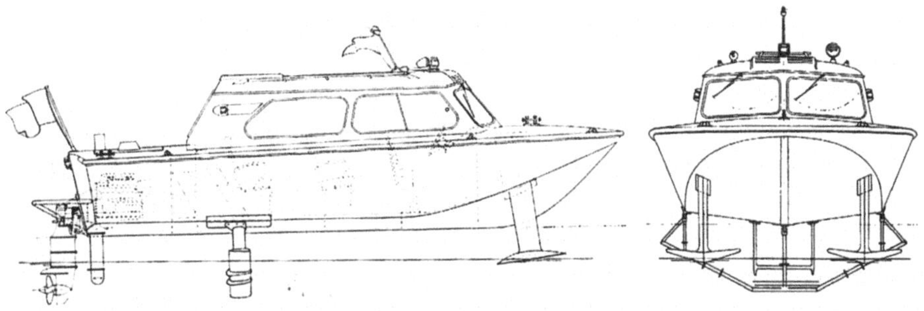

Fig. 10.14 Aquavion general arrangement

longitudinal stability and seaworthiness, so was used for many craft from Supramar and subsequently Rodriquez.

(c) Canard Arrangement

In this arrangement, the bow foil is significantly smaller than the stern foil, so it is impossible to keep greater stiffness of bow foil than stern foil. A canard foil arrangement with fixed foils requires a surface piercing bow foil to achieve higher stiffness of the forward foil, while the aft foil may be fully submerged flat or similar foil with smaller stiffness.

An example of the canard system is the design of Aquavion [6]. The load on the stern foil is about 85 % ~ 90% of the total load, with largely submerged foils. In order to satisfy transverse stability, a submerged stern foil with a small piercing area was applied, as shown in Fig. 10.14. This main lifting foil was located some way forward of the stern. At the transom, a small submerged flat foil was mounted to assist with pitch stability. Primary trim stability was from two bow planing buoyant bodies giving lift and higher pitch stiffness.

A prism shape surface piercing bow foil and fully submerged mid foil with inclined side foils were applied on the Canadian Bras D'Or hydrofoil to increase as far as possible the bow foil stiffness and to decrease the stern foil stiffness, see Figs. 7.28 and 7.29. Air ventilation was applied to the foils. To protect against bow foil plow-in air ventilation was shut off increasing the lift, to ensure sufficient bow-up moment.

Summing up, longitudinal stability has to be considered carefully for the foil arrangement of hydrofoil craft with fixed foils. Longitudinal stability is a fundamental basis for investigating seaworthiness. This will be discussed in the next section.

Separately, for a craft with a single hydrofoil and aft planing surface (see Fig. 7.11), higher stiffness is available with the hull planing surface so the forward foil should have sufficient stiffness to provide seaworthiness, and a fully submerged mid-foil with surface piercing side foils may be applied to minimize craft drag.

10.3.3 Transverse Stability of Hydrofoil Craft

Similar to vertical stability, transverse stability does not exist on a deep submerged foil. For such a foil in roll motion, the foil lift has to be adjusted by control surfaces to generate restoring moment.

On a fixed shallow submerged foil or surface piercing foil, lift forces change due to varying immersion and submerged foil area that generate restoring forces.

Shallow Fully Submerged Foil

In case of a transverse disturbance on a foil, as the craft rolls the submerged depth increases, so lift is increased, but the submerged depth and lift of the other side of the foil is decreased, so the resultant lift shifts to the heeling side and generates a restoring moment.

In general, this restoring moment is positive, and similar to a displacement ship, the initial transverse stability can be obtained based on moments around the metacenter. Assuming the lift distribution before heeling is uniform along the span, and with a trapezoid shape after the heeling, as shown in Fig. 10.15, then the submerged depth at both sides after heeling should be:

$$h_1 = h_0 \cos\theta + \frac{l}{2}\sin\theta \approx h_0 + \frac{l}{2}\theta$$

$$h_2 = h_0 \cos\theta - \frac{l}{2}\sin\theta \approx h_0 - \frac{l}{2}\theta$$

where:

l foil span length
θ foil inclination (heeling plus dihedral) angle

Then the restoring moment can be calculated as follows. The force acting on the foil can be divided into two parts; a rectangular and a superimposed triangular part. The center of action of the rectangular part is located at mid-span, point 0, and does not cause a moment around the craft CG. The lift of the triangle part is:

$$\Delta L = \frac{dC_L}{dh} \cdot \frac{\Delta h}{c} \cdot \frac{1}{2}\rho V^2 lc$$

This is caused by the difference of submerged depth, so the lift coefficient C_L for the side being submerged should change to reflect $\overline{h}$ for that side and similar for the up-going side. The difference between both sides' submerged depth is:

$$\Delta h = hl - h2 = l sin\theta \approx l\theta$$

Putting Δh into the expression, we have:

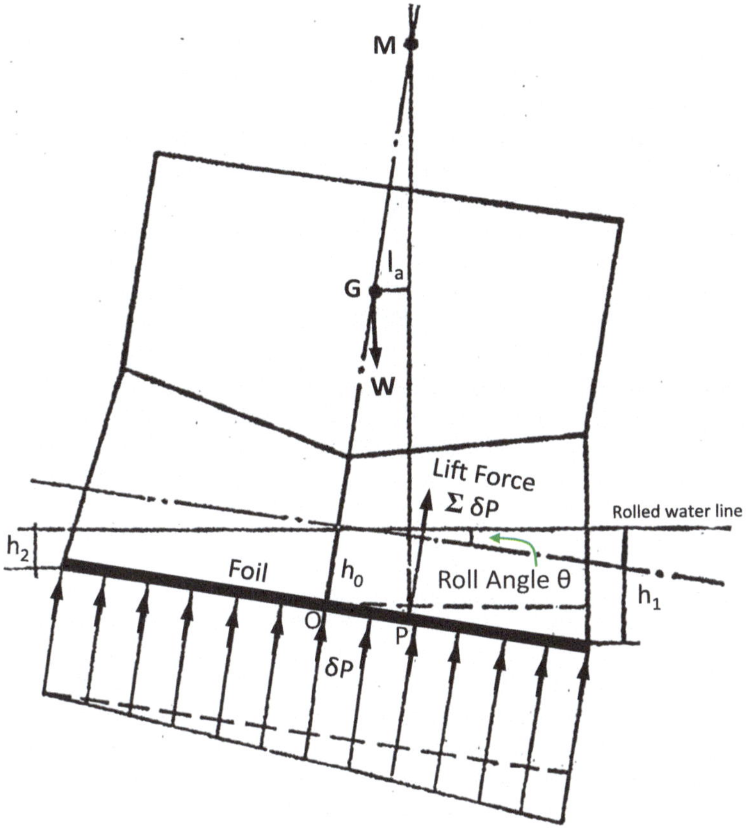

Fig. 10.15 Transverse stability of shallow submerged flat foil

$$\Delta L = \frac{dC_L}{dh}\frac{1}{2}\rho V^2 l^2 \theta$$

This acts on the centroid of the triangle, so the moment arm is:

$$\frac{2l}{3} - \frac{l}{2} = \frac{l}{6}$$

The restoring moment is therefore:

$$m_\theta = \frac{dC_L}{dh}\frac{1}{12}\rho V^2 l^3 \theta \qquad (10.26)$$

Assuming the total lift operates at point P in Fig. 10.15, the restoring moment generated by the total lift is equal to this moment and:

$$C_L \frac{1}{2} \rho V^2 c l \overline{OP} = \frac{dC_L}{d\bar{h}} \frac{1}{12} \rho V^2 l^3 \theta \tag{10.27}$$

Similar to displacement stability, from point P a straight line perpendicular to the water plane meets point M as the intersection point of the metacenter, and the metacenter height is $h_p = \overline{GM}$ so that $\overline{OP} = \left(\overline{OG} + \overline{GM}\right) \tan\theta = \left(\overline{OG} + h_p\right) \tan\theta$.

Putting this into (10.27), then for small θ:

$$\overline{OG} + h_p = \frac{l^2}{12c} \frac{\frac{dC_L}{d\bar{h}}}{C_L} \quad \text{and so} \quad h_p = \frac{l^2}{12c} \frac{\frac{dC_L}{d\bar{h}}}{C_L} - \overline{OG} \tag{10.28}$$

Equation (10.28) is an expression for nominal initial metacenter height of hydrofoil craft, actually from $h_p + \overline{OG} = \overline{GM}$, the restoring moment can be obtained directly.

$$m_\theta = \Delta \cdot \overline{OM} \cdot \theta \tag{10.29}$$

where:

Δ displacement of craft.

In (10.29), the $\overline{O}M$ is the equivalent restoring moment arm to the metacenter on a displacement ship or planing hull measured from the keel line (KM).

From (10.28) one can see initial stability is in proportion to square power of the span, and in linear power to shallow submergence effect. So increasing span is the most effective means to improve transverse stability. It is therefore likely the airplane or canard foil arrangement is more favorable compared to a tandem arrangement, from the point of view of transverse stability.

The transverse stability of a shallow submerged hydrofoil craft during take-off is rather difficult due to its unstable transverse situation and even capsizes have happened to early craft during this phase.

At low-speed transverse stability is provided by buoyancy, and at high speed provided by shallow submergence effect of foils, however, during takeoff, when the craft is rising up, but still not lifted off from water with small buoyancy and also small submerged effect due to deep submergence and low speed, the foils cannot provide enough restoring moment.

To solve this problem on the first Russian shallow submerged hydrofoil craft "Raketa," various takeoff aids were installed. Two side plates were added to forward foils a little bit higher than the main foils. When the craft heeled the side plates operated in planing mode to provide additional lift. Later the craft was modified to add small ladder foils at the outside wall of the struts. The arrangement for Raketa is shown in Fig. 10.16.

The auxiliary foils and bow takeoff planing surface are above water surface contact at design speed, so they do not influence high-speed performance; however, as they are shallow submerged foils during takeoff, the foil installation angle (fixed angle of attack relative to baseline) is about 2° ~ 3° larger than the main foil.

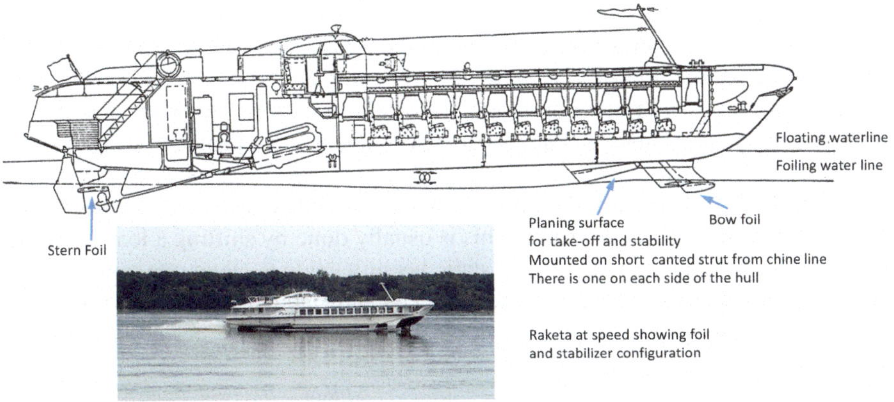

Fig. 10.16 Raketa bow foils with takeoff aids and planing stabilizer foils

Fig. 10.17 Transverse stability of surface piercing foil

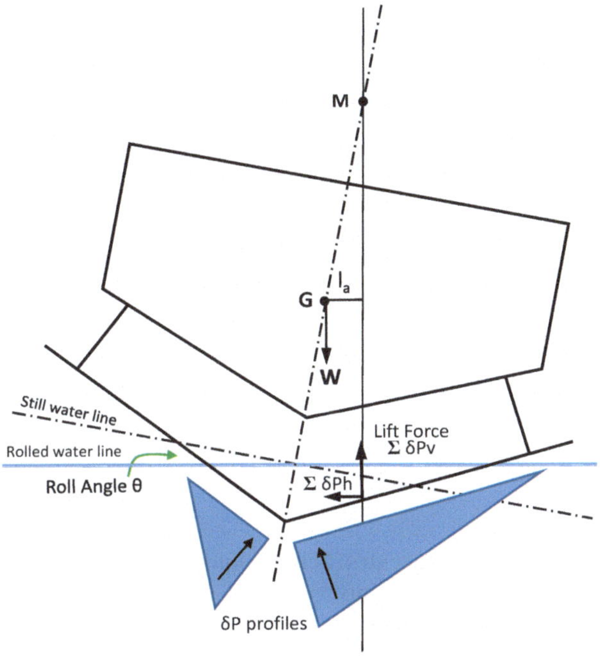

Surface Piercing Foil (Surface Piercing Foil)

During heeling for a V-type surface piercing foil, the restoring moment is provided mainly by the change of foil area. Meanwhile, due to the change of dihedral angle the lift of the heeling down foil will be increased, so the lift force resultant generates restoring moment about CG, as shown in Fig. 10.17.

It must be noted that for a very high-speed design, the submerged area and foil width may be very small, so the restoring moment may be negative. So in general, a limited speed for the hydrofoil craft with V-type foil should be applied for keeping positive transverse stability, that is, the craft stability is positive below this speed; otherwise, instability may occur. Alternatively, active control surfaces may be installed.

Similar to a planing hull, the initial transverse stability also can be obtained by model tests or prototype craft trials. This is usually done by shifting a load on board to make a heeling moment and initial static heeling angle θ_O, then one measures the correspondent heeling angle θ at various speeds and calculates the initial transverse metacenter height h_p.

$$\overline{OM} = \overline{OG} + h_p = \frac{m_\theta}{\Delta \cdot \theta} = \frac{\Delta h \theta_0}{\Delta \cdot \theta} = \frac{\theta_0}{\theta} h$$

$$h_p = \frac{\theta_0}{\theta} h - \overline{OG}$$

where:

m_θ heeling moment, $m_\theta = \Delta h \theta_0$
Δ Displacement
θ measured heeling angle at speed V
h metacenter height at static buoyancy,
$\overline{OG}$ CG height from foil baseline.

It is seen that the initial metacentric height of a craft at speed increases inversely with heeling angle, and so correspondent h_p versus θ can be obtained.

Reference [7] presented an approximate equation for initial metacenter height of a shallow submerged hydrofoil craft as:

$$h_y = 0.2bA^2 f(\overline{h}) - (2g + H_K) \tag{10.30}$$

where:

h_y initial metacentric height (equiv h_p above)
A foil aspect ratio
H_K running altitude of keel when foil borne
$f(\overline{h})$ is function of lift of shallow submerged foil with respect to change of submerged depth. Generally:

$$f(\overline{h}) = f - 0.5 \cdot 10^{-(\overline{h})} \tag{10.31}$$

where $\overline{h} = h/c$ relative submerged depth.

For a V-type surface piercing foil, according to [8], the initial metracentric height can be estimated as:

$$h_y = \frac{a_F l_F}{2 \tan \beta_F} + \frac{a_a l_a}{2 \tan \beta_a} - (Z_g + H_K) \tag{10.32}$$

where:

a_F, a_a load of forward and aft foil over the displacement,
l_F, l_a span of forward and aft foils,
β_F, β_a deadrise angle at span direction of forward and aft foils,
z_g CG height.

For transverse stability at large heeling angles, the calculation is similar to that on a displacement ship. However the foil lift is extremely difficult to calculate due to the three-dimensional effects, so the calculation is very complicated and with poor precision.

Suffice it to say that hydrofoil operation should be controlled within environmental conditions where the hydrofoil craft can maintain dynamic stability in roll, and pitch. If forces are such that high angles may occur then either the foil configuration would need redesign, or the craft allowable operating envelope would have to be reduced.[1]

10.4 Maneuverability of Hydrofoil Craft

The maneuverability of a fixed hydrofoil craft is similar to the planing hull and is also implemented by the installation of a rudder or multiple rudders. The maneuverability of a hydrofoil craft is related to the lateral underwater projected area, its distribution, and center of horizontal rotation. For a canard foil arrangement, course stability is good, however with poor turning performance due to the underwater area concentrated at the stern.

The airplane foil-type turning performance is much better, due to the distribution of underwater appendages. Due to the high speed of hydrofoil craft, their turning diameter is large so optimization is important.

Also due to the high speed, rudder efficiency is high, so small area rudders can be applied aft of the propeller, or aft and connected to a strut. When using water jet propulsion, thrust steerable nozzles can be applied, while still using water rudders installed on aft struts.

[1] See the IMO High Speed Craft rules for guidance on stability guidelines and assessment for hydrofoils: International Code of Safety for High Speed Craft, IMO, publication IA-185E, ISBN 92789 28,014 2402, 2000. Amendments and resolutions after 2000 are available on IMO web site IMO.org

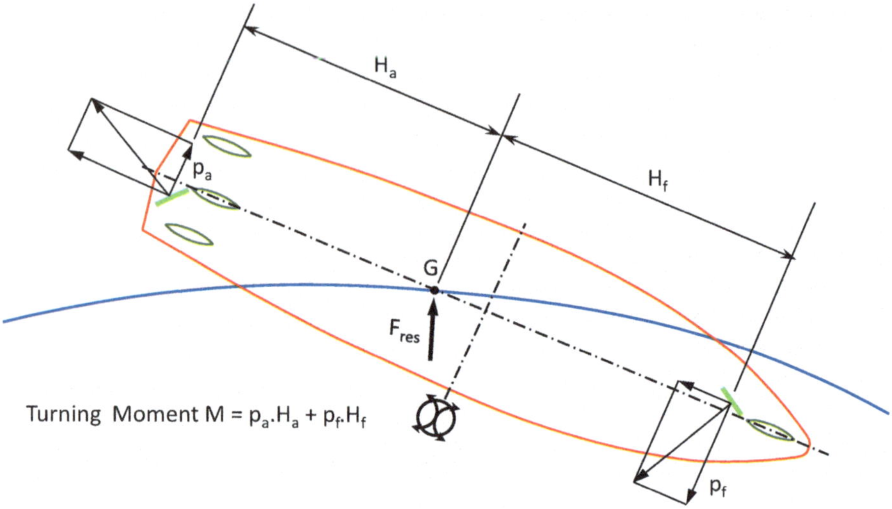

Fig. 10.18 Forces acting on the rudders during turning of hydrofoil craft

For shallow fully submerged foils, due to shallow draft in foil borne operation, the rudder is generally mounted aft of the propeller. This is the normal arrangement on Russian inland water hydrofoil craft.

For stern foils with deeper submerged depth, a rudder is usually difficult to be mounted aft of the propeller, so it can be mounted on the aft strut.

Studies have demonstrated that during the turning of hydrofoil craft, as the craft CG is aft of amidships the lever arm is shortest to a stern-mounted rudder. If rudders are mounted on the bow foil, the turning effect is significantly greater, as shown in Fig. 10.18. In addition, due to the presence of the struts, the rudder efficiency of an aft rudder is less than a bow rudder.

So purely from the point of view of hydrodynamics, it is more optimal to mount the rudder at the bow, and perhaps the best option is using a bow strut-rudder system on a hydrofoil craft with canard foil arrangement. This was been adopted for *Bras D'Or*, and later by Boeing on Jetfoil.

When calculating the turning performance of hydrofoil craft, the hydrodynamic forces acting on the craft in swaying and heeling motion have to be considered during the turn. For the turning performance of hydrofoil craft with fixed foils, the following two points may be noted:

1. At high speed, inward banking has to be created as far as possible if tight radius turns are to be executed. In turn, the centrifugal force will roll the craft outwards. Since the CG is a considerable distance from the roll center at the foil elevation the stabilizing moment is limited, and once the craft starts to roll outward the situation could easily become unstable. Dynamic stabilizing moments can be generated with surface piercing hydrofoils and V-type foils. When applying a yawing angle β, the lateral component of incoming flow $V\sin\beta$ generates an

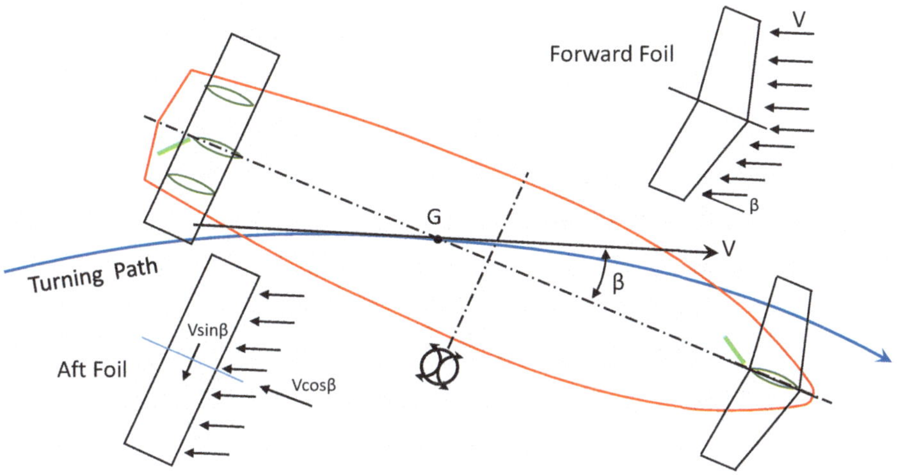

Fig. 10.19 Foil velocity vectors in a turn

inward heeling moment, as shown in Fig. 10.19. There is no such inward heeling moment on a fully submerged flat foil. However if a swept foil is used an inward heeling moment is available, as shown in Fig. 10.19 at the forward foil. The flow direction during turning causes greater lift acting on the foil external side and smaller lift on the inner side of the foil, so as to create the inward heeling moment. A swept main foil is applied on most hydrofoil craft with shallow submerged flat foils to provide inward heeling moment, and to defer cavitation. In addition, such craft usually have surface piercing components that have dihedral and variable chord to tune the forces and moments for maneuvers.

2. The yawing angle during turning causes support struts to be located in oblique flow possibly with a high angle of attack. This causes high negative pressure on the strut back, which can induce cavitation and air ventilation, as the strut extends through the water surface. The air venting may reach the lifting foil back decreasing lift, and finally significant reduction in a turn. Thus fences should be mounted on the strut to arrest the progress of ventilation to the lifting foil.

For hydrofoil craft with fully submerged foils and active control surfaces, the rudders or/and flaps can be controlled to generate inward banked turns (referred to here as "coordinated" turns) or turns where the craft stays upright (referred to here as "platform" turns).

The coordinated turn is favored by crews as it minimizes perceived transverse acceleration to the body, but platforming turning can have advantages for some instrument and control systems, and would normally be applied for gentle turns where crew or passengers may feel more comfortable without the roll motion.

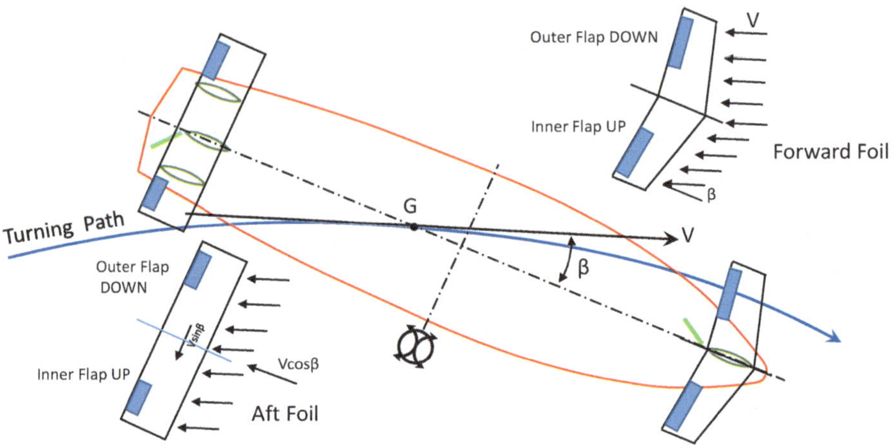

Fig. 10.20 Operation of flaps for coordinated turning

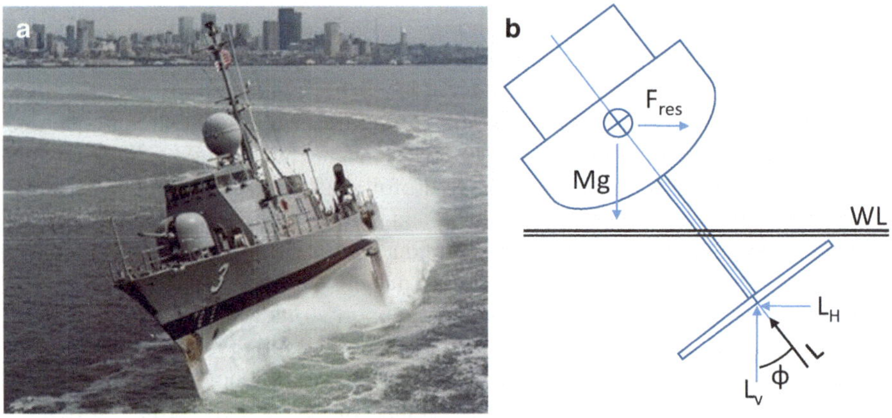

Fig. 10.21 Coordinated turning of PHM craft

In general, for sharper turns, coordinated turns with differential action of flaps is used, as shown in Fig. 10.20. For example, these were used on hydrofoil craft *Tucumcari* and US Navy guided missile hydrofoil craft, the PHM. However, a water rudder was still installed on the craft for hull borne operation. In addition, in order to avoid air ventilation of struts during turning a rotating bow strut was installed enabling it to maintain zero angle of attack with incoming flow during turns. Figure 7.4 shows the arrangement of a PHM hydrofoil craft with canard foil arrangement and side flaps, which enable inward banked turning, and with the rotating bow single strut. Figure 10.21 presents the banked turning of PHM, and the forces acting on the foils during the coordinated turns [9].

10.5 Seaworthiness of Hydrofoil Craft

Assessment of hydrofoil craft seaworthiness includes its response in waves, slamming acceleration, speed loss in waves, spray, and green water on decks, dynamic stability, and maneuverability.

The motion response in waves is a basic characteristic of craft seaworthiness. We will discuss mainly the longitudinal motion response, and it is not difficult to analyze the transverse response based on the same principle. According to the different wave sizes, there are three typical motion modes:

10.6 Platforming Operation

This is a motion mode with the CG moving on horizontal straight level, without vertical motion disturbed by waves, as shown in Fig. 10.22. Such motion is an ideal motion mode, minimizing accelerations on equipment.

For fixed foil hydrofoil craft (FF hydrofoil craft), no matter whether shallow submerged flat foil or surface piercing foil, this is only possible on "calm water"; however, for automated control deep submerged foil hydrofoil craft (ACS hydrofoil craft), this mode can be maintained in moderate sea states.

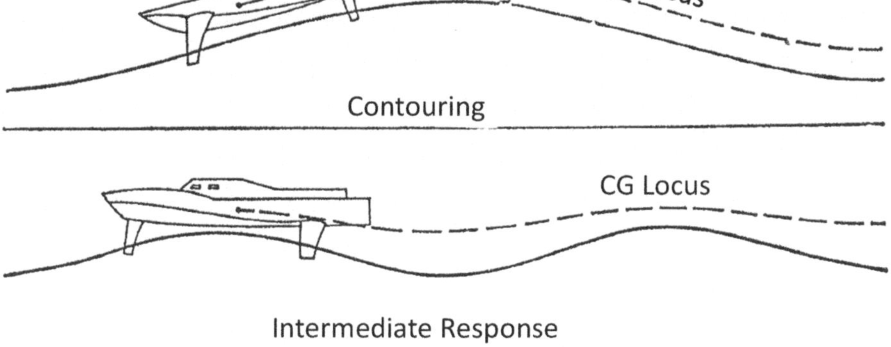

Fig. 10.22 Motion response of hydrofoil craft in waves

10.7 Motion Following the Wave Profile

This motion type is the CG following the wave surface contour, as shown in Fig. 10.22 middle diagram.

It is impossible for a fixed foil hydrofoil craft to maintain "platforming" motion, and in general, such a craft will pitch and heave following the wave profile of longer waves in a sea state. An ACS hydrofoil craft also cannot maintain platforming operation in rough seas and so will have a pitch and heaving motion, though much less than following the wave profile, as in the middle diagram of Fig. 10.22.

10.8 Intermediate Response Motion

In this condition, craft motion is between the two motion modes above, as shown in Fig. 10.22 lower diagram. In rough seas ACS hydrofoil craft cannot reach platforming operation, nor when in following seas. To avoid large motions the control system should be set to respond to the intermediate-length waves. The craft will then pitch and heave following the longer-length waves (swell). FF hydrofoil craft will profile wave swells and respond to the shorter waves with damping depending on the relation to the natural period of heave and pitch. The design for motion damping is therefore important.

It is required to avoid the hull touching the water, and especially foils breaking the water surface as far as possible, no matter what type of foil arrangement is used. The safe operational limit will be before such extreme motion would occur.

For ideal seaworthiness an automated control system is the best choice, so that the lift of foils in waves can be adjusted continually, responding to different waves, so as to optimize the motion response.

For FF hydrofoil craft, the motion response is fixed by the foil geometry and cannot reach an ideal motion response; however, for a well-designed foil system, the seaworthiness also can be maximized within a realistic sea state envelope using simple equipment and lower cost compared to ACS hydrofoil craft.

In order to understand the wave response of hydrofoil craft, we discuss first the wave response of a single foil in waves. We will follow that by discussing the response of a hydrofoil craft.

Wave Response of Single Foil

The wave response of a single foil is similar to that discussed above in the stability section. The difference is the perturbation for stability analysis is assumed small and quasi-static with displacement but no velocity of perturbation. However, a foil moving in waves has to consider perturbation displacement, velocity, and even acceleration. The perturbation acceleration can be considered as an impacting force acting on the foils and hull, and the perturbation itself as a limited value.

Vertical Response of Foil—"Stiffness" and Its Influence

The foil response in terms of perturbation displacement has been discussed in the above "Stability" section, as shown in Eq. (10.24).

Vertical Velocity Response of Hydrofoil—"Damping" and Its Influence

The hydrofoil velocity relative to the water surface is induced by two factors. The first is the heaving, pitching, and rolling motion of the craft in waves. The vertical velocity caused by heaving and pitching is uniform along the span. Motion caused by rolling is a variable value along the span. A second factor is the vertical velocity component of water molecules due to the circular motion of water molecules in a wave. So the vertical velocity of water acting on the foil is the superimposition of these two parts.

The influence of velocity on a foil corresponds to the change of values and direction of incoming flow. The foil is under the action of the resultant from incoming flow velocity and vertical motion velocity of the foil. Due to the high speed, the vertical motion velocity is a small value compared with craft speed, so it can be neglected in the resultant absolute velocity value. However, the change of angle of attack $\Delta\alpha$ caused by the change of flow direction cannot be neglected, since it is the same magnitude as the original angle of attack α_0, as shown in Fig. 10.23. We show the circular motion of water molecules in a wave as an example in Fig. 10.24.

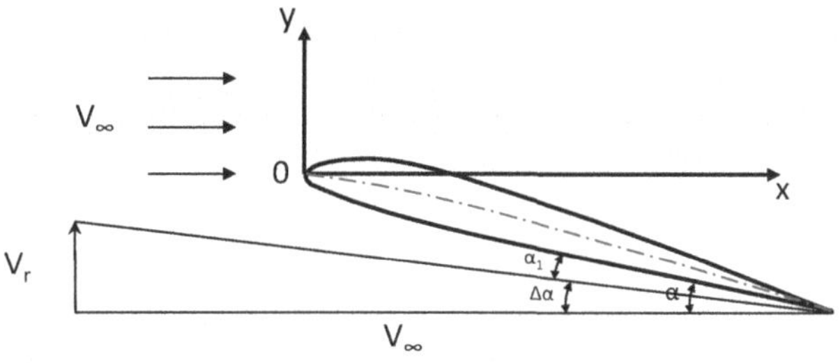

Fig. 10.23 Influence of vertical velocity

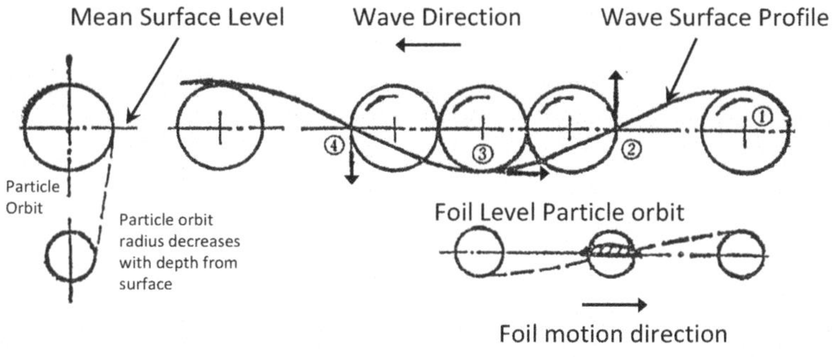

Fig. 10.24 Velocity caused by wave circular oscillatory motion

At the wave peak and trough, the direction of circular oscillatory velocity on the water surface is parallel to incoming flow, and superimposes it, with no change to the angle of attack. However, at the wave frontal and back surface, the direction of water molecule circular oscillatory velocity is vertical. Assuming craft speed is $Vs = 50$ knots, operating in lower limit of sea state 4, with wave height $H = 1.25$ m, wavelength is 20 times wave height, that is, $\lambda = 25$ m, the radius of circular oscillatory motion at the wave surface is $r_0 = H/2 = 0.635$ m, and the tangential velocity is:

$$v_o = \omega r_o = \frac{2\pi}{T} r_o$$

According to small amplitude deep water wave theory, the relation of wave period and wavelength is:

$T = \sqrt{\frac{2\pi\lambda}{g}} \approx 0.8\sqrt{\lambda}$ For sea state 4, $T = 0.8\sqrt{25} = 4$ s, and so $v_o = \frac{2\pi}{4} 0.625 \approx 1$ m/s.

The circular oscillatory velocity is therefore a small value compared with craft speed, (which is $V = 50 \times 0.5144 = 25.72$ m/s); however, the angle of attack on the wave frontal and back surface changes to:

$$\Delta\alpha = \frac{v_o}{V} \approx \frac{1}{25.72} \approx 0.039 = 2.2^\circ$$

This is the same order as the operational angle of attack of the hydrofoils. In sea state 6 the $\Delta\alpha \approx \pm 6^\circ$. The influence of vertical motion velocity is the same. So the vertical velocity will influence the lift via the change of angle of attack, as a damping force or damping coefficient as follows:

$$\frac{dL}{dv_V} = \frac{dL}{V_\infty d\alpha} = \frac{1}{V_\infty} \frac{d}{d\alpha}\left[C_L \frac{\rho}{2} V_\infty^2 S\right] = \frac{\rho}{2} V_\infty S \frac{dC_L}{d\alpha} \tag{10.33}$$

where v_v is the vertical velocity component and $\frac{dL}{dv_V} \propto \frac{dL}{d\alpha}$, so $\frac{dL}{d\alpha}$, the lift derivative with respect to the angle of attack, is the foil damping.

From (10.33) one can see that as the lift loading is reduced, that is, foil area is increased for a fixed load, damping is increased. Damping increases in proportion with $\frac{dC_L}{d\alpha}$, and for a three-dimensional thin foil considering the influence of the free water surface, we have (see also Eq. 9.40):

$$\frac{dC_L}{d\alpha} = \frac{2\pi\eta\gamma}{\frac{1}{\cos\beta} + \frac{2\eta\gamma\varsigma}{A}(1+\tau)K_e} \tag{10.34}$$

Here the influence of aspect ratio, upward dihedral angle, and submerged depth h is apparent. The first two items of (10.34) reflect that damping increases with aspect ratio A and varies inversely with dihedral angle. The influence of submerged depth is via coefficients γ and ζ. In general, damping decreases with relative submergence $\overline{h}$. Therefore, to decrease damping, foils with a smaller aspect ratio, and surface piercing foils with shallow submerged depth can be applied. Larger foil damping comes with a high aspect ratio and large submerged depth.

Motion Response of Hydrofoil Craft in Waves

We consider here how the geometric configuration of foils on a craft will affect its motion response, and considerations to minimize the motions.

(1) Requirements for Bow and Stern Foils to Reduce Craft Motion in Waves

The requirements for stiffness of bow and stern foils have been discussed in the section of longitudinal stability. Regarding seakeeping quality, the requirements for stiffness are similar to that for longitudinal stability, that is, hopefully, a certain stiffness is available for both bow and stern foils and the bow foil stiffness is higher than the stern foil, to ensure bow up trimming is maintained. For hydrofoil craft it is most important to avoid plough in of the craft bow, losing foil operation mode. So improving bow foil stiffness is an important step for avoiding the bow plough in. It may be noted that the stiffness required is in relation to the load carried by the foil.

Regarding the damping value, the bow foil should have small damping for maintaining longitudinal stability and seakeeping. The bow foil will be less sensitive to the circular oscillatory motion of water particles in waves, that is, with small damping in waves. For a craft running in head waves, as shown in Fig. 10.25, and bow foil located at wave frontal surface, the upwards circular velocity causes increasing angle of attack of the bow foil, and lift causes bow up motion, so the craft follows the wave profile.

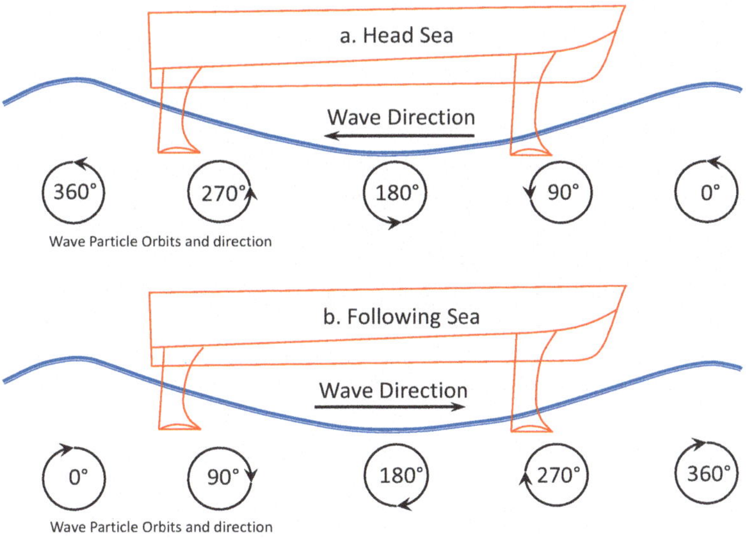

Fig. 10.25 Hydrofoil craft running in head and following seas

For a craft running in following seas, the case is contrary. When the bow is located on a wave back, the circular motion of waves decreases bow foil lift and causes bow down motion possibly with bow ploughing into the incoming wave peak, as shown in Fig. 10.25b. When the bow foil is located at a wavefront (behind the wave trough due to craft direction); a bow-up moment occurs taking the bow foil out of the water.

Therefore, in order to improve hydrofoil craft seakeeping quality running in following seas, the bow foil should not only have higher stiffness but also small damping, so as to be less sensitive to the circular oscillatory motion of wave particles.

For the stern foil, it is located in the wake of the bow foil, so the effect of wave-particle motion is not so sensitive, so larger damping can decrease the motion amplitude. In addition, high damping ($\frac{dL}{d\alpha}$ great, or $\frac{dC_L}{d\alpha}$ great) usually gives high efficiency to the hydrofoil.

To sum up, the requirements of seakeeping quality suggest the bow foil should have high stiffness and small damping, and stern foil with small stiffness and high damping.

(2) Motion Response of Hydrofoil Craft with Various Foil Arrangements in Waves

A hydrofoil with *airplane geometry fully submerged foils* can keep both high stiffness and damping of bow foil with good seakeeping quality running in head waves and following the wave profile with a damped motion. However, plough in and foil egress out of the water may still occur in following seas. Furthermore, heavy rolling may occur while a hydrofoil is out of the water, with serious consequences.

Since the bow foil of an airplane foil arrangement will be closest to the CG, and the hull forward of the foil is perhaps half craft length, large impact loads may occur when pitching down. Impact may also cause yaw and roll due to wave surface geometry resulting in further unstable motions dangerous to dynamic stability.

The higher stiffness and lower damping of hydrofoil craft with an *airplane geometry and surface piercing foil system* may give better seaworthiness than a fully submerged shallow foil system; however, the problem concerned with the length of bow hull extended forward still has not been solved.

The forward hull problem on *tandem foil craft* can be reduced a little bit with lower stiffness of bow foils so it is suggested to apply bow surface piercing together with shallow submerged flat stern foils. This will keep higher stiffness and low damping of the bow foil, and low stiffness as well as larger damping of the stern foil.

Lowest damping of bow foil with higher damping of stern foil of *canard foil system*, and no long bow ahead of CG, are the main seakeeping quality advantages of a canard foil system as implemented on the Canadian anti-submarine patrol hydrofoil craft "Bras D'Or" [10], with bow foil loading of 10% and stern foil load 90%. It had a fully submerged flat main stern foil, upward and downward dihedral side foils with sub-cavitating foil section with high damping and appropriate stiffness.

The camber and installation angle of the outer dihedral foils is greater than the flat central foil, to give rolling stiffness and lift during takeoff. The profile of the main

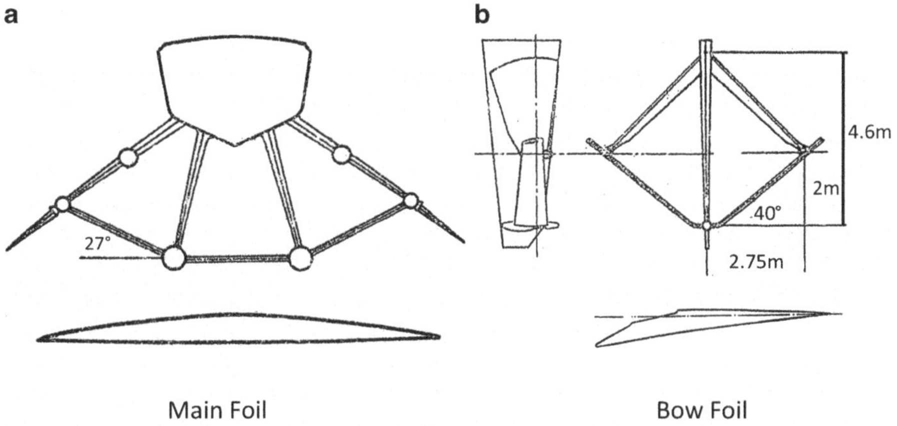

Fig. 10.26 Main and bow foils of hydrofoil craft Bras D'or

foil is as shown in Fig. 10.26a. The bow foil is composed of both upward and downward air-ventilated foils with prismatic cross section geometry and Tulin two-parameter foil section. Thus variable stiffness is available, as well as area change, and in the case of craft plough-in with deep submergence the air ventilation is cut off at the bow foil, causing rapid increased lift and stiffness. The bow foil air ventilation decreases the sensitivity of lift to the angle of attack and decreases damping in normal operation. Figure 10.26b shows the bow foil and foil section profiles.

Suitable Sea State for ACS Hydrofoil Craft

The adjustment of lift for fixed foil systems is through the change of submerged depth and vertical velocity of foils. The active adjustment of foil lift is a better control to adjust the lift, just as for an airplane.

At the initial stage of hydrofoil craft development manual adjustment of foil lift was used, to demonstrate the possibility of manual handling of craft for takeoff and touching down in practice. However, it is extremely difficult to react to the requirement of lift adjustment on time, as the density of salt water is 820 times greater than air thus the fluctuation and frequency are high, and it is difficult to be handled by manual operation. It is this that has led to automated controls on hydrofoil craft.

In general, an automated control system can be applied on hydrofoil craft with any foil arrangement successfully so long as the operating forces for the control surfaces are analyzed correctly and system power is therefore adequate. The arrangement and type of foil system should be considered together with the range of operations, that is, hull borne operation, the retractable capability of foil system, etc.

Automatic foil system controls can be successfully applied on hydrofoil craft, however, with definite limitations in terms of wave height or sea state. There are three motion modes of operating in waves, that is, platforming, following the wave profile, and intermediate mode as discussed above.

A hydrofoil craft with an automated control foil system will have maximum ability for platforming operation mode in waves and the ability to maintain intermediate mode in higher sea state, compared with other vessels with the same displacement. However, in rough seas, the motions will follow the wave profile similar to a planing craft though damped somewhat. Once the sea state is such that impacts with the hull are regular, speed has to be reduced and the vessel will have greatly increased motions and accelerations. Generally, for an ACS hydrofoil craft there are four grades of sea state to consider:

(a) *"Calm" water:* No wave or small waves, where craft inertia and high foil damping can prevent high motions. Due to the small wave amplitude and high frequency the hydrofoil is less responsive to the waves and maintains platforming operation. In such sea state, it is not necessary to use an automated control system. A fixed surface piercing foil system may cause motions due to its higher sensitivity to the waves, so a shallow submerged flat foil system is most suitable to be applied in such sea state.
(b) *Moderate sea state:* Strictly speaking, the moderate sea state implies a wave height smaller than the clearance between the bottom of hull and water surface, so the wave will not touch the hull. In such a sea state, an automatic control foil system can reach an ideal platforming operation.

    Recent experimental investigations demonstrate that even in case of wave height greater than hull bottom clearance, craft with automated control foil systems can be operated in platforming mode so long as the wave height being not greater than 1.5–2.0 times hull clearance. The design requirements for the control system will be higher, and the craft may experience "wave piercing" situation from time to time. Nevertheless, the craft motion and the force acting on it still will be smaller than that in the intermediate operation mode.

    In a moderate sea state, good seaworthiness for craft with surface piercing foils with appropriate stiffness and damping can be obtained; however, it is impossible to maintain platforming operation and vertical acceleration will be 2–4 times higher than that of ACS hydrofoil craft.
(c) *Higher sea state:* This is wave height larger than $1.5 \sim 2.0$ times hull clearance from the calm water surface. In such a sea state, it is necessary to use the intermediate operation mode for the ACS hydrofoil craft. Performance is degraded due to being unable to reach platforming operation mode. A well-designed surface piercing foil craft can achieve similar seaworthiness. For a craft operating in "motion following waves" mode, wave height is not the main factor influencing craft motion, rather it is the wave slope.
(d) *Extreme high sea state:* In this case, wave height is higher than three times hull clearance. Meanwhile, the craft is difficult to keep in foil borne mode. Surface piercing foil craft with V-type foil have the highest ability to maintain foil borne mode, due to their high lift margin, particularly V-type foils with a ladder arrangement. It is apparent ACS hydrofoil design is not necessary in such cases.

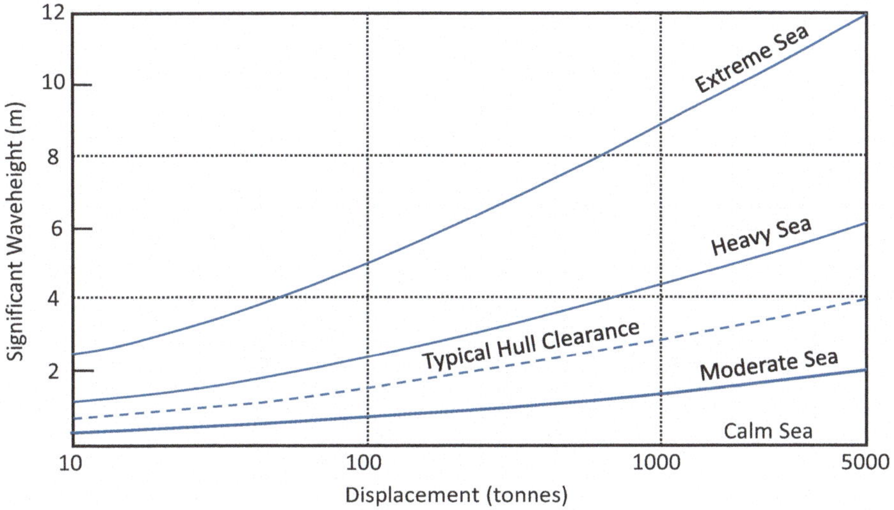

Fig. 10.27 Platforming motion range for various craft displacements

Thus use of ACS hydrofoil design is dependent on the design sea state. In a high sea state, where mainly platforming operation can be achieved, together with some intermediate mode operation, the ACS hydrofoil craft has a large advantage. However, for craft operating in "motion following with waves," there is no advantage with such foil system, particularly in extreme rough seas.

The four sea states mentioned above are relative to the craft displacement, and the absolute wave height for craft with various displacements is different. Reference [10] gives sea state classes for different craft displacement, as shown in Fig. 10.27, taken from statistical data of hull bottom clearance.

For ACS hydrofoil craft, the most important is the boundary between the "Moderate sea state" and "Higher sea state." This is because an automatic control system is effective below the higher sea state boundary to keep the craft heave and pitch very low, in platforming mode. The limits can be defined based on the vessel geometry combined with statistical data of waves in the operational sea region, so as to guide specification of the foil control surfaces and the control system powering and response.

As an example, Fig. 10.27 presents statistical data for the boundary line of moderate, heavy, and extreme sea states for different craft displacements. When defining the design project of a hydrofoil craft, designers should judge the application of automated control systems according to the environmental data for the operational sea region for design and drawing such boundary curves.

In Fig. 10.28a, boundary lines for operability (P%) are plotted according to the significant wave height (1) and maximum wave height (2) for the North Atlantic Ocean for different sizes of craft based on the performance capability in Fig. 10.27. Thus, a 100-ton craft may expect an operability of around 70% based on the

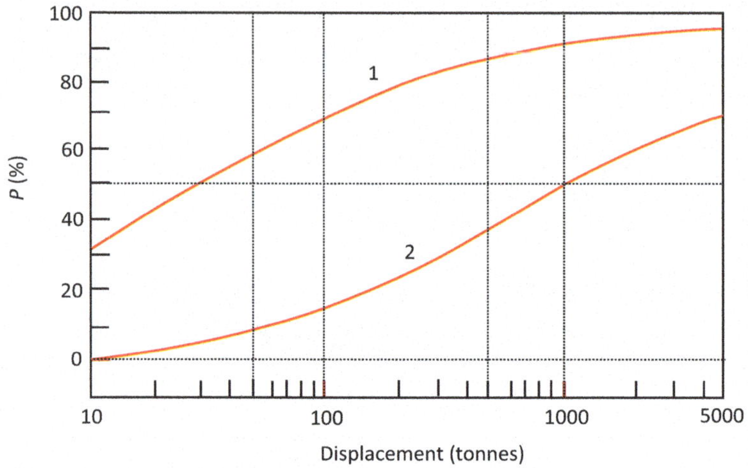

Fig. 10.28a Sea state in North Atlantic Ocean
1—based upon significant wave height;
2—upon maximum wave height

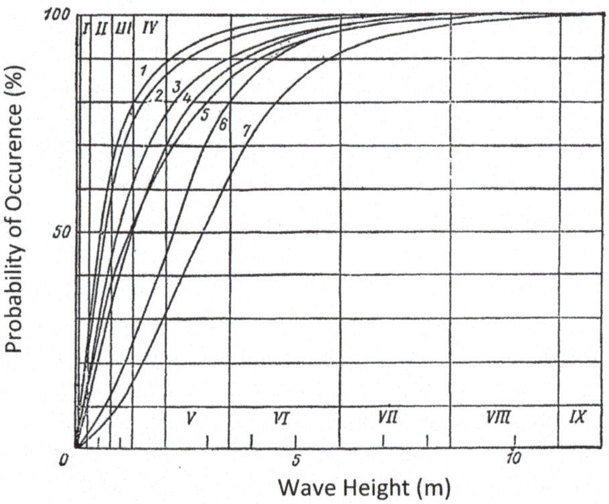

Fig. 10.28b Average annual wave height occurrence for the sea regions around Russia
Key to figure:
1—Black Sea
2—Baltic Sea
3—Sea of Japan
4—Sea of Okhotsk
5—Barents sea
6—Northwestern part of the Pacific Ocean
7—North Atlantic
The bands I to IX show the Beaufort scale bands equivalent to the wave height)

significant wave annual statistics, while if considering limiting sea states for the craft the operability may be as low as 15% to stay on foils.

Figure 10.28b presents the statistical probability of the annual average wave height of sea regions around Russia, where wave height is with 3% occurrence. These areas are much less aggressive than the North Atlantic placing lower demands on foil clearance and motion control systems.

Vertical Acceleration and Speed Loss of Hydrofoils in Waves

The dynamic response of hydrofoil craft in waves is much better than a planing hull with the same displacement, due to the low stiffness of the hydrofoils, so causing low

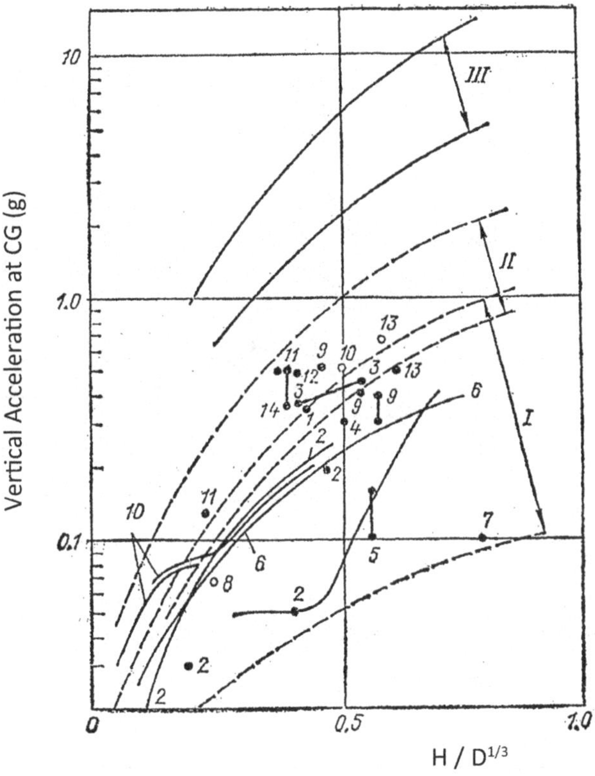

Fig. 10.29 Vertical acceleration of hydrofoil craft at CG
Key to figure
I Hydrofoils with automated controls
1—Seaflight H57
2—*Dolphin*
3—*High Point*
4—*Tucumcari*
5—*Pegasus*
6—P-420 *Swordfish*
7—Boeing Jetfoil
8—*Flipper*
9—Supramar PT-150
II Surface Piercing Hydrofoils without stabilization:
10—*Denison*
11—PT-50
12—PT-20
13—*Bras D'Or*
14—*Meteor*
III Planing vessels

vertical acceleration. Among the different hydrofoil configurations, the fully submerged foil is less stiff than a surface piercing foil system and both have significantly lower accelerations than slower displacement ships. Reference [1] presents the statistical data of vertical acceleration at CG, as shown in Fig. 10.29 showing planing hulls and hydrofoil craft with and without automatic control foil systems. The ordinate denotes the ratio of vertical acceleration at CG with acceleration due to gravity a_g/g. The abscissa denotes the ratio of wave height with 1/3 power of craft displacement, from which one can see the vertical acceleration of hydrofoil craft is smaller than a planing hull, particularly for automatic control foil system craft. Surface piercing hydrofoil craft have responses clustered around 0.5g. Craft with submerged hydrofoils such as 2, 6, and 7 have much reduced response. The response of planing vessels is much higher as indicated by III. Note that waveheight is the 3% occurrence haveheight, not significant waveheight, in this figure.

Figure 10.30 presents a comparison of vertical acceleration of a hydrofoil catamaran, the Westermarin 2900 Foilcat, which has an automated control foil system, with a conventional catamaran with the same length, and comparison of seaworthiness based on the ISO standard for the judgment of work degradation and fatigue of personnel [11].

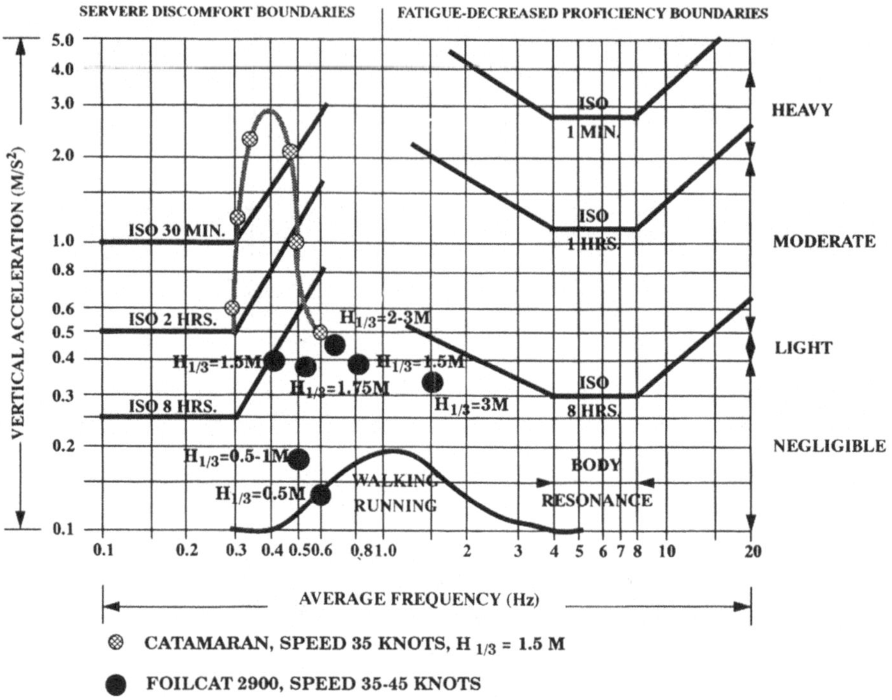

Fig. 10.30 Comparison of comfort between hydrofoil catamaran with automated ride control system with conventional catamarans in various sea states

Fig. 10.31 Vertical acceleration a_g/g versus wave height for the hydrofoil *Dolphin*
1—Projected from wave piercing performance
2—Projected from wave contouring performance
3—Full scale data at 45–50 knots
A: Zone where wave height exceeds foil clearance

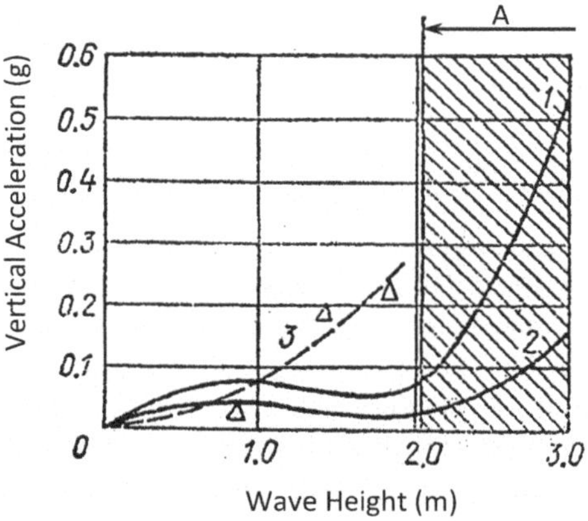

The vertical acceleration of hydrofoil craft is relative to the wave direction. It is maximum in head seas, minimum in following seas, and intermediate in beam seas. Figure 10.31 shows vertical acceleration of a 60 t prototype hydrofoil craft, the Grumman "Dolphin" in foil borne operation mode at different wave directions where

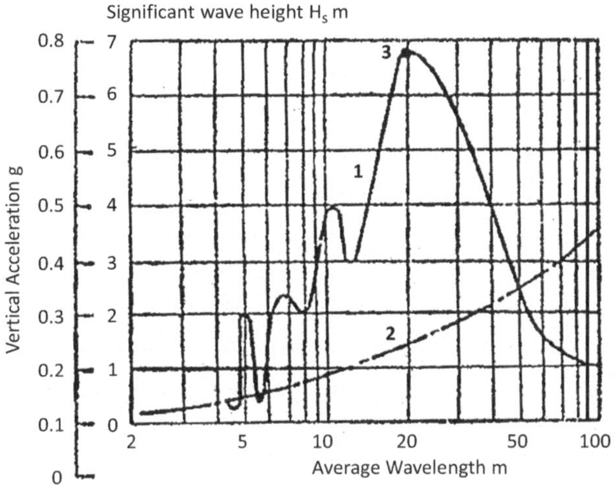

Fig. 10.32 Influence of wavelength on vertical acceleration according to research by Supramar for the PT-50 vessel (LBP = 20 m, *V* 46 knots, in regular waves)
Key to diagram
1—Vertical acceleration
2—Wave height
3—Greatest acceleration occurs at wavelength 20 m, i.e., similar to LBP

1 is head seas, 2 is beam seas, and 3 is following seas. The hatched area shows where the waves would impact the hull.

Similar to other ships, vertical acceleration of all ships is relative to the ratio of wavelength and craft length, and for hydrofoil craft the ratio of wavelength with spacing of foils. If the foil spacing is a round number of wavelengths, the vertical acceleration is at peak value. Figure 10.32 shows test data of vertical acceleration at the CG of hydrofoil craft PT-50 at different wavelengths and heights [1]. The waveheight variation with length is shown as dotted curve 2, while the acceleration measured is the solid line 1. From the figure, one can see vertical acceleration will be maximum when wavelength is equal to foil spacing labelled as 3 and it will decrease quickly after wavelength greater than foil spacing.

Reference [12] presents a comparison of vertical acceleration for various marine craft, at a similar speed in increasing waveheight, and one can see the vertical acceleration of hydrofoil craft is lowest, as shown in Fig. 10.33.

The vertical acceleration of hydrofoils is far smaller than a planing hull, so the impact load in waves is also far smaller. As the response of hydrofoil craft is smaller than other fast craft, the corresponding drag increment and speed loss in waves are also smaller, especially for automated control hydrofoil craft. Reference [1] presents an empirical equation for drag increment of hydrofoil craft with a nonautomatic control foil system in waves:

In transit operation mode:

$$\Delta R/_R = 1.2\left(V/_{V_0}\right)^3 \frac{h_w^2}{h_{st}\Delta^{1/3}} \quad (10.35)$$

In foil borne mode:

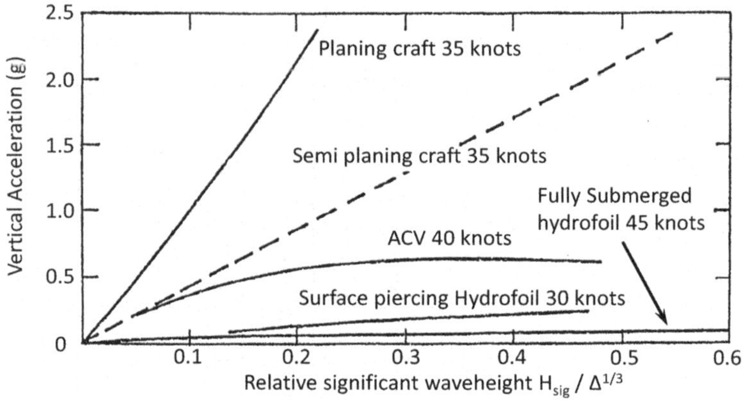

Fig. 10.33 Comparison of vertical acceleration for various craft

$$\Delta R/R = \left[0.35 + \frac{0.85}{\left(V/V_0\right)^2}\right] \frac{h_w^2}{h_{st}\Delta^{1/3}} \tag{10.36}$$

where:

ΔR drag increment in waves.
R drag on calm water.
V speed.
V_O take-off speed of hydrofoil craft.
h_w wave height in 3% occurrence
h_{st} strut height
Δ craft displacement.

Figure 10.34 shows the relation of drag increment with speed in terms of Eqs. (10.35) and (10.36), and comparison with the data from some existing craft. One can see that the empirical equations seem reliable.

The speed loss of hydrofoil craft in waves is also lower than other craft, as shown in Fig. 10.35 [12], where the ordinate is the speed ratio of speed in waves with speed on calm water V/V_O. The abscissa is the ratio of wave-significant height with 1/3 power of displacement $H_{1/3}/\Delta^{1/3}$. The fully submerged hydrofoil craft are ACS hydrofoils, and one can until $H_{1/3}/\Delta^{1/3} \leq 0.55$ these craft are best. However, for rougher waves, the speed loss may be worse than surface piercing foil hydrofoil craft.

Reference [1] presents statistical data of speed loss in waves of hydrofoil craft and some other high-speed craft, as shown in Fig. 10.36. From the figure, one can see the speed loss of hydrofoil craft in waves (areas I to IV in the diagram) is far smaller than other high-speed craft, especially for the ACS hydrofoil craft.

Reference [12] presents the statistics of power for various craft in waves, as shown in Fig. 10.37, where the hydrofoil craft are ACS hydrofoil craft. The ordinate is the ratio of required power times displacement and speed, $P/\Delta V$. The abscissa is

Fig. 10.34 Drag increment of hydrofoil craft in waves versus craft speed
Key to diagram
1—Model test data (triangles)
2—Transient mode operation (solid line)
3—*Bras D'Or*
4—*Raketa*
5—PT-20
6—VS-8
7—*Arrow*
8—*Meteor*
9—Flying mode (solid line)
Full-scale data points

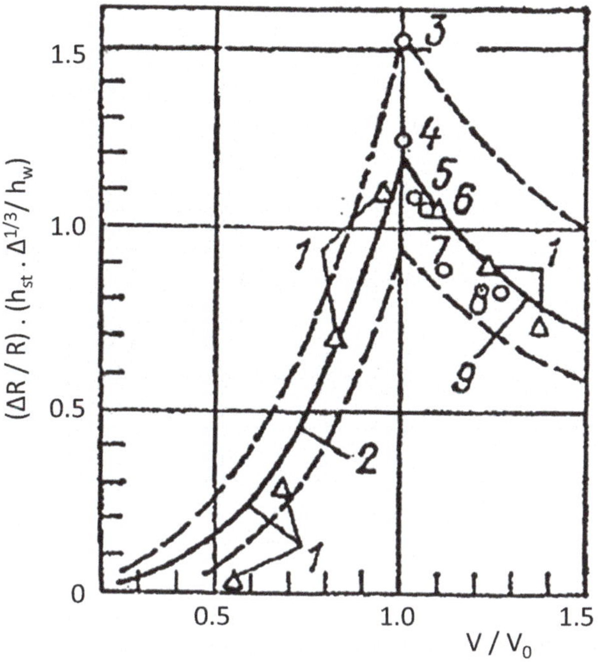

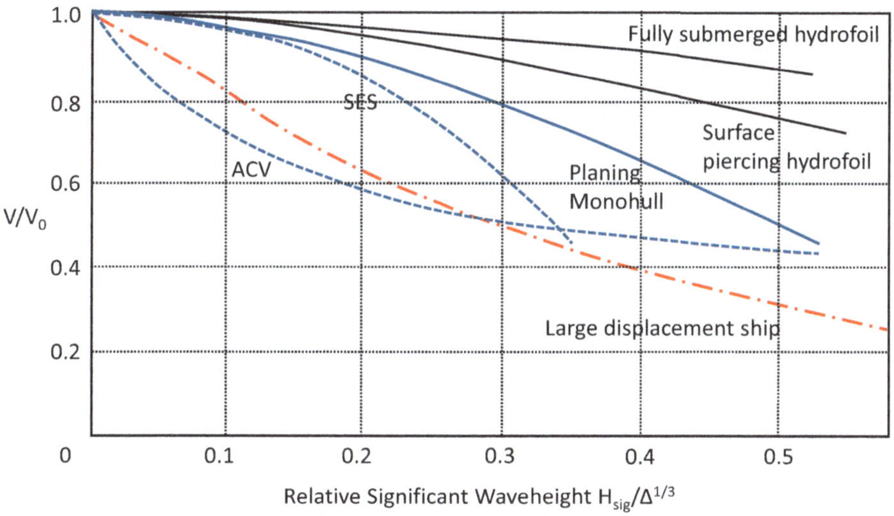

Fig. 10.35 Comparison of speed loss in waves for various craft

the Froude number $F_V = V/\Delta^{1/3}$. Data is shown for calm water and at moderate sea state represented by $H/\Delta^{1/3}$. No matter if in calm water or in waves, the required power for hydrofoil craft is far smaller than planing craft, particularly in waves. When compared with an ACV the hydrofoil is a little bit worse on calm water, but better in waves, especially in moderate sea states.

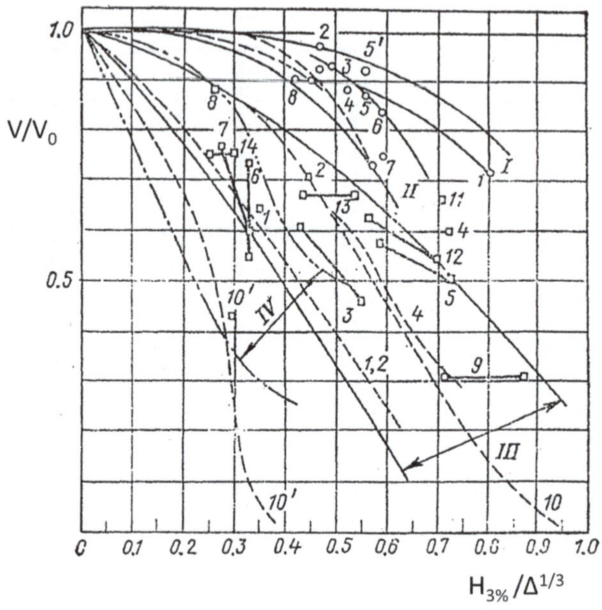

Fig. 10.36 Speed loss in waves of various high-speed craft

Key to figure:

I (circles) Hydrofoils with automated ride control 1—*Dolphin*, 2—*Typhoon*, 5' PT-50 with air stabilization system

II (circles) Hydrofoils with no automated ride control 3—PT-20, 4—*Arrow*, 5—PT-50, 6—*Bras D'Or*, 7—*Denison*, 8VS-8

III (squares) Hovercraft with flexible skirts: 1—SR N2, 2—SR N3, 3—SR.N4, 4—SR N5, 5—SR N6, 6—BH7, 7—BH8, 8—HM-2 (SES), 9—VT-1, 10—SKMR-1, 11—SK-5, 12—SK-6, 13—SK-9, 14—SK-10

IV Commercial vessels with a displacement of 10,000–20,000 tons

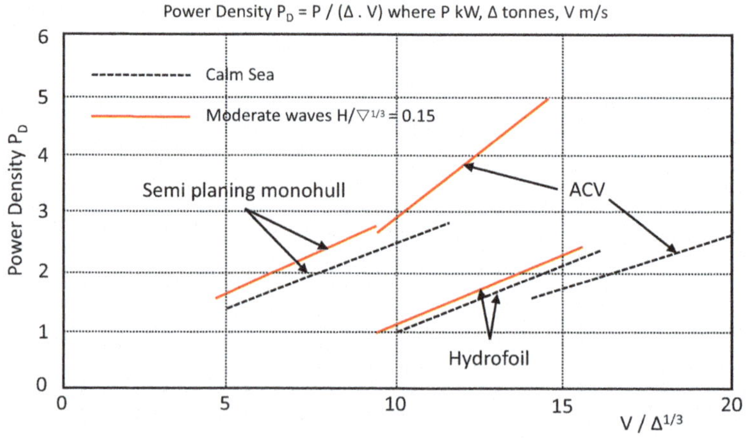

Fig. 10.37 Power requirements for various craft in waves

10.9 Automated Control of Hydrofoil Motions

The ACS hydrofoil craft is the best foil system for seakeeping. It can be operated in platforming mode in a "Moderate sea state," with "transitional motion response" in high seas, and wave following motion in "Extreme rough seas," using different control modes of the automated control system. The classification of sea state in terms of craft displacement can be found in Fig. 10.27. The seaworthiness of hydrofoil craft is related to the capability of the automatic foil control system, which also influences the other performance parameters, such as speed performance, stability, and maneuverability.

There are several different automatic foil control systems for hydrofoil craft, including mechanical, air ventilation, and electronic control systems.

The *mechanical control* system is an early devised equipment that works best for small craft and small prototypes. It is a simple low-cost approach whereby a small buoyant surface tracker is mounted on a long feeler arm at the bow. The arm in turn controls the bow foil flap angle or foil angle itself through a linkage that is set up to maintain the desired running depth of the foil. The system has worked best where there are two separate bow foils as that can also help with coordinating turns.

The simple equipment and high reliability of *air ventilation control* systems is an advantage, however with a penalty of reducing hydrofoil hydrodynamic efficiency. It has been applied on the bow foil of Canadian hydrofoil craft "Bras D'Or." On other surface piercing hydrofoil craft, such as Rodriquez, it is used as a passive stabilizing system. It is one means also to stabilize foil performance as a craft accelerates through the semi-cavitating regime to service speed for high-speed hydrofoils. This same steadying of performance is helpful in higher sea states.

Electronic automatic foil control requires an advanced automated control system with high reliability and fast response. It has the disadvantages of complex components and high cost. Nevertheless, such systems have been very successful and reliable on craft such as the Boeing and Kawasaki Jetfoils. The basics for an automated control system for deep-submerged craft are in the block diagram below.

Simplified block chart for electronic automatic control installation.

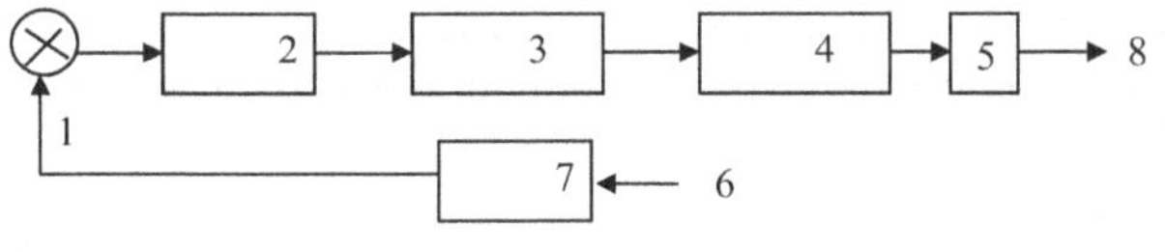

where 1—instruction to program; 2—computer; 3—actuator; 4—force generator; 5—craft foil system and hydrodynamic control surfaces; 6—wave forcing; 7—sensors; 8—motion response

In the case of waves acting on the craft, sensors generate signals that are input to a control computer as parameters. The computer interprets the input via a program that outputs signals to a powered actuator system to move the actuators and so the control flaps and change the hydrofoil lift. Depending on the parameters, the control actuation may be to adjust quasi-static trim or to continuously adjust foil lift to respond to wave forces that heave, pitch, or roll the craft.

In order to optimize running attitude it is necessary to respond not only to wave amplitude, but also to craft motions, including the amplitude, velocity, and acceleration of heaving, pitching, rolling motions, as well as course control. Thus instrumentation has to be installed to monitor all these parameters in parallel.

The second requirement is sufficient reliability due to the complexity, so a reserve system (parallel backup) is necessary. In modern ACS hydrofoil craft, one or even two independent reserve systems are available. A monitoring system is also necessary, and resetting if a malfunction occurs, whether to the control system or to the craft. If the craft has a malfunction or perhaps an event such as a plough in the controls need to be able to respond and maintain safe operation.

Component Overview

(a) Sensors

The functions of sensors are transmitting the motion and relative position signals of the craft to the computer. Different sensors (transducers) are available for the different position and motion parameters.

Height sensor	The sensor measuring the wave surface relative to the craft, some different types are available as follows:
Electric height sensor	The height can be measured by the change of electric resistance of seawater wetted resistor, which may be not precise due to the influence of spray in waves.
Ultrasound height sensor	The function is as same as sonar, using the ultrasound wave to measure the height of water surface. This is applied often at present but has the disadvantage of the effect of sound interference.
Radar height measurer	Using electromagnetic waves to measure the height. This is the preferred instrument currently due to the high transmission speed of electromagnetic wave and high precision without interference.

In addition, mechanical and laser height instruments are available, however these have not been frequently applied to date.

In general, the height transducer is installed on the hull bottom for measuring the distance between the hull bottom and wave surface, and sending the height signal to the computer for process. In order to maintain the craft platforming an averaged value of height over a certain time period should be used to avoid wave following operation of craft.

This will also maintain vertical stability. In rough seas not suitable for platforming, the time interval can be shortened and craft motions adjusted to intermediate response operation. When the wave is exceeding the platforming limitation, the craft can operate in wave following motion, according to the height output.

(b) Computer

Angle transducer	Measuring the pitching and rolling angle. Normally, an aviation vertical gyro is applied for measuring the pitching and rolling angle, and horizontal gyro for course deviation angle
Angular velocity transducer	An angular velocity gyro is used to measure the angular velocity of roll, pitch, and yawing, and also vertical or horizontal gyros can be used for measuring the angular displacement, and the angular velocity can be obtained by derivation through integration of the sampled signals using fast fourier transformation
Vertical acceleration transducer	Measures heaving acceleration, by means of servo-accelerator

The function of the computer is to process the signals from the transducers, transforming time step data and through the use of preprogrammed response operators to output the required control positions to the actuator system to respond to the input forcing.

Through the development period for hydrofoils, computers, microelectronics and sensors have advanced rapidly, and so at the present time, much more precise control is possible.

(c) Actuator system

Generally, hydraulic systems are applied right now. These can transfer the electric signal from the computer to mechanical motion via the actuator, which comprises two parts normally as the actuating force applied to the control surface is normally via a hydraulic actuator:

Electric-hydraulic servomechanism	Transferring the electric signal from computer to a hydraulic signal and using an electric liquid servo-valve to transfer the electric signal to hydraulic oil flow
Hydraulic actuator	Hydraulic actuator and connecting rod as well as the control surface (force generator) rod

It is possible to use electric-actuators, that is, the servomotor can drive the control surface rod directly via electric cylinder, according to the signal from the computer. The precision and reliability of electric actuators are high and avoid a hydraulic oil circuit, so lowering maintenance. On the other hand, higher force is available from hydraulic systems, so there is a tradeoff for the designer to make.

(d) Force generator

The force generator is the hydrofoil lift control system. In aviation there are many types for controlling the lift; however, on hydrofoil craft, only two types are generally used, that is, rotating a whole hydrofoil directly to change the angle of attack and flaps that change camber and angle of attack. Application of rotating flap can save power consumption in the control system up to 75% compared with rotating the whole foil, maximizing control efficiency. It may be noted that flaps on a

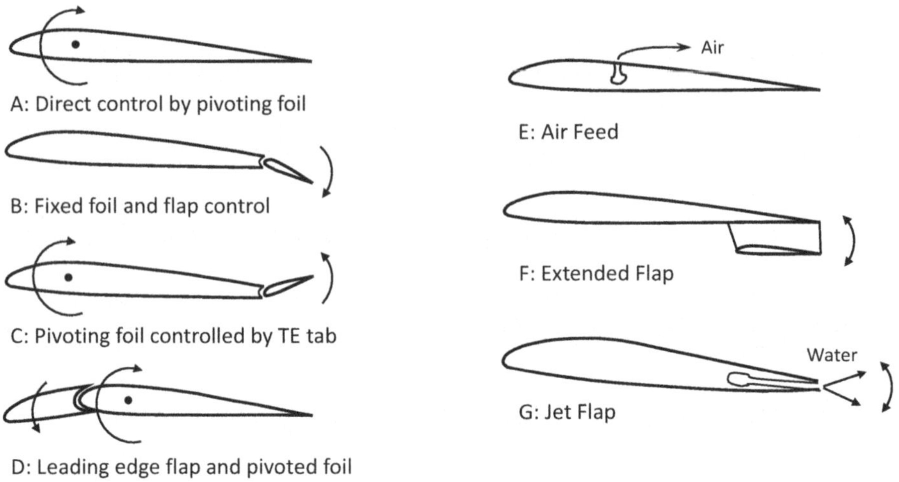

Fig. 10.38 Various lift control types (and required powers)

hydrofoil combine the function carried out by both flaps and elevators on an airplane. They are used for trim control and then varying positions for motion control.

Figure 10.38 shows the range of different potential controls. The relative power requirements compared to a rotating foil (A:1.0) to achieve similar turning moment on equivalent foils are as follows B:0.25, C:0.02, D:0.75, E:2.0, F:0.5 to 1.0, G:2.0.

Even though the power consumption for rotating a whole foil is great, the precision requirement is lower than a rotating flap. So it has also been a choice used on hydrofoil craft, such as "Flagstaff" (PGH-1), manufactured by Grumman Aircraft Co, USA, with displacement of $\Delta = 68$ tons, and power consumption of 100 shp for the foil control in design sea state.

In general, control mechanism power is in proportion to the 2/3 power of vessel displacement, so in case of displacement being 1000 tons the power for controlling the hydrofoils is up to 6000 hp. This does need to be taken in context since the main power for such a very large hydrofoil craft would be in the region of 120,000 shp.

Another hydrofoil craft "Plainview" (AGEH-1) also manufactured by Grumman, with displacement of $\Delta = 320$ tons used a whole foil control system for design sea state 6 and a speed of 50 knots. The foil pivot moment is about 138 ton meters, the total flow of hydraulic pump 5 $\mathrm{m^3/min}$, pressure 200 $\mathrm{kg/cm^2}$, so it must be noted that the power consumption of such a whole foil control system is significant.

Even if using a flap control system on a large hydrofoil craft the power consumption is significant, so an air ventilation control system may be able to minimize the problem; however, due to the lower efficiency the power consumption is still great. The advantage of the air ventilation is the simple and reliable mechanism.

In addition, the regulating tab and extending flap control, as applied on aircraft may be an opportunity for optimization for hydrofoil craft. In the first case, the moment for rotating a whole foil can be provided by the hydrodynamic force from extending a tab. In the latter case, the rotating moment is reduced due to the flap being located in a high-pressure zone, which is useful to avoid cavitation. Table 10.1 presents several hydrofoil automated control systems deployed on US hydrofoils.

Feedback Circuits of an Automatic Control System
A craft control system can be divided into two circuits, as follows:

Longitudinal motion circuit: including pitching and heaving.
Transverse motion circuit: including rolling and yawing; the latter is rudder control.

Strictly speaking, longitudinal and transverse motion interact with each other, however, such interaction is small so these two circuits can be considered separate from each other. The longitudinal motion circuit can include forward and aft foils circuits for controlling the pitching and heaving. On most recent hydrofoil craft these two circuits can act independently, that is, the forward foil controls motion attitude, and aft foil controls trimming. In this case both right and left systems control synchronously. In transverse control the circuits perform left/right differential action of main foil controls to operate rolling and yawing (or turning), and the latter also can be realized by rotating strut flaps or rudders. One can therefore see that while the control logic may be considered separately, they operate on the same control surfaces and use different combinations of input from the same sensors and signal conditioning in Fig. 10.39. It is primarily a matter of parallel computation and also generally provision of duplicated circuitry and backup processing to provide failure safety.

The key response coefficients that have to be incorporated into the ACS electronic computer algorithms for longitudinal motion are as follows:

K_z operator coefficient for heaving signal
K_ψ operator coefficient for pitching signal
$K_{\dot{z}}$ operator coefficient for heaving signal
$K_{\dot{\psi}}$ operator coefficient for pitching angular velocity signal
$K_{\ddot{z}_f}$ operator coefficient of heaving acceleration for forward foil
K_{ZT} operator coefficient for attitude balance
$K_{\ddot{z}_a}$ operator coefficient of heaving acceleration for aft foil
$K_{\psi T}$ operator coefficient for pitching balance.

The key response coefficients that have to be incorporated into the ACS computer algorithms for transverse motion are as follows:

Where:

K_{θ_A} operator coefficient for rolling signal due to flap
$K_{\ddot{y}_R}$ operator coefficient for rolling acceleration due to rudder system
$K_{\dot{\theta}_A}$ operator coefficient for rolling angular signal due to flap
$K_{\ddot{y}_A}$ operator coefficient for swaying acceleration due to flap
K_{θ_R} operator coefficient for rolling signal due to rudder system
$K_{\dot{\theta}_R}$ operator coefficient for swaying acceleration signal due to rudder system
K_{C_A} operator coefficient for course instruction due to flap
K_{C_R} operator coefficient for course instruction due to rudder system.

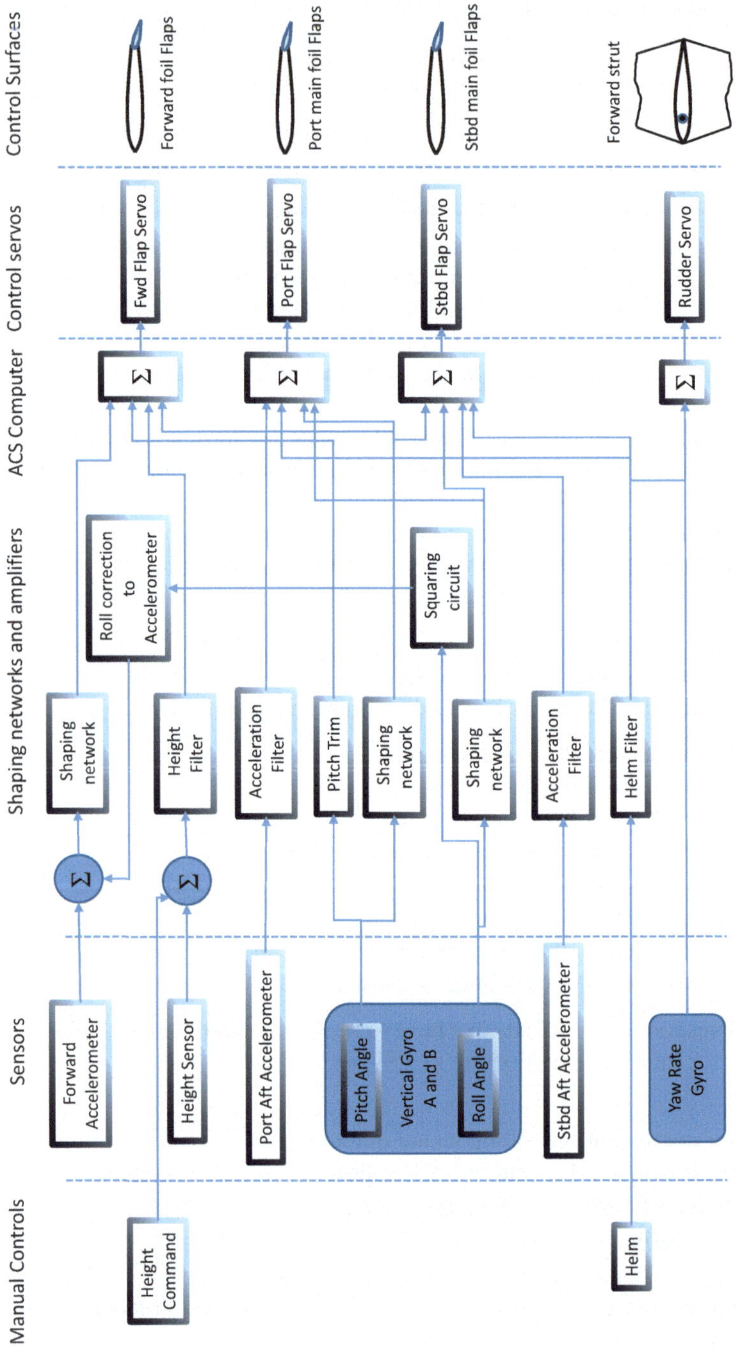

Fig. 10.39 Logic block chart for automated controls system

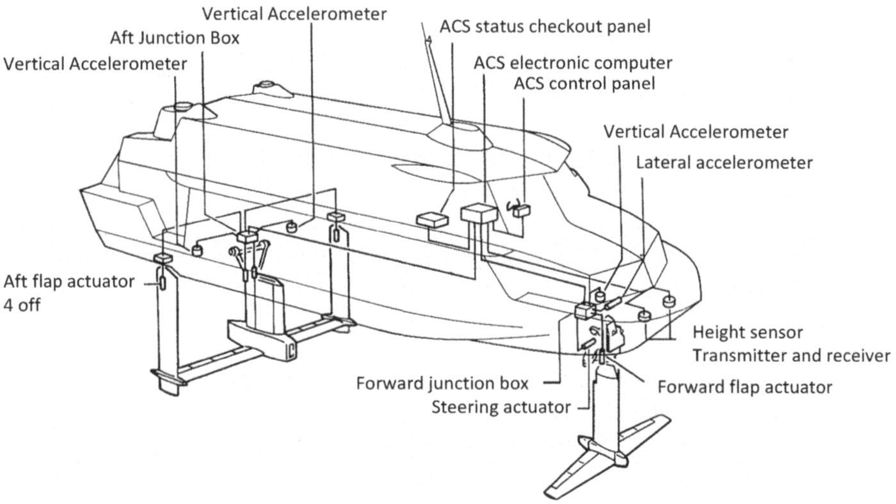

Fig. 10.40 Control system components and arrangement of *Jetfoil*

The computer then operates on inputs from the manual controls and sensors, after filtration (smoothing and feedback compensation) and application of the above coefficients to determine the correct movement of the specific control surface, as shown in logic diagram Fig. 10.39 [13]. This illustrates the system developed for the Jetfoil. A diagrammatic of the control system in the vessel is shown in Fig. 10.40.

Determination of the coefficients requires steady flow computer hydrodynamic analysis, model testing, and eventually simulation or/and prototype testing to refine the various parameters and coefficients employed in the control system.

The operation of controls for motion attenuation, flight height, and maneuvering may then be summarized as follows:

(1) Attitude control

The attitude transducer signal combined with pilot's instruction is used to obtain the foil submerged depth difference signal, which drives the flap angle to control the submerged depth of forward hydrofoil via a compensated feedback circuit. Meanwhile, the trimming change signal, from trimming gyro, drives the flap angle of the aft hydrofoil, to adjust the trimming angle and the flying height of aft foil. Continuous monitoring of signals of the flying height of both forward and aft foils can be obtained. Simultaneously, the vertical acceleration sensors, mounted on the forward strut (one piece), and aft strut (two pieces), output the signals, via the feedback compensated circuit of each, to the servos of left/right/inner/outer flaps, so as to improve the altitude adjustment and avoid sudden change of the ride height altitude.

(2) Pitch control

From the average pitch signal from duplicated vertical gyros on the craft, and via the compensated feedback circuit, the flaps of both forward and aft foils are used to provide pitch restoring moment. In addition, the pitch angular signal produced from the compensated feedback circuit (differential of pitch angle) increases the pitch damping. From the vertical acceleration signals from one sensor on the forward strut and two on aft struts via each compensated feedback circuit, the signals are sent to servo-valves of forward foil and both inner/outer flaps of aft foil for improving the pitch performance.

(3) Roll control

Using two gyros to produce the rolling signal, and via three electronic channels to control four flaps on the aft foil, differential action of the inner and outer left and right flaps is used to provide rolling moment and improving rolling performance. Meanwhile one vertical gyro, via a compensated feedback circuit, is used to control left outer flap, and another gyro signal controls right outer flap, and the average signal of which controls the inner flaps of right/left flaps, and each signal of the compensated circuit is independent. The rolling angular velocity, in a compensated circuit can also be determined via the differential of rolling angle signal, to increase the roll damping. The signals from two vertical accelerometers on left/right struts, via each compensated circuit, can be output to the servo-valves of inner/outer left and right flaps, for improving the roll performance. The "inspection and signal interruption" feedback circuit is also added in the output channel of the roll signal, which can automatically interrupt the signal in case of rolling signal failure from gyros to protect the craft safety in operation.

(4) Course and yaw control

From the deviation signal of course angle and course instruction, together with the yawing angular velocity signal from the yaw angular velocity gyro and lateral acceleration signal from the lateral accelerometer mounted on the forward strut, each compensated circuit transmits these signals to the servo-valves on the forward strut, to drive the rotation of forward strut and control the course and yawing of the craft. The course can be changed but only in the range of 5°, greater change is completed by turning the control from the rudder or water jet nozzle rotation.

(5) Turning control

From the rudder signal, and via a compensated feedback circuit, differential action of inner/outer of right/left flaps and heeling of the craft are used for coordinated turns. Turning is also accomplished by the horizontal component of foil lift. Meanwhile, after integration of the rolling signal, yawing angular velocity, lateral acceleration, the system transmits these signals via a servo-valve on the forward strut. Rotation at a small angle of the forward strut enables coordinated turning with smooth and inward banked turning, including by the horizontal component of foil lift and transverse lateral force (rudder force).

(6) Takeoff control

Using takeoff control mode, the trimming angle can be decreased via flap positioning on the forward foil, to takeoff smoothly with speed self-adapting control to avoid too large a trimming angle.

(7) Speed self-adapting control

This system functions during speed change, via the adjustment of trimming angle and angle of attack to adjust the foil lift. The system also responds to change to the drag from change of strut draft, to maintain or adjust the speed. In case of changed CG and craft weight, the lift of forward and aft foils can be adjusted by the same system.

(8) Running attitude control

Two operation modes, "Platforming" and "Wave following" operations are set up on the control panel. When the wave height is close to strut height while using "platforming" operation mode in rough seas the signals of vertical acceleration gain are reduced, and the system moves into the "wave following" operation mode to avoid the hydrofoil takeoff from the water surface and wave impacting on the hull bottom.

In order to maximize the reliability of hydraulic systems two independent hydraulic actuator systems are installed. In normal condition, system A drives the two aft outer flaps and forward flap while system B drives two aft inner flaps and forward strut. If one system fails, the systems can be switched between each other, to maintain control of forward flaps, forward strut and a pair of aft outer flaps or control of a pair of aft inner flaps, and safe operation of craft.

Although ACS hydrofoil craft such as Jetfoil have fine operational performance, no passive stability is available so the requirement for the reliability of the flight control systems is extremely high, rather like an aircraft. The speed in operation is usually limited to below 60 knots due to cavitation.

Scaling of this type of hydrofoil is limited, because the speed cannot be enhanced. The increase of foil size can't compensate the increase of displacement, as foil size is in proportion of square power of linear size, and displacement is increased with cubic power of linear size, so in case of a given speed, a larger foil area is necessary than a linear scale, so reducing the hydrodynamic efficiency. Thus, a craft of size close to that of Jetfoil is optimal for the concept.

10.10 Review of Foil Arrangements on Hydrofoil Craft

There are three configurations commonly applied to the hydrofoil craft, the shallow submerged flat foil with auxiliary foils, the surface piercing type and the deep submerged foil with automatic controls. The three configurations have been deployed in environments of increasing severity—rivers and lakes, coasts with mild sea states, and more exposed environments.

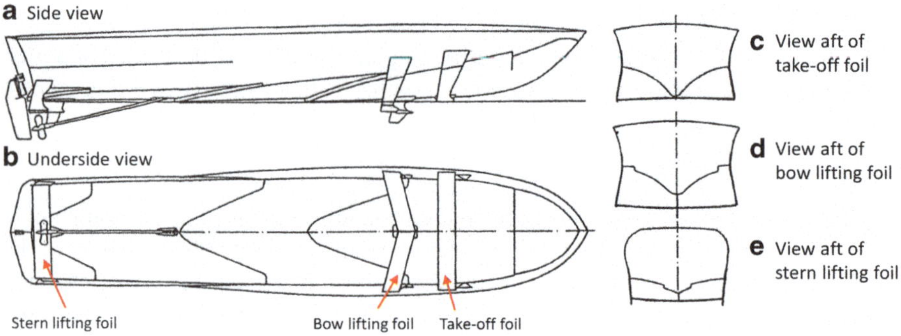

Fig. 10.41 Foil arrangement of hydrofoil craft type *Volga*

Shallow Submerged Hydrofoil Craft with Passive Stability
For the hydrofoil craft operating on sheltered water, the fixed shallow submerged foil system is the best choice. Alekseyev's foil system has been widely applied in inland rivers and lakes in Russia. The "Volga" shown Fig. 10.41 is a prototype hydrofoil craft with displacement of 1.9 t, and tandem foil system, with forward and aft shallow submerged flat foils. A takeoff assisting foil is mounted before the forward foil providing additional lift before takeoff. Such hydrofoil craft have limited seakeeping quality, and so are not suitable for operation in an exposed seaway.

For the hydrofoil craft operating in large lakes, this hydrofoil system can be modified using side piercing foils to improve seakeeping. The side foil dihedral angle should be sufficient to provide roll stiffness, as shown in Fig. 7.2c. The above waterline area should be sufficient to prevent bow plough in and improve take-off. These craft tend to have high L/B and remain rather sensitive when in following seas.

In Russia and China, craft have been built with just a bow foil, as shown in Fig. 7.24. The aft hull bottom operates as a planing hull. These so-called foil-planing craft run in following waves more like a planing craft. When the forward foil is located at a wave back and forward foil lift is dropping down, the aft planing surface located at a wave peak is not pushed higher as a rear submerged foil would be, and so there is less plough-in tendency.

On vessels with shallow foils it is also possible to install a middle foil, arranged over the main foil, for increased vertical, transverse, and longitudinal stability. Figure 10.42 shows the Russian hydrofoil craft type Meteor, with a displacement of 55 ton. It has a shallow fully submerged forward foil with slight V.

The takeoff aid foil is installed above the main foil, also in V form, and another take-off aid foil is mounted at the craft amidships close to the keel, with shallow V form. Both bow and mid-takeoff-aid foils also provide additional restoring force and moment in waves. The seaworthiness is limited by the hull clearance. This is set by the design environment of river or lake service.

The advantages of these craft are simple construction, efficient hydrodynamic properties operating in sub-cavitation regime, with speed between 40 and 50 knots.

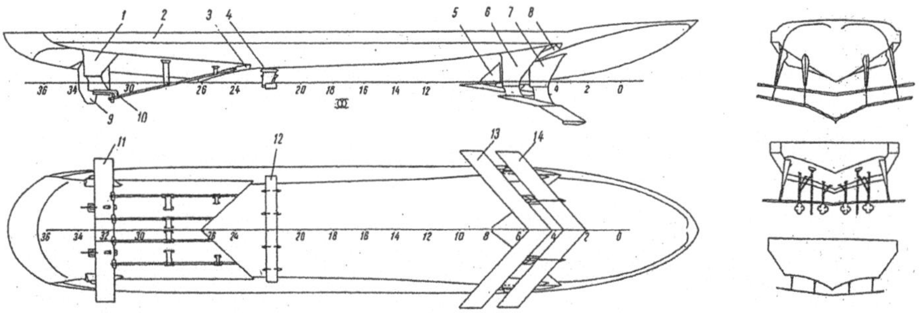

Fig. 10.42 Hydrofoil arrangement of Russian hydrofoil *Meteor*

The main application constraints are the limited sea-keeping quality, particularly in following waves with plough-in sensitivity. They have nevertheless had wide application on inland waterways in Europe and Russia.

Hydrofoil Craft with Surface Piercing Foils

This is the second hydrofoil system widely used in the world so far, based around the geometry developed by Supramar. It is a surface piercing foil system with tandem or airplane foil arrangement. Its seakeeping quality is higher than a shallow submerged flat foil system suitable to operate along coastal areas. The PT series is a typical hydrofoil craft of this configuration manufactured by Supramar Co and later by Rodriquez as the RHS series hydrofoils. The RHS series modified the configuration of the stern foil to a deep submerged flat foil, with simple automatic control system. Figure 10.43 shows the general arrangement of passenger surface piercing automatic control hydrofoil craft type PT-50 MK-II. This has a load distribution for forward and aft foils at 65% and 35%, respectively.

Figure 10.44 shows the general arrangement of Italian passenger hydrofoil craft type RHS-150, where the load for forward and aft foils is 66% and 34%, respectively. This has the simple automatic control foil system at the stern foil and is referred to as "automatic self-stability" hydrofoil craft. There are many hydrofoil craft with surface piercing bow foil systems using such simple automatic control system at the stern foils for improving seakeeping quality.

Fully Automated Control Deep Submerged Foil System (ACS)

The hydrofoil craft with ACS systems manufactured by Grumman and Boeing Co, USA, are the typical of these kinds around the world. The Grumman hydrofoil has airplane foil arrangement with full foil rotation for controlling the lift, and water

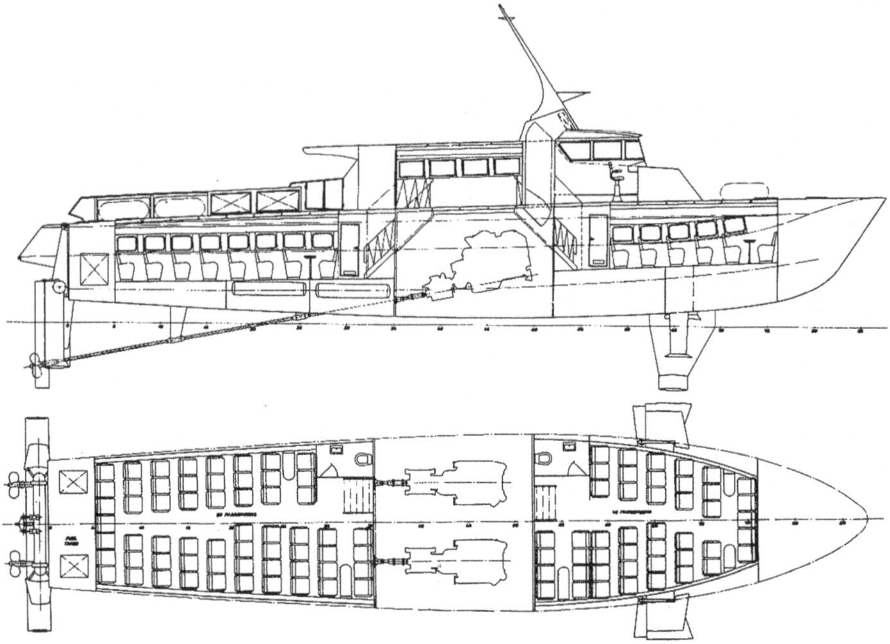

Fig. 10.43 General arrangement of PT-50 MK-II

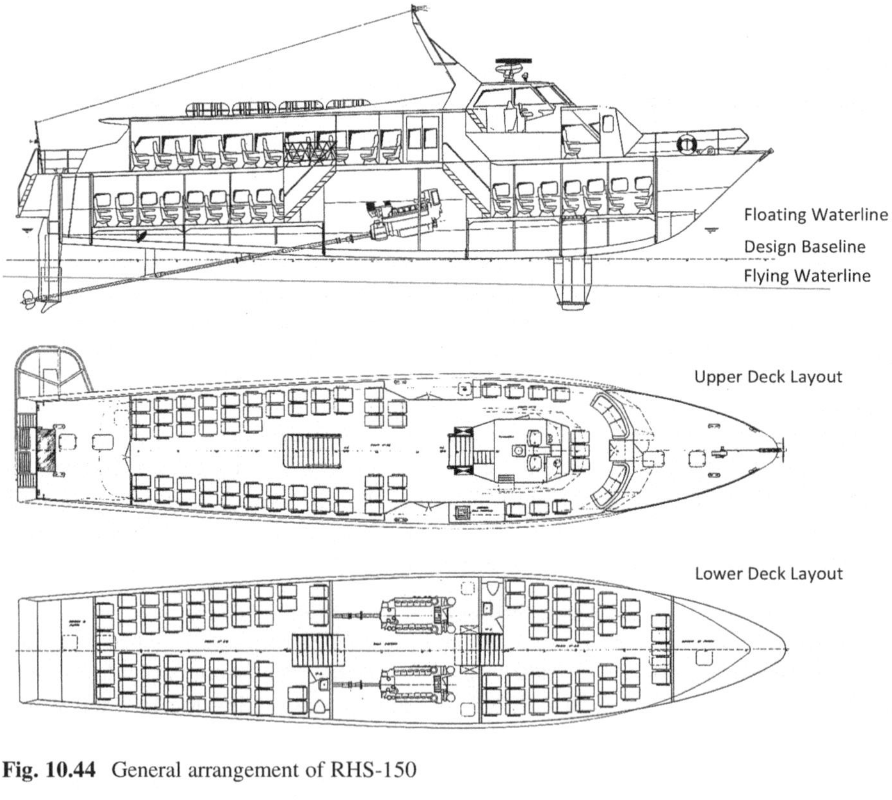

Fig. 10.44 General arrangement of RHS-150

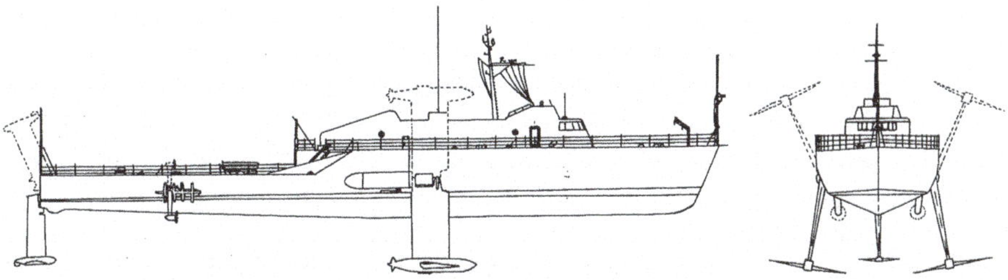

Fig. 10.45 Foil arrangement of hydrofoil craft AGEH-1 "Plainview"

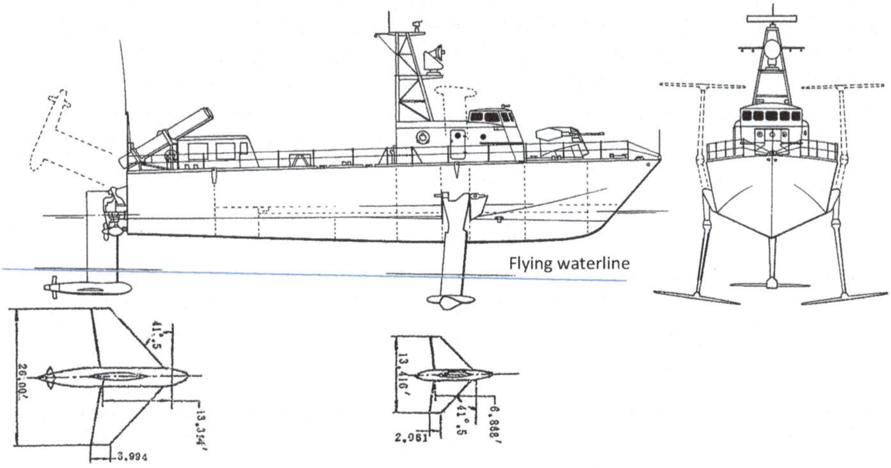

Fig. 10.46 Foil arrangement of hydrofoil craft PGH-1 *Flagstaff*

propeller with Z drive power transmission. The Boeing craft have a canard foil arrangement with flaps to control foil lift and water jet propulsion.

The load share between forward and aft foils of hydrofoil craft of Grumman Co, AGEH-1 (Plainview) are 90% and 10% respectively. This type features split forward foils, to avoid hydrodynamic interference between the foils, and the ability for retracting the foils on deck and using a swept foil plan profile. The foil arrangement is shown in Fig. 10.45. This is a large craft with displacement 328 ton, craft length 64.6 m, beam 12.3 m, draft on foils 3.1 m, maximum speed 50 knots, with two GE-240 gas-turbines, each 12,350 kW driving Z drive power transmission for propellers plus two diesels for hull borne operation, with power of 2 x 370 kW.

PGH-1 (Flagstaff) is another hydrofoil craft with ACS system manufactured by Grumman Co., with airplane foil arrangement, fully rotating foil control system, vertical Z drive power transmission, and the loads for forward and aft foil are 70% and 30% respectively. The craft is shown in Fig. 10.46. The full displacement is 58 t, and other main parameters are craft length 27.1 m, width 6.5 m, foil borne draft

1.7 m, main engine one gas-turbine type MK-621 with power output 3300 kW, and one diesel with 120 kW for hull borne operation.

The hydrofoil craft PCH-1 (Highpoint) of Boeing Co. has a canard foil arrangement, and load distribution for forward and aft foils are 30% and 70%, respectively, so the reverse of Grumman. The foil struts can be retracted around the hull with a hydraulic system for reducing the hull borne draft. In practice, it was found that the reliability of this type of retraction was poor.

Lift is controlled by hinged flaps, with forward foil controlling attitude. The aft inner flaps control trimming and the outer flaps control rolling. It was found that the moment of the outer flaps on their own was unsatisfactory for roll control, and so the system rearranged to engage both external and inner flaps to control the rolling.

Initially, the forward strut rudder was used to control the course but this was found a weak control, and later modified to full forward strut rotation plus the differential action of the foil flaps to control the course in coordinated turning. The foil arrangement is as shown in Fig. 10.47. The displacement is 110 t, craft length 35.4 m, beam 9.84 m, foil borne draft 2.0 m, with two gas turbines 2 × 3100kw, plus a diesel, 1 × 440 kW for hull borne operation.

Boeing's PGH-2 (*Tucumcari*), Fig. 10.48, was developed in competition with Grumman's design PGH-1 testing the attributes of a canard geometry against the aeroplane with split forward foils of the Grumman craft, as we have outlined in Chap. 7. The craft have the same speed and payload. The differences regarding the foil configuration are:

1. Canard foil arrangement on *Tucumcari*, with loading on forward foil 30%, aft foil 70%, with downward dihedral for both forward and aft foils, and both forward and aft foils can be retracted using split aft foils.
2. Water jet propulsion (compared with propellers used on *Flagstaff*), with one GT, 1 × 3100kw, plus one diesel, 120 kW for hull borne operation.
3. Flap control foil system for adjusting lift rather than all moving foils.
4. Using coordinated turning, with differential action of left/right main foil flaps.
5. Use of forward strut rotation for maintaining coordination and avoiding strut air ventilation and cavitation during turning.

The design of the canard foil arrangement and water jet propulsion for PGH-2 resulted in the pump power in foil borne and hull borne operation to be lower than power levels in PGH-1. The seakeeping quality was also better than PGH-1, and the coordinated turning appreciated by the users. Based upon the successful experience of this craft the guided missile hydrofoil craft type PHM-1 (*Pegasus*) and passenger hydrofoil craft "Jetfoil" were developed.

PHM-1 also has a canard foil system, with forward foil loading 32%, aft foil 68%, downward dihedral aft foils, and integrated aft struts and water jet intakes. Her prototype, PGH-2, had a split aft foil which was sideways retractable, and had a flat forward foil, as shown in Fig. 7.5, as well as the general arrangement in Fig. 10.48. The PHM-1 had a simplified development with a single aft foil that rotated back over the stern, as shown in Figs. 7.4 and 7.21. This arrangement was further simplified to a flat aft foil arrangement on the Jetfoil.

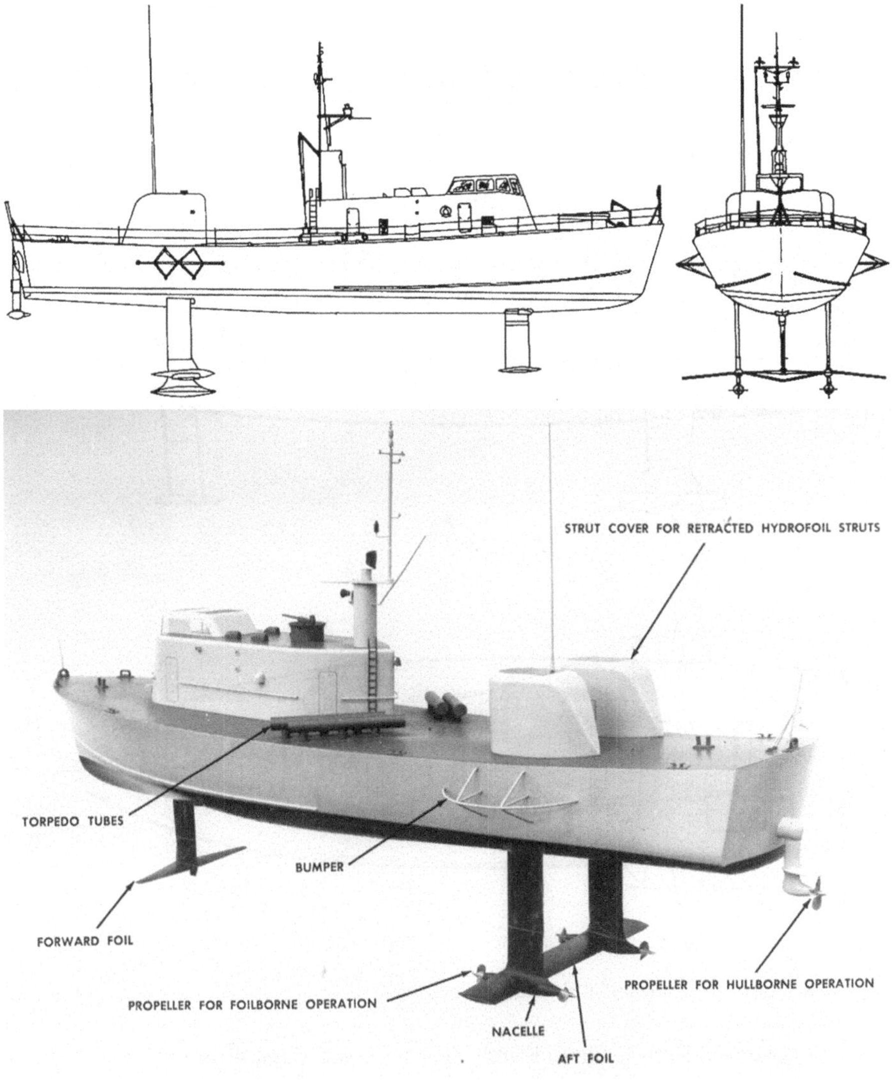

Fig. 10.47 PCH-1 "Highpoint"

The PHM displacement is 238 t, craft length 40.5 m, width 8.9 m, foil borne draft 2.8 m, main engine for water jet propulsion is one GT type LM2500, power 20,000 kW, plus two diesels for hull borne propulsion with the power of 2 × 590kw.

Taking PGH-2 and PHM design as a prototype, the passenger hydrofoil craft design series "Jetfoil" was a successful development with very good seaworthiness, as a passenger ferry for 250 passengers operating at maximum speed 50 knots and normal operation speed 45 knots. This capability has been successfully deployed for services between Hong Kong and Macau for over 30 years, and also for a number of

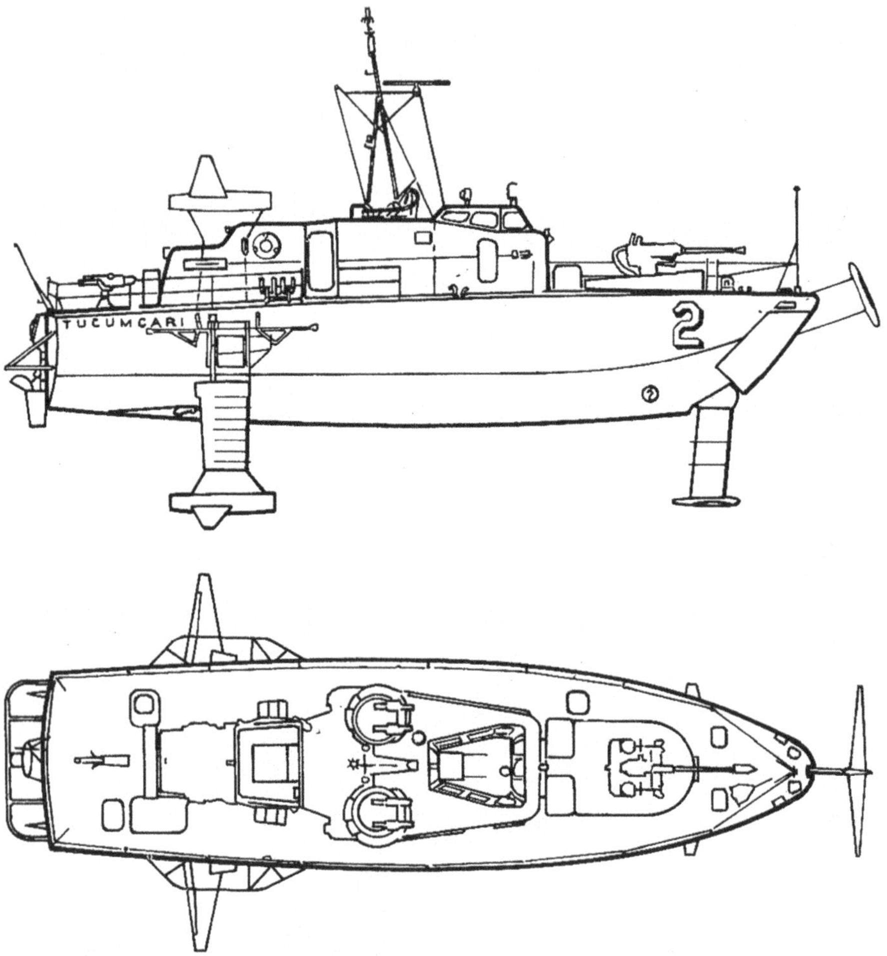

Fig. 10.48 PGH-2 "Tucumcari" foil arrangement

routes in Japan and from Japan to Korea. As a high-speed maritime city center to city center commuter ferry, it is ideal, avoiding the need for transits to airports, etc., which short flights involve.

References

1. B.A. Kozealov et al., *Handbook for Design of Dynamic Supported Ship* (Leningrad (shipbuilding), 1980) (in Russian)
2. R.J. Johnston, *'Chapter 5, hydrofoils', Naval Engineers Journal, February* (US Society of Naval Engineers, 1985), pp. 143–199
3. S F Hoerner, 'Fluid Dynamic Lift', 1975, Published by the author

4. J.B. Hadler, The prediction of power performance on Planing craft. Transactions SNAME **74** (1966)
5. V I Blumin et al., 'Transport hydrofoil craft', (in Russian)
6. The Aquavion System—International Hydrofoil Society (yumpu.com)
7. A.M. Vagohov, *Design of High Speed Craft* (Shipbuilding, 1978) (in Russian)
8. International Code of Safety for High Speed Craft, IMO, publication IA-185E, ISBN 92789 28014 2402, 2000. Amendments and resolutions after 2000 are available on IMO web site IMO.org
9. R.J. Johnson, Hydrofoils. Nav. Eng. J. **2**, 142–199 (1997)
10. M.C. Eames, T.G. Drummond, H.M.C.S bras d'Or: Sea trials and future prospects. Transactions Royal Institution of Naval Architects, UK, Vol **115**, 69–87 (1973)
11. K.J. Minsås, Design and development of hydrofoil catamarans in Norway, in *Proceedings. FAST93 Conference*, vol. 1, (Tokyo, 1993), pp. 83–99
12. Baron H Von Schertel, 'The Design and Application of Hydrofoils, and Their Future Prospects', Institute of Marine Engineers Transactions, Volume 86, 1973, pp 53–64, ISSN: 0309-3948
13. R J Johnston, G W O'Neil, 'The development of automatic control systems for hydrofoil craft' Hovering Craft & Hydrofoil International Conference, May 13–16, *Brighton* (Kalerghi Publications, England, 1974)

Chapter 11
Some Special Hydrofoil Craft

11.1 Introduction

The advantages of a hydrofoil are:

- High lift/drag ratio, particularly for foils with high aspect ratio.
- Effective control force for improving seakeeping quality, particularly in case of high speed.
- By using foil-flap control and a low-power hydraulic system, to reduce motions and the seasickness rate of both passengers and crews on craft in a seaway.

Foils have been applied as a useful part of ride control systems (RCS) on various HPMV, for example, on SWATH, semi-SWATH, and high-speed catamarans, in recent years. However, these foil systems do not play a role in supporting the craft weight and will not be involved in this chapter of the present book.

Hybrid hydrofoil and planing hull configurations where the foils support most or all of the craft weight, forming novel craft types, are introduced in this section.

It is well known that the disadvantages of a monohull hydrofoil craft are low cabin space and deck area for accommodating passenger seats or cargo. They have poor stability during take-off, particularly for deep submersible hydrofoils, where almost no passive transverse stability is available, so a ride control system is necessary for safety. This increases the cost. In addition, in order to improve the seaworthiness of Deep Submersible Automatic Control hydrofoils, high struts from the foils have to be available, and this increases the draft in hull-borne operation, then a retractable foil system has to be mounted on such craft and, thus, further increase the construction and maintenance cost again.

All of these challenges have limited the development of the monohull hydrofoil. However, the slender hull catamaran has the potential to ameliorate some of the disadvantages, since they have wide deck and cabin space, good transverse stability and low hull drag. The twin hull slenderness also offers the possibility to minimize ride height as interaction with higher waves may not be so critical as a monohull

Z.-S. Dong et al., *High-Speed Monohull and Hydrofoil Craft*,
https://doi.org/10.1007/978-3-031-62760-6_11

hydrofoil. Combining a hydrofoil and a catamaran to form a hybrid configuration had potential, and so was developed by several companies through the 1990s.

In this section, we will introduce two types of hydrofoil-catamaran:

- *Foil-assisted catamaran*: Here, about 50% craft weight is supported by bow-located hydrofoils, and another 50% weight is supported by twin hulls.
- *Foil catamaran*: In case of high speed, most part of the craft weight will be supported by hydrofoil, and only a little bit part of the weight supported by foil struts; however, in case of lower speed, that is, before take-off, and at hull-borne operation, most part of the weight is supported by twin hulls.

11.2 Foil-Assisted Catamaran

The design concept we introduce here is a hybrid craft [1–3], typical of which is type "Superfoil-40," developed by Marine Technology Development Corporation, (MTD), Russia, operated in the Baltic Sea, between Helsinki, Finland, and Tallinn, Estonia, and also in Shanghai (after 2005) between "Yang San" harbor and Shanghai.

The craft is composed of twin hulls and two bow foils, each mounted under the twin hulls, thus mixing the advantages of both hydrofoil and catamaran, with automated RCS for automatically controlling the bow foils and stern interceptors for improving the takeoff ability and seaworthiness, as shown in Figs. 11.1 and 11.2.

The craft is constructed of aluminum alloy, with twin hull and hull-borne operation at low speed before take-off, as shown in Fig. 11.1b, and each hull is of hard chine planing configuration for high-speed operation. At high speed after take-off, the craft is supported both by bow foils and short planing surfaces of twin hulls at the stern, with interceptors at the transoms for augmenting the lift and planing on the water surface, as shown in Fig. 11.1a, achieving a speed of up to 55 knots.

The main features and technical characteristics of the craft are:

- Deep submersible bow foil, offering efficient hydrodynamics, as $h/C > 1$.
- Twin hull configuration, with fine passive transverse stability at low speed, and also positive passive stability in foil borne operation even if the RCS fails.
- The retractable foil system between the twin hulls, as shown in Fig. 11.1b, ensures the foils are safe, easy to maintain, repair, and the vessel to dock.
- Shallow draft of 1.2 m when in foil borne operation using water jet propulsion.
- Skegs and interceptors at transom stern so the hydrodynamic efficiency of the twin hull planing surfaces is high.
- With the aid of RCS for controlling the bow foils and stern interceptors, the seaworthiness is high.

The general arrangement of the craft is shown in Fig. 11.2. The leading particulars are:

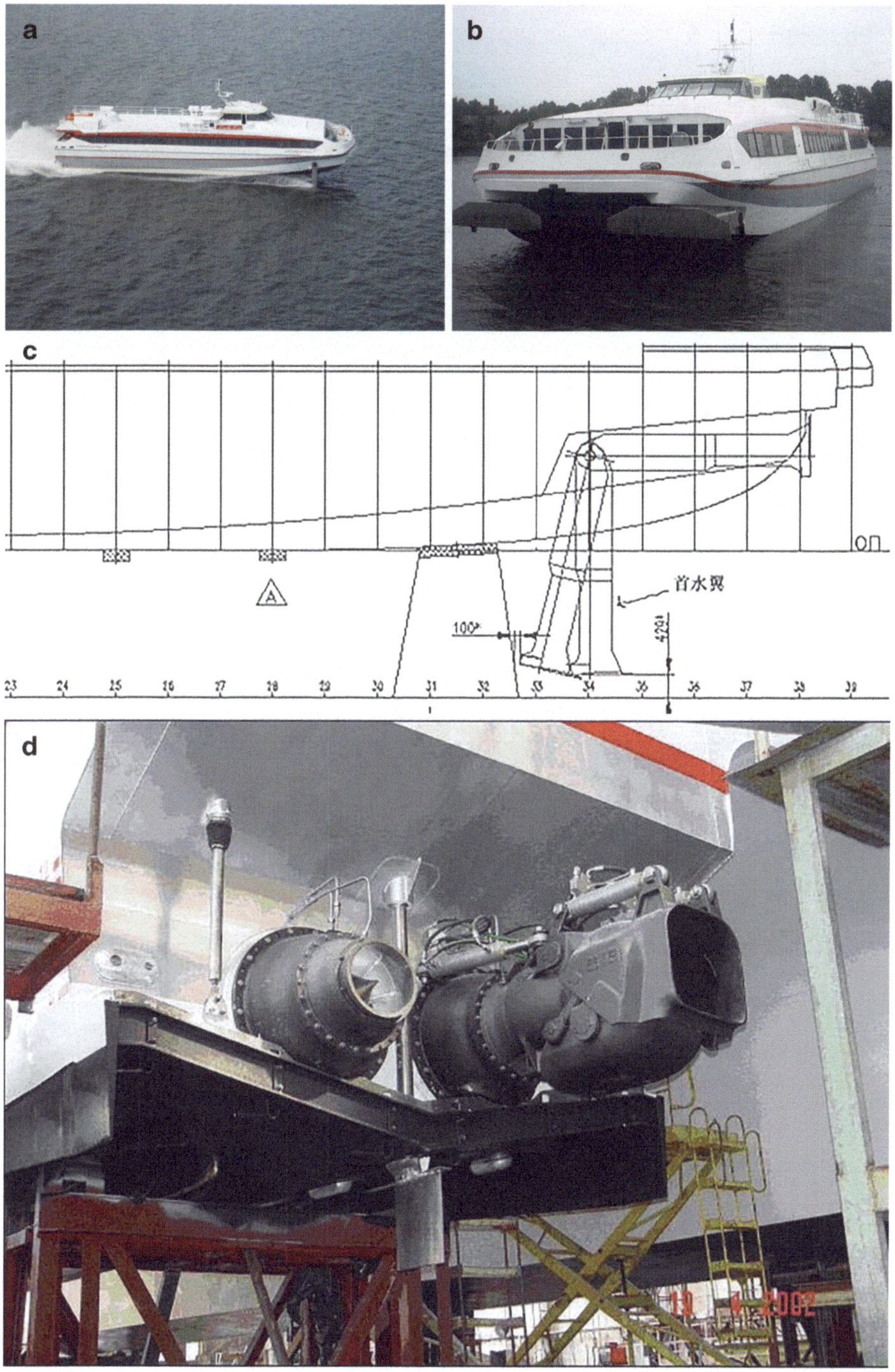

Fig. 11.1 (**a**) Superfoil-40 at speed (**b**) Superfoil-40 bow view afloat with raised bow foils (**c**) Geometry of Superfoil-40 foil system deployed and raised (**d**) Stern view of one demi hull showing one fixed and one steerable propulsion waterjet

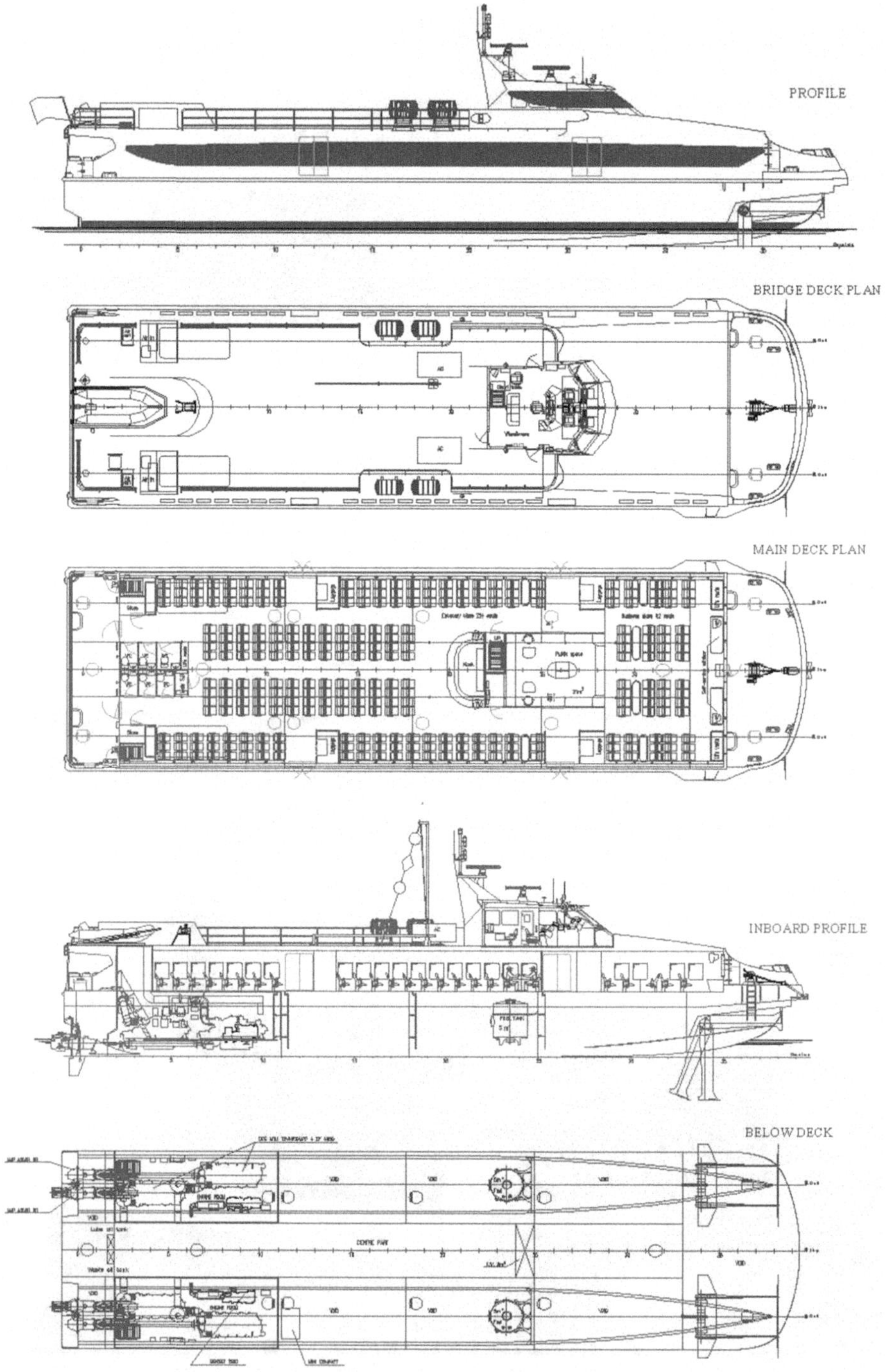

Fig. 11.2 FACAT 40 general arrangement

Loa	41.00 m
Lbp	36.36 m
Lwl	40.28 m
Boa	12.40 m
B, molded	11.00 m
Depth, molded	3.55 m
Draft, full loaded, foil borne	1.20 m
Draft, static, fully loaded, foil extended	3.40 m
Draft, static, fully loaded, foil retracted	2.10 m
Payload, total	33.00 tonnes
Passengers, economy class	224
Passengers, business class	62
Power plant and propulsion	Four MTU 12 V 4000 M70, 1740 kw/2400 rpm driving two MJP J ISSO RDD water jets and two MJP JSSO BDB water jets
Range	140 Nautical miles at service speed 45 knots, SS4
Directional control	Two steerable water jets, plus two auxiliary, semi-balanced rudders, one on each hull for improving both maneuverability and course keeping ability
Bow foils	One foil located under each hull at bow, constructed of high-strength titanium alloy, $h/c > 1$ Foils are deep submersed type, no flap, with foil incidence control. Figure 11.1c Shows a drawing of the bow area with craft on maintenance chocks with foil in lowered and raised positions.
Interceptors	One interceptor mounted at each transom (Fig. 11.1d)

The interceptors are controlled via a hydraulic system, for upward and downward movement, to generate lift at the stern and to improve the seakeeping quality of the craft, with bow foils.

The stern skegs (Fig. 11.1d) act together with the interceptors to generate lift that increases with speed. Since the submersion depth for the interceptor is small, just 30–40 mm (0.08–0.1% of L), and within the boundary layer, the added drag is small, so the hydrodynamic efficiency is high, improving the craft hydrodynamics and the take-off to foiling operation. The skeg length is almost equal to the wetted length of the demi-hull at service speed.

Hydrodynamic Performance

The hydrodynamic performance should be high, as mentioned above and shown in Fig. 11.3. From the figure, one can see that the resistance/lift ratio of the craft at the speed of 50 knots, Fr_l 1.5, can be as low as 0.108.

Seaworthiness

Figure 11.4a shows the vertical acceleration (root mean square value) of the craft running at the speed of 40–45 knots, at various wave directions, in seas of significant wave height 2.0 m, with wave period 5.2 s.

The vertical acceleration value curve is also compared with ISO 2631/3 to show the seasickness and discomfort rate of crew and passengers on the craft, where line 1 is the vertical acceleration limit for the seasickness, while lines 2, 3, 4, 5, and

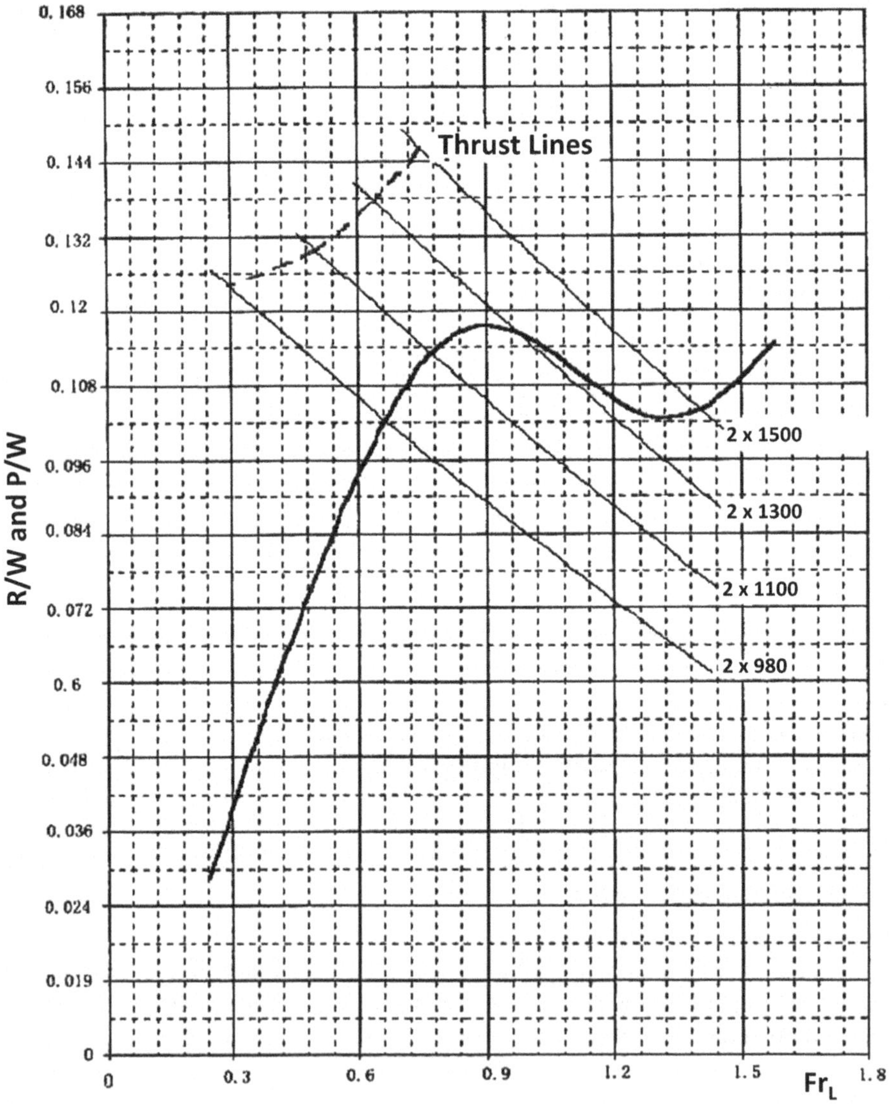

Fig. 11.3 FACAT 40 performance

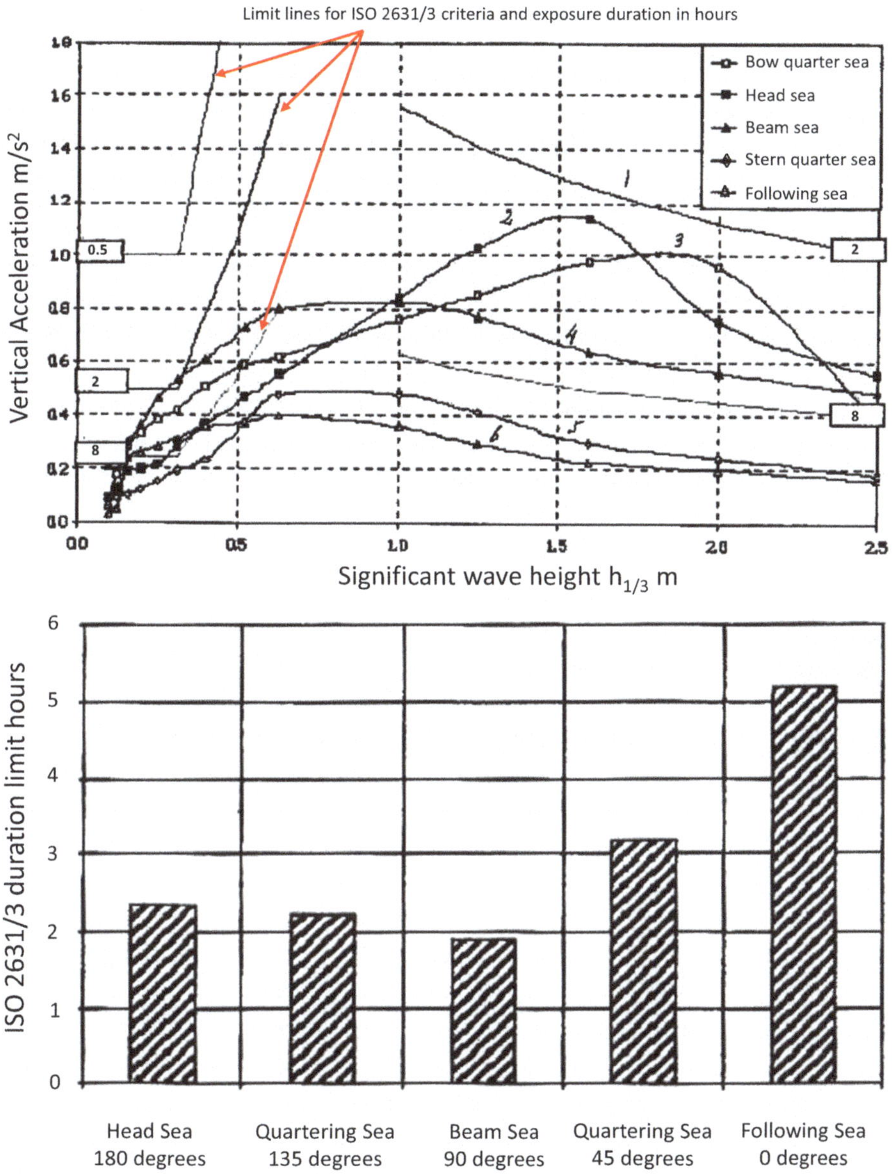

Fig. 11.4 (**a**) FACAT 40 vertical acceleration performance in a seaway (**b**) FACAT 40 motion comfort limits

6 show the vertical acceleration in the wave directions of head sea, bow quartering, beam sea, stern quartering, and following sea respectively.

Figure 11.4b shows the time limit (hours) for passengers to avoid serious discomfort on the craft, at 40–45 knots in 2.0 m significant wave height, at various wave directions. One can see that seasickness is not likely to happen for passengers until about 2 h, or even 5.1 h for following sea, and 3.1 h for stern quartering sea.

Figure 11.5 shows the Block Diagram for the foil/interceptor ride control system (RCS).

The RCS enables banked and coordinated turns, with smaller turning diameter and course angular speed rate > 0.5°/s. Meanwhile, quicker take-off capability is also available by adjusting the height of interceptor depth below the transom gradually with craft speed.

The key to numbered items in the diagram, Fig. 11.5, is as follows: 1 Gyrocompass, 2 GPS, 3 motion sensors, 4 FCI basic control panel, 5 PC2 reserve control panel, 6 MMP12–1 system basic monitor screen, 7 FC3 central control panel, 8 PP-UBC2 altitude transmitter and receiver, 9 A052–2 Altitude antenna, 11 FC4 linear accelerometers for the directions of X,Y,Z, 10 FC6 power system control,

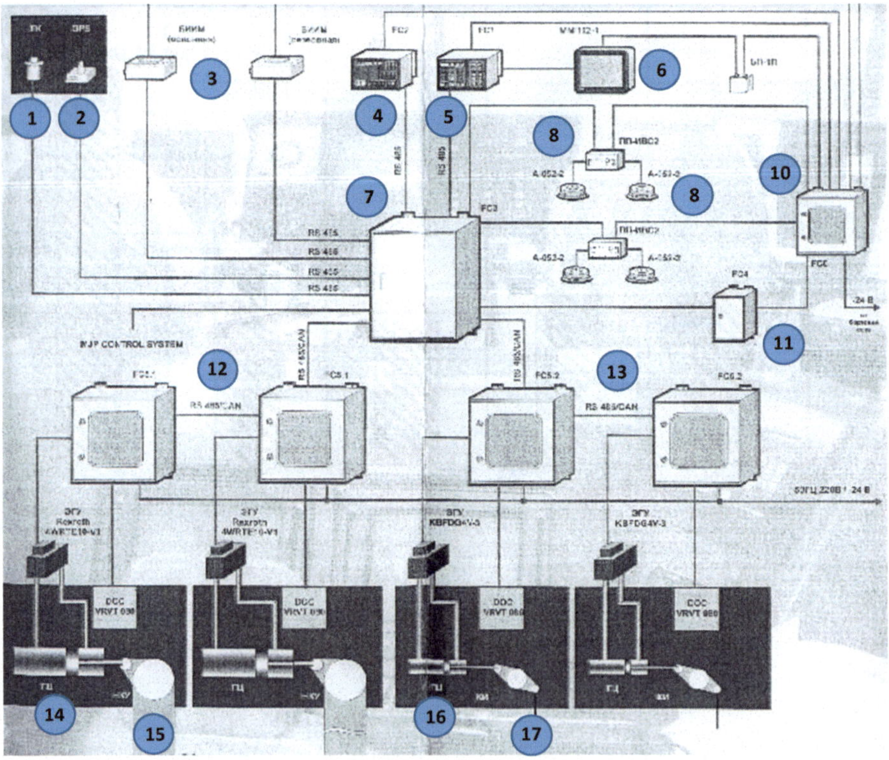

Fig. 11.5 FACAT 40 foil control system

12 FC5.1 mechanical control for bow hydrofoils, 13 FC5.2 mechanical control for stern interceptors, 14 hydraulic cylinder, 15 bow hydrofoils, 16 electric hydraulic cylinder, and 17 interceptors.

HYSUCRAFT Foil-Assisted Catamaran

Reference [4] introduces another type of foil-assisted catamaran, with the foils located between demi-hulls and lift between 30% and 40% of craft weight; Fig. 11.6a, b show the principle and a typical general arrangement for this type of craft. The aim of the foils between the catamaran demi-hulls is to provide dynamic lift using a much more efficient surface than a planing hull bottom.

The study of hydrofoil-supported catamarans has been a focus of Prof K.G.W Hoppe and his team at the Marine Engineering Department of the University of Stellenbosch in South Africa since the late 1980s. His work has focused on configurations similar to the HPC described above, while he has applied the principles to the design and construction of a significant list of projects. Prof Hoppe has investigated various configurations of the main lifting foil and support stern foils through analysis and many model tests. Following a systematic initial series of model tests, it was possible to identify configurations of foils that would reduce the drag of a planing catamaran by 30–40% compared with the "bare hull" catamaran. In addition, the interaction of the foils, primarily the main lifting foil as it passes through waves, was found to stabilize heave motions, giving a smoother ride than the bare hull.

How does the improvement occur? The reason is that the ratio of drag/lift for a hydrofoil is typically 0.03–0.05, while the same ratio for the planing hulls of a catamaran is 0.25–0.3, a difference of 10 times.

The lifting foil cross section may be a thin "airfoil" section or a c-type section with a sharp leading edge. Since the foils operate close under the water surface, the suction pressure over the upper surface will lead to cavitation or air entrainment. The sharp leading edge ensures a stable operation when in the cavitation regime. It may be noted that for effective lift, the immersion of the main foil H_w should be greater than 20–30% of the foil chord depending on the exact foil profile, as when immersion is decreased toward the surface, the lift force rapidly declines. This favors a deep vee hull section as the foil needs to be at or above the keel line for safety.

Due to the complexity of combining the elements of a catamaran and the foils, a combination of analysis and model testing is used. Initially, the catamaran base design is prepared to conform to the payload and other mission requirements and static stability. Following a choice of potential lift support from foils in the range of 20–40%, the foil geometry is selected and overlaid on the hull geometry.

Foil-supported catamarans to Professor Hoppe's design have been built as passenger ferries, utility vessels, and recreational vessels. An example is shown in Fig. 11.6c, that shows the Nordblitz 115 passenger fast ferry. The vessel is 22 m LOA, service speed 34 knots, with tandem foil system front foil forward of CG and main foil aft of CG. With Servogear propeller propulsion system. Information on the HYSUCAT can be found in chapter 10 of [5] and at http://fastcc.hysucraft.com.

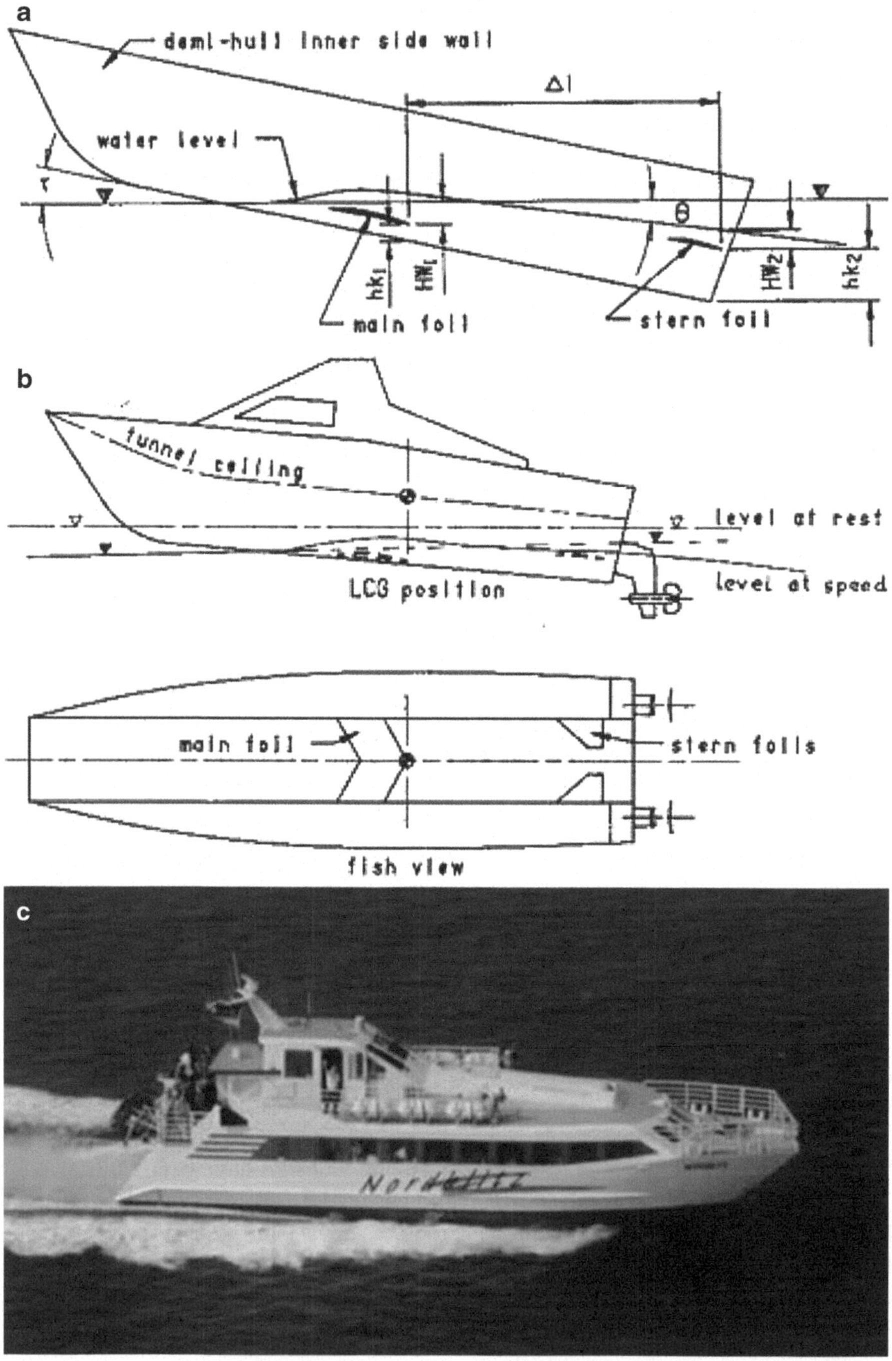

Fig. 11.6 The HYSUCAT foil assisted catamaran (**a**) The principle (**b**) Typical general arrangement (**c**) Nordblitz 115 passenger fast ferry

11.3 Foilcat

The difference between the foilcat [6, 7] and the foil-assisted catamaran is that on the Foilcat, the weight is fully supported by foils that are located under the twin hulls, while on a foil-assisted catamaran, only about 30–50% of craft weight is supported by the foils located at the bow (FACAT) or between the hulls (HYSUCAT).

During the 1990s, Kvaerner Fjellstand in Norway developed the Foilcat FC40. In parallel, Westamarin in Mandal developed their own design of Foilcat, the W2900.

Figure 11.7 shows the profile of a Foilcat FC40 from Kværner Fjellstrand. The twin hulls widen the deck and cabin space and improve transverse stability during hull borne and take-off operation. There are twin foils at the bow and a single foil across the stern with water jet intakes incorporated into the two main support struts. The flying waterline is a little below the keel line of the hulls rather than the substantial height of the Jetfoil. The support by twin hulls with slim body plan and high deadrise angle is aimed at low and progressive forces in waves when the hulls interact.

Figure 11.8 shows the forward foil (right side) and rear foil (left side) arrangement of this craft. One can see the forward foils are two small foils located separately under the twin hulls. The stern foil is a single one extended across under twin hulls transversely with a high aspect ratio, giving efficient hydrodynamic properties and supporting the major part of the craft weight.

Before take-off, since the speed is small, the craft weight is mostly supported by hull buoyancy. However, as speed increases, the craft weight is transferred gradually to the foils until the twin hulls clear from the water surface, and almost the weight of the craft is supported by hydrofoils. Figure 11.9 shows the thickened stern foil strut that accommodates the water jet inlet and the propulsion arrangement. The propulsion efficiency is high due to a short inlet length, reducing the hydrodynamic loss in the duct.

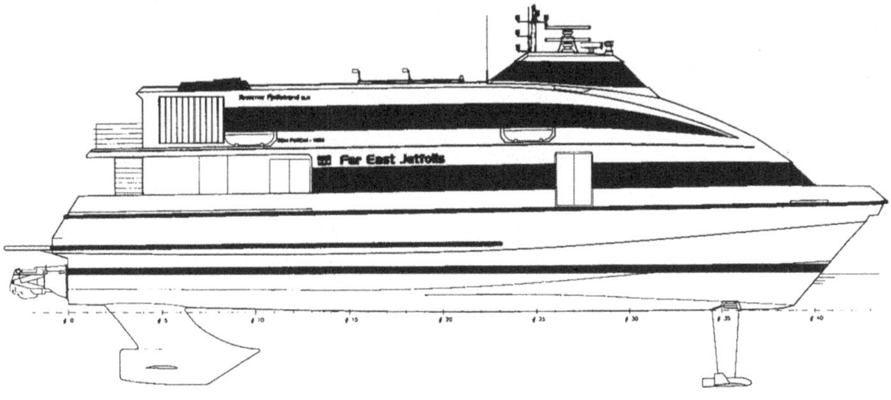

Fig. 11.7 Kværner Fjellstrand Foilcat FC40 profile

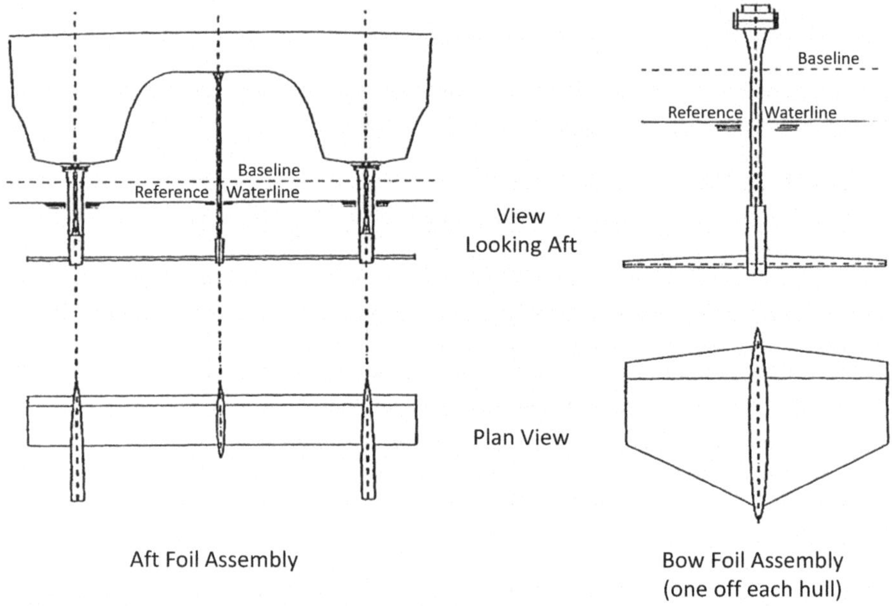

Fig. 11.8 Foilcat foil arrangement

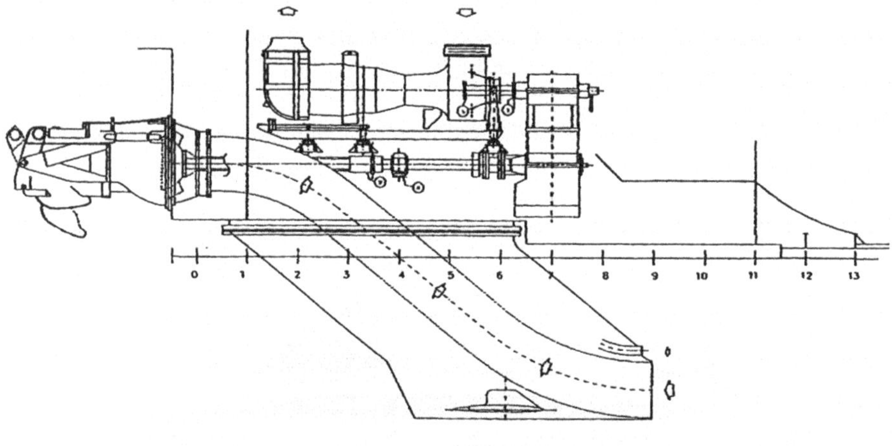

Fig. 11.9 Foilcat propulsion system

Wave making resistance in hull borne operation is minimzed due to the high slenderness of the catamaran demihulls. Figure 11.10 shows the percent of drag for hulls, foils, and strut at various speeds.

From the figure, one can see that the foil drag increases rapidly after a speed of 10 knots as a consequence of gaining lift; meanwhile, hull drag increases slowly as a consequence of being lifted out of the water. The craft takes off gradually, and the

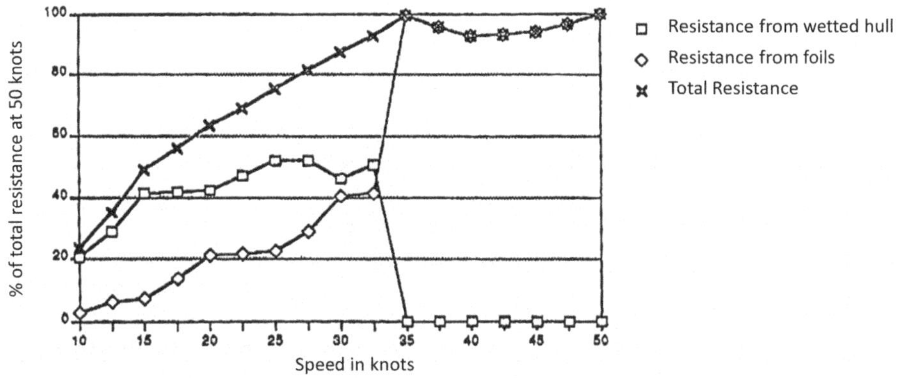

Fig. 11.10 Foilcat resistance components from hull and foils with speed

drag builds progressively rather than with a sharp drag hump until at a speed of 35 knots take-off is complete and the drag of foils becomes the whole craft drag.

Power buildup of the engines is also smooth, which is seldom experienced in the family of dynamically supported craft, such as ACV, SES, WIG, and planing hulls.

The advantages of Foilcat can be summarized as:

- Smooth drag curve and without drag hump.
- High propulsion efficiency.
- High turning angular velocity and small turning diameter, due to the bow foil's ability to rotate horizontally (skeg can be turned) and allowing coordinated turning operation.
- High seaworthiness with the aid of automatic control flaps. These are controlled by a number of instruments mounted in the craft, as shown in Fig. 11.11.
- Shallow draft due to the shorter struts, to allow a fixed main foil system. It may be noted that a retractable foil system causes high complexity and cost. However, this shallower draft concept causes increased impacting force of hulls in waves as they fly with a low clearance gap compared to hydrofoils such as the Jetfoil. Nevertheless, the slender twin hulls with high deadrise angle should minimize slamming to a satisfactory level.
- High safety, due to the positive passive transverse stability of twin hulls. In addition, the limited wave clearance reduces the crash stop accelerations in case of an RCS system failure.

Figure 11.12 shows one of the two Turbojet Hong Kong ferries FC40 Foilcat craft at speed between Hong Kong and Macao.

Research similar to that in Kvaerner Fjellstrand was also carried out in Mitsubishi, Japan, to develop the Super-Shuttle 400 "Rainbow" [8], with a foil arrangement similar to FC 40 in Norway. The "Rainbow" had a larger single foil across the bow as well as a large fixed foil crossing transversely the whole width at the stern. The bow foils have strut-mounted rudders for turning as well as flaps on

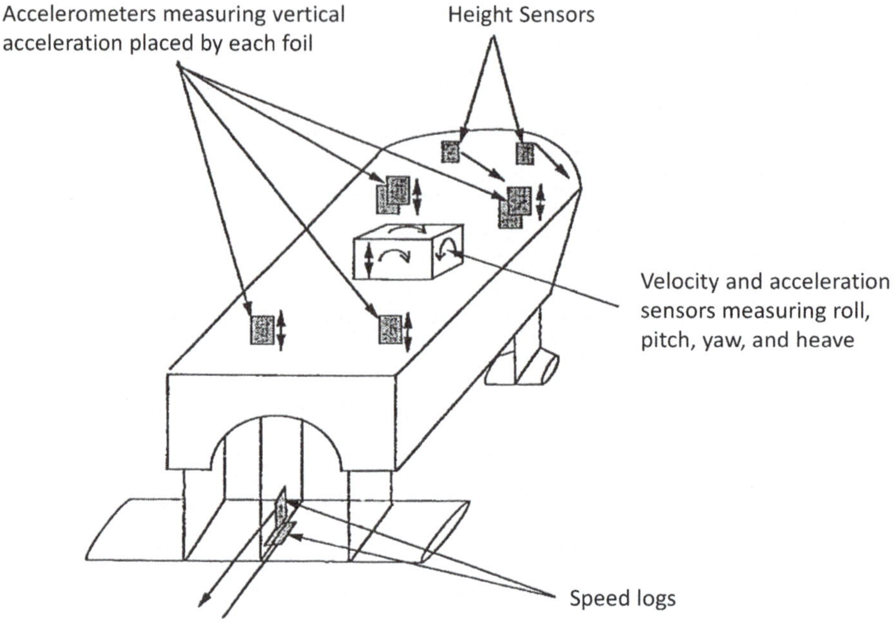

Fig. 11.11 Foilcat instrumentation to allow active automated ride control

Fig. 11.12 Turbojet Hong Kong ferries Fjellstrand foilcat at speed

both forward and aft foils. A full RCS system was installed. The Super-Shuttle 400 Rainbow is shown in Fig. 11.13.

Rainbow entered service on a route between Honshu and the Oki Islands in Japan in 1993. The ride control system was developed further to optimize motion response after construction of the second craft, initially named Rainbow 2 in 1998 [9], using a sea trials program to refine the system parameters.

Fig. 11.13 Mitsubishi Super-Shuttle 400 "Rainbow"

The leading particulars of "Rainbow" are:

Loa	33.24 m
Lwl	28.50 m
Boa	11.00 m
Depth, molded	4.20 m
Draft, hull borne	4.50 m
Foil borne	2.10 m
Gross tonnage (Japanese)	302 t
Passengers	341
Dead weight	35 t
Main engines	Mitsubishi S16R-MTK-S x 4 (2850 shp Ea)
Service speed	38 knots
Maximum speed	45 knots

Later, a Super Shuttle 400 "Dream" also entered service between Fukuoka in Japan and Busan in South Korea for Daea Express Shipping. The craft is shown in Fig. 11.14.

While Kværner Fjellstrand developed the FC40, Westamarin in Mandal developed a smaller hydrofoil-supported catamaran, the FC 2900. The Super Shuttle 400 is propelled by water jets, as are the two FC 40 craft, while the FC 2900 employed Z drive podded tractor propellers for its propulsion. A comparison of the leading particulars is as follows:

Item	FC40	FC2900
Full displacement, t	190	125
Passengers	300–400	150–200
Loa, m	40.55	29.25
Beam, m	12.00	8.35

(continued)

Fig. 11.14 Daea Express Shipping Super Shuttle 400 "Dream"

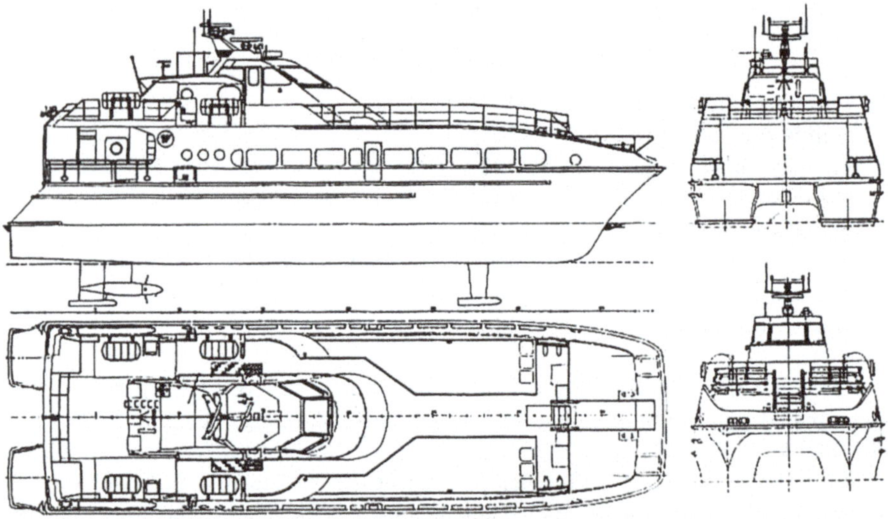

Fig. 11.15 Westamarin Foilcat FC2900 general arrangement

Item	FC40	FC2900
Max foil draft, m	4.00	3.70
Span of bow foils, m	3.00	2.50
Flying height, m	0.80	0.35

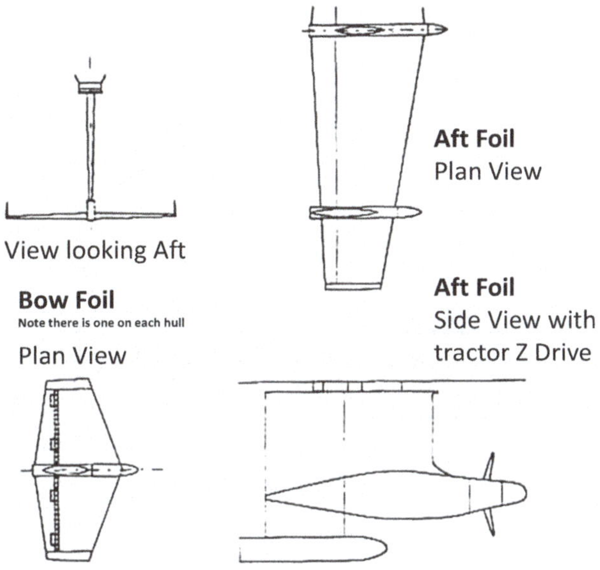

Fig. 11.16 Foil and propulsor details Westamarin FC2900

Fig. 11.17 Westamarin 2900 in service

Figure 11.15 shows the general arrangement of Foilcat 2900. Figure 11.16 shows for Foilcat 2900. Figure 11.17 shows the craft in service, and the foil arrangement. Figure 11.18 shows the seakeeping quality performance for both the Foilcat 2900 and a conventional catamaran based on trials data. This shows that the vertical

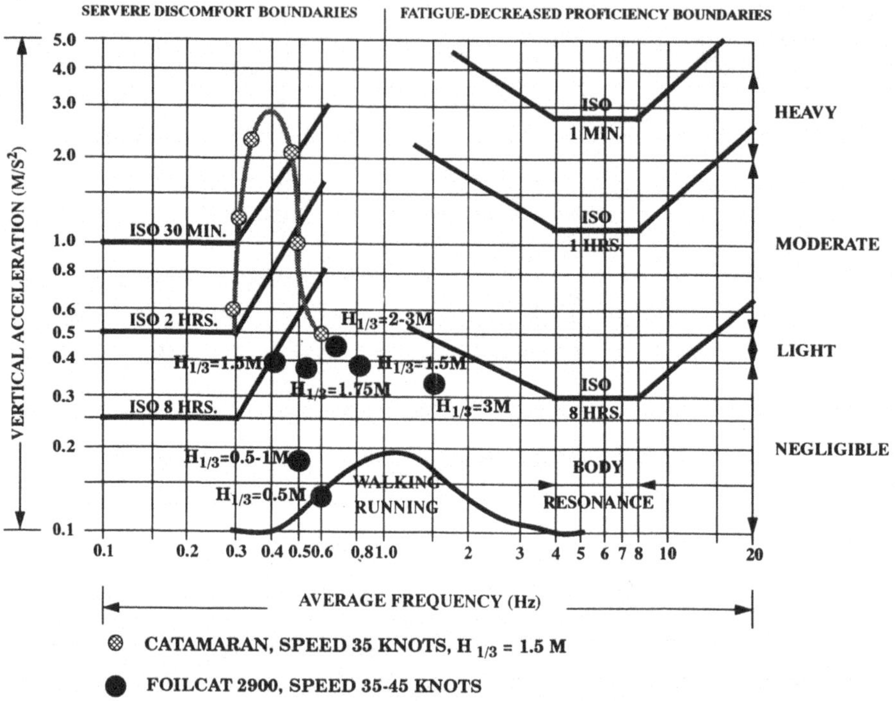

Fig. 11.18 Motions response data for Catamaran and Westamarin FC2900

acceleration and seasickness rate are significantly lower for the Foilcat than a conventional catamaran.

The two foilcat craft from Fjellstrand have operated for Turbojet in Hong Kong on the Macau route for many years. The FC2900 began service in Denmark and later was sold for service between the Greek Islands. The Super Shuttle 400 began services in Japan in 1993, and a craft was later put into service on a route between Busan in Korea and Fukuoka in Japan in 1998. The craft remained in operation until 2021.

Both the FACAT and Foilcat examples here have demonstrated significant technical advances in terms of efficiency and ride quality on open sea service routes while providing spacious accommodation for passengers.

Due to the high power density required for a vessel traveling at 50 knots and more, the concept has not been a stepping stone to a much larger craft that was envisaged as a goal by technologists back in the 1970s. The technical achievement is nevertheless a very significant one!

References

1. S.P. Pavlov, L. Yun, G. Mo, A Brief introduction to a high speed passenger craft-hydrofoil assisted catamaran "Superfoil 40", in *Proceedings, HPMV Conference, April 9–11, 2005, Shanghai*, (2005). Published also as selected paper in textbook 'Performance, Technology and Application of High Performance Marine Vessels Volume One', Chapter 3 paper 3.3.2, published by Cambridge Scholars Publishing, 2018, 819 pp ISBN13: 978-1-5275-0356-4
2. L. Yun, *Brief Introduction to the Hydrofoil & Hydrofoil Assisted Catamaran*, Science and technical report (Flying dragon science and technology Ltd, Shanghai branch, Hong Kong, 2004)
3. F.H. Wang, Y.J. Li, Experimental investigations on the performance of hydrofoil catamaran, in *Proceedings, HPMV Conference, Shanghai*, (1999)
4. K.G. Hoppe, Optimization of foil supported catamarans, in *Proceedings of 3rd International Conference on Fast Sea Transportation (FAST '95), Travemunde, Germany, September 1995*, vol. 1, (1995), pp. 301–308
5. L. Yun, A. Bliault, H.Z. Rong, *Chapter 10: High Speed Catamarans and Multihulls* (Springer Publications, 2018), pp. 463–467. ISBN 978-1-4939-7889-2
6. J.-H. Jorde, The development of a 50 knot 40m FoilCAT, in *Proceedings, FAST'91, Trondheim, Norway*, vol. 2, (1991), pp. 1093–1111. ISBN 82-519-0962-7
7. K.J. Minsæs, Design & development of hydrofoil catamaran in Norway, in *Proceedings, FAST'93*,
8. K. Kihara et al., Diesel driven-Fully submerged hydrofoil catamaran, in *Mitsubishi Super Shuttle 400, the "Rainbow", Proceedings, FAST'93*,
9. T. Kitamura, A problem and solution of the motion control system, in *IFAC Control Applications in Marine Systems Conference, Glasgow, Scotland, 2001, Proceedings*, (2001), pp. 473–478

Chapter 12
Postscript

This is the last book in the series instigated by Professor Yun Liang. It has been a task spanning nearly four decades. Liang was passionate about the technology of high-speed craft and spent much of his career in their research. This same period has been one of rapid advance, both in hydrodynamics and in the machinery and structures suitable for such vessels. It has been an exciting time for a maritime engineer!

The work at MARIC, based in Shanghai, together with his colleagues, gave him both challenge and satisfaction. MARIC designed and participated in many fast craft both for R&D and for commercial use. This has included ACV, SES, and WIG craft, as well as more conventional vessels.

An informal partnership with the undersigned has resulted in the successful preparation of a textbook series covering all of these craft. We together have drawn upon the research of many well-known engineers across the world, both East and West, including several of Liang's colleagues. This technology has been the result of a global endeavour and the volumes in the series reflect that.

This volume has focussed on monohull vessels and their extension with hydrofoils. In common with other volumes in the series, we have assembled work from Russia and China as well as USA, Europe, Japan, and more, giving a slightly wider view than often presented.

At the time of publication, hydrodynamic analysis using computer-based modeling for geometry, hull structure, stability, and hydrodynamics through finite element-based fluid modeling (CFD) has become a basic tool in larger commercial and military organizations. A considerable amount of university-based research has also been carried out. A sample of reference papers on CFD research is given in Chap. 14—Resources.

There are two limitations with such models. The first is that for accuracy element, numbers in the 100,000s are needed, making iterative solution of time step simulation a significant task. A consequence of the number of elements required is that timescale and costs may be significant, particularly for response in a seaway, and thus, it may be as simple to carry out model tests.

Z.-S. Dong et al., *High-Speed Monohull and Hydrofoil Craft*,
https://doi.org/10.1007/978-3-031-62760-6_12

The second is that unbounded multiphase flow modeling is limited to higher-end programs and is very difficult to calibrate against physical tests and observations. The multiphase issue affects modeling in the surface boundary for planing craft, as well as ACV, SES, WIG in the transition phase, and hydrofoils.

CFD, therefore, needs to be a tool in the toolkit rather than a replacement for model testing, for example, as discussed by Stefano Brizzolara in [1]. If a series of craft is envisaged based on a geometry, CFD can also be used to look at the effect of geometric developments. By its nature, a CFD model will also determine pressure patterns across a hull surface allowing optimization and assisting structural design.

Prior to computer fluid modeling, designers developed designs by interpretation from existing model test series. Description of the most well-known series has formed the core of this volume for monohull fast craft. Regression analysis of many of these has allowed predictive models to be prepared that can provide an initial estimate for drag and trim in the speed range up to full planing. The hull shapes available as examples are wide enough now to provide a useful start point.

The work carried out in the CFD references in Chap. 14 seems to suggest that the Savitsky semi-empirical model is accurate in the planing regime. Below this, the regression models discussed in this book form a good basis for initial design for all shapes of monohull craft. We have developed a number of spreadsheet models that may be useful to a student. These largely complement the models developed by Professor Radojcic and his colleagues [2]. This reference is recommended for a detailed explanation of the mathematical modeling approach.

Where a hull form is of the hard chine type, once above about $Fn_\nabla = 2.5$ or so, the empirical equations and approach developed by Daniel Savitsky provide accurate first estimates of drag and trim. This has been validated in research by others in the years since Savitsky's work was first published in 1964. We include spreadsheets for the Savitsky model and its extension to a single-step hull carried out by Svahn [3] in the electronic materials associated with Chap. 4.

There is much to learn from the fundamentals of "wing sections" operating in the water surface interface and close under it, and this is the reason for working through the fundamentals for planing surfaces in Chap. 3 and hydrofoils in Chap. 8, providing input from research in many countries.

Readers will understand from the chapters on seakeeping that the bow form is important for motions and accelerations and that there is a trade-off between motions and minimized drag when designing a new vessel.

The success of high-speed craft has led to their use in increasingly exposed environments and for longer endurance. This has led to the development of craft with a stern form with hard chines and a bow form that is slender and with a vertical stem, giving a form of wave piercing. We describe this in Chap. 6.

For the monohull then, it is possible to make a choice of geometry that should give the best combination of speed and ride for the intended service, whether recreational runabout, military patrol vessel, or high-speed ferry.

Our treatment of hydrofoil craft is based mainly on the US R&D programs and subsequent military and commercial Jetfoil craft, together with the programs carried out in the USSR. The two programs were targeted at very different environments for the most part. In the USA, the target was a much faster offshore patrol craft, while in the USSR, the mission was for medium- to long-distance ferry service along the rivers that were the main logistics arteries until recent times. In Europe and Scandinavia, fast passenger hydrofoils became very popular until traffic increased to a level that was only able to be serviced by the newer high-speed catamarans, as the hydrofoils could not be "scaled up" economically to accommodate vehicles as well as passengers.

Since the beginning of the twenty-first century, CFD has been used to develop hydrofoil geometries for America's cup competition (as well as the sails!). The racing sailboats in a sense formed a full-scale testbed, and so this technology has advanced such that there are small test passenger hydrofoils built on this experience. While there may be some improvements in efficiency, the basic arrangement of the Jetfoil is unlikely to be radically improved for commercial or military hydrofoils. Perhaps the opportunity will be if a zero-emission alternative to the gas turbine propulsion system were to be possible.

It is perhaps an interesting thought that commercial marine craft have a wide range of geometries, in contrast to the limited range used for commercial airliners. Some of this may be attributed to the variety of local route conditions and terminal interfaces; after all, runways are essentially all the same, passenger egress has been standardized worldwide, and the ariel environment is steadier than the ocean surface once at cruising altitude.

Perhaps that can be looked at as one of the joys of marine design. While a vessel may be designed for a "limited run" build, most projects are likely to be "bespoke." It is fortunate therefore that there is such a useful body of knowledge available in the public domain, as we have described here!

Going back to CFD as a tool, there are open access programs—some are listed in Chap. 14, and the power available on a current PC or MacBook may be sufficient for simulation by a student in a stand-alone manner or perhaps used for meshing and linked to a university computer for the simulation itself. The main thing is that new tools are becoming available. Hopefully, in the next decade or so, we will see new "Standard Series" created through the joint use of physical tests and CFD to provide a finer grid for parameter variation and implementation of new regression models for the lower end of the speed range.

Why would we need that?

The greatest challenge now upon us is the energy transition and the need for zero-emission powering. The higher cost of the fuel employed, whether batteries, hydrogen, or synfuel, means that optimization must rise to a new level. Hydrodynamics is just part of the story for the design of a fast marine craft, and it must be linked with structure and outfit, as well as a new approach to propulsion that compact electrical motors bring in. Electronic control systems are now reaching the stage where active optimization, including in-line adjustments through "artificial intelligence," is possible, and this should assist power drawdown during operation in direct response to the external environment.

It will be most satisfying for an engineer and designer to shape a hull and its power system to perform efficiently on its route or mission and at the same time provide an advance in the energy transition through automation.

This book, as with others in the series, is one to use as a reference book, as well as following the "storyline," and we offer it to readers as that. Please remember though that it reflects our understanding—which is a personal "snapshot." Continued work at the various model test facilities and universities will give further insight into this topic, so it is important to follow this, whether through the learned societies or the internet and open-source publications.

Alan Bliault, FRINA
Sola
October2023

References

1. S. Brizzolara, F. Serra, Accuracy of CFD Codes in the prediction of planing surfaces hydrodynamic characteristics. University of Genoa, Department of Naval Architecture and Marine Technology, 2nd international conference on marine research and transportation, Ischia, Naples, Italy, June 2007, available at http://www.icmrt07.unina.it/
2. D. Radojcic, M. Kalajdzic, A. Simic, *Power Prediction Modelling of Conventional High-Speed Craft* (Springer Nature, 2019) ISBN 978-3-030-30606-0
3. D. Svahn, Performance Prediction of hulls with transverse steps. Master's Thesis at Royal Institute of Technology, KTH, Centre for Naval Architecture, Stockholm, Sweden, 2009

Chapter 13
Appendix 1: Analytical Models

13.1 Introduction

The development of analytical models by regression analysis of parametric model series was initiated to help define basic hull characteristics for a new vessel prior to expensive and time-consuming model tests, for hull shapes that were close to previous designs.

We have introduced the approach and some models in Sect. 2.3.4 for round bilge hulls and Sect. 4.5 for hard chine hulls. Our intent with this appendix is to provide further discussion of models developed so far and reference to the data needed to assemble such a model as a spreadsheet both for a round bilge craft and for a hard chine craft.

Readers will find several of the models introduced in Chaps. 2 and 4 detailed by Radojcic and colleagues in [1] together with spreadsheets available in electronic form at the Springer site page for their book. Radojcic's book explains in detail the approach taken for regression analysis modeling and also the more recent "artificial neural network" models (ANN) that he and his colleagues have prepared over the last three decades. In this book, we have been able to present some other models including those from China. These models have also been prepared as spreadsheets accessible from the electronic materials on the Springer page for this book. In this appendix we will give a short overview of our selected models and make note of where a model is dealt with in detail in [1] which is heartily recommended reading.

S C Fung [2] has summarized and reviewed early models developed before presenting his own model development based on that primarily of M Aughey carried out at the beginning of the 1980s for the US Navy. Aughey's work is not available publicly, but S C Fung worked for US department of the Navy, NAVSEA, and so

---

Supplementary Information The online version contains supplementary material available at https://doi.org/10.1007/978-3-031-62760-6_13.

Z.-S. Dong et al., *High-Speed Monohull and Hydrofoil Craft*,
https://doi.org/10.1007/978-3-031-62760-6_13

was able to access the report and data to assist developing his own. The models discussed were as follows:

- Morrison 1954, 42 cargo ship tests
- Doust 1959, 130 trawler tests at NPL, UK, and 200 cargo ships 1962-4
- Todd 1966, 233 tests. Analysis extended by Neal 1966
- Sabit 1972, BSRA series 60 and SSPA Cargo Liner series
- Mercier and Savitsky 1973, 118 model tests, round bilge, and hard chine fast craft
- Jin 1980 model developed from Mercier and Savitsky with substitution of data from Series 63, Series 64, and SSPA high-speed round bilge tests so that round bilge data predominates
- Aughey 1983 and Lin 1984 prepared models based on Surface Combatant model tests for the US Navy

Fung prepared his regression model based on his analysis of the accuracy of residuary resistance prediction of these earlier models to select parameters for a regression equation. For round bilge naval combatants with a transom stern that were his focus, he had data of 426 hulls including test series such as the NPL round bilge series by David Bailey, the Davidson Series 63, DTMB series 64, and several others. His data set included both round bilge and hard chine craft, as did the work of Mercier and Savitsky which we have presented in Chap. 2, Sect. 2.3.1.

In Chap. 2, we have presented the models developed by MARIC based on Series 63/NPL/Nordstrom/DeGroot and their tests in Sect. 2.3.2, and the Shanghai University model in Sect. 2.3.3 in some detail, so this Appendix 1 we will discuss their application.

Additional models related to round bilge hull forms developed from model test results presented in Chap. 2 are the VTT model. The Radojcic SKLAD model, the Bogovic AMECRC model and additional models specific to the NPL and extended NPL hull forms series.

Below we will discuss each of the models and their application referring to the model tests presented in Chap. 2. It is important to note that readers will need to refer back to the reference materials documented in the chapter. Reference [1] presents round-bilge based models for NPL, SKLAD, VTT, and NTUA double chine forms.

In Chap. 4, we have introduced the hard chine models developed by Delft University for Series 62 and DSDS hull forms, the US Navy Academy model based on their test series, and the Naples Series model developed by Naples University. Models prepared by Prof Radojcic and his colleagues for the TUN series, Series 62, and the Naples warped chine series have been mentioned. In addition, material for a Series 50 model can be found in [1]. Delft University has also developed regression models based on their own model tests and series 62 that can be found in reference [3].

It should be noted that the hard chine craft are intended for service in the planing regime beyond $Fn_\nabla = 2.5 - 3.0$. Resistance for the planing region is reliably predicted using the Savitsky or equivalent approach as presented in Sect. 4.3.2. The regression models are concerned with the lower part of the speed range for these type of craft.

We will discuss first the models related to round bilge craft and then follow with the hard chine planing vessels. Where we have prepared spreadsheet models from material published by original authors, we will indicate that. Where a spreadsheet model is available from [1], we will indicate that. In what follows, the aim is to present the models and the relevant parameter space. Readers will find it useful to refer back to Chaps. 2 and 4 where we have discussed the effect of key parameters on performance. These need to be considered when a designer is carrying out initial definition studies, and wishes to set parameters for using a regression model for initial performance prediction.

There are two types of models that have been developed—*speed-independent* where the key equation coefficients are singular through the range and *speed-dependent* models where a different table of coefficients is developed for each speed. As S C Fung has discussed in [2], the speed-dependent models have been shown to better identify the humps and hollows in the residual drag curve. High-speed vessels do need to define this drag curve carefully as the primary peak at around $Fn_L = 0.5$ may well define the required powering in calm water particularly for planing craft. Prof Radojcic and his colleagues have taken a further step in developing neural network models that aim for improved predictive capability. A number of these can be found in [1].

When considering a model to use, it is recommended first to check the general hull shape against the examples discussed below, determine the principal form parameters, and check against the range defined by the model tests (Tables 2.2 and 4.2). For the models presented by Radojcic and his colleagues, the applicability check is included in the relevant spreadsheet model within a VBA routine.

13.2 Round Bilge and Semi Planing Craft

13.2.1 S C Fung

In [2], S C Fung presents the equation he developed together with the coefficient tables for $V/L_{wL}^{0.5}$ between 0.6 and 2.3 (Fn_L approx. 0.2–0.7). These have been collated to a spreadsheet model by the present authors. The model is developed from model tests of fast displacement combatants with a transom stern.

Residuary Resistance Equation
The equation has the following form:

$$C_R = f(\mathrm{D/L, Bx/Tx, Cp, Cx, IE, TA, TW, TT, BA, CWS})$$

Hull Form Parameters

D/L	Displacement Δ/L_{WL}, tons/ft
Bx/Tx	Breadth/draft at maximum section
C_P	Prismatic coefficient

Cx	Maximum section area coefficient (Ax/(Bx.Tx))
I_E	Half entrance angle, degrees
TA	Transom area ratio A_{20}/Ax where A_{20} = section area at ordinate 20
TW	Transom width ratio B_{20}/Bx where B_{20} = width at ordinate 20
TT	Transom depth ratio T_{20}/Tx
BA	Bow area ratio A_0/Ax where A_0 = section area at ordinate 0
FB	Lcp/L_{WL} (L_{cp} is length between perpendiculars, ordinates 0 and 20)
CWS	Wetted surface coefficient, where:
CWS	$A_{WS}/(L_{WL}.\ \Delta)^{0.5}$ where given by:

$$CWS = [6.223 - 0.94D/L + 16.0209/(L_{wL}/Bx) + 0.9207(Bx/Tx) + 5.4630/(Bx/Tx) + 9.8528Cx + 7.1592/Cx + 0.0201I_E + 2.1857/I_E + 1.0359TW + 2.4925BA + 3.7181FB]$$

Please note that dimensions are ft and long tons, with A_{WS} in square feet.

Fung formulated a regression equation with ten terms as follows:

$$CR_1 = C1 + a1^* D/L + a2/(D/L)$$
$$CR_2 = C2 + a3^*(Bx/Tx) + a4/(Bx/Tx)$$
$$CR_3 = C3 + a5^* Cp + a6/Cp$$
$$CR_4 = C4 + a7^* Cx + a8/Cx$$
$$CR_5 = C5 + a9^* I_E + a10/I_E$$
$$CR_6 = C6 + a11^* TA + a12^* TA^2$$
$$CR_7 = C7 + a13^* TW + a14^* TW^2$$
$$CR_8 = C8 + a15^* TT$$
$$CR_9 = C9 + a16^* BA$$
$$CR_{10} = C10 + a17^* CWS$$

Following regression and correlation analyses on the data sets he had compiled, he eventually focused on the data for transom stern vessels. He then produced tables for the ten terms for the speed range enabling the residual resistance to be computed from the simple relation:

$C_R = \sum (C_R)_n$ where n = 1 to 10, and
$C_R = R_R/(\rho V^2/2^* A_{ws})$ for the given speed
$R_T = R_R + R_F$ where
$R_F = C_F{}^* \rho V^2/2^* A_{WS}$

The tables for $(C_R)_1$ to $(C_R)_{10}$ against $V/L_{WL}{}^{0.5}$ and the relevant parameter can be found in the spreadsheet "Fung model" included in the electronic materials to this appendix.

The most significant parameters controlling residual drag are D/L, thus slender hulls will be advantageous, and following this I_E and linked to it Cp with slender entry giving lower drag at speeds up to $V/L_{WL}^{0.5}$ around 1.4 or Fn_L between 0.4 and 0.5. Above this, wider entry becomes more favorable as planing speeds are approached. Fung gave an estimation for I_E appropriate for this type of vessel if lines are not available as follows:

$$I_E = [120.0192 + 69.2808/(D/L) + 0.1131^*LwL/Bx + 104\text{-}8640/(LwL/Bx) \\ - 28.5611^*Cp - 26.5287/Cp - 24.8166^*Cx - 16.2033/Cx \\ + 2.6398^*TW - 49.6570^*FB]$$

From his analysis using his equation, Fung prepared two plots of suggested parameters that would give favorable residual resistance, see Figs. 13.1 and 13.2, depending on the speed ($V/L^{\wedge 0.5}$ or Fn_L).

His data set did include that analyzed by Mercier and Savitsky and so the model may be used as a means of comparison, the Fung model being mainly round bilge hulls, while the Mercier and Savitsky being more weighted to planing hull geometry. Both are wide data sets rather than specific to a hull parameter set from a test series.

Having determined the residual drag curve for a craft, the frictional drag has to be determined based on the Reynolds number and so friction coefficient and the wetted area. This will then give the total hydrodynamic drag. Aerodynamic drag can be determined from the above waterline presented area and shape.

Fig. 13.1 Suggested Bx/Tx and IE

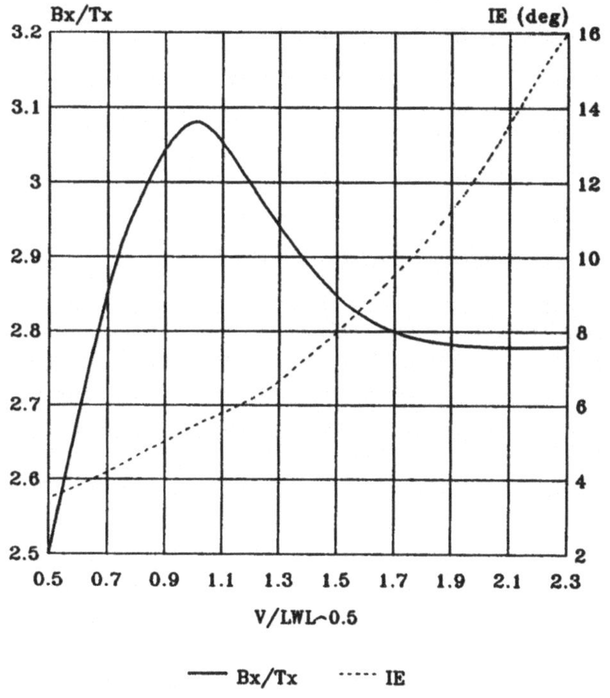

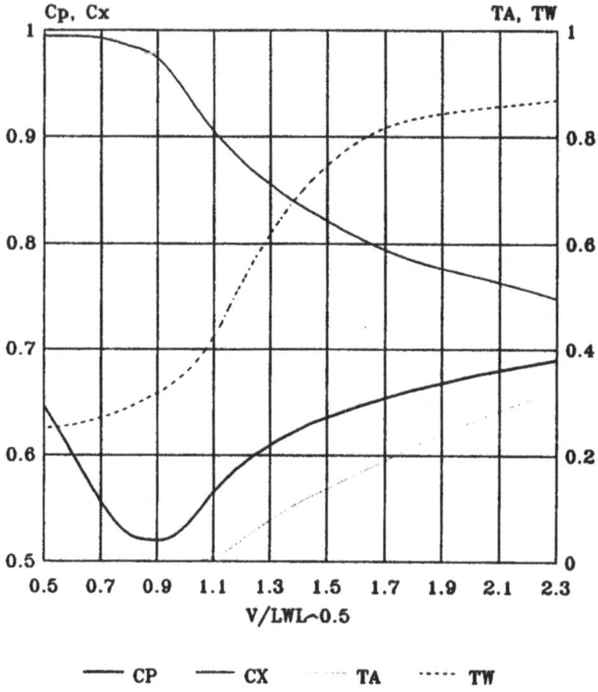

Fig. 13.2 Suggested Cp, Cx, TA, and TW

Fung has included initial assessments of appendage drag and powering prediction. The present authors have taken a more direct approach to estimating these as the possible choice of propulsion (traditional propellers; tunnel, surface, and podded propellers; and water jets) needs individual attention once the main hull parameters have been selected.

13.2.2 Mercier and Savitsky

This analysis is presented in Sect. 2.3.1. The total resistance was analyzed by regression normalized to a standard hull size at 100,000 lbs. displacement. Prediction of resistance for craft at different displacements and so hull dimensions is determined by scaling and adjusting the frictional resistance to account for the change according to the relation given in Eq. 2.8.

The model was built initially using a total of 27 terms including linear, cross coupling, and quadratic terms with associated coefficients that were determined for the speed range of Fn_∇ from 1.0 to 2.0 (approx. Fn_L 0.25–0.9). This was reduced to 14 terms in the final model after evaluation of impact on results.

The key hull form parameters used were L, Bx, Ax, A_T, I_E, displacement Δ, and displaced volume ∇.

While forming a part of Fung's data set, this work was aimed at craft that were intended to operate in the higher speed planing region and to provide a better estimate of drag at sub-planing speeds. It nevertheless does include data from round bilge craft and so may be considered useful for both hard chine and short chine craft for estimating drag at low speeds.

The hull form parameters applicable space is given in [4] as a number of plots combining the model test data. A simpler way to approach this is to consider the model test series data included that have shape closest to the new design and stay within these parameters (see Table 2.1 in Chap. 2 referring to NPL, Nordström, DeGroot, SSPA, and series 62, 63, and 64).

Mercier and Savitsky's model determine R_T/Δ and so the friction resistance component will be extracted by estimating the wetted surface area using Eq. 2.9 in Chap. 2 and using C_f based on the proposed craft dimensions and related Reynolds number. It has to be remembered that the regression model data is for a basic hull and so appendage drag has to be calculated separately before considering powering.

A model based on this analysis is presented as "Mercier and Savitsky 500t correlation" in the electronic materials to this appendix.

13.2.3 MARIC

MARIC developed a regression model based on their model test results and NPL series, De Groot series, Nordstrom series, and Series 63. This was expected to give more reliable results for semi planing round bilge craft than the more general Mercier and Savitsky model. The MARIC model is detailed in Sect. 2.3.2. MARIC chose to determine residual drag and estimated wetted area from which friction resistance would be determined separately and added to give R_T.

An equation for the residual drag coefficient is given in the regression formula as a function of volumetric displacement/length coefficient $C_\nabla = \nabla /(L/10)^3$, the prismatic coefficient C_p, the relative transom area A_T/A_X, the longitudinal location of C_B, the position of the center of buoyancy aft of amidships as %L, X_b, the half entrance angle of the bow water line i_S, and the length Froude number Fn_L (Fn). MARIC presents the five coefficients in plots against Fn_L in Figs. 2.29–2.33.

The wetted area estimate is given in Eqs. 2.12 and 2.13.

The model is proposed to be useful for the following range of parameters:

$C_\nabla = 1.5–5.7$ for $Fn = 0.9 - 1.0$
$C_\nabla = 1.5–11.0$ for $Fn = 0.4 - 0.9$
$A_T/A_X = 0.0–0.74$
$C_P = 0.573–0.764$
$X_b = 0.0–6.4\%$ aft of amidships
$i_S = 4–12°$ ($=I_E$)

It is useful to compare these parameters with the model test summary in Chap. 2, Table 2.1.

A spreadsheet model for this series was prepared by the present authors and compared with the NPL model discussed below. The additional hull models included in the MARIC data set increased the C_R and R_T/Δ slightly compared with the NPL specific model and the more general Mercier and Savitsky model. It was concluded that it is recommended to simply refer to those models as a start point.

13.2.4 Shanghai University

Shanghai University developed their regression model from their own model test series of 38 craft, as described in Sect. 2.3.3.

The regression equation is for residual drag as for MARIC and is based on three key hull parameters, slenderness $L/\nabla^{1/3}$, longitudinal position of CB (X_b), and block coefficient C_B. There are eight terms including linear, quadratic, and cross coupling items.

Rather than graphic presentation for the coefficients, Shanghai University has prepared the data as two tables, for calculation using either Fn_L or using Fn_∇.

A spreadsheet model for this series is presented as "Shanghai JT University 500t correlation" in the electronic materials to this appendix. This includes the calculation of wetted area according to the equation in Sect. 2.3.3 and friction drag.

The parameter range appropriate for this model is as follows:

Slenderness Coefficient $L/\nabla^{1/3} = 5.16–8.15$
Longitudinal position of CB: $X_b/L = 1.25–6.4\%$ aft of midship section
Block coefficient $C_B = 0.39–0.51$
Beam draft ratio $B/T = 3.00–4.26$

13.2.5 VTT

VTT carried out regression analysis of the data from NPL, SSPA, and the VTT series yielding 65 sets of model data and developed equations to extend from the work by Mercier and Savitsky for prediction purposes.

Data correlation testing was carried out resulting in conclusions that the original Mercier and Savitsky model was reliable for the data set used below $Fn_\nabla = 1.8$ while a new equation was developed for the speed range up to $Fn_\nabla = 3.2$. The new equations are for total resistance, consistent with the Mercier and Savitsky model.

VTT have developed "*speed-independent*" models for both the round bilge and separately the hard chine model test data that they gathered that are applicable in the speed range $Fn_\nabla = 1.8 - 3.2$ so extending through the transition to planing. The Savitsky planing model is recommended for the planing regime above $Fn_\nabla = 2.5 - 3.2$ depending on hull form.

The limits of applicability of parameters for the round bilge VTT model are

$L/\nabla^{1/3} = 4.47–8.13$
$L/B = 3.33–8.21$
$B^3/\nabla = 0.68–7.76$
$B/T = 1.72–10.24$

$A_T/A_X = 0.16–0.82$
$C_X = 0.567–0.888$

The limits of applicability of parameters for the hard chine VTT model are

$L/\nabla^{1/3} = 4.49–6.81$
$L/B = 2.73–5.43$
$B/T = 3.75–7.54$
$A_T/A_X = 0.43–0.995$

The same approach as Mercier and Savitsky is used, thus basing on a 100,000lbs or 45.36 tonnes hull.

The equation for round bilge craft is as follows:

$$\frac{R_T}{\Delta_{100000}} = A_0 + \sum_{i=1}^{6} A_i.P_i + \left(\sum_{i=7}^{10} A_i.P_i\right) Fn_\nabla + \left(\sum_{i=11}^{13} A_i.P_i\right) Fn_\nabla^2 + \left(\sum_{i=14}^{16} A_i.P_i\right) Fn_\nabla^3 + \left(\sum_{i=17}^{24} A_i.P_i\right) Fn_\nabla^4$$

The tables for parameters P_i and related coefficients A_i are listed in spreadsheet model "VTT Radojcic 500t correlation" in the electronic materials to this appendix. The equation above is valid for $Fn_\nabla = 1.8$ up to 3.2. It is intended that the Mercier and Savitsky model is used in the range up to 2.0, and in the region $Fn_\nabla = 1.5$ up to 2 that a weighted average value be prepared to graduate between the two models.

The equation for hard chine craft is as follows:

$$\frac{R_T}{\Delta_{100000}} = A_0 + A_1.P_1 + \left(\sum_{i=2}^{3} A_i.P_i\right) Fn_\nabla + \left(\sum_{i=4}^{6} A_i.P_i\right) Fn_\nabla^2$$

The tables for parameters P_i and related coefficients A_i are listed in spreadsheet model "VTT Radojcic 500t correlation" in the electronic materials to this appendix. It should be noted that the hard chine model is assembled from just 13 vessels and so is specific to such craft in the interim regime up to planing in the range $Fn_\nabla = 1.8$ up to 3.3. A weighted average is proposed prepared for the speed range $Fn_\nabla = 1.5$ up to 2 as for the round bilge craft.

Since the model is based on a craft with displacement of 45.36 tonnes, a correction of frictional resistance is needed for varying craft size. This is given as:

$$\left(\frac{R_T}{\Delta}\right)_{corr} = \left(\frac{R_T}{\Delta}\right)_{100,000} + \{(C_F + C_A) - C_{F100,000}\}0.5\frac{S}{\nabla^{1/3}} Fn_\nabla^2$$

Where:

$(R_T/\Delta)_{corr}$	corrected total resistance/displacement ratio
$(R_T/\Delta)_{100,000}$	total resistance/displacement ratio for 100,000 lbs. craft
$C_{F100,000}$	frictional resistance coefficient for the 100,000 lbs. craft
C_F	frictional resistance coefficient for the new craft
C_A	incremental resistance coefficient for ship/model correlation

The wetted area S, volumetric displacement ∇, and Fn_∇ all relate to the new craft. The total resistance can then be determined for the new craft at the actual displacement Δ.

Both of these VTT models are also described in detail, and a spreadsheet model is presented in Section 8.2.2 of [1] by Radojcic et al. The present authors have developed a model also from the VTT data, it is essentially the same as Radojcic's model.

The following models are presented based on data published by Professor Radojcic [1]—Sklad and NPL, and P K Sahoo et al [5]—AMECRC and S-NPL.

13.2.6 Sklad

A regression model was developed by Radojcic in 1997 [6] and further detailed in [1] based on his equations for C_R, dynamic wetted area coefficient CS, and dynamic slenderness (M).

The approach is a stepwise regression procedure for prediction of residual drag coefficient C_R based on the four key parameters of L/B, B/T, $L/\nabla^{1/3}$, and C_B at a specific speed adding variables and significance testing at each step until all significant independent variables, including cross coupling, are inside the model. Additionally, if variables lose significance after a new variable is added, it may be removed and the model retested. The equation is built, and coefficients determined at each speed in the range. The final *speed-dependent* regression equation and coefficients are then tested against model test results. The expression for the SKLAD test series with a total of 48 potential terms of which 21 are used is as follows:

$$C_R = a_0 + a_1 x_2{}^2 x_3 + \cdots a_{21} x_1{}^{1/3} x_2{}^{1/3} x_3^2$$

Radojcic and colleagues have developed a dynamic wetted area regression based on their model tests, so that first a dynamic slenderness (M) is calculated and from this the appropriate wetted area at the specific speed. This is then used to calculate the friction drag coefficient and subsequently the total drag.

The speed range is between $Fn_\nabla = 1.0$ and 3.0 in steps of 0.25. The complete equations for C_R, C_S, and (M) and coefficient tables are contained in spreadsheet model "SKLAD Radojcic 500t correlation" in the electronic materials to this appendix.

For reference the key parameter ranges for SKLAD are:

$L/\nabla^{1/3}$	4.5–8.5
L/B	4.0–8.0
B/T	3.0–5.0
C_B	0.35–0.55

13.2.7 HSDHF/AMECRC

Sahoo et al. have prepared a model for the AMECRC series that succeeded the HSDHF program. The same parameters are used for x_1, x_2, x_3, and x_4 as for the SKLAD. The regression equation is of similar form, while having 48 potential terms, of which 30 are used

$$R_R/\Delta = a_1 + a_1x_1 + a_2x_2 + a_3x_3 + a_4x_1^2 + a_5x_2^2 \cdots a_{48}x_3^2x_4^2$$

The complete equation and coefficient tables are contained in spreadsheet model "AMECRC Radojcic 500t correlation" in the electronic materials to this appendix.

Associated with this is a model for wetted area developed by P Bogovic at the Australian Maritime College in 1998 giving the wetted area coefficient $C_S = S/\nabla^{2/3}$. The equation for AMECRC series is:

$$C_S = C_0 + C_1\left(\frac{L}{B}\right)^{2/3}(C_B)^{-2/3} + C_2\left(\frac{B}{T}\right)(C_B)^{-2/3} + C_3\left(\frac{L}{B}\right)^{2/3}\left(\frac{B}{T}\right)(C_B)^{-1} + C_4\left(\frac{L}{B}\right)^{-4/3}(C_B)^{-1} + C_5\left(\frac{B}{T}\right)^{1/3}(C_B)^{-1/3}$$

The coefficients and wetted surface calculation are included in the spreadsheet.

For reference the key parameter ranges for AMECRC are:

$L/\nabla^{1/3}$	4.3–8.7
L/B	4.0–8.0
B/T	2.5–4.0
C_B	0.4–0.5

13.2.8 NPL

The model developed for the NPL series in Sect. 2.2.1 is from work prepared by Radojcic [6] using the same approach. The same parameters x_1, x_2, and x_3 as used for SKLAD were employed, and an equation with 27 terms, of which 14 are used in the final model, taking the same form:

$$R_R/\Delta = a_1 + a_1x_1 + a_2x_2 + a_3x_3 + a_4x_1^2 + a_5x_2^2 \ldots a_{27}x_2^2x_3^2$$

In this case, the analysis is between $Fn_\nabla = 0.8$ and 3.0 in steps of 0.2. The complete equation and coefficient tables are contained in spreadsheet model "NPL

Radojcic 500t correlation" in the electronic materials to this appendix. A wetted surface calculation has been prepared by Bogovic with the following form:

$$C_S = C_0 + C_1\left(\frac{L}{B}\right)^{2/3}\left(\frac{B}{T}\right) + C_2\left(\frac{L}{B}\right)^{-2}\left(\frac{B}{T}\right) + C_3\left(\frac{B}{T}\right)^{-1}$$
$$+ C_4\left(\frac{L}{B}\right)^{-2}\left(\frac{B}{T}\right)^{2/3} + C_5\left(\frac{B}{T}\right)^{2/3} + C_6\left(\frac{L}{B}\right)^{2}\left(\frac{B}{T}\right)$$
$$+ C_7\left(\frac{L}{B}\right)^{2/3}\left(\frac{B}{T}\right)^{2/3}$$

The coefficients and wetted surface calculation are included in the spreadsheet.

Radojcic in [1] has prepared both a *speed-dependent model* and a *speed-independent* model for R_T/Δ, and a model for wetted surface S. The dynamic trim τ is also provided as output to these models.

For reference the key parameter ranges for NPL are:

$L/\nabla^{1/3}$	4.47–8.3
L/B	3.33–7.5
B/T	1.5–2,5
C_B	0.397

13.2.9 Southampton University—High L/B NPL Form

Building on their experience with the Radojcic NPL model, Sahoo et al have completed an equivalent model for the extended NPL geometry model tests carried out by Southampton University [7]. This test series was concerned with the hulls of slender catamarans and included a test series for the individual demi-hulls. This effectively extended the parameter space from the original series at NPL in Feltham as shown in Chap. 2, Fig. 2.4.

The parameters applied were as follows:

$$x_1 = \frac{\left(\frac{L}{B} - \frac{\left(\frac{L}{B}\right)_{\min} + \left(\frac{L}{B}\right)_{\max}}{2}\right)}{\frac{\left(\frac{L}{B}\right)_{\max} - \left(\frac{L}{B}\right)_{\min}}{2}}, x_2 = \frac{\left(\frac{B}{T} - \frac{\left(\frac{B}{T}\right)_{\min} + \left(\frac{B}{T}\right)_{\max}}{2}\right)}{\frac{\left(\frac{B}{T}\right)_{\max} - \left(\frac{B}{T}\right)_{\min}}{2}},$$

$$x_3 = \frac{\left(\frac{L}{\nabla^{1/3}} - \frac{\left(\frac{L}{\nabla^{1/3}}\right)_{\min} + \left(\frac{L}{\nabla^{1/3}}\right)_{\max}}{2}\right)}{\frac{\left(\frac{L}{\nabla^{1/3}}\right)_{\max} - \left(\frac{L}{\nabla^{1/3}}\right)_{\min}}{2}}$$

The regression equation of the Sahoo S-NPL model is of the form with a total of 27 possible terms of which 19 are used in the final model:

$$R_R/\Delta = a_1 + a_1x_1 + a_2x_2 + a_3x_3 + a_4x_1^2 + a_5x_2^2 \ldots a_{27}{x_2}^2{x_3}^2$$

The complete equation and coefficient tables are contained in spreadsheet model “S-NPL model” in the electronic materials to this appendix. The analysis for S-NPL has been carried out between $Fn_\nabla = 1.0$ and 2.5 in steps of 0.25.

The same wetted surface equation developed for NPL series is applicable to the S-NPL series. This is included in the spreadsheet “S_NPL 500t simulation.”

For reference the key parameter ranges for S-NPL are:

$L/\nabla^{1/3}$	6.3–9.5
L/B	7.0–15.1
B/T	1.5–2.5
C_B	0.397

13.2.10 US Naval Academy Series (Round Bilge)

The US Naval Academy Hydrodynamic Laboratory prepared a relatively simple regression model applicable to craft with geometry similar to their hulls. The model is described in Sect. 4.4.5. The Naval Academy tests were with 3 round bilge and 3 hard chine shapes partly aimed at looking at the difference between their drag performance at relatively low speeds. As discussed in Chaps. 4 and 6, a hard chine craft tends to have higher drag than a round bilge craft at low speed. The benefits of the hard chine form is higher stability particularly at high speed.

The regression equation developed for residual drag coefficient $C_R = R_R/\Delta$ is as follows:

$$C_R \times 10^3 = \mathbf{A} + \mathbf{B}\,(\text{Lpp}/\mathbf{B}) + \mathbf{C}\,(\Delta/\text{Lpp}) + \mathbf{D}\,(100.\text{LCG}/\text{Lpp})$$

The table of coefficients A, B, C, and D is given in Table 4.4 for each of the round bilge and hard chine hull forms.

The wetted surface area for the models is given in Table 4.4. On the basis that resistance estimates will be for craft within the parameters for this model series an estimate for wetted surface and friction drag is included in the NAHL Model spreadsheet for both hull types.

For reference the key parameter ranges for NAHL are:

$L/\nabla_{1/3}$	6.3–9.5
L/B	3.9–5.2
B/T	3.0–5.0
C_B	0.432–0.452
X_b	8% aft of amidships approx.
LCG	5–15% aft of amidships

The NAHL model tests were all in the displacement speed range, and so a spreadsheet has not been included here. Preparation is possible from the data presented in Chap. 4.

13.3 Hard Chine Planing Vessels

13.3.1 Mercier and Savitsky

The Mercier and Savitsky model is discussed above in Sect. 13.2.2 and a spreadsheet model is presented in the electronic materials to this appendix. Its application for hard chine planing craft in the displacement and semi planing speed regime may give best results when the new design parameters are within those of the Series 62 or Series 65 (see Chap. 4 Table 4.1 and model test summaries in Sect. 4.4).

13.3.2 Series 62/Series 65

Radojcic and his colleagues have carried out a regression analysis for these two series that is presented in [1] to prepare a *speed-dependent* model following the approach introduced in Sect. 13.2.6. The speed range covered is $Fn_\nabla = 1.0$ up to 3.5 in steps of 0.25.

The model is fully described in Section 8.3.1 of [1] and an electronic spreadsheet model included in the electronic materials to [1]. Output from the model is $(RT/\Delta)_{100,\ 000}$, L/L_{pp}, Trim τ, and wetted area coefficient $S/\nabla^{2/3}$.

This model therefore covers craft with deadrise between 12.5° and 25°, a considerable L/B range and a bow shape that is fuller than today's racing craft. Readers are referred to [1] for the model and further discussion.

A further two models for the Series 62 and Delft University series with deadrise 12.5°, 25°, and 30° have been prepared by Radojcic. The first is a "'simple" regression model and the second a more complex analytical neural network or ANN model. This can be found in Section 8.3.4 of [1].

13.3.3 Delft DSDS

Delft University has prepared model test data and materials for generating prediction models in [3]. Once an account to the repository has been granted by application to the university, this data and models can be accessed. Within the parameter space covered by the DSDS series, an estimate can be made for resistance, trim, and CG heave.

The DSDS series was developed from the US Series 62 described in Sect. 4.4.2 and extended to a range of deadrise, and twist/rocker variations, described in [3] and Sect. 4.4.10.

13.3.4 US Naval Academy Series (Hard Chine)

Reference is made to Sect. 13.2.10 above and Sect. 4.4.5

13.3.5 VTT

The model for hard chine craft developed from their model tests is discussed in Sect. 13.2.5. The tables for parameters P_i and related coefficients A_i are listed in spreadsheet model "VTT Radojcic 500t correlation" in the electronic materials to this appendix.

13.3.6 USCG and TUN

Radojcic and his colleagues have carried out an artificial neural network (ANN) analysis for these two series combined that is presented in [1]. Speed-dependent models following the approach introduced in Sect. 13.2.6 have been developed. The speed range covered is $Fn_\nabla = 0.6$ up to 3.5 in steps of 0.25.

The model is fully described in Section 8.3.2 of [1] and spreadsheet models are included in the electronic materials to [1] for both a simple and a complex model. Output from the model is $(R_T/\Delta)_{100,\ 000}$, L_K/L, and wetted area coefficient $S/\nabla^{2/3}$.

These models cover a hull shape with essentially constant deadrise and beam aft of amidships and so relate to many smaller planing craft for fishing, coastal interdiction, and general recreation. Radojcic and his colleagues propose that the simple model is useful for initial study, and the more complex model for design analysis perhaps for comparison with CFD models or model tests. The test programs have been described in Sects. 4.4.6 and 4.4.7.

13.3.7 Series 50

M Morabito [8] carried out a reanalysis of the US Series 50 model test series to provide data in the modern format, see Sect. 4.4.1.

Subsequently Radojcic and his colleagues have completed ANN analysis using this updated material to prepare a prediction model using 65 terms in the main expression. The model is given in Section 8.3.3 of [1]. It provides R_R/Δ, $L_m/L_p = ((L_k - L_c)/2)/L_p$ where L_p is projected chine length at SWL; L_k and L_c are wetted lengths at keel and chine; trim angle from baseline $\tau_{BL} = \tau_0 + \tau$; and over the speed range between $Fn_\nabla = 1.0$ to 5.5.

The Series 50 form is rather full in the bow and the bottom has significant concave curvature along the length, as well as twist. One might therefore expect differences to occur in drag and trim compared with a prediction from Savitsky in the planing region above about $Fn_\nabla = 2.5$ and so it is useful to have this model to compare.

While the overall hull form has been considered "outdated" by some in recent times, this shape is still used for high-speed daughter craft in Europe and the Mediterranean.

13.3.8 Naples University Warped Hard Chine Series

Fabio De Luca and C Pensa published results from a planing model test series carried out at the University of Naples "Federico II," Italy, in 2017. A full description of the tests is given in Sect. 4.4.9. Following the tests, they prepared a 20-term polynomial equation from the test data, documented in [9]. This was used to predict residual resistance, running wetted surface, and waterline length as a function of $L_P/\nabla^{1/3}$ and Fn_L over the speed range $Fn_\nabla = 1.1$ to 4.2 ($Fn_L = 0.5 - 1.6$). A polynomial equation was developed that related to each of the five models tested, which was therefore useful for looking at parameter adjustments to an individual hull shape. This is a little different to the remainder of the models discussed in this appendix.

Radojcic has subsequently taken the model test data and performed analysis to generate an ANN-based model that generates $(R_T/\Delta)_{100,\ 000}$, L_{WL}/L_p, trim τ, and wetted area coefficient $S/\nabla^{2/3}$ in [1]. This can be used for example to compare with results obtained from the models for USCG/TUN and VTT to consider the effect of geometry changes in the speed range up into planing. Above this, the Savitsky model is a starting point followed by review of the potential impact of parameters as discussed in Sect. 2.4 for performance below planing as these issues will be common.

13.4 General Discussion

We have presented here an introduction to nine round bilge and eight hard chine models for drag estimation. A selection of the models are made available directly as spreadsheets to the reader, and reference is given to the valuable source of rather more sophisticated models provided with [1] with thanks to Professor Radojcic and colleagues.

Between the physical descriptions and body plans in Chaps. 2 and 4 together with the model summaries here, readers should be able to select one or more models to use for initial estimates for their designs together with investigation of consequences of parameter changes.

It does need to be reiterated that when used for planing craft, these statistically based models need to be faired in to drag and trim estimates for the planing regime made using Savitsky or similar models.

At low speeds in the range $Fn_\nabla = 0.5 - 1.5$, wave generation can cause rapidly changing forces on a hull. Equally this is the same speed range that vortices are generated behind the transom, and it is gradually "dried" above around $Fn_\nabla = 1.2$. This is the reason most of the testing and these models start around $Fn_\nabla = 1.0$. Below this, the overall drag value is still low, the key interest being the humps that occur up to $Fn_\nabla = 2.5$ before full planing is reached.

In this middle range, the flow around the hull is sensitive to changes in hull geometry. At lower speed, changes to the surface curvature and position of center of buoyancy as well as B/T and fineness will be important. As speed increases toward the planing regime, geometry changes in the afterbody can cause turbulent flow and increasing boundary layer which is detrimental to performance. This is the drive for investigations such as that in [9].

The choice of propulsor has a significant impact on selection of the hull shape aft of amidships, whether for a hard chine or round bilge craft. Fortunately, now the designer has options available to him including water jets and podded propulsors that are less restrictive on afterbody geometry than a traditional inclined shaft propeller drive. Initial definition for a high-speed semi planing or planing craft nevertheless does need powering options to be an integral part of the early studies.

Given this approach, the models here together with the discussion of parameters in Chaps. 2 and 4 may be useful to give some confidence in initial specification of the hull geometry, and perhaps using the more complex models provided in [1], a designer can go a little farther around the design spiral before committing to model tests, or detailed CFD modeling.

The reader should note that for convenience, no roughness allowance has been applied to the friction coefficient C_f and no correlation allowance for correlation between model and full-scale data. This is consistent with the examples given in [1]. An exception to this is that the S C Fung modeling does include roughness allowance.

13.5 Spreadsheet Models Provided

Spreadsheet models referred to in Sect. 13.2.1 to 1.9 have been prepared and placed in the electronic materials on the Springer site for this book. In addition, a summary sheet has been prepared with plots of C_R and R_T/Δ for the model data tested. The intent has been to see how the different regression models predict resistance for a hull with similar displacement and main geometry, one that also can be checked against a prediction by Prof Radojcic in [1].

Main conclusions from this work are that for a round bilge geometry, the lines of Sklad or NPL provide a good starting point where the configuration is toward the center of the parameter range as shown in Figs. 13.3 and 13.4. The AMECRC hull was aimed at improving response in a seaway, so while the resistance is higher, one needs to look at this aspect in relation to that geometry. This is discussed in Chaps. 2–6.

Models have also been built for the S C Fung regression analysis [2] and using data provided by Professor Sahoo [5] for the S-NPL series carried out at Southampton University. The Fung model is targeted at frigates and destroyers with a rather

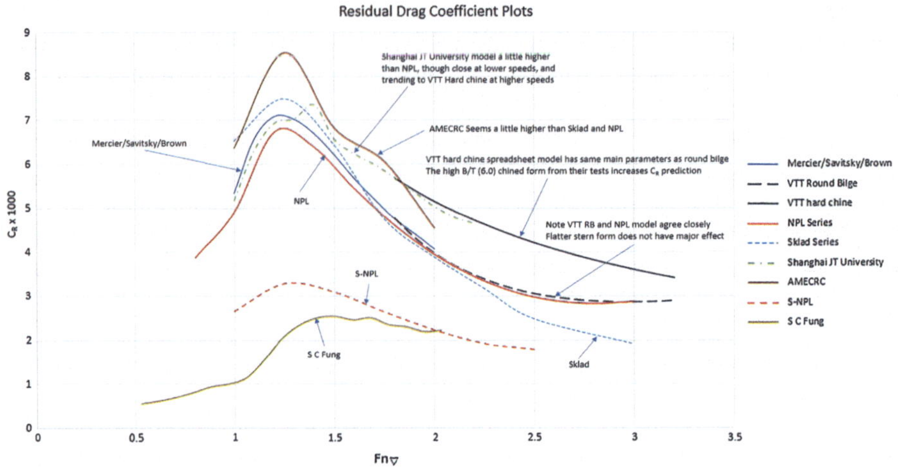

Fig. 13.3 C_R Comparative plot

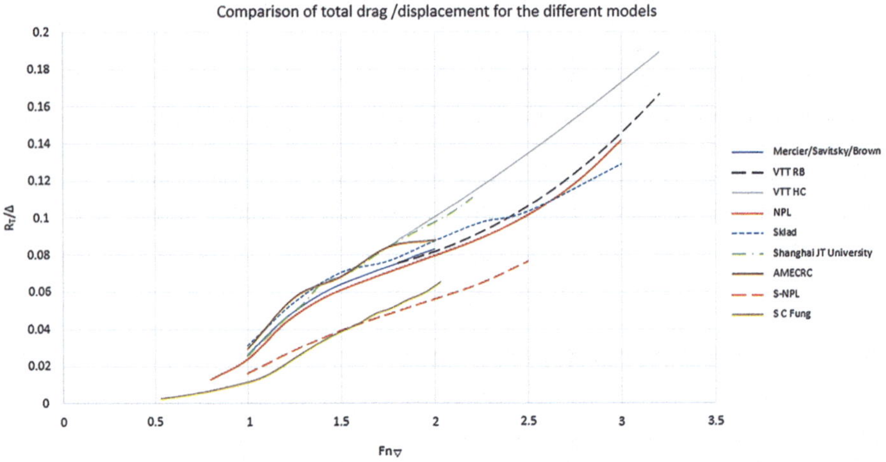

Fig. 13.4 RT/Δ Comparative plot

high B/T, while the S-NPL data is derived from high L/B hulls suitable for catamarans or slender patrol vessels. These have much lower B/T and so these two models cannot be compared directly. The Fung model has been tested against data provided in the appendix to [2], and so there is confidence that it works as intended. The S-NPL model seems to correlate with results presented in [7] and so may be used with care for initial assessment of slender high L/B patrol craft.

The Mercier/Savitsky/Brown regression model applies to both round bilge and hard chine craft, though much of the data used is from round bilge forms. Prof Radojcic has prepared a number of regression models from the data for the hard chine US Series 50, 62, and 65 as well as the TUNS and USCG data sets [1]. It is left to the reader to build their own models from this reference work for drag prediction below planing speed. A start point can always be set by using the Mercier/Savitsky/Brown model up to $Fn_\nabla = 1.8$ and the VTT RB model up to around $Fn_\nabla = 2.5$, and run the Savitsky planing model (see electronic materials for Chap. 4) above this speed. The VTT HC model seems to be a special case due to its low L/B in comparison to the round bilge hulls. This might be investigated against predictions from hard chine models derived from [1].

References

1. D. Radojcic, M. Kalajdzic, A. Simic, *Power Prediction Modelling of Conventional High-Speed Craft* (Springer Nature, 2019) ISBN 978-3-030-30606-0
2. S.C. Fung, Resistance and powering prediction for transom stern forms during early-stage ship design. SNAME Trans. **99**, 29–84 (1991) ISBN 0-939773-07-04
3. J.A. Keuning, L. Hillege, The results of the Delft Systematic Deadrise Series. FAST 2017 Transactions Session B1, pp 97–106, Data is available at www.dsds.tudelft.nl once you set up an account for access, the paper is via www.fast2017.com
4. D. Savitsky, P. Ward Brown, Procedures for the hydrodynamic evaluation of planing hulls in smooth and rough water. SNAME Mar. Technol. **13**(4), 381–400 (1976)
5. P.K. Sahoo, H. Peng, J. Won, D. Sangarasigamany, Re-evaluation of resistance prediction for high-speed round bilge hull forms. 11th international conference on fast sea transportation, FAST 2011, Honolulu, Hawaii, USA September 2011
6. D. Radojcic, T. Rodic, N. Kostic, Resistance and trim predictions for the NPL high speed round bilge displacement hull series. RINA symposium, power, performance and operability of small craft, Southampton, UK, September 15–16, 1997
7. A.F. Molland, J.F. Wellicome, P.R. Couser, Resistance experiments on a systematic series of high-speed displacement catamaran forms: Variation of length-Displacement ratio and breadth-draught ratio. Ship Science Report 71, University of Southampton, 1994
8. M. Morabito, *Re-Analysis of Series 50 Tests of V-Bottom Motor Boats*, vol 121 (SNAME, New York., Transactions, 2015), pp. 190–223
9. F. De Luca, C. Pensa, The Naples warped hard chine hulls systematic series. Ocean Eng. **139**, 205–236 (2017) An extensive test series at University of Naples, Italy, including effect of transom interceptors for trim control. Open access article published by Elsevier under creative commons. It can be found at: www.dx.doi.org/10.1016/j.oceaneng.2017.04.038

Chapter 14
Appendix 2: Resources and General Reference Material

Our intent with this appendix is to provide readers a direct reference to internet locations where a reader may find additional information, together with a listing of key reference books and reports relevant to high-speed monohull vessels and hydrofoils. There are two parts to this appendix—the internet locations of organizations and a listing of links and reference to conferences, textbooks, reports, and academic papers.

The material presented here should be viewed as a sample that is relevant to high-speed craft analysis and design. We include material related to regulations, power and propulsion systems, and outfitting, in addition to the hydrodynamics focused on in this book, as a designer needs data from all these areas to assemble a successful design.

14.1 Internet Resources

Listed below are internet locations for organizations gathered in subject areas for ease of initial reference.

Please note that the listing is not exhaustive, and we tend to focus on smaller vessel applications. Many of the commercial organizations supply equipment or services to the marine industry for fast vessels and these are similarly applicable generally. The idea is to give you a good start. We have also given additional direction to pages that are directly useful as many of the sites for large companies or groups have different ways of presenting their information and finding the area for information relevant to our subject of fast multihull vessels is not obvious.

It is important to note that companies do change, get absorbed into larger organizations, and some more specialist organizations often pass further on to become part of different larger organizations. Our advice is that if the link doesn't work for you, try searching on the main name and see if parts of the address have been changed, perhaps ".co.uk" to ".com" for example. For much basic information,

Z.-S. Dong et al., *High-Speed Monohull and Hydrofoil Craft*,
https://doi.org/10.1007/978-3-031-62760-6_14

it is helpful also to look at what exists in Wikipedia. Please note that Wiki also exists in various local forms, so Wikipedia in German is not the same as the UK or US version. Please do be careful to check that a site is genuine if searching on keywords, as there are many fakes out there. It is important to have virus protection working for you. Happy hunting!

Technical Societies
The Societies that best form a start point are RINA and SNAME as these publish research papers and technical journals. There are many other societies and special interest clubs that can provide useful information if a key word search is used. University electronic library sites are now becoming more common, with access once an account is registered. Sites such as www.scribd.com also have some useful papers available.

www.RINA.org.uk	Royal Institution of Naval Architects (London, UK)
www.SNAME.org	Society of Naval Architects and Marine Engineers (NY, USA)

Universities, Marine Institutes, and Industry Organizations

https://www.tudelft.nl/en/library/	Location for Delft University of Technology library
www.NTNU.no	Site for Trondheim University and Test Basin
www.engineering.unsw.edu.au	University of New South Wales at /mechanical-engineering/Naval Architecture. Prof Laurie Doctors and his book on hydrodynamics of High-Performance Marine Vessels (see books below). Degree now suspended.
www.utas.edu.au/	University of Tasmania, Marine Technology and Naval Architecture, with extensive list of research papers available on request at www.eprints.utas.edu.au.
www.ucl.ac.uk	search under /mecheng, /our-courses, /postgraduate/naval-architecture for courses, research and contacts
www.southampton.ac.uk/	search under engineering maritime engineering, ship science, naval architecture for Naval Architecture course and contacts.
www.wumtia.soton.ac.uk	/about-us/published-papers/high-speed-craft-code-review for High-Speed Craft

	reviews by Southampton University including wind heeling moment research
www.theses.gla.ac.uk	University of Glasgow library and repository for theses, look under section v for Naval and Naval Architecture
https://www.stevens.edu/davidson-laboratory	Located at Stevens Institute of Technology,1 Castle Point Terrace, Hoboken, NJ, USA. Site of many of Savitsky's test programs
https://www.navsea.navy.mil/Home/Warfare-Centers/NSWC-Carderock/	Location of the David W Taylor Naval Ship Research and Development Center, earlier the David Taylor Ship Model Basin
www.wegemt.com	Site for European Universities summer school papers. Includes HPMV
www.MARIN.nl	Marine Research Institution Netherlands, at Wageningen
www.SSPA.se	Swedish Ship Model Basin, Goteborg, Sweden
www.HSVA.de	Hamburg Ship Model Basin, Hamburg, Germany
www.en.wikipedia.org/wiki/David_Taylor_Model_Basin	for DTMB basic info
www.dome.mit.edu/handle/1721.3/48001	for DTMB reports
https://interferry.com/	Interferry industry organization for all types of ferries

Research and Development

www.Cordis.europa.eu	EU R&D sites
https://www.epo.org/index.html	European Patent Office
	Patents search at

https://www.epo.org/searching-for-patents/legal/register.html#tab-1
https://worldwide.espacenet.com/advancedSearch?locale=en_EP

	Google for patents
https://patents.google.com/patent/US5570650	
http://krylov-centre.ru/en/	Krylov State Research Centre, St Petersburg
http://offshore.cssc.net.cn/en/index.php	MARIC
http://www.cssrc.com/	CSSRC

https://en.wikipedia.org/wiki/PLA_Naval_University_of_Engineering	China Naval Engineering University (CNEU)
http://www.sssri-marin-jv.com/	SSSRI
http://www.sssri-marin-jv.com/about-us/sssri/	
https://en.wikipedia.org/wiki/Rostislav_Alexeyev	
http://www.ckbspk.ru/en/	Fast Ships Bureau, Alexeev Design Institute, Nizhny Novgorod
http://krylov-centre.ru/en/	Krylov Ship Research Institute (KSRI)
http://www.nas.gov.ua/EN/STRUCTURE/Pages/default.aspx	Ukraine Scientific Academy Hydrodynamic Research Institute (UHRI)
http://www.tsagi.com/	CAHI Central Aerohydrodynamic research institute (N Zhukovsky Institute, Zhukovsky, Moscow Region, 140180, Russian Federation)
https://www.en.mas.bg.ac.rs/confluence-belgrade-golden-team-of-naval-architecture-students/	
https://novimar.eu/partners/university-of-belgrade/	University of Belgrade, Mechanical Engineering Faculty, Naval Architecture

Designers

www.invincibleboats.com	Ventilated stepped vee hulls as from "mpyd"
www.aeromarineresearch.com	tunnel boats—HS cats, design books, and software
www.revenger.co.uk	Stepped hull RIB's
www.lornecampbelldesign.com	Planing and racing craft—ref presentation on steps
www.cdicorp.com	/engineering/government-services/naval-architecture-advanced-ship-design/ including Band Lavis group working on ACV and catamarans, etc.

Boat and Shipbuilders

There are thousands of fast boat and shipbuilders on a worldwide basis. A student may start by searching for yards or companies in his own country, or countries of interest. A publication such as Janes' High-Speed Marine Transportation may also be a good start (published up to 2012, search on Amazon, or book sellers sites)

http://www.tuco.dk/index.php/en/cases/78-air-supported-vessel	Tuco Marine, Denmark
https://www.beneteau.com/en/innovate	Beneteau, France
www.damen.com	Damen, Holland (and shipyards globally)
https://en.wikipedia.org/wiki/List_of_ships_built_by_Fincantieri	Italy, Fincantieri MDV 3000, Check MDV 3000 Jupiter class
http://www.zdship.ru/en/	Zelenodolsk Shipyard, Gorky, Russia Ref A145 monohull and various hydrofoils before that

Rules and Regulations

www.lr.org	Lloyds Register Classification Society
www.dnvgl.com	DNV Classification Society—High Speed Service Craft Rules
www.eagle.org	ABS rules at /rules-and-resources/rules-and-guides.html
www.turkloydu.org/en-us	Lloyds Turku Classification Society Home page, go to /publications/turk-loydu-rules.aspx# for regulations, part C High Speed Crafts
www.krs.co.kr	Korean Register of Shipping Classification Society technical rules listing—GB11 for HSC, GC06 and 07 for recreation, and WIG. Rules are at: http://krsusa.cloudapp.net/Files/KRRules/KRRules2016/KRRulesE.html
www.ccs.org.cn	China Classification Society, go to /ccswzen/font/fontAction!moudleIndex.do for high speed craft rules
www.gov.uk/	UK Maritime and Coastguard Agency Go to: guidance/high-speed-craft-construction-and-maintenance-standards#construction-standards-for-high-speed-craft
www.amsa.gov.au	Australian Marine Standards Association

www.sjofartsdir.no/en	Norwegian maritime authority go to /shipping/legislation/#regulations
www.bureauveritas.com www.veristar.com/portal	Home page at /home/our-services/classification/ /veristarinfo/detail/generalinfo/giRulesRegulations/bvRules/rulenotes for full rules listing including HSC, etc., items in red can be downloaded. Also link to erules. Erules loads a popup application also. Note NR 396 are joint between BV, GL, and RIN from 2002, and so link also to DNV rules

http://www.imo.org/en/OurWork/Safety/Regulations/Pages/HSC.aspx
http://www.imo.org/en/OurWork/Safety/StabilityAndSubdivision/Pages/Default.aspx

IMO for High Speed Craft Rules, etc.

International Organizations

www.ittc.info/	International Towing Tank Conference Home Page—source for guidance on model testing and hydrodynamic analysis including CFD for vessel and propulsors. Go to the publications list for PDFs of key procedures and guidelines. Full reports of each ITTC are available under the downloads section.
www.imo.org/en	Home page for the International Maritime Organization—HSC Code and updates—HSC Code can buy paper or electronic.

Software

www.bentley.com	Maxsurf is at extension: /en/products/brands/Maxsurf. Hydromax is also documented on the site. Maxsurf is for hull modeling and statics, and Hydromax for wave generation and drag
www.dnvgl.com/services	go to /services/global-fe-analysis-software-shipload-18522 for DNVGL shipload, or /linear-and-non-linear-hydrodynamic-analysis-of-vessels-including-forward-speed-wasim-2413 for linear and nonlinear hydrodynamic analysis or /hydrodynamic-analysis-and-stability-analysis-sesam-hydrod-2410 for

	stability and linear hydrodynamic analysis or /marine-project-management-efficient-collaboration-in-ship-design-shipbuilding-and-aftermarket-synergi-project-18373 for Synergi Project Management
www.proteusengineering.com/	Fastship / for hull dev / Visual SMP ship motions prog from USN using strip theory. US Based with Alion
www.alionscience.com	owners of Proteus—use Navcad resistance and powering (from Hydrocomp), GHS for stability and hydrostatics (from Creative Systems), and Visual SMP for seakeeping
www.hydrocompinc.com	Navcad speed and power, using 2D theory based on volume rather than surface ordinates
www.ghsport.com	Creative Systems Inc, General Hydrostatics programs, used by Navatek and Damen
www.autoship.com	Detail design software
www.aerohydro.com	Multisurf 3D design and interface to WAMIT—note modelling is body form for hydrostatics and dynamics and depends on WAMIT for motions prediction.
www.boatdesign.net	Information network for boat design. Useful site and software links
www.aeromarineresearch.com	Information site for power boat design including tunnel hull catamaran planing craft
www.hawaii-marine.com/templates/	Various spreadsheets for hydrostatics and planing Hull resistance calculations including Savitsky (outside EU only to individuals)
www.aveva.com	Aveva Marine comprehensive Naval Architecture for ship projects, hull, and outfitting design including project workflow, initial design of hull form, structure, hydrostatics and dynamics, followed by materials, drafting, PDMS, systems and fabrication setup as used by Hyundai, etc.

www.delftship.net	Hull modeler and hydrostatics, free and professional at eur150, plus extensions ref Danish Yachts
www.wumtia.soton.ac.uk/software	Southampton University Wolfson Unit Marine Design Software—free and to purchase. The free shipshape program can output to DXF as well as their own format for further processing.
www.rhino3d.com/	Rhino 3D modeler and rendering software available for PC and MAC. Euro 995 for full software or Euro 1700 for whole package. Works with NURBS surfaces, etc.
www.orca3d.com	Builds on rhino modeling and provides hull design and fairing, hydro and stability, speed and power, and weight/cost tracking for around 3000 USD
www.autodesk.com/products	Autodesk site for CAD products and integration with Solidworks and Nastran FEM
www.autodesk.com/navisworks	Site for Autodesk CAD and Navisworks viewer and project management
www.solidworks.com/	CAD modeling suite
www.SSI-corporate.com	Ship Constructor 2017 based on AutoCAD and Navisworks, linking to modules mechanical, P&ID, Plant 3D, use for detailed systems design
www.napa.fi	Finnish company that supports major shipyards with design software and Autocad drafting
www.ptc.com	Software design site
www.adina.com/	FEM suite, German origin, with nonlinear analysis
www.plm.automation.siemens.com/en_us/	Siemens PLM home page with access to all products
www.mdx.plm.automation.siemens.com/star-ccm-plus	Siemens Star CD and CCM+ etc. fluid simulation software for internal and general turbulent flows, earlier was CDadapco Star CD and CCM+. Now integrated into Siemens design automation solutions

www.paramarine.qinetiq.com/products/paramarine/Pages/default.aspx	Qinetiq Paramarine software builds on Siemens PLM parasolid modeling. Seakeeping via 2D Rankine source approach for Freq Domain response, also has structural model aimed at Naval projects. Also used by Adhoc Marine and Keel Marine for wind farm catamaran design
www.mscsoftware.com	Home for Finite Element–based software including Nastran Structural design FEM software linear, fatigue and nonlinear, and multiple linked structures.
www.ansys.com	AQWA at: /Products/Structures/ANSYS-Aqwa Diffraction based software for wave loads and link to ANSYS ASAS suite for structural analysis
www.ansys.com	ANSYS-Fluent at: /Products/Fluids/ANSYS-Fluent CFD software tools including Fluent and CFX. Fluent is general modeling while CFX covers turbomachinery. Note that this has been sued by Navatek to design their SWATH vessels
www.openfoam.com/	Open source CFD software—Also has visual CFD.
https://www.cfd-online.com	go to /Wiki/Main Page for lots of info, sources both open and commercial including mesh generation, visualization, etc.
http://www.lincolnproject.eu/lincosim/	Opensource web-based CFD software that links with openfoam and automates the meshing and can link with larger machines for simulation
www.sunrise-sys.com	Suppliers of pipe-net piping system modeling for flow analysis including fire-water sprinklers, go to /index.asp
www.swan.tudelft.nl/	SWAN software from TUI Delft freely available for wave generation modeling in coastal areas and inland waters, so useful for wave wash

www.ptc.com	Site for Mathcad software, also links to Solidworks
www.mathworks.com	Site for Matlab for solving matrix-based problems
www.strand7.com/	FE software with useful images in gallery showing twisted cat. Aimed at smaller company
www.reliasoft.com/products.htm	Reliability and FMEA software tools
www.fmea-fmeca.com	Information site on FMEA/FMECA and industry standards for FMEA

Propulsion Waterjets

www.wartsila.com	Wartsila water jets in range 4500–26,000 kW incorporates LIPS from earlier, look under propulsion products for water jets; are in Holland.
www.marinejetpower.com	Successor to MJP at /index2.php including Ultrajet range, see history
www.rolls-royce.com/marine/	Rolls Royce subsidiary KaMeWa water jets in power range from 100 kW to 40 mW (under /propulsors/waterjets)
www.hamiltonjet.com	Hamiltonjet water jets in range to approx 4000 kW
www.castoldijet.it/en	Castoldijet water jets up to 1987 kW at /waterjet_en/applications_en.html
www.namjet.com/	Site for North American Marine Jet, suppliers of waterjets from 287 to 1016 mm diameter (294 to 588 kW) for utility and military craft. Axial pump design, and electrical controls not hydraulic for improved reliability
www.scottwaterjet.com	at /products/index.html New Zealand supplier for smaller jet units in range 50–2000 shp
www.rbbi.com	at /links/drives/waterjet.htm list of waterjet manufacturers worldwide with links
www.berkeleyjet.com/	US manufacturer for planing craft in power range 205–430 shp
www.americanturbine.com/	Another supplier similar to Berkleyjet for jetboats, mixed flow jets including inducers and aluminum intakes can be welded into aluminum hulls.
www.doen.com/	Dutch manufacturer for water jest in range 100–4000 kW

Propellers

www.servogear.no/	propellers classic for fast ferries, in hull partial tunnel (Norway)
www.wartsila.com	at /products/marine-oil-gas/propulsors-gears, though oriented to large ferries and ships (Finland)
www.rolls-royce.com/marine	KaMeWa are now part of Rolls Royce Marine under propulsion and continue to supply propellers for fast vessels including CP and supercavitating
www.elicheradice.com	at /page.php?pageid=PHOME001 home page for propeller, shaft and skeg supplier in Italy including surface drives (they say)
www.piening-propeller.de/en/	Propellers and propulsion packages (gearbox, shaft, etc.) also high speed
www.qmarine.co.nz	at /products/propulsion-systems inc surface drives and market MJP waterjets
www.tuprop.com/	tunnel propeller system, mounted on transom to 1000 shp for utility and fast boats. Is the site for MSA Marine systems gmbh
http://www.andritz.com	at /products-and-services/pf-detail?productid=9659, for EscherWyss CP high speed propeller systems
www.amartech.nl/products	Propeller and transmission designer and manufacturer
www.bruntons-propellers.com/	Smaller propeller manufacturer has built props for many high speed craft, builds up to 3000 KW while sister company Stone Marine builds larger propellers
www.teignbridge.co.uk/	Propeller manufacturer including high-speed propellers and surface drive propellers
www.miwheel.com/	Michigan Wheel—various propeller series for outboards, inboards, and speeds in range up to Fn about 0.7 I think. They are big.
www.volvopenta.com	Search on IPS system for their integrated engine and rotatable contra-rotating puller propeller drives in range to 740 kW, with control system

Surface Drives

www.Rolla-propellers.ch	Rolla propellers (also analytical consultants) for surface drive and fully submerged high-speed propellers (Italy)
www.Arneson-industries.com	Arneson surface drives—the complete drive system (USA)
www.zf.com	ZF Searex surface drive
www.francehelices.fr	Surface drive system and propeller manufacture
www.q-spd.com/	surface drives manufacturer supplied by Qmarine above
www.levidrives.com	Levi surface drives and propellers. Drive unit has fixed prop and cover which doubles as rudder mechanism
http://msa-marine-systems.com/	Tunnel Surface Drives in range to around 2000kW. Drive unit hinged for both trim and turning

Engines

www.deutz.com	engines up to around 600 shp air and water cooled, still independent
www.mtu-online.com	Go to products/engine-program/diesel engines. MTU provide main propulsion for passenger ships, ferries and fast ferries. Thier power range is up to 9000 kW. MTU are owned by Rolls Royce Holdings and Daimler AG. Detroit Diesel is also part of the group.
www.cat.com	go to /en_US/by-industry/marine.html. Main basis is Germany, which is descendant of MWM (now Caterpillar Energy Solutions gmbh). Used by LCS project, etc.
www.marine.man.eu	MAN engines, go to /applications/ferries
www.cumminsengines.com	Main site for marine, go to resources to download data sheets, etc.
www.rolls-royce.com	go to /products-and-services/marine/product-finder/diesel-and-gas-engines.aspx#section-featured-product too locate med speed (Bergen) diesels, propulsion pods, etc., as well as KaMeWa waterjets and CP Propellers
www.energy.siemens.com/hq/en	go to /fossil-power-generation/gas-turbines. Siemens use RR as core drivers 4–50 mW

www.scania.com/global/en	at /global/products-and-services/engines/our-range/marine-engines.html for marine engines up to 1150 bhp. Eight Di13 engines being used in ferries on Potomac, etc., by Metal Shark
www.geaviation.com/marine/	Gas Turbines from 4.5 to 42 MW are produced. GE diesels for marine service are medium speed and heavy, aimed at traditional displacement vessels.
www.dieselturbo.man.eu/	The MAN B&W internet site covers mainly large slow speed diesel engines for displacement ships.
www.volvopenta.com	Engines and integrated propulsion sterndrives
www.yanmarmarine.com	Powerboat and commercial marine diesels up to 4480kW
www.mercurymarine.com	Outboards and sterndrives to 550 bhp
www.suzukimarine.com	Outboards up to 350 bhp
www.evinrude.com	Outboards up to 300 bhp

Intake Filtration

www.sulzer.com/en	go to /Products-and-Services/Separation-Technology/Separators for knitmesh filters
www.knitted-mesh.com	Chinese supplier of knitted mesh products including demister materials

Service Suppliers and Marine Equipment Suppliers

www.frydenbo-industri.no	go to /eng/engines/deutz/deutz-marine-engines and others in Norway
www.european-diesels.co.uk	go to ../engines for their service and spares range. This includes engines from the ranges of Ruston, Bergen Diesel, Dorman, Perkins and English Electric. These have been used in semi displacement vessels since the 1950's. Ruston sold to Siemens, but diesels seem to have stopped, so only spares now
www.vetus.com/	Suppliers of equipment and outfitting for boats and smaller vessels ranging from engines and ancillaries to windows to fire retardant insulation materials (under engines and around)

Gearboxes and Transmission

www.reintjes-gears.de	Reintjes, also propulsor plus pdfs of vessels
www.zf.com	go to /corporate/en_de/products/further_product_ranges/marine/index.html for marine gearboxes and fast ferries, etc. Also do fixed pitch propellers and tunnel thrusters.
www.twindisc.com	at /marine-products/ for gearboxes and transmission, trim tabs, propellers, and other products. Parent to Rolla and Arneson.
www.prm-newage.com	at /c1-marine-gearbox for marine gearboxes at smaller power end of market
www.renksystems.com	marine at /renk-marine-gears.php., have supplied also US Navy LCS
www.renk.biz/home-en.html	Renk main site, go to vehicle transmissions and products to find marine transmissions and download brochure
www.renk-maag.ch/en/company/	Renk Swiss subsidiary providing high-performance gearboxes
www.regalpts.com	Jaure s.a. specialist in marine transmission shafts and couplings at /industries/marine/Pages/marine.aspx.
	Note that Rolls Royce, Wartsila, and MAN also provide transmission components or an integrated system

Stabilizers and Interceptors

www.humphree.com/	Interceptors and stabilizers, electric
www.naiaddynamics.com/	Successor to Maritime Dynamics, designers of stabilization systems for fast marine vessels including forward T foils and stern tabs, with active control systems

Rolls Royce and Servo gear also deliver stabilizers, and check on this site for other potential suppliers:
www.nauticexpo.com/boat-manufacturer/stabilizer-19818.html

Safety Outfitting, etc.

www.surviteczodiac.com	Survivetec group main company in France
www.rfd.co.nz	Survivetec—NZ—evacuation and survival gear
www.Viking-life.com	Evacuation and life rafts
www.lsames.com	Evacuation and life rafts
www.actionair.co.uk	HVAC duct marine fire dampers for rectangular or round ducting

Rubber Mountings

www.mackayrubber.com.au	Rubber mounts for top superstructures for Incat, Austal, etc., as well as complete industrial range of vibration isolation
http://isoflextech.com	Suppliers of isolation rubber mounts for machinery, superstructures, and wheelhouses for commercial vessels
www.nauticexpo.com	Site for Ships and Yacht windows and various other marine outfitting equipment

Marine Fire and Sound Insulation

www.glava.no	go to /marine-offshore/solutions for insulation, fire protection, etc.
www.promat-marine.com/en	marine fire and noise insulation

Marine Architectural Panels Including Suspended Ceilings

www.duflex.com.au	/duflex2/products/featherlight for FRP panels
http://www.lautex.com	marine ceilings with sound absorption
http://dampa.com	/marine/products/ marine ceilings panels and design—supply Austal and are on 126m Trimaran
www.ceilingworks.com.au	Ferry internal suspended ceilings, etc.
www.altrofloors.com	Maritime safety flooring at /floors/transport-floors/Maritime

Seat Manufacturers

www.eknes.no	Eknes classic seating for fast ferries
www.beurteaux.com/	Supplier of fast ferry seating for more than 900 vessels
www.westmekan.com/no	Suppliers of ferry seats and propulsion gear out of Nordfjordeid
www.pacificmarine.net	/marine-deck/marine-seats/ferry-passenger-seats.htm. US company for seats and other internals for ferries
www.deltafurniture.com	at /passenger transportation/ferry seating Ferry seating and passenger transport specialist—Canada location
www.ferryseating.com/	China supplier
www.ferryseat.com/	Sanhui, another Chinese factory with IMO HSC certified seats
www.springfieldmarine.co.uk	UK supplier of ferry seats, crew transfer, and suspension seats for crew, etc.
www.grammer.co.uk/home.php	Grammer seats for suspended marine and all sorts of transport, etc. (not pax seats)

Marine Interior Design

www.speargreen.com.au/	interior design specialist for fast ferries, out of Sydney since 1993. Also do Naval Architecture. Worked with AMD and Austal for Oman, etc.
www.arosmarine.com/en	marine outfitting contractor does installation—interesting is "proven brands" banner bottom left on page
www.kaefer.com	/Accommodation.html another contractor for installation especially for cruise ships, etc., with range of own brand equipment also
www.alusys.com.sg/	Singapore-based internal outfitting contractor including for ferries

14.2 General References and Internet Links

We list below a selection of basic references that can form a starting point for technical search by designers as well as the links given above. The material referenced includes conference proceedings, textbooks, technical reports, and published papers and is intended as a general addition to the direct references given in each chapter.

Conference Materials

HIPER Conferences	www.hiper-conf.info	details from 2017
FAST Conferences	http://fast2017.com/	

Digital Reports and Document Libraries

University of North Texas Digital Library, NACA Technical Memoranda Reports, Available at https://digital.library.unt.edu/ark:/67531/metadc64973/

University of Michigan Digital Library, David Taylor Model Basin Reports, Available at https://www.lib.umich.edu/database/david-taylor-model-basin-reports

Technische Universitiet Delft (TUDelft), Holland

Research documentation is available at https://repository.tudelft.nl/ including early work on polymer injection for drag reduction and more recent work on air lubrication and Air Cavities, including:

1. J A Keuning, J L Gelling, 'The influence of the bow shape on the operability of a fast ship in a seaway', HIPER 2006, session A pp 219–234.
2. O Sverkhovskyi, 'Ship drag reduction by air cavities', PhD dissertation 2014, ISBN 9789461086389
3. J Westerweel, T J C van Terwisga, O Zverkhovskyi, 'A flat bottomed vessel and a method for controlling the length of at least one air cavity', European Patent under WIPO, WO 2015/080574 A1

Naval Architecture Texts

1. John P Comstock, editor, Principles of Naval Architecture, Revised edition 1967, further updates on regular basis to the present, now available as a series of books via www.SNAME.org, Society of Naval Architects and Marine Engineers, New York, NY 10048, USA.
2. Ship Hydrostatics and Stability, A. B. Biran, Butterworth Heinemann, 2003, ISBN 978-0-7506-4988-9. (BH is an impression of Elsevier, ref Elsevier.com)
3. Architecture Navale, Connaissance et pratique, D Presles, D Paulet, Editions de la Villette, 2005, ISBN 2-915456-14-3. (info at www.paris-lavillette.archi.fr)

4. Resistance et Propulsion du navire (Resistance and propulsion of ships - French text), Doutreleau,Y, Laurens JM, Jodet L, Technosup ENSTA Bretagne, Paris, 2011 ISBN 978-2-7298-6490-3
5. K J Rawson, E C Tupper, 'Basic Ship Theory', 2 volumes or one combined volume, Butterworth Heinemann, 5th Edition 2001, ISBN 978-0750653961/78 and 978-0750653981
6. E C Tupper, 'Introduction to Naval Architecture', Butterworth Heinemann, 5th Edition 2013, 490pp, ISBN 978-0080982373
7. P A Wilson, 'Basic Naval Architecture – Ship Stability', Springer, New York, 2018, 228pp, ISBN 978-3319728049
8. E M Lewandowski, 'The dynamics of marine craft – Maneuvering and seakeeping', Advanced series on Ocean Engineering, Vol 22, World Scientific Publishing Co Pte Ltd, Singapore, 2004, ISBN 981-02-4755-9, 424 pp. A useful overview, with a chapter 6 on the dynamics of marine craft.

Textbooks on Hydrodynamics of High-Speed Marine Vessels

1. Hydrodynamics of High-Performance Marine Vessels, Professor Laurence J Doctors, two volumes, available at www.Amazon.com in 2 volumes, 888pp, 2018, ISBN-13 978-15112244717
2. Performance by Design – Hydrodynamics for high-speed vessels, D J Blount, published by Donald J Blount and Associates, 342pp, 2014, ISBN-10-0989083713
3. Practical Design of Advanced Marine Vehicles, Chris B McKesson, 392pp, 2014, ISBN 13 978-1497396890
4. Hydrodynamics of High-Speed Marine Vehicles, Professor O M Faltinsen, NTNU Trondheim, published by Cambridge University Press, 2005, 454pp, ISBN 978-0-521-84568-7

Hydrodynamics and Aerodynamics Classic Texts

1. Fundamentals of Hydro- and Aeromechanics, based on lectures of L Prandtl, Ph. D., O.G. Tietjens, Ph. D., Dover Publications, New York, 1957, 270pp, ISBN-13 978-0486603742
2. Applied Hydro- and Aeromechanics, based on lectures of L Prandtl, Ph. D., O.G. Tietjens, Ph. D., Dover Publications, New York, 1957, 306pp, ISBN-13 978-0486603759
3. Aerodynamic Theory, ed. William F Durand in six volumes, Dover Publications, New York 1963, Volume VI, Section S, Hydrodynamics of Boats and Floats, Library of Congress No 63-19489

MARIN Research Reports (www.MARIN.nl)

The Marine Research Institute in the Netherlands at Wageningen has carried out significant research into air lubrication and air cavity physics as well as fast craft as referenced elsewhere in this book. It is necessary to register an account first and then access is available to a range of R&D papers, including the following documents:

1. G Rotte, M Kerkvleit, T van Terwisga, 'On the Turbulence Modelling for an Air Cavity Interface', NuTTS conference 2017
2. O Zverkhovskyi, M Kerkvleit, A Lampe, G Vaz, T van Terwisga, 'Numerical Study on Air Cavity Flows', NuTTS conference 2015
3. E J Foeth, R eggers, F H H A Quadvlieg, 'The efficacy of air bubble lubrication for decreasing friction resistance', International Conference on ship drag reduction, SMOOTH Ships, Istanbul, Turkey, 20–21 May 2010
4. G Rotte, O Zwerkhovskyi, M Kerkvliet, T van Terwisga, 'On the physical mechanisms for the numerical modelling of flows around air lubricated ships', International Conference on Hydrodynamics, ICHD 2016. Modelling using MARIN CFD code ReFRESCO including looking at re-entrant flows at tail of cavity and wave pinch-off and effect of real fluid turbulence. Paper also at:

   http://www.ichd2016.nl/onlineproc/proceedings/documents/15.pdf
5. C Thill, S Toxopeus, F van Walree, 'Project Energy-saving air-lubricated ships (PELS)', 2nd International Symposium on Seawater Drag Reduction, Busan, Korea, 23–26 May 2005.

Summary of PELS Joint Industry Project model testing of a segmented instrumented barge model with cavities, and parallel simulation analysis

University of Southampton Reports

Available from www.eprints.soton.ac.uk

Use refined search and author name, sort on year to get easiest usable result. The reports below are a sample, as the R&D has been quite extensive over the 1990's and 2000's and to date.

1. Ship Science Report No. 71, 1994, A F Molland, J F Wellicome, P R Couser, 'Resistance experiments on a systematic series of high-speed displacement catamaran forms: variation of length-displacement ratio and breadth-draught ratio'.
2. Ship Science Report No. 106, 1999, J F Wellicome, A F Molland, J Cie, D J Taunton, 'Resistance experiments on a high-speed displacement catamaran of Series 64 form'.
3. Ship Science Report No. 122, 1999, A F Molland, P A Wilson, D J Taunton, 'A systematic series of experimental wash wave measurements for high-speed displacement monohull and catamaran forms in shallow water'.
4. Ship Science Report No. 125, November 2002, A F Molland, P A Wilson, and D J Taunton, 'Theoretical prediction of the characteristics of ship generated near field wash waves'.
5. Ship Science Report No. 127, 2003, A F Molland, P A Wilson, and D J Taunton, 'Resistance experiments on a series of high-speed displacement monohull and catamaran forms in shallow water'.

Flying Boat and Seaplane Hydrodynamics

Note that much of the early work on planing was concerned with the hulls and floats of seaplanes in the early part of the twentieth century, much of this work is available either through textbooks as sampled below, or the NACA and DTMB report series from the 1930s to 1960s

1. Aerodynamic Theory, ed. William F Durand in six volumes, Dover Publications, New York 1963, Volume VI, Section S, Hydrodynamics of Boats, and Floats, Library of Congress No 63-19489
2. W Sottorf, 'Systematic Model Research on the stability limits of the DVL Series of Float Designs', NACA TM 1254 December 1949
3. N A Sokolov, 'Hydrodynamic properties of planing surfaces and flying boats', NACA TM 1246, October 1950; Translation from CAHI Report 149, 1932
4. Walter S Deil, 'The application of basic data on planing surfaces to the design of flying boat hulls', NACA Report 694, 1940
5. E Jablonski, 'Sea Wings – The romance of the flying boats, An illustrated history', Doubleday and Company, Garden City, New York, 1972, Library of Congress Card Number 72-76173, 259pp

High Speed Stability

1. Y Ikeda, T Katayama, 'Stability of High-Speed Craft', in 'Contemporary ideas on Ship Stability', Elsevier Science Ltd, 1st Edition 2000, pp 401–409, ISBN 978008054715, www.elsevier.com
2. T Cellano, 'The prediction of porpoising inception for modern planing craft' SNAME Transactions, Vol 106, 1998, pp 269 to 292, ISBN 0.939773-25-2.
3. M Simeone, 'Hydrodynamic lift effect on the stability and on the banking angle of fast craft', FAST'91, Trondheim, Norway, Volume 1, pp 331 to 346, ISBN 82-519-0962-7
4. S Tavakoli, A Dashtimanesh, S Mancini, 'A theoretical method to explore the influence of free roll motion on the behavior of a high-speed planing vessel through steady yawed motion',

RINA Transactions, Vol 160, Part B1, International Journal of Small Craft Technology, January–June 2018, pp B67–77, ISSN 0035-8967. Paper looks at two planing hulls in roll fixed or roll free behavior when yawed in calm water to assess resistance and motion response.

5. D L Blount, L T Codega, 'Dynamic Stability of Planing Boats', Marine Technology Vol 29, No 1, January 1992, pp 4–12, SNAME, ISSN 0025-3316
6. A W Troesch, 'On the hydrodynamics of vertically oscillating planing hulls', Journal of Ship Research, Vol 36, No 4, December 1992, pp 317–331, SNAME, ISSN 0022-4502
7. A W Troesch, 'Dynamics and Hydrodynamics of High-Speed Craft', PASI 2010, Dynamics and Control of Manned and Unmanned vehicles, June 29th, 2010, Barranquilla, Colombia. Presentation including outline of R&D carried out at University of Michigan. A very useful overview of steady and unsteady dynamics.
8. M Martin, 'Theoretical prediction of motions of high-speed planing boats in waves', SNAME Journal of Ship Research, Vol 22, No 3, September 1978, pp 140–169, ISSN 0022-4502. Linearized response of constant deadrise planing boats in head and following waves compared with model tests by Fridsma at Davidson Laboratory of Stevens Institute of Technology, Report R-1275, November 1969.
9. M Martin, 'Theoretical determination of porpoising instability of High-speed Planing Boats', SNAME Journal of Ship Research, Vol 22, No 1, March 1978, pp 32–53, ISSN 0022-4502, Theoretical method developed to estimate the natural frequencies and damping for prismatic hulls and resultant motions and instability boundaries. Comparison with testing by Day and Haag.
10. P R Payne, 'On the high-speed porpoising instability of a prismatic hull', SNAME Journal of Ship Research, Vol 28 No 2, June 1984, pp 77–89, ISSN 0022-4502. Analytical modelling of a 'chines dry' prismatic hull with deadrise.
11. P De Jong, F van Waltree, 'Hydrodynamic lift in a time-domain panel method for the seakeeping of fast ships', 6th International Conference on High-Performance Marin Vehicles, Naples, Italy 18–10 September 2008, pp 95–105
12. J A Keuning, G L Visch, 'Use of a vertical fin for the combined yaw and roll stabilization of a fast patrol boat', 6th International Conference on High-Performance Marin Vehicles, Naples, Italy 18–10 September 2008, pp 107–116

Systematic Series

These references are intended as a general addition to the direct references used in Chaps. 2 and 4.

1. E P Clement, D L Blount, 'Resistance tests of a systematic series of planing hull forms', SNAME Transactions 1963. Outlines the Series 62 programme and results.
2. J B Hadler, E N Hubble, 'Prediction of the power performance of the Series 62 Planing Hull forms' Transactions SNAME 1971, pp 366 to 404.
3. S M Calisal, D McGreer, 'A resistance study on a systematic series of low L/B vessels', SNAME Marine Technology Vol 30, No 4, Oct 1993, pp 286–296, ISSN 0025-3316. A series at University of British Columbia, Vancouver, including single chine, double chine and round bilge craft for fishing, with algorithms for resistance and wetted surface. Displacement speed regime.
4. D J Taunton, D A Hudson, R A Shenoi, 'Characteristics of a series of high-speed hard chine planing hulls. Part I: Performance in calm water, Transactions of the Royal Institution of Naval Architects, Vol 152 Part B2 2011, ISSN 1740-0694. A systematic series programme at Southampton University
5. D J Taunton, D A Hudson, R A Shenoi, 'Characteristics of a series of high-speed hard chine planing hulls. Part II: Performance in waves, Transactions of the Royal Institution of Naval Architects, Vol 153 Part B1 2010, ISSN 1740-0694

6. F De Luca, C Pensa, 'The Naples warped hard chine hulls systematic series', Ocean Engineering 139 (2017) pp 205–236, www.dx.doi.org/10.1016/j.oceaneng.2017.04.038, Open access article published by Elsevier under creative commons. An extensive test series at University of Naples, Italy, including effect of transom interceptors for trim control.
7. D J Kim, S Y Kim, Y J You, K P Rhee, S H Kim, Y G Kim, 'Design of high-speed planing hulls for the improvement of resistance and seakeeping performance', International Journal of Naval Architecture and Ocean Engineering 5 (2013) pp161–177, www.dx.doi.org/10.2478/IJNAOE-2013-0124. A useful series of three deep vee planing and wave piercing forms at Seoul University and Korea Institute of Ocean Science
8. G J Grigoropoulos, T A Loukakis, 'Resistance and Seakeeping Characteristics of a systematic series in the pre-planing condition (Part 1)' SNAME Transactions Vol 110, 2002, pp 77 to 113, ISBN 0.939773-37-6
9. J A Keuning, L Hillege, 'The results of the Delft Systematic Deadrise Series', FAST 2017 Transactions Session B1, pp 97–106, Data is available at www.dsds.tudelft.nl once you set up an account for access, or via www.fast2017.com
10. D Radojcic, M Kalajdzic, 'Resistance and trim modelling of the Naples hard chine systematic series', RINA Transactions, Vol 160, Part B1, International Journal of Small Craft Technology, January–June 2018, pp B31–41, ISSN 0035-8967
11. D Radojcic, G Grigoropoulos, D P Damala, T Rodic, T Kuvelic, 'The resistance and trim of semi-displacement, double-chine, transom stern Hull series', 6th International Conference on Fast Sea Transportation (FAST'01), Southampton, U.K. The NTUA double chine model test series.
12. L J Doctors, 'Optimization of marine vessels on the basis of model test series', Journal of Marine Science and Technology, 2020, 25:887–900, https://doi.org/10.1007/s00773-019-00687-4

General Planing Hydrodynamics Papers

1. N Santoro, E Bergovic, C Bertorello, A Bove, S De Rosa, F Franco, 'Experimental study of the hydrodynamic loads on high speed planing craft', International symposium on dynamic response and failure of composite materials, Draf2014, Procedia Engineering 88 (2014) pp 186–193, www.creativecommons.org/license/by-nc-nd/3.0/ available at ScienceDirect.com
2. M Ricciardo, Planing Hull Hydrodynamics, Study of the Effects Caused by Variation of the Thrust Line Due to Displacement, Series 62 Model No. 4667-1, Research Paper, University of Michigan, Naval Architecture and Marine Engineering, March 2010, 32 pp.
3. S Pennino, A Scamardella, 'Dynamic equilibrium and resistance assessment for warped hulls bymeans of total pressure distribution', RINA Transactions, Vol 160, Part B1, International Journal of Small Craft Technology, January–June 2018, pp B91–100, ISSN 0035-8967. Looks at pressure distribution using Morabito method published in Journal of Ship research in 2014 to improve understanding for warped hulls.
4. D Villa, M Viviani, E Ferri, 'Application of CFD Calculations for the improvement of planing crafts manoeuvrability mathematical models' 12th International Conference on Hydrodynamics, Egmond an Zee, Holland, 13 to 23 September 2016, Paper 33. Go to www.ichd2016.nl, transactions and search on author, click on tracking (paper) number for full text as pdf.
5. A R Kohansal, H Ghassemi, M Ghaisi, 'Hydrodynamic Characteristics of high-speed planing hulls, including trim effects' Transactions of the Turkish Journal of Engineering and Environmental Science, 34 (2010) pp 155–170.
6. D J Kim, S Y Kim, Y J You, K P Rhee, S H Kim, Y G Kim, 'Design of high speed planing hulls for the improvement of resistance and seakeeping performance', International Journal of Naval Architecture and Marine Engineering (2013) 5:161–177, www.dx.doi.org/10.2478/IJNAOE-2013-0124.
7. M Iacono, 'Hydrodynamics of planing hull by CFD', Masters' Thesis at University of Naples 'Federico II', presented academic year 2014/2015. STAR CCM+ used to study one

monohedral and one warped deadrise hull including comparisons with experiment, and Savitsky and Marabito's empirical formulations.

8. M G Morabito, 'Reanalysis of William Froude's studies of planing craft', Transactions of the Royal Institution of Naval Architects, Vol 157, 2015, Part B1, pp B33–B47, ISSN 1740-0694. A review, retest and discussion around the Ramus stepped hull 'Rocket-Ram' proposal originally tested by William Froude in the Torquay Test Tank compared to Series 62 and a flat planing plate.
9. C Lai, A W Troesch, 'Modelling issues related to the hydrodynamics of three-dimensional steady planing', SNAME Journal of Ship Research, Vol 39, No 1, March 1995, pp 1–24, ISSN 0022-4502. Development of a three-dimensional numerical model for lift, Centre of pressure and pressure profile, and comparison with some experimental results showing general agreement with Savitsky data.
10. D Radojcic, 'An engineering approach to predicting hydrodynamic performance of planing craft using computer techniques', Transactions of the Royal Institution of Naval Architects, Vol 133, 1991, pp 251–267, ISSN 0035-8967. Presentation of Prof Radojcic's regression technique applied to model tests of planing craft for prediction of resistance, powering and seakeeping with models for resistance and trim in the speed range $Fn_{\nabla} = 1.0$ to 4.0. The model presented is based on Series 62 and Series 65 data.
11. D Savitsky, J L Gore, 'Re-evaluation of the planing hull form', Journal of Hydronautics, Vol 14, No 2, Article No 79-2028R, April 1980. Presented as paper 79-2028 at AIAA/SNAME Advanced Marine Vehicle Conference, Baltimore, Maryland Oct 2–4, 1979.
12. G Grigoropoulos, 'Recent Advances in the Hydrodynamic Design of Fast Monohulls', Ship Technology Research, January 2005, DOI: 10.1179/str.2005.52.1.003

Resistance Prediction

1. M De Vos, P K Sahoo, 'A critical Analysis of resistance prediction using regression methods for high-speed hull forms', 1995, available via www.Researchgate.com.
2. P K Sahoo, H Peng, J Won, D Sangarasigamany, 'Re-evaluation of resistance prediction for high-speed round bilge hull forms', 11th International Conference on Fast Sea Transportation, FAST 2011, Honolulu, Hawaii, USA September 2011.
3. S Ozum, B Sener, K Unlugencoglu, 'Resistance prediction of a high-speed craft by using CFD', Ovidius University Anals of Mechanical and Marine Engineering, Vol 10, Part 1 2010, ISSN 1223-7221
4. S C Fung, 'High speed displacement vessels parametric studies and calm water resistance predictions – state of the art', Association of Scientists and Engineers of the Naval Sea Systems Command, 23rd Annual Symposium, 1986, NSC Report AD-A174 027, April 1986.
5. S S Meyer, P K Sahoo, 'A numerical wave resistance prediction method for high-speed round bilge hull forms', Ninth International Conference on High-Performance Marine Vehicles, Athens, Greece, 3–5 December 2014. ANSYS FLUENT simulations compared with AMECRC model test data showing agreement within 10%
6. Y-H Lin, C-W Lin, 'Numerical simulation of seakeeping performance on the preliminary design of a semi-planing craft', Journal of Marine Science and Engineering, 2019, 7, 199; doi:10.3390/jmse7070199. Linear motions spectral analysis and comparison with model tests on a series 62 geometry hard chine hull at 7, 13 and 33 knots full scale in 1.5m significant sea state. The study included all six degrees of freedom motions and directions in spread sea states.

Spray

1. R Latorre, 'Study of prismatic planing model spray and resistance components', SNAME journal of Ship Research, vol 27, No 3, September 1983, pp 187 to 196.
2. D Savitsky, M DeLorme, R Datla, 'Inclusion of "Whisker Spray" Drag in Performance Prediction Method for High-Speed Planing Hulls', Davidson Laboratory, Stevens Institute of Technology; Hoboken, New Jersey, USA, Technical Report SIT-DL-06-9-2845, March 2006

3. S Pennino, S Mancini, A Scamardella, 'Dynamic equilibrium and resistance evaluation for warped planing hulls', Proceedings of FAST 2017 pp 338–346. Uses Star CCM+ to look at prediction including whisker spray.
4. R Latorre, 'Study of prismatic planing model spray and resistance components', SNAME Journal of Ship Research, Vol 27, No 3, September 1983, pp 187–196, ISSN 0022-4502. Model tests carried out at University of New Orleans on four prismatic models in acrylic so that wetted, and spray area could be studied in detail.

Spray Rails

1. Randolph Ashton, 'Effect of spray strips on various power boat designs', Stevens Institute of Technology TM 99, February 1949
2. G J Grigoropoulos, T A Loukakis, 'Effect of spray rails on the resistance of planing hulls', FAST 1995, Lubeck- Travemunde, Germany, Transactions Volume 1, pp 33 to 44. Looked at chine located horizontal spray rails on models of the series 62 form, a deep vee, double chine and rounded bilge form. Results mixed while positive in reducing resistance at higher speeds.
3. Jeonghwa Seo, Hak-Kyu Choi, Uh-Cheul Jeong, Dong Kun Lee, Shin Hyung Rhee, Chul-Min Jung, Jaehoon Yoo, 'Model tests on resistance and seakeeping performance of wave-piercing high-speed vessel with spray rails', International Journal of Naval Architecture and Ocean Engineering 8 (2016) pp 442 to 455, at Science Direct.com
4. Dae-Hyuk Kim, Kye-pyo Rhee, Nakwan Kim, Jin-Hyung Ahn, 'Experimental Study of the effect of spray rails for a V-shaped high speed Planing hull', International Conference on Hydrodynamics, ICHD 2016, paper 120. Studied effect of position, geometry and number of rails and determined effect on resistance of planing hull.

Transom Sterns

1. W H Cheng, 'Computation of 3D Transom stern flows', Proceedings of 5th Conference on Numerical Ship Hydrodynamics, 1990, pp 581 to 589, 24–28 September 1989 Hiroshima, Japan, ISBN 0-309-+4241-0
2. L J Doctors, 'A Numerical Study of the Resistance of Transom-Stern Monohulls', 5th International Conference on High Performance Marine Vehicles, 8–10 November, 2006
3. L J Doctors, 'Influence of the Transom-Hollow Length on Wave Resistance', available at http://iwwwfb.org/Abstracts/iwwwfb21/iwwwfb21_10.pdf from International Workshop on waves and floating bodies, www.iwwwfb.org session 21, 2nd–5th April 2006, Loughborough, UK
4. M Orych, L Larsson, 'Hydrodynamic aspects of transom stern optimisation', 5th High Performance Yacht Design Conference, Auckland, 10–12 March 2015, pp 247 to 256
5. T C. Fu, A M. Fullerton, T Ratcliffe, L Minnick, D Walker, M L Pence, and K Anderson, 'A Detailed Study of Transom Breaking Waves', Naval Surface Warfare Centre, Hydromechanics Department Report NSWCCD-50-TR-2009/025 May 2009
6. K J Maki, 'Transom stern hydrodynamics' PhD dissertation University of Michigan Department of Naval Architecture and Marine Engineering, 2006
7. M Haase, G Thomas, J R Binns, N Bose, 'Wave Piercing Catamaran Transom Stern Ventilation Process', Ship Technology Research, April 2016, available at www.ResearchGate.net DOI: https://doi.org/10.1080/09377255.2015.1119922

Stepped Hulls

1. A De Marco, S Mancini, S Miranda, R Scognamiglio, L Vitiello, 'Experimental and numerical hydrodynamics analysis of a stepped planing hull', Applied Ocean Research 64 (2017) pp 135–154, available at https://doi.org/10.1016/j.apor.2017.02.004. An experimental and parallel CFD programme improving understanding of flow behind a step on a deep vee hull carried out at University of Naples, Italy
2. D Savitsky, M Morabito, 'Surface wave Contours associated with the forebody wake of stepped planing hulls', presentation to New York Metropolitan Section of SNAME, March 10, 2009
3. Kobus Potgeiter 'Understanding design and performance of stepped hulls', article, at Kobus Naval Designs, South Africa, www.navaldesign.co.za/articles/Stepped Hulls- Feb07.pdf

4. M G Morabito, M E Pavkov, 'Experiments with stepped planing hulls for special operations craft', Transactions of the Royal Institution of Naval Architects, vol 156, 2014, Part B2, pp B87–B95, ISSN 1740-0694. Study at the US Naval Academy. Discusses step configuration for moderate and high speeds, different step heights and porpoising stability
5. S Hajizadeh, M Fathi Kazerooni, A Mohammadi, 'RANS simulation of stepped planing hull hydrodynamic performance in calm water and waves, Transactions of the Royal Institution of Naval Architects, Vol 161, 2019, Part B2, pp B31–B37, ISSN 1740-0694. The authors from Persian Gulf University, Iran, used RANS based CFD to evaluate Taunton and Hudson's stepped model C1. Results suggested CFD can calibrate with model test data for resistance and wave induced motions.

Computational Fluid Dynamics

1. H M. Bakhtiari, S.T.G. Veysi, H. Ghassemi, 'Numerical Modeling of the Stepped Planing Hull in Calm Water', International Journal of Engineering (IJE), Transactions B: Applications Vol. 29, No. 2, (February 2016)
2. Flavio Di Caterino, Rasul Niazmand Bilandi, Simone Mancini, Abbas Dashtimanesh, Maria De Carlini, 'A Numerical Way for a Stepped Planing Hull Design and Optimization', Technology and Science for the Ships of the Future, IOS Press, 2018, https://doi.org/10.3233/978-1-61499-870-9-220
3. Hassan Ghassemi, Mojtaba Kamarlouei, Sajad Taj Golah Veysi, 'A Hydrodynamic Methodology and CFD analysis for performance prediction of stepped planing hulls', Polish Maritime Research 2(86) 2015 Vol. 22; pp. 23–31, https://doi.org/10.1515/pomr-2015-0014
4. Mario Caponnetto, 'Practical CFD simulations for planing hulls', Rolla Research, Balerna, Switzerland, the paper is available from the author at the following location: https://caponnetto-hueber.com/wp-content/uploads/Practical_CFD_simulations_for_planing_hulls_HIPER_2001.pdf
5. Simone Mancini, Fabio De Luca, Anna Ramolini, 'Towards CFD guidelines for planing hull simulations based on the Naples Systematic Series', VII International Conference on Computational Methods in Marine Engineering, MARINE 2017, https://upcommons.upc.edu/bitstream/handle/2117/332155/Marine-2017-89_Towards%20CFD%20guidelines.pdf
6. F Pacuraru, A Mandru, A Bekhit, 'CFD Study on Hydrodynamic Performances of a Planing Hull', Journal of Marine Science Engineering 2022, 10, 1523. https://doi.org/10.3390/jmse10101523
7. J Guo, Z Hu, Z Zhu, Z Chen, L Cui, G Li, 'Numerical calculation and analysis of resistance performance of planing craft combining Savitsky method and overset grid technology', Chinese Journal of Ship Research, 2022, 17(3). http://www.ship-research.com/en/article/doi/10.19693/j.issn.1673-3185.02417
8. N V R Nimmagadda, L R Polisetty, A S V Iyer, 'Simulation of Air–Water Interface Effects for High-speed Planing Hull', Journal of Marine Science and Application, 2020 19:398–414, https://doi.org/10.1007/s11804-020-00172-0
9. J Townsend, 'The application of computational fluid dynamics to the modelling and design of high speed boats', PhD Thesis, College of Engineering, Swansea University, 2020, available from the university: https://cronfa.swan.ac.uk/Record/cronfa58462/Download/58462__21395__af965f14c1504c1db720e4cca3363a43.pdf
10. M J Crosby, 'A CFD Study on the Performance of High Speed Planing Hulls', (2019). Masters Thesis at Grand Valley State University, available at https://scholarworks.gvsu.edu/theses/966
11. Raffaele Ponzini, Francesco Salvadore, Ermina Begovic, Carlo Bertorello, 'Automatic CFD analysis of planing hulls by means of a new web-based application: Usage, experimental data comparison and opportunities', Ocean Engineering, Volume 210, Aug 2020, https://doi.org/10.1016/j.oceaneng.2020.107387

   This paper describes use of Openfoam and Lincosim open source CFD modelling.

12. Elin Lindbergh, Felicia Ahlstrand, 'Methods to Predict Hull Resistance in the Process of Designing Electric Boats', Masters' Degree Project at KTH Royal Institute of Technology, School of Engineering Sciences, Stockholm, Sweden, June 2020. https://www.diva-portal.org/smash/get/diva2:1465549/FULLTEXT01.pdf

Interceptors

1. T Hansvik, 'Use of interceptors and stepped hull to improve performance of high speed planing', High Speed Craft – ACV's WIG's and Hydrofoils Conference at RINA 31 October–1 November 2006

Waterjets

1. The specialist Committee on Waterjets, Final Report and Recommendations to the 22nd International Towing Tank Conference, 1999 available at www.ITTC.info. Procedures for testing waterjets.
2. N W H Bulten, R Verbeek, B P M van Esch, 'CFD simulations of the flow through a waterjet installation', Transactions of the Royal Institution of Naval Architects, vol 148, 2006, pp 141–151, ISSN 0035-8967

   A paper verifying the ability to accurately model waterjet installations with CFD with 43 useful references.
3. D C Murrin, N Bose, 'Waterjet propulsion system tested in a wind tunnel and compared with numerical simulations', Transactions of the Royal Institution of Naval Architects, vol 148, 2006, pp 305–313, ISSN 0035-8967

Hydrofoils

1. S Brizzolara, A Federici, 'Super-cavitating profiles for ultra-high-speed hydrofoils and a hybrid CFD design approach', 9th High Speed Marine Vessels Conference, Naples, 25–27 May 2011

Index

A
Air resistance, 167, 226–231
Analytical equations, 207–209
Appendages, 6, 34, 38, 81–82, 95, 103, 110–119, 167, 210, 211, 226–241, 255, 268–272, 319, 505, 515–516, 529
Axe bow, 338–341

C
Cavitation, 92, 157, 229, 255, 354, 389, 437, 493, 575
Computer fluid dynamics, 588
Concept development, 355
Configurations, 4, 6–8, 12, 16, 17, 54, 86–88, 90, 112, 123, 131, 162, 164–166, 169, 187, 198, 226, 228, 253, 257, 285, 304–311, 320, 327, 336, 347, 349, 350, 352–355, 357, 358, 361, 363, 375, 378, 389, 408, 414–421, 493, 494, 496, 514, 529, 537, 543, 557, 559, 562, 567, 568
Control, 5, 104, 154, 209, 269, 345, 492, 493, 567, 589

D
Deep vee, 9, 16, 17, 23, 28, 29, 225, 306, 312, 315–341, 356
Drag, 3, 33, 123, 161, 251, 315, 345, 394, 446, 495, 567
Dynamic stability, 29, 51, 91, 123, 252, 255, 263, 264, 529, 533, 538

E
Empirical methods, 135, 167–179, 199, 451
Experimental data, 52, 151, 169, 246

F
Foil arrangements, 349, 353, 354, 493, 496, 498, 506, 517, 522, 523, 526, 529, 530, 532, 534, 538, 539, 557–564, 579, 583
Foil-assisted catamaran, 568–575, 577
Foilcat, 382, 543, 577–584
Free surface effect, 441–442, 508

H
Hydrofoil arrangement, 513
Hydrofoil development, 354–375
Hydrofoil theory, 401, 437–492

I
Initial designs, 68, 102, 163, 211, 294, 318, 323, 494, 588
Introduction, 3–32, 39, 107, 161–164, 345–379, 428, 493

M
Model tests, 5, 9, 16, 28, 30, 33, 34, 36, 38–70, 78, 81, 84, 91, 93–95, 97, 99, 100, 102–106, 113, 115, 116, 118, 162, 163, 172, 178–180, 182–184, 187, 189,

Z.-S. Dong et al., *High-Speed Monohull and Hydrofoil Craft*,
https://doi.org/10.1007/978-3-031-62760-6

206–216, 218, 227, 232, 255, 261, 273, 276, 278, 286, 293, 294, 299, 300, 303, 304, 308, 324–327, 337, 341, 459–461, 468, 516, 528, 575, 587, 588
Model test series, 35, 36, 39, 42, 56, 58, 70, 82, 90, 99, 125, 161, 162, 178, 179, 207, 209, 588
Monohull, 3–119, 197, 380–382, 500, 567, 587, 588, 611, 615

P
Parameters, 9, 29–33, 36, 38, 39, 41–43, 47, 50, 51, 55, 56, 58–60, 62, 63, 65, 68, 70, 71, 78, 81–91, 95, 99, 100, 102, 106–108, 110, 115, 125, 131, 161–163, 179, 180, 184, 186–191, 193, 196, 198, 200, 202, 203, 205, 208, 209, 218, 222, 244, 255, 257, 259, 261, 265, 267, 285, 291, 295, 304–311, 320, 422, 455, 465, 494, 495, 498, 499, 504, 549, 550, 555, 561, 580, 589, 592–607
Performances, 14, 17, 19, 29, 30, 32, 36, 39, 45, 46, 50, 51, 60, 71, 81, 87–90, 106, 107, 110–119, 151–158, 162, 165, 179, 184, 187, 193, 197, 204, 206, 208, 232, 234, 243, 244, 263, 273, 274, 283, 304, 306, 308, 311–312, 315, 316, 319, 320, 323, 324, 331, 334, 336, 341, 349, 352, 354, 356, 371, 375, 377, 389, 414–421, 437, 454, 459–462, 468, 486, 490, 491, 493, 503, 526, 529, 530, 537, 540, 541, 549, 556, 557, 571, 583
Planing, 5, 35, 123, 161, 251, 315, 345, 403, 439, 504, 568
Planing hulls with steps, 11, 165, 216–226, 363
Planning, 5

R
Resistance, 9, 16, 30, 38–82, 84, 85, 90, 97, 100, 130, 132, 134–137, 141, 143, 144, 147, 148, 151–154, 161–247, 251, 273, 288, 311, 345, 353, 426, 550, 571, 578
Round bilge, 4, 8, 9, 17, 35–37, 42, 45, 49, 52, 54, 66, 69, 70, 73, 77, 81–88, 91–119, 179, 188, 189, 207, 283, 291, 316–320, 323–328, 330, 331, 333–336

S
Sample data, 379
Seakeeping, 9, 11, 12, 29, 36, 39, 42, 46, 51, 57, 59, 60, 62, 63, 68, 86, 88, 93–108, 110, 113, 116, 117, 123, 184, 191, 203, 204, 208, 251–312, 315–317, 323, 325, 327, 332–333, 336–339, 341, 352, 353, 356, 357, 361, 372, 375, 377, 379, 381, 382, 385, 493, 510, 537, 538, 549, 558, 559, 562, 567, 571, 583, 588
Seaworthiness, 9, 17, 35, 36, 88, 106, 251, 304–312, 346, 353, 358, 365, 377, 518, 523, 533, 534, 538–540, 543, 549, 558, 563, 567, 568, 572, 579
Section profiles, 88–90, 228, 279, 315, 389, 423, 426, 486–492, 494, 499, 514, 522, 539
Semi-displacement, 7, 8, 17, 33–119, 161, 189, 316
Semi-planing, 5, 17, 28, 29, 32, 35, 55, 81–82, 88, 90, 93, 106–108, 161, 162, 179, 197, 200, 204, 315–341
Special designs, 210
Stability, 8, 42, 123, 251, 316, 346, 419, 468, 493, 567

T
Theories, 93, 97–100, 116, 131, 133, 135, 137–147, 162, 168, 174, 213, 233, 235, 243, 264, 267, 268, 292–294, 301, 302, 389–405, 409, 412, 414–416, 419, 432, 433, 437, 440, 442, 448–451, 455, 464, 472, 473, 475, 476, 482, 487, 493, 505, 536
Three-dimensions, 94, 125, 134, 135, 148, 243, 301, 404, 420, 536

V
Viscous flow, 147–151, 210, 212, 216, 432
Vortex models, 394–396, 405, 417, 472

W
Wing theory, 116, 128, 129, 134, 137–142, 169, 177, 255, 389, 396, 412, 427, 437–492